DYNAMICS OF TWO-PHASE FLOWS

Edited by
Owen C. Jones
Director for Multiphase Research
Rensselaer Polytechnic Institute
Troy, New York

Itaru Michiyoshi
Department of Nuclear Engineering
Kyoto University
Kyoto, Japan

begell
house

CRC Press
Boca Raton   Ann Arbor   London   Tokyo

Library of Congress Cataloging-in-Publication Data

Catalog record is available from the Library of Congress.

Developed by Begell House, Inc.

This book represents information obtained from authentic and highly regarded sources. Reprinted material is quoted with permission, and sources are indicated. A wide variety of references are listed. Every reasonable effort has been made to give reliable data and information, but the author and the publisher cannot assume responsibility for the validity of all materials or for the consequences of their use.

Neither this book nor any part may be reproduced or transmitted in any form or by any means, electronic or mechanical, including photocopying, microfilming, and recording, or by any information storage and retrieval system, without permission in writing from the publisher.

Direct all inquiries to CRC Press, Inc., 2000 Corporate Blvd., N.W., Boca Raton, Florida, 33431.

© 1992 by CRC Press, Inc.

International Standard Book Number 0-8493-9925-4

Printed in the United States   1 2 3 4 5 6 7 8 9 0

Printed on acid-free paper

Contents

Preface .... *vii*

Fundamental Equations and Closure Law

Discrete Modeling Considerations in Multiphase Fluid Dynamics 3
  V.H. Ransom, and J.D. Ramshaw

Closure Problems and Basic Equations for Correlation Functions in
  Two-Phase Flow 27
  Isao Kataoka

Some Tests of Two-Fluid Models for Two-Phase Flow 47
  Graham B. Wallis

Estimation of Volumetric Fractions of Each Phase in Gas-Liquid-Solid
  Three-Phase Slug Flow in Vertical Pipes 61
  T. Sakaguchi, H. Minagawa, T. Saibe, and K. Sahara

Flow Regime Modeling and Dynamics

Experimental Study on the Vibration Force of a Two-Phase Impingement Jet 89
  N. Nakamori, K. Kawanishi, M. Isono, J. Kasahara, and T. Kamiya

Fundamental Study of Interfacial Waves in Stratified Flow 103
  T. Sakaguchi, M. Ozawa, Y. Shiomi, S. Inoue, and Y. Murai

A Study of Air-Water Flow in a Narrow Rectangular Duct Using
  an Image Processing Technique 141
  *K. Mishima, S. Fujine, K. Yoneda, K. Yonebayashi, K. Kanda, and
  H. Nishihara*

The Effect of a Flat-Plate-Type Obstacle on a Thin Liquid Film Flow 161
  T. Fukano, T. Tanaka, K. Kutaragi, and M. Kanamori

Disturbance Wave in Boiling Flow 185
  S. Nakanishi, S. Yamauchi, and T. Sawai

Two-Equation Turbulence Modeling of an Upper-Plenum Water Pool above
  a Horizontal Perforated Plate with Steam Condensation 213
  S.G. Bankoff, and C.L. Chen

Bubble Coalescence and Transition from Wall Void Peaking to Core Void
  Peaking in Turbulent Bubbly Flow 233
  I. Zun, I. Kljenak, and A. Serizawa

Phase Separation and Distribution Phenomena

Effects of Droplet-Size Distribution and Flow-Blockage upon Inertia Collection of Droplets
  by Horizontal Cylinders in Downward Flow of Gas-Liquid Mist 253
  T. Aihara, W.S. Fu, and Y. Suzuki

Void Fraction Distribution in Two-Phase Single-Component Condensing Flow 275
    S. Toda, and Y. Hori

Countercurrent Gas-Liquid Flow in a Boiling Water Reactor Core during
    Postulated Loss-of-Coolant Accidents 295
    M. Murase, H. Suzuki, Y. Katoaoka, and S. Hatamiya

The Analysis of Phase Separation Phenomena in Branching Conduits 313
    R.T. Lahey, Jr.

Role of Bubble Behavior in Turbulence Structure of Vertical Bubbly Flow-
    Simultaneous Measurement of Bubble Size, Bubble Velocity and Liquid
    Velocity Using Phase Doppler Method 347
    K. Ohba, and J. Isoda

Characteristic Structure of Upward Bubble Flow 359
    G. Matsui

Bubble Size Effect on Phase Distribution 379
    A. Serizawa, I. Michiyuoshi, I. Kataoka, and I. Zun

Two-Phase Gas-Liquid Flow Distributions in Multiple Channels 401
    Y. Sato, and M. Sadatomi

Measurement of Phase Distribution Phenomena in High Pressure
    Steam-Water Two Phase Flow 423
    T. Narabayashi, S. Morooka, T. Kagawa, T. Ishizuka, A. Hoshide, and
    T. Ishiyama

Wave and Shock Phenomena and Critical Flow

Flow Pattern Transition Due to Instability of Voidage Wave 441
    S. Morioka, F. Jousselin, and H. Monji

An Analysis of Wave Propagation Phenomena in Two-Phase Flow 453
    R.T. Lahey, Jr.

Stability of Numerical Analysis on Void Waves in Stratified Two-Phase Flow 485
    T. Watanabe, and M. Hirano

Flow Instabilities in Parallel Channels with Nucleate Boiling and Film Boiling 503
    K. Akagawa, N. Takenaka, T. Fujii, and Y. Oku

Heating Limits and Instability of Downward Two-Phase Flow 537
    K. Fukuda, T. Kondoh, H. Sakai, and S. Hasegawa

A Study on Critical Two-Phase Flow through a Labyrinth 553
    K. Hijikata, T. Nagasaki, and T. Kanaya

Critical Flashing Flow in Pipes and Cracks 571
    V.E. Schrock

Nucleation and Flashing of Initially Subcooled Liquids in Nozzles:
    1. A Distributed Nucleation Model 603
    T.S. Shin, and O.C. Jones, Jr.

Contents

Nucleation and Flashing of Initially Subcooled Liquids Flowing in Nozzles:
2. Comparison with Experiments Using a 5-Equation Model for Vapor
Void Development 637
    V.N. Blinkov, and O.C. Jones, Jr.

Forced Convective and Post-Dryout Heat Transfer

An Analogy between Heat Transfer and Pressure Drop in Forced Convective
Boiling Flow 669
    K. Sekoguchi, H. Zhen-Xing, M. Kaji, T. Imasaka, and Y. Sumiyoshi

Critical Heat Flux and Flow Characteristics of Subcooled Flow Boiling with
Water in Narrow Tubes 689
    H. Nariai, and F. Inasaka

Interfacial Instability and Modeling of Wave Effect on Heat Transfer in
Annular Two-Phase Flow 709
    K. Suzuki, and Y. Hagiwara

Critical Power Characteristics in a Rod Bundle with Narrow Gap 739
    K. Arai, S. Tsunoyama, S. Yokobori, and K. Yoshimura

Thermo-Hydraulic Behavior of Inverted Annular Flow (Effects of Flow
Direction and Channel Diameter) 753
    M. Aritomi, and A. Inoue

Preface

Research in two-phase flow and heat transfer with change of phase currently emphasizes the dynamics of two-phase flows, flow boiling, and flow condensation. This work is motivated in large part by the great need to understand better the physical phenomena taking place in various conventional and nuclear energy conversion/production systems. This research is directed to developing the capability to improve thermal efficiency and to optimize the design of such systems, and more precisely evaluate the safety aspects of these systems. Major research programs are being conducted in government, industry, and university laboratories in both Japan and the United States. The Japanese work is characterized as rather fundamental whereas the U.S. workers have a tendency to deal with more applied work.

The importance and value of a joint Japan-U.S. seminar was recognized in 1979 when the first seminar on Two-Phase Flow Dynamics, cosponsored by the Japan Society for the Promotion of Science and the U.S. National Science Foundation, was held at the University Seminar House in Kansai. Five years later, the second seminar similarly cosponsored was held in Lake Placid, New York. During these intensive interaction periods, a valuable exchange of information occurred in the technology of multiphase dynamics. As a result of these seminars, considerable collaborative research and much interchange between researchers in the two countries was initiated.

The research field on two-phase flow had grown in the four intervening years to the time when this Third Japan-U.S. Seminar on Two-Phase Flow Dynamics was held in July of 1988 at Lake Biwa, Japan. This seminar, too, was jointly sponsored by the JSPS and the USNSF. Emphasis had been on developing a better understanding of the physics of the phenomena and improvements in dynamic analysis of normal transient or accident situations. Therefore, the holding of this third seminar was especially appropriate and beneficial to all parties. This volume, therefore, represents a distillation of the research reported during this meeting.

The seminar featured a broad review of the status of research relating to two-phase flow dynamics both with and without phase change. The papers fall into several categories which formed the natural grouping for the seminar, and for this volume, These major groupings are:

1. Fundamental equations and closure laws;
2. Flow regime modeling and dynamics;
3. Phase separation and distribution phenomena;
4. Wave and shock phenomena and critical flows;
5. Forced convective and post-dryout heat transfer.

In the final analysis, the value of any gathering such as this can only be judged in relation to the lasting effects as evidenced in the direction and quality of research which follows. The exchange of information by these two active nations is expected to substantially enhance progress in many areas of energy conversion including power plant efficiency, reliability, and safety. Ultimately, it is hoped that the seminar will lead to an improved understanding of process and system behavior, which in turn will lead to improved safety, longevity, and optimization of operational efficiencies of future energy conversion systems which utilize multiphase flows.

Finally, the organizers of the Japan-U.S. Seminar on Two-Phase Flow Dynamics would be remiss if they did not acknowledge the vision and foresight of Professor Arthur E. Bergles, Dean of Engineering at Rensselaer Polytechnic Institute, and Professor Seikan Ishigai, Professor Emeritus of Osaka University who jointly conceived and organized the first seminar a decade ago. It was through their efforts that this has become a successful series of quadrennial events. The second seminar in 1984 was jointly organized by Professor Koji Akagawa, Professor Emeritus of Kobe University, and Professor Owen C. Jones of Rensselaer Polytechnic Institute. In addition, the editors of this volume would also be negligent if they did not recognize the kind support of the Japanese Society for the Promotion of Science and the United States National Science Foundation, and for all institutions and companies which provided support for members of their organizations to participate in the seminar. This support provides extremely high leverage for continued collaboration and progress in the field.

In closing, the editors would like to express thanks to members of their respective organizations who provided valuable assistance and support in the planning, preparation, and conduct of the seminar. This support was the key to conducting a successful seminar and developing what we hope will be a useful compendium of research papers in the field.

Owen C. Jones
Troy, New York

Itaru Michiyoshi
Kyoto, Japan

FUNDAMENTAL EQUATIONS AND CLOSURE LAW

DISCRETE MODELING CONSIDERATIONS IN MULTIPHASE FLUID DYNAMICS*

V.H. Ransom and J.D. Ramshaw
Idaho National Engineering Laboratory
Idaho Falls, ID 83415, USA

ABSTRACT

A discussion is given of discrete modeling considerations in multiphase fluid dynamics and related areas. By the term "discrete modeling" we refer to a collection of ideas and concepts which we hope will ultimately provide a philosophical basis for a more systematic approach to the solution of practical engineering problems using digital computers. Our conception of discrete modeling is still evolving and has not yet led to useful results; thus the present paper is of the nature of a preliminary report on work in progress, and its primary purpose is to stimulate further thought and discussion. As presently constituted, the main ingredients in our discrete modeling Weltanschauung are the following considerations: (1) Any physical model must eventually be cast into discrete form in order to be solved on a digital computer. (2) The usual approach of formulating models in differential form and then discretizing them is an indirect route to a discrete model. It is also potentially hazardous: the length and time scales of the discretization may not be compatible with those represented in the model. It may therefore be preferable to formulate the model in discrete terms from the outset. (3) Computer time and storage constraints limit the resolution that can be employed in practical calculations. These limits effectively define the physical phenomena, length scales, and time scales which cannot be directly represented in the calculation and therefore must be modeled. This information should be injected into the model formulation process at an early stage. (4) Practical resolution limits are generally so coarse that traditional convergence and truncation-error analyses become irrelevant. (5) A discrete model constitutes a reduced description of a physical system, from which fine-scale details are eliminated. This elimination creates a closure problem, which has an inherently statistical character due to uncertainty about the missing details.

---

*Research supported by DOE Office of Basic Energy Sciences under DOE Contract DOE-AC07-76ID01570.

Methods from statistical physics may therefore be useful in the formulation of discrete models. In the present paper we elaborate on these themes and illustrate them with simple examples.

INTRODUCTION

Multiphase flow plays a fundamental role in a wide variety of technologically important processes, systems, and devices, including nuclear reactors and other energy systems. Consequently there is great interest in analytical models for describing multiphase flow behavior. Just as in single phase flow, there is a spectrum of modeling approaches that can be taken depending upon the application. These include the atomistic simulation of behavior at the molecular level, local instant continuum modeling, and averaged macroscale modeling. It is pertinent to this discussion to note that the evolution of these modeling approaches starts with a particle formulation in terms of ordinary differential equations at the atomistic level. Then, by averaging over systems of particles, a local instant continuum model in the form of partial differential equations (PDEs) is obtained. The macroscale model is obtained by still another averaging of the local instant continuum model. These modeling approaches applied to single phase fluid dynamics yield the equations of molecular kinetics, the Navier-Stokes equations, and the Reynolds averaged Navier-Stokes equations respectively. The primary application that we address in this paper is system modeling, and the only approach that is numerically tractable for this case, considering the current state of computer capability, is the averaged macroscale model.

The local instant or microscopic behavior of multiphase flow is accurately modeled by the Navier-Stokes formulation with the addition of interfaces across which appropriate jump conditions hold. A particularly elegant formulation of this problem has been developed by Kataoka (1986) to obtain a local instant formulation for multiphase flow analogous to the Navier-Stokes formulation for single phase flow. The Kataoka formulation can be solved in principle to obtain a microscopic description of a multiphase flow process. Unfortunately, just as in the case of the Navier-Stokes formulation, the application to large-scale engineering problems is numerically intractable. Therefore, to obtain macroscale solutions we must resort to further averaging and seek solutions for the average values of the system dependent variables. This second averaging operation introduces the need for additional closure models, and the details of the averaging methods have been the subject of debate. Nevertheless, the multiphase system models in use today are based on this approach and research efforts to improve these models are continuing.

In spite of significant research efforts within the past decade, existing macroscale multiphase flow models cannot yet be regarded as satisfactory in terms of either accuracy or predictive capability. They rely heavily on empiricism, often of an <u>ad hoc</u> variety, important physical effects are sometimes neglected altogether, and the interplay between model and numerical solution scheme is not well characterized. This reflects, of course, the immense complexity of multiphase flow, in which all of the difficulties of single phase flow and turbulence occur in conjunction with the additional complications of interfacial dynamics, flow regime transitions, interphase transport, phase transitions, and so on.

In view of the complexity of the problem, it would be unrealistic to hope that a truly comprehensive and predictive macroscale multiphase flow model will emerge in the near future, although incremental progress toward this goal can and should be expected. Unfortunately, the rate of such progress seems to have slowed in recent years. The purpose of the present paper is to discuss a somewhat different approach which we feel holds promise for further progress in macroscale modeling.

At the Japan-U.S. Seminar on Two-Phase Flow Dynamics, Ransom and Trapp (1984) discussed the importance of well-posedness and stability of two-phase flow models with respect to constructing numerical solutions. Tacit in that discussion was the assumption that a multiphase flow model for the macroscale behavior could be formulated as a system of PDEs. That being the case, obtaining a well-posed model becomes tantamount to including appropriate and sufficient physics in differential form; numerical solutions could then be obtained in a straightforward manner. In the present paper we examine the basic premise of whether it is reasonable to expect that a sufficiently complete PDE model exists for the macroscale description and whether improved modeling can be achieved by seeking missing differential terms representing additional physics of the flow. In this context, even the well-posed models that were discussed by Ransom and Trapp (1984) were not claimed to be complete physical descriptions in the limit of short-wavelength effects, but rather merely to have benign behavior at short wavelengths. These models are not equivalent to the local instant formulation, and it is not clear that further refinement of the macroscale model within the PDE framework would ever achieve such a result. We will presently argue the premise that it may be more productive to recognize practical modeling and computational limitations from the outset and construct discrete models that are compatible with the inherently discrete numerical solution process.

Our expository task is made more difficult by the fact that our views are still evolving and have not yet resulted in a specific model. Nevertheless, our contention is that insufficient consideration has been given to certain serious and fundamental difficulties with current approaches to multiphase flow modeling and computation, and that an alternative view and approach is perhaps better suited to the construction of models suitable for implementation and solution on digital computers. For want of a better term, we use "discrete modeling" to refer to the collection of ideas and concepts of which our current view is comprised and which constitute the subject of the present discussion.

The basic motivations for discrete modeling have already been summarized in the Abstract, and they will be amplified upon and the approach outlined in the subsequent sections. It is convenient to divide this discussion into four main parts. First we discuss continuum formulations and attempt to highlight some perspectives that are frequently lost sight of. Next we discuss numerical considerations, and then we outline the development of a discrete model. Finally, statistical concepts are introduced. This subdivision is somewhat artificial as there are numerous points of interconnection between the different sections. We shall attempt to call attention to these connections at appropriate places.

CONTINUUM FORMULATIONS

Within the framework of classical mechanics, the basic dynamical laws at the molecular level may be expressed as a system of ordinary differential equations for the coordinates and velocities of the molecules. A useful mathematical model is obtained by formulating the average space-time behavior of the system properties in terms of continuous functions. With this approach the physical laws then appear as PDEs that govern the behavior of the continuum. The dynamical description then consists of the PDEs, a specified initial state, and appropriate boundary conditions. This practice is commonplace in fluid mechanics and results in highly useful formulations. Only in extreme situations, such as highly rarified gases, is it necessary to resort to more fundamental particulate descriptions.

As a result of the averaging process used to obtain a continuum approximation, macroscopic closure laws are required. In the case of fluid mechanics, Newton's law of viscosity, Fourier's law of conduction, and Fick's law of diffusion are such closure laws for momentum, energy, and mass.

Continuum Limitations

It is apparent that such partial differential formulations are not exact; they are merely very useful approximations. For example, Fourier's law of heat conduction is very accurate for most applications, yet it predicts that thermal signals propagate with infinite speed, which is clearly unphysical. This is an illustration of the very general fact that <u>partial differential descriptions typically break down at very short length and time scales</u>. In particular, the familiar Navier-Stokes equations, with which we all feel so comfortable as a description of ordinary single-phase fluid dynamics, become inaccurate at sufficiently short lengths and times; the goal of generalized hydrodynamics (Mountain, 1977; Alder and Alley, 1984) is to remedy this deficiency.

In general, the length and time scales below which continuum descriptions become inaccurate are on the order of the characteristic lengths and times associated with the microstructure of the medium. For example, the Navier-Stokes equations break down, and generalized hydrodynamics is called into play, when length and time scales approach those of a molecular magnitude. Only for lengths and times much larger than characteristic molecular lengths and times do the Navier-Stokes equations provide a quantitatively correct description of fluid dynamical behavior. Fortunately, this range of lengths and times is very wide and encompasses essentially all macroscopic fluid flow phenomena. That is to say, the differential description is useful because there is a <u>wide separation of scales</u> between the smallest macroscopic scales of interest and the microscopic scales associated with the internal structure of the fluid.

If the microscales were always of a molecular magnitude then questions of scale separation would seldom arise. But in many of the descriptions employed for engineering purposes, the characteristic scales of the internal structure being modeled are themselves macroscopic in nature. For example, the microstructure in a porous medium consists of the detailed geometry of the individual pores. Flow in porous media is usually described by equations in which the effects of this microstructure are modeled in terms of distributed porosity, frictional drag forces, etc. (Whitaker, 1986). The resulting PDEs are clearly valid only on length scales much greater than a characteristic pore size, even though they in principle possess solutions on all length scales. In such situations the desired separation between the calculated and modeled scales is much less clear cut, and one must be careful not to attribute quantitative significance to any predicted solution features with scales comparable to the internal microscales.

A word is in order about the nature of the inaccuracies that occur when a continuum description is pushed too far; i.e., applied on scales too small. Ideally one hopes that such inaccuracies, while necessarily quantitative, are not qualitative or catastrophic. For example, the Navier-Stokes equations predict profiles for strong shock waves which are qualitatively reasonable even though quantitatively inaccurate. If, however, viscosity is neglected while thermal conductivity is retained, the profiles acquire a qualitatively unphysical discontinuity (Zel'dovich and Raizer, 1967).

There are also situations in which the inaccuracies of a continuum model are catastrophic in nature. In the present context, the relevant example is the one-dimensional two-phase flow equations with both viscosity and surface tension neglected, which constitute an improperly posed problem characterized by violent and unbounded instabilities at short wavelengths (Ramshaw and Trapp, 1978). These instabilities make it impossible to calculate even longer wavelength components of the solution (unless they are controlled by introducing artificial diffusion, as is often done in practice). Thus for a continuum model to be useful at all, even over a restricted range of length and time scales, it is necessary that its inaccuracies at smaller scales be benign in nature, and that it exhibit at least qualitatively reasonable behavior in spite of the inaccuracies.

In nonlinear problems there may be even more stringent requirements concerned with the necessity to prevent errors at small scales from contaminating larger scales. The obvious example is the rate of energy dissipation, which must be correctly represented in both shock waves and turbulence in spite of any other inaccuracies which may be present at small scales. Both artificial shock viscosities of the Von Neumann-Richtmyer type and subgrid scale (large eddy simulation) eddy viscosities for turbulence, when implemented conservatively, possess this essential property (Von Neumann and Richtmyer, 1950; Ramshaw, 1979). In both cases, the effect of the artificial viscosity is to artificially expand the microscale on which the energy dissipation occurs until it becomes large enough to resolve in the calculation. This expansion is done in such a way as to preserve the rate at which energy is dissipated. It seems likely that similar considerations will apply to discrete modeling in more general contexts as well.

It should be emphasized that the breakdown in continuum descriptions at small scales does not imply that the various continuous variables in the problem, such as velocities, densities, etc., become in any sense ill-defined. The implication is rather than when such variables vary too rapidly in space and/or time they no longer satisfy PDEs of a simple structure. The further implication is that it may not in fact be useful to define and deal with quantities which cannot be calculated in any simple way. That is to say, continuous quantities may not be the most natural description of the system at small length and time scales. The statistical mechanics literature provides additional support for this view. Statistical theories of continuous systems frequently encounter short-wavelength divergences (sometimes referred to as ultraviolet catastrophes by analogy to quantum electrodynamics), particularly in the treatment of thermal fluctuations of continuous quantities. These difficulties are commonly circumvented by adopting some sort of coarse-graining procedure such as dividing the system into little cells (Bedeaux et al., 1982; Gunton and Droz, 1983; Visscher, 1978, 1985), which is tantamount to replacing a continuum model by a discrete one. We shall presently argue more generally that this is indeed the simplest and most natural approach to avoid difficulties with continuum models at short scales (cf. Visscher, 1978).

In this regard, it is interesting to note that there are increasingly frequent speculations to the effect that the fundamental laws of nature on the most basic level, and perhaps even space and time themselves, may be discrete in nature (Feynman, 1982; Namsrai, 1985; Lee, 1987). However, we shall resist any temptation to draw support for our discrete modeling ideas from this source, as our motivation stems more from the scale separation considerations discussed above and from the computational considerations discussed below. For our purposes we are entirely content to accept Newton's laws for molecular motion and the Navier-Stokes equations for fine-scale fluid motion as starting points.

Multiphase Formulations

We now consider the implications of the above considerations for multifluid descriptions of multiphase flow, in which each phase is described by its own fluid dynamical variables and equations coupled appropriately to those of the other phases (Banerjee, 1986; Drew, 1983; Bedford and Drumheller, 1983; Stewart and Wendroff, 1984; Drew and Wood, 1985; Ishii, 1975). The basic idea is that these variables and equations refer to some sort of appropriate averages over the detailed fine-scale structure of the multiphase flow. For example, in a bubbly flow one may have a cloud of bubbles of various sizes and shapes, each with its own position, orientation, and velocity, and one seeks a reduced description in which only a mean velocity, void fraction, and perhaps a few bubble size distribution parameters are retained. The central point now is that the length and time scales associated with the detailed flow structures are not always widely separated from the macroscopic length and time scales on which one would like to predict the average behavior. That is to say, the desired separation of scales upon which the validity of a continuum description depends rarely obtains in multiphase flow, where the scales of the internal structures over which we desire to average may be comparable to the macroscopic scales over which the desired averaged quantities vary. This is precisely the sort of situation in which a simple partial differential description may be expected to break down. Thus it is somewhat doubtful whether PDEs for the averaged flow variables, however arrived at, can realistically be expected to accurately predict the variations of those variables on the length and time scales of interest.

A related concern emerges from a consideration of how these types of averaged multiphase flow equations are usually obtained. The usual procedure is to apply space and/or time averaging to the local instantaneous equations (Banerjee, 1986; Drew, 1983; Bedford and Drumheller, 1983; Stewart and Wendroff, 1984; Drew and Wood, 1985; Ishii, 1975). These types of averaging are subject to certain fundamental objections, which will be discussed below, but this is not the point at issue here. (A related concern is that when all the dust settles the averaging scales seem frequently to have disappeared from the model, whereas the model must in fact still depend on which scales have been modeled and which are represented directly.) The point is rather that the characteristic lengths and times over which these averages are performed clearly define lower limits for the scales over which the equations accurately represent differential variations. Yet this point is often lost sight of and it is tacitly assumed that the continuous solutions of the resulting equations are significant even at smaller scales. Of course, the hope is that if the equations are properly formulated they will not predict any structure at scales less than the averaging scales; i.e., that the solutions will in fact be smooth on those scales. Unfortunately, one has no real assurance that this in the case. The converse danger also exists, namely that the model will be solved with resolution lengths and times that <u>exceed</u> the averaging scales, so that the physical scales are insufficiently represented in the model.

A simple illustration of these considerations is provided by the separated or stratified flow of two immiscible incompressible fluids between parallel plates (Ramshaw and Trapp, 1978). Viscosity will be neglected so that potential flow theory can be applied, while surface tension will be retained to make the problem well posed. Attention is further restricted to small perturbations about a uniform steady-state solution, so that the problem may be linearized. Within these restrictions the exact dispersion relations for the two-dimensional problem are determined by (Ramshaw and Trapp, 1978)

$$\alpha_1 \rho_2 F(\alpha_2 kH)(ku_2-\omega)^2 + \alpha_2 \rho_1 F(\alpha_1 kH)(ku_1-\omega)^2$$
$$- \alpha_1 \alpha_2 \sigma H k^4 = 0 \qquad (1)$$

where α_i, ρ_i, and u_i (i=1,2) are respectively the volume fraction, density, and velocity of phase i, σ is the surface tension, H is the plate spacing, k is the wave number, ω is the angular frequency, and $F(z) = z \coth z$. This problem may also be described using the one-dimensional area-averaged two-phase flow equations (Ramshaw and Trapp, 1978), and in this description the dispersion relations are determined by

$$\alpha_1 \rho_2 (ku_2-\omega)^2 + \alpha_2 \rho_1 (ku_1-\omega)^2 - \alpha_1 \alpha_2 \sigma H k^4 = 0 \qquad (2)$$

Equations (1) and (2) are essentially equivalent provided that $kH \ll 1$; i.e., that the wavelength of the disturbance is long compared to the plate spacing. (In this limit $F(\alpha_1 kH)$ and $F(\alpha_2 kH)$ both tend to unity.) The plate spacing here plays the role of an internal characteristic length scale. When the wavelength becomes comparable to the plate spacing, the one-dimensional description becomes quantitatively inaccurate, in spite of the fact that the equations may possess solutions with structure on that scale. Because the full problem is nonlinear, there is also a danger that these inaccurate small-scale structures will feed back into and contaminate the longer wavelength components of the solution. Thus the mere fact that a differential formulation is able to generate continuous solutions on all scales does not imply that all scales are represented with equal

MULTIPHASE FLUID DYNAMICS

fidelity, or indeed that inaccuracies on small scales will not eventually contaminate the solution on larger scales as well. The smallest length scales in the problem are effectively determined by the value of the surface tension. For physical values these length scales are much shorter than the plate spacing. Thus we have a situation where the solution does in fact contain components on length scales below the internal scale, and these components are inaccurate. This can be circumvented by artificially increasing the surface tension (Ramshaw and Trapp, 1978), but again one must be concerned with the effect this may have on the accuracy of the longer wavelengths as well.

In summary, the absence in multiphase flow of a clear separation between the scales associated with the microstructure and the scales of macroscopic interest makes it somewhat doubtful that the latter can be accurately described by means of PDEs which implicitly assume that such a separation exists. In such a situation it may be more appropriate to formulate the model directly in discrete terms based on the averaging scales that one wishes to employ, e.g., by constructing control volumes of a given size and considering their contents as the discrete variables in the description. We will later elaborate on this suggestion, and argue that the size of such volumes should be defined with respect to the resolution employed in a practical numerical calculation.

COMPUTATIONAL CONSIDERATIONS

A further and purely pragmatic argument for formulating models in discrete terms from the outset is that it will eventually be necessary to cast or convert the models into discrete form in any case so that they can be solved on digital computers. From a practical point of view, what is wanted is a discrete model corresponding to the resolution that can be employed in practice for the solution of a given problem, and one would like this model to be as faithful as possible to the physics within those constraints. Formulating a differential model and then discretizing it is a rather indirect route to such a model, and it is fraught with pitfalls. Perhaps the main danger is that the length and time scales of the eventual discretization may not be compatible with those implicit in the averaging procedure on which the model was based. A further difficulty is that the inevitable discretization errors are not taken into consideration during the process of model formulation, which might conceivably have been done in such a way as to reduce them. It therefore seems sensible to eliminate the continuum middleman, so to speak, and to attempt to formulate such discrete models directly, in a single step, rather than by discretizing a continuum model whose fidelity is questionable even if it were solved exactly.

One might argue that discretization errors can be separately analyzed and dealt with by conventional techniques for doing so, such as convergence and truncation-error analyses

(Roache, 1972). The problem is that these traditional methods of assessing the size and effects of discretization errors are largely irrelevant in practical calculations because of the coarse resolution limits imposed by computer time and storage constraints. Such analyses are based on the low-order terms in Taylor series expansions in the space and time increments, and when these increments are not small the low-order terms bear little or no relation to the overall errors. Indeed, high-order difference schemes can yield highly inaccurate solutions in problems with large gradients, and conversely schemes that are formally of zeroth-order accuracy (i.e., inconsistent schemes) can nevertheless yield accurate results under certain conditions (Pike, 1987; Levermore et al., 1987). In practice one is hardly ever in a position to keep refining the resolution until the solution no longer changes, and one must seek other means of obtaining the most accurate results possible under the circumstances. In any case, there is clearly no point in examining the behavior of the system in the limit of infinitely fine resolution when the models themselves do not apply on scales less than those over which the averaging is performed in their derivations.

DISCRETE MODEL FORMULATION

Our research has not progressed to the point of a complete formulation, let alone numerical experimentation. However, based on the preceding discussion, we can now outline the approach that is envisioned. A direct derivation, or perhaps construction, of a discrete model might proceed along the following outline. Begin by dividing the system of interest into a number of computational elements (possibly control volumes), the number of which is to be determined by computer time and storage considerations. Let the discrete variables in the description be appropriate integral parameters associated with the computational elements (for example, the total mass, momentum, and energy of each phase within each control volume). Then postulate and/or derive a time advancement procedure by which the values of these discrete variables on each time level may be calculated from those on the previous time level. Ideally this procedure would be developed on the basis of appropriate accuracy- or error-based criteria to attain the maximum accuracy possible within the constraints of the discrete description adopted and for a given level of algorithmic complexity. Statistical methods might be employed for this purpose, as will be discussed below. In practice a more heuristic approach will usually be required, based on such considerations as postulated fluxes of conserved quantities between computational elements, the length scales of the various physical processes and whether they are larger or smaller than the resolution length, and so on. For example, in a bubbly flow various interphase exchange effects can be expressed in terms of the bubble size distribution, and the portion of that distribution corresponding to bubble diameters less than the resolution length would need to be modeled while larger bubbles could be represented directly.

Cascade phenomena (e.g., drop or bubble breakup), in which larger length scales evolve into scales too small to be further resolved, would clearly require special consideration. The effects of unresolved scales back on resolved ones would also require consideration and formulation. The way in which such effects are represented will depend strongly on the physics; no general statements appear possible. For example, coalescence of subgrid scale drops or bubbles may produce drops or bubbles large enough to be resolved directly, which would then need to be properly introduced into the calculation. In other circumstances a stochastic approach analogous to "eddy noise" in subgrid scale turbulence modeling (Rose, 1977; Yakhot et al., 1985) might be more appropriate. The general hope is that by focusing directly on the discrete quantities that are actually being computed, one can formulate models in which the relevant physics is represented within the constraints of the available resolution in a more harmonious manner.

It will be noted that this proposed approach bears more than a passing resemblance to the control-volume approach to deriving difference equations (Roache, 1972). Because of this resemblance, the existing one-dimensional two-phase thermal-hydraulic systems codes already possess something of the character that we think should be associated with discrete models, although many of the elements that such models should possess are simply ignored or omitted. Some of the mathematical manipulations will also bear a close resemblance to those of the conventional volume-averaging approach, in spite of the fact that there are irreconcilable philosophical differences between the two approaches. One essential difference is that in conventional volume averaging, the control volume is not considered as a fixed finite region corresponding to a discretization element, but rather as a sliding control volume which can be centered at any point in the region. Another is that ensemble averaging plays an essential role in the approach proposed here, as will be discussed in the next section.

The general philosophy underlying the approach proposed above is very much in line with that of the subgrid scale or large eddy simulation turbulence models (Ramshaw, 1979; Rose, 1977; Yakhot et al., 1985; Ferziger, 1983), the basic idea of which is that scales large enough to resolve are directly calculated while scales below the resolution are necessarily modeled. In this way minimum demands are placed upon the model, and the less that we ask of it the more likely it is to measure up to our expectations. The modeling process then largely consists of formulating subgrid scale models for physical processes occurring on scales below the resolution scales, and coupling them to the computed scales in a consistent and computationally efficient manner.

STATISTICAL CONSIDERATIONS

Averaging

It is commonly recognized that most multiphase fluid flows have a statistical character, and that the macroscopic equations describing them should be based on some appropriate averaging procedure. In spite of this, explicitly statistical considerations have played almost no role in the modeling of the various fluctuation terms that arise from the averaging. This may be symptomatic of the fact that as a general rule, there has been a tendency to become overly preoccupied with formal manipulations while neglecting the conceptual foundations of the averaging procedures themselves. In particular, the usual running space and/or time averaging procedures are not in our view well founded; the appropriate averaging procedure for general use is ensemble averaging. To be sure, lip service is sometimes paid to the idea that space and time averages are really only substitutes for ensemble averaging, to which they are hopefully equivalent by virtue of some ergodic theorem. But any such equivalence is possible only for a system which is statistically homogeneous and steady in time. In transient inhomogeneous systems, the usual running space and time averages over finite regions are not really averages at all, as they are not superpositions or weighted combinations of different realizations or possible outcomes. It is therefore not really sensible to attempt to model the fluctuation terms that thereby arise in terms of the averaged variables, for this presupposes a relation between quantities that are inherently unrelated. The values of the fluctuation terms clearly depend on information about the fine-scale details of the flow. To model these terms as functions of the averaged variables is to assume a closure where none can possibly exist--there can be no closure within a single realization. If closure occurs at all, it can occur only "on the average;" i.e., only after performing an appropriately weighted ensemble average over all possible realizations of the flow. Thus the various closure relationships which have been proposed and studied become sensible only when an underlying ensemble average is regarded as having been implicitly performed. It then becomes questionable what the space and/or time averaging originally adopted is contributing to the desired final formulation, a question that has been almost universally ignored.

Indeed, in contrast to the common view, it is not even correct to think of the usual sliding space- and time-averaging operations as discarding any information about the small-scale flow features. Such averages <u>attenuate</u> short-wavelength and high-frequency components (i.e., the amplitude of those components is reduced) but do not eliminate them (see Appendix A). The situation is analogous to early attempts to derive the Boltzmann equation from statistical mechanics using a running time average, where it was eventually recognized that the formal time smoothing operation was not having the assumed and intended effect

(Zwanzig, 1967). For an averaging operation to do its job some information must be discarded, and this is not done by sliding averages.

Yet the <u>intent</u> of the usual averaging procedures is basically sound. The intent is to somehow lump together the effects of detailed dynamical processes taking place on length and time scales too small to resolve, and this must surely be done in any practical multiphase flow model. Our contention here is that the simplest and most direct way to do this is to integrate over fixed nonsliding control volumes, and to let the ensemble averages of the integrated contents of these control volumes be the discrete variables of which the description is comprised. In contrast to a running average, an integration over fixed regions <u>does</u> discard information, and the problem acquires a legitimately statistical character because of uncertainty about the missing details. It is these missing fine-scale details over which ensemble averaging averages.

The Closure Problem

Once a conceptually well-founded averaging procedure is adopted one can confront the closure problem in a proper setting. There are two fundamental aspects to the closure problem. The first is to decide what variables are to be included in the desired macroscopic description. For example, suppose we have divided a two-phase system up into control volumes of a size corresponding to the resolution we wish to employ in a practical calculation. Based on past experience, we will probably want to include the mass, momentum, energy, and volume fraction of each phase within each control volume in our set of macroscopic variables. But this will rarely be sufficient; other parameters will be necessary to specify information about the flow regime. In a bubbly flow, for example, information equivalent to a number density of bubbles will be required to determine whether a given void fraction corresponds to many small bubbles or a few large bubbles. Information about the distribution of bubble sizes will probably also be required. Interfacial area has been suggested as an additional variable (Ishii, 1975; Ishii et al., 1982), but while it is clearly a critical parameter whose value must be known to predict interphase transfer rates, it is not clear whether it is a fundamental or a derived variable. The latter seems more likely for bubbly flow, where the bubble number density seems more fundamental.

The difficulty in identifying a complete set of macroscopic variables is forcefully illustrated by the observation that such a set of variables would permit the prediction of flow regime transitions, and this is clearly a very difficult proposition. We are usually in the less satisfactory position of dealing with a less complete set of variables which suffices to describe a restricted class of flow situations (e.g., bubbly flow). Unfortunately, to our knowledge there is no general theoretical approach to the determination of a complete set of macroscopic variables in any particular situation. This has been and remains a question on which the present statistical theories are silent, although certain insights and general guidelines have evolved. As a practical matter, one must simply introduce variables corresponding to one's conception of what is needed to characterize the important physics on the small-scale level and proceed to work out the consequences.

The second aspect of the closure problem is simpler to deal with in principle, but in practice the complications are so great that the required procedure can rarely be explicitly carried out. Once the macroscopic variables are identified, it is necessary to derive or postulate a closed system of equations which these variables satisfy. The usual approach to this problem has been the hierarchy approach (Leslie, 1973), in which one systematically derives equations for the moments of fluctuating quantities that appear in the equations for the variables of interest. These equations then contain unknown higher moments of the fluctuating quantities, for which further equations can be derived, and so on ad infinitum. This process must be truncated at a low order to obtain equations sufficiently simple to be useful, and this truncation has been an arbitrary and unsystematic process.

An alternative and more systematic approach to this aspect of the closure problem is provided by a non-hierarchial closure method based on information theory; i.e., the maximum entropy formalism (Levine and Tribus, 1979; Rosenkrantz, 1983). This formalism provides a very general way of assigning probability distributions over microstates subject to constraints embodying known macroscopic information (see Appendix B). This approach can in principle be used to derive closed reduced descriptions in a single step, without the necessity of generating or truncating an infinite hierarchy of moment equations. It has recently been used to give a formal derivation of closed nonlinear dissipative evolution equations in Hamiltonian statistical mechanics (Ramshaw, 1986a). However, further theoretical developments are needed before this approach will be applicable to multiphase flow phenomena. The Navier-Stokes equations are dissipative rather than Hamiltonian in structure, and if they are adopted as the starting point the theory will need to be generalized to a non-Hamiltonian microscopic dynamics. This will entail, among other things, addressing the well-known ambiguity in defining the appropriate volume element in state space (Ramshaw, 1986b,c). This difficulty can be circumvented by adopting the Hamiltonian equations of motion for the constituent molecules as the microscopic starting point, but one has the feeling that it should not be necessary to go all the way down to the molecular level to find a sound foundation for the theory. And even if this were done, it is not yet clear how to properly identify the "parameter of slowness" appearing in the general theory in the case of a spatially distributed system.

Work to overcome these theoretical obstacles is in progress. Even if it is fully successful, however, there will be practical limitations to the use of the results. The resulting closed descriptions will contain complex formal expressions involving the detailed microscopic dynamics in a rather intractable way; they will be essentially multiphase analogs of the familiar Green-Kubo time-correlation function expressions for molecular transport coefficients (Ramshaw, 1986a). Such expressions can yield valuable information about the structure of the theory and what variables appear in the constitutive relations, but they will seldom be amenable to quantitative evaluation. The situation is analogous to that of the ordinary molecular transport coefficients of a dense fluid, which can in principle be calculated from the intermolecular force law by means of the Green-Kubo expressions, but which in practice must be measured experimentally. Such theories will therefore never completely eliminate the need for empirical data; their proper function is rather to provide a suitable framework within which experimental data can be compiled, interpreted, and utilized, and to provide a well-founded starting point for approximations.

Stochastic Models

Our discussion so far has centered on deterministic models for the determination of ensemble averaged quantities. An alternative approach is to formulate equivalent stochastic models, in which only a single (hopefully representative) realization of the ensemble is simulated. In this approach the evolution equations for the macroscopic variables of interest contain stochastic terms, which in the present context would represent part or all of the effects of the unresolved small-scale motions back on the resolved larger-scale motions. The evolution equations then assume the character of nonlinear Langevin equations, which may be solved by stochastic simulation methods (Suzuki, 1981; Murthy, 1983; Gerling, 1984; Tartaglia and Chen, 1984; Heermann, 1986). The stochastic approach has the potential major advantage that it automatically accounts for fluctuation-renormalization effects (Zwanzig, 1980; Rodriquez and Pena-Auerbach, 1984) without the necessity for explicitly including the nonlocal terms to which such effects typically give rise. When fluctuations are large, as is frequently the case in multiphase flow, the equivalent deterministic models for ensemble averaged quantities may be expected to contain terms which are nonlocal in both space and time, in the sense that constitutive relations depend on information from an entire region of space and a finite time interval rather than just on information from the immediate neighborhood of the space-time point in question (Zwanzig, 1980; Rodriquez and Pena-Auerbach, 1984). Such nonlocal terms are very inconvenient to deal with in practical calculations, and if they turn out to be essential for accuracy then stochastic models may well be preferable to deterministic models.

CONCLUSION

As previously emphasized, our ideas on discrete modeling are still in a formative stage, and the present discussion is not intended to be the last word on the subject. We recognize that the picture we have presented is still incomplete, as we do not yet have a well-defined universal procedure that one can follow to generate models of the type we advocate. Indeed, no such procedure may exist; it may always be necessary to adopt different approaches tailored to different physical circumstances. Needless to say, at this stage we do not recommend that existing models be discarded, nor that attempts to improve them by the conventional approaches be terminated. Progress should proceed on a wide front by a variety of approaches. What we do hope is to promote a greater awareness of some of the difficulties with the current approaches and with some of the issues we have discussed, many of which do not seem to be very widely appreciated. And while we do not expect to be joined on the discrete modeling bandwagon by a mass exodus of passengers from more conventional modes of transportation, we would be pleased to hear from and interact with occasional more adventurous travelers who might consider themselves kindred spirits.

APPENDIX A

REMARKS ON SLIDING VOLUME AND TIME AVERAGES

Our purpose here is to call attention to some properties of sliding volume and time averages that are sometimes lost sight of. For brevity the mathematical development will be restricted to volume averaging, but entirely analogous results and statements hold for time averaging.

A general sliding volume average of an arbitrary function of position $f(\underline{r})$ may be defined by

$$\bar{f}(\underline{r}) = \int d\underline{r}'\, W(\underline{r}-\underline{r}')\, f(\underline{r}') \tag{A-1}$$

where $W(\underline{r})$ is a weighting function. The function $W(\underline{r})$ is assumed to be everywhere positive, to be localized in the vicinity of $\underline{r} = 0$ (and therefore to tend toward zero for large $|\underline{r}|$), to have a characteristic width on the order of the desired spatial averaging length L, and to satisfy

$$\int d\underline{r}\, W(\underline{r}) = 1 \tag{A-2}$$

The latter property ensures that when $f(\underline{r})$ is uniform

MULTIPHASE FLUID DYNAMICS 19

its value is not changed by the averaging. Frequently $W(\underline{r})$ is taken to have the simple box or step function form $W(\underline{r}) = (1/L^3) \, B(x)B(y)B(z)$, where $B(s)$ is unity for $-L/2 < s < L/2$ and zero otherwise.

Since Eq. (A-1) is of convolution form, it may be Fourier transformed to yield

$$\hat{\bar{f}}(\underline{k}) = \hat{W}(\underline{k}) \, \hat{f}(\underline{k}) \tag{A-3}$$

where $\hat{F}(\underline{k})$ denotes the Fourier transform of $F(\underline{r})$. As is well known, $\hat{W}(\underline{k})$ will have a characteristic width of order $1/L$. Now even if $W(\underline{r})$ is taken to have the simple box form given above, its Fourier transform $\hat{W}(\underline{k})$ will in general be nonzero for all finite values of $\underline{k}$, although it must of course tend to zero as $|\underline{k}|$ tends to infinity. Equation (A-3) may therefore be solved for $\hat{f}(\underline{k})$ from a knowledge of $\hat{\bar{f}}(\underline{k})$, so that $f(\underline{r})$ may be determined from a knowledge of its spatial average $\bar{f}(\underline{r})$. The sliding spatial averaging of Eq. (A-1) therefore does not discard or suppress any information contained in $f(\underline{r})$. In particular, it does not remove or destroy the short-wavelength components; it merely attenuates them. The averaged function $\bar{f}(\underline{r})$ contains exactly the same information as the unaveraged function $f(\underline{r})$. It therefore cannot be expected to satisfy equations of any simpler structure than $f(\underline{r})$, and moreover it clearly cannot be assumed to be smooth on scales less than the averaging length L unless the original function $f(\underline{r})$ was already smooth on those scales. For an averaging procedure to have the desired effect it must discard some information, and since the present procedure does not do so it cannot be regarded as a satisfactory basis for multiphase flow modeling.

Two remedies for this difficulty suggest themselves. The first is to define $W(\underline{r})$ in such a way that its Fourier transform is a box function in $\underline{k}$-space; e.g., $\hat{W}(\underline{k}) = (\text{const.})D(k_x)D(k_y)D(k_z)$, where $D(s)$ is unity for $-2/L < s < 2/L$ and zero otherwise. Equation (A-3) then shows that the short-wavelength components of $f(\underline{r})$ are actually removed by the averaging, and that $f(\underline{r})$ can no longer be reconstructed from a knowledge of its average $\bar{f}(\underline{r})$. But now a new difficulty appears; when the inverse transform of this $\hat{W}(\underline{k})$ is taken it is found to have negative regions, so it is not acceptable as a weighting function.

The second remedy is to no longer permit the volumes over which the averaging is performed to slide continuously around, but rather to fix them in space; e.g., to identify them with the cells in a fixed grid. The spatial averaging then no longer has the convolution form of Eq. (A-1) and the above objections no longer apply. In particular, integration over such fixed cells clearly discards information about the fine-scale structure contained within those cells, as we would like an averaging procedure to do.

Nevertheless, spatial averaging over fixed cells is still not sufficient in itself for the reasons already discussed. It is still not really an averaging in the sense that it is applied only to a single realization of the flow and is not a weighted combination of different possible realizations or outcomes. It is therefore more appropriate to think of it as a lumping procedure; i.e., an operational procedure for generating integral variables corresponding to the continuous variables of the fine-scale differential description. As discussed in the text, it is still necessary to ensemble average these integral variables before one can have any hope that they constitute a closed macroscopic description of the system.

APPENDIX B

OUTLINE OF MAXIMUM ENTROPY CLOSURE

Here we wish to indicate how the maximum entropy formalism (Levine and Tribus, 1979; Rosenkrantz, 1983) provides a very simple route to closure in terms of whatever macroscopic variables are utilized to describe a physical system. We consider a system whose microstate is denoted by $\underline{X}$ and whose relevant macroscopic variables are $\underline{A}(\underline{X}) = (A_1(\underline{X}), A_2(\underline{X}), ...)$. The ensemble average values of the variables $\underline{A}(\underline{X})$ are presumed to be known and are denoted by $\underline{a} = (a_1, a_2, ...)$. The maximum entropy formalism supplies the least biased or maximally noncommittal (in the sense of information theory) probability distribution in $\underline{X}$ consistent with the specified values of $\underline{a}$. This distribution is of the generalized canonical form

$$\rho(\underline{X}; \underline{a}) = Q^{-1}(\underline{a}) \, g(\underline{X}) \, \exp[\underline{\gamma}(\underline{a}) \cdot \underline{A}(\underline{X})] \qquad (B-1)$$

where the partition function $Q(\underline{a})$ is given by

$$Q(\underline{a}) = \int d\underline{X} \, g(\underline{X}) \, \exp[\underline{\gamma}(\underline{a}) \cdot \underline{A}(\underline{X})] \qquad (B-2)$$

and the Lagrange multipliers $\underline{\gamma}(\underline{a})$ are functions of $\underline{a}$ implicitly determined by the requirement that

$$\int d\underline{X} \, \rho(\underline{X}; \underline{a}) \, \underline{A}(\underline{X}) = \underline{a} \qquad (B-3)$$

MULTIPHASE FLUID DYNAMICS

Here $g(\underline{X})$ is the probability distribution of complete ignorance (Levine and Tribus, 1979; Rosenkrantz, 1983); i.e., the distribution that would be assigned by an unbiased observer with no knowledge whatever of $\underline{X}$. It is not always clear how to determine $g(\underline{X})$, but in many cases it may be found by group-invariance arguments (Levine and Tribus, 1979; Rosenkrantz, 1983), the rough idea of which is that truly complete ignorance should remain complete under changes of viewpoint. The distribution of Eq. (B-1) maximizes the information entropy

$$S = - \int d\underline{X}\, \rho(\underline{X};\underline{a})\, \ln\, [\rho(\underline{X};\underline{a})/g(\underline{X})] \tag{B-4}$$

subject to the constraints of Eq. (B-3).

Since $\rho(\underline{X};\underline{a})$ depends parametrically on $\underline{a}$, so will any average evaluated using it; i.e.,

$$\langle F(\underline{X}) \rangle = \int d\underline{X}\, \rho(\underline{X};\underline{a})\, F(\underline{X}) \equiv f(\underline{a}) \tag{B-5}$$

The distribution $\rho(\underline{X};\underline{a})$ therefore provides a method of closure; i.e., for expressing the average f of any desired microscopic quantity $F(\underline{X})$ as a function of the relevant macroscopic variables $\underline{a}$. Further, this closure is effected in a single step, without the necessity of forming an infinite hierarchy of moment equations and arbitrarily truncating it.

This procedure is appealingly simple in concept, but it is by no means a straightforward matter to apply it to two-phase flow modeling. There is as yet no satisfactory general procedure for identifying $g(\underline{X})$, and there are mathematical difficulties in dealing with microstates described in terms of continuous fields. Moreover, in dynamical problems one does not strictly have the freedom to apply the maximum entropy formalism at different times during a transient (although it may nevertheless be a useful approximation to do so), because the time evolution of the probability distribution $\rho(\underline{X})$ is governed by a Liouville equation (Ramshaw, 1986a). The maximum entropy distribution may be invoked only as an initial condition, following which the evolution proceeds according to the microscopic dynamics of the system. This circumstance is in fact the essential reason why nonequilibrium statistical mechanics is more difficult than, and differs in structure from, equilibrium statistical mechanics.

In spite of these highly nontrivial difficulties, the fact that the maximum entropy formalism effects a systematic non-hierarchial closure in a single step makes it an appealing alternative to more traditional hierarchy procedures. It therefore seems worthy of further consideration in connection with multiphase flow modeling.

NOMENCLATURE

English

$\underset{\sim}{A}$	=	macroscopic system variables
$\underset{\sim}{a}$	=	ensemble average of macrostate variables
$B(x)$	=	box function in position space
$D(k)$	=	box function in Fourier transform space
$F(z)$	=	z coth z, or arbitrary function
$f(x)$	=	arbitrary function
$g(\underset{\sim}{X})$	=	probability distribution of complete ignorance
H	=	plate spacing
k	=	wave number
L	=	spatial averaging length
Q	=	partition function
$\underset{\sim}{r}$	=	position variable
S	=	information entropy
s	=	dummy variable
u	=	velocity
$W(\underset{\sim}{r})$	=	weighting function
$\underset{\sim}{X}$	=	microscopic variables
x	=	spatial coordinate
y	=	spatial coordinate
z	=	dummy variable, or spatial coordinate

Greek

α	=	volume fraction
$\underset{\sim}{\gamma}$	=	Lagrangian multipliers
ρ	=	density or probability distribution
σ	=	surface tension
ω	=	complex angular frequency

Subscripts

i	=	phase index: liquid=1, vapor=2
x,y,z	=	denotes components of wave vector

Overscore

$-$	=	spatial average
$\wedge$	=	Fourier transform

Underscore

$\sim$	=	vector quantity

Other

$<\ >$	=	ensemble average over microstates

BIBLIOGRAPHY

Alder, B. J. and Alley, W. E., 1984, *Physics Today*, January, p. 56.

Banerjee, S., 1986, "Multifield Methods for Nuclear Thermalhydraulics Problems," Proceedings of the Second International Conference on Simulation Methods in Nuclear Engineering (Canadian Nuclear Society, Montreal, 14-16 October), p. 824.

Bedeaux, D., Mazur, P., and Van Saarloos, W., 1982, Physica $\underline{112A}$, 514.

Bedford, A. and Drumheller, D. S., 1983, Int. J. Engng. Sci. $\underline{21}$, 863.

Drew, D. A., 1983, Ann. Rev. Fluid Mech. $\underline{15}$, 261.

Drew, D. A. and Wood, R. T., 1985, "Overview and Taxonomy of Models and Methods for Workshop on Two-Phase Flow Fundamentals," International Workshop on Two-Phase Flow Fundamentals, National Bureau of Standards, Gaithersburg, MD (September).

Ferziger, J. H., 1983, "Higher-Level Simulations of Turbulent Flows, in *Computational Methods for Turbulent, Transonic, and Viscous Flows*, edited by J. A. Essers (Hemisphere, Washington), and references cited therein.

Feynman, R. P., 1982, Int. J. Theo. Phys. $\underline{21}$, 467.

Gerling, R. W., 1984, Z. Phys. B $\underline{55}$, 263.

Gunton, J. D. and Droz, M., 1983, *Introduction to the Theory of Metastable and Unstable States* (Springer-Verlag, Lecture Notes in Physics, Vol. 183, Berlin).

Heermann, D. W., 1986, *Computer Simulation Methods* (Springer-Verlag, Berlin).

Ishii, M., 1975, *Thermo-Fluid Dynamic Theory of Two-Phase Flow* (Eyrolles, Paris).

Ishii, M., Mishima, K., Kataoka, I. and Kocamustafaogullari, G., 1982, "Two-Fluid Model and Importance of the Interfacial Area in Two-Phase Flow Analysis," Proceedings of the Ninth U. S. National Congress of Applied Mechanics (ASME, New York, ca.), p. 73.

Kataoka, I., 1986, Int. J. Multiphase Flow $\underline{12}$, 745.

Lee, T. D., 1987, J. Stat. Phys. $\underline{46}$, 843.

Leslie, D. C., 1973, *Developments in the Theory of Turbulence* (Oxford University Press).

Levermore, C. D., Manteuffel, T. A. and White, A. B., Jr., 1987, "Numerical Solution of Partial Differential Equations on Irregular Grids," International Symposium on Computational Fluid Dynamics and International Conference on Computational Techniques and Applications, Sydney, Australia, 24-27 August; Los Alamos National Laboratory Unclassified Release LA-UR-87-2842.

Levine, R. D. and Tribus, M., eds., (1979) *The Maximum Entropy Formalism* (MIT Press, Cambridge, MA).

Mountain, R. D., 1977, Advan. Mol. Relax. Proc. **9**, 225.

Murthy, K. P. N., 1983, in *Stochastic Processes--Formalism and Applications* (Springer-Verlag, Lecture Notes in Physics, Vol. 184, Berlin), p. 116.

Namsrai, Kh., 1985, Int. J. Theo. Phys. **24**, 741.

Pike, J., 1987, J. Comput. Phys. **71**, 194.

Ramshaw, J. D. and Trapp, J. A., 1978, Nucl. Sci. Eng. **66**, 93.

Ramshaw, J. D., 1979, "Alternate Interpretation of the Subgrid Scale Eddy Viscosity," Los Alamos Scientific Laboratory Report LA-7955-MS (August).

Ramshaw, J. D., 1986a, J. Stat. Phys. **45**, 983.

Ramshaw, J. D., 1986b, Phys. Lett. A **116**, 110.

Ramshaw, J. D., 1986c, Phys. Lett. A **117**, 172.

Ransom, V. H. and Trapp, J. A., 1984, "Well-Posedness and Stability Characteristics of Multi-Phase Models," Proceedings of the Japan-U.S. Seminar on Two-Phase Flow Dynamics, Lake Placid, NY, USA, July 29 to August 3.

Roache, P. J., 1972, *Computational Fluid Dynamics* (Hermosa, Albuquerque).

Rodriguez, R. F. and Pena-Auerbach, L. de la, 1984, Physica **123A**, 609.

Rose, H. A., 1977, J. Fluid Mech. **81**, 719.

Rosenkrantz, R. D., ed., 1983, *E. T. Jaynes: Papers on Probability, Statistics and Statistical Physics* (Reidel, Dordrecht).

Stewart, H. B. and Wendroff, B., 1984, J. Comput. Phys. **56**, 363.

Suzuki, M., 1981, Advan. Chem. Phys. **46**, 195.

Tartaglia, P. and Chen, S.-H., 1984, J. Stat. Phys. <u>35</u>, 171.

Visscher, P. B., J. Stat. (1978) Phys. <u>18</u>, 59.

Visscher, P. B., 1985, J. Stat. Phys. <u>38</u>, 989.

Von Neumann, J. and Richtmyer, R. D., 1950, J. Appl. Phys. <u>21</u>, 232.

Whitaker, S., 1986, Transp. Porous Media <u>1</u>, 3.

Yakhot, V., Orszag, S. A. and Pelz, R. B., 1985, "Renormalization Group-Based Subgrid Scale Turbulence Closures," in <u>Ninth International Conference on Numerical Methods in Fluid Dynamics</u> (Springer-Verlag, Lecture Notes in Physics, Vol. 218, Berlin), p. 592.

Zel'dovich, Ya. B. and Raizer, Yu. P., 1967, <u>Physics of Shock Waves and High-Temperature Hydrodynamic Phenomena</u>, Vol. II (Academic, New York).

Zwanzig, R. W., editor, 1967, <u>Selected Topics in Statistical Mechanics</u> (John Gamble Kirkwood Collected Works, Gordon and Breach, New York), pp. xiv-xv.

Zwanzig, R., 1980, in <u>Systems Far from Equilibrium</u>, L. Garrido, ed., (Springer-Verlag, Lecture Notes in Physics, Vol. 132, Berlin), p. 198.

CLOSURE PROBLEMS AND BASIC EQUATIONS FOR CORRELATION FUNCTIONS IN TWO-PHASE FLOW

ISAO KATAOKA
Institute of Atomic Energy
Kyoto University
Uji, Kyoto 611 Japan

ABSTRACT

Closure problems in two-phase basic equations are discussed for local instant and averaged formulations. For the local instant formulation, the equation of interfacial movement is necessary to close the formulation. The averaged formulation is obtained by directly averaging the local instant formulation. As the result of averaging, several correlative terms arise that are composed of turbulence terms, interfacial momentum transfer terms and the correlative terms between characteristic function and velocity fluctuations. Since the constitutive equations of first two terms are known, special attention is paid to the last terms. In order to close the averaged formulation, the conservation equations for the last terms are derived, based on the conservation equation of momentum fluctuation.

INTRODUCTION

In analyzing two-phase flow phenomena, it is indispensable to accurately formulate the conservation equations of two-phase flow. A number of formulations have been developed so far. Among these, the most detailed one is the two-fluid model (or separate flow model) formulation where the conservation equations of each phase are considered separately (Wallis, 1969; Ishii, 1975; Delhaye, 1968).

This formulation is given in appropriately averaged forms. As a result of averaging, it is not closed, which means the number of unknown parameters exceeds the number of equations. In order to close the formulation, constitutive equations for turbulence transport terms and interfacial transport terms are necessary. However, when these constitutive equations are given, the formulation is still not closed. Usually, the assumed relation between the averaged pressure of each phase is given in order to finally close the formulation.

This relation for pressure has a different nature from the constitutive equations for turbulence and interfacial momentum transports. The turbulence and interfacial transport terms come directly from averaging. On one hand, it is natural to need constitutive equations. On the other hand, the pressure relation is not naturally required by the averaging process. It is arbitrarily given in order to close the formulation.

Due to this arbitrariness, the relation for pressure has certain problems. One problem is that in some cases the pressure relation makes the formulation ill-posed, from which may arise the serious problem of the existence of stable solutions. The other problem is that it is somewhat difficult to make a reliable relation for the pressures based on physical models, particularly in the case of multi-dimensional and/or microscopic analysis of two-phase flow. This is because the pressure of each phase is complicatedly related to average velocity, turbulent velocity, their gradients, and interfacial movements. One method which avoids the above-mentioned problems is to derive the averaged formulation which does not require the pressure relation. This is the subject of this paper.

Starting from the local instant formulation, the averaged basic equations which do not require the pressure relation have been derived. These equations include the conservation equations for correlation functions between characteristic function and velocity fluctuations.

For the convenience of discussion, the most simple case is treated here, where in both phases are incompressible and there are no phase changes and no heat exchange between phases. Therefore, only mass and momentum equations are considered.

LOCAL INSTANT FORMULATION AND ITS CLOSURE PROBLEM

There are two methods to derive the averaged formulation of two-phase flow. One is to consider a certain control volume which includes gas liquid interfaces and to derive the averaged conservation equations with help of integral theorems. The other is to directly average the local instant (deterministic) conservation equations of two-phase flow.

In this paper, the latter method is adopted because this method makes it more convenient to consider closure problems and to derive the conservation equations of correlation functions. In order to make direct averaging possible, the local instant formulation must be given in the form of field equations, which can be defined in all the time and space domains that are under consideration. Such field equations can be given in terms of the characteristic function of each phase and the conservation equations within each phase (Kataoka, 1986a).

CORRELATION FUNCTIONS

(mass)
$$\phi_k \frac{\partial v_{k\alpha}}{\partial x_\alpha} = 0 \quad (k=1,2),\tag{1}$$

(momentum)
$$\phi_k \frac{\partial v_{k\alpha}}{\partial t} + \phi_k \frac{\partial}{\partial x_\beta}(v_{k\alpha} v_{k\beta}) = -\phi_k \frac{1}{\rho_k}\frac{\partial P_k}{\partial x_\alpha} + \phi_k \frac{1}{\rho_k}\frac{\partial \tau_{k\alpha\beta}}{\partial x_\beta} + \phi_k F_{k\alpha} \quad (k=1,2).\tag{2}$$

Here, suffix k denotes each phase and suffixes α,β,γ denote the components of a three dimensional Cartesian coordinate. For suffixes α,β,γ, the Einstein summation rule is applied whereas for suffix k it is not applied.

ϕ_k is the characteristic function of each phase. It takes value of unity within phase k and otherwise takes the null value zero (Kataoka, 1986a). It is related to the configuration of the gas-liquid interface and is governed by the equation of interfacial movement which is given by (Kataoka, 1986a; Kataoka & Serizawa, 1987)

$$\frac{\partial \phi_k}{\partial t} + v_{k\alpha}\frac{\partial \phi_k}{\partial x_\alpha} = 0 \quad (k=1,2).\tag{3}$$

Here, the following relations are used under the assumption of no phase change.

$$v_{k\alpha}\frac{\partial \phi_k}{\partial x_\alpha} = v_{k\alpha i}\frac{\partial \phi_k}{\partial x_\alpha} \quad (k=1,2),\tag{4}$$

$$v_{1\alpha i} = v_{2\alpha i} = v_{i\alpha}.\tag{5}$$

Here, $v_{k\alpha i}$ represents the velocity of phase k at the interface and $v_{i\alpha}$ represents the velocity of the interface. Since characteristic functions satisfy

$$\phi_1 + \phi_2 = 1,\tag{6}$$

the two equations in Eq. (3)(k=1,2) are dependent on each other[1].

External force $F_{k\alpha}$ is usually given as a known field quantity and stress tensor $\tau_{k\alpha\beta}$ is given in terms of velocity gradients for Newtonian fluids. As already assumed, the density of each phase ρ_k is a constant and known quantity.

Under these circumstance, the local instant basic equations mentioned above are closed, because the number of independent unknown quantities is five ($\phi_1, v_{1\alpha}, v_{2\alpha}, P_1, P_2$) and the number of independent equations is five (Eq.(1) for k=1,2;

Eq.(2) for k=1,2; and Eq.(3) for k=1).

Using Eq. (1), Eq. (3) can be rewritten by

$$\frac{\partial \phi_k}{\partial t} + \frac{\partial}{\partial x_\alpha}(\phi_k v_{k\alpha}) = 0 \quad (k=1,2). \tag{7}$$

There are four equations in Eqs. (1) and (7). However, since one of them is dependent upon the others (footnote 1), only three equations in Eqs. (1) and (7) are independent. As easily shown (see Appendix A), any combination of the three independent equations is possible.

With some modifications of momentum equations using Eqs. (4) and (7), one obtains the following set of local instant conservation equations.

$$\phi_k \frac{\partial v_{k\alpha}}{\partial x_\alpha} = 0 \quad (k=1,2), \tag{8}$$

$$\frac{\partial}{\partial t}(\phi_k v_{k\alpha}) + \frac{\partial}{\partial x_\beta}(\phi_k v_{k\alpha} v_{k\beta}) = -\phi_k \frac{1}{\rho_k} \frac{\partial P_k}{\partial x_\alpha} + \phi_k \frac{1}{\rho_k} \frac{\partial \tau_{k\alpha\beta}}{\partial x_\beta}$$

$$+ \phi_k F_{k\alpha} \quad (k=1,2), \tag{9}$$

$$\frac{\partial \phi_k}{\partial t} + \frac{\partial}{\partial x_\alpha}(\phi_k v_{k\alpha}) = 0 \quad (k=1,2). \tag{10}$$

It should be noted again that the number of independent equations in this formulation is five (one equation is dependent upon others).

AVERAGED FORMULATION AND ITS CLOSURE PROBLEM

Averaged formulation can be obtained directly by averaging the local instant formulation, Eqs. (8) through (10). Three types of averaged terms appear, depending upon the relation between the characteristic function and the derivatives.

---

[1]footnote

$$\frac{\partial \phi_1}{\partial t} + v_{1\alpha} \frac{\partial \phi_1}{\partial x_\alpha} = \frac{\partial \phi_1}{\partial t} + v_{1\alpha i} \frac{\partial \phi_1}{\partial x_\alpha} \quad \text{(from Eq.(4))}$$

$$= \frac{\partial \phi_1}{\partial t} + v_{2\alpha i} \frac{\partial \phi_1}{\partial x_\alpha} \quad \text{(from Eq.(5))}$$

$$= -\frac{\partial \phi_2}{\partial t} - v_{2\alpha i} \frac{\partial \phi_2}{\partial x_\alpha} \quad \text{(from Eq.(6))}$$

$$= -(\frac{\partial \phi_2}{\partial t} + v_{2\alpha} \frac{\partial \phi_2}{\partial x_\alpha}) \quad \text{(from Eq.(4))}$$

CORRELATION FUNCTIONS

They are

$$\overline{\frac{\partial}{\partial t}(\phi_k A_k)}, \quad \overline{\frac{\partial}{\partial x_\alpha}(\phi_k A_k)}, \quad \text{and} \quad \overline{\phi_k \frac{\partial}{\partial x_\alpha} A_k}.$$

In the first two-terms, it can be easily shown that averaging and differentiation are exchangeable(Kataoka & Serizawa, 1987). That is

$$\overline{\frac{\partial}{\partial t}(\phi_k A_k)} = \frac{\partial}{\partial t}\overline{(\phi_k A_k)} = \frac{\partial}{\partial t}(\overline{\phi_k}\,\overline{\bar{A}_k}), \tag{11}$$

$$\overline{\frac{\partial}{\partial x_\alpha}(\phi_k A_k)} = \frac{\partial}{\partial x_\alpha}\overline{(\phi_k A_k)} = \frac{\partial}{\partial x_\alpha}(\overline{\phi_k}\bar{A}_k). \tag{12}$$

Here, $\bar{\bar{A}}_k$ is the phase weighted average of A_k which is defined by

$$\bar{\bar{A}}_k = \overline{\phi_k A_k} / \overline{\phi_k}. \tag{13}$$

The last term is given by (see Appendix B)

$$\overline{\phi_k \frac{\partial A_k}{\partial x_\alpha}} = \overline{\phi_k}\frac{\partial \bar{\bar{A}}_k}{\partial x_\alpha} + \frac{\partial}{\partial r_\alpha} R(\phi_k', A_k')\Big|_{\mathbf{r}=0}, \tag{14}$$

$$R(\phi_k', A_k') = \overline{\phi_k'(\mathbf{x})\phi_k(\mathbf{x+r})A_k'(\mathbf{x+r})} - \overline{\phi_k(\mathbf{x})\phi_k'(\mathbf{x+r})A_k'(\mathbf{x})}. \tag{15}$$

Here, ϕ_k' and A_k' are the fluctuating terms and they are defined by

$$\phi_k' = \phi_k - \overline{\phi_k}, \tag{16}$$

$$A_k' = A_k - \bar{\bar{A}}_k. \tag{17}$$

Note that ϕ_k' is the field quantity which is defined in all the time and space domains under consideration, whereas A_k' is not. To make A_k' a field quantity, it is multiplied by ϕ_k, as in

$$\phi_k A_k'$$

This situation is quite different from single phase flow where all fluctuating variables are field quantities.

Using the relations, Eqs. (11) through (15), one can average the local instant formulation, Eqs. (8) through (10). The result is

$$\overline{\phi_k}\frac{\partial \bar{\bar{v}}_{k\alpha}}{\partial x_\alpha} = -\frac{\partial}{\partial r_\alpha} R(\phi_k', v_{k\alpha}')\Big|_{\mathbf{r}=0} \quad (k=1,2), \tag{18}$$

$$\frac{\partial}{\partial t}(\overline{\phi_k \bar{\bar{v}}_{k\alpha}}) + \frac{\partial}{\partial x_\beta}(\overline{\phi_k \bar{\bar{v}}_{k\alpha} \bar{\bar{v}}_{k\beta}}) = -\overline{\phi_k \frac{1}{\rho_k}} \frac{\partial \bar{\bar{P}}_k}{\partial x_\alpha} + \overline{\phi_k \frac{1}{\rho_k}} \frac{\partial \bar{\bar{\tau}}_{k\alpha\beta}}{\partial x_\beta} + \overline{\phi_k \bar{\bar{F}}_{k\alpha}}$$

$$- \frac{\partial}{\partial x_\beta}(\overline{\phi_k \overline{v'_{k\alpha} v'_{k\beta}}}) - \frac{1}{\rho_k} \frac{\partial}{\partial x_\alpha} R(\phi'_k, P'_k)\big|_{\mathbf{r}=0}$$

$$+ \frac{1}{\rho_k} \frac{\partial}{\partial r_\beta} R(\phi'_k, \tau'_{k\alpha\beta})\big|_{\mathbf{r}=0} \qquad (k=1,2) \qquad (19)$$

$$\frac{\partial \overline{\phi_k}}{\partial t} + \frac{\partial}{\partial x_\alpha}(\overline{\phi_k \bar{\bar{v}}_{k\alpha}}) = 0 \qquad (k=1,2) . \qquad (20)$$

This formulation includes five independent equations (one equation is dependent upon the others)[2].

The independent unknown quantities are $\overline{\phi}_1$ (in view of Eq. (6)), $\bar{\bar{v}}_{1\alpha}$, $\bar{\bar{v}}_{2\alpha}$, $\bar{\bar{P}}_1$ and $\bar{\bar{P}}_2$. In addition to them, as a result of averaging, the correlative terms

$$\overline{v'_{k\alpha} v'_{k\beta}} \quad (k=1,2), \quad R(\phi'_k, v'_{k\alpha}) \quad (k=1,2), \quad R(\phi'_k, P'_k) \quad (k=1,2),$$

and $\quad R(\phi'_k, P'_k) \quad (k=1,2)$

appear as unknown quantities. Therefore, in order to close the averaged formulation, the constitutive equations for above-mentioned correlative terms are necessary. It should be noted that here the pressure relation is not needed to close the formulation and the requirements for constitutive equations naturally come from the averaging process.

Among the correlative terms mentioned above, Reynolds stress

$$\overline{v'_{k\alpha} v'_{k\beta}} \quad (k=1,2)$$

can be treated in a similar way to single phase flow. There have been several constitutive equations based on diffusion models (Sato & Sekoguchi, 1975). A more precise method to determine this term is based on the conservation equations of turbulence. Recently, the basic equations of turbulent energy and dissipation have been derived (Kataoka & Serizawa, 1987; Serizawa & kataoka, 1987; Lahey, 1987) and a numerical analysis was carried out (Lahey, 1987).

----

[2] Averaging Eq.(E-1)(Appendix A), one obtains

$$\frac{\partial \overline{\phi_1}}{\partial t} + \frac{\partial}{\partial x_\alpha}(\overline{\phi_1 \bar{\bar{v}}_{1\alpha}}) = -\left\{ \frac{\partial \overline{\phi_2}}{\partial t} + \frac{\partial}{\partial x_\alpha}(\overline{\phi_2 \bar{\bar{v}}_{2\alpha}}) \right\} .$$

Therefore, two equations in Eq. (20)(k=1,2) are dependent upon each other.

CORRELATION FUNCTIONS

The correlation functions between fluctuations of characteristic function, pressure and stress tensor

$$\frac{\partial}{\partial r_\alpha} R(\phi_k', P_k')\Big|_{\mathbf{r}=0} \quad (k=1,2), \quad \text{and} \quad \frac{\partial}{\partial r_\beta} R(\phi_k', \tau_{k\alpha\beta}')\Big|_{\mathbf{r}=0} \quad (k=1,2)$$

can be related to interfacial momentum transfer terms (see Appendix C). Reliable constitutive equations for these terms have already been developed based on experiments and physical modeling (Wallis, 1969; Ishii, 1975; Delhaye, 1968; Delhaye, 1984; Ishii & Chawla, 1979).

The remaining correlative terms to be determined are the correlative terms given by

$$\frac{\partial}{\partial r_\alpha} R(\phi_k', v_{k\alpha}')\Big|_{\mathbf{r}=0} \quad (k=1,2).$$

These terms are considered to represent phase exchange due to velocity fluctuation. The constitutive equations for these terms have not been developed yet. Since these quantities are related to minute structures of interfacial movement and turbulence, it seems somewhat difficult to make a measurement of these terms and to construct a simple physical model for them with current knowledge. Therefore, an attempt will be made here to develop governing equations for these correlative terms.

Since these terms are composed of correlation functions between fluctuating terms of velocity and characteristic functions, the governing equations of them can be derived from the conservation equations of momentum and interfacial movements (Eqs. (9) and (19)) for fluctuating parts.

CONSERVATION EQUATIONS OF FLUCTUATING PARTS AND THEIR CORRELATION FUNCTIONS

Conservation equations of mass, momentum and interfacial movement for fluctuating parts can be obtained by subtracting the averaged equations (Eqs. (18) through (20)) from the local instant equations (Eqs. (8) through (10)). With some modifications (see Appendix D), they are finally given by

$$\overline{\phi_k \frac{\partial v_{k\alpha}'}{\partial x_\alpha}} = \frac{\overline{\phi_k}}{\overline{\phi_k}} \frac{\partial}{\partial r_\alpha} R(\phi_k', v_{k\alpha}')\Big|_{\mathbf{r}=0} \quad (k=1,2) \tag{21}$$

$$\phi_k \frac{\partial v'_{k\alpha}}{\partial x_\alpha} + \phi_k \frac{\partial}{\partial x_\beta}(v'_{k\alpha}\bar{\bar{v}}_{k\beta} + \bar{\bar{v}}_{k\alpha}v'_{k\beta} + v'_{k\alpha}v'_{k\beta})$$

$$= -\phi_k \frac{1}{\rho_k}\frac{\partial P'_k}{\partial x_\alpha} + \phi_k \frac{1}{\rho_k}\frac{\partial \tau'_{k\alpha\beta}}{\partial x_\beta} + \phi'_k F'_{k\alpha} + \bar{\bar{v}}_{k\alpha}\frac{\phi_k}{\phi_k}\frac{\partial}{\partial x_\beta}R(\phi'_k, v'_{k\beta})\big|_{r=0}$$

$$+ \frac{\phi_k}{\phi_k}\frac{\partial}{\partial x_\beta}(\overline{\phi'_k\, v'_{k\alpha}v'_{k\beta}}) + \frac{\phi_k}{\phi_k}\frac{1}{\rho_k}\frac{\partial}{\partial r_\alpha}R(\phi'_k, P'_k)\big|_{r=0}$$

$$- \frac{\phi_k}{\phi_k}\frac{1}{\rho_k}\frac{\partial}{\partial x_\beta}R(\phi'_k, \tau'_{k\alpha\beta})\big|_{r=0} \quad (k=1,2) \tag{22}$$

$$\frac{\partial \phi'_k}{\partial t} + \frac{\partial}{\partial x_\alpha}(\phi_k v'_{k\alpha} + \phi'_k \bar{\bar{v}}_{k\alpha}) = 0 \quad (k=1,2). \tag{23}$$

Equations (21) through (23) are the conservation equations at point **x**. Therefore, multiplying Eqs. (21) through (23) by $\phi'_k(\mathbf{x+r})$ and averaging them, one obtains conservation equations for correlation functions.

As discussed in the previous section, what is needed here are the equations which determine

$$\frac{\partial}{\partial r_\alpha}R(\phi'_k, v'_{k\alpha})\big|_{r=0} \quad (k=1,2).$$

These two terms are dependent each upon other because averaging Eq. (A-2) and in view of Eq. (14), one obtains

$$\frac{\partial}{\partial r_\alpha}R(\phi'_2, v'_{2\alpha})\big|_{r=0} = \frac{\partial}{\partial r_\alpha}R(\phi'_1, v'_{1\alpha})\big|_{r=0} + \overline{\phi_1}\frac{\partial \bar{\bar{v}}_{1\alpha}}{\partial x_\alpha} - \overline{\phi_2}\frac{\partial \bar{\bar{v}}_{2\alpha}}{\partial x_\alpha}. \tag{24}$$

Therefore, it is sufficient to derive the conservation equation for only one of the terms. Here the conservation equation for phase 1

$$\frac{\partial}{\partial r_\alpha}R(\phi'_1, v'_{1\alpha})\big|_{r=0}$$

will be derived.

In view of Eq. (15), one needs to derive the conservation equations for

$$\overline{\phi'_1(\mathbf{x})\phi'_1(\mathbf{x+r})v'_{1\alpha}(\mathbf{x})} \quad \text{and} \quad \overline{\phi'_1(\mathbf{x})\phi'_1(\mathbf{x+r})v'_{1\alpha}(\mathbf{x+r})}.$$

The conservation equation for the first term can be obtained from Eq. (22) multiplied by $\phi'_1(\mathbf{x+r})$. With some modification, one obtains

CORRELATION FUNCTIONS

$$\frac{\partial}{\partial t}(\phi'_{1\mathbf{r}}\phi_1 v'_{1\alpha}) + \frac{\partial}{\partial x_\beta}(\phi'_{1\mathbf{r}}\phi_1 v'_{1\alpha}\bar{\bar{v}}_{1\beta} + \phi'_{1\mathbf{r}}\phi_1 v'_{1\alpha}v'_{1\beta})$$

$$- v'_{1\alpha}\{\frac{\partial}{\partial t}(\phi'_{1\mathbf{r}}\phi_1) + v_{1\beta}\frac{\partial}{\partial x_\beta}(\phi'_{1\mathbf{r}}\phi_1)\} + \bar{\bar{v}}_{1\alpha}\frac{\phi'_{1\mathbf{r}}\phi_1}{\bar{\phi}_1}\frac{\partial v'_{1\beta}}{\partial x_\beta} + \phi'_{1\mathbf{r}}\phi_1 v'_{1\beta}\frac{\partial \bar{\bar{v}}_{1\alpha}}{\partial x_\beta}$$

$$= -\frac{1}{\rho_1}\phi'_{1\mathbf{r}}\phi_1\frac{\partial P'_1}{\partial x_\alpha} + \frac{1}{\rho_1}\phi'_{1\mathbf{r}}\phi_1\frac{\partial \tau'_{1\alpha\beta}}{\partial x_\beta} + \phi'_{1\mathbf{r}}\phi_1 F'_{1\alpha} + \bar{\bar{v}}_{1\alpha}\frac{\phi'_{1\mathbf{r}}\phi_1}{\bar{\phi}_1}\frac{\partial}{\partial r_\beta}R(\phi'_1,v'_{1\beta})|_{\mathbf{r}=0}$$

$$+ \frac{\phi'_{1\mathbf{r}}\phi_1}{\bar{\phi}_1}\frac{\partial}{\partial x_\beta}(\overline{\phi_1 v'_{1\alpha}v'_{1\beta}}) + \frac{1}{\rho_1}\frac{\phi'_{1\mathbf{r}}\phi_1}{\bar{\phi}_1}\frac{\partial}{\partial r_\alpha}R(\phi'_1,P'_1)|_{\mathbf{r}=0}$$

$$- \frac{1}{\rho_1}\frac{\phi'_{1\mathbf{r}}\phi_1}{\bar{\phi}_1}\frac{\partial}{\partial r_\beta}R(\phi'_1,\tau'_{1\alpha\beta})|_{\mathbf{r}=0} \quad . \tag{25}$$

Using Eq. (21) and the conservation equation for $\phi'_{1\mathbf{r}}\phi_1$ (see Appendix E),

$$\frac{\partial}{\partial t}(\phi'_{1\mathbf{r}}\phi_1) + v_{1\beta}\frac{\partial}{\partial x_\beta}(\phi'_{1\mathbf{r}}\phi_1) = \frac{\partial}{\partial r_\beta}(\phi'_{1\mathbf{r}}\phi_1 v'_{1\beta} - \phi_1\phi_{1\mathbf{r}}v'_{1\beta\mathbf{r}})$$

$$+ \frac{\partial}{\partial r_\beta}\{\phi'_{1\mathbf{r}}\phi_1(\bar{\bar{v}}_{1\beta} - \bar{\bar{v}}_{1\beta\mathbf{r}})\} \tag{26}$$

and averaging, one finally obtains

$$\frac{\partial}{\partial t}(\overline{\phi'_{1\mathbf{r}}\phi_1 v'_{1\alpha}}) + \frac{\partial}{\partial x_\beta}(\overline{\phi'_{1\mathbf{r}}\phi_1 v'_{1\alpha}}\bar{\bar{v}}_{1\beta} + \overline{\phi'_{1\mathbf{r}}\phi_1 v'_{1\alpha}v'_{1\beta}})$$

$$= -\frac{1}{\rho_1}\overline{\phi'_{1\mathbf{r}}\phi_1\frac{\partial P'_1}{\partial x_\alpha}} + \frac{1}{\rho_1}\overline{\phi'_{1\mathbf{r}}\phi_1\frac{\partial \tau'_{1\alpha\beta}}{\partial x_\beta}} - \overline{\phi'_{1\mathbf{r}}\phi_1 F'_{1\alpha}} + \frac{\overline{\phi'_{1\mathbf{r}}\phi_1}}{\bar{\phi}_1}\frac{\partial}{\partial x_\beta}(\overline{\phi_1 v'_{1\alpha}v'_{1\beta}})$$

$$+ \frac{1}{\rho_1}\frac{\overline{\phi'_{1\mathbf{r}}\phi_1}}{\bar{\phi}_1}\frac{\partial}{\partial r_\alpha}R(\phi'_1,P'_1)|_{\mathbf{r}=0} - \frac{1}{1}\frac{\overline{\phi'_{1\mathbf{r}}\phi_1}}{\bar{\phi}_1}\frac{\partial}{\partial r_\beta}R(\phi'_1,\tau'_{1\alpha\beta})|_{\mathbf{r}=0}$$

$$- \overline{\phi'_{1\mathbf{r}}\phi_1 v'_{1\beta}}\frac{\partial \bar{\bar{v}}_{1\alpha}}{\partial x_\beta} + \frac{\partial}{\partial r_\beta}(\overline{\phi'_{1\mathbf{r}}\phi_1 v'_{1\alpha}v'_{1\beta}} - \overline{\phi_1\phi_{1\mathbf{r}}v'_{1\beta\mathbf{r}}v'_{1\alpha}})$$

$$+ \frac{\partial}{\partial r_\beta}\{\overline{\phi'_{1\mathbf{r}}\phi_1 v'_{1\alpha}}(\bar{\bar{v}}_{1\beta} - \bar{\bar{v}}_{1\beta\mathbf{r}})\} \quad . \tag{27}$$

Here, for the convenience of notation, the quantities at $\mathbf{x} + \mathbf{r}$ are denoted by suffix $\mathbf{r}$, or

$$\phi_{1\mathbf{r}} \equiv \phi_1(\mathbf{x}+\mathbf{r}) \tag{28}$$

$$v_{1\alpha r} \equiv v_{1\alpha}(\mathbf{x+r}) \ . \tag{29}$$

Similarly, the conservation equation for the second term, $\overline{\phi_1' \phi_{1r} v_{1\alpha r}'}$ can be obtained from the momentum conservation equation of fluctuating parts at $\mathbf{x + r}$, which is, in view of Eq. (22), given by

$$\phi_{1r}\frac{\partial v_{1\alpha r}'}{\partial t} + \phi_{1r}\frac{\partial}{\partial x_\beta}(v_{1\alpha r}'\bar{\bar{v}}_{1\beta r} + \bar{\bar{v}}_{1\alpha r}v_{1\beta r}' + v_{1\alpha r}'v_{1\beta r}')$$

$$= -\phi_{1r}\frac{1}{\rho_1}\frac{\partial P_{1r}'}{\partial x_\alpha} + \phi_{1r}\frac{1}{\rho_1}\frac{\partial \tau_{1\alpha\beta r}'}{\partial x_\beta} + \phi_{1r}F_{1\alpha r}' + \bar{\bar{v}}_{1\alpha r}\frac{\phi_{1r}}{\overline{\phi_{1r}}}\frac{\partial}{\partial r_\beta'}R(\phi_{1r}', v_{1\beta r}')\big|_{\mathbf{r'}=0}$$

$$+ \frac{\phi_{1r}}{\overline{\phi_{1r}}}\frac{\partial}{\partial x_\beta}(\overline{\phi_{1r}' v_{1\alpha r}'v_{1\beta r}'}) + \frac{\phi_{1r}}{\overline{\phi_{1r}}}\frac{1}{\rho_1}\frac{\partial}{\partial r_\alpha'}R(\phi_{1r}', P_{1r}')\big|_{\mathbf{r'}=0}$$

$$- \frac{\phi_{1r}}{\overline{\phi_{1r}}}\frac{1}{\rho_1}\frac{\partial}{\partial r_\beta'}R(\phi_{1r}', \tau_{1\alpha\beta r}')\big|_{\mathbf{r'}=0} \ . \tag{30}$$

Here, the correlation function at $\mathbf{x + r}$, is defined by

$$R(\phi_{1r}', A_{1r}') = \overline{\phi_1'(\mathbf{x+r})\phi_1(\mathbf{x+r+r'})A_1'(\mathbf{x+r+r'})} - \overline{\phi_1(\mathbf{x+r})\phi_1'(\mathbf{x+r+r'})A_1'(\mathbf{x+r})}. \tag{31}$$

Multiplied by ϕ_1', Eq. (30) can be rewritten by

$$\frac{\partial}{\partial t}(\phi_1'\phi_{1r}v_{1\alpha r}') + \frac{\partial}{\partial x_\beta}(\phi_1'\phi_{1r}v_{1\alpha r}'\bar{\bar{v}}_{1\beta r} + \phi_1'\phi_{1r}v_{1\alpha r}'v_{1\beta r}')$$

$$- v_{1\alpha r}'\{\frac{\partial}{\partial t}(\phi_1'\phi_{1r}) + v_{1\beta r}\frac{\partial}{\partial x_\beta}(\phi_1'\phi_{1r})\} + \bar{\bar{v}}_{1\alpha r}\frac{\phi_1'\phi_{1r}}{\overline{\phi_{1r}}}\frac{\partial v_{1\beta r}'}{\partial x_\beta} + \phi_1'\phi_{1r}v_{1\beta r}'\frac{\partial \bar{\bar{v}}_{1\alpha r}}{\partial x_\beta}$$

$$= -\frac{1}{\rho_1}\phi_1'\phi_{1r}\frac{\partial P_{1r}'}{\partial x_\alpha} + \frac{1}{\rho_1}\phi_1'\phi_{1r}\frac{\partial \tau_{1\alpha\beta r}'}{\partial x_\beta} + \phi_1'\phi_{1r}F_{1\alpha r}'$$

$$+ \bar{\bar{v}}_{1\alpha r}\frac{\phi_1'\phi_{1r}}{\overline{\phi_{1r}}}\frac{\partial}{\partial r_\beta'}R(\phi_{1r}', v_{1\beta r}')\big|_{\mathbf{r'}=0} + \frac{\phi_1'\phi_{1r}}{\overline{\phi_{1r}}}\frac{\partial}{\partial x_\beta}(\overline{\phi_{1r}' v_{1\alpha r}'v_{1\beta r}'})$$

$$+ \frac{\phi_1'\phi_{1r}}{\overline{\phi_{1r}}}\frac{1}{\rho_1}\frac{\partial}{\partial r_\alpha'}R(\phi_{1r}', P_{1r}')\big|_{\mathbf{r'}=0} - \frac{\phi_1'\phi_{1r}}{\overline{\phi_{1r}}}\frac{1}{\rho_1}\frac{\partial}{\partial r_\beta'}R(\phi_{1r}', \tau_{1\alpha\beta r}')\big|_{\mathbf{r'}=0}. \tag{32}$$

Using the relations obtained in view of Eq. (21)

CORRELATION FUNCTIONS

$$\phi_{1\mathbf{r}}\frac{\partial v'_{1\beta\mathbf{r}}}{\partial x_\beta} = \frac{\phi_{1\mathbf{r}}}{\phi_{1\mathbf{r}}}\frac{\partial}{\partial r'_\beta}R(\phi'_{1\mathbf{r}},v'_{1\beta\mathbf{r}})\Big|_{\mathbf{r}'=0} \tag{33}$$

$$\frac{\partial}{\partial t}(\phi'_1\phi_{1\mathbf{r}}) + v_{1\beta\mathbf{r}}\frac{\partial}{\partial x_\beta}(\phi'_1\phi_{1\mathbf{r}}) = v_{1\beta\mathbf{r}}\phi_{1\mathbf{r}}\frac{\partial \phi'_1}{\partial x_\beta} - \phi_{1\mathbf{r}}\frac{\partial}{\partial x_\beta}(\phi'_1 v'_{1\beta} + \phi'_1 \bar{\bar{v}}_{1\beta}). \tag{34}$$

(See Appendix E),
and averaging, Eq. (32) becomes

$$\frac{\partial}{\partial t}(\overline{\phi'_1\phi_{1\mathbf{r}}v'_{1\alpha\mathbf{r}}}) + \frac{\partial}{\partial x_\beta}(\overline{\phi'_1\phi_{1\mathbf{r}}v'_{1\alpha\mathbf{r}}}\bar{\bar{v}}_{1\beta} + \overline{\phi'_1\phi_{1\mathbf{r}}v'_{1\alpha\mathbf{r}}v'_{1\beta\mathbf{r}}})$$

$$= -\frac{1}{\rho_1}\overline{\phi'_1\phi_{1\mathbf{r}}\frac{\partial P'_{1\mathbf{r}}}{\partial x_\alpha}} + \frac{1}{\rho_1}\overline{\phi'_1\phi_{1\mathbf{r}}\frac{\partial \tau'_{1\alpha\beta\mathbf{r}}}{\partial x_\beta}} + \overline{\phi'_1\phi_{1\mathbf{r}}F'_{1\alpha\mathbf{r}}} + \frac{\overline{\phi'_1\phi_{1\mathbf{r}}}}{\overline{\phi_{1\mathbf{r}}}}\frac{\partial}{\partial x_\beta}(\overline{\phi_{1\mathbf{r}}}\overline{v'_{1\alpha\mathbf{r}}v'_{1\beta\mathbf{r}}})$$

$$+ \frac{\overline{\phi'_1\phi_{1\mathbf{r}}}}{\overline{\phi_{1\mathbf{r}}}}\frac{1}{\rho_1}\frac{\partial}{\partial r'_\alpha}R(\phi'_{1\mathbf{r}},P'_{1\mathbf{r}})\Big|_{\mathbf{r}'=0} - \frac{\overline{\phi'_1\phi_{1\mathbf{r}}}}{\overline{\phi_{1\mathbf{r}}}}\frac{1}{\rho_1}\frac{\partial}{\partial r'_\alpha}R(\phi'_{1\mathbf{r}},\tau'_{1\alpha\beta\mathbf{r}})\Big|_{\mathbf{r}'=0}$$

$$- \overline{\phi'_1\phi_{1\mathbf{r}}v'_{1\beta\mathbf{r}}}\frac{\partial \bar{\bar{v}}_{1\alpha\mathbf{r}}}{\partial x_\beta} - \frac{\partial}{\partial x_\beta}(\overline{\phi'_1\phi_{1\mathbf{r}}v'_{1\alpha\mathbf{r}}v'_{1\beta\mathbf{r}}} - \overline{\phi'_1\phi_{1\mathbf{r}}v'_{1\alpha\mathbf{r}}v'_{1\beta\mathbf{r}}})$$

$$- \frac{\partial}{\partial r_\beta}\{\overline{\phi'_1\phi_{1\mathbf{r}}v'_{1\alpha\mathbf{r}}}(\bar{\bar{v}}_{1\beta} - \bar{\bar{v}}_{1\beta\mathbf{r}})\}. \tag{35}$$

Subtracting Eq. (27) from (35), differentiating with regard to r_γ and taking a limit of $\mathbf{r} \to 0$, one finally obtains

$$\frac{\partial}{\partial t}(\frac{\partial}{\partial r_\gamma}R(\phi'_1,v'_{1\alpha})\Big|_{\mathbf{r}=0}) + \frac{\partial}{\partial x_\beta}(\bar{\bar{v}}_{1\beta}\frac{\partial}{\partial r_\gamma}R(\phi'_1,v'_{1\alpha})\Big|_{\mathbf{r}=0})$$

$$= -\frac{1}{\rho_1}\frac{\partial}{\partial r_\gamma}(\overline{\phi'_1\phi_{1\mathbf{r}}\frac{\partial P'_{1\mathbf{r}}}{\partial x_\alpha}} - \overline{\phi'_{1\mathbf{r}}\phi_1\frac{\partial P'_1}{\partial x_\alpha}})\Big|_{\mathbf{r}=0} + \frac{1}{\rho_1}\frac{\partial}{\partial r_\gamma}(\overline{\phi'_1\phi_{1\mathbf{r}}\frac{\partial \tau'_{1\alpha\beta\mathbf{r}}}{\partial x_\beta}} - \overline{\phi'_{1\mathbf{r}}\phi_1\frac{\partial \tau'_{1\alpha\beta}}{\partial x_\beta}})\Big|_{\mathbf{r}=0}$$

$$+ \frac{\partial}{\partial r_\gamma}(\overline{\phi'_1\phi_{1\mathbf{r}}F'_{1\alpha\mathbf{r}}} - \overline{\phi'_{1\mathbf{r}}\phi_1 F'_{1\alpha}})\Big|_{\mathbf{r}=0} - \frac{\partial}{\partial x_\beta}\{\frac{\partial}{\partial r_\gamma}(\overline{\phi'_1\phi_{1\mathbf{r}}v'_{1\beta}v'_{1\alpha\mathbf{r}}} - \overline{\phi'_{1\mathbf{r}}\phi_1 v'_{1\alpha}v'_{1\beta}})\Big|_{\mathbf{r}=0}\}$$

$$+ \overline{\phi_1(1-\overline{\phi_1})}\frac{\partial}{\partial x_\gamma}\{\frac{1}{\overline{\phi_1}}\frac{\partial}{\partial x_\gamma}(\overline{\phi_1}\overline{v'_{1\alpha}v'_{1\beta}})\} + \overline{\phi_1(1-\overline{\phi_1})}\frac{\partial}{\partial x_\gamma}\{\frac{1}{\overline{\phi_1}}\frac{\partial}{\partial r_\alpha}R(\phi'_1,P'_1)\Big|_{\mathbf{r}=0}\}$$

$$- \overline{\phi_1(1-\overline{\phi_1})}\frac{\partial}{\partial x_\gamma}\{\frac{1}{\overline{\phi_1}}\frac{\partial}{\partial r_\beta}R(\phi'_1,\tau'_{1\alpha\beta})\Big|_{\mathbf{r}=0} - \frac{\partial}{\partial r_\gamma}R(\phi'_1,v'_{1\beta})\Big|_{\mathbf{r}=0}\frac{\partial \bar{\bar{v}}_{1\alpha}}{\partial x_\beta}$$

$$- \frac{\partial}{\partial r_\beta}R(\phi'_1,v'_{1\alpha})\Big|_{\mathbf{r}=0}\frac{\partial \bar{\bar{v}}_{1\beta}}{\partial x_\gamma} - \frac{\partial}{\partial r_\gamma}R(\phi'_1,v'_{1\alpha})\Big|_{\mathbf{r}=0}\frac{\partial \bar{\bar{v}}_{1\beta}}{\partial x_\beta} \tag{36}$$

$$+ \frac{\partial^2}{\partial r_\gamma \partial r_\beta}\{(\overline{\phi'_1\phi_{1\mathbf{r}}v'_{1\beta}v'_{1\alpha\mathbf{r}}} - \overline{\phi'_{1\mathbf{r}}\phi_1 v'_{1\alpha}v'_{1\beta}}) + (\overline{\phi'_1\phi_{1\mathbf{r}}v'_{1\beta\mathbf{r}}v'_{1\alpha}} - \overline{\phi'_1\phi_{1\mathbf{r}}v'_{1\alpha\mathbf{r}}v'_{1\beta\mathbf{r}}})\}\Big|_{\mathbf{r}=0}.$$

In this equation, many correlative terms appear. Among them, the term

$$(\overline{\phi_1' \phi_{1r}' F_{1\alpha r}'} - \overline{\phi_{1r}' \phi_1 F_{1\alpha}'})$$

is the correlative term of external force and usually is considered to be a given quantity.

The turbulent and interfacial momentum transfer terms

$$\overline{\phi_1} \; \overline{v_{1\alpha}' v_{1\beta}'} \;, \quad R(\phi_1', P_1') \;, \quad \text{and} \quad R(\phi_1', \tau_{1\alpha\beta}')$$

were already discussed in Sec. 3 and the constitutive equations for these terms can be given.

Besides these terms, Eq. (36) includes the higher order correlative terms as

$$(\overline{\phi_1' \phi_{1r} \frac{\partial P_{1r}'}{\partial x_\alpha}} - \overline{\phi_{1r}' \phi_1 \frac{\partial P_1'}{\partial x_\alpha}}) \;, \quad (\overline{\phi_1' \phi_{1r} \frac{\partial \tau_{1\alpha\beta r}'}{\partial x_\beta}} - \overline{\phi_{1r}' \phi_1 \frac{\partial \tau_{1\alpha\beta}'}{\partial x_\beta}}) \;,$$

$$(\overline{\phi_1 \phi_{1r} v_{1\beta}' v_{1\alpha r}'} - \overline{\phi_{1r}' \phi_1 v_{1\alpha}' v_{1\beta}'}) \quad \text{and} \quad (\overline{\phi_1 \phi_{1r} v_{1\beta r}' v_{1\alpha}'} - \overline{\phi_1' \phi_{1r} v_{1\alpha r}' v_{1\beta r}'}).$$

At present, nothing is known for constitutive equations for these higher order terms. However, analogous to single phase turbulent modeling, it may be possible to relate the higher order terms to the derivatives of correlation function $R(\phi_1', v_{1\alpha}')$ and/or other averaged quantities.

If the constitutive equations for these higher order correlative terms can be given, Eq. (36) determines the nine quantities

$$\frac{\partial}{\partial r_\gamma} R(\phi_1', v_{1\alpha}')\Big|_{\mathbf{r}=0} \quad (\alpha=1 \sim 3, \; \gamma=1 \sim 3),$$

because Eq. (36) includes nine equations for

$$\alpha = 1 \sim 3 \;, \quad \gamma = 1 \sim 3 \;.$$

Then the source terms in Eq. (18)

$$\frac{\partial}{\partial r_\alpha} R(\phi_1', v_{1\alpha}')\Big|_{\mathbf{r}=0}$$

can be determined and the averaged formulation of two-phase flow, Eqs. (18) through (20) is closed.

CONCLUSIONS

Closure problems in basic formulations of two-phase flow are considered. Based on local instant formulation, the averaged conservation equations for mass, momentum and interfacial movement are obtained (Eqs. (18) through (20)). In these equations, correlative terms appear as source terms. They are composed of turbulent stress terms, interfacial

CORRELATION FUNCTIONS

momentum transfer terms and the correlative terms between characteristic function and velocity fluctuations.

In order to close the averaged formulations, the conservation equations are derived to determine the correlative terms between characteristic function and velocity fluctuations (Eq. (35)). With appropriate constitutive equations for higher order correlative terms, the averaged formulation, Eqs. (18) through (29) and (22), becomes a closed equations system. In this formulation, there is no need for relations between the averaged pressures of both phases. Modeling and development of reliable constitutive equations for higher order correlative terms are desirable projects for future research.

Appendix A: Dependent Equations in Eqs. (1) and (7).

Assuming that the two equations in Eq. (1) (k=1,2) are independent, the two equations in Eq. (7) (k=1,2) become dependent upon each other as shown below.

$$\frac{\partial \phi_1}{\partial t} + \frac{\partial}{\partial x_\alpha}(\phi_1 v_{1\alpha}) = \frac{\partial \phi_1}{\partial t} + v_{1\alpha}\frac{\partial \phi_1}{\partial x_\alpha} + \phi_1 \frac{\partial v_{1\alpha}}{\partial x_\alpha}$$

$$= \frac{\partial \phi_1}{\partial t} + v_{1\alpha}\frac{\partial \phi_1}{\partial x_\alpha} \qquad \text{(from Eq.(1))}$$

$$= -(\frac{\partial \phi_2}{\partial t} + v_{2\alpha}\frac{\partial \phi_2}{\partial x_\alpha}) \qquad \text{(from Footnote 1)}$$

$$= -(\frac{\partial \phi_2}{\partial t} + v_{2\alpha}\frac{\partial \phi_2}{\partial x_\alpha} + \phi_2 \frac{\partial v_{2\alpha}}{\partial x_\alpha}) \qquad \text{(from Eq.(1))}$$

$$= -\{\frac{\partial \phi_2}{\partial t} + \frac{\partial}{\partial x_\alpha}(\phi_2 v_{2\alpha})\} \quad . \qquad (A-1)$$

Conversely, assuming that the two equations in Eq. (7) (k=1,2) are independent, the two equations in Eq. (1) (k=1,2) become dependent as follows.

$$\phi_1 \frac{\partial v_{1\alpha}}{\partial x_\alpha} = \frac{\partial \phi_1}{\partial t} + \frac{\partial}{\partial x_\alpha}(\phi_1 v_{1\alpha}) - (\frac{\partial \phi_1}{\partial t} + v_{1\alpha}\frac{\partial \phi_1}{\partial x_\alpha})$$

$$= -(\frac{\partial \phi_1}{\partial t} + v_{1\alpha}\frac{\partial \phi_1}{\partial x_\alpha}) \qquad \text{(from Eq.(7))}$$

$$= \frac{\partial \phi_2}{\partial t} + v_{2\alpha}\frac{\partial \phi_2}{\partial x_\alpha} \qquad \text{(from Footnote 1)}$$

$$= \frac{\partial \phi_2}{\partial t} + v_{2\alpha}\frac{\partial \phi_2}{\partial x_\alpha} - \{\frac{\partial \phi_2}{\partial t} + \frac{\partial}{\partial x_\alpha}(\phi_2 v_{2\alpha})\} \qquad \text{(from Eq.(7))}$$

$$= -\phi_2 \frac{\partial v_{2\alpha}}{\partial x_\alpha} \quad . \tag{A-2}$$

Therefore, one of the four equations in Eqs. (1) and (7) (k=1,2) can be derived from the other three equations. In other words, any three equations in Eqs. (1) and (7) (k=1,2) are independent.

Appendix B: Derivation of Eq. (14)

From the definition of the averaged derivative of two-phase flow, one obtains

$$\overline{\phi_k \frac{\partial A_k}{\partial x_\alpha}} = \lim_{\Delta x_\alpha \to 0} \overline{\frac{1}{\Delta x_\alpha} \phi_k(x_\alpha)\phi_k(x_\alpha + \Delta x_\alpha)\{A_k(x_\alpha + \Delta x_\alpha) - A_k(x_\alpha)\}}$$

$$= \lim_{\Delta x_\alpha \to 0} \frac{1}{\Delta x_\alpha}(\overline{\phi_k(x_\alpha)\phi_k(x_\alpha + \Delta x_\alpha)A_k(x_\alpha + \Delta x_\alpha)} - \overline{\phi_k(x_\alpha)A_k(x_\alpha)})$$

$$- \lim_{\Delta x_\alpha \to 0} \frac{1}{\Delta x_\alpha}(\overline{\phi_k(x_\alpha)\phi_k(x_\alpha + \Delta x_\alpha)A_k(x_\alpha)} - \overline{\phi_k(x_\alpha)A_k(x_\alpha)})$$

$$= \frac{\partial}{\partial r_\alpha}(\overline{\phi_k(x)\phi_k(x+r)A_k(x+r)})\Big|_{r=0} - \frac{\partial}{\partial r_\alpha}(\overline{\phi_k(x)\phi_k(x+r)A_k(x)})\Big|_{r=0}. \tag{B-1}$$

Correlation functions in Eq. (B-1) can be divided into averaged terms and fluctuating terms by

$$\overline{\phi_k(x)\phi_k(x+r)A_k(x+r)} = \overline{\phi_k(x)\phi_k(x+r)}\,\overline{A_k(x+r)} + \overline{\phi_k'(x)\phi_k(x+r)A_k'(x+r)}, \tag{B-2}$$

$$\overline{\phi_k(x)\phi_k(x+r)A_k(x)} = \overline{\phi_k(x)\phi_k(x+r)}\,\overline{A_k(x)} + \overline{\phi_k(x)\phi_k'(x+r)A_k'(x)} . \tag{B-3}$$

Substituting Eqs. (B-2) and (B-3) into Eq. (B-1), one finally obtains

$$\overline{\phi_k \frac{\partial A_k}{\partial x_\alpha}} = \phi_k(x)\frac{\partial}{\partial x_\alpha}\overline{\overline{A_k(x)}} + \overline{\overline{A_k(x)}}\frac{\partial}{\partial r_\alpha}\overline{\phi_k(x)\phi_k(x+r)}\Big|_{r=0} + \frac{\partial}{\partial r_\alpha}\overline{\phi_k'(x)\phi_k(x+r)A_k'(x+r)}\Big|_{r=0}$$

$$- \overline{\overline{A_k(x)}}\frac{\partial}{\partial r_\alpha}\overline{\phi_k(x)\phi_k(x+r)}\Big|_{r=0} - \frac{\partial}{\partial r_\alpha}\overline{\phi_k(x)\phi_k'(x+r)A_k'(x)}\Big|_{r=0}$$

$$= \overline{\phi_k}\frac{\partial}{\partial x_\alpha}\overline{\overline{A_k}} + \frac{\partial}{\partial r_\alpha}R(\phi_k', A_k')\Big|_{r=0} \quad . \tag{B-4}$$

Here, the following relations are used.

$$\frac{\partial}{\partial r_\alpha}\overline{\overline{A_k(x+r)}}\Big|_{r=0} = \frac{\partial}{\partial x_\alpha}\overline{\overline{A_k(x)}} \quad , \tag{B-5}$$

$$\overline{\phi_k(x)\phi_k(x+r)}\Big|_{r=0} = \overline{\phi_k(x)} \quad . \tag{B-6}$$

Appendix C: Relation Between Correlation Terms and Interfacial Momentum Transfer Terms

The averaged momentum equation (Eq. (19)) can be equivalently rewritten using interfacial transfer terms by

$$\frac{\partial}{\partial t}(\overline{\phi_k}\overline{\overline{v}}_{k\alpha}) + \frac{\partial}{\partial x_\beta}(\overline{\phi_k}\overline{\overline{v}}_{k\alpha}\overline{\overline{v}}_{k\beta}) = -\frac{1}{\rho_k}\frac{\partial}{\partial x_\alpha}(\overline{\phi_k}\overline{\overline{P}}_k) + \frac{1}{\rho_k}\frac{\partial}{\partial x_\beta}(\overline{\phi_k}\overline{\overline{\tau}}_{k\alpha\beta})$$

$$- \frac{\partial}{\partial x_\beta}(\overline{\phi_k}\overline{v_{k\alpha}' v_{k\beta}'}) + \overline{\phi_k}\overline{\overline{F}}_{k\alpha} + \frac{1}{\rho_k}M_{ki\alpha} \quad (k=1,2) \quad . \tag{C-1}$$

Here, M_{ki} is the interfacial momentum transfer term given by

$$M_{ki\alpha} = \overline{P_{ki}n_{ki\alpha}a_i + \tau_{k\alpha\beta}'n_{ki\beta}a_i} \quad (k=1,2) \quad , \tag{C-2}$$

where n_{ki} and a_i are the normal outward vector of phase k at interface and local instant interfacial concentration (Kataoka, 1986a).

Comparing Eq. (19) with Eq. (C-1), one obtains

$$-\frac{\partial}{\partial r_\alpha}R(\phi_k', P_k')\Big|_{r=0} + \frac{\partial}{\partial r_\beta}R(\phi_k', \tau_{k\alpha\beta}')\Big|_{r=0} = M_{ki\alpha} - \overline{\overline{P}}_k\frac{\partial\overline{\phi_k}}{\partial x_\alpha} + \overline{\overline{\tau}}_{k\alpha\beta}\frac{\partial\overline{\phi_k}}{\partial x_\beta} \quad (k=1,2). \tag{C-3}$$

Appendix D: Derivation of Eqs. (21) through (23)

Equation (21) is obtained by subtracting Eq. (18) multiplied by $\phi_k'/\overline{\phi}_k$ from Eq. (8).

Equation (19) can be rewritten in view of Eqs. (18) and (20) by

$$\overline{\phi}_k \frac{\partial \overline{\overline{v}}_{k\alpha}}{\partial t} + \overline{\phi}_k \frac{\partial}{\partial x_\beta}(\overline{\overline{v}}_{k\alpha} \overline{\overline{v}}_{k\beta}) = -\overline{\phi}_k \frac{1}{\rho_k} \frac{\partial \overline{\overline{P}}_k}{\partial x_\alpha} + \overline{\phi}_k \frac{1}{\rho_k} \frac{\partial}{\partial x_\beta} \overline{\overline{\tau}}_{k\alpha\beta} + \overline{\phi}_k \overline{\overline{F}}_{k\alpha}$$

$$- \frac{\partial}{\partial x_\alpha}(\overline{\phi}_k \overline{v'_{k\alpha} v'_{k\beta}}) - \frac{1}{\rho_k} \frac{\partial}{\partial r_\alpha} R(\phi_k', P_k')\big|_{\mathbf{r}=0} + \frac{1}{\rho_k} \frac{\partial}{\partial r_\beta} R(\phi_k', \tau_{k\alpha\beta}')\big|_{\mathbf{r}=0}$$

$$- \overline{\overline{v}}_{k\alpha} \frac{\partial}{\partial r_\beta} R(\phi_k', v_{k\beta}')\big|_{\mathbf{r}=0} \quad (k=1,2) \quad . \tag{D-1}$$

Subtracting Eq. (D-1) multiplied by $\phi_k'/\overline{\phi}_k$ from Eq. (9), one obtains Eq. (22).

Equation (23) is directly obtained by subtracting Eq. (20) from Eq. (10). Here, $\phi_k v_{k\alpha}$ is divided into the averaged part and a fluctuating part by

$$\phi_k v_{k\alpha} = \overline{\phi}_k \overline{\overline{v}}_{k\alpha} + \phi_k v_{k\alpha}' + \phi_k' \overline{\overline{v}}_{k\alpha} \quad (k=1,2) \quad . \tag{D-2}$$

Note that terms such as

$\phi_k' v_{k\alpha}' \quad (k=1,2)$

can not be defined as field quantities.

Appendix E: Derivation of Eqs. (26) and (34)

Equation (26):
One starts from Eq. (10) for k=1.

$$\frac{\partial \phi_1}{\partial t} + \frac{\partial}{\partial x_\alpha}(\phi_1 v_{1\alpha}) = 0 \quad . \tag{E-1}$$

Multiplied by $\phi_{1\mathbf{r}}'$ and modified, it becomes

$$\phi_{1\mathbf{r}}' \frac{\partial \phi_1}{\partial t} + \frac{\partial}{\partial x_\alpha}(\phi_{1\mathbf{r}}' \phi_1 v_{1\alpha}) - \phi_1 v_{1\alpha} \frac{\partial \phi_{1\mathbf{r}}'}{\partial x_\alpha} = 0 \quad . \tag{E-2}$$

Using the relation

$$\frac{\partial \phi_{1\mathbf{r}}'}{\partial x_\alpha} = \frac{\partial \phi_{1\mathbf{r}}'}{\partial r_\alpha} \quad , \tag{E-3}$$

Eq. (E-2) can be rewritten by

$$\phi_{1\mathbf{r}}' \frac{\partial \phi_1}{\partial t} + \frac{\partial}{\partial x_\alpha}(\phi_{1\mathbf{r}}' \phi_1 v_{1\alpha}) - \frac{\partial}{\partial r_\alpha}(\phi_{1\mathbf{r}}' \phi_1 v_{1\alpha}) = 0 \quad . \tag{E-4}$$

CORRELATION FUNCTIONS

On the other hand, in view of Eq. (23), the conservation equation of fluctuation term of $\phi_{1\mathbf{r}}$ can be given by

$$\frac{\partial \phi'_{1\mathbf{r}}}{\partial t} + \frac{\partial}{\partial x_\alpha}(\phi_{1\mathbf{r}} v'_{1\alpha\mathbf{r}} + \phi'_{1\mathbf{r}} \bar{\bar{v}}_{1\alpha\mathbf{r}}) = 0 \ . \tag{E-5}$$

In this equation, $\frac{\partial}{\partial x_\alpha}$ can be exchanged for $\frac{\partial}{\partial r_\alpha}$, and then

$$\frac{\partial \phi'_{1\mathbf{r}}}{\partial t} + \frac{\partial}{\partial r_\alpha}(\phi_{1\mathbf{r}} v'_{1\alpha\mathbf{r}} + \phi'_{1\mathbf{r}} \bar{\bar{v}}_{1\alpha\mathbf{r}}) = 0 \ . \tag{E-6}$$

Multiplied by ϕ_1, this equation becomes

$$\phi_1 \frac{\partial \phi'_{1\mathbf{r}}}{\partial t} + \frac{\partial}{\partial r_\alpha}(\phi_{1\mathbf{r}} v'_{1\alpha\mathbf{r}} + \phi_1 \phi'_{1\mathbf{r}} \bar{\bar{v}}_{1\alpha\mathbf{r}}) = 0 \ . \tag{E-7}$$

Adding Eqs. (E-4) and (E-7) and using Eq. (8), one obtains

$$\frac{\partial}{\partial t}(\phi'_{1\mathbf{r}} \phi_1) + v_{1\alpha} \frac{\partial}{\partial x_\alpha}(\phi'_{1\mathbf{r}} \phi_1) = \frac{\partial}{\partial r_\alpha}(\phi'_{1\mathbf{r}} \phi_1 v'_{1\alpha} - \phi_1 \phi_{1\mathbf{r}} v'_{1\alpha\mathbf{r}})$$

$$+ \frac{\partial}{\partial r_\alpha}\{\phi'_{1\mathbf{r}} \phi_1 (\bar{\bar{v}}_{1\alpha} - \bar{\bar{v}}_{1\alpha\mathbf{r}})\} \ . \tag{E-8}$$

Since suffixes are arbitrary, Eq. (E-8) is identical to Eq. (26).

Equation (34):
In view of Eq. (3), the conservation equation of $\phi_{1\mathbf{r}}$ can be given by

$$\frac{\partial \phi_{1\mathbf{r}}}{\partial t} + v_{1\alpha\mathbf{r}} \frac{\partial \phi_{1\mathbf{r}}}{\partial x_\alpha} = 0 \ . \tag{E-9}$$

Multiplied by ϕ'_1 and modified, one obtains

$$\phi'_1 \frac{\partial \phi_{1\mathbf{r}}}{\partial t} + v_{1\alpha\mathbf{r}} \frac{\partial}{\partial x_\alpha}(\phi'_1 \phi_{1\mathbf{r}}) - v_{1\alpha\mathbf{r}} \phi_{1\mathbf{r}} \frac{\partial \phi'_1}{\partial x_\alpha} = 0 \ . \tag{E-10}$$

On the other hand, Eq. (23) for k=1 gives

$$\frac{\partial \phi'_1}{\partial t} + \frac{\partial}{\partial x_\alpha}(\phi_1 v'_{1\alpha} + \phi'_1 \bar{\bar{v}}_{1\alpha}) = 0 \ . \tag{E-11}$$

Multiplied by $\phi_{1\mathbf{r}}$ and adding Eq. (E-10), one obtains

$$\frac{\partial}{\partial t}(\phi'_1 \phi_{1\mathbf{r}}) + v_{1\alpha\mathbf{r}} \frac{\partial}{\partial x_\alpha}(\phi'_1 \phi_{1\mathbf{r}}) = v_{1\alpha\mathbf{r}} \phi_{1\mathbf{r}} \frac{\partial \phi'_1}{\partial x_\alpha} - \phi_{1\mathbf{r}} \frac{\partial}{\partial x_\alpha}(\phi_1 v'_{1\alpha} + \phi'_1 \bar{\bar{v}}_{1\alpha}) \ . \tag{E-12}$$

Due to the arbitrariness of the suffix, Eq. (E-12) is identical to Eq. (34).

It should be noted that quantities $\phi'_{1\mathbf{r}} \phi_1$ and $\phi'_1 \phi_{1\mathbf{r}}$ are related to the interfacial area concentration (Kataoka & Serizawa, 1987; Kataoka, 1986b) by

$$a_i = -\frac{1}{2\pi}\int_0^{2\pi}\int_0^{\pi}\{\frac{\partial \phi_{1\mathbf{r}}}{\partial \mathbf{r}}\Big|_{\mathbf{r}=0} - 2\frac{\partial}{\partial \mathbf{r}}(\phi_1 \phi_{1\mathbf{r}})\Big|_{\mathbf{r}=0}\}\frac{d\mathbf{r}}{|\mathbf{r}|} \quad , \tag{E-13}$$

where $\frac{\partial}{\partial \mathbf{r}}$ denotes directional differentiation.

NOMENCLATURE

English

a_i	Interfacial area concentration
A	Physical variable
F	External force
M	Interfacial momentum transfer term
P	Pressure
r	Coordinate of displacement vector
$\mathbf{r}, \mathbf{r'}$	Displacement vectors
$R(\phi_k', A_k')$	Correlation function defined by Eq. (15)
t	time
v	velocity
x	Space coordinate
$\mathbf{x}$	Position vector

Greeks

Δx	Increment of x
ρ	Density
τ	Stress tensor
ϕ	Characteristic function

Subscripts and others

i	Value at interface
k	Phase k (k=1,2)
$\mathbf{r}$	value at $\mathbf{x} + \mathbf{r}$
α, β, γ	Components of Cartesian coordinate
$-$	Average
$=$	Phase weighted average defined by Eq. (13)

BIBLIOGRAPHY

Delhaye, J. M., 1968, "Equation Fondamentales des Ecoulment Diphasiques, I. II.", CEA-R-3429.

Delhaye, J.M., 1984, "Jump Conditions and Entropy Sources in Two-Phase Flow Systems, Local Instant Formulation", Int. J. Multiphase Flow, **9**, 227-236.

Ishii, M., 1975, Thermo-Fluid Dynamic Theory of Two-Phase Flow, Eyrolles, Paris.

Ishii, M. and Chawla, T.C., 1979, "Local Drag Laws in Dispersed Two-Phase Flow", ANL 79-105, NUREG/CR1230.

Kataoka, I., 1986a, "Local Instant Formulation of two-Phase

Flow", Int. J. Multiphase Flow, **12**, 745-758.

Kataoka, I., 1986b, "Interfacial Area Concentration of Two-Phase Flow Based on Statistical Averaging", Proc. 23rd Nat. Heat Transfer Symp. Japan, 259-261, Sapporo, Japan, May 27-29, (in Japanese).

Kataoka, I. and Serizawa, A., 1987, "Interfacial Area Concentration and its Roles in Local Instant Formulations of Two-Phase Flow", Proc. of 1987 ICHMT Int. Seminar on Transient Phenomena in Multiphase Flow, Dubrovnik, Yugoslavia, May 24-30.

Lahey, Jr., R.T., 1987, "Turbulence and Phase Distribution Phenomena in Two-Phase Flow", Proc. of 1987 ICHMT Int. Seminar on Transient Phenomena in Multiphase Flow, Dubrovnik, Yugoslavia, May 24-30.

Sato, Y. and Sekoguchi, 1975, K., "Liquid Velocity Distribution in Two-Phase Bubbly Flow", Int. J. Multiphase Flow, **2**, 79-95.

Serizawa, A. and Kataoka,I., 1987, "Phase Distribution in Two-Phase Flow", Proc. of 1987 ICHMT Int. Seminar on Transient Phenomena in Multiphase Flow, Dubrovnik, Yugoslavia, May 24-30.

Wallis, G.B., 1969, One dimensional Two-Phase Flow, McGraw Hill, New York.

SOME TESTS OF TWO-FLUID MODELS FOR TWO-PHASE FLOW

GRAHAM B. WALLIS
Thayer School of Engineering
Dartmouth College
Hanover, NH 03755

ABSTRACT

Solutions of simple well-defined potential two-phase flows are presented. A coefficient of added mass, the pressure difference between phases and the equation of motion of each phase are derived, for these cases, in terms of a single function, β, that depends only on the void fraction. Three specific examples are used to test recent formulations of the "two-fluid" model. Only Geurst's equations, derived from a variational aproach, pass these tests.

INTRODUCTION

Many recent approaches to two-phase flow problems have been based on the "two-fluid" or "separated" flow model. Conservation laws are written separately for each phase and interactions between the phases are expressed as appropriate terms appearing in both sets of equations. These interaction terms are sometimes justified on an empirical or ad hoc basis and sometimes, as in the case of inertial coupling, by an idealized analysis assuming inviscid or potential flow.

It is very difficult to make enough measurements with sufficient accuracy over a broad enough range of parameters to obtain confident verification of such "general" equations and correlations for the coefficients in the terms representing interactions between the phases. As a filter for postulated equations and coefficients, it is therefore useful to have available a set of solutions to well-defined, relatively simple problems that are recognized as correct and perhaps even "exact." There are very few such examples.

The purpose of this paper is to present a few of these "standard solutions" and to compare them with the predictions of some recent formulations of the two-fluid model.

STANDARD SOLUTIONS

A Stationary Sphere in an Accelerating Steady Flow

G.I. Taylor (1928) showed that the force per unit volume, f_2, necessary to restrain an object held in an inviscid fluid with velocity v_1 varying in the x-direction is

$$f_2 = -\rho_1(1+C)v_1 \frac{dv_1}{dx} \qquad (1)$$

where C is a "coefficient of added mass" that is equal to 1/2 for a sphere in a dilute suspension. The force is a factor of $(1+C)$ greater than would be calculated by multiplying the volume of the sphere by the pressure gradient in the fluid:

$$\frac{dp_1}{dx} = -\rho_1 v_1 \frac{dv_1}{dx} \qquad (2)$$

We can also show, by integrating the pressure around a sphere held in a uniform flow with speed v_1, that the average pressure on the sphere is

$$p_2 = p_1 - \frac{1}{2}\rho v_1^2 C \qquad (3)$$

a result that can also be deduced for objects of a more general shape (Wallis, 1989).

Equations (1), (2) and (3) can be combined to give the simple result:

$$f_2 = \frac{dp_2}{dx} \qquad (4)$$

which shows that one has to be careful in defining what is meant by "pressure" in a two-phase flow.

These results are valid when the velocity does not change much over the length of the suspended object. They can be confirmed, subject to the previous condition, for special cases such as a sphere in the field of a single source (Wallis, 1989; Milne-Thomson, 1949; Pauchon & Banerjee, 1986) or a cylinder in the field of two equal and opposite vortices that have images in the cylinder.

Suspensions of Spheres

Consider potential flow through a bed of stationary spheres occupying a volumetric fraction α_2. The average potential gradient associated with a fluid volumetric flux, $\mathbf{j}_0$, will exceed the potential gradient for the same flux in the absence of the spheres by the factor "β" which depends on α_2, i.e.

$$-\nabla \phi = \beta \mathbf{j}_0 \qquad (5)$$

For the analogous problem of electrical current flow past a matrix of non-conducting spheres, Maxwell (1881) showed that

$$\beta = \frac{1 + \alpha_2/2}{1 - \alpha_2} \tag{6}$$

a result confirmed by Turner (1976) for fluidized beds.

The kinetic energy of the fluid, per unit total volume, is

$$K_1 = -\frac{1}{2}\rho_1 \mathbf{j}_0 \cdot \nabla \phi = \frac{1}{2}\rho_1 j_0^2 \beta \tag{7}$$

The kinetic energy of the fluid per unit volume of fluid is then

$$k_1 = \frac{K_1}{\alpha_1} = \frac{1}{2}\rho_1 v_1^2 \alpha_1 \beta \tag{8}$$

where

$$\mathbf{v}_1 = \frac{\mathbf{j}_0}{\alpha_1} \tag{9}$$

is the average fluid velocity.

In the general case in which both phases move it is easy to show (Wallis, 1989) that their total kinetic energy per unit of total volume is

$$k = \alpha_1 k_1 + \alpha_2 k_2 = \frac{1}{2}\alpha_1 \rho_1 v_1^2 + \frac{1}{2}\alpha_2 \rho_2 v_2^2 + \frac{1}{2}\alpha_1(\alpha_1\beta - 1)\rho_1 w^2 \tag{10}$$

where

$$\mathbf{w} = \mathbf{v}_1 - \mathbf{v}_2 \tag{11}$$

is the relative velocity. Equation (10) will not be needed for the simple results presented in this paper, but it forms the basis for numerous further derivations, for example, the variational approach used by Geurst (1985a, 1985b, 1986) to derive the general momentum equations in the two-fluid model of an inviscid suspension.

Returning to the case in which the particles are at rest, we have for every streamline through the matrix that comes from a common source with stagnation pressure p_0,

$$p_0 = p + \frac{1}{2}\rho_1 u_1^2 \tag{12}$$

u_1 is the local fluid velocity, distinguished from the average velocity v_1. If we average (12) throughout the fluid volume, we get

$$<p_0>_1 = <p>_1 + <\frac{1}{2}\rho_1 u_1^2>_1 \tag{13}$$

where $<>_1$ denotes the operation

$$<a>_1 = \left(\int_1 a \, dV\right)/V_1 \tag{14}$$

i.e. the integral of a variable "a" over the volume occupied by phase 1, divided by the total volume of phase 1.

We now define the macroscopic pressure of phase 1, the fluid, to be

$$p_1 = <p>_1 \tag{15}$$

Since p_0 is uniform, it is identical with its average.

The final term in (13) is the average kinetic energy of the fluid per unit volume expressed by (8), therefore, (13) can be rewritten as

$$p_0 = p_1 + \frac{1}{2}\rho_1 v_1^2 \alpha_1 \beta \tag{16}$$

or as

$$p_0 = p_1 + \frac{1}{2}\rho_1 v_1^2 + \frac{1}{2}\rho_1 v_1^2(\alpha_1 \beta - 1) \tag{17}$$

to emphasize that the final term plays the role of a correction to Bernoulli's equation resulting from the relative motion and its accompanying three-dimensional flow field.

Any candidate for the two-fluid conservation laws must be compatible with (17) in the special case of steady inviscid flow through a porous medium along a macroscopic streamline.

The average pressure at the interface between the phases (i.e. on the surface of the spheres) may be obtained by using an energy method and considering the work done by the interphase pressure when each sphere grows uniformly (Wallis, 1989). The result is

$$p_i = p_0 + \frac{1}{2}\rho_1 j_0^2 \frac{d\beta}{d\alpha_1} \tag{18}$$

In the absence of surface tension, or similar effects, it is reasonable to set the average pressure in phase 2 equal to the interphase pressure,

$$p_2 = p_i \tag{19}$$

Using (9) and (19) in (18) and combining with (16) we obtain, for the case where the particles are at rest,

$$p_2 - p_1 = \frac{1}{2}\rho_1 v_1^2 \alpha_1 \frac{d(\alpha_1 \beta - 1)}{d\alpha_1} \tag{20}$$

or

$$p_2 = p_1 - \xi \rho_1 v_1^2 \tag{21}$$

with ξ defined as

$$\xi = -\frac{\alpha_1}{2} \frac{d(\alpha_1 \beta - 1)}{d\alpha_1} \tag{22}$$

The reason for using the combination $(\alpha_1 \beta - 1)$ in (20) is that it is the same coefficient that appears in (10) and (17) and plays a fundamental role in describing inertial coupling or "added mass" effects. It has been called the "exertia" (Wallis, 1989). For a Maxwellian suspension of spheres we have, from (6), since $\alpha_1 = (1 - \alpha_2)$,

$$(\alpha_1 \beta - 1) = \frac{\alpha_2}{2} \tag{23}$$

and, therefore,

$$\xi = \frac{\alpha_1}{4} \tag{24}$$

For a dilute suspension $\alpha_1 \to 1$, $\alpha_2 \to 0$, $\xi \to 1/4$.

When both phases are moving, v_1 should be replaced by the relative velocity w in (20) and (21).

The momentum per unit overall volume is, by straightforward averaging,

$$\mathbf{m} = \alpha_1 \rho_1 \mathbf{v}_1 + \alpha_2 \rho_2 \mathbf{v}_2 \tag{25}$$

By conserving both energy and momentum for the uniform acceleration of a uniform suspension in a one-dimensional uniform duct, (10) and (25) can be used to derive effective equations of motion for each phase (Wallis, 1989),

$$\rho_1 \frac{dv_1}{dt} + \rho_1(\alpha_1 \beta - 1)\left(\frac{dv_1}{dt} - \frac{dv_2}{dt}\right) = -\frac{\partial P}{\partial x} + b_1 + f_1 \tag{26}$$

$$\rho_2 \frac{dv_2}{dt} - \rho_1 \frac{\alpha_1}{\alpha_2}(\alpha_1 \beta - 1)\left(\frac{dv_1}{dt} - \frac{dv_2}{dt}\right) = -\frac{\partial P}{\partial x} + b_2 + f_2 \tag{27}$$

$\partial P/\partial x$ is a macroscopic pressure gradient (it is the same for both phases since the relative motion is uniform) and b and f are body forces and "other forces per unit volume of that phase" respectively. Equations (26) and (27) relate $(\alpha_1 \beta - 1)$ to certain forms of "added mass coefficient."

A final result that can be deduced from (10) by an energy method is the external force per unit volume necessary to restrain a particle in a void fraction gradient in steady one-dimensional flow in a uniform duct (Wallis, 1989). It is simply

$$f_2 = -\frac{\rho_1 j_1^2}{2} \frac{d^2 \beta}{d\alpha_2^2} \frac{d\alpha_2}{dx} \tag{28}$$

Now, since j_1 is constant and equal to $\alpha_1 v_1$, we have

$$\frac{d\alpha_2}{dx} = -\frac{d\alpha_1}{dx} = \frac{\alpha_1}{v_1} \frac{dv_1}{dx} \tag{29}$$

Therefore (28) can also be expressed as

$$f_2 = -\frac{1}{2} \rho_1 v_1 \frac{dv_1}{dx} \alpha_1^3 \frac{d^2 \beta}{d\alpha_2^2} \tag{30}$$

TWO-FLUID MODELS

All of the results derived above are available for "testing" proposed conservation equations for two-phase flow.

COMPARISON WITH HYPOTHESIZED TWO-FLUID EQUATIONS

Many equations have been suggested to represent the conservation laws for the two-fluid model. There is usually agreement on the form of the continuity equations because they are simple and averaging is straightforward. The momentum equations, however, have proliferated in wonderful variety. For the purposes of this paper a few of the more recent one-dimensional formulations will be selected. Moreover, attention will be confined to the influences of inertial coupling and pressure in inviscid incompressible flows, ignoring effects such as drag and phase change.

The momentum equation for the continuous phase is written in the form adopted by Drew and Wood (1985), using the present nomenclature and correcting what appears to be an error in sign in the final term:

$$\rho_1 \left(\frac{\partial v_1}{\partial t} + v_1 \frac{\partial v_1}{\partial x}\right) = -\frac{\partial p_1}{\partial x} - \xi \rho_1 \frac{w^2}{\alpha_1} \frac{\partial \alpha_1}{\partial x} - \frac{\alpha_2}{\alpha_1} c_{vm} \rho_1 \left[\left(\frac{\partial v_1}{\partial t} + v_1 \frac{\partial v_1}{\partial x}\right) \right.$$
$$\left. - \left(\frac{\partial v_2}{\partial t} + v_2 \frac{\partial v_2}{\partial x}\right)\right] + \frac{1}{\alpha_1} \frac{\partial}{\partial x}\left[\alpha_1 \alpha_2 (a_1 + b_1) \rho_1 w^2\right] \quad (31)$$

with $a_1 = -3/20$, $b_1 = -1/20$ for a dilute suspension.

This equation resembles one due to Pauchon and Banerjee (1986), who omit the final term and set $\xi = 1/4$ and $c_{vm} = 1/2$, the values for a dilute suspension of spheres.

c_{vm} is a "coefficient of virtual mass" that is consistent with (26) if it is defined as

$$c_{vm} = \frac{\alpha_1}{\alpha_2}(\alpha_1 \beta - 1) \quad (32)$$

From (6), $c_{vm} = \alpha_1/2$ for a Maxwellian suspension.

For the equation of motion of the continuous phase we choose the Pauchon-Banerjee form with the added mass term in the form given by Drew and Wood,

$$\rho_2 \left(\frac{\partial v_2}{\partial t} + v_2 \frac{\partial v_2}{\partial x}\right) = -\frac{\partial p_2}{\partial x} + c_{vm} \rho_1 \left[\left(\frac{\partial v_1}{\partial t} + v_1 \frac{\partial v_1}{\partial x}\right) - \left(\frac{\partial v_2}{\partial t} + v_2 \frac{\partial v_2}{\partial x}\right)\right] \quad (33)$$

Drew and Wood have essentially the same equation but with p_2 replaced by

$$p_2 = p_1 - \xi \rho_1 w^2 \quad (34)$$

with ξ regarded as constant. This is the correct form of (21) when both phases move. Equation (33) is consistent with (27) when (32) is adopted.

Test 1 — A Sphere in an Accelerating Steady Flow

In this case, in order to hold the sphere in place, we must introduce an external force per unit volume of phase 2, f_2. We have the conditions

$$v_2 = 0, \quad \frac{\partial}{\partial t} = 0, \quad \alpha_2 = 0, \quad \frac{\partial \alpha_1}{\partial x} = 0$$

and therefore (31) reduces to the usual single-phase flow equation for the fluid,

$$\rho_1 v_1 \frac{dv_1}{dx} = -\frac{dp_1}{dx} \tag{35}$$

(33) becomes

$$0 = -\frac{dp_2}{dx} + c_{vm} \rho_1 v_1 \frac{dv_1}{dx} + f_2 \tag{36}$$

Use of (34) and (35) in (36) gives the external force on the sphere as

$$f_2 = -\rho_1 v_1 \frac{dv_1}{dx} (1 + c_{vm} + 2\xi) \tag{37}$$

Since $c_{vm} = 1/2$ and $\xi = 1/4$ in the limit of the dilute suspension, (37) has the effect of counting the virtual mass force twice, compared with (1). Alternatively, we may compare (36) with (4) and conclude that the term involving added mass should either not be there or should be cancelled by an additional term that is absent from the formulation.

Test 2 — Steady Inviscid Flow Through a Stationary Porous Medium

In this case, (17) is valid and may be differentiated to give

$$\rho_1 v_1 \frac{dv_1}{dx} = -\frac{dp_1}{dx} - \frac{d}{dx}\left[\frac{1}{2}\rho_1 v_1^2(\alpha_1 \beta - 1)\right] \tag{38}$$

Expanding the differential in square brackets and using (22) and (32) we obtain

$$\rho_1 v_1 \frac{dv_1}{dx} = -\frac{dp_1}{dx} - \rho_1 v_1 \frac{dv_1}{dx} c_{vm} \frac{\alpha_2}{\alpha_1} + \frac{1}{2}\rho_1 v_1^2 \frac{\xi}{\alpha_1} \frac{d\alpha_1}{dx} \tag{39}$$

TWO-FLUID MODELS

The term involving c_{vm} is the same as appears in (31). However, the term involving ξ has the opposite sign from the one appearing in (31) and it is not clear how the final term in (31) can resolve the disagreement.

Test 3 — Force on the Stationary Dispersed Phase in a Void Fraction Gradient in a Uniform Duct

In this case we need to compare the predictions of the two-fluid models with (30). Clearly, the only way the second derivative of β can emerge from (31) or (33) is by way of the differential of ξ; therefore, we cannot regard ξ as constant.

From (38) and (29) we have

$$\frac{dp_1}{dx} = -\rho_1 v_1 \frac{dv_1}{dx}(\alpha_1 \beta + \xi) \qquad (40)$$

and, from (34) and (29), since $v_2 = 0$,

$$\frac{dp_2}{dx} = \frac{dp_1}{dx} - \rho_1 v_1 \frac{dv_1}{dx}\left(2\xi - \alpha_1 \frac{d\xi}{d\alpha_1}\right) \qquad (41)$$

Combining (40) and (41) gives

$$\frac{dp_2}{dx} = -\rho_1 v_1 \frac{dv_1}{dx}\left(3\xi + \alpha_1 \beta - \alpha_1 \frac{d\xi}{d\alpha_1}\right) \qquad (42)$$

The terms in parentheses in (42) are readily shown, by using (22), to be equivalent to $\alpha_1^3 (d^2\beta/d\alpha_1^2)/2$. Therefore, (42) becomes

$$\frac{dp_2}{dx} = -\frac{1}{2}\rho_1 v_1 \frac{dv_1}{dx} \alpha_1^3 \frac{d^2\beta}{d\alpha_1^2} = -\frac{1}{2}\rho_1 v_1 \frac{dv_1}{dx} \alpha_1^3 \frac{d^2\beta}{d\alpha_2^2} \qquad (43)$$

Comparing (43) and (30) we have the very simple result

$$0 = -\frac{dp_2}{dx} + f_2 \qquad (44)$$

which indicates either that the added mass term does not belong in (33) or that it should be balanced by some other term that has not been included - the same conclusion that was reached for Test 1.

GEURST'S EQUATIONS

Geurst (1985a, 1985b, 1986) developed a variational approach to two-phase flow, starting from the equivalent of (10). He used a parameter "m," which in the present notation is equivalent to

$$m = \alpha_1(\alpha_1\beta - 1) = c_{vm}\alpha_2 \tag{45}$$

He used the symbol m' for $dm/d\alpha_2$ and we may easily show that

$$m + \alpha_1 m' = 2\alpha_1 \xi \tag{46}$$

Geurst (1985b) presents several equations of motion for each phase. His equation (2.24), the one-dimensional equation of motion for phase 1, in the present notation for incompressible phases, is

$$\left(\frac{\partial}{\partial t} + v_1\frac{\partial}{\partial x}\right)\left(v_1 - \frac{m}{\alpha_1}(v_2 - v_1)\right) - \frac{m}{\alpha_1}(v_2 - v_1)\frac{\partial v_1}{\partial x}$$

$$+ \frac{\partial}{\partial x}\left\{\frac{p_2}{\rho_1} + \frac{1}{2}m'(v_2 - v_1)^2\right\} = 0 \tag{47}$$

Using (45) and (46), (47) may be transformed to

$$\frac{\partial v_1}{\partial t} + v_1\frac{\partial v_1}{\partial x} = \frac{\partial}{\partial t}\left\{c_{vm}\frac{\alpha_2}{\alpha_1}(v_2 - v_1)\right\} + v_1\frac{\partial}{\partial x}\left\{c_{vm}\frac{\alpha_2}{\alpha_1}(v_2 - v_1)\right\}$$

$$+ c_{vm}\frac{\alpha_2}{\alpha_1}(v_2 - v_1)\frac{\partial v_1}{\partial x} - \frac{\partial}{\partial x}\left\{\left(\xi - \frac{c_{vm}\alpha_2}{2\alpha_1}\right)(v_2 - v_1)^2\right\} - \frac{1}{\rho_1}\frac{\partial p_2}{\partial x} \tag{48}$$

or, combining terms and using (22),

$$\frac{\partial v_1}{\partial t} + v_1\frac{\partial v_1}{\partial x} = -\frac{1}{\rho_1}\frac{\partial}{\partial x}(p_2 + \xi\rho_1 w^2) - w^2\frac{\xi}{\alpha_1}\frac{\partial \alpha_1}{\partial x} - c_{vm}\frac{\alpha_2}{\alpha_1}\left(\frac{\partial v_1}{\partial t} - \frac{\partial v_2}{\partial t}\right.$$

$$\left. + v_1\frac{\partial v_1}{\partial t} - v_2\frac{\partial v_2}{\partial x}\right) + (v_2 - v_1)\left(\frac{\partial}{\partial t} + v_1\frac{\partial}{\partial x}\right)\left(c_{vm}\frac{\alpha_2}{\alpha_1}\right) \tag{49}$$

In view of (34), (49) is identical with (31) in all but the final term, which can also be written, using (22) and (45), as

$$-2(v_2 - v_1)\frac{\xi}{\alpha_1}\left(\frac{\partial \alpha_1}{\partial t} + v_1\frac{\partial \alpha_1}{\partial x}\right)$$

With this final term, (49) passes Test 2.

TWO-FLUID MODELS

Geurst's equation (2.25), the one-dimensional equation of motion for the discontinuous phase in incompressible flow, is, adding an external force per unit volume of phase 2, f_2

$$\left(\frac{\partial}{\partial t}+v_2\frac{\partial}{\partial x}\right)\left\{v_2+\frac{\rho_1 m}{\rho_2 \alpha_2}(v_2-v_1)\right\}+\frac{\rho_1 m}{\rho_2 \alpha_2}(v_2-v_1)\frac{\partial v_2}{\partial x}+\frac{1}{\rho_2}\frac{\partial p_2}{\partial x}=\frac{f_2}{\rho_2} \qquad (50)$$

which can be rearranged to

$$\rho_2\left(\frac{\partial}{\partial t}+v_2\frac{\partial}{\partial x}\right)v_2 = -\frac{\partial p_2}{\partial x}+f_2+\rho_1 c_{vm}\left(\frac{\partial v_1}{\partial t}-\frac{\partial v_2}{\partial t}+v_1\frac{\partial v_1}{\partial x}-v_2\frac{\partial v_2}{\partial x}\right)$$

$$-\rho_1(v_2-v_1)\frac{\partial}{\partial t}c_{vm}-\rho_1 c_{vm}(v_2-v_1)\frac{\partial}{\partial x}(v_2-v_1) \qquad (51)$$

The final term in (51) enables it to pass <u>both Test 1 and Test 3</u>.

Although Geurst's equations presented here pass the three tests, they may not be universally applicable to one-dimensional flows as they were derived for flow in a constant area duct. It might be necessary to reduce the vector equations presented by Geurst (1986) to one-dimensional form to describe flows in nozzles, for instance.

CONCLUSIONS

Solutions to five simple potential two-phase flows have been presented.

In the first, the difference in mean pressure between the phases is related to the relative motion, in agreement with previous formulations. In addition, the coefficient ξ in the resulting expression is found to be more generally related to the derivative with respect to void fraction of a function, $(\alpha_1 \beta - 1)$, resembling an "added mass coefficient."

The second example, in which both phases are uniformly accelerated, leads to equations of motion containing "added mass coefficients," allowing several formulations to be compared. The result is a more general formulation of the added mass coefficient, valid for all void fractions and consistent with other formulations in the limit of a dilute suspension.

The three further situations chosen to test general "two-fluid" equations of motion suggested in the literature were:

a) The force on a single sphere in an accelerated steady flow.

b) The averaged Bernoulli equation for steady ideal incompressible flow through a stationary porous medium.

c) The force on a sphere in a void fraction gradient for steady flow in a uniform duct.

The Pauchon-Banerjee and Drew-Wood formulations, that are typical of recent hypotheses, failed all three tests, indicating either that some terms in the equations are incorrectly formulated or that additional terms are required.

There is a need for more such "standard" solutions that can be used as further tests of suggested general equations and can provide a filter for picking out possible errors.

NOMENCLATURE

a	Any variable
b	Body force per unit volume of a particular phase
a_1, b_1	Coefficients in (31)
C, c_{vm}	Coefficients of added mass
f	Force per unit volume of a particular phase
j	Volumetric flux
j	One-dimensional component of **j**
k	Kinetic energy per unit volume
K	Kinetic energy
m	Coefficient of added kinetic energy (see (45))
m	Momentum density
p	Pressure
P	Macroscopic pressure
t	Time
x	One-dimensional coordinate
u	Local velocity
v	Average velocity
v	One-dimensional component of **v**
V	Volume
w	Relative velocity $\mathbf{v}_1 - \mathbf{v}_2$
w	One-dimensional component of **w**

Greek

α	Volumetric fraction
β	Defined in (5)
ξ	Defined in (22)
ρ	Density
ϕ	Potential

Subscripts

0	With particles at rest, stagnation conditions
1	Phase 1 (continuous)
2	Phase 2 (discontinuous)
i	Interface

REFERENCES

Drew, D.A. and Wood, R.T. (1985) Overview and Taxonomy of Models and Methods, International Workshop on Two-Phase Flow Fundamentals, National Bureau of Standards, Maryland.

Geurst, J.A. (1985a) Virtual Mass in Two-Phase Bubbly Flow, *Physica*, vol. 129A, p. 233.

Geurst, J.A. (1985b) Two-Fluid Hydrodynamics of Bubbly Liquid/Vapour Mixture Including Phase Change, *Philips J. Res.*, vol. 40, p. 352.

Geurst, J.A. (1986) Variational Principles and Two-Fluid Hydrodynamics of Bubbly Liquid/Gas Mixtures, *Physica*, vol. 135A, p. 455.

Maxwell, J.C. (1881) *A Treatise on Electricity and Magnetism*, 2nd ed., vol. 1, p. 398, Clarendon Press, Oxford.

Milne-Thomson, L.M. (1949) *Theoretical Hydrodynamics*, 2nd ed., MacMillan, London, (a) page 420, paragraph 15-43 - note error in denominator in book; (b) page 417; (c) page 433, problem 39; (d) page 92.

Pauchon, C. and Banerjee, S. (1986) Interphase Momentum Interaction Effects in the Averaged Multifluid Model, Part 1, Void Propagation in Bubbly Flow, *Int. J. Multiphase Flow*, vol. 12, p. 559.

Taylor, G.I. (1928) The Forces on a Body Placed in a Curved or Converging Stream of Fluid, *Proc. Roy. Soc.*, vol. A120, pp. 260-283.

Turner, J.C.R. (1976) Two-Phase Conductivity: The Electrical Conductance of Liquid-Fluidized Beds of Spheres, *Chem. Eng. Sci.*, vol. 31, p. 487.

Wallis, G.B. (1989) Inertial Coupling in Two-Phase Flows: Macroscopic Properties of Suspensions in an Inviscid Fluid, *Multiphase Science & Technology*, vol. 5.

ESTIMATION OF VOLUMETRIC FRACTIONS OF EACH PHASE IN GAS-LIQUID-SOLID THREE-PHASE SLUG FLOW IN VERTICAL PIPES

T. SAKAGUCHI and H. MINAGAWA
Kobe University
Rokkodai, Nada, Kobe, 657, Japan

K. SAHARA
Sumitomo Electric Industries Co. Ltd.
Kitahama, Higashi, Osaka, 541, Japan

T. SAIBE
Mitsubishi Kasei Co. Ltd.
Mizushima Factory
Ushio-dori, Kurashiki, 712
Japan

ABSTRACT

A framework has been proposed to estimate the volumetric fractions of each phase in gas-liquid-solid three-phase flows in vertical pipes(Sakaguchi,1987a;Sakaguchi,1987b). The framework is applied to the gas-liquid-solid three-phase slug flow, and two main equations are solved with many correlations to estimate the volumetric fractions. The necessary correlations are mainly derived by the extension of those used in the gas-liquid two-phase slug flow to the three-phase slug flow and by the experimental results. The estimated values are compared with the experimental results of air-water-solid particles three-phase slug flow in vertical pipes and we conclude that this method is useful.

INTRODUCTION

Gas-liquid-solid three-phase flows are encountered in a lift pipe of an air lift pump which is used to mine the manganese nodules from deep sea beds, and in preheaters and reactors of coal liquefaction plants. In order to design such facilities, it is important to have an estimate of the volumetric fractions of each phase.

In previous papers(Sakaguchi,1987a;Sakaguchi,1987b), the framework of a method has been presented to estimate the volumetric fractions of each phase in gas-liquid-solid three-phase flows in vertical pipes. In this paper, the volumetric fractions in the gas-liquid-solid three-phase slug flow are estimated on the basis of this method and compared with the experimental results.

The two main equations are based principally on the drift flux model(Zuber,1965;Giot,1982). The terminal rising velocity of a large gas bubble is used in the equation for the gas phase. The suspension volumetric flux of the liquid phase, by which the solid particles are suspended or hovering in the gas-liquid two-phase flow, is used for the solid phase. These two main equations must be solved with many correlations which express the relation between the rising velocity of the large gas bubble and volumetric fluxes, the volumetric fractions of each phase in the large gas bubble part and in the liquid slug part and the volumetric fractions in the slug unit, etc. Since no such correlations have been proposed for the three-phase slug flow, the correlations for the gas-liquid two-phase slug flow are extended to the three-phase slug flow, considering the existence of the solid particles in the liquid phase. Experiments are carried out and their results are used to obtain several constants and coefficients. The experimental results are well correlated by this method and the equations. It is concluded from these results that the proposed framework of the estimation of the volumetric fractions is useful if the solid phase exists within the liquid phase.

EXPERIMENTAL APPARATUS AND INSTRUMENTATION

The volumetric fractions of each phase in gas-liquid-solid three-phase slug flows are measured in vertical pipes of 20.9 mm and 30.6 mm inside diameter which are made of transparent acrylic resin. The characteristics of solid particles used in this experiment are shown in Table 1. Water and air are used as the liquid and as the gas phases, respectively, at room temperature and at atmospheric pressure.

Figure 1 is a schematic diagram of the experimental apparatus. It consists mainly of six parts: the test section, the solid and water supply line, the water supply line, the water drainage, the air supply line, and instruments.

The solid particles accumulated in the hopper are supplied to the water tank with the electromagnetic feeder, and then fed to the test section with water through the solid-water inlet by the first Mohno pump. A large amount of water is required to supply the solid particles without any choking or burning accident. Additional water is supplied from the bottom of the test section by the second Mohno pump to lift the solid particles and prevent them from settling and piling in the bottom. Excess water is drained away through the water drainage by the third Mohno pump. The flow rate of the solid particles is controlled by the vibration frequency of the electromagnetic feeder and that of water by the combination of the rotating speed of the three Mohno pumps. The flow rate of the solid particles and of water is measured, respectively, by the graduated cylinders. Air from the compressor is fed to the test section through the cooler, the filters, the regulator valve, and the critical flow nozzle. The flow rate of air is controlled and measured by the pressure at the upstream side of the critical flow nozzle. The relation between the flow rate of air and the pressure is calibrated in advance.

Air, water, and the solid particles flow up in the test section as the gas-liquid-solid three-phase cocurrent upflow. Air is separated from the solid particles and water in the air separator. The solid particles are separated from water in the solid separator. The solid particles and water return to the hopper and the water tank, respectively.

TABLE 1. Characteristics of solid particles

Material	Aluminium	Ceramics	Aluminium
Shape	Sphere	Sphere	Sphere
Mean diameter d_s mm	2.57	4.17	2.96
Density ρ_s kg/m³	2380	2400	2640
Free settling velocity V_{st} m/s †	0.306	0.432	0.375
Drag coefficient C_D†	0.498	0.409	0.453

†T_ℓ = 20°C

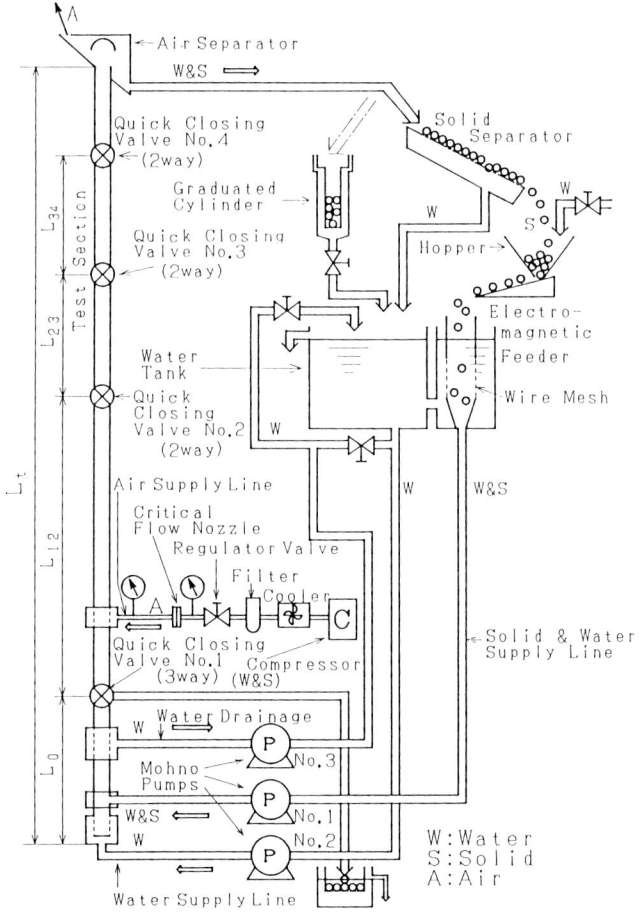

	L_t (mm)	L_0 (mm)	L_{12} (mm)	L_{23} (mm)	L_{34} (mm)
D=20.9mm	9820	599	3853	2002	2020
D=30.6mm	9965	716	3841	2003	2002

FIGURE 1. Schematic diagram of experimental apparatus

Seven pressure taps are installed on each test pipe to measure the static pressure distribution along the test section. Four valves are installed on each test section to measure the volumetric fractions of each phase. Three of them are the two-way valves, and one is the three-way valve. The volumetric fractions of each phase are measured by the quick-closing valve method. Therefore, the obtained volumetric fractions are not the local values but the average values between the two valves. The volumes of the solid particles and the gas phase trapped between the two valves are obtained from their heights by using the relations between their heights and volumes calibrated in advance for each test section and each solid particles. The volume of the liquid phase is calculated by subtracting both volumes of the solid and the gas phase from the whole volume between two valves. The volumetric fractions of each phase are obtained by dividing the volume of each phase by the whole volume. Two measured values of the volumetric fractions are obtained in the volumes between valves No.2 and No.3 and between valves No.3 and No.4 by one closing in this experimental apparatus. It is confirmed that there is no difference in qualitative characteristics between these two measured values of the volumetric fractions. This fact proves that the flow is fully developed. Thus, the two values are adopted for the experimental data bases.

The volumetric flux of the gas phase is corrected to the value corresponding to the static pressure at the center of the two valves, supposing that the gas phase isothermally experiences volume changes.

The experiments for the volumetric fraction in gas-liquid-solid three-phase flows are carried out in the slug flow region. The ranges of the volumetric fluxes are $<j_g>$ = 0.282-0.893 m/s, $<j_\ell>$ = 0.473-0.978 m/s, and $<j_s>$ = 0.00429-0.0635 m/s. Here, $<\ >$ denotes the average value of a quantity in the cross-sectional area as defined in Eq.(1)(Zuber,1965).

$$<j_i> = \frac{1}{A} \int_A j_i(r) dA \qquad (1)$$

where $j_i(r)$ is the local value of j_i, and the subscript i (=g,ℓ,s) denotes gas, liquid, and solid, respectively.

The volumetric flux of the liquid phase by which the solid particles are suspended or hovering in the gas-liquid two-phase flow is also measured by this experimental apparatus. This volumetric flux is henceforth referred to as the suspension volumetric flux of the liquid phase. Two sheets of wire mesh are installed on valves No.2 and No.3 to prevent the solid particles from flowing out of the test section. The solid particles whose volume is measured beforehand are accumulated in the test section between these valves. Water and air is supplied to the test section until the solid particles are suspended or hovering uniformly throughout the test section. Then, the volumetric flux of the liquid phase is measured by the graduated cylinders, that of the gas phase is measured by the pressure at the upstream side of the critical flow nozzle, and the volumetric fraction of the gas

SLUG FLOW IN VERTICAL PIPES

phase is measured by the quick-closing valve method. This experiment is carried out for the slug flow region.

EXPERIMENTAL RESULTS OF VOLUMETRIC FRACTION

The volumetric fractions of each phase in the gas-liquid-solid three-phase slug flow in vertical pipes are measured for six experimental conditions created by the combination of the two test pipes with three kinds of the solid particles. The precision index(ASME,1985) of the measured values is 0.0147 for $<\alpha_g>$, 0.0149 for $<\alpha_\ell>$, 0.00212 for $<\alpha_s>$ in the range $0.01 < <\alpha_s> < 0.02$ and 0.00522 in the range $0.05 < <\alpha_s> < 0.06$.

All of the experimental results of the volumetric fractions of each phase are shown versus the volumetric flux $<j_s>$ of the solid phase in Fig.2. The values of $<j_\ell>$ and $<j_g>$ in the figures are the mean values of their experimental values which are held in the range of ± 3 %, respectively. In every case, the volumetric fraction $<\alpha_s>$ of the solid phase increases with $<j_s>$, and decreases with increasing $<j_g>$ or $<j_\ell>$. The value of $<\alpha_g>$ increases with $<j_g>$, and decreases with increasing $<j_\ell>$ or $<j_s>$. In the same manner, the value of $<\alpha_\ell>$ increases with $<j_\ell>$, and decreases with increasing $<j_g>$ or $<j_s>$. That is, in this experimental range, the volumetric fraction of a certain phase increases with the volumetric flux of the corresponding phase, and decreases with the increase of the volumetric fluxes of the other phases. Furthermore, in this experimental range, the value of $<\alpha_s>$ increases with the mean solid diameter d_s and slightly increases with the density ρ_s of the solid phase. The effects of d_s and ρ_s on $<\alpha_g>$ or $<\alpha_\ell>$ are not obvious. The effect of the pipe diameter D on the volumetric fractions of each phase is not obvious, either. The solid lines in Fig.2 are drawn by the proposed estimation described later.

ESTIMATION OF VOLUMETRIC FRACTIONS OF EACH PHASE

In the previous papers(Sakaguchi,1987a;Sakaguchi,1987b), the framework of a method has been proposed to estimate the volumetric fractions of each phase in gas-liquid-solid three-phase flows in vertical pipes.

The volumetric fraction $<\alpha_g>$ of the gas phase may be estimated by the same equation as proposed by Zuber-Findlay (1965) and Giot(1982):

$$<\alpha_g> = \frac{<j_g>}{\bar{V}_g} = \frac{<j_g>}{C_g<j_T>+K_g} \qquad (2)$$

where $<j_T>$ and $<j_g>$: the volumetric flux of the mixture and of the gas phase, $\bar{V}_g$: the volumetric fraction weighted mean velocity of the gas phase, C_g: distribution parameter of the gas phase, K_g: the volumetric fraction weighted mean drift velocity of the gas phase to the mixture. The volumetric fraction weighted mean velocity is defined in Eq.(3)(Zuber,

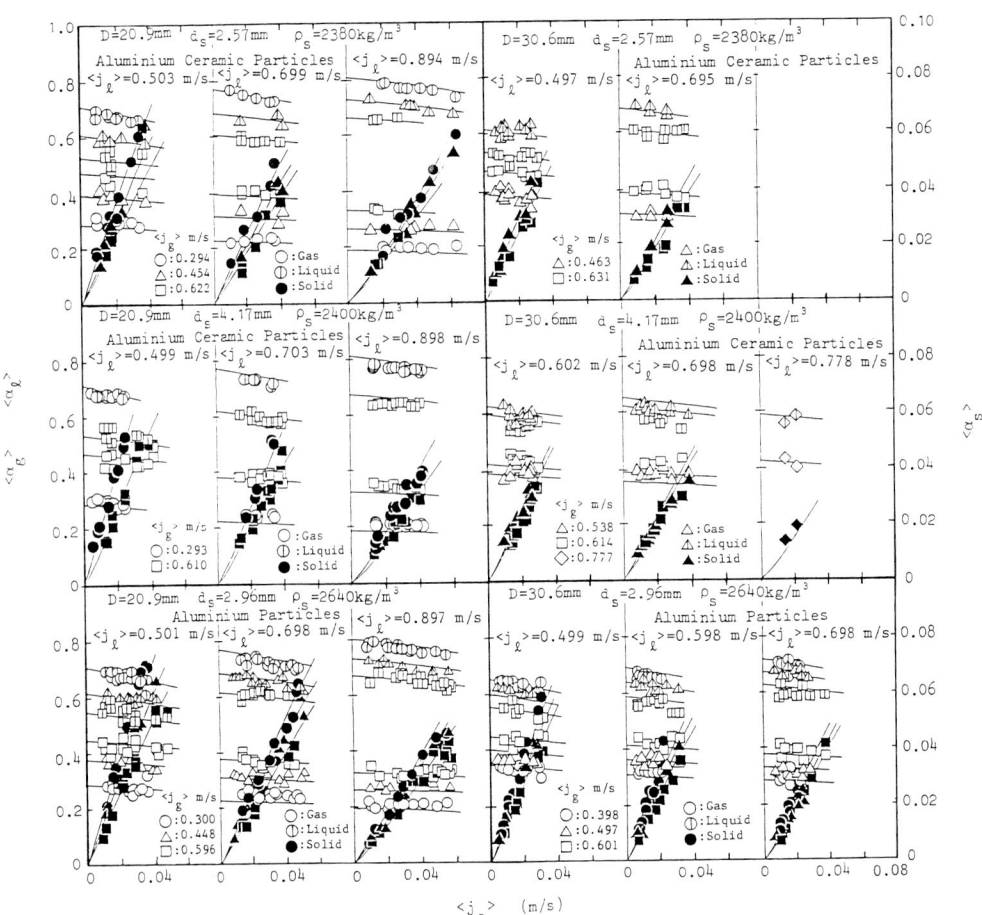

FIGURE 2. Volumetric fractions of each phase in gas-liquid-solid three-phase slug flow (———: Estimated by the proposed method)

1965).

$$\overline{V}_i = \frac{\frac{1}{A}\int_A \alpha_i(r)V_i(r)dA}{\frac{1}{A}\int_A \alpha_i(r)dA} = \frac{<\alpha_i V_i>}{<\alpha_i>} \qquad (3)$$

The volumetric fraction $<\alpha_s>$ of the solid phase may be estimated by Eq.(4) which was derived from the drift flux model(Sakaguchi,1987a;Sakaguchi,1987b).

$$<\alpha_s>^* = \frac{<j_s>^*}{\overline{V}_s^*} = \frac{<j_s>^*}{C_s^*(<j_\ell+j_s>^* - <j_\ell>_f^*)} \qquad (4)$$

SLUG FLOW IN VERTICAL PIPES

where

$$<\alpha_s>^* = \frac{<\alpha_s>}{1 - <\alpha_g>} \qquad (5)$$

$$<j_\ell>^* = \frac{<j_\ell>}{1 - <\alpha_g>} \qquad (6)$$

$$<j_s>^* = \frac{<j_s>}{1 - <\alpha_g>} \qquad (7)$$

$$<j_\ell + j_s>^* = \frac{<j_\ell + j_s>}{1 - <\alpha_g>} \qquad (8)$$

$$<j_\ell>_f^* = \frac{<j_\ell>_f}{1 - <\alpha_g>} = \frac{Q_{\ell f}}{(1 - <\alpha_g>)A} \qquad (9)$$

$\bar{V}_s^*$: the volumetric fraction weighted mean velocity of the solid phase in the cross-sectional area not occupied by the gas phase, C_s^*: a coefficient, $<j_\ell>_f^*$, $<j_\ell>_f$: the suspension volumetric flux of the liquid phase for the cross-sectional area that the gas phase does not occupy, and for the total cross-sectional area, respectively, $Q_{\ell f}$: the volume flow rate of the liquid phase for the condition that the solid particles are suspended in the gas-liquid two-phase flow, and A: total cross-sectional area of the pipe. Thus, * denotes the value defined in the region not occupied by the gas phase. Equation (4) has been derived by applying the method of estimation of the volumetric fraction of the solid phase in liquid-solid two-phase flow(Sakaguchi,1987a;Sakaguchi,1987b) to the part of the liquid-solid mixture, assuming that this mixture flows in the cross-sectional area that the gas phase does not occupy.

The suspension volumetric flux $<j_\ell>_f^*$ of the liquid phase has been used in this method as another basic point to correlate the $\bar{V}_s$-$<j_T>$ line(Sakaguchi,1987a;Sakaguchi,1987b) in addition to the terminal settling velocity of the solid particles in the stationary liquid which had been usually related to the volumetric fraction weighted mean drift velocity of the solid phase to the mixture(Zuber,1965).

In order to obtain the volumetric fractions of each phase from Eqs. (2) and (4) under the given quantities, namely, physical properties, flow conditions and geometries (ρ_g, ρ_ℓ, ρ_s, μ_g, μ_ℓ, σ, P, T, $<j_g>$, $<j_\ell>$, $<j_s>$, D and d_s), it is necessary to prepare correlations for C_g, K_g, C_s^* and $<j_\ell>_f^*$.

In the previous papers(Sakaguchi,1987a;Sakaguchi,1987b), the values of $<\alpha_g>$ and $<\alpha_s>$ were estimated for air-water-solid particles three-phase slug flow at room temperature and at atmospheric pressure. In those studies, it was assumed that the experimental equation(10) of $\bar{V}_{g2}$ in air-water two-phase slug flow can be applied to air-water-solid particles three-phase slug flow when the volumetric fraction of the solid phase is less than the value of 0.1 by replacing the term

$\langle j_{g2}+j_{\ell 2}\rangle$ with the term $\langle j_g+j_\ell+j_s\rangle$ as shown in Eq.(11),

$$\overline{V}_{g2} = \frac{\langle j_{g2}\rangle}{\langle \alpha_{g2}\rangle} = 1.13 \langle j_{g2} + j_{\ell 2}\rangle + 0.15 \tag{10}$$

$$\overline{V}_g = \frac{\langle j_g\rangle}{\langle \alpha_g\rangle} = 1.13 \langle j_g + j_\ell + j_s\rangle + 0.15 \tag{11}$$

where the subscript 2 denotes two-phase flow. That is, the values of 1.13 and 0.15 obtained from the experimental results of the air-water two-phase slug flow were used for the value of C_g and of K_g in Eq.(2), respectively.

The values of C_s^* and $\langle j_\ell\rangle_f^*$ were given by Eqs.(12) and (13) which were obtained from the experimental results of the air-water-aluminium particles three-phase slug flow.

$$C_s^* = \frac{0.66}{\left(\frac{\langle \alpha_s\rangle^*}{0.021}\right)^{1.6} + 1} + 1 \tag{12}$$

$$\langle j_{\ell f}\rangle^* = 0.311(1 - \langle \alpha_s\rangle^*)\exp\left(-2.52 \frac{\langle \alpha_s\rangle^*}{1 + \langle \alpha_s\rangle^*}\right) \tag{13}$$

Then, the volumetric fractions of each phase were estimated by solving simultaneous Eqs.(2),(4)-(13). The volumetric fraction $\langle \alpha_\ell\rangle$ of the liquid phase were obtained by Eq.(14).

$$\langle \alpha_\ell\rangle = 1 - (\langle \alpha_g\rangle + \langle \alpha_s\rangle) \tag{14}$$

Since it has been the main object in the previous papers to propose the framework to estimate the volumetric fractions of each phase in the gas-liquid-solid three-phase flow, the empirical equations obtained from the limited experimental data were used. So, in this paper, more general correlations for C_g, K_g, C_s^* and $\langle j_\ell\rangle_f^*$ are derived by applying the correlations used in the gas-liquid two-phase slug flow to the three-phase slug flow and by using the experimental data obtained from the six experimental conditions.

As mentioned above, the correlations in gas-liquid two-phase flows are applied to three-phase flows while considering the existence of the solid particles in the liquid phase. In order to apply them, a one-dimensional three-phase slug flow model shown in Fig.3 is considered. The model consists of two parts: the large gas bubble part and the liquid slug part. Their lengths are L_g and L_ℓ, respectively. The volumetric fraction weighted mean velocities $\overline{V}_{ij}$ and the volumetric fractions $\langle \alpha_{ij}\rangle$ of each phase in each part are defined as shown in the figure. Here, the subscript $j(=g,\ell)$ denotes the large gas bubble part and the liquid slug part, respectively.

In the large gas bubble part, there is a large gas bubble which almost fills the total cross-sectional area, and there

SLUG FLOW IN VERTICAL PIPES

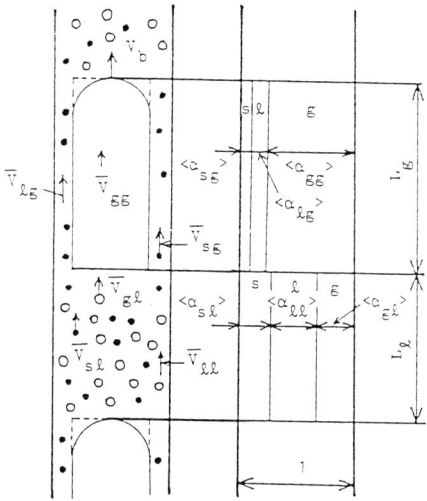

FIGURE 3. One-dimensional three-phase slug flow model

are no solid particles in it. The liquid film containing a small amount of solid particles and no small gas bubbles surrounds the large gas bubble.

In the liquid slug part, small gas bubbles and solid particles are dispersed in the liquid phase. This flow condition is supposed to be the gas-liquid-solid three-phase bubbly flow in this study.

In the first part of this section, the estimation of the volumetric fraction of the gas phase will be described by making a few assumptions, and by using some correlations for gas-liquid two-phase flows. In the second part, the estimation of the volumetric fraction of the solid phase will be discussed.

Correlations to Estimate Volumetric Fraction of Gas Phase

The volumetric faction $\langle\alpha_g\rangle$ of the gas phase is expressed in Eq.(2) by the drift flux model. In order to calculate $\langle\alpha_g\rangle$ by Eq.(2) for given physical properties, flow conditions and geometries, the value of the distribution parameter C_g and the volumetric fraction weighted mean drift velocity K_g of the gas phase must be obtained.

Now, the correlations for C_g and K_g will be derived. Following three assumptions are made on the three-phase slug flow model.
Assumption:
(1) The flow is isothermal and one-dimensional. There is no heat and mass transfer between phases.
(2) Both the rising velocity V_b of the large gas bubble and the mean gas velocity $\bar{V}_{g\ell}$ in the liquid slug part have

the same value as the volumetric fraction weighted mean gas velocity $\overline{V}_g$.

$$V_b = \overline{V}_{g\ell} = \overline{V}_g \qquad (15)$$

(3) The solid particles uniformly distribute in the liquid phase only.

Fukano et al.(1980) proposed Eq.(16) for the rising velocity V_{b2} of a large gas bubble in the gas-liquid two-phase slug flow whose liquid slug part contains small gas bubbles.

$$V_{b2} = C_{12} \overline{V}_{\ell\ell 2} + V_{bt2} \qquad (16)$$

Namely, the rising velocity of the large gas bubble is expressed by the sum of the terminal rising velocity V_{bt2} of a large gas bubble in a stationary liquid and the product of the liquid velocity $\overline{V}_{\ell\ell 2}$ in the liquid slug and a coefficient C_{12}. The value $\overline{V}_{\ell\ell 2}$ is estimated by Eq.(17) which was derived by the volume balance in the liquid slug part considering the existence of small bubbles whose volumetric fraction is expressed by $<\alpha_{g\ell 2}>$.

$$\overline{V}_{\ell\ell 2} = \frac{<j_{T2}> - <\alpha_{g\ell 2}>\overline{V}_{g2}}{1 - <\alpha_{g\ell 2}>} \qquad (17)$$

Substituting Eq.(17) into Eq.(16), V_{b2} can be expressed in Eq.(18).

$$V_{b2} = \frac{C_{12}}{1 + (C_{12}-1)<\alpha_{g\ell 2}>}<j_{T2}> + \frac{(1 - <\alpha_{g\ell 2}>)V_{bt2}}{1 + (C_{12}-1)<\alpha_{g\ell 2}>} \qquad (18)$$

By the analogy of this procedure, equations of the rising velocity V_b of a large gas bubble in the three-phase slug flow is supposed to be expressed by Eq.(19).

$$V_b = C_1 \overline{V}_{\ell\ell} + V_{bt} \qquad (19)$$

Here, $\overline{V}_{\ell\ell}$ is the liquid velocity in the liquid slug part, V_{bt} is the terminal rising velocity of a large gas bubble in the condition that the total volumetric flux $<j_T>$ is zero, and C_1 is a coefficient. According to the volume balance in the liquid slug part, the following relation is obtained.

$$A<j_T> = A(\overline{V}_{g\ell}<\alpha_{g\ell}> + \overline{V}_{\ell\ell}<\alpha_{\ell\ell}> + \overline{V}_{s\ell}<\alpha_{s\ell}>) \qquad (20)$$

Eliminating $\overline{V}_{\ell\ell}$ from Eqs.(19) and (20), and using Eq.(15), $\overline{V}_g$ is expressed by the next equation.

$$\overline{V}_g = \left\{ \frac{C_1}{1 + (C_1 - 1)<\alpha_{g\ell}> - <\alpha_{s\ell}>} \right\}<j_T>$$

$$+ \left\{ \frac{(1 - <\alpha_{g\ell}> - <\alpha_{s\ell}>)V_{bt} - C_1<\alpha_{s\ell}>\overline{V}_{s\ell}}{1 + (C_1 - 1)<\alpha_{g\ell}> - <\alpha_{s\ell}>} \right\} \qquad (21)$$

SLUG FLOW IN VERTICAL PIPES

The coefficient of $\langle j_T \rangle$ corresponds to C_g, and the second term corresponds to K_g in Eq.(2). Correlations which relates the unknown quantities $\langle \alpha_{g\ell} \rangle$, $\langle \alpha_{s\ell} \rangle$, $\bar{V}_{s\ell}$, C_1 and V_{bt} in Eq.(21) to the physical properties, flow conditions and geometries must be obtained to calculate $\langle \alpha_g \rangle$ by Eqs.(2) and (21). However, such correlations have not been obtained yet. Therefore, the correlations of the unknown quantities in Eq.(21) are derived by making use of not only the physical properties, flow conditions and geometries but also the volumetric fractions $\langle \alpha_g \rangle$, $\langle \alpha_\ell \rangle$ and $\langle \alpha_s \rangle$. By giving the trial values to these volumetric fractions which should be finally estimated, the iterative method is performed to estimate them until all the equations are satisfied.

First, the correlations for $\langle \alpha_{g\ell} \rangle$ and $\langle \alpha_{s\ell} \rangle$ will be derived. The assumption (3) gives the following relation.

$$\frac{\langle \alpha_{sg} \rangle}{\langle \alpha_{\ell g} \rangle} = \frac{\langle \alpha_{s\ell} \rangle}{\langle \alpha_{\ell \ell} \rangle} \tag{22}$$

Since the sum of the volumetric fractions in any cross-sectional area is unity, the following equations are obtained.
For the large gas bubble part:

$$\langle \alpha_{gg} \rangle + \langle \alpha_{\ell g} \rangle + \langle \alpha_{sg} \rangle = 1 \tag{23}$$

For the liquid slug part:

$$\langle \alpha_{g\ell} \rangle + \langle \alpha_{\ell \ell} \rangle + \langle \alpha_{s\ell} \rangle = 1 \tag{24}$$

By these equations, $\langle \alpha_{s\ell} \rangle$ can be expressed as Eq.(25).

$$\langle \alpha_{s\ell} \rangle = \frac{1 - \langle \alpha_{g\ell} \rangle}{1 - \langle \alpha_{gg} \rangle} \langle \alpha_{sg} \rangle \tag{25}$$

In order to relate these volumetric fractions of each phase in each part to $\langle \alpha_g \rangle$ and $\langle \alpha_s \rangle$, the length L_g of the large gas bubble part and the length L_ℓ of the liquid slug part are introduced. The volumetric fraction of the gas phase and that of the solid phase are related to them by Eqs.(26) and (27), respectively.

$$\langle \alpha_g \rangle = \frac{L_g \langle \alpha_{gg} \rangle + L_\ell \langle \alpha_{g\ell} \rangle}{L_g + L_\ell} \tag{26}$$

$$\langle \alpha_s \rangle = \frac{L_g \langle \alpha_{sg} \rangle + L_\ell \langle \alpha_{s\ell} \rangle}{L_g + L_\ell} \tag{27}$$

Substituting Eqs.(26) and (27) into Eq.(25) to eliminate $\langle \alpha_{gg} \rangle$ and $\langle \alpha_{sg} \rangle$, and rearranging it to obtain the equation of $\langle \alpha_{s\ell} \rangle$.

$$\langle \alpha_{s\ell} \rangle = \frac{(1 - \langle \alpha_\ell \rangle)\langle \alpha_s \rangle}{1 - \langle \alpha_g \rangle} \qquad (28)$$

As for the correlation of $\langle \alpha_{g\ell} \rangle$, Akagawa et al.(1965) proposed the expression $\langle \alpha_{g\ell 2} \rangle = \langle \alpha_{g2} \rangle^z$ (z=1.8) for the gas-liquid two-phase slug flow. This type of the formula is supposed to be able to be extended to the gas-liquid-solid three-phase slug flow by considering the existence of the solid phase as follows. The total cross-sectional area A is assumed to be reduced to $A(1-\langle \alpha_{s\ell} \rangle)$ in the liquid slug part, and to $A(1-\langle \alpha_s \rangle)$ in a slug unit. Then, this formula is extended to the three-phase slug flow as shown by Eq.(29).

$$\frac{A\langle \alpha_{g\ell} \rangle}{A(1 - \langle \alpha_{s\ell} \rangle)} = \left(\frac{A\langle \alpha_g \rangle}{A(1 - \langle \alpha_s \rangle)} \right)^z \qquad (29)$$

Rearranging Eq.(29), the next formula for $\langle \alpha_{g\ell} \rangle$ is obtained.

$$\langle \alpha_{g\ell} \rangle = \left(\frac{\langle \alpha_g \rangle}{1 - \langle \alpha_s \rangle} \right)^z (1 - \langle \alpha_{s\ell} \rangle) \qquad (30)$$

By solving Eqs.(28) and (30), unknown quantities in Eq.(21), $\langle \alpha_{g\ell} \rangle$ and $\langle \alpha_{s\ell} \rangle$, are obtained and they are expressed as the function of only $\langle \alpha_g \rangle$ and $\langle \alpha_s \rangle$ as shown in Eqs.(31) and (32), respectively.

$$\langle \alpha_{g\ell} \rangle = \left(\frac{\langle \alpha_g \rangle}{1 - \langle \alpha_s \rangle} \right)^z \frac{1 - \langle \alpha_g \rangle - \langle \alpha_s \rangle}{1 - \langle \alpha_g \rangle - \left(\frac{\langle \alpha_g \rangle}{1 - \langle \alpha_s \rangle} \right)^z \langle \alpha_s \rangle} \qquad (31)$$

$$\langle \alpha_{s\ell} \rangle = \frac{\left(1 - \frac{\langle \alpha_g \rangle}{(1 - \langle \alpha_s \rangle)} \right)^z \langle \alpha_s \rangle}{1 - \langle \alpha_g \rangle - \left(\frac{\langle \alpha_g \rangle}{1 - \langle \alpha_s \rangle} \right)^z \langle \alpha_s \rangle} \qquad (32)$$

The value of the exponent z must be evaluated by the experimental results.

Secondly, the correlation of $\bar{V}_{s\ell}$ in Eq.(21) will be derived. $\bar{V}_{s\ell}$ is the mean velocity of the solid phase in the liquid slug part. There is also no correlations to estimate it from only the physical properties, flow conditions and geometries. As mentioned above, the flow situation in the liquid slug part of the three-phase slug flow is supposed to be that of the three-phase bubbly flow. Then, the value of the velocity $\bar{V}_{s\ell}$ can be calculated by the equations for the volumetric fraction weighted mean solid velocity of the three-phase bubbly flow.

The volumetric fraction weighted mean solid velocity in three-phase flows can be generally expressed by Eq.(33) from

SLUG FLOW IN VERTICAL PIPES

Eq.(4).

$$\overline{V}_s^* = C_s^* (<j_\ell + j_s>^* - <j_\ell>_f^*) \tag{33}$$

where * denotes the value in the region not occupied by the gas phase as mentioned above. However, for the value of the volumetric fraction weighted mean velocity of the solid phase, $\overline{V}_s^*$ can be replaced by $\overline{V}_s$ because $\overline{V}_s^*$ is equal to $\overline{V}_s$ as shown in Eq.(34).

$$\overline{V}_s^* = \frac{<j_s>^*}{<\alpha_s>^*} = \frac{<j_s>/(1-<\alpha_g>)}{<\alpha_s>/(1-<\alpha_g>)} = \frac{<j_s>}{<\alpha_s>} = \overline{V}_s \tag{34}$$

Therefore, Eq.(33) can be written in Eq.(35).

$$\overline{V}_s = C_s^* (<j_\ell + j_s>^* - <j_\ell>_f^*) \tag{35}$$

$<j_\ell>_f^*$ is the suspension volumetric flux of the liquid phase in the region not occupied by the gas phase when solid particles are suspended or hovering in the gas-liquid two-phase flow.

Now, Eq.(35) should be modified to the equation by which the volumetric fraction weighted mean solid velocity $\overline{V}_{s\ell}$ in the liquid slug part can be estimated. The term $<j_\ell + j_s>^*$ in Eq.(35) should be replaced by the sum of the volumetric fluxes of the liquid and the solid phases $<j_{\ell\ell} + j_{s\ell}>^*$ in the region not occupied by the gas phase in the liquid slug part. This is expressed by Eq.(36).

$$<j_{\ell\ell} + j_{s\ell}>^* = \frac{<\alpha_{\ell\ell}>\overline{V}_{\ell\ell}}{1 - <\alpha_{g\ell}>} + \frac{<\alpha_{s\ell}>\overline{V}_{s\ell}}{1 - <\alpha_{g\ell}>} \tag{36}$$

Substituting Eqs.(2),(15),(20) and (36) into Eq.(35), and rearranging, the next equation is obtained.

$$\overline{V}_{s\ell} = C_{sb}^* \left(\frac{<j_T> - <\alpha_{g\ell}><j_g>/<\alpha_g>}{1 - <\alpha_{g\ell}>} - <j_\ell>_{fb}^* \right) \tag{37}$$

where, the subscript b denotes the gas-liquid-solid three-phase bubbly flow. In order to solve this equation by the given quantities, the equations for C_{sb}^* and $<j_\ell>_{fb}^*$ must be obtained. Since no theoretical or empirical equations have been proposed, following equations are obtained as empirical ones from the experimental results of the gas-liquid-solid three-phase bubbly flow(Sakaguchi,1989). The details will be presented in separate papers.

$$C_{sb}^* = 1.14 \tag{38}$$

$$<j_\ell>_{fb}^* = V_{st}(1-<\alpha_s>^*)^{M_b} / <\alpha_s>^{*N_b} \tag{39}$$

where,

$$M_b = \left[(10.5 \frac{d_s}{D} - 6.40)<\alpha_g> - 0.457(\frac{d_s}{D})^{0.7} + 1.31 \right] / <\alpha_s>^{*1.20} \tag{40}$$

$$N_b = \left[10.0\frac{d_s}{D} - 4.96\right]<\alpha_g> - 0.797\frac{d_s}{D} + 0.732 \tag{41}$$

and V_{st} is the free settling velocity of a single solid particle in a stationary liquid. The values $<\alpha_s>*$ and $<\alpha_g>$ in Eqs.(39)-(41) are evaluated by $<\alpha_{s\ell}>*$ defined in Eq.(42) and by $<\alpha_{g\ell}>$, respectively.

$$<\alpha_{s\ell}>* = \frac{<\alpha_{s\ell}>}{1 - <\alpha_{g\ell}>} \tag{42}$$

Therfore, $\overline{V}_{s\ell}$ is calculated by the following equations.

$$\overline{V}_{s\ell} = 1.14\left(\frac{<j_T> - <\alpha_{g\ell}><j_g>/<\alpha_g>}{1 - <\alpha_{g\ell}>} - V_{st}\frac{\{1 - <\alpha_{s\ell}>/(1 - <\alpha_{g\ell}>)\}^{M_b}}{\{<\alpha_{s\ell}>/(1 - <\alpha_{g\ell}>)\}^{N_b}}\right) \tag{43}$$

where,

$$M_b = \left[(10.5\frac{d_s}{D} - 6.40)<\alpha_{g\ell}> - 0.457(\frac{d_s}{D})^{0.7} + 1.31\right] / \left[\frac{<\alpha_{s\ell}>}{1 - <\alpha_{g\ell}>}\right]^{1.20} \tag{44}$$

$$N_b = \left[10.0\frac{d_s}{D} - 4.96\right]<\alpha_{g\ell}> - 0.797\frac{d_s}{D} + 0.732 \tag{45}$$

Lastly, correlations of the rest two unknown quantities C_1 and V_{bt} in Eq.(21) will be derived. For gas-liquid two-phase slug flows, C_{12} and V_{bt2} have been expressed by the function of the Reynolds number ($Re_2 = \rho_\ell D <j_{T2}>/\mu_\ell$) and the Bond number ($Bo_2 = \rho_\ell g D^2/\sigma$), respectively(Sakaguchi,1985;Sakaguchi, 1987a).

$$C_{12} = 1.2 + 0.8 \left[1 + \{(Re_2 - 600)/585\}^{1.1}\right]^{-1} \tag{46}$$

$$V_{bt2} = \left[0.35 - \frac{0.25}{\{(\sqrt{Bo_2} - 1.9)/2.12\}^{2.67} + 1}\right]\sqrt{gD(\rho_\ell - \rho_g)/\rho_\ell} \tag{47}$$

These equations are extended to the three-phase slug flows supposing that the gas and the liquid phase flow in the reduced cross-sectional area $A(1 - <\alpha_s>)$ by the existence of the solid phase. Therefore, the pipe diameter D is reduced to $D\sqrt{1 - <\alpha_s>}$. The gas and the liquid phase are supposed to flow in this narrowed pipe. The total volumetric flux of the gas and the liquid phase flowing in this narrowed cross-sectional area is expressed by $<j_g + j_\ell>/(1 - <\alpha_s>)$. Then, the Reynolds number in Eq.(46) is extended as Eq.(48).

$$Re = \rho_\ell <j_g + j_\ell> \frac{1}{\mu_\ell} \frac{D}{\sqrt{1 - <\alpha_s>}} \tag{48}$$

It is assumed that Eq.(46) is applied to the gas-liquid-solid three-phase slug flow by substituting the Reynolds number

SLUG FLOW IN VERTICAL PIPES

expressed by Eq.(48) into Eq.(46), namely,

$$C_1 = 1.2 + 0.8 \, [1 + \{(Re - 600)/585\}^{1.1}]^{-1} \tag{49}$$

Equation(47) and the Bond number in the equation are extended as follows.

$$V_{bt} = \left[0.35 - \frac{0.25}{\{(\sqrt{Bo}-1.9)/2.12\}^{2.67}+1}\right] \sqrt{gD \sqrt{1-<\alpha_s>} \, \frac{\rho_\ell - \rho_g}{\rho_\ell}} \tag{50}$$

$$Bo = \rho_\ell g D^2 (1 - <\alpha_s>) / \sigma \tag{51}$$

Thus, all the unknown quantities in Eq.(21) were derived and expressed by Eqs.(31),(32),(43),(44),(45),(48),(49),(50) and (51). Therefore, C_g and K_g in Eq.(2) can be calculated. The flow chart for the calculation of the value of $<\alpha_g>$ is given in the first half of Fig.4. A new value of the volumetric fraction $<\alpha_g>$ of the gas phase is calculated by these equations under the given trial value of the volumetric fractions of the gas and the solid phase, if the value of z is given. The iterative method should be performed until the convergence is reached within a desired value of accuracy.

Correlations to obtain the value of $<\alpha_s>$ will be derived in the next section.

<u>Correlations to Estimate Volumetric Fraction of Solid Phase</u>

The main equation to estimate the volumetric fraction $<\alpha_s>$ of the solid phase is Eq.(4). Substituting Eq.(35) into (4), the following equation is obtained.

$$<\alpha_s> = \frac{<j_s>}{C_s^*(<j_\ell + j_s>^* - <j_\ell>_f^*)} \tag{52}$$

In order to solve this equation, correlations of $<j_\ell>_f^*$ and C_s^* are necessary. Both correlations are given by empirical equations in this study because no theoretical ones have been proposed for them.

The suspension volumetric flux $<j_\ell>_f^*$ of the liquid phase in three-phase slug flows will be described. The experimental results of $<j_\ell>_f^*$ are shown in Fig.5 versus the volumetric fraction $<\alpha_s>^*$ of the solid phase in the region not occupied by the gas phase. The precision index(ASME,1985) of the measured values of $<j_\ell>_f^*$ is 0.00910 m/s.

The value of $<j_\ell>_f^*$ decreases with increasing value of $<\alpha_s>^*$, and with decreasing mean particle diameter. The effect of the pipe diameter is not obvious. The effect of the volumetric fraction $<\alpha_g>$ of the gas phase is hardly recognized when the ratio d_s/D is small, however, slightly recognized when d_s/D is large. The experimental results in this experimental conditions are correlated by the next type of equation (Dedegil,1982).

$$<j_\ell>_f^* = V_{st}(1 - <\alpha_s>^*)^M / <\alpha_s>^{*N} \tag{53}$$

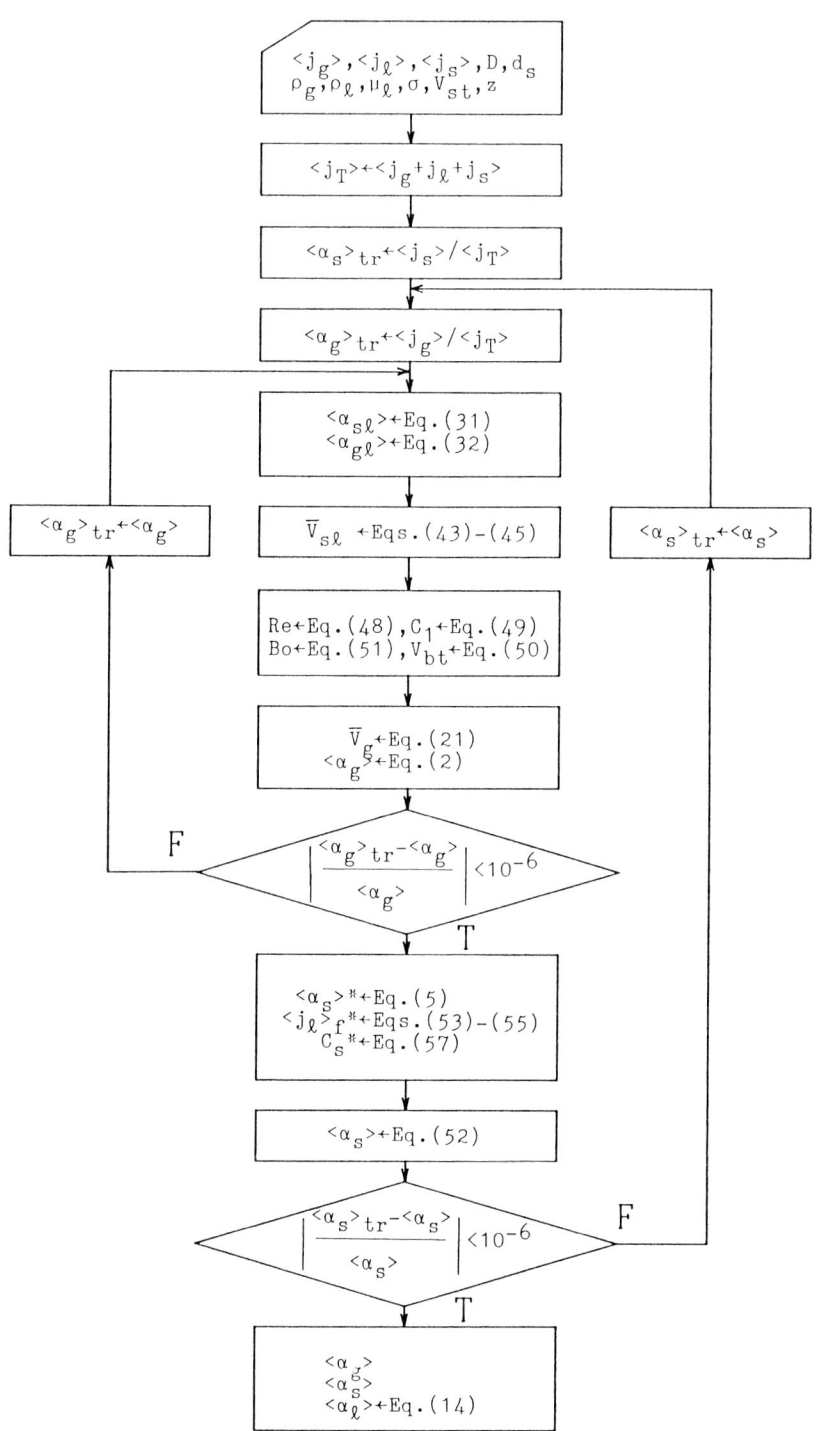

FIGURE 4. Flow chart for estimation of volumetric fractions of each phase

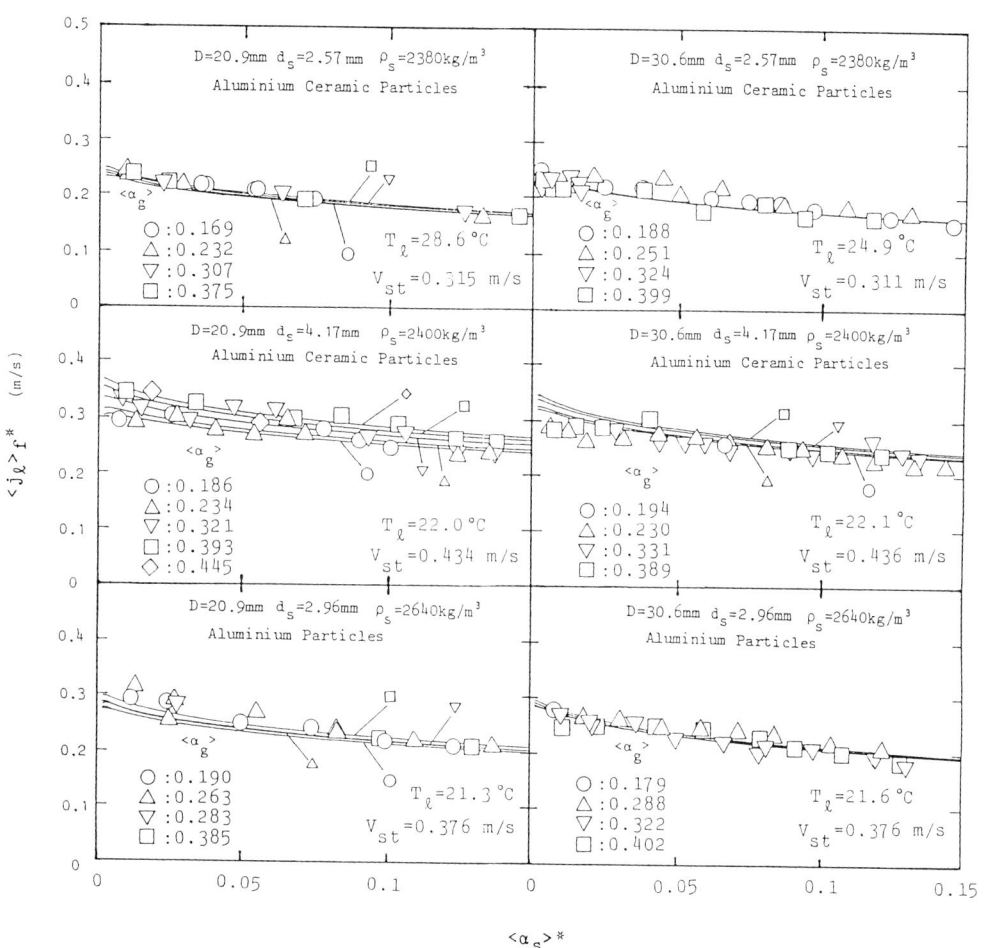

FIGURE 5. Suspension volumetric flux $\langle j_\ell \rangle_f^*$ of liquid phase for slug flow region (———— : Calculated by Eqs.(53)-(55))

where,

$$M = \left\{ (-2.34 \frac{d_s}{D} + 0.155)\langle \alpha_g \rangle - 0.916 \frac{d_s}{D} + 0.948 \right\} / \langle \alpha_s \rangle^{*1.14} \quad (54)$$

$$N = -0.642 \frac{d_s}{D} + 0.342 \quad (55)$$

In this study, the effects by the terms $\langle \alpha_g \rangle$ and d_s/D are also considered. The solid lines in Fig.5 are the correlated results by Eqs.(53)-(55). The mean value of the ratio of the calculated results by these equations and the experimental values is 0.989, and the standard deviation SD(%) around the mean value defined in Eq.(56) is 6.94 %.

$$SD = \sqrt{\frac{1}{n} \Sigma \left(\frac{X_{cal}}{X_{exp}} - \frac{1}{n} \Sigma \left(\frac{X_{cal}}{X_{exp}} \right) \right)^2} \times 100 \; (\%) \qquad (56)$$

Here, n is the number of data, X_{exp}: experimental result, and X_{cal}: calculated value of a quantity X.

The coefficient C_s^* defined by Eq.(4) is an important factor for this estimation, too. It corresponds to the distribution parameter of the solid phase for gas-liquid-solid three-phase flow.

For the three-phase slug flow, the coefficient C_s^* is calculated by the substitution of the experimental results of $<\alpha_s>$, $<\alpha_g>$, $<j_\ell>$, $<j_s>$, and $<j_\ell>_f^*$ into Eqs.(4)-(8). It is observed that C_s^* depends on the volumetric flow fraction β_s^* of the solid phase in the area not occupied by the gas phase ($=<j_s>^*/<j_\ell+j_s>^*$), $<j_\ell+j_s>^*$, $<j_\ell>_f^*$, d_s and D in this experimental range. Then, the relationship between C_s^* and these quantities are shown in Figs.6 and 7. The effect of $<j_\ell>_f^*/<j_\ell+j_s>^*$ on C_s^* is shown in Fig.6, and that of d_s/D is shown in Fig.7. The value of C_s^* increases with d_s/D, and with decreasing $<j_\ell>_f^*/<j_\ell+j_s>^*$ and β_s^*. When β_s^* increases, the value of C_s^* seems to approach unity gradually. Considering these characteristics of C_s^*, the empirical equation(57) is obtained for all the experimental conditions.

$$C_s^* = \frac{-1.02 \frac{<j_\ell>_f^*}{<j_\ell+j_s>^*} + 0.856 \frac{d_s}{D} + 0.414}{\left(\frac{\beta_s^*}{0.03}\right)^2 + 1} + 1 \qquad (57)$$

The solid lines in Figs.6 and 7 are the calculated values by Eq.(57). The mean value of the ratio of these calculated values and the experimental results is 1.03, and SD(%) in Eq.(56) is 12.4 %.

ESTIMATION PROCEDURE

The flow chart for the estimation of the volumetric fractions of each phase in gas-liquid-solid three-phase slug flow is shown in Fig.4. There are two loops in this flow chart. The input parameters are the volumetric fluxes $<j_g>$, $<j_\ell>$, and $<j_s>$, the pipe diameter D, the density ρ_g of the gas phase, that of the liquid phase ρ_ℓ, the free settling velocity V_{st} of the solid particle in a stationary liquid, the mean solid diameter d_s, the surface tension σ and the exponent z in Eq.(29). Then, trial values of $<\alpha_s>$ and $<\alpha_g>$ which are denoted by $<\alpha_s>_{tr}$ and $<\alpha_g>_{tr}$ are given. In this estimation, the volumetric flow fraction $<j_s>/<j_T>$ of the solid phase and that of the gas phase $<j_g>/<j_T>$ are used for them as the trial values, respectively. By setting these values, an iteration calculation estimates $<\alpha_g>$ until the relative error of $<\alpha_g>$ reaches a desired value of accuracy. In the estimation, the value is set at 10^{-6}. If the relative error

SLUG FLOW IN VERTICAL PIPES

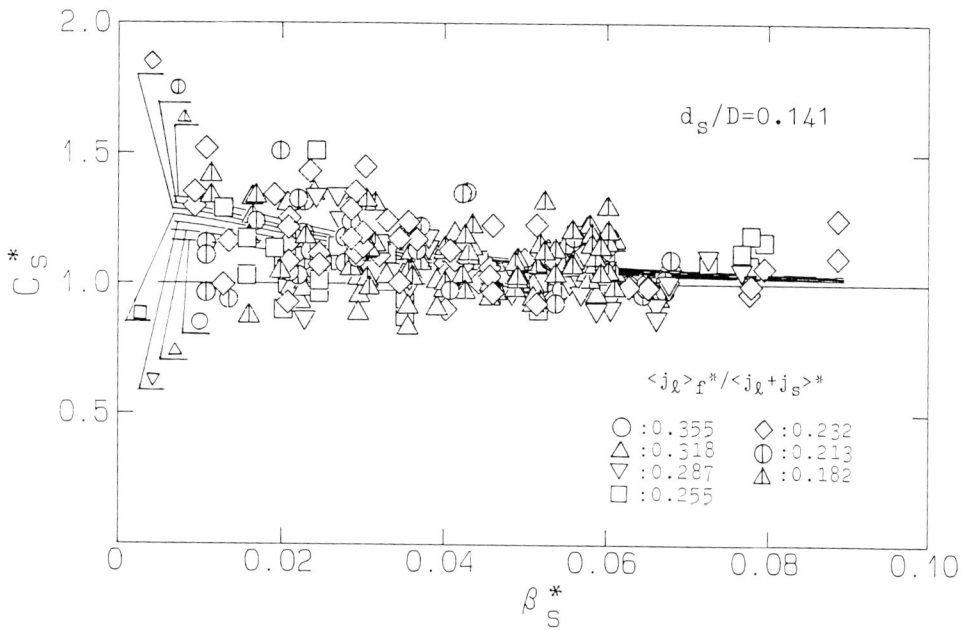

FIGURE 6. Relation between C_s^* and β_s^* (parameter: $\langle j_\ell \rangle_f^* / \langle j_\ell + j_s \rangle^*$; ——— : Calculated by Eq.(57))

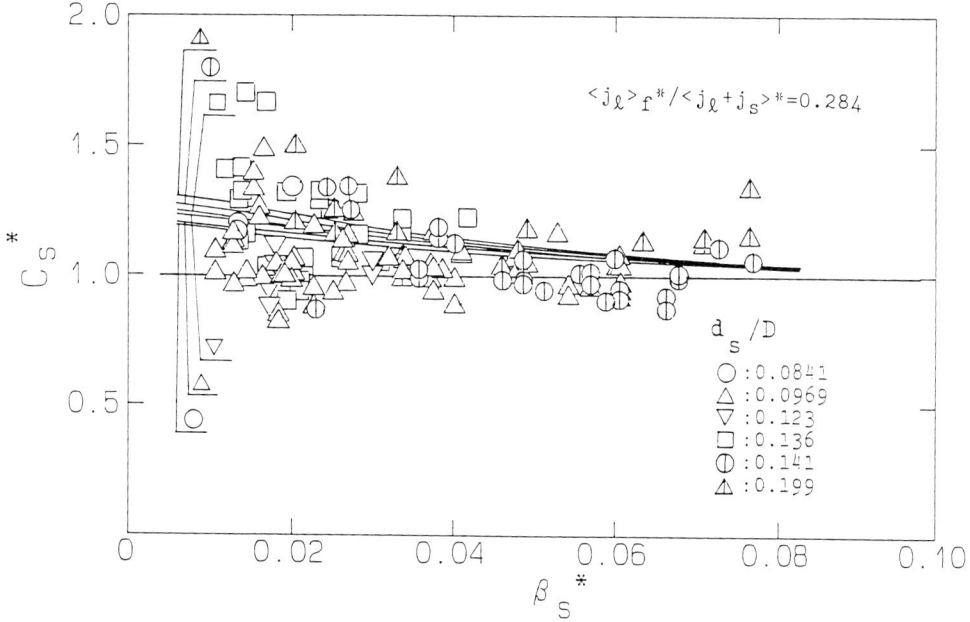

FIGURE 7. Relation between C_s^* and β_s^* (parameter: d_s/D ; ——— : Calculated by Eq.(57))

is larger than the value, $<\alpha_g>_{tr}$ is replaced by the estimated value of $<\alpha_g>$, and the estimation of $<\alpha_g>$ is repeated.

Next, $<\alpha_s>$ is estimated, and again its relative error is checked. If it is larger than the desired value of accuracy, the value of $<\alpha_s>_{tr}$ is replaced by the estimated value of $<\alpha_s>$, and the iteration calculation for both $<\alpha_g>$ and $<\alpha_s>$ is continued. If the relative error for $<\alpha_s>$ reaches the desired value of accuracy, the values of $<\alpha_g>$ and $<\alpha_s>$ are the final values to be obtained as the volumetric fractions of the gas and the solid phases of the gas-liquid-solid three-phase slug flow. That of the liquid phase is obtained by Eq.(14).

COMPARISON OF ESTIMATED VALUES WITH EXPERIMENTAL RESULTS

The estimated results for the volumetric fractions of each phase in gas-liquid-solid three-phase slug flows calculated according to the flow chart in Fig.4 are shown in Fig.2 by the solid lines. Here the value of z = 1.05 is used for the estimation. These lines agree well with experimental results. The mean value of the ratio of the estimated results and the experimental results is 0.993 for $<\alpha_g>$, 1.002 for $<\alpha_\ell>$, and 0.987 for $<\alpha_s>$, and the value of SD(%) is 6.25 % for $<\alpha_g>$, 3.53 % for $<\alpha_\ell>$, and 11.9 % for $<\alpha_s>$. Through these values and the relation between the solid lines and the experimental results in Fig.2, it is recognized that the proposed estimation can estimate well both qualitatively and quantitatively the volumetric fractions of each phase in gas-liquid-solid three-phase slug flows.

In Fig.8, examples of the estimated results are shown on $\overline{V}_g$, $\overline{V}_s$-$<j_T>$ diagram with experimental results. The plotted symbols are the experimental results of the air-water-aluminium particles three-phase slug flow for $<j_g>$ = 0.633 m/s and in the pipe of D = 20.9mm. The values of $<j_g>$ and $<j_s>$ in the figure are the mean values of their experimental values held in the range of ± 5 % and ± 20 %, respectively. The solid lines in the figure are the estimated values of $\overline{V}_g$ and $\overline{V}_s$ for each value of $<j_s>$. Both $\overline{V}_g$ and $\overline{V}_s$ increase with total volumetric flux $<j_T>$. In the cases of small values of $<j_s>$, the value of $\overline{V}_s$ increases steeply with increasing value of $<j_T>$. Almost all the experimental data of $\overline{V}_g$ distribute around a single straight line, whereas those of $\overline{V}_s$ disperse widely in the relatively large region of $<j_T>$. The value of $\overline{V}_s$ increases with decreasing $<j_s>$. The estimated lines predict well the characteristics of the experimental data.

In Fig.9, $\overline{V}_g$,$\overline{V}_s$-$<j_T>$ diagram is shown with $<\alpha_s>$ as parameter. The experimental results of $\overline{V}_g$ and $\overline{V}_s$ at the same experimental conditions given in Fig.8 are plotted in the figure in the range of $<\alpha_g>$=0.362 ± 5 %. The values of $<\alpha_s>$ are the mean values of the experimental results held in the range of ± 20 %. In the relatively large region of $<j_T>$, the experimental data of $\overline{V}_s$ disperse widely, and the value of $\overline{V}_s$ increases with decreasing value of $<\alpha_s>$. One experimental data, in this figure, is larger than the value of $<j_T>$. These characteristics of $\overline{V}_s$ are also observed for liquid-solid two-phase flows(Newitt,1961;Toda,1967;Sakaguchi,1988).

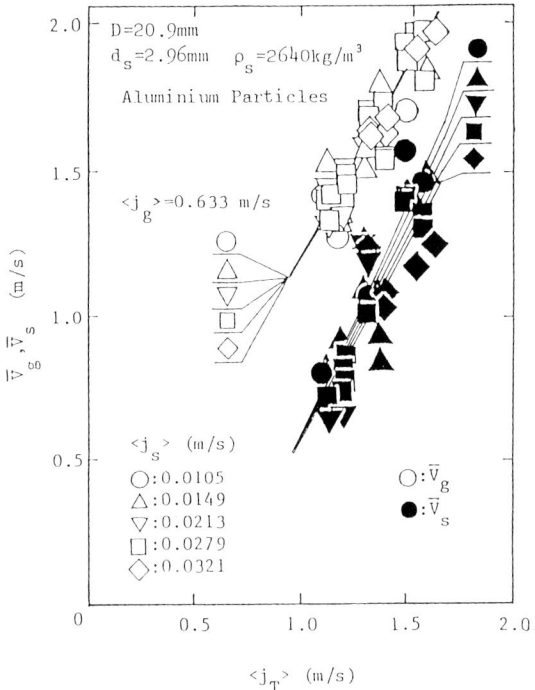

FIGURE 8. $\bar{V}_g, \bar{V}_s$-$<j_T>$ diagram (parameter : $<j_s>$; ——— : Estimated by the proposed method)

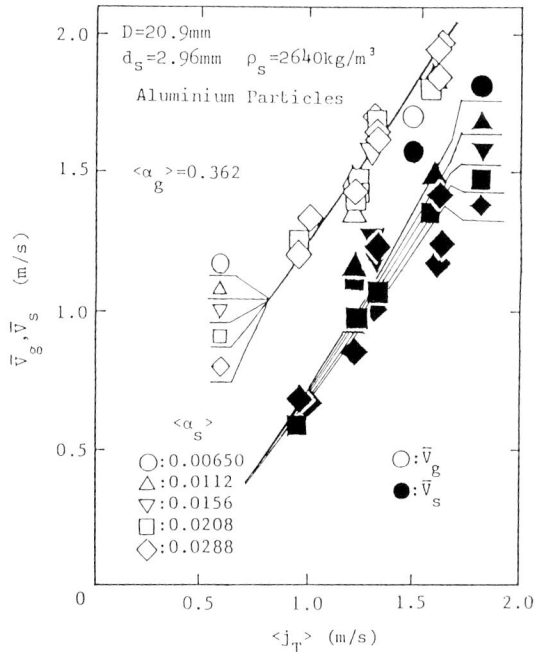

FIGURE 9. $\bar{V}_g, \bar{V}_s$-$<j_T>$ diagram (parameter : $<\alpha_s>$; ——— : Estimated by the proposed method)

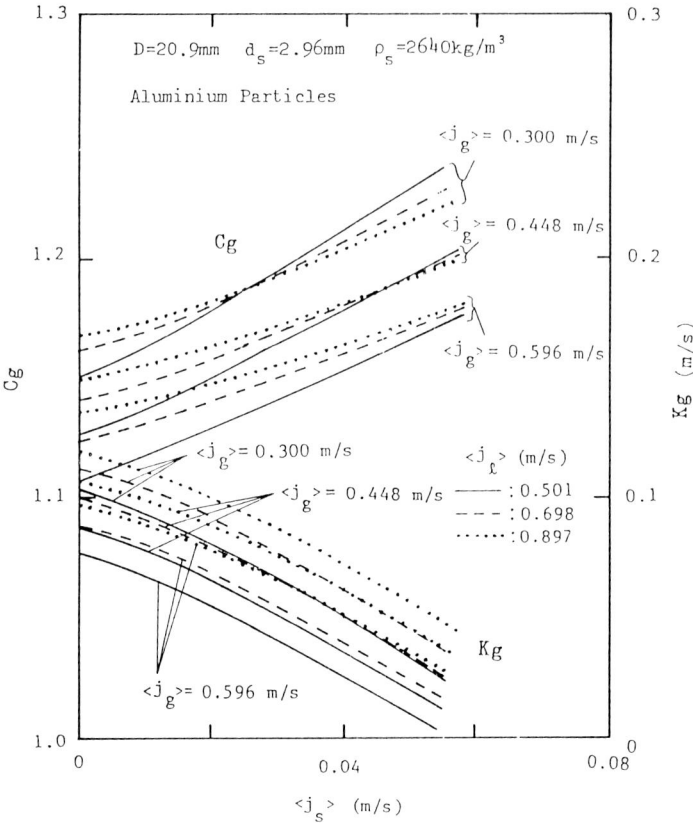

FIGURE 10. Estimated results of C_g and K_g

They are explained by the fact that solid particles concentrate in the center of the pipe. The solid lines in the figure are the estimated results. They predict these characteristics qualitatively.

Finally, the estimated results of C_g and K_g in Eq.(2) are descrived. These values are obtained by Eq.(21) as the coefficient of $<j_T>$ and the second term on the right hand side, respectively. In Fig.10, the examples of the estimated values of C_g and K_g are shown as curves for air-water-aluminium particles three-phase slug flows. The parameters in the figure are set at the same values as in the corresponding part of Fig.2. The value of C_g increases with increasing $<j_s>$, with decreasing $<j_g>$, with increasing $<j_\ell>$ in the range of smaller value of $<j_s>$ and with decreasing $<j_\ell>$ in the range of larger value of $<j_\ell>$. The value of K_g increases with decreasing $<j_s>$, with increasing $<j_\ell>$ and with decreasing $<j_g>$. As the experimental values of these parameters C_g and K_g which must be calculated from local values of $\alpha_g(r)$, $j_T(r)$ and $V_g(r)$(Zuber,1965), have not been obtained in the three-phase slug flow, the comparison can not be made between the estimated and the experimental values.

SLUG FLOW IN VERTICAL PIPES

CONCLUSIONS

The volumetric fractions of each phase in gas-liquid-solid three-phase slug flow are estimated by two main equations which are principally derived from the drift flux model. The terminal rising velocity of a large gas bubble is related to the volumetric fraction weighted mean drift velocity of the gas phase to the mixture and the suspension volumetric flux of the liquid phase is used as another basic point to express the main equation for the solid phase.

Correlations are derived to solve these two main equations by extending the correlations in the gas-liquid two-phase slug flow to those of the gas-liquid-solid three-phase slug flow. By this method of estimation, the experimental data are well correlated.

It is recognized, in the case that the solid phase may exist within the liquid phase, that the correlations in the gas-liquid two-phase flow may be applicable to those in the gas-liquid-solid three-phase flow, if they are modified slightly by considering the existence of the solid phase.

At the present stage, the application range of this estimation is confined to the experimental range as the some of the correlations are obtained experimentally from the limited experimental results. These general correlations must be derived theoretically in future.

ACKNOWLEDGEMENT

The authors wish to acknowledge Messrs. H. Shakutsui, K. Hashimoto, K. Yamakoshi, L. Dang, S. Inoue, S. Kitani, H. Takahashi, H. Minoyama, M. Ushio, S. Koike and K. Baba who performed some of the experimental work or data reduction.

NOMENCLATURE

English

A : Cross-sectional area (m²)
Bo : Bond number (-)
C : Distribution parameter (-)
C_1 : Coefficient used in Eq.(19)
D : Pipe diameter (m)
d_s : Mean particle diameter (m)
g : Gravitational acceleration (m/s²)
j : Volumetric flux (m³/m²s)
K : Volumetric fraction weighted mean drift velocity (m/s)
L : Length (m)
M : Constant in Eq.(39) (-)
M_b : Constant in Eq.(53) (-)
N : Constant in Eq.(39) (-)
N_b : Constant in Eq.(53) (-)
P : Pressure (Pa)
Q : Volume flow rate (m³/s)

Re : Reynolds number (-)
SD : Standard deviation defined in Eq.(56) (%)
T : Temperature (°C)
V : Velocity (m/s)
z : Exponent used in Eq.(29) (-)

Greek and Other

α : Volumetric fraction (-)
β : Volumetric flow fraction (-)
μ : Viscosity (Pa·s)
ρ : Density (kg/m³)
σ : Surface tension (N/m)
$\langle\ \rangle$: Cross-sectional mean value defined in Eq.(1)
$\overline{}$: Volumetric fraction weighted mean value defined in Eq.(3)

Subscripts

b : large gas bubble, bubbly flow region
cal : calculated value
exp : experimental result
f : suspended condition
g : gas phase, large gas bubble part
ℓ : liquid phase, liquid slug part
s : solid phase
T : total
t : terminal condition
tr : trial
2 : two-phase flow

Superscripts

* : value defined in the region not occupied by the gas phase

BIBLIOGRAPHY

Akagawa, K., and Sakaguchi, T., 1965, On the Characteristics of Void Fluctuations in Gas-Liquid Two-Phase Flow, *Trans. JSME*, vol. 31, no. 224, pp. 594-600. (in Japanese)

ASME, 1985, Measurement Uncertainty, *ASME Performance Test Codes*, ANSI/ASME PTC 19.1-1985, translated by JSME, 1987. (in Japanese)

Dedegil, M. Y., 1982, Neuere Untersuchungen zum Lufthebeverfahren, *Verfahrenstechnik*, vol. 16, no. 4, pp. 229-232.

Fukano, T., Matsumura, K., Kawakami, Y., and Sekoguchi, K., 1980, Unsteady Phenomena of Slug Flow, *Trans. JSME*, vol. 46, no. 412, pp. 2412-2419. (in Japanese)

Giot, M., 1982, in *Handbook of Multiphase Systems*, ed. Hetsroni, G., pp. 7-29-7-52, McGraw-Hill, New York.

Newitt, D. M., Richardson, J. F. and Gliddon, B. J., 1961, Hydraulic Conveying of Solids in Vertical Pipes, *Trans. Instn.*

Chem. Engrs., vol. 39, pp. 93-100.

Sakaguchi, T., Shakutsui, H., Hamaguchi, H., Ono, M., and Minagawa, H., 1985, Basic Study for the Conversion from Gas-Liquid Two-Phase Slug Flow to Bubbly Flow, *Prepr. JSME*, vol. 85-4, no. 2, pp. 74-76. (in Japanese)

Sakaguchi, T., Minagawa, H., Kato, Y., Kuroda, N., Matsumoto, T., and Sahara, K., 1987a, Estimation of In-Situ Volume Fraction of Each Phase in Gas-Liquid-Solid Three-Phase Flow, *Trans. JSME*, vol. 53, no. 487, pp. 1040-1046. (in Japanese)

Sakaguchi, T., Minagawa, H., Sahara, K., Kato, Y., Kuroda, N., and Matsumoto, T., 1987b, Estimation of Volumetric Fraction of Each Phase in Gas-Liquid-Solid Three-Phase Flow, *Proc. 1987 ASME-JSME Thermal Engng. Joint Conf.*, vol. 5, pp. 373-380.

Sakaguchi, T., Minagawa, H., Shakutsui, H., Sahara, K., Saibe, T., Hashimoto, K., and Dang, L., 1988, Volumetric Fraction of Solid Phase and Particle Velocity of Liquid-Solid Two-Phase Flow in Vertical Pipes, in *Experimental Heat Transf., Fluid Mechs., and Thermodynamics 1988*, ed. Shah, R. K., Ganić, E. N., and Yang, K. T., pp. 1353-1360, Elsevier, New York.

Sakaguchi, T., Shakutsui, H., Minagawa, H., Tomiyama, A, and Takahashi, H., 1989, Characteristics of Volumetric Fraction in Gas-Liquid-Solid Three-Phase Bubbly Flow in Vertical Pipe, 1989, *Prepr. of JSME*, no. 894-2, pp. 47-48. (in Japanese)

Toda, M., Konno, H., Saito, S., and Maeda, S., 1967, Hydraulic Conveying of Solids through Horizontal and Vertical Pipes, *Kagaku Kogaku (Chem. Engng.)*, vol. 33, pp. 67-73.(in Japanese)

Zuber, N., and Findlay, J. A., 1965, Average Volumetric Concentration in Two-Phase Flow Systems, *Trans. ASME, J. Heat Transf.*, Ser. C, vol. 87, no. 4, pp. 453-468.

FLOW REGIME MODELING AND DYNAMICS

EXPERIMENTAL STUDY ON THE VIBRATION FORCE OF A TWO-PHASE IMPINGEMENT JET

N. NAKAMORI, K. KAWANISHI, M. ISONO, and J. KASAHARA
Mitsubishi Heavy Industries, Ltd.
Takasago Research & Development Center
Takasago, Hyogo Prefecture, Japan
T. KAMIYA
Mitsubishi Heavy Industries, Ltd.
Takasago Machinery Works
Takasago, Hyogo Prefecture, Japan

ABSTRACT

The dynamic and static forces which a two-phase jet gave to an impingement plate were studied using two-phase air and water test apparatus under the conditions of atmospheric pressure and room-temperature. As a result, the dynamic and static characteristics of the two-phase jet were well understood and the calculation method for the dynamic and static forces was established. This calculation method can be useful for the design of a shell and tube type heat exchanger, so the improvement of its reliability is expected.

INTRODUCTION

In the shell and tube heat exchanger, the impingement plate is usually installed at the inlet portion of shell-side fluid to prevent the jet from the inlet nozzle from impinging directly on the tube bundle. This indicates that the impingement plate is directly exposed to the jet, so its design is very important from the viewpoint of the heat exchanger reliability. Especially, because the two-phase jet may give a larger vibration force to the impingement plate than the single-phase jet does, the fatigue strength evaluation of the impingement plate becomes very important.

However, few data on the vibration force of the two-phase impingement jet have been published thus far. Therefore, designers must evaluate roughly on the basis of the turbulence of the single-phase jet. On the other hand, it has been reported (Nakamura, 1985) already that the vibration force of the two-phase cross-flow in the tube bundle is much larger than that of the single-phase cross-flow. This fact indicates that the vibration force of the two-phase impingement jet may be much larger than that of the single-phase impingement jet, therefore this study was planned to establish the calculation method for the vibration force of the two-phase impingement jet.

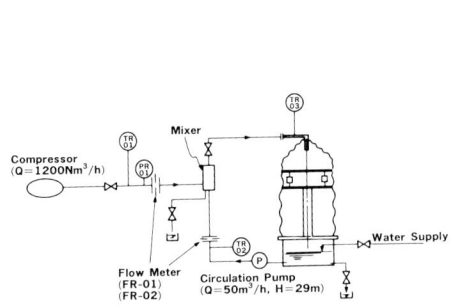

Fig. 1 System of Test Apparatus

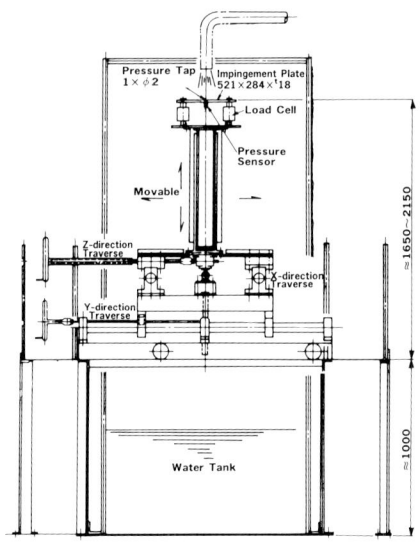

Fig. 2 Schematic View of Traversing Apparatus

TEST APPARATUS AND TEST CONDITIONS

Test Apparatus

<u>System of test apparatus</u> The system of the test apparatus is shown in Fig. 1. The air and water were mixed in the mixer after their flow rate was measured individually. After that, the air and water mixture was discharged from the discharge pipe to the impingement plate, and only the water was circulated.

<u>Traversing apparatus</u> A schematic view of the traversing apparatus is shown in Fig. 2. The traversing apparatus was able to traverse the impingement plate three-directionally. Also, it was stiffly constructed to increase its natural frequency and was connected with flexible pipes to isolate the mechanical vibration of the pump. Furthermore, the circulation pump was separated from the pipes and the base floor by the rubber.

<u>Discharge pipe and impingement plate</u> The configurations of the discharge pipe and the impingement plate, which simulate those of the representative heat exchanger, are shown in Fig. 3. The impingement plate, which was made of aluminum and the natural frequency of which was about 180Hz, was supported by three load cells and could be traversed three-directionally by the traversing apparatus.

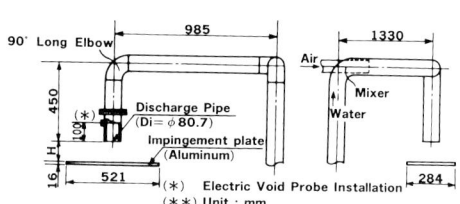

Fig. 3 Configurations of Discharge Pipe and Impingment Plate

Table 1 Measured Items and Instruments

Measured Items	Instruments
Static and Dynamic Forces Acting on Impingement Plate	Load Cell
Static Pressure and Pressure Fluctuation on Impingement Plate	Strain Gauge Type Pressure Sensor
Air and Water Flow Rates	Orifice Flow Meter

<u>Measurements</u> Table 1 summarizes the measured items and their instruments. The load of the impingement plate, which was given by the jet, was obtained by adding the outputs of the three load cells. Furthermore, it was analyzed to obtain the static load and the dynamic load (root mean square value and peak to peak value).

<u>Test Conditions</u>

Tests were conducted under specified room-temperature and atmospheric pressure conditions. The superficial air and water velocities (j_g, j_ℓ), and the ratio of the distance between the outlet of discharge pipe and the impingement plate (H) to the inner diameter of discharge pipe (D_i) were varied below to cover the conditions of the representative heat exchangers.

$$10 \leq j_g \leq 71 \text{ (m/s)}, \quad 0.6 \leq j_\ell \leq 2.5 \text{ (m/s)} \quad (1)$$

$$0.47 \leq H/D_i \leq 2.9 \quad (2)$$

These velocity conditions indicate that the flow pattern in the discharge pipe was the annular flow or the dispersed annular flow due to the Golan's flow pattern map (Golan, 1969∼1970) as shown in Fig. 4.

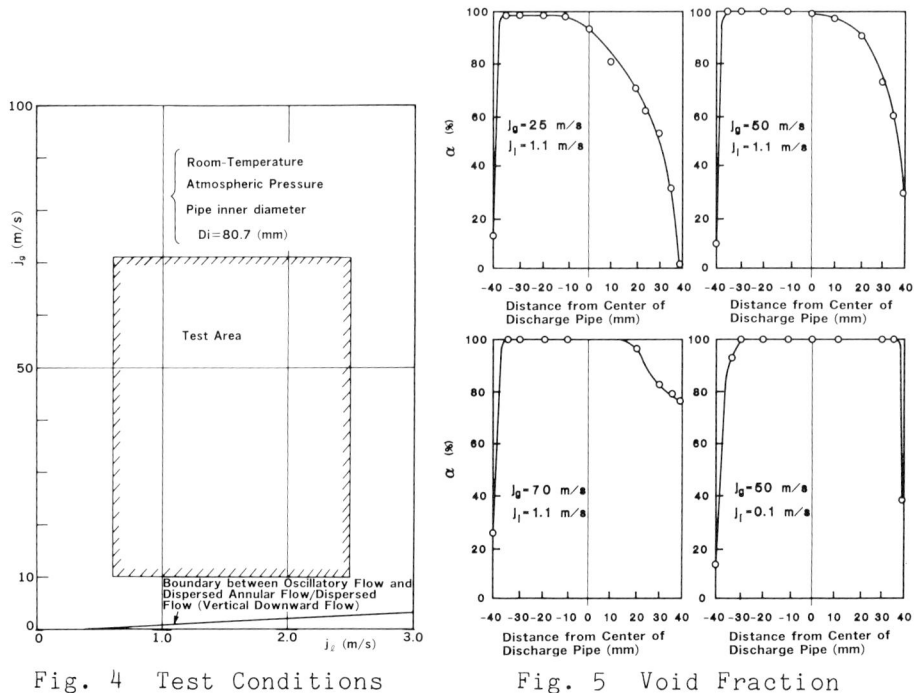

Fig. 4 Test Conditions

Fig. 5 Void Fraction Distribution

TEST RESULTS

Void Fraction Distribution

The void fraction distributions 100mm upstream from the outlet of discharge pipe were measured by an electric void probe. The representative measured results are shown in Fig. 5. These figures indicate that most of the water flowed on the wall of the discharge pipe as a water film and the water film on the outer side of elbow was thicker than that on the inner side of the elbow due to centrifugal force. This tendency was remarkable at the high water flow rate ($j_\ell = 1.1$ m/s). On the other hand, the water film thickness was comparatively uniform at the low water flow rate ($j_\ell = 0.1$ m/s).

Static and Dynamic Forces Acting on Impingement Plate

The relationship between the static force (F) and dynamic forces (ΔF_{RMS}, ΔF_{PP}) acting on the impingement plate and the jet force (J_{HM}) calculated with the homogeneous flow model are shown in Figs. 6 to 8. There was no remarkable effect of H/D_i on the static and dynamic forces. This was due to the fact that the jet did not expand at the maximum H/D_i of 2.94. The measured static force was smaller than the jet force calculated with the homogeneous flow model. Also, the

TWO-PHASE IMPINGEMENT JET

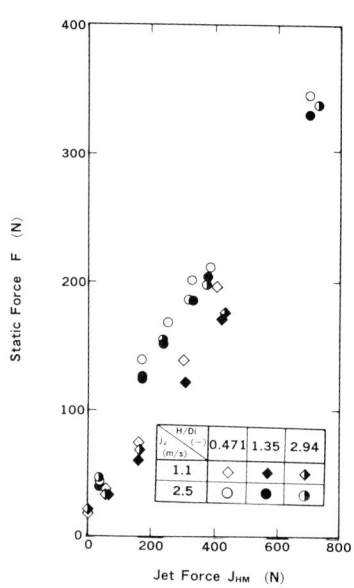

Fig. 6 Relationship between Static Force and Jet Force

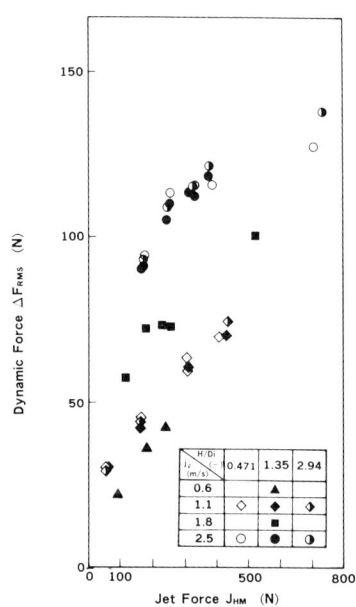

Fig. 7 Relationship between Dynamic Force (RMS) and Jet Force

static and dynamic forces vary with the superficial water velocity. These facts indicated that the actual water velocity may be slower than that calculated with the homogeneous flow model. Therefore, data reduction using the homogeneous flow model may be unavailable. This problem is discussed later in detail.

Relationship between ΔF_{PP} and ΔF_{RMS}

The relationship between the peak to peak value (ΔF_{PP}) and the root mean square value (ΔF_{RMS}) of the dynamic force is shown in Fig. 9. As shown in the figure, the ΔF_{PP} is well correlated with the ΔF_{RMS} and the relationship between the ΔF_{PP} and the ΔF_{RMS} is as follows.

$$\Delta F_{PP} \cong 7 \times \Delta F_{RMS} \tag{3}$$

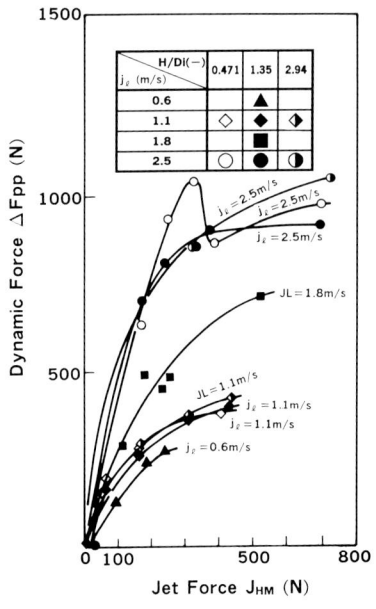

Fig. 8 Relationship between Dynamic Force (Peak to Peak) and Jet Force

Fig. 9 Relationship between ΔF_{PP} and ΔF_{RMS}

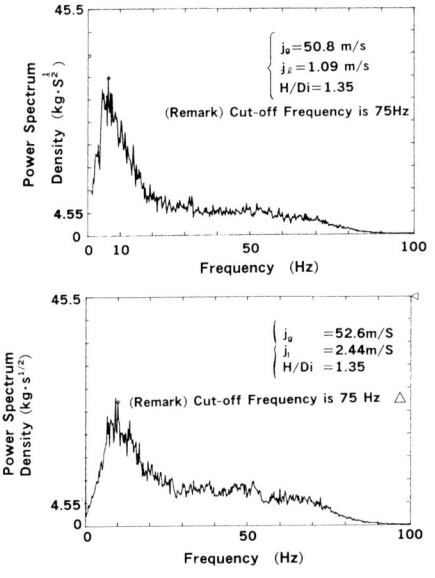

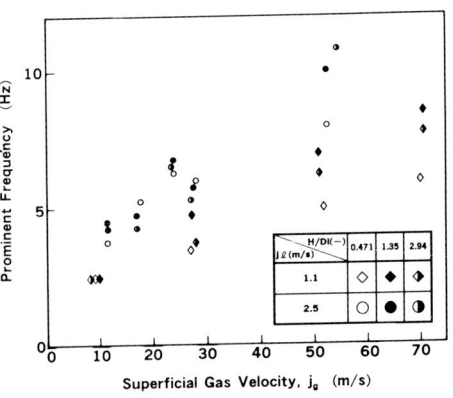

Fig. 10 Typical Power Spectrum Density

Fig. 11 Prominent Frequency of Dynamic Force

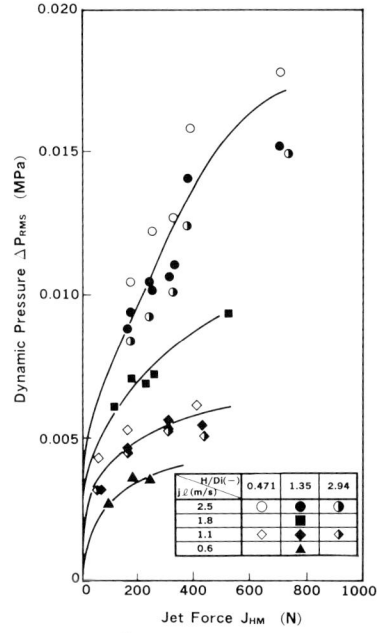
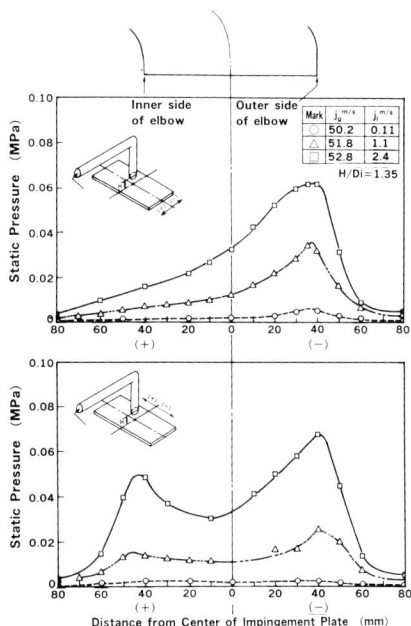

Fig. 12 Relationship between Dynamic Pressure (RMS) and Jet Force

Fig. 13 Static Pressure Distribution on Impingement Plate

Power Spectrum Density of Dynamic Force

The power spectrum density of the dynamic force was obtained after the load cell signal was processed through the low-pass filter, the cut-off frequency of which was 75 Hz. Most of the measured power spectrum densities have a similar characteristic, the representative results of which are shown in Fig. 10. As shown in the figure, the dynamic force has the prominent frequency and the measured prominent frequencies are summarized in Fig. 11. This figure indicates that the prominent frequency is lower than about 10 Hz. This result was similar to that of the two-phase cross-flow in the tube bundle. Also, the prominent frequency was the same order as the slug transit frequency in the slug flow or the froth flow.

Pressure Fluctuation Acting on Impingement Plate

The relationship between the pressure fluctuation on the impingement plate (peak to peak value, ΔP_{pp}) and the jet force (J_{HM}) is shown in Fig. 12. This figure indicates that the characteristic of the pressure fluctuation is similar to that of the dynamic force, therefore the pressure fluctuation was fully consistent with the dynamic force.

The representative static pressure distribution on the impingement plate, which was measured traversing the impingement plate, is shown in Fig. 13. This figure indicates that the static pressure on the outer side of the

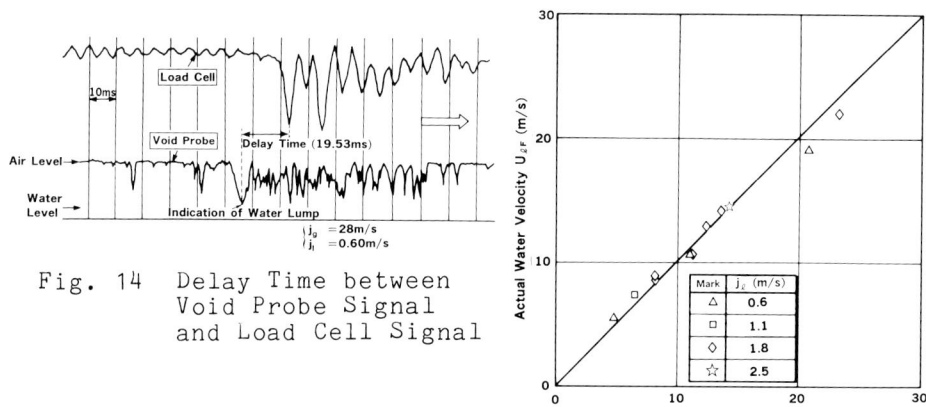

Fig. 14 Delay Time between Void Probe Signal and Load Cell Signal

Fig. 15 Comparison of $U_{\ell F}$ with $U_{\ell DT}$

elbow was greater than that of the other region. This result was fully consistent with the result of the void fraction distribution and indicated that the pressure fluctuation on the impingement plate, which was consistent with the dynamic force acting on the impingement plate, was induced by the water film behavior, that is, the variations of the actual water velocity and the water film thickness.

Actual Water Velocity

The actual water velocity ($U_{\ell DT}$) can be calculated with a delay time (Sekoguchi, 1985) between the void probe signal and the load cell signal. Figure 14 shows the representative example of the void probe signal and the load cell signal. The figure indicates that there was certainly the delay time between the void probe signal to indicate the arrival of a water lump and the variation of load acting on the impingement plate. The actual water velocities obtained by the above method are shown in Fig. 15 and these results are discussed later in detail.

DISCUSSION

Availability of Separated Flow Model

The measured static and dynamic forces indicate that the homogeneous flow model may be unavailable as explained previously. Therefore, a separated flow model is discussed below as a new flow model to describe the flow in detail. The schematic view of the separated flow model is shown in Fig. 16. and the jet force (J_{SM}) calculated with the separated flow model is as follows.

$$J_{SM} = (\rho_\ell \times \frac{j_\ell^2}{1-\alpha} + \rho_g \times \frac{j_g^2}{\alpha}) \qquad (4)$$

TWO-PHASE IMPINGEMENT JET

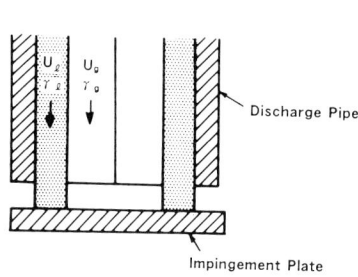

Fig. 16 Separated Flow Model

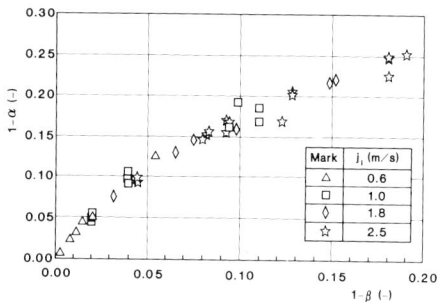

Fig. 17 Relationship between $(1-\alpha)$ and $(1-\beta)$

From Eq. (4), the void fraction (α) and the actual water velocity (U_ℓ) are given as

$$\alpha = \frac{(-\rho_\ell j_\ell^2 + \rho_g j_g^2 + \frac{J_{SM}}{A}) + \sqrt{(\rho_\ell j_\ell^2 - \rho_g j_g^2 - \frac{J_{SM}}{A})^2 - 4\frac{J_{SM}}{A}\cdot\rho_\ell j_g^2}}{2\frac{J_{SM}}{A}} \quad (5)$$

$$U_\ell = \frac{j_\ell}{1-\alpha} \quad (6)$$

In Eq. (5), equating the jet force (J_{SM}) and the measured static force (F), the void fraction (α) can be calculated on the basis of the test data. Also, from Eq. (6), the actual water velocity ($U_{\ell F}$) can be calculated.

In order to verify the availability of the separated flow model, the actual water velocity ($U_{\ell F}$) calculated from the above equation was compared with the actual water velocity ($U_{\ell DT}$) obtained by the delay time measurement between the void probe signal and the load cell signal. This result is shown in Fig. 15. The $U_{\ell F}$ agrees with the $U_{\ell DT}$, therefore it is considered that the separated flow model was available within the conditions of this test.

Calculation Method of Void Fraction

In order to calculate the actual water velocity (U_ℓ), the void fraction (α) must be calculated. Therefore, the calculation method of the void fraction is discussed below. Generally speaking, a void fraction (α) is correlated with a volume flow quality (β) using a flow parameter (K) (Armand, 1946)

$$\alpha = K\beta \quad (7)$$

In a vertical upward flow, the K is about 0.7.

On the other hand, the relationship between the $(1-\alpha)$ and the $(1-\beta)$, which are calculated with the flow conditions (j_g, j_ℓ) and the measured static force (F) using the Eq. (5), is shown in Fig. 17. This figure indicates that these test results are different from that of the vertical

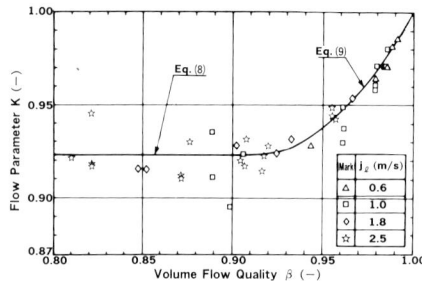
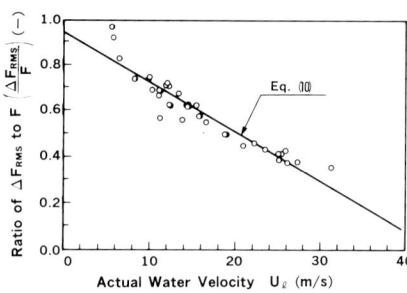

Fig. 18 Change of Flow Parameter with Volume Flow Quality

Fig. 19 Relationship between $\Delta F_{RMS}/F$ and U_ℓ

upward flow. Therefore, the flow parameter (K) is discussed below on the basis of this test data.

The change of flow parameter (K) with the volume flow quality (β) is shown in Fig. 18. The flow parameter (K) is well correlated with the volume flow quality (β).

$$K = 0.923 \quad (0.8 < \beta \leq 0.9) \tag{8}$$

$$K = 8.445 - 16.578\beta + 9.133\beta^2 \quad (1 > \beta \geq 0.9) \tag{9}$$

The correlations of the flow parameter (K) with the volume flow quality (β) consisted of the two equations, Eq. (8) and Eq. (9), and it was considered that the boundary between the two equations may have been the transition point from the annular flow to the dispersed flow.

Ratio of Dynamic Force to Static Force

The measured pressure fluctuation on the impingement plate indicated that the water film behavior may play an important role in the dynamic forces acting on the impingement plate. Therefore, the effect of the measured actual water velocity on the measured ratio of the dynamic force (ΔF_{RMS}) to the static force (F) was examined. This result is shown in Fig. 19 and the $\Delta F_{RMS}/F$ correlates with the actual water velocity (U_ℓ) as.

$$\frac{\Delta F_{RMS}}{F} = 0.936 - 0.0215 \times U_\ell \quad (U_\ell: m/s) \tag{10}$$

$$\text{where} \quad 0.6 < j_\ell < 2.5 \text{ (m/s)} \tag{11}$$
$$0.8 < \beta < 1$$

Also, as indicated in Eq. (10), the $\Delta F_{RMS}/F$ became close to 1.0 at the low actual water velocity and this value was much greater than that of the single-phase flow. On the other hand, Eq. (10) also indicated that the faster the actual water velocity, the smaller the $\Delta F_{RMS}/F$ became. This result may have been due to the fact that the two-phase flow became nearly a homogeneous condition, that is, the single-phase flow.

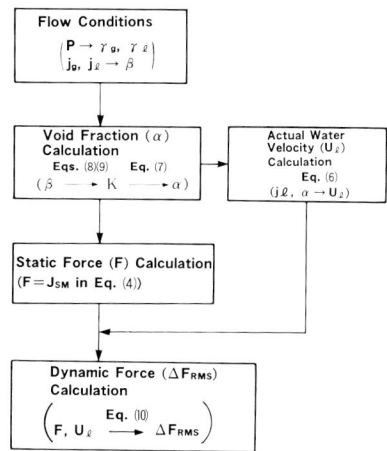

Fig. 20 Calculation Procedure

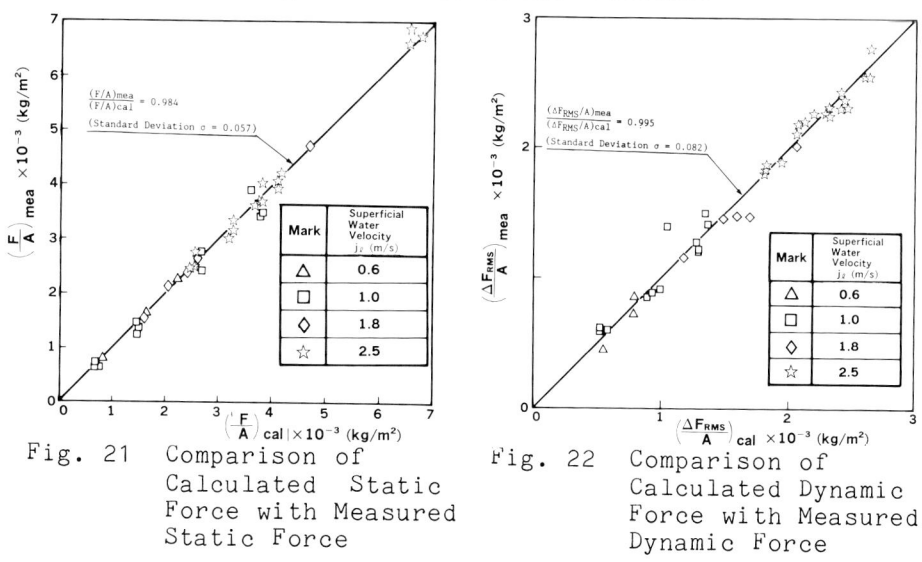

Fig. 21 Comparison of Calculated Static Force with Measured Static Force

Fig. 22 Comparison of Calculated Dynamic Force with Measured Dynamic Force

SUMMARY OF CALCULATION PROCEDURE

Calculation Procedure of Static and Dynamic Forces

The whole calculation procedure of the static and dynamic forces is shown in Fig. 20 and the summary of the calculation procedure is as follows.

(1) The flow conditions were given.
 (i) The fluid pressure (P) gave the fluid densities (ρ_g, ρ_ℓ).
 (ii) The superficial velocities (j_g, j_ℓ) gave the volume flow quality (β).
(2) The void fraction could be calculated. From Eq. (8) or Eq. (9), the flow parameter (K) could be calculated

with the β. From Eq. (7), the void fraction (α) can be calculated with the K and the β.
(3-1) Equating the jet force (J_{SM}) and the static force (F) in Eq. (4), the F could be calculated.
(3-2) From Eq. (6), the actual water velocity (U_ℓ) could be calculated with the j_ℓ and the α.
(4) From Eq. (10), the dynamic force (root mean square value, ΔF_{RMS}) could be calculated with the F and the U_ℓ. If the maximum dynamic force (ΔF_{PP}) was necessary, the ΔF_{PP} could be calculated with the ΔF_{RMS} from Eq. (3).

Comparison of Calculated Forces and Measured Forces

The comparisons of the calculated forces $(F/A)cal$, $(\Delta F_{RMS}/A)cal$ with the measured forces $(F/A)mea$, $(\Delta F_{RMS}/A)mea$ are shown in Figs. 21 and 22. The calculated forces agreed well with the measured forces and each least square lines are as follows.

$$\frac{(F/A)mea}{(F/A)cal} = 0.984 \qquad (12)$$

(Standard Deviation $\sigma = 0.057$)

$$\frac{(\Delta F_{RMS}/A)mea}{(\Delta F_{RMS}/A)cal} = 0.995 \qquad (13)$$

(Standard Deviation $\sigma = 0.082$)

In the design of the actual heat exchanger, the design margin must be added to Eq. (12) and Eq. (13), considering the standard deviation to calculate conservatively.

CONCLUSION

In order to obtain the fundamental data for the fatigue strength evaluation of the impingement plate of actual heat exchanger, the two-phase impingement jet was experimentally studied using two-phase air and water test apparatus under the conditions of room-temperature and atmospheric pressure. The major conclusions are as follows.
(1) The two-phase impingement jet could be treated by the separated flow model and the static force acting on the impingement plate could be calculated accurately using the separated flow model.
(2) The ratio of the dynamic force to the static force acting on the impingement plate correlated well with the actual water velocity. Using this correlation and the calculated static force, the dynamic force could be calculated accurately.
(3) Most of the measured power spectrum densities of the dynamic force had the similar characteristic of prominent frequency, which was lower than about 10 Hz.
(4) The fatigue strength of the impingement plate could be more accurately evaluated using the measured dynamic force and its power spectrum density.

(5) In the near future, a wider study in this field is expected to greatly extend this calculation method to high pressure conditions and various discharge pipe configurations.

NOMENCLATURE

A	= flow area of discharge pipe ($=\frac{\pi}{4} \times 0.0807^2 = 5.11 \times 10^{-3}$ m^2)
D_i	= inner diameter of discharge pipe (=80.7mm)
F	= static force acting on impingement plate in N
ΔF_{RMS}	= dynamic force (root mean square value) actin on impingement plate in N
ΔF_{PP}	= dynamic force (pean to peak value) acting on impingement plate in N
H	= distance between end of discharge pipe and impingement plate in m
J_{HM}	= jet force calculated with homogeneous flow model in N
J_{SM}	= jet force calculated with separated flow model in N
P	= static pressure acting on impingement plate in MPa
ΔP_{RMS}	= pressure fluctuation acting on impingement plate in MPa
Q_g	= air flow rate in m^3/s
Q_ℓ	= water flow rate in m^3/s
U_ℓ	= actual water velocity in m/s
$U_{\ell F}$	= actual water velocity obtained with static force acting on impingement plate in m/s
$U_{\ell DT}$	= actual water velocity obtained with delay time between void probe signal and load cell signal in m/s
α	= void fraction
β	= volume flow quality ($= \frac{j_g}{j_g + j_\ell}$ or $\frac{Q_g}{Q_g + Q_\ell}$)
j_g	= superficial air velocity in m/s
j_ℓ	= superficial water velocity in m/s
ρ_g	= air density in kg/m^3
ρ_ℓ	= water density in kg/m^3

REFERENCES

(1) Nakamura, T., et al., Study on flow Induced Vibration of a Tube Array by a Two-Phase Flow (1st Report: Large Amplitude Vibration by Air-water Flow), JSME No.85-0283A, 1985.
(2) Golan, L. P., Stenning, A. H., Proc. Ind. Mech. Engrs., 184, pt. 3C, 108, 1969~1970.
(3) Sekoguchi, O., et al., Proceedings of 4th Multi-phase Flow Symposium, Japan, 12, 4/5, 1985.
(4) Armand, A. A., Izvest. V Teplotech. Inst., No.1, 16, 1946, No.4, 1, 1947, No.2, 1950.

FUNDAMENTAL STUDY OF INTERFACIAL WAVES IN STRATIFIED FLOW

T. SAKAGUCHI, M. OZAWA, Y. SHIOMI[1], S. INOUE[2], and Y. MURAI[3]
Faculty of Engineering
Kobe University
Rokkodai-cho, Nada-ku, Kobe 657, Japan

ABSTRACT

Fundamental aspects of the behavior of fully developed solitary waves in two-phase stratified and wavy flows are discussed based on experimental results and results of numerical experiments. The wave-propagation velocity of fully developed solitary waves in two-phase stratified and wavy flows is expressed as the sum of the velocity of solitary waves on stationary liquid and the mean liquid velocity in stratified flow. The wave profile is given approximately by the wave profile on stationary liquid. The liquid velocity is given by the sum of the velocity induced by the wave on stationary liquid and the velocity of the stratified liquid flow. The static pressure in the liquid phase is given approximately by the sum of the static pressure of a solitary wave on stationary liquid and the gas pressure.

These simple relationships hold for flow conditions of fully developed solitary waves with relatively low velocities of liquid and gas. When the mean liquid velocity becomes large, the width of the wave profile decreases. The effect of the mean gas velocity on the solitary wave was rather small in the author's range of experiments up to 3.0 m/s. The applicability of the above-mentioned relationships was examined for the developing solitary wave, which led to the onset of slugging. The typical behavior of instability is discussed based on the data of the static pressure in liquid phase.

INTRODUCTION

An abrupt increase in the gas flow rate in a horizontal two-phase stratified or wavy flow induces slugging under certain

---

[1]Faculty of Science and Technology, Ryukoku University, Seta, Otsu 520-21, Japan
[2]Nippon Denso Co. Ltd., Kariya, Aichi 448, Japan
[3]Mitsubishi Heavy Industry Co. Ltd., Uzumasa-tatsumi-cho, Ukyo-ku, Kyoto 616, Japan

operating conditions. After the liquid slug flows out of the channel, the flow pattern returns to the initial separated flow. This flow phenomenon is referred to as a transient slug flow (Sakaguchi et al., 1973; 1977; 1978; 1980; 1987).

The onset of slugging in such transient conditions is closely related to the stability of liquid run-up or some kinds of finite amplitude wave induced by the abrupt increase in the gas flow. In a case of flow pattern transition from wavy to slug flow under conditions with constant flow rates, the mechanism of the onset of slugging has been usually explained by means of the Kelvin-Helmholtz instability theory or its extension to finite amplitude waves (Kordyban et al., 1970; Taitel et al., 1976; Mishima et al., 1980; Hihara et al., 1983). Thus the onset of slugging is closely related to the stability of the finite amplitude wave, such as the "solitary wave" (Taitel et al., 1976; Hihara et al., 1983) or the "most dangerous wave" (Mishima et al., 1980) in the wavy flow regime. Based on such stability theory, various criteria of the onset of slugging have been proposed by many researchers.

Although the dynamic behavior of finite amplitude waves in two-phase stratified and wavy flows plays an important role in the onset of slugging, such dynamic behavior has not been clearly investigated in experiments to date. Furthermore, in most of the analyses on slugging criteria, the velocity profile of liquid under the finite amplitude wave is given simply by superposition of the uniform velocity profile of the undisturbed liquid flow and the profile induced by the wave itself. This assumption has not been verified through experiments. This is mainly due to the difficulty in the identification of developing finite amplitude waves in experiments and analyses.

Thus the authors aimed to look insight fully into the flow behavior of such a finite amplitude wave in two-phase stratified and wavy flow regimes by imposing a well-defined solitary wave as an input into a horizontal two-phase flow system. The authors conducted laboratory experiments and numerical experiments successively on the following items: the flow behavior of solitary waves on stationary liquid, the flow behavior of solitary waves on flowing liquid but without gas flow above it, and the flow behavior of solitary waves in two-phase stratified and wavy flows. Finally, the dynamic behavior of solitary waves, which led to the onset of slugging, was investigated in two-phase wavy flow. Through these series of experiments, the authors investigated the effects of liquid and gas flows on the behavior of solitary waves and derived simple relationships which expressed these effects on solitary waves. In this paper, the authors present the experimental data on the behavior of solitary waves in various conditions of flow and discuss the above-mentioned effects on solitary waves.

The solitary wave is a typical example of finite amplitude waves. As for solitary waves on stationary liquid, investigations have been reported by Scott-Russel (see Wiegel, 1964), McCowan (1891), Laitone (1960), Daily and Stephan

(1953), Sakaguchi et al. (1986a) and others. Sakaguchi et al. (1986a) investigated the wave-propagation velocity, the wave profile, and the liquid-velocity profile of the solitary wave. Based on their experimental results, the applicability of McCowan's and Laitone's theories and also the numerical experiment which used the modified SOLA-SURF computer code to analyze the solitary wave on stationary liquid was verified. Thus the behavior of solitary waves on stationary liquid can be predicted accurately at this stage of research.

Benjamin (1967) analyzed the solitary wave on flowing liquid. However, the applicability of the result of Benjamin's study has not yet been confirmed experimentally. Moreover, the behavior of the solitary wave, such as the wave-propagation velocity, the wave profile, the velocity profile and the static-pressure profile in two-phase stratified and wavy flows has seldom been reported. Thus, the experimental data presented in this paper proves useful in examining the applicability of theories and simulations.

The authors describe briefly the behavior of solitary waves on stationary liquid based on Sakaguchi's work (1986a). Further, the behavior of solitary waves in two-phase stratified and wavy flow regimes, including fully developed and developing conditions is discussed based on the experimental results and the results of the numerical experiment.

EXPERIMENT

The experimental apparatus used in this investigation is shown schematically in Fig. 1. The main parts of the apparatus were a horizontal test channel of rectangular cross section, a water supply system, and an air supply system. The test channel was made of transparent acrylic resin plates, and was 100 mm in width, 82.5 mm in height and 10.8 m in

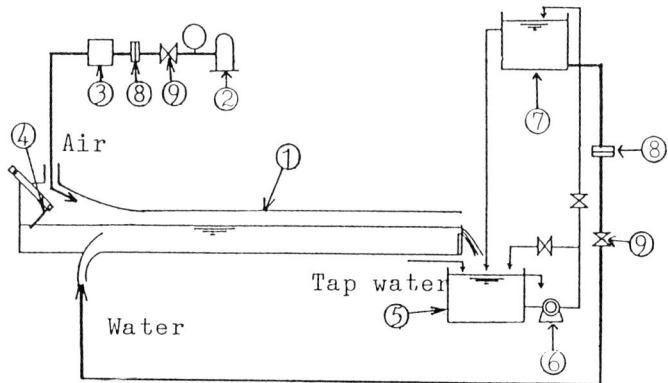

1: Test channel   2: Blower   3: Seeding generator
4: Wave generator   5: Water reservoir   6: Pump
7: Overflowing tank 8: Orifice 9: Flow control valve

Fig. 1 Experimental apparatus

length. At the end of the test channel, a wave generator was mounted in order to generate the solitary wave propagating concurrently with two phases. The wave generator was made of a flat plate which could be turned over at one edge of the plate.

Water and air were used as test fliuds. The water was supplied from an overflowing tank located at about 4 m above the test channel, and was introduced smoothly into the test channel. Then, the water flowed horizontally through the channel and flowed out over a weir at the end of the test channel. From a water reservoir at the end of the channel, the water was partly recirculated to the overflowing tank. Adjusting the height of the weir, an initial steady state value of the liquid depth was set to a predetermined value. The air was supplied smoothly into the test channel from a blower through a seeding generator. In seeding generator, scattering particles for LDV measurement were entrained into the air flow. Since tap water was used, extra scattering particles for LDV measurement were not needed because of the contamination of water.

In the first stage of experiments, inital steady state velocities of the liquid phase u_{lb} and of the gas phase u_{gb} were measured for predetermined initial liquid height h_i, volumetric fluxes of the liquid phase j_{lb}, and of gas phase j_{gb}. Then, the transient values of the liquid height h and the velocity components of two phases u_l, v_l, u_g and v_g were measured during the propagation of the solitary wave with a predetermined wave height h_a.

Two series of experiments were carried out with reference to the wave height. The first series of experiments were those for relatively small values of h_a. In this case the flow relating to the solitary wave was fully developed at the measuring position and did not change downstream of that position except for the damping of wave height owing to fluid friction. This type of solitary wave is referred to as a fully developed solitary wave. The second series of experiments were those for relatively large values of h_a. In this case the wave height increased significantly along the channel owing to the effect of gas flow above the liquid surface. In addition, the slugging took place at a certain position along the channel. This type of solitary wave is referred to as a developing solitary wave.

The liquid height was measured by conductance probes (Sakaguchi et al., 1980). From the results of the liquid height measurement, the wave profile and the wave propagation velocity C were determined. The wave profile was obtained through the transformation of the liquid height versus time curve into the liquid height versus the horizontal-length curve. This transformation was carried out by multiplying the time by the wave-propagation velocity because the changes of wave-propagation velocity and of wave profile were negligible during the residence time of the wave at the measuring position of the liquid height. Indeed, the wave profile obtained through such transformation almost coincided with the profile obtained simultaneously by means of several conductance probes.

The wave-propagation velocity C was defined as an average velocity of the wave crest in a certain distance along the channel. This distance was set to 0.7 m. All these quantities are shown schematically in Fig. 2 together with the half-value distance x_m.

The velocities of both phases were measured by using a He-Ne LDV system. Owing to the limitation of the LDV system and to the mutual relationship of the arrangement of the conductance probes and the laser beam, the horizontal and the vertical components u and v were obtained separately in the following manner. At first the horizontal component u of the velocity was measured, and next the velocities V_{30} and V_{330} were measured in the planes ±30 degrees inclined from the horizon, respectively. The geometrical relationships among these quantities are shown in Fig. 3. The vertical component v was calculated by using these two velocities V_{30} and V_{330} and Eqs. (1) to (3):

$$v = V \sin\theta \tag{1}$$

$$V = \frac{V_{30}}{\cos(\pi/6 - \theta)} \tag{2}$$

$$\theta = \tan^{-1}\left\{ \frac{V_{30} - V_{330}}{(V_{30} + V_{330})\tan(\pi/6)} \right\} \tag{3}$$

where V represents the absolute value of the velocity vector, and θ the inclination angle of the velocity vector from the horizon.

It was impossible to measure all the velocities in the flow area at the same time. Thus, the velocity profiles were obtained by traversing the LDV system in the vertical direction. The authors confirmed that generation of the solitary wave could be reproduced and that the dispersion of the wave height was kept within the range of ±0.3 mm.

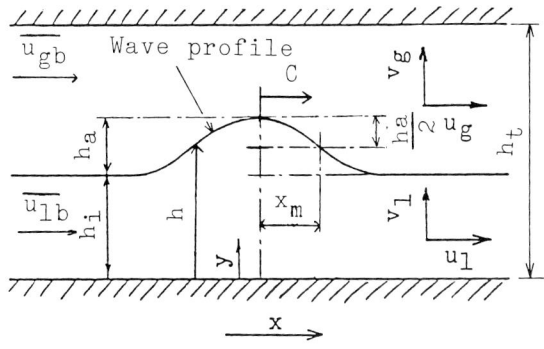

Fig. 2 Definition of variables

The static pressures of the gas and liquid phases were measured by using reactance-type pressure transducers (Sokken Model P7D) located at the top wall and the side wall at the height of 30 mm from the bottom of the channel, respectively.

In this experiment, the error in the measurement of the liquid depth was about ±0.1mm. The dispersion of the vertical component v was about ±0.02 m/s when the horizontal component u was about 0.4 m/s. This dispersion included the error in estimating the value v by means of the procedure mentioned above and also the dispersion due to the fluctuation of the flow condition. It was difficult to distinguish these two factors. The error in the pressure measurement was about ±2 Pa. The time delays of the response of the instrumentations for the liquid-depth measurement and for the pressure measurement were calibrated in pre-examinations, and were 0.04 s. and 0.005 s., respectively. Thus these time delays were compensated for the data analysis. The time response of the LDV system was fast enough for the experiment and the compensation was not needed.

Measurements of liquid height, velocities, and static pressures were carried out at the same position of $x=5.2$ m from the inlet of the channel. In the pre-examination, it was confirmed that the flow and the solitary wave were well-developed at the measuring position mentioned above when the wave height was relatively low. The velocity profiles of the liquid and the gas phases in the transverse direction of the channel were also measured at several positions along the channel, and the development of the flow along the channel was confirmed.

NUMERICAL EXPERIMENT

Numerical experiments for the solitary wave on flowing liquid were carried out using the modified SOLA-SURF code. This computer code was developed for the finite difference analysis of the Navier-Stokes equation and the continuity equation for two-dimensional flow with free surface. The details of the code can be found elswhere (Sakaguchi et al., 1986a; Hirt

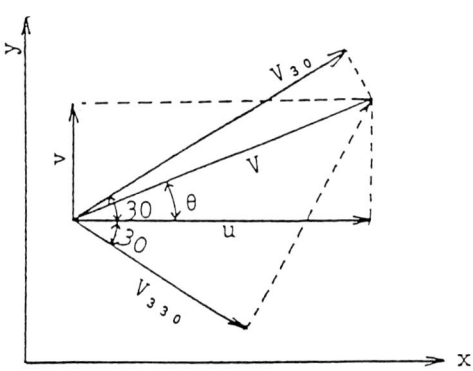

Fig. 3 Geometrical relationship among velocity components

INTERFACIAL WAVES IN STRATIFIED FLOW

et al., 1975). The applicability of this code to solitary waves has been confirmed (Sakaguchi et al., 1986a; 1986b). The effect of the gas flow on the motion of the solitary wave was not taken into account in the numerical experiment. However, in this experiment the gas velocity was rather low and the interfacial shear stress and the frictional pressure drop in the gas phase was very small. Therefore, only the liquid flow had dominant effects on the solitary wave in a certain range with relatively low gas velocity. It was expected that the numerical experiment would not differ significantly from the experimental results.

The following flow model was used in the numerical experiment:
(1) The flow field is two-dimensional, and the liquid is incompressible.
(2) Mass and momentum transfers through the gas-liquid interface do not exist.

The fundamental equations used in the code are given by

$$\frac{\partial u_1}{\partial x} + \frac{\partial v_1}{\partial y} = 0 \tag{4}$$

$$\frac{\partial u_1}{\partial t} + \frac{\partial u_1^2}{\partial x} + \frac{\partial u_1 v_1}{\partial y} = -\frac{1}{\rho_1}\frac{\partial p}{\partial x} + \nu_1 \left(\frac{\partial^2 u_1}{\partial x^2} + \frac{\partial^2 u_1}{\partial y^2}\right) \tag{5}$$

$$\frac{\partial v_1}{\partial t} + \frac{\partial u_1 v_1}{\partial x} + \frac{\partial v_1^2}{\partial y} = -\frac{1}{\rho_1}\frac{\partial p}{\partial y} + \nu_1 \left(\frac{\partial^2 v_1}{\partial x^2} + \frac{\partial^2 v_1}{\partial y^2}\right) - g \tag{6}$$

where g is the gravitational acceleration, p is a pressure, and ν_1 is a kinetic viscosity of liquid.

Boundary conditions at the free surface are given as follows:
(1) The surface is continuous,

$$\frac{\partial h}{\partial t} + u_1 \frac{\partial h}{\partial x} = v_1 \tag{7}$$

(2) The shear stress is equal to zero,

$$\frac{\partial u_1}{\partial y} + \frac{\partial v_1}{\partial x} = 0 \tag{8}$$

(3) The normal stress is equal to zero at the horizontal surface,

$$p_s = 2\nu_1 \frac{\partial v_1}{\partial y} \tag{9}$$

where p_s represents the pressure above the free surface, and is constant all over the flow field.

A staggered-mesh system was used in the finite difference approximation. The forward difference scheme was applied to the unsteady terms and the pressure terms, the centered difference scheme to the viscous terms, and the second upwind scheme was applied to the convective terms. The boundary condition at the bottom wall was set to the no-slip condition.

The initial condition for the velocity profile of liquid was given by superposition of the liquid velocity u_{1o} given by McCowan's equation (10) (McCowan, 1891) of the solitary wave on stationary liquid upon the liquid velocity u_{1b} of the initial flow in Fig. 4. The initial condition for the vertical component of liquid velocity was directly given by McCowan's equation (11).

$$u_{1o} = C_o N \frac{1 + \cos(\frac{My}{h_i})\cosh(\frac{Mx}{h_i})}{\{\cos(\frac{My}{h_i}) + \cosh(\frac{Mx}{h_i})\}^2} \tag{10}$$

$$v_{1o} = C_o N \frac{1 + \sin(\frac{My}{h_i})\sinh(\frac{Mx}{h_i})}{\{\cos(\frac{My}{h_i}) + \cosh(\frac{Mx}{h_i})\}^2} \tag{11}$$

where C_o is the propagation velocity of solitary wave on stationary liquid and is given by Boussinesq's equation (12) (Wiegel, 1964):

$$C_o = \sqrt{g(h_i + h_a)} \tag{12}$$

The constants M and N in Eqs. (10) and (11) are given by

$$\frac{h_a}{h_i} = \frac{N}{M} \tan\{\frac{M}{2}(1 + \frac{h_a}{h_i})\} \tag{13}$$

$$N = \frac{2}{3}\sin^2\{M(1 + \frac{2h_a}{3h_i})\} \tag{14}$$

The initial wave profile was given by McCowan's equation (15) (McCowan, 1891):

$$h = h_i + h_a \operatorname{sech}^2\{\sqrt{\frac{3h_a}{4(h_i + 19h_a/12)}}\frac{x}{h_i}\} . \tag{15}$$

In Eqs. (10) to (15), the origin of the horizontal coordinate x is set at the position corresponding to the wave crest, and that of the vertical coordinate is set at the bottom of the channel. The applicability of Eqs.(10) to (15) to the solitary wave on stationary liquid has been verified by Sakaguchi et al. (1986a) as described in the next section.

INTERFACIAL WAVES IN STRATIFIED FLOW

INITIAL STEADY STRATIFIED FLOW AND SOLITARY WAVE

The flow behavior of the initial two-phase stratified and wavy flows before the generation of the solitary wave in the channel were considered to be a basis for discussing the interaction between the flow and the wave. The authors describe here the velocity profile of this initial steady flow.

In this experiment, the liquid height was kept constant, h_i=50 mm. The void fraction of the initial flow was 0.394. Experiments were carried out in the range of the volumetric fluxes of liquid phase j_{lb}=0 - 0.102 m/s and of gas phase j_{gb}=0 - 1.18 m/s. The mean velocity of liquid phase was $\overline{u_{lb}}$=0 - 0.168 m/s, and the mean velocity of gas phase was $\overline{u_{gb}}$=0 - 3.0 m/s. In cases of the mean gas velocity being relatively low, the gas-liquid interface was smooth enough, and the flow was in stratified flow regime. In cases of the mean gas velocity being relatively high, for example, $\overline{u_{gb}} \geqq 2.5$ m/s, the gas-liquid interface became wavy.

The liquid flow induced by the solitary wave was approximately two-dimensional, except in the region near the side walls of the channel (Sakaguchi et al., 1986a). On the other hand, the initial flow in the rectangular channel is three-dimensional owing to the existence of the secondary flow (Schlichting, 1968). At present, it is too difficult to distinguish between the bulk flow along the channel and the secondary flow in the measured values of velocity. Thus, the discussion in this paper is carried out based on the velocity profile measured in the center of the channel width, and the effect of secondary flow on the solitary wave is beyond the scope of this discussion.

Examples of the velocity profiles of the initial flow are shown in Fig. 4. In the case of $\overline{u_{gb}}$=0, the maximum point of the liquid velocity did not appear at the free surface, which

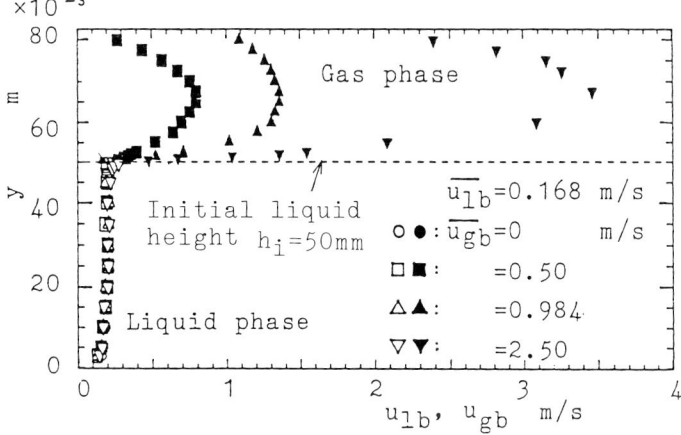

Fig. 4 Velocity profile in initial steady stratified flow

may have been due to the existence of the secondary flow (Schlichting, 1968). When the mean gas velocity became large enough, the liquid velocity became large only in the narrow region adjacent to the free surface due to the shear stress induced by the gas flow. In other words, the effect of gas flow on the liquid velocity was limited in such a very narrow region adjacent to the free surface, and the effect of gas flow was very small in a bulk of liquid flow.

Next, the authors describe briefly the behavior of wave-propagation velocity, wave profile, liquid-velocity profile and static-pressure profile of solitary waves on stationary liquid. These behaviors represent the fundamental characteristics of the solitary wave which was introduced into the channel as an input. Detailed description of the behavior of solitary waves on stationary liquid can be found elsewhere (Sakaguchi et al., 1986a).

The propagation velocity C_o of solitary wave on stationary liquid is, in general, uniquely determined for given h_i and h_a as shown in Fig. 5 (Sakaguchi et al., 1986a). The experimental data including the data of Daily and Stephan (1953) are correlated by using h_i+h_a, which is a linear function of $\sqrt{h_i+h_a}$. The solid line given by the Boussinesq's equation (12) agrees with the experimental data.

The wave profile of the solitary wave on stationary liquid is uniquely determined for h_a and h_i (Sakaguchi et al., 1986a). The wave profiles obtained in the experiment for stationary liquid are shown in Fig. 6, where the origin of x-coordinate is set to the position of the wave crest and the region $x>0$ corresponds to the wave front. Experimental results of the wave profile were obtained through the transformation of sampled data on the liquid depth versus the time curve. Thus the experimental results and the results of numerical experiments of wave profiles are represented by plotted data

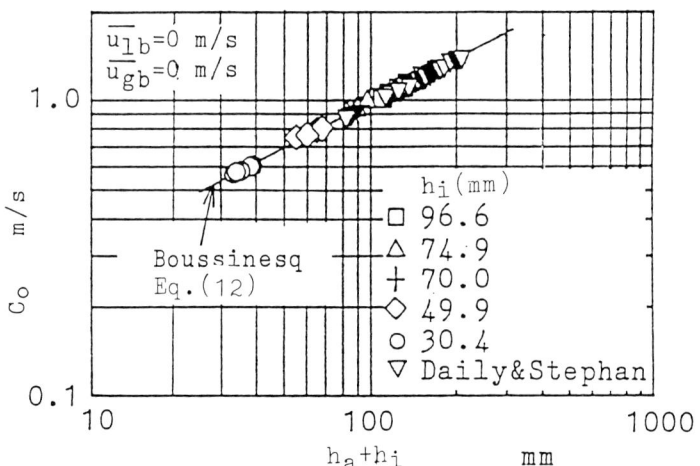

Fig. 5 Propagation velocity of solitary wave on stationary liquid

points. The experimental results agreed with the curve obtained by using McCowan's equation (15) and also with the results obtained in the numerical experiment (Sakaguchi et al., 1986a).

When the solitary wave propagates on stationary liquid, the liquid rises upward at the front side of the wave, and sinks downward at the back side as shown in Fig. 7. The experimental results of liquid velocity profile are shown in Fig. 8. In Figs. 8(a) and 8(b), broken lines represent the origins of $u_1=0$ and $v_1=0$ at every elevation, respectively. The

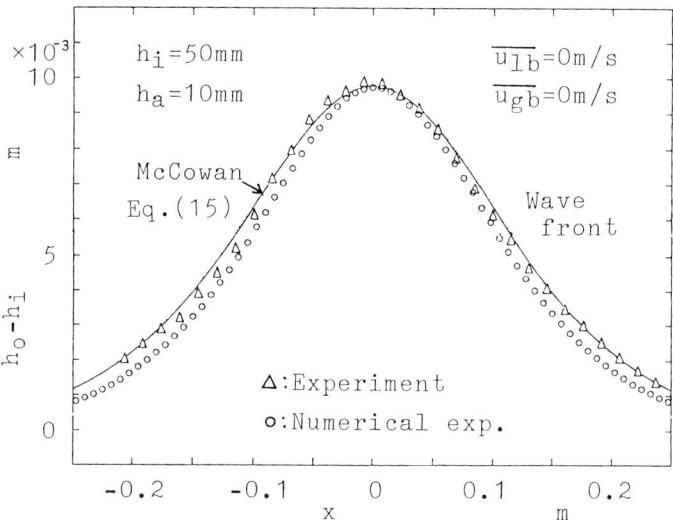

Fig. 6 Wave profile of solitary wave on stationary liquid

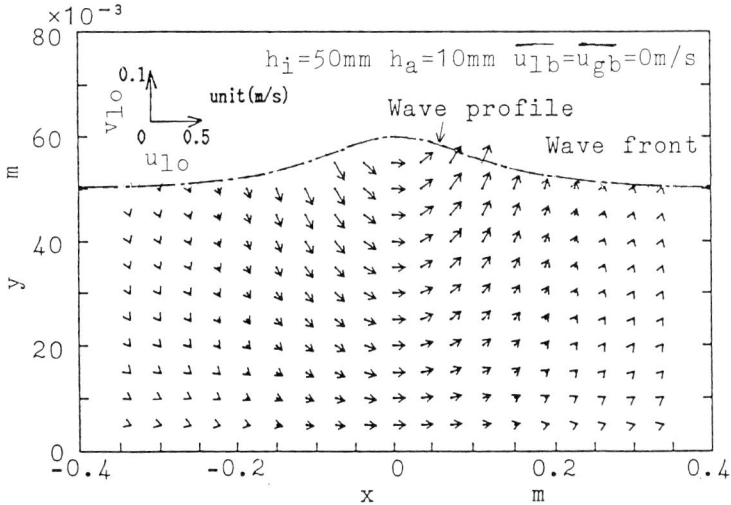

Fig. 7 Velocity vectors of solitary wave on stationary liquid (numerical experiment)

intersection points of the vertical coordinates with broken lines represent the height of the measuring position, y=15 mm, 25 mm, 35 mm, 45 mm, 50 mm and 55 mm in the case of Fig. 8. The data for y=55 mm represent the velocity profile in the liquid run-up part of the wave. The horizontal component u_l increases along the horizontal coordinate x, reaches a maximum at the wave crest, and then decreases. This tendency does not change for any values of y. The profile of u_l is approximately symmetric for front and back sides of the wave. On the other hand, the vertical component v_l is positive, and has a maximum at the front side; and v_l is negative, and has a minimum at the back side of the wave. The value v_l becomes zero at the position of the wave crest. The velocity profiles

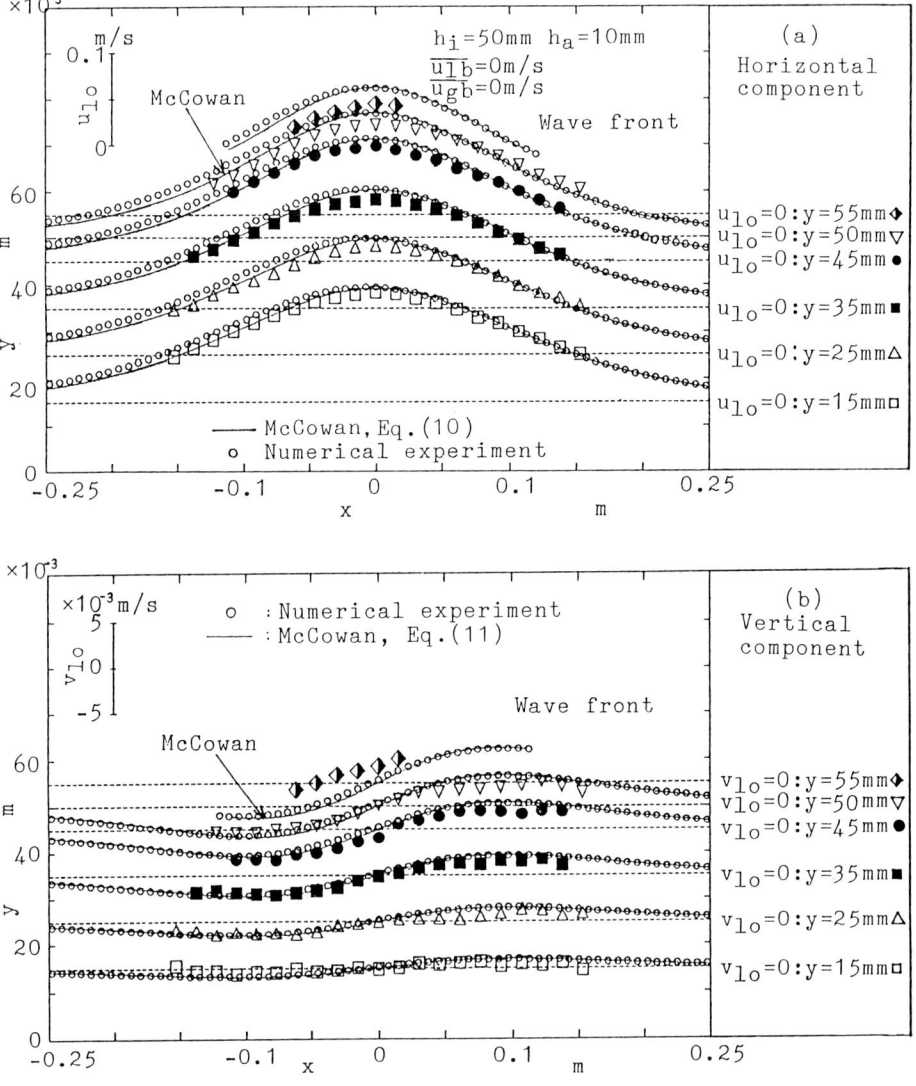

Fig. 8 Velocity profile of solitary wave on stationary liquid

obtained by the numerical experiment and McCowan's equations (10) and (11) agreed with the experimental results (Sakaguchi et al., 1986a).

Figure 9 shows the static-pressure profile ($p_{hi}-p_a$) at the initial liquid height, $y=h_i$, where p_a is the atmospheric pressure. The pressure p_{hi} was estimated by subtracting the static head $\rho_l g(h_i-y)$ from the measured value of static pressure at $y=30$ mm. It was confirmed that under the horizontal surface the static head of liquid $\rho_l g(h_i-y)$ at $y=30$ mm coincides with the pressure difference $(p(y=30 \text{ mm})-p_{hi})$. The static-pressure profile is similar in shape to the wave profile, because the dominant factor contributing to the pressure profile is the static head of liquid above the initial liquid height. The result in the numerical experiment was in agreement with the experimental one. The gas pressure (p_g-p_a) at the top of the channel was almost zero along the horizontal coordinate. This indicated that the pressure change induced by the wave motion was very small in the gas phase. A broken line represents the static pressure profile ($p_{hi}'-p_a$) obtained by the sum of the gas pressure (p_g-p_a) at the top of the channel, static head of gas phase and the static head of liquid in the liquid run-up of the wave, that is,

$$p_{hi}'-p_a \equiv p_g-p_a+\rho_g g(h_t-h)+\rho_l g(h-h_i). \tag{16}$$

The value of ($p_{hi}-p_a$) is about 10 % smaller than that of ($p_{hi}'-p_a$). This pressure difference between p_{hi} and p_{hi}' was caused by the curved stream lines in the wave.

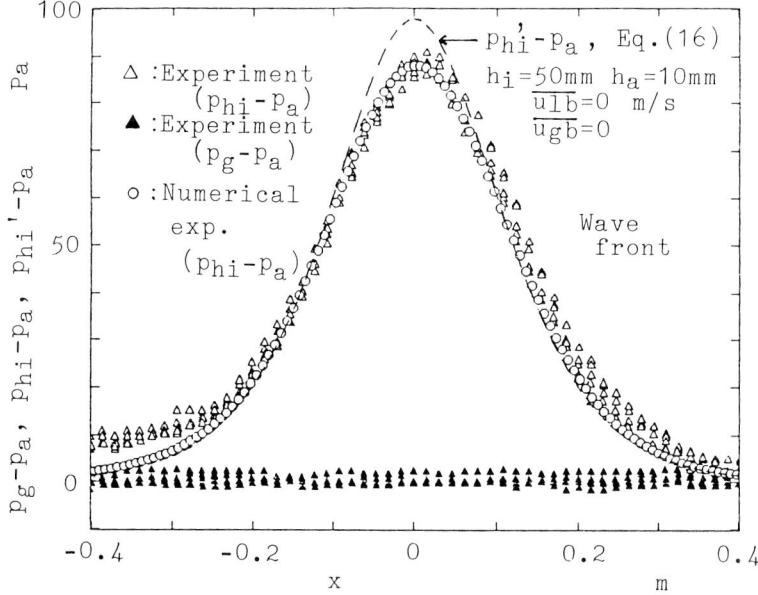

Fig. 9 Static-pressure profile of solitary wave on stationary liquid

In the next section, the authors describe the experimental results and the results in the numerical experiment of the solitary wave on the gas-liquid interface in stratified and wavy flow regimes.

BEHAVIOR OF FULLY DEVELOPED SOLITARY WAVES IN STRATIFIED AND WAVY FLOWS

When the wave height of solitary waves introduced into two-phase stratified and wavy flows is relatively low, the wave height retains an almost constant value during the residence time in the measuring section of the liquid height, velocities and static pressure, and the wave profile retains almost the same shape during the residence time. In this section, the authors describe the flow characteristics of such fully developed solitary waves, that is, the behavior of wave-propagation velocity, of wave profile, of velocity profile and of static-pressure profile. In all items, at first the behavior of solitary waves under the conditions of $\overline{u_{1b}} \neq 0$ and $\overline{u_{gb}} = 0$ is described and then the behavior of solitary wave under the conditions of $\overline{u_{1b}} \neq 0$ and $\overline{u_{gb}} \neq 0$, that is, in two-phase stratified and wavy flows, is described.

Propagation Velocity of Solitary Wave

Benjamin (1967) proposed the equation to give the propagation velocity of solitary wave on flowing liquid:

$$C = \overline{u_{1b}} + \sqrt{g(h_i + h_a) + gh_i F_{ri}^2 \Delta(3 + 4\frac{h_a}{h_i})} \qquad (17)$$

where F_{ri} represents the Froude number of the initial flow of liquid, and is defined by

$$F_{ri} \equiv \frac{u_{1b}}{\sqrt{gh_i}} \; . \qquad (18)$$

The momentum correction factor Δ is defined by

$$\Delta \equiv \frac{\int_0^{h_i} \{[u_{1b}(y) - \overline{u_{1b}}]/\overline{u_{1b}}\}^2 dy}{h_i} \qquad (19)$$

In Eq. (17), the first term on the right side corresponds to the contribution of the volumetric flux of liquid to the wave-propagation velocity, the first term in the root represents the wave-propagation velocity on stationary liquid and the second term represents the effect of the non-uniform velocity profile on the inertia force. The second term in the root is relatively small as described below. Then the wave-propagation velocity depends mainly on the mean liquid velocity for given values of the initial liquid height and of the wave height.

Thus experimental data of the wave-propagation velocity for given h_i and h_a under the conditions of $\overline{u_{gb}}$=0 m/s are plotted against $\overline{u_{1b}}$ in Fig. 10. The propagation velocity increases with the increase in the mean liquid velocity $\overline{u_{1b}}$. The relationship between the wave-propagation velocity and the mean liquid velocity of the initial flow is expressed approximately by a linear function of $\overline{u_{1b}}$ for given h_i and h_a. The results of the numerical experiment are also plotted in Fig. 10, and almost coincide with the experimental data. The solid line in Fig. 10 is drawn by using Benjamin's equations (17) to (19), and it agrees with the present data.

In general, the momentum correction factor Δ is small, for instance Δ=0.016 when the velocity profile is given by the 1/7-power law. The Froude number F_{ri} is at most 0.5 in these experiments. Thus, the term $F_{ri}^2\Delta$ becomes 0.004 and the second term in the root in Eq. (17) is so small as to be neglected in comparison with the first term in the root. Then, Eq. (17) reduces to

$$C = C_o + \overline{u_{1b}} \,. \tag{20}$$

The relationship C versus $\overline{u_{1b}}$ obtained by using Eq. (20) almost coincides with the solid line in Fig. 10. Thus, the wave propagation velocity on flowing liquid is expressed as the sum of that on stationary liquid C_o and the mean liquid velocity $\overline{u_{1b}}$. This means that the solitary wave propagates with its own propagation velocity in the frame of reference moving with the bulk motion of the flowing liquid, that is, $\overline{u_{1b}}$.

The propagation velocity of the solitary wave in two-phase stratified and wavy flows, that is, in cases $\overline{u_{1b}}\neq 0$ and $\overline{u_{gb}}\neq 0$, are also plotted in Fig. 10. In spite of the mean gas velocities being different, all data coincide with each other, and fall on the solid line. This fact indicates that

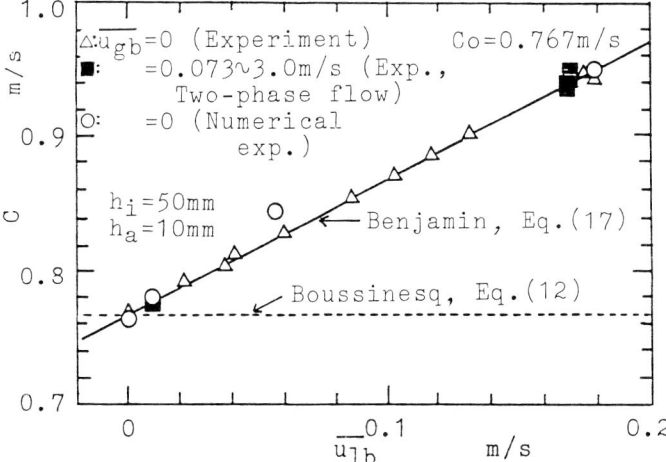

Fig. 10 Wave-propagation velocity of fully developed solitary wave in stratified flow

the effect of gas flow on the wave propagation velocity is so small as to be neglected in the range of this experiment. Thus the wave-propagation velocity is approximately expressed by Eqs. (17) and (20) in cases of fully developed solitary wave in the range of these experiments.

Wave Profile

Benjamin (1967) proposed the equation to give the wave profile of solitary wave on flowing liquid:

$$h = h_i + a \cdot \text{sech}^2\left(\frac{x}{b}\right) \tag{21}$$

where

$$\left.\begin{aligned}
a &= h_i \frac{g^{-1} - I_2(1)}{g I_4(1)} \\
b &= 2h_i \left\{ \frac{I_2(1)\beta(1) - \alpha(1)}{g^{-1} - I_2(1)} \right\} \\
I_2(1) &= \int_0^1 \frac{dZ}{W^2(Z)} \\
I_4(1) &= \int_0^1 \frac{dZ}{W^4(Z)} \\
\alpha(1) &= \int_0^1 \int_0^X \int_0^Y \frac{W^2(Y)}{W^2(X)W^2(Z)} dZ dY dX \\
\beta(1) &= \int_0^1 \int_0^Y \frac{W^2(Y)}{W^2(Z)} dZ dY \\
W(y) &= u_{1b}(y) - C.
\end{aligned}\right\} \tag{22}$$

Based on these equations, he reported that the width of the wave profile becomes smaller with the increase in the mean liquid velocity.

The experimental results of the wave profile on flowing liquid under the condition of $\overline{u_{gb}}$=0 m/s are shown in Fig. 11. The wave profiles are similar to that for stationary liquid in Fig. 6. The calculated result for $\overline{u_{1b}}$=0.168 m/s by means of Eq.(21) agreed with the experimental results. Although it can be seen in the results of Benjamin that the width of the wave profile decreases slightly with the increase in the mean liquid velocity $\overline{u_{1b}}$, such effect of liquid flow on the wave profile was not remarkable in this range of experiments, and almost all data coincided with each other independently on the mean liquid velocity. The effect of liquid flow on the wave profile is discussed by referring to the numerical experiment later. The wave profile obtained using McCowan's equation (15) is shown by solid line in Fig. 11 and agrees with the experimental result except a certain part of the back side of the wave. The liquid height at the back side of the wave was higher than that of McCowan's curve which was symmetric at the front and back sides. This difference in the liquid height at the back side was mainly due to the secondary wave which followed the primary wave. Thus the discussion of the wave profile was limited only at the front side of the wave here. The wave profiles obtained in the numerical experiments are also shown in Fig. 11. The results of the numerical experiments showed that the wave profile became small in its width with the increase in the mean liquid velocity. Although the width of the wave profile in the numerical experiment was slightly small compared with the experimental results, the result obtained in the numerical experiment approximately agreed with the experimental results.

Owing to the limitation of the experimental range, the effect of liquid flow on the wave profile was not clear in the experiment. Thus, the effect of liquid flow on the wave profile was investigated in the numerical experiment. The width of the wave profile was characterized by means of the half-value distance x_m. The half-value distance x_m of the wave profile was defined as a horizontal distance between the wave crest and the position where the liquid height reaches the value $(h_i+h_a/2)$ as shown in Fig. 2. In the numerical experiment where the secondary wave which followed the primary wave was also observed, the wave profile was not symmetric at the front and back sides with respect to the point of wave crest. Thus in this case, the authors discuss only the half-value distance at the front side of the wave.

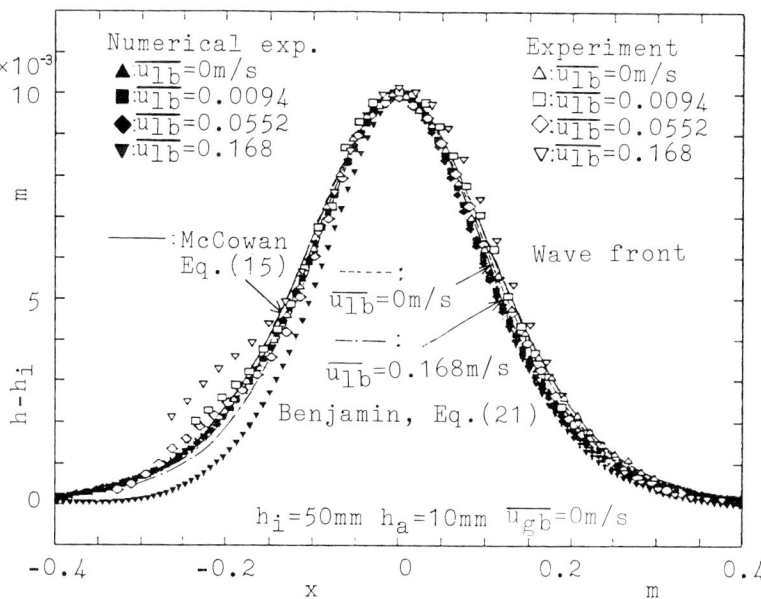

Fig. 11 Wave profile of fully developed solitary wave on flowing liquid

In general, the wave profile of a solitary wave was closely related to the wave height h_a and the initial liquid height h_i. The wave-propagation velocity C was <u>also</u> closely related to h_i and h_a and was proportional to $\sqrt{h_i+h_a}$. Thus the wave profile was closely related to the wave propagation velocity C. Then, the normalized half-value distance x_m/x_{mo} is plotted against the Froude number F_{rc} of the wave propagation velocity in Fig. 12, where x_{mo} represents the half-value distance of the wave profile on stationary liquid and the Froude number F_{rc} is defined by

$$F_{rc} \equiv \frac{C}{\sqrt{g(h_i+h_a)}} \quad . \tag{23}$$

Substituting Eqs.(12) and (20) into Eq.(23), Eq.(23) is rewritten as

$$F_{rc} = \frac{\overline{u_{1b}}}{\sqrt{g(h_i+h_a)}} + 1 \quad . \tag{24}$$

The normalized half-value distance x_m/x_{mo} decreases approximately linearly with increasing F_{rc}. The relationship between x_m/x_{mo} and F_{rc} is approximately expressed by

$$x_m/x_{mo} = 1.22 - 0.22 F_{rc} \quad . \tag{25}$$

When u_{1b} becomes the same order of magnitude of C_o, the width of solitary wave decreases about 20 % from that for $\overline{u_{1b}}=0$ m/s. The normalized half-value distance obtained by means of Benjamin's equation (21) is also plotted in Fig. 12. The results of Benjamin almost coincide with those of the numerical experiment.

When $\overline{u_{1b}}$ is relatively low, the difference between x_m and x_{mo} is small. Thus the following simple relationship approxi-

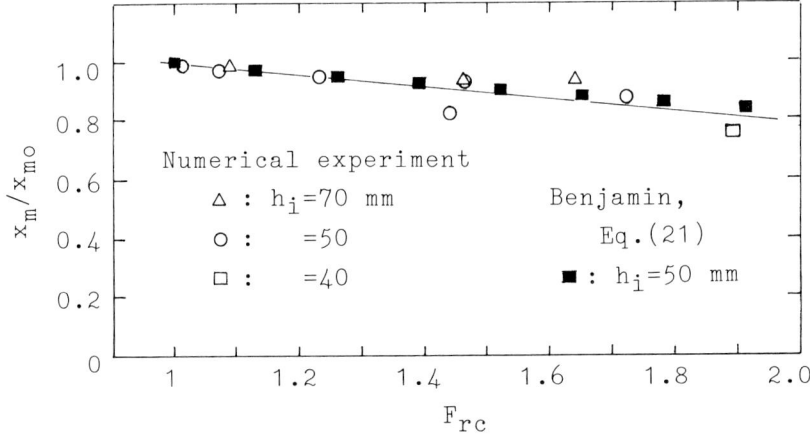

Fig. 12 Half-value distance of wave profile on flowing liquid

INTERFACIAL WAVES IN STRATIFIED FLOW

mately holds for given initial liquid height and wave height:

$$x_m = x_{mo}. \tag{26}$$

In the next part of this section, the wave profile of the solitary wave in two-phase flow will be discribed.

The experimental results of the wave profile in two-phase stratified and wavy flows, that is, in cases of $\overline{u_{1b}} \neq 0$ and $\overline{u_{gb}} \neq 0$, are shown in Fig. 13. The wave profiles in two-phase flow are similar to those in cases of $\overline{u_{gb}} = 0$ m/s in Fig. 11. It should be noted here that small waves whose amplitude was about 0.5 mm were superposed on the solitary wave in cases of relatively large value of $\overline{u_{gb}}$, that is, $\overline{u_{gb}} \geq 2.5$ m/s, and that the fluctuation of the liquid height for $\overline{u_{gb}} \geq 2.5$ m/s was larger than that for relatively small value of $\overline{u_{gb}}$. Therefore the wave profiles for $\overline{u_{gb}} \geq 2.5$ m/s are averaged profiles in several runs. The difference of widths of waves was actually very small in this experimental data and the effect of gas flow on the wave profile is not clear in Fig. 13.

The wave profiles obtained by using Eqs. (15) and (21) and that obtained in the numerical experiment are also shown in Fig. 13. All curves approximately agrees with the experimental result, although the effects of liquid flow and/or gas flow were not taken into account in the calculation. This fact indicates that in cases of relatively low values of the initial liquid and gas velocities the wave profile in two-phase stratified and wavy flows can be approximately estimated by using Benjamin's equation (21) for flowing liquid, by using this numerical experiment and more simply by using McCowan's equation (15) for stationary liquid. Thus the simple relationship of Eq.(26) approximately holds not only in cases of $\overline{u_{gb}} = 0$ m/s but also in two-phase stratified and

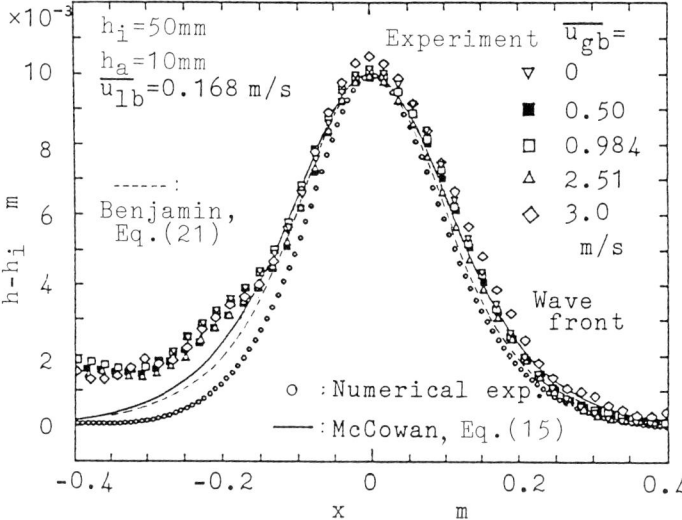

Fig. 13 Wave profile of fully developed solitary wave in two-phase flow

MODELING AND DYNAMICS

wavy flows in this experimental range.

Velocity Profile

In this section, the authors first describe general features of the velocity profile of the solitary wave on flowing liquid under the condition of $\overline{u_{gb}}=0$ m/s. The description is based on the experimental results and on the results of the numerical experiments.

Figure 14 shows one of the examples of the velocity profiles obtained in the experiment. The broken lines represent the origin of the velocity, that is, $u_l=0$ m/s in Fig. 14(a) and $v_l=0$ m/s in Fig. 14(b), respectively. The intersections with the vertical axis represent the height of the measuring position from the bottom. The curves represented by small

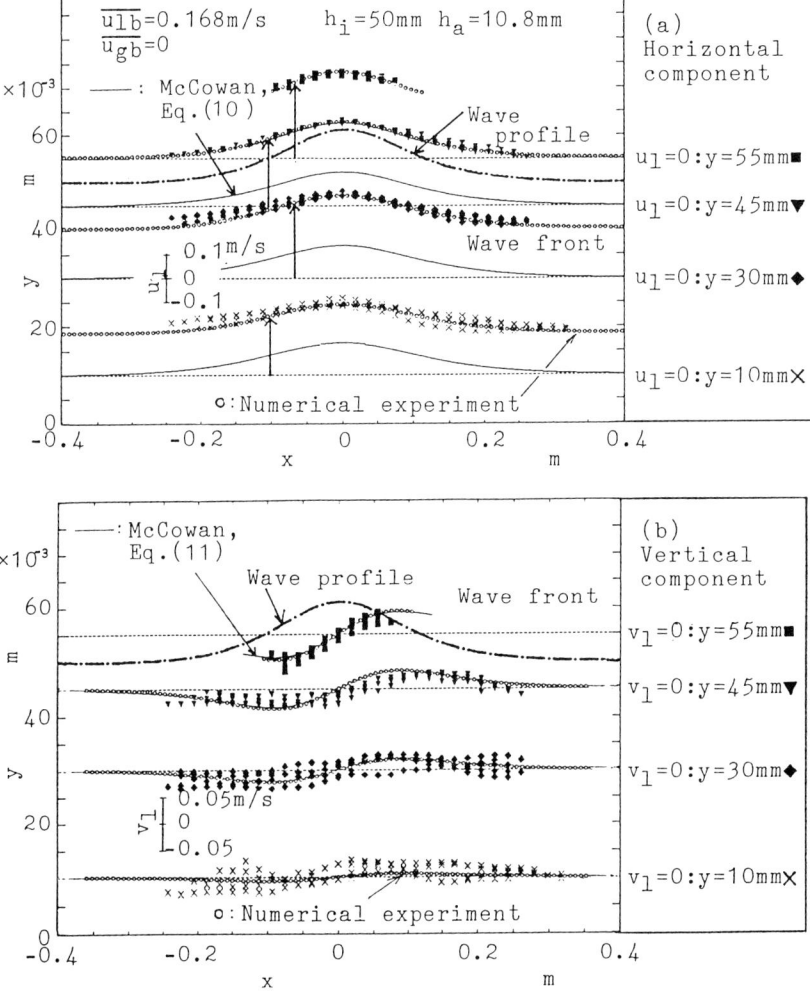

Fig. 14 Velocity profile of fully developed solitary wave on flowing liquid

circles are obtained in the numerical experiment and the dot-dash lines represent the wave profile. Solid lines represent the velocity profiles ontained by using McCowan's equations (10) and (11) for solitary waves on stationary liquid. Plotted data were obtained in several runs of experiments. Although the absolute value of the horizontal component u_1 in Fig. 14(a) is larger than that for stationary liquid in Fig. 8 as can be seen from the comparison between the experimental data and the solid-line curves, u_1 shows a tendency similar to that for stationary liquid throughout the experimental results. The dispersion of the experimental results of the vertical component v_1 is relatively large compared with that of u_1 owing to the method of measurement applied to this experiment. However, it is clear that the data of vertical component v_1 have also a tendency similar to those for stationary liquid in Fig. 8. They approximately agree with the solid-line curves of McCowan's equation for stationary liquid. The velocity profiles obtained in the numerical experiment agree with the experimental results.

Next the authors discuss the relationship between the liquid velocities of the solitary wave and of the initial liquid flow. In the previous section, the authors described that the solitary wave propagated in its own velocity in the frame of reference moving with the mean liquid velocity of the initial liquid flow. Thus it may be possible to relate the liquid velocity of the solitary wave to the mean liquid velocity. However, the initial liquid velocity and also the liquid velocity induced by the solitary wave were not uniform in the vertical direction. Thus in this case, it is assumed that every fluid particle moves in its own velocity induced by the solitary wave in the frame of reference moving with the same velocity as the fluid particles in the initial liquid flow. In other words, it is assumed that the liquid velocity of the solitary wave on flowing liquid is expressed as a sum of that on the stationary liquid u_{1o} and the initial liquid velocity u_{1b} at each height:

$$u_1 = u_{1b} + u_{1o} . \qquad (27)$$

The vertical velocity component of the initial liquid flow will be very small and is assumed to be zero in this case. Then the vertical component v_1 is approximately given by

$$v_1 = v_{1o} . \qquad (28)$$

In the part of the wave above the initial liquid height, the liquid velocity u_{1b} is assumed to be equal to that at the interface ($y=h_i$), because there is not initial flow of liquid in that region.

Fukano et al. (1985) analyzed the stability of wave motion in thin liquid-film flow under the assumption that the liquid velocity profile of wave was similar to those of the base film in shape. By using the present notions, Fukano's assumption can be expressed by

$$u_1(y) = \frac{h_i+h_a}{h_i} u_{1b}(\frac{h_i}{h_i+h_a}y) \qquad (29)$$

where $u_1(y)$ represents the liquid velocity of wave at elevation y, $u_{1b}(h_i y/[h_i+h_a])$ represents the liquid velocity of initial flow at elevation $h_i y/[h_i+h_a]$. This method may be applicable to the estimation of the initial flow of liquid in the part of the wave above the initial liquid height. However, this assumption results in that the initial flow of liquid involves the effect of the existence of wave on the velocity profile, and that it is not easy to discuss the relationship between the initial flow without a wave above it and the solitary wave. Thus the simple assumption described before is used.

One of the examples of the (u_1-u_{1b}) profile is shown in Fig. 15. The profile of (u_1-u_{1b}) in the experiment approximately coincides with the curve of McCowan's equation (10) and with the results of the numerical experiment for stationary liquid. As is described above, the experimental results of vertical component v_1 approximately agreed with the curve of McCowan's equation (11). These facts show that the relationships of Eqs.(27) and (28) approximately hold in this range of experiments.

As the wave profile was slightly affected by the initial liquid flow, the liquid velocity profile was also affected by the initial liquid flow. The effect of the initial liquid flow on the velocity profile is discussed here using the (u_1-u_{1b}) versus x and v_1 versus x planes. Assuming that the above-mentioned relationships in Eqs. (27) and (28) hold for large values of $\overline{u_{1b}}$, for instance $\overline{u_{1b}}$=0.35 m/s (F_{ri}=0.5), the profiles of (u_1-u_{1b}) and v_1 obtained in the numerical

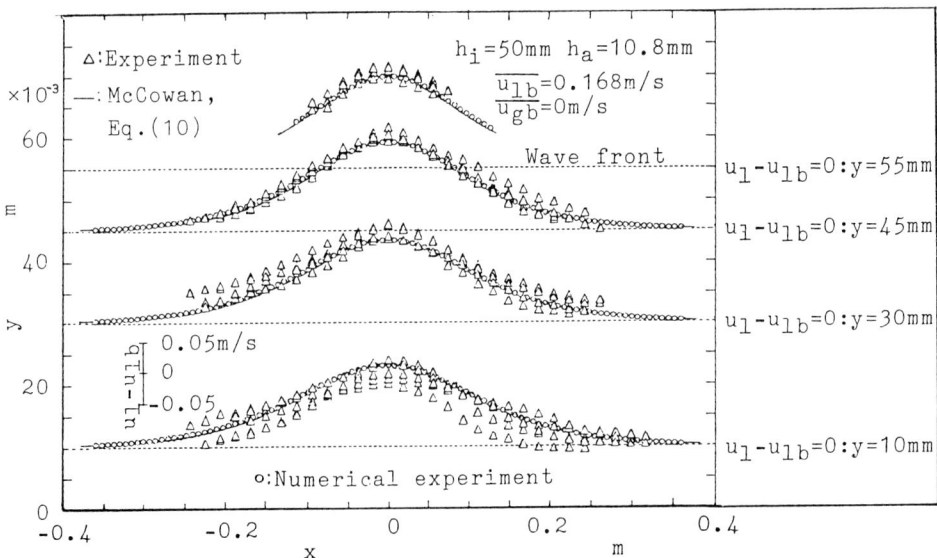

Fig. 15 Liquid velocity of fully developed solitary wave relative to initial liquid flow

INTERFACIAL WAVES IN STRATIFIED FLOW 125

experiment in such case are compared with those for stationary liquid, that is, $\overline{u_{1b}}=0$ m/s, in Fig. 16. With respect to the velocity component (u_1-u_{1b}) at the same distance from the wave crest, the value of (u_1-u_{1b}) for the solitary wave on flowing liquid becomes slightly smaller than that of the solitary wave on stationary liquid. As to the vertical component, the horizontal distance between the wave crest and the maximum position of v_1 of the solitary wave on flowing liquid is slightly smaller than that on stationary liquid. These facts correspond to the difference in the wave profiles between the solitary waves on flowing liquid and on stationary liquid. Thus it is concluded in the first part of this section that the dominant effect of the initial liquid flow on the liquid velocity of the solitary wave on flowing liquid can be approximately expressed by using linear relationships

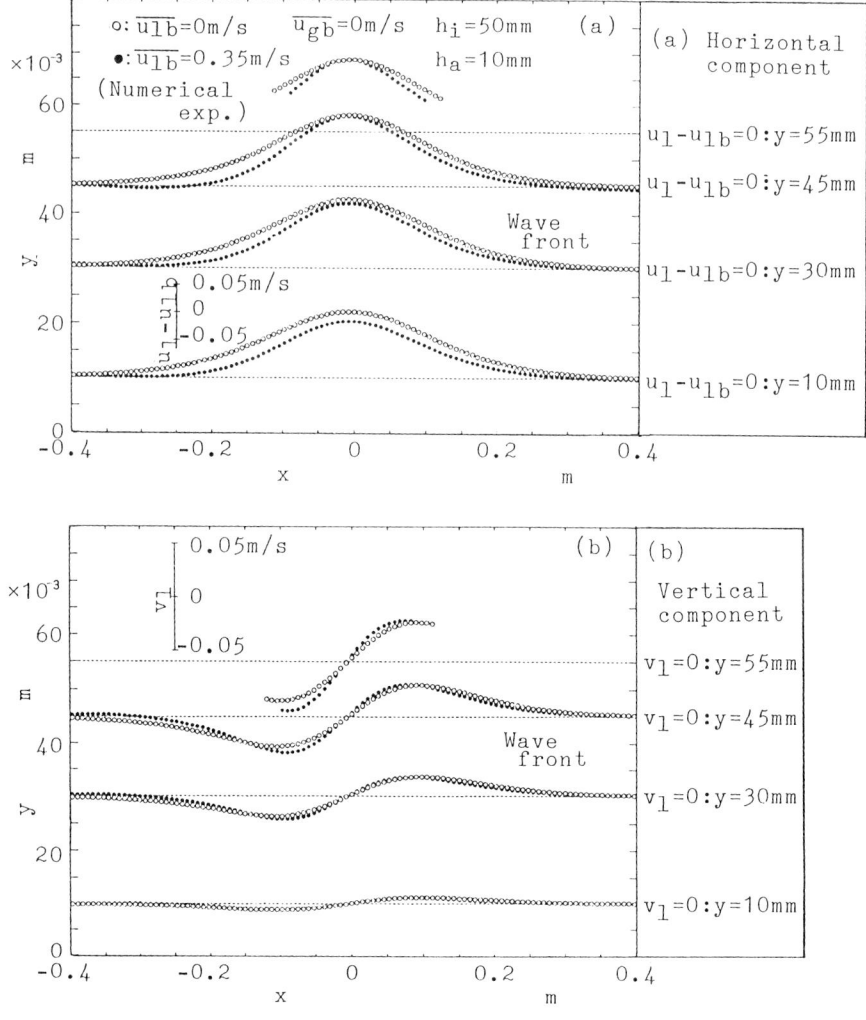

Fig. 16 Liquid velocity of fully developed solitary wave relative to initial liquid flow (numerical experiment)

in Eqs. (27) and (28) and that secondary effect of the initial liquid flow was observed in the width of the velocity profile.

Next the authors discuss the velocity profile of the solitary wave in two-phase stratified and wavy flows, that is, in cases of $\overline{u_{lb}} \neq 0$ and $\overline{u_{gb}} \neq 0$. The velocity profiles of the solitary wave are shown in Figs. 17 and 18. The broken lines represent the origin of the velocity component at each height as shown in Fig. 14. The curves represented by small circles are results of the numerical experiment which do not take into account the effect of gas flow. The horizontal component u_l and also the vertical component v_l in liquid phase show tendencies similar to those in the case of $\overline{u_{gb}}=0$ m/s in Fig. 14. The liquid-velocity profile obtained in the numerical

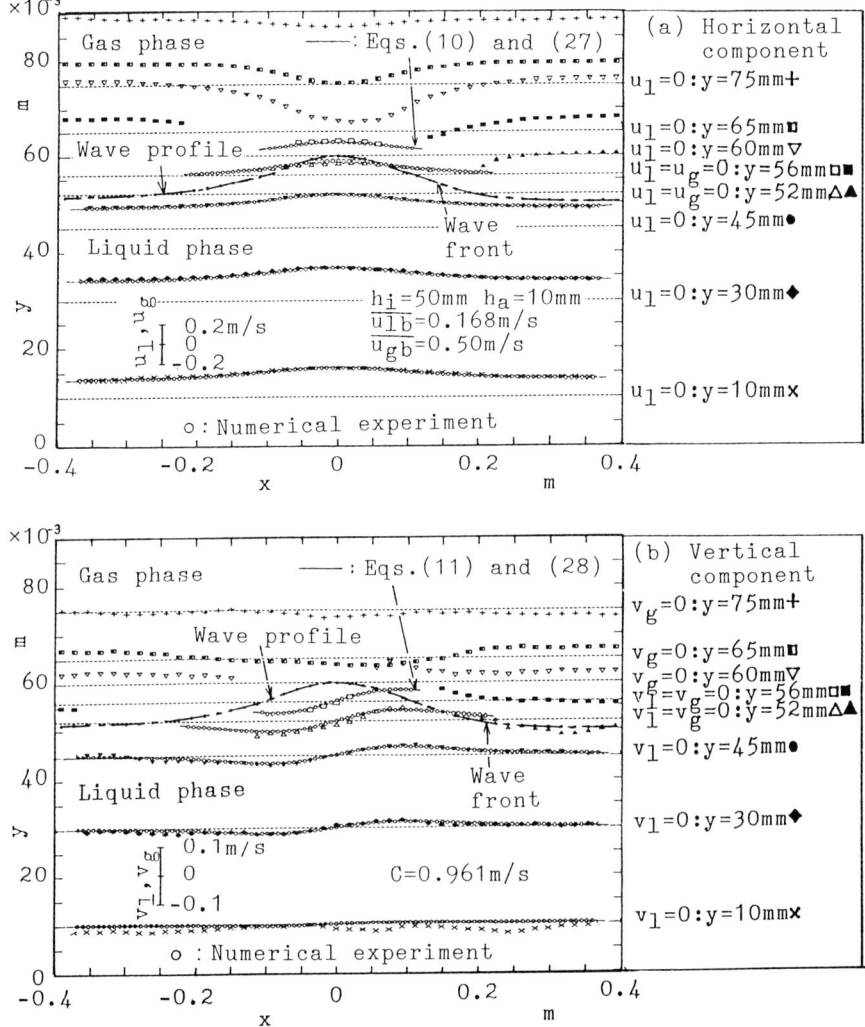

Fig. 17 Velocity profile of fully developed solitary wave in two-phase flow ($\overline{u_{gb}} < C$)

INTERFACIAL WAVES IN STRATIFIED FLOW

experiment almost coincides with that in experiments in two-phase stratified and wavy flows. Solid lines represent the liquid-velocity profiles which are estimated by means of McCowan's equations (10) and (11), Eqs.(27) and (28) and the initial liquid velocity u_{lb} in Fig. 4. Solid lines almost coincide with the experimental results. This means that the effect of gas flow on the liquid-velocity profile of the solitary wave is rather small in this range of experiments. Furthermore, the simple relationships of Eqs.(27) and (28) approximately hold in two-phase stratified and wavy flows.

In the case of the mean gas velocity $\overline{u_{gb}}$=0.50 m/s, the horizontal component of the gas velocity, u_g, decreases on approaching the wave crest, and has a minimum value at the wave crest. The vertical component, v_g, is rather small

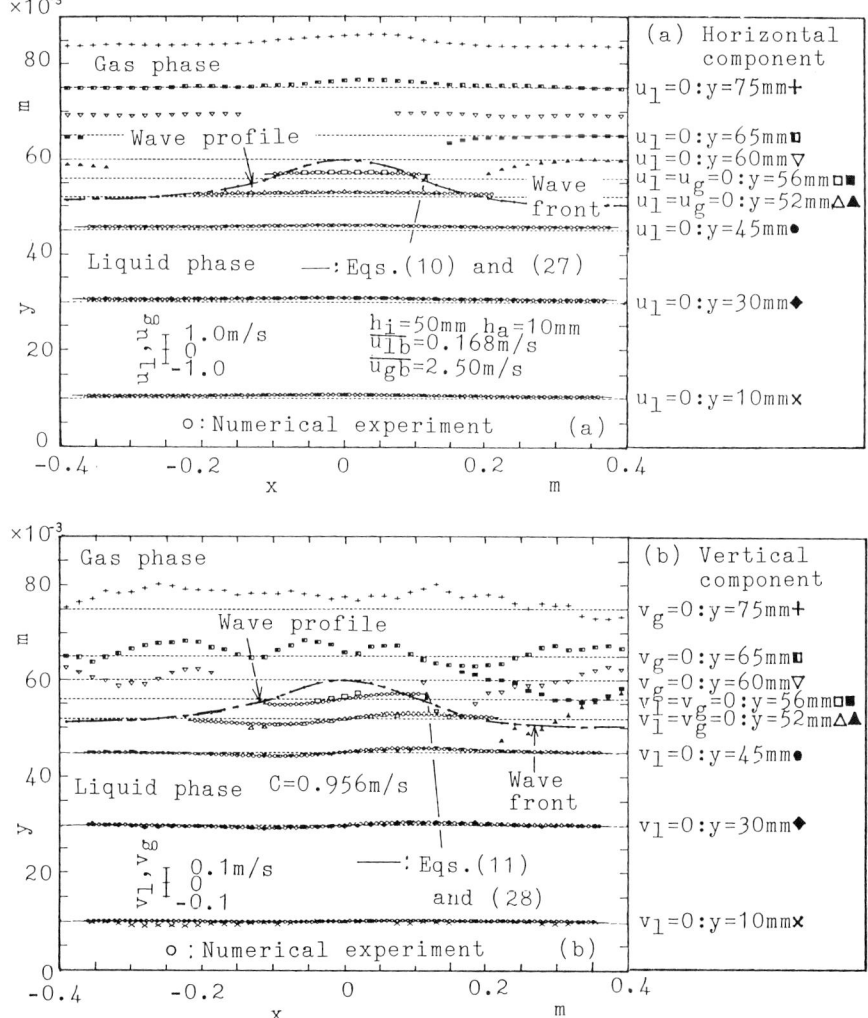

Fig. 18 Velocity profile of fully developed solitary wave in two-phase flow ($\overline{u_{gb}} > C$)

compared with the horizontal component and it is difficult to find general tendencies from this example. In the case of $\overline{u_{gb}}$=2.5 m/s, u_g increases on approaching the wave crest and has a maximum value at the wave crest. The profile of u_g shows the opposite tendency to that in the former case. The vertical component v_g is very small compared with the horizontal component and shows a certain magnitude of fluctuation. Thus it is also difficult to find general tendencies in these data on the vertical velocity component of gas phase.

The behavior of u_g is explained in the following simplified model. The frame of reference is set so as to move with the solitary wave. Then, the relative mean gas velocity with respect to the wave can be given by $\overline{u_{gb}}-C$ above the initial liquid height. At a certain position away from the solitary wave, the flow area of the gas is given by (h_t-h_i) for unit width of the channel, and that above an arbitrary liquid height is given by (h_t-h). Assuming the constant density of gas phase, the mass balance of gas phase is given by

$$\rho_g(\overline{u_{gb}} - C)(h_t - h_i) = \rho_g(\overline{u_{gb}}' - C)(h_t - h) \tag{30}$$

where $\overline{u_{gb}}'$ is the mean gas velocity above the liquid height h. Rearranging Eq.(30) with respect to $\overline{u_{gb}}'$, one obtains

$$\overline{u_{gb}}' = \overline{u_{gb}} + \frac{h}{h_t - h}(\overline{u_{gb}} - C) . \tag{31}$$

Equation (30) represents the relationship between the mean gas velocity at the position away from the wave and that above the wave. As the velocity profile of the gas phase is not uniform in the vertical direction, this equation does not hold in all regions of the gas flow. For example, data from the horizontal components u_g at $y=52$ mm and 56 mm show different tendencies from those at elevation $y \geq 60$ mm. Thus these components are not valid in the regions near the wave nor the channel walls, but are valid in the bulk region of the flow.

In the case of $\overline{u_{gb}}<C$, the second term in the right side in Eq. (31) becomes negative. This means that the horizontal component of gas velocity decreases on approaching the wave crest and has a minimum value at the wave crest as can be seen from the data at $y=65$ mm and 75 mm in Fig. 17. In the case of $C<\overline{u_{gb}}$, the second term in the right side in Eq. (31) becomes positive. The gas velocity increases on approaching the wave crest and has a maximum at the wave crest as can be seen from the data at $y=65$ mm and 75 mm in Fig. 18.

The general behavior of v_g can be anticipated using the frame of reference moving with the wave. The horizontal component of gas velocity had, in the case of Fig. 17, a negative direction of the x-axis with respect to the frame of reference moving with the wave. Then, the gas flow at the wave front could be considered as convergent flow in the frame of reference moving with the wave. This means that the value v_g had a positive value at the wave front. The gas flow at the back of the wave became divergent flow, and v_g had a negative value. In the case of Fig. 18, the gas flow at the wave front became divergent flow in the frame of reference moving with the wave. Then, v_g became negative at the wave front. The value v_g became positive at the back of the wave. However, these expected tendencies of v_g were not clear in the experimental results in Figs. 17 and 18. This was due to the fluctuation of v_g and/or the secondary flow of gas phase, but it was not easy to estimate the order of magnitude of the secondary flow and its effect on the velocity profile. Further discussion on the behavior of v_g will be reported in a subsequent paper.

The order of magnitude of the vertical component was, for example, in the case of Fig. 18, about one tenth of the horizontal component. Therefore, the effect of the vertical component v_g on the behavior of solitary wave was not so significant. Thus it is concluded that the liquid velocity profile can be approximately estimated by means of this numerical experiment and more simply by means of Eqs.(27) and (28) without taking into account the effect of gas flow under relatively low values of the mean liquid and gas velocities as in this range of the experiment.

Static-Pressure Profile

This section presents the static-pressure profiles of solitary waves, at first, on flowing liquid under the condition of $\overline{u_{gb}}$=0 m/s and then, in two-phase stratified and wavy

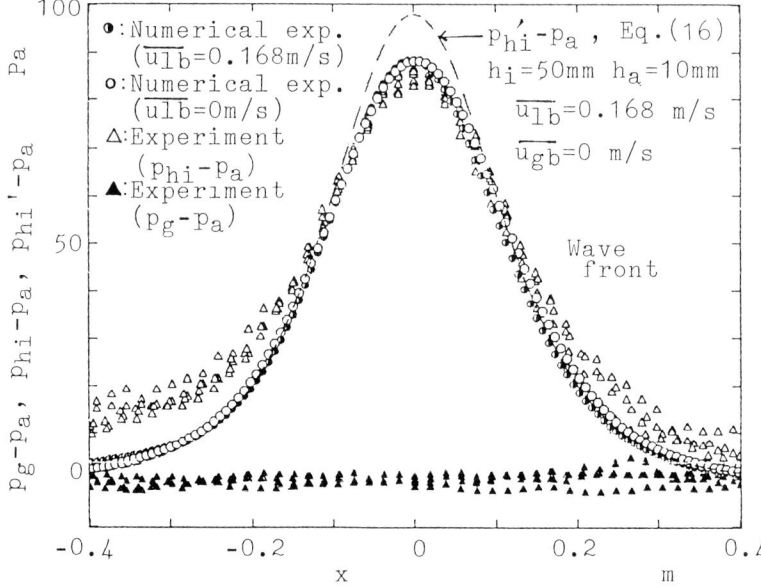

Fig. 19 Static-pressure profile of fully developed solitary wave on flowing liquid

flows.

Figure 19 shows experimental results of the static pressure ($p_{hi}-p_a$) in the liquid phase at the initial liquid height, $y=h_i$, in the case of $\overline{u_{gb}}=0$ m/s. The static pressure profile showed a similar shape to that observed in cases of solitary waves on stationary liquid shown in Fig. 9. The broken line shows the calculated results by using Eq.(16) of the pressure profile ($p_{hi}'-p_a$) owing to the static heads of liquid and gas above the initial liquid height. The static pressure ($p_{hi}-p_a$) was about 10% smaller than the value of ($p_{hi}'-p_a$). Almost the same characteristics were observed in this case as were observed in the case of Fig. 9. The change of gas pressure p_g was very small and the value of (p_g-p_a) was almost zero. This was mainly due to the fact that the mean gas velocity $\overline{u_{gb}}$ was zero in this case and thus the pressure drop in the gas phase is almost zero. The result of the numerical experiment for $\overline{u_{1b}}=0.168$ m/s and $\overline{u_{gb}}=0$ m/s is also shown in the figure and is in agreement with the experimental results. In the figure, the static-pressure profile obtained in the numerical experiment for $\overline{u_{1b}}=0.168$ m/s and $\overline{u_{gb}}=0$ m/s is compared with that for $\overline{u_{1b}}=0$ m/s and $\overline{u_{gb}}=0$ m/s. The former was slightly lower than the latter, but the maximum values at the wave crest almost coincided between the two cases. The effect of liquid flow on the static-pressure profile was mainly observed in the width of the profile, which corresponded to those in the wave profile and in the velocity profile. Though this effect became significant when the mean liquid velocity was high, the effect of liquid flow on the static pressure was small for relatively low mean liquid velocity and the static pressure p_l in liquid phase of the solitary wave on flowing liquid was approximately given by

$$p_l = p_g + (p_{lo}-p_{go}) \tag{32}$$

where the static pressure p_{lo} of the solitary wave on stationary liquid could be predicted by means of this numerical experiment as shown in Fig. 9. When the values of (p_g-p_a) and ($p_{go}-p_a$) were very small compared with ($p_{lo}-p_a$), equation (32) reduced to

$$p_l = p_{lo} . \tag{33}$$

Of course, the static pressure p_{lo} can be approximately estimated only by taking into account the static head of liquid. However, this simple estimation causes a relatively large error mainly at the wave crest as previously discussed.

The static-pressure profiles in two-phase stratified and wavy flows are shown in Fig. 20. The static pressure p_{hi} at $y=h_i$ in liquid phase showed a profile similar to that for $\overline{u_{1b}}=0.168$ m/s and $\overline{u_{gb}}=0$ m/s in Fig. 19. The static pressure profile for $\overline{u_{gb}}=0.5$ m/s was quite similar to that for $\overline{u_{gb}}=2.5$ m/s, and thus the effect of gas flow on the static pressure in liquid phase was very small in this range of the experiment. The gas pressures p_g are also shown in Fig. 20. The pressure drop in the channel was very small and was at most a few Pa in both cases of $\overline{u_{gb}}$ in Fig. 20. The change of

p_g was also very small along the solitary wave. The results of the numerical experiment for $\overline{u_{gb}}$=0 m/s are shown in Fig. 20, and approximately agree with the experimental results in two-phase stratified and wavy flows. These facts indicated that the static pressure in liquid phase could be approximately predicted using Eq.(32) or Eq.(33). These simple relationships were applicable only in the cases of the fully developed solitary wave with relatively low mean velocities of liquid and gas. In the case of the developing solitary wave, the static pressure in gas and liquid phases were significantly affected by the gas flow as will be described in the next section.

Based on the preceding discussion, the behavior of solitary waves in two-phase stratified and wavy flows can be summarized as follows: The wave-propagation velocity could be predicted by the sum of the mean liquid velocity and the propagation velocity of the solitary wave on stationary liquid (Eq. (20)). The wave profile was approximately equal to that of stationary liquid (Eq. (26)). The liquid velocity could be predicted by the sum of the initial liquid velocity and the liquid velocity induced by the solitary wave on stationary liquid (Eqs. (27) and (28)). And finally, the static-pressure in liquid phase could be predicted by the sum of the gas pressure and the pressure in liquid phase of the solitary wave on staionary liquid (Eq. (32)). These simple relationships were applicable to the fully developed solitary wave in two-phase stratified and wavy flows.

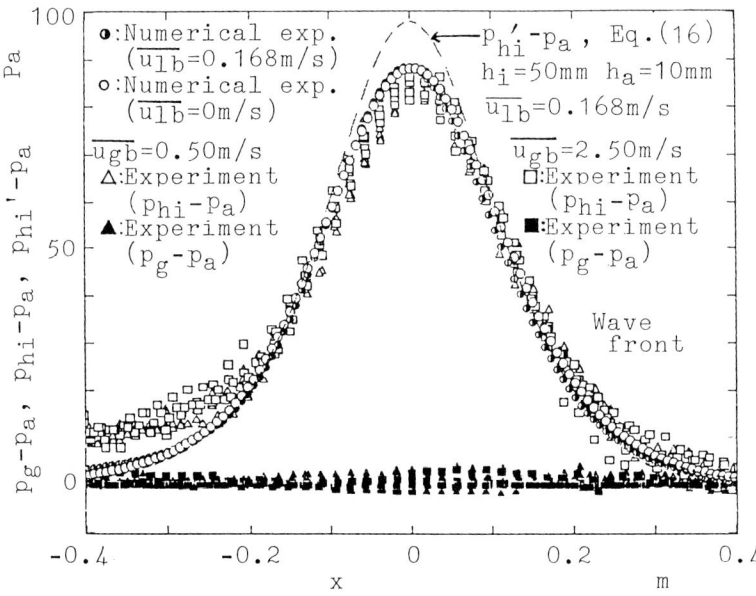

Fig. 20 Static-pressure profile of fully developed solitary wave in two-phase flow

FLOW BEHAVIOR OF DEVELOPING SOLITARY WAVE

When the wave height of the solitary wave introduced into two-phase stratified and wavy flows was high enough beyond a certain limit, the solitary wave grew in its wave height along the channel, the wave crest reached the upper wall of the channel at a certain position along the channel and the slugging took place. In this section, the flow characteristics of such developing solitary wave in two-phase stratified and wavy flows are discussed. For example, the behavior of wave-propagation velocity, of liquid height, of liquid and gas velocities and the behavior of static pressure are described.

The behavior of developing solitary wave was observed by means of high-speed VTR (Kodak EKTA-PRO-1000). The wave height increased gradually and approximately linearly along the channel, and then it increased rapidly. This rapid increase in the wave height caused the onset of slugging. The behavior of the developing solitary wave in the state of gradual increase in the wave height is discussed here. The process of the rapid increase in the wave height was out of scope because measurements in such rapid transients were quite difficult.

The wave-propagation velocities in the state when the wave height increased gradually are plotted in Fig. 21 against the mean value of (h_a+h_i) in the distance of 0.6 m where the wave-propagation velocity was measured. In this case, the wave height grew at the speed of about 6 mm/s. Thus the wave height changed about 4 mm during the residence time of the wave in the distance of 0.6 m. The solid line represents the wave-propagation velocity obtained by using Benjamin's equation (17) for fully developed solitary waves on flowing

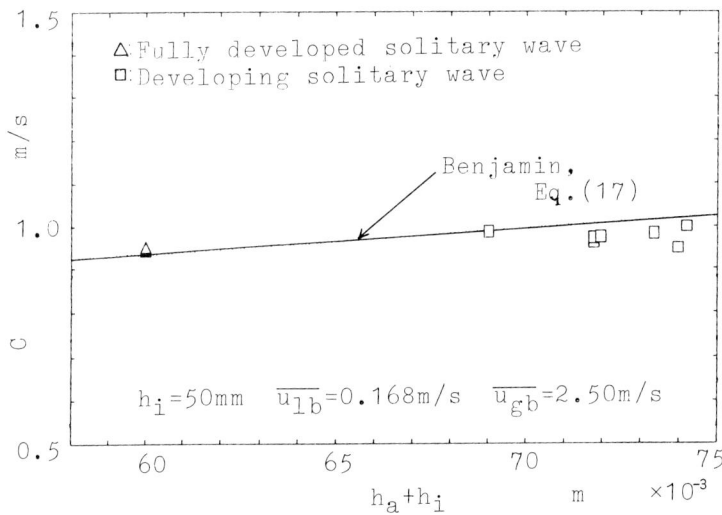

Fig. 21 Wave-propagation velocity of developing solitary wave in two-phase flow

INTERFACIAL WAVES IN STRATIFIED FLOW

liquid. Although in the case of the developing solitary wave experimental results of the wave-propagation velocity were slightly smaller than those on this solid line, the wave-propagation velocity could be approximately predicted by Benjamin's equation (17) and more simply by Eq. (20).

Figure 22 shows the liquid height versus time curve obtained using one conductance probe when the developing solitary wave passed through the position of the measurement. The measuring position was about 0.5 m upstream of the position of the onset of slugging. As the wave height increased with time, the wave profile changed at every moment in Fig. 22. Thus, the transformation of the liquid height versus time curve to the liquid height versus horizontal length curve was impossible in this case. The origin of the time scale was set to the time when the wave crest reached the measuring position of the liquid height. The feature of the liquid height versus time curve was similar to that for the fully developed solitary wave.

In principle, the wave profile or the liquid height versus time curve in this case was different from those for the fully developed solitary wave with constant wave height. However, if it is assumed that the maximum value of the wave height in Fig. 22 is a mean value of the wave height during the measuring period of 1.2 sec. which corresponded to the horizontal distance of about 1.2 m, the liquid height versus time curve can be compared with that for a fully developed solitary wave with the same wave height as the above-mentioned mean value of wave height. The values h_a in Figs.

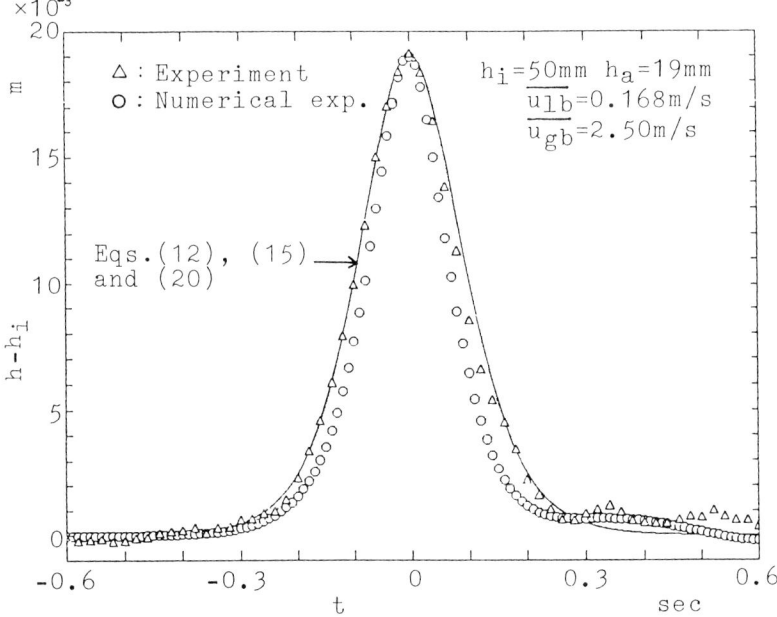

Fig. 22 Liquid height versus time curve of developing solitary wave in two-phase flow

22 to 24 have such mean values of wave height. The liquid height versus time curve obtained by McCowan's equation (15) and Eqs.(12) and (20) and that obtained in this numerical experiment are shown by solid line and small circles, respectively, in Fig. 22. Both curves agreed with the experimental results. Thus, if the wave height was given by some means, the liquid height versus the time curve could be estimated by this numerical experiment, or more simply by McCowan's equation (15) and Eqs.(12) and (20). This relationship holds only in the case when the increase in the wave height per unit period was relatively small.

As the position of the onset of slugging scatters widely along the channel, the vertical components v_l and v_g of the liquid and the gas velocities were not obtained in our experiments. Thus only the horizontal componets u_l and u_g are considered in this part. Figure 23 shows experimental results of the behavior of liquid velocity u_l and gas velocity u_g against time. The origin of the time scale was set to the time when the wave crest reached the measuring position. The liquid velocity u_l versus time curve was similar to that observed for the fully developed solitary wave, that is, u_l increased on approaching the wave crest, had a maximum and then decreased. The gas velocity u_g versus time curve showed a tendency similar to that for the fully developed solitary wave in cases of $\overline{u_{gb}} > C$. That is, u_g increased on approaching the wave crest, had a maximum and then decreased.

In the same manner as in Fig. 22, it is assumed that the maximum value of the liquid height versus time curve was the mean value of the wave height during the measuring period. The velocity versus time curves were compared with those for the fully developed solitary wave with the same wave height as this mean value of wave height. The results of this numerical experiment are plotted in Fig. 23. Comparing both the experimental and the calculated results of u_l, the values

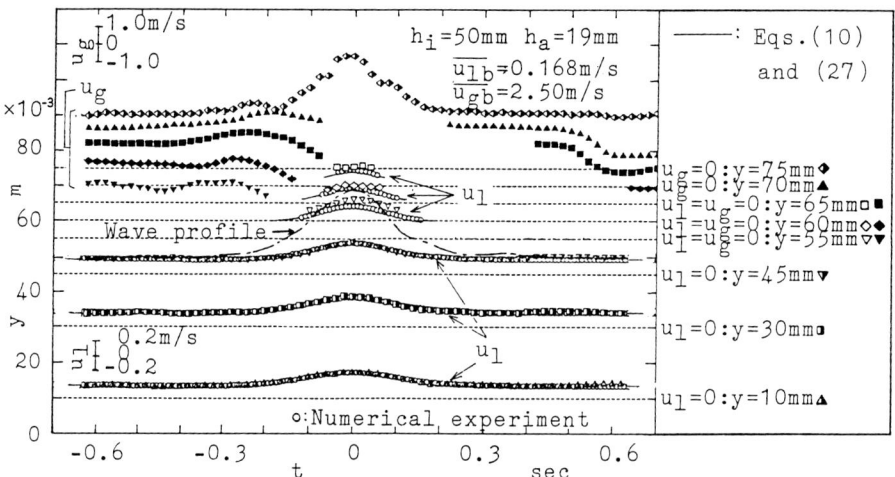

Fig. 23 Horizontal-velocity component of developing solitary wave in two-phase flow

of u_l in the experiment were slightly higher than those in the numerical experiment, especially in the liquid run-up above the initial liquid height. As the difference in u_l between both curves was relatively small compared with their absolute values, the horizontal component of the liquid velocity could be approximately predicted by means of this numerical experiment if the wave height was given. The solid lines represent the estimated liquid velocity obtained using McCowan's equation (10), Eqs.(12), (20) and (27) for the fully developed solitary wave and the initial liquid velocity profile in Fig. 4. The solid lines almost coicided with the experimental results. Thus the simple relationship of Eq.(27) held approximately in this case of the developing solitary wave.

The vertical component of the liquid velocity v_l of the solitary wave with the wave height corresponding to the maximum liquid height in Figs. 22 and 23 may be predicted by this numerical experiment or by McCowan's equation (11) and Eq.(28) under the same assumption of the mean wave height as in the case of the horizontal velocity component. However, in this developing solitary wave the wave height increased gradually, and this increase in the wave height meant that the amount of liquid rising upward at the wave front was larger than that sinking downward at the back of the wave. Thus the absolute values of v_l were different between those at the wave front and at the back of the wave. The estimation using the relationships for the fully developed solitary wave may cause a relatively large error in this case of the developing solitary wave.

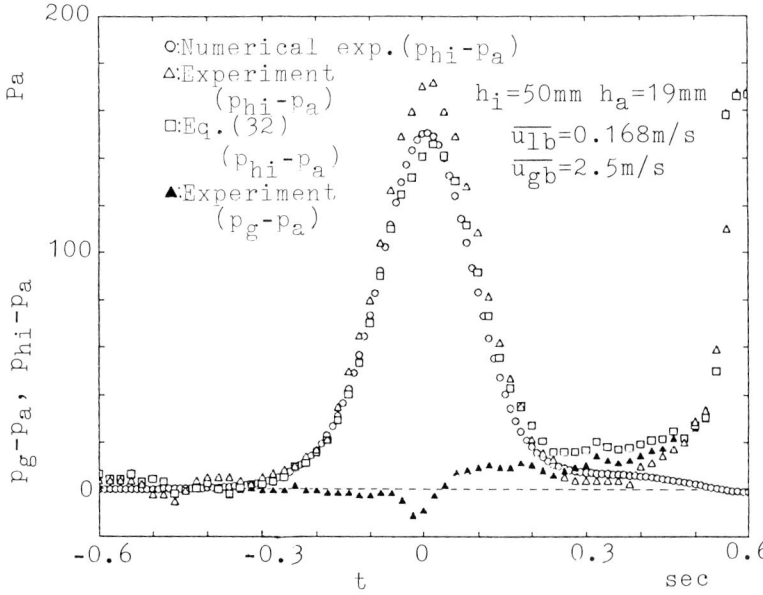

Fig. 24 Static pressure of developing solitary wave in two-phase flow

Figure 24 shows the behavior of the static pressure ($p_{hi}-p_a$) in liquid phase at the initial liquid height and the static pressure (p_g-p_a) in gas phase at the top of the channel. The origin of the time scale was set in the same manner as in the case of Figs. 22 and 23. The general features of the static pressure in liquid phase were similar to those observed for the fully developed solitary waves.

The authors set the same assumption of the mean wave height as that in Figs. 22 and 23 and compared the experimental results with the results of the numerical experiment for the fully developed solitary wave with the wave height corresponding to the maximum value of the liquid height versus time curve. The result of this numerical experiment is shown by small circles. The calculated results of the static pressure ($p_{hi}-p_a$) were about 20 Pa smaller than the experimental results at the wave crest. Of course, small-scale surface waves existed on the gas-liquid interface. But the amplitudes of these waves were rather small, for example, 0.5 mm in height, and did not cause such a large pressure difference. Moreover, the applicability of this numerical experiment was verified in the experiments for the large-amplitude solitary waves on stationary liquid. Thus this difference of 20 Pa between the experiment and the numerical experiment can be explained by other flow mechanisms.

In the previous section, it was shown that the relationship of Eq.(32) held for the fully developed solitary wave in two-phase stratified and wavy flows. This equation (32) was applied to the case of a developing solitary wave in two-phase wavy flow. The value of ($p_{lo}-p_{go}$) at $y=h_i$ was given by the numerical experiment for the mean wave height, and the static pressure of the gas phase can be given by the experimental results of (p_g-p_a) in Fig. 24. Then the static pressure (p_l-p_a) at $y=h_i$, that is, ($p_{hi}-p_a$), was obtained from Eq.(32) and the values mentioned above. Such estimated pressure ($p_{hi}-p_a$) is plotted in Fig. 24. Comparing the estimated pressure with the experimental results, the experimental results of ($p_{hi}-p_a$) were much higher than the estimated values of ($p_{hi}-p_a$) found by using Eq.(32) at the wave crest. Thus the pressure difference between the estimated value and the experimental value at the wave front and at the wave crest corresponded to the driving force for growing up the wave height owing to a certain flow mechanism, for example, the well-known Kelvin-Helmholtz instability or its extension to finite amplitude waves for the onset of slugging.

On the other hand, the gas pressure (p_g-p_a) decreased gradually with time, had a minimum of a negative value at the wave crest ($t=0$ sec.) and then increased. The magnitude of the change of this static pressure in gas phase was very large compared with that for the fully developed solitary wave in Fig. 20. Assuming one-dimensional flow of gas, the difference in the static pressures of gas phase between that above the initial liquid height and that above the wave crest of the mean wave height was roughly estimated by applying Bernoulli's equation and continuity equation, and became about -7 Pa. This value was of the same order of magnitude with the minimum value of about -10 Pa of the static pressure ($p_g - p_a$) in Fig. 24.

In sum, the behavior of the developing solitary wave in two-phase wavy flow was as follows: the wave-propagation velocity, the liquid height versus time curve and the horizontal velocity component of liquid could be approximately estimated using the relationships for the fully developed solitary wave in two-phase stratified and wavy flows when the wave height was given suitably by some means. The change in the static pressure of gas phase is explained by Bernoulli's theorem. On the other hand, the change of the static pressure in liquid phase was much larger than that predicted using Eq. (32) for the fully developed solitary wave. This was probably due to the flow instability mechanism, like the Kelvin-Helmholtz instability.

CONCLUSION

In order to look insight fully into the behavior of the solitary wave in two-phase stratified and wavy flows, experiments and numerical experiments were carried out successively for the following conditions: fully developed solitary waves on stationary liquid; on flowing liquid without gas flow above it; fully developed solitary waves in two-phase stratified and wavy flows; and, the developing solitary wave under the condition of the onset of slugging. The experimental data of the wave-propagation velocity, the wave profile, the liquid- and gas-velocity profiles and the static pressure profile were presented for the above-mentioned flow conditions. Simple relationships which express the effects of liquid and gas flows on the behavior of fully developed solitary wave were derived based on the experimental results and the results of the numerical experiments. Based on the experimental results of the developing solitary wave, the applicability of these simple relationships was examined.

The authors consider this investigation as a first step in their research on wave phenomena in two-phase flow. The simple relationships derived here will be useful in analyses of the onset of slugging. Moreover, this data will also be useful for the examination of theories and simulations for wave phenomena in two-phase stratified and wavy flows.

ACKNOWLEDGEMENT

The authors wish to express their thanks to Messrs. T. Higuchi, K. Miyazaki, Y. Asao and S. Ono for their assistance in carrying out the experiments. This study is partly supported by a Grant-in-Aid of Scientific Research of the Ministry of Education, Science and Culture of Japan.

NOMENCLATURE

C	:	Wave-propagation velocity
F_{rc}	:	Froude number with respect to wave-propagation velocity
F_{ri}	:	Froude number with respect to initial liquid flow
h	:	Liquid height
h_a	:	Wave height
h_i	:	Initial liquid height
h_t	:	Channel height
V_{30}	:	Velocity component in the plane 30 degrees inclined from the horizon
V_{330}	:	Velocity component in the plane -30 degrees inclined from the horizon
g	:	Gravitational acceleration
p	:	Pressure
p_a	:	Atmospheric pressure
p_{hi}	:	Liquid pressure at $y=h_i$
t	:	Time
u	:	Horizontal component of velocity
$\bar{u}$	:	Mean velocity in each phase
v	:	Vertical component of velocity
x	:	Horizontal coordinate
x_m	:	Half-value distance of solitary wave profile
y	:	Vertical coordinate

Greek

Δ	:	Momentum correction factor
θ	:	Inclination angle
ν	:	Kinetic viscosity

Subscripts

o	:	Value for stationary liquid
b	:	Value in initial flow
g	:	Value of gas phase
l	:	Value of liquid phase

REFERENCES

Benjamin, T.B., 1967 The Solitary Wave on a Stream with an Arbitrary Distribution of Vorticity, *J. Fluid Mech.*, 12, pp.97-116.

Daily, J.W. and Stephan, S.C., Jr., 1953 Characteristics of the Solitary Wave, *Trans. ASCE*, 18, pp.575-587.

Fukano, T., Itoh, A. and Ousaka, A., 1985 Breakdown of a Liquid Film Flowing Concurrently with Gas in Horizontal Line, *Physico Chemical Hydrodynamics*, Vol.6, No.1/2, pp.23-47.

Hihara, E. and Saito, T., 1983 Slug Flow Transition in a Horizontal Tube, *Proc. ASME-JSME Thermal Eng. Joint Conf., Hawaii*, Vol.1, pp.19-24.

Hirt, C.W., Nichols, B.D. and Romero, N.C., 1975 SOLA-A Numerical Solution Algorithm for Transient Fluid Flows, L.A.5852, Los Alamos Sci. Lab.

Kordyban, E.S. and Ranov, T., 1970 Mechanism of Slug Formation in Horizontal Two-Phase Flow, *Trans. ASME, J. Basic Eng.*, 92, pp.857-864.

Laitone, E.V., 1960 The Second Approximation to Cnoidal and Solitary Waves, *J. Fluid Mech.*, 9-3, pp.430-444.

McCowan, J., 1891 On the Solitary Wave, *Phil. Mag.*, Ser.5, 32-7, pp.45-48.

Mishima, K. and Ishii, M., 1980 Theoretical Prediction of Onset of Horizontal Slug Flow, *Trans. ASME, J. Fluids Eng.*, 102-4, pp.441-445.

Sakaguchi, T., Akagawa, K., Hamaguchi, H. and Ashiwake, N., 1973 Transient Behavior of Air-Water Two-Phase Flow in a Horizontal Tube, *ASME Paper 73-WA/HT-21*, pp.1-15.

Sakaguchi, T., Akagawa, K., Hamaguchi, H., Arima, H. and Takaoka, T., 1977 Water-Air Two-Phase Slug Flow in Horizontal Tubes, *Proc. 17th Internat. Congress of the IAHR, Baden-Baden*, Vol. 1, A49, pp.387-394.

Sakaguchi, T. Akagawa, K. and Hamaguchi H., 1978 Transient Behavior of Flow Pattern for Air-Water Two-Phase Flow in Horizontal Tubes, *Theoretical and Applied Mechanics*, 26, pp.445-459.

Sakaguchi, T., Akagawa, K., Hamaguchi, H. and Amano, T., 1980 Developing Steady and Transient Air-Water Two-Phase Flow in Horizontal Tubes, in *Multiphase Transport, Fundamentals, Reactor Safety*, Vol.1, ed. S. Kakac and T.N. Veziroglu, Hemisphere Pub., Washington D.C., pp.45-74.

Sakaguchi, T, Ozawa, M., Takahashi, R. and Shiomi, Y., 1986a Liquid Velocity Measurement of Solitary Wave by LDV, *Memoirs of the Faculty of Engineering, Kobe University*, No. 33, pp.33-62.

Sakaguchi, T, Ozawa, M., Shiomi, Y., Uchida, S., Harima, T. and Takahashi, R., 1986b Weak and Strong Interaction Between Soliatry Waves on Liquid Surface, *Proc. the 3rd Asian Congress of Fluid Mechanics, Tokyo*, pp.475-478.

Sakaguchi, T, Ozawa, M., Hamaguchi, H., Nishiwaki, F. and Fujii, E., 1987 Analysis of the Impact Force by a Transient Liquid Slug Flowing Out of a Horizontal Pipe, *Nuclear Eng. and Design*, 99, pp.63-71.

Schlichting, H., 1968 *Boundary Layer Theory*, 6th Edition, McGraw-Hill, New York, pp.575-578.

Taitel, Y. and Dukler, A.E., 1976 A Model for Predicting Flow Regime Transitions in Horizontal and Near Horizontal Gas-Liquid Flow, *AIChE J.*, 22-1, pp.47-55.

Wiegel, R.L., 1964 *Oceanographical Engineering*, Prentice-Hall, Englewood Cliffs.

A STUDY OF AIR-WATER FLOW IN A NARROW RECTANGULAR DUCT USING AN IMAGE PROCESSING TECHNIQUE

K. MISHIMA, S. FUJINE, K. YONEDA, K. YONEBAYASHI*,
K. KANDA, and H. NISHIHARA
Research Reactor Institute
Kyoto University
Kumatori-cho, Sennan-gun, Osaka 590-04, Japan

ABSTRACT

The purpose of this study is to observe an air-water flow in a narrow rectangular duct and to measure the void fraction and the interfacial area by applying an image processing technique. Video images of a two-phase flow were taken either by an optical method or by a neutron radiography technique. Flow regimes were determined by observing the video images.

Observed flow regimes were slug, froth and annular flows. Since the transition of flow regimes occurred rather gradually, the boundary between them was expressed with a band. Data of void fraction obtained with the image processing technique were compared to those obtained with the conventional conductance probe technique. The agreement was good and the results were well-reproduced by the drift flux correlation. A strong correlation was observed between the interfacial area concentration and the void fraction.

INTRODUCTION

The behavior of gas-liquid two-phase flow in a narrow channel has recently attracted considerable attention because of the importance of such situations relative to tight lattice core in a high conversion light water reactor (HCLWR). The purpose of this study was to observe the behavior of an air-water two-phase flow in a narrow rectangular duct in which two parallel walls restrain the bubble motion, and to measure its characteristics.

The first step to studying the behavior of a two-phase flow is to visualize it. A number of techniques have been proposed for flow visualization (Asanuma, 1986). When the duct is transparent, the most popular way is to use an optical method such as photography, high-speed camera and television. When the duct is opaque, however, some special device is needed, such as a radiography technique.

---

*Currently at Takahama Nuclear Power Plant, Kansai Electric Power Co., 1 Tanoura, Takahama-cho, Ooi-gun, Fukui 919-23.

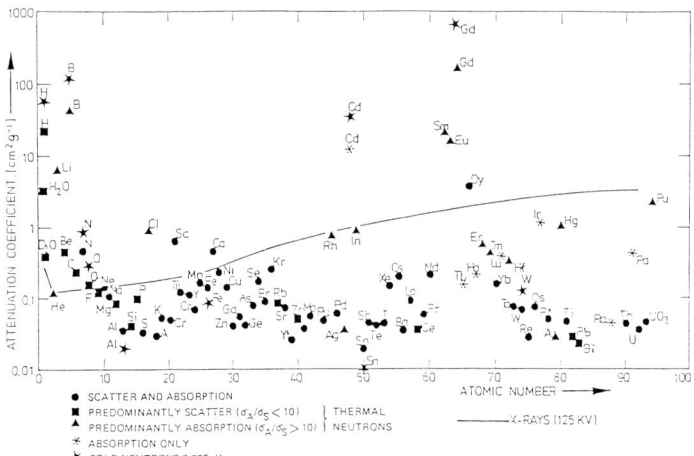

FIGURE 1. Neutron and X-ray mass attenuation coefficients for the elements (von der Hardt, 1981).

Neutron radiography (NRG) is a radiography technique which makes use of the difference in attenuation characteristics of radiation in materials (von der Hardt, 1981). Figure 1 shows the mass attenuation coefficient of thermal neutron and X-ray in the elements (von der Hardt, 1981). The attenuation coefficient of X-rays increases monotonically with the atomic number of the element. On the other hand, thermal neutrons easily penetrate heavy materials like steel, while they are attenuated in light materials such as water, hydrocarbons and boron. Therefore, it is clear that NRG is more suitable for observing the fluid behavior in a metal duct.

A neutron television (TV) system, which is an applied technique of NRG, has been developed to observe real-time images of the dynamic behavior of a hydrogenous material in a metal vessel (Stewart, 1983; Robinson, 1983). Thus some attempts have been made to observe gas-liquid two-phase flow by neutron TV (Robinson, 1983; Fujine, 1985; Harris, 1986; Tamaki, 1986). This technique was also used here to observe a two-phase flow in a metal duct.

Provided that the video images of the two-phase flow have been taken either by optical method or by NRG, they may be processed to calculate some flow characteristics. Therefore, some attempts were made here to obtain the void fraction and the interfacial area between two phases. In order to use the same image processing technique, the optical method was so designed that it simulated the NRG method.

EXPERIMENT

Basic Idea of Flow Visualization

Figure 2 shows the schematic diagram of the flow visualization

FLOW IN A NARROW RECTANGULAR DUCT

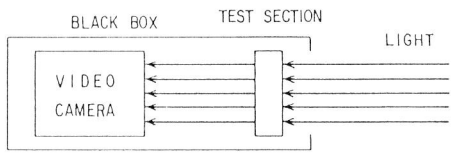

OPTICAL METHOD
FOR AIR-GRAY WATER TWO-PHASE FLOW

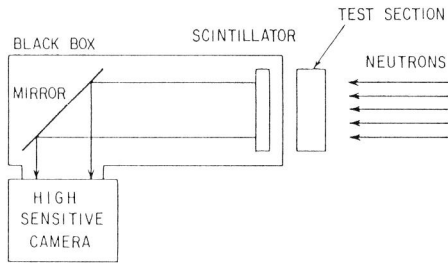

NEUTRON RADIOGRAPHIC METHOD
FOR AIR-WATER TWO-PHASE FLOW

FIGURE 2. Basic concept of the NRG method and the optical method.

methods. In the NRG method, the neutron beam attenuated by water in the test section made the scintillator emit visible light, producing an image of two-phase flow which was taken by a high-sensitive camera. In the optical method, the neutron attenuation was simulated by light attenuation in gray water which was tinted with black ink; the light was detected directly by a camera. Thus a neutron beam attenuated by water was simulated by a light beam attenuated by gray water.

Optical Method

Experimental apparatus. The experimental apparatus consisted of two sections. One was a two-phase flow loop and the other was a measuring system. The two-phase flow loop is shown schematically in Fig. 3.

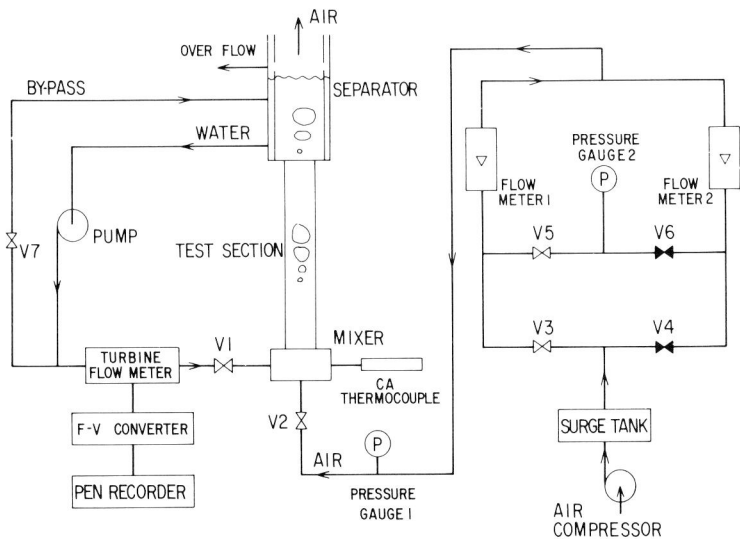

FIGURE 3. Schematic diagram of the two-phase flow loop.

The test section was a transparent rectangular duct made of acrylic resin. Its gap, width and height were 2.4 mm, 50 mm and 600 mm, respectively. Gray water and air flowed upward through the test section. The gray water was circulated in the loop by a centrifugal pump. Its flow rate was measured by a turbine flow meter installed just upstream of the inlet

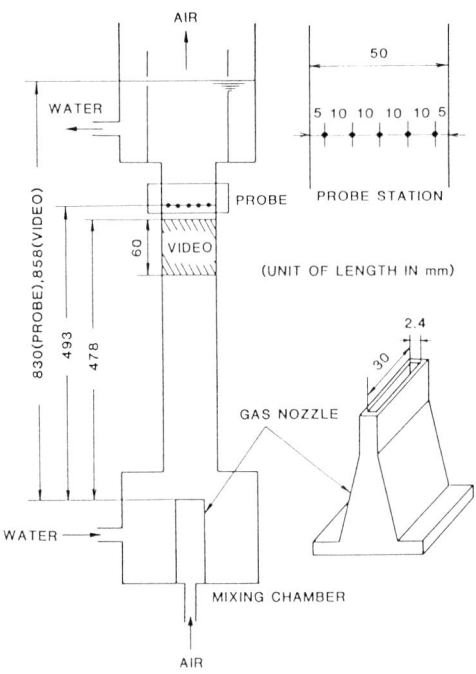

FIGURE 4. Measured portion and the gas nozzle.

valve V1. The air was supplied by an air compressor, introduced into the mixing chamber through a narrow slit as shown in Fig. 4, and was finally released into the atmosphere after flowing through the test section. The flow rate of the air was measured by a float type flow meter. The air pressure was measured at the inlet of the flow meter and just upstream of the inlet valve V2. The temperature is measured by a chromel-alumel thermocouple at the entrance of the test section.

The measuring portion in the test section is shown in Fig. 4 together with the sketch of the gas inlet nozzle. The region of interest in image processing begins at 418 mm downstream from the gas inlet and the length of the region along the stream is 60 mm.

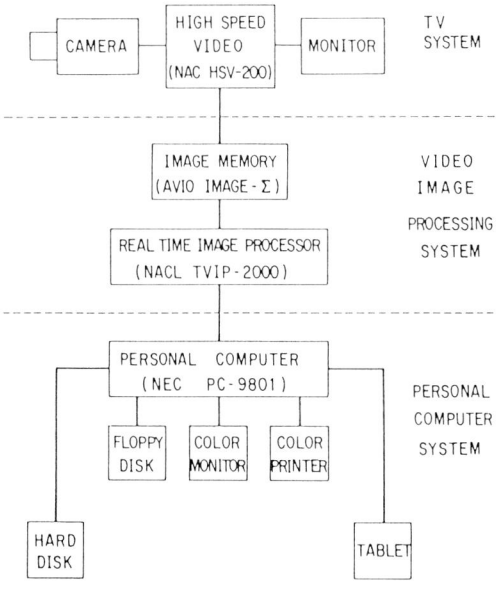

FIGURE 5. Block diagram of the image processing system.

The measuring system consisted of a TV system, a video image processing system and personal computer system, as shown in Fig. 5. The image processing system consists of an image memory (AVIO IMAGE-Σ) and a processor (NACL TVIP-2000). The personal computer system, consisting of a personal computer (NEC PC-9801), a hard disk unit and some input/output devices, processes the digitized images to calculate the mean void fraction and the interfacial area concentration using FORTRAN software. The arrangement of the TV system is shown in Fig. 6. The test section and a high speed camera were installed in a black box. The light beam from a strobe was scattered by a translucent plastic plate to light the test section uniformly.

FLOW IN A NARROW RECTANGULAR DUCT

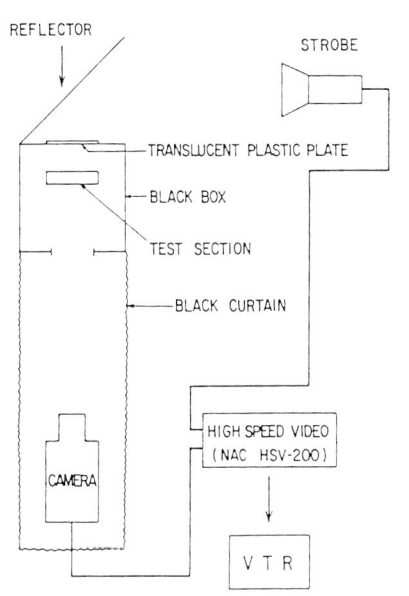

FIGURE 6. Block diagram of the optical system.

The high-speed video system (NAC HSV-200) took 200 frames of pictures per second.

Void fractions were also measured using the conventional conductance probe technique to compare them with the image processing technique. The probe station was placed at 15 mm downstream from the end of the video region. Five probes were installed there at equal intervals across the stream.

<u>Data processing</u>. Image memory IMAGE-Σ was a digital image processing system which divided the video images into up to 480 × 640 pixels and recorded them into the memory. It can integrate up to 999 video images. In this study, at least five images were taken at random by IMAGE-Σ.

TVIP-2000 was a digital image processing system which divided one image frame into 512 × 480 pixels and specified the brightness in each pixel into a gray level from 0 to 255. TVIP-2000 was connected to a computer (NEC PC-9801) and operated by a software which was developed by combining an existing software IMAGE COMMAND 98 (NACL) and a subroutine library TVIP-2000 handler (NACL). The program was written in MS-FORTRAN.

The thickness of the gray water layer through which the light beam passed was calculated from the gray level in each pixel, knowing the relation between them. The calibration sample was made of acrylic resin and had six gaps of an already-known thickness filled with gray water. The gap thicknesses were 0, 0.3, 1.1, 1.9, 2.7, and 3.5 mm. The concentration of the gray water was determined to be 7 ml black ink per liter of water so that the best calibration curve was obtained. With this concentration, the surface tension of gray water was almost the same as that of water.

The basic concept of calculating the void fraction and the interfacial area is depicted in Fig. 7. Using the calibration data, the thickness of the gray water layer in the pixel was calculated. Then, the mean void fraction in the pixel, namely the pixel-mean void fraction, was obtained from the thickness of the gray water layer divided by the thickness of the duct. Finally, all the pixel-mean void fractions in the region of interest were averaged to obtain the overall-mean void fraction.

In calculating the interfacial area concentration, it was

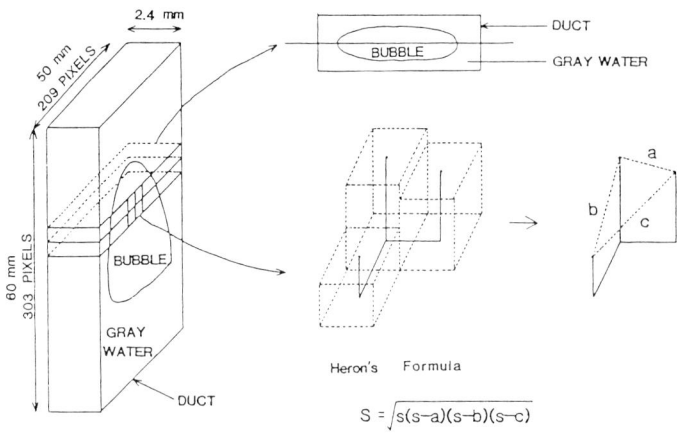

FIGURE 7. Basic idea of image processing for the void fraction and the interfacial area.

assumed that the flow was symmetrical with respect to the center of the gap because the gap was narrow. A single bubble was assumed to be in the center of the gap, and the thickness of the liquid film along the bubble was just a half of that calculated from the gray level as shown in Fig. 7. Using the film thickness in three neighboring pixels, the interfacial area in the region surrounded by them was approximated by the area of the triangle which was calculated using Heron's formula. Interfacial areas so obtained were summed up over the volume of interest and divided by the volume to obtain the interfacial area concentration.

Neutron Radiography

Experimental facility. When the test section was opaque, the flow inside was visualized by using the NRG technique. The two-phase flow loop was the same as that employed in the optical method, as shown in Fig. 3 except that the test section was made of aluminum alloy. The gap, width and height of the test section were 2.4, 40 and 1400 mm, respectively. The sight of the neutron beam was in the circle centered at 600 mm downstream of the entrance of the test section and its diameter was 160 mm.

The calibration sample made of aluminum alloy had five gaps whose thicknesses were 0.6, 1.2, 1.8, 2.4 and 3.0 mm. The gaps were filled with the same water that flows in the test section. The sample was mounted just beside the test section.

Experiments were performed using the NRG facility at the E-2 experimental port of the Kyoto University Research Reactor (KUR). Figure 8 shows the NRG facility of the KUR whose characteristics are presented in Table 1. Experiments were performed also at the Nuclear Safety Research Reactor (NSRR) of the Japan Atomic Energy Research Institute (JAERI). Although

FLOW IN A NARROW RECTANGULAR DUCT

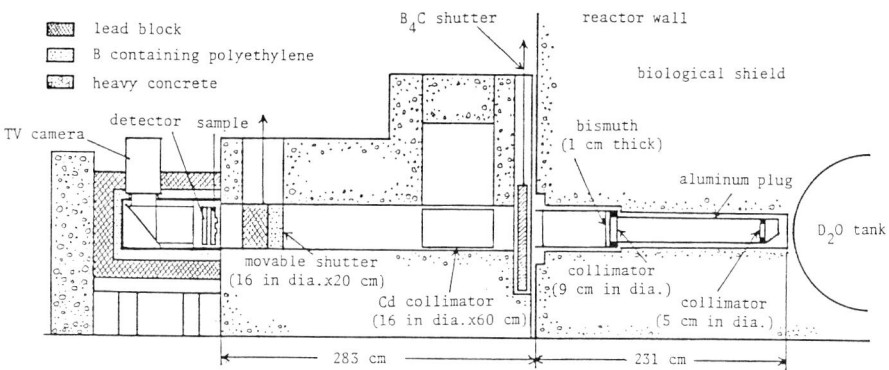

FIGURE 8. NRG facility at the KUR.

TABLE 1. Characteristics of the NRG Facility of the KUR

Reactor / Power	KUR / 5000kW
Peak Thermal Neutron Flux in Core	6×10^{13} n/cm²/s
Thermal Neutron Flux at the Collimator	9×10^{10} n/cm²/s
Range of L_c	500 cm
Standard L_c/D_c	100
Thermal Neutron Flux at the Film	1.2×10^6 n/cm²/s
Gamma Dose Rate	4.2 R/h
Cadmium Ratio	400
n/γ Ratio	1.1×10^6 n/cm²/s
Film Size Available	16 cm in diameter
ASTM-75 Specification	85-12-11

detailed characteristics of the NRG facility of the NSRR were not known, it was expected that its pulsed neutron beam produced high-quality frozen images of a two-phase flow since the half width of a pulse was as short as several milliseconds and the neutron fluence was estimated to be higher than 10^9 nvt.

The measuring system was the same as shown in Fig. 6 except that the TV system was replaced by a neutron TV system. The arrangement of the neutron TV system was shown in Fig. 9. Two imaging systems were employed in the experiment. One was a HITACHI XTV-10A consisting of a fluorescent converter SAKURA KH ($Gd_2O_2S(Tb)$) and a 3 in. image-orthicon camera (HS-131). The other was a combination of a Nuclear Enterprise NE426 scintillator (LiF+ZnS(Ag)) and a Tokyo Densi-Kogyo VC-7000 silicon intensifier tube.

Data processing. The same procedures were followed for image processing as in the optical method. The neutron beam attenu-

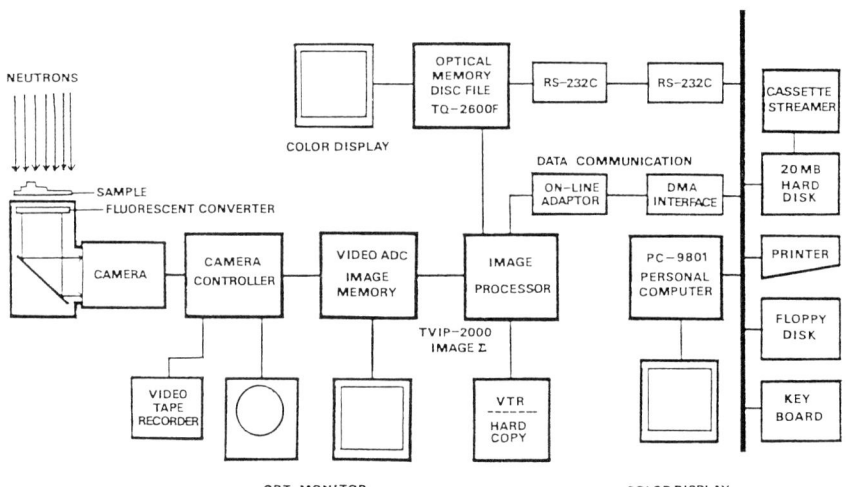

FIGURE 9. Block diagram of the neutron TV system.

ated in the sample hit the conversion screen to form the image of the sample on it. The video image of each gap was processed and converted into a gray level from which a calibration curve for the relation between the gray level and the water layer thickness was obtained.

Since the original image quality by the NRG method was not enough to calculate the void fraction, the overall-mean void fraction was calculated after 255 images were integrated using IMAGE-Σ to obtain sufficient statistics. This integral process indicated time-averaging of the images. Therefore, the overall mean void fraction, in this case, was obtained by the space-averaging of a time-averaged image, which was proved to be equivalent to the averaging process in the optical method. In most cases, five time-averaged images, therefore 255 × 5 original video images, were enough to calculate the overall-mean void fraction with a standard error of 5%.

Conductance Probe Method

In comparison with the image processing method, the void fraction was measured using the conductance probe method as well. A probe consisted of a point electrode placed in the stream and a ground electrode mounted on the duct wall. An enamel-covered copper wire (0.2 mm in diameter) was uncovered at the tip to make a point electrode.

A two-phase flow measuring system KANOMAX 7931 was used for measurement. The existence of an air-bubble at the tip was detected by the change in the electric conductance between the tip and the ground electrode. Thus the time-averaged void fraction at the tip was calculated as the probability that the tip touched air bubbles. The overall mean void fraction was calculated by averaging the void fraction at the tips.

FLOW IN A NARROW RECTANGULAR DUCT

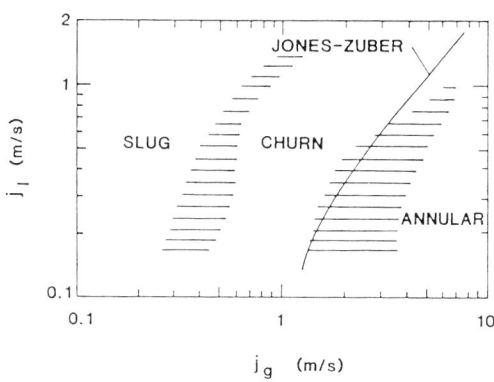

FIGURE 10. Flow regime map obtained by optical method.

RESULTS

Flow Regimes

Flow regimes were determined by observing the video images. Both the optical method and the NRG produced video images with enough quality to delineate flow regimes. The standard sample for the optical method is shown in Photo 1(a) and (b). The gray levels in Photo 1(a) were translated into color tones in Photo 1(b), which is called here a pseudo-color image. Photos 2 through 7 are examples of flow regimes. Photos denoted by (a) are original video images and those denoted by (b) are pseudo-color images.

Flow regimes observed by the NRG method at the KUR are shown in Photos 8 through 13. The images of the standard sample are also shown. The thermal neutron flux at the scintillator was about 10^6 n/cm^2/s. With this neutron-beam intensity, the image quality was good enough to observe flow regimes but not sufficient to calculate the interfacial area. Photos 12 and 13 demonstrate an original image and an integrated one, respectively, of a slug bubble which was standing still in the liquid downflow.

Better image quality was attained by pulsed NRG using NSRR. Pictures taken by the NRG facility are shown in Photos 14 through 17. In those cases, photos were taken by mounting the film just behind the scintillator instead of using TV camera. Each neutron radiograph was taken by a pulse of neutron beam whose half width was about 7.5 msec and whose neutron fluence was estimated to be at least 10^9 nvt. Since the smallest hole in the standard sample was clearly imaged, the spatial resolution was better than 0.25 mm.

The flow map obtained from the video images was plotted in terms of the superficial liquid velocity j_l vs. the superficial gas velocity j_g as shown in Fig. 10. The areas shaded with horizontal lines are transition regions, indicating that a flow regime changed gradually into another and the boundary was not obvious. In this case, bubbly flow was not observed because of the geometry of the gas inlet nozzle.

In Fig. 10, the prediction by the Jones-Zuber correlation (Jones, 1979) is also shown for the boundary between slug and annular flows. Since they did not discriminate between those flow regimes, the agreement between the present result and Jones-Zuber's appears to be reasonable.

150   **MODELING AND DYNAMICS**

(a) Original image   (b) Pseudo-color image

PHOTO 1. Video image of the calibration sample taken by the optical method.*

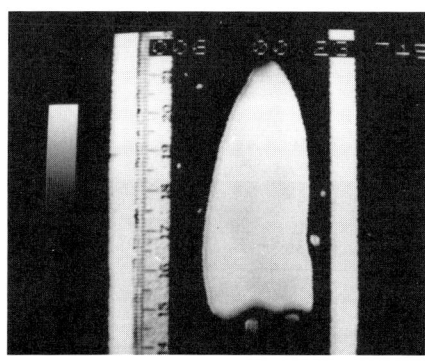 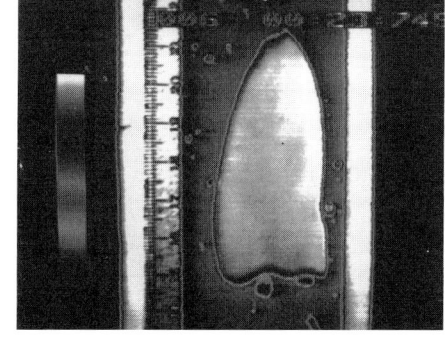

(a) Original image   (b) Pseudo-color image

PHOTO 2. Slug flow image taken by the optical method.

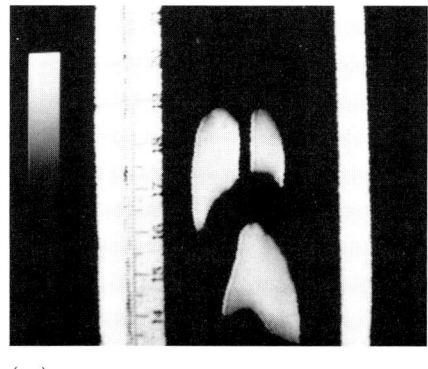 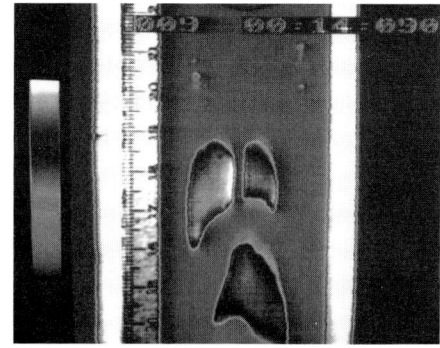

(a) Original image   (b) Pseudo-color image

PHOTO 3. Slug flow image taken by the optical method.

* Photos 1 through 17 appear in color following page 312.

FLOW IN A NARROW RECTANGULAR DUCT

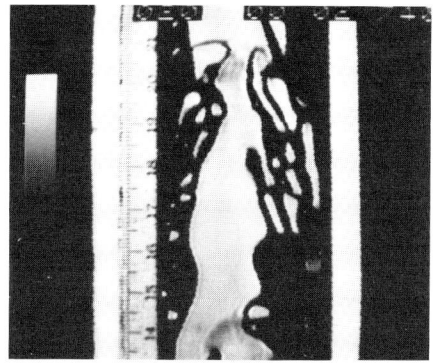
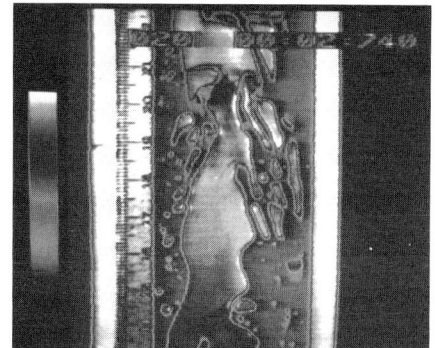

(a) Original image   (b) Pseudo-color image

PHOTO 4.   Froth flow image taken by the optical method.

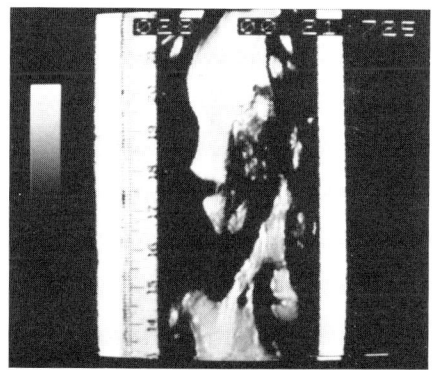

(a) Original image   (b) Pseudo-color image

PHOTO 5.   Froth flow image taken by the optical method.

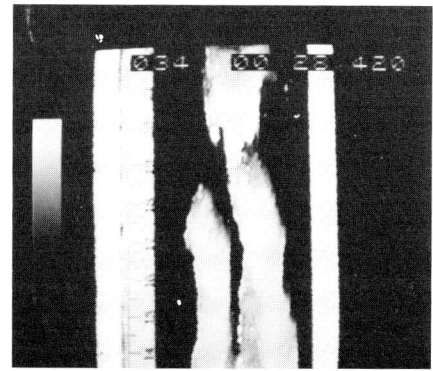
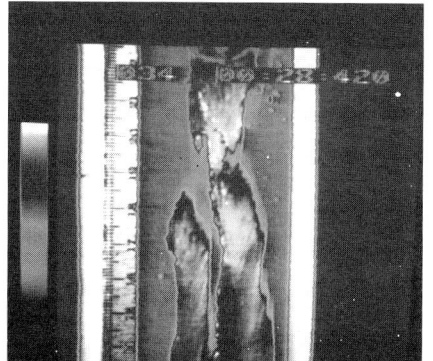

(a) Original image   (b) Pseudo-color image

PHOTO 6.   Annular flow image taken by the optical method.

152 MODELING AND DYNAMICS

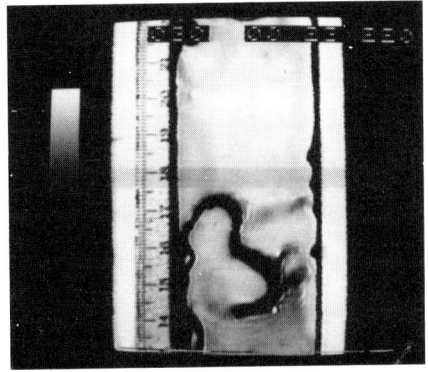 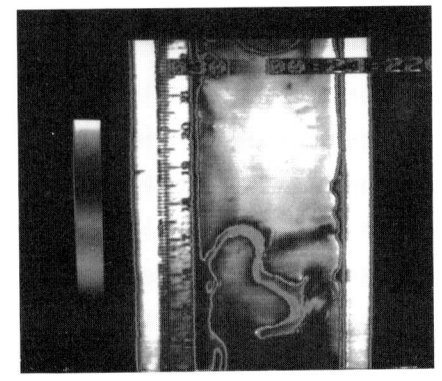

(a) Original image   (b) Pseudo-color image

PHOTO 7. Annular flow image taken by the optical method.

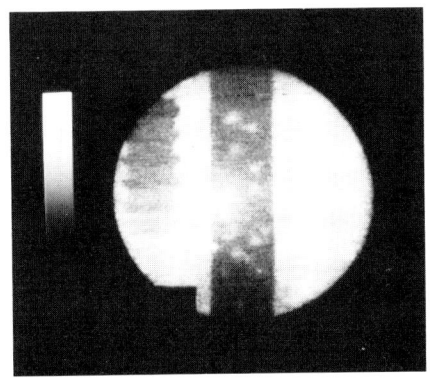 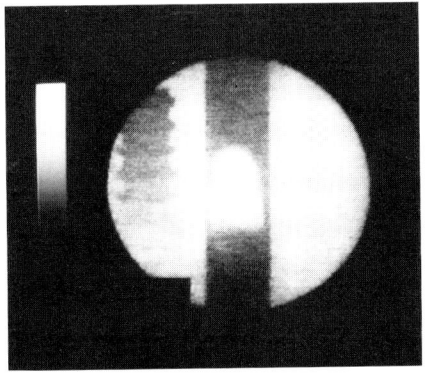

PHOTO 8. Bubbly flow image taken by NRG.

PHOTO 9. Slug flow image taken by NRG.

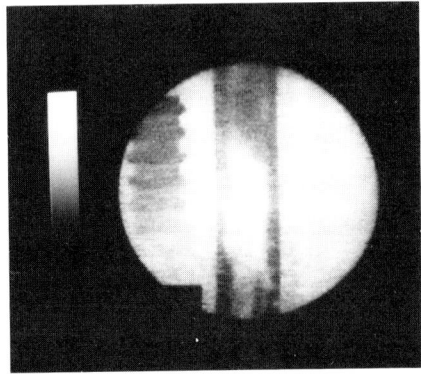 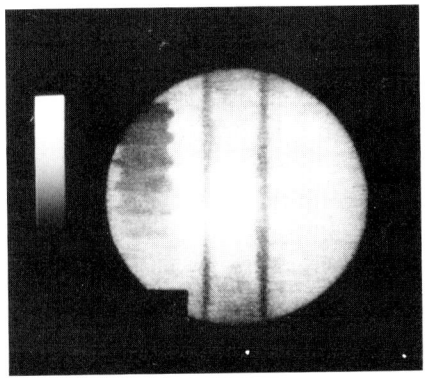

PHOTO 10. Froth flow image taken by NRG.

PHOTO 11. Annular flow image taken by NRG.

FLOW IN A NARROW RECTANGULAR DUCT 153

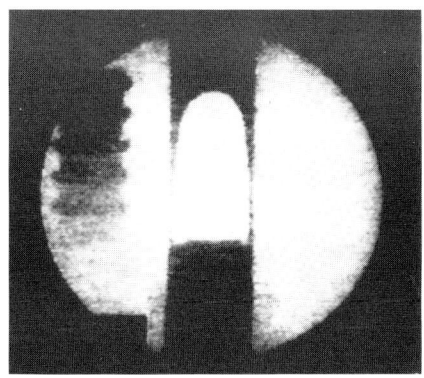

PHOTO 12. Original image of a stagnant slug bubble taken by NRG.

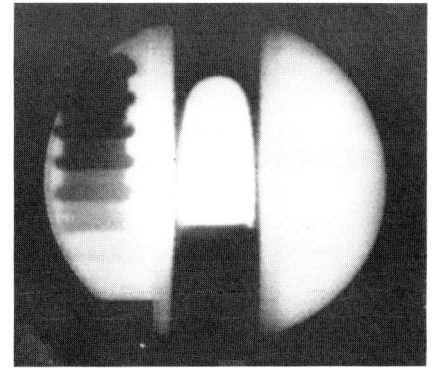

PHOTO 13. Integrated image of a stagnant slug bubble taken by NRG.

PHOTO 14. Deformed-bubble image taken by pulsed NRG.

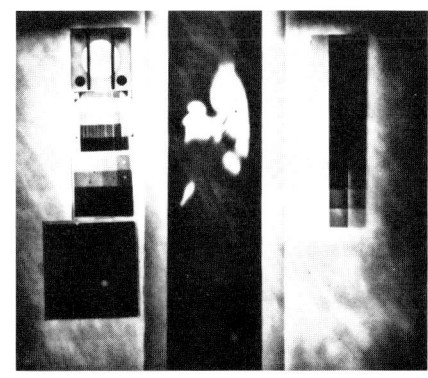

PHOTO 15. Slug flow image taken by pulsed NRG.

PHOTO 16. Froth flow image taken by pulsed NRG.

PHOTO 17. Annular flow image taken by pulsed NRG.

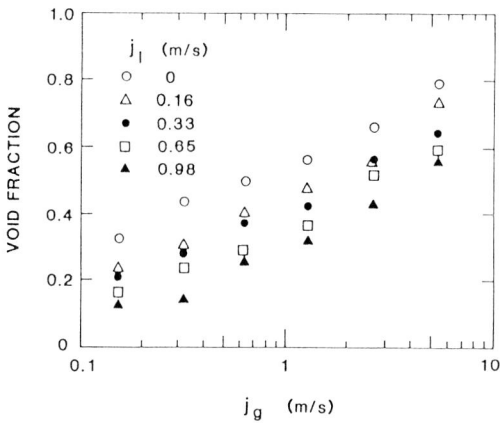

FIGURE 11. Overall mean void fraction as a function of the superficial velocities, taken by the optical method.

Overall Mean Void Fraction

Overall mean void fraction was plotted against the superficial gas velocity using the superficial liquid velocity as a parameter as shown in Fig. 11. The drift flux model was used to correlate the mean void fraction. The equation is as follows:

$$v_g = j_g/\alpha = C_0 j + V_{gj} \quad , \quad (1)$$

The drift velocity for rectangular ducts has been obtained by Griffith (1963) and is given by the following equation:

$$V_{gj} = (0.23 + 0.13 s/w)\sqrt{\Delta\rho g w/\rho_l} \quad , \quad (2)$$

As for the distribution parameter, Jones-Zuber (1979) used the value of 1.2. Ishii (1977), however, deduced the value of distribution parameter from several sources and presented a correlation as follows:

$$C_0 = 1.35 - 0.35\sqrt{\rho_g/\rho_l} \quad . \quad (3)$$

The gas velocity v_g in Eq. (1) can be calculated based upon the superficial gas velocity and measured overall mean void fraction.

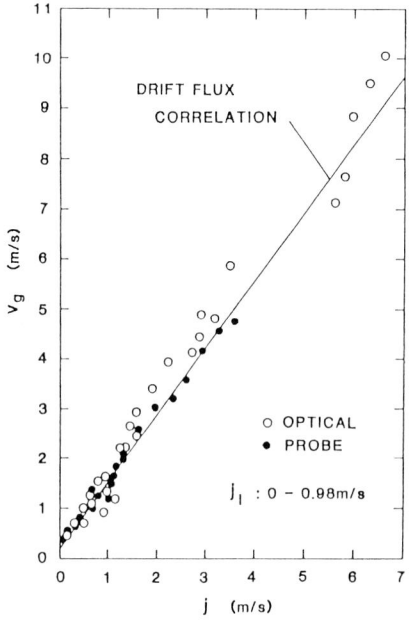

FIGURE 12. Drift flux correlation of the void data taken by the optical method and the probe method in comparison with Eq. (1).

Thus the results obtained by the image processing method and the conductance probe method were compared to the drift flux correlation in Fig. 12, where the video images were taken by the optical method. It turned out that the value of C_0 as given by Eq. (3) correlated the present data well. The figure indicates that the scatter in the data points is smaller by the conductance probe method

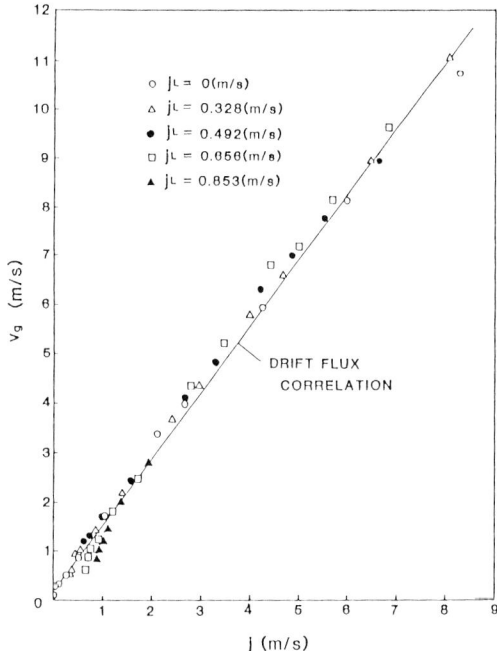

FIGURE 13. Drift flux correlation of the void data taken by the NRG method using integral images.

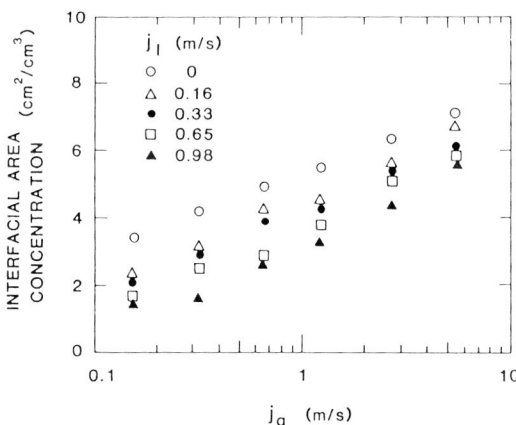

FIGURE 14. Interfacial area concentration as a function of the superficial velocities, taken by the optical method.

than by the image processing method. The range of the measurement, however, extended to higher volumetric flux by the optical method than by the probe method. This is because the narrow gap of the duct limited the thickness of the probes and they could not maintain enough stiffness at high velocities. In any case, the agreement with the drift flux correlation appeared to be good.

The original image quality by the NRG method was found to be not so good as that by the optical method. Therefore, the original video images were integrated for better statistics, since the value of the mean void fraction would not be affected by integration. The integrated results by the NRG method are shown in Fig. 13 where 255 frames of a sample of original images were integrated and the average of five randomly-chosen samples is plotted. The scatter in the data points was much less than in Fig. 12.

Interfacial Area

Since the image processing technique required an instantaneous image of two-phase interface to calculate the interfacial area, good image quality was required for interfacial area. It turned out that sufficient image quality was obtained by the optical method, but not by the NRG method.

The result of the interfacial area concentration a_i obtained by the optical method is shown in Fig. 14. Here, the interfacial area concentration is defined as the area of the interface between two phases per unit volume of the mixture. As can be seen by comparing Figs. 11 and 14, the general trends of the data of

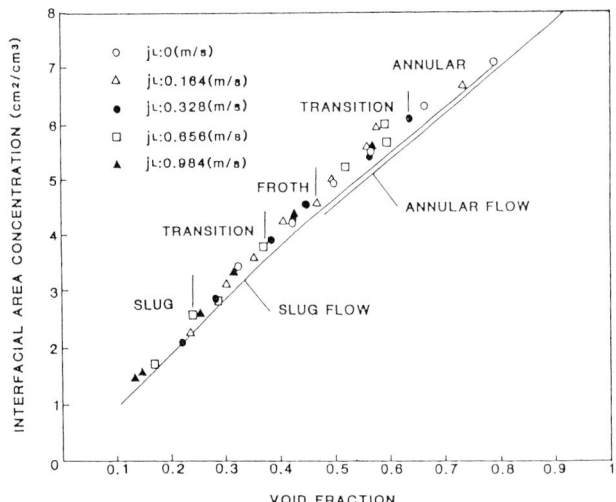

FIGURE 15. Interfacial area concentration as a function of the overall mean void fraction.

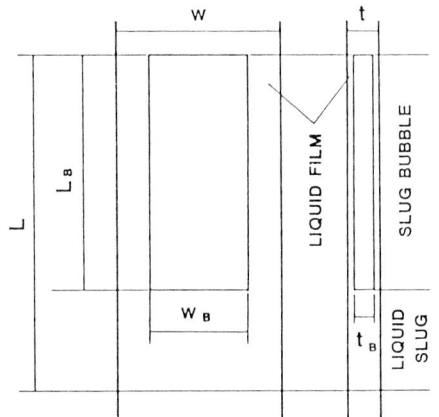

FIGURE 16. Slug flow model for narrow rectangular duct.

the interfacial area concentration and the overall mean void fraction were very similar to each other, indicating that there is a strong correlation between them.

Therefore, the interfacial area concentration was plotted against the overall mean void fraction with the superficial liquid velocity as a parameter as shown in Fig.15. Approximate locations of the flow regimes were also shown in the figure. The figure indicates that the interfacial area concentration in a rectangular duct with a large aspect ratio was correlated only with the overall mean void fraction. Empirical equations which depend on flow regime are given as follows:

slug flow: $a_i = 9.52\alpha^{0.937}$, (4)

froth flow: $a_i = 9.42\alpha^{0.92}$, (5)

and annular flow:

$a_i = 8.78\alpha^{0.813}$. (6)

Equations (4) through (6) are independent of the liquid superficial velocity. The above-mentioned trend of the interfacial area concentration can be reproduced by a simple model as shown in Fig. 16. Based upon the observation, one can consider a slug bubble to be like a tablet, where a rectangular nose is assumed for simplicity. From a simple geometrical consider-ation, one obtains an equation for the interfacial area as follows:

$$a_i = \frac{2\alpha}{t_B} + \frac{2}{wL}(L_B + w_B) \ . \qquad (7)$$

FLOW IN A NARROW RECTANGULAR DUCT

As for the width of a slug bubble, experimental observation indicated that w_B was in the range between $0.6w$ and $0.7w$. Akagawa (1966) has pointed out also that the void fraction in the slug bubble section was about 0.65 for slug flow in a round tube. Therefore, we assume here that

$$w_B = 0.65w . \tag{8}$$

The relation between the slug bubble length L_B and the void fraction has been obtained by Griffith (1961) as follows:

$$1 - \alpha = 0.087 + 0.526 \frac{w}{L_B} . \tag{9}$$

Although the above equation was derived from round tube data, it is assumed that one may use it approximately for a narrow rectangular channel taking the width of the duct as the characteristic length. Finally, the relation among the void fraction, the slug bubble length and the length of a unit slug bubble and a liquid slug is given by:

$$\alpha = \frac{w_B L_B}{wL} . \tag{10}$$

Using the above equations, one obtains

$$a_i = 9.64\alpha - 0.76\alpha^2 , \tag{11}$$

where t_B is approximated by t.

The equation for annular flow can be obtained if one assumes that $1/L_B$ is neglected compared to $1/t$ and $L_B/L \simeq 1$. Thus, the equation for annular flow becomes:

$$a_i = 8.33\alpha + 0.4 . \tag{12}$$

The predictions by Eqs. (11) and (12) are shown in Fig. 15 with solid lines. It can be seen that the experimental data was larger than the prediction. This difference may be attributed to the roughness of wavy interface, the existence of small bubbles in the liquid and droplet entrainment.

It should be noted here also that in general the interfacial area concentration depends upon the liquid superficial velocity as well as the void fraction. Serizawa (1983) proposed a correlation for a round tube (inner diam.: 60 mm) and an annulus (outer diam. of the inner tube : 12 mm, and inner diam. of the outer tube : 52 mm) which was proportional to $(j_l)^{-0.5}$. The exponent of the void fraction is 0.85 which lies in between these results for annular flow and the other flow regimes.

CONCLUSIONS

Video images of a two-phase flow in a narrow rectangular duct were taken either by the optical method or the NRG method. Those video images were reproduced and processed to calculate the overall mean void fraction and the interfacial area concentration. The results are summarized as follows:

a. The neutron television system demonstrated its usefulness in observing dynamic behavior of two-phase flow, while NRG by pulsed neutron beam was found to be effective in taking frozen images of two-phase flow to observe the structure of phase interface.

b. Image quality was good enough to delineate two-phase flow regimes, which were slug, froth and annular flows. A flow regime map was obtained.

c. With sufficient image quality, the image processing technique can be used to calculate the void fraction and the interfacial area which were consistent with those obtained by using the conductance probe technique as well as the drift flux correlation.

ACKNOWLEDGMENT

The authors wish to express their thanks to Dr. M. Sobajima, Mr. S. Ohtomo and other staff of Nuclear Fuel Safety Division of Japan Atomic Energy Research Institute for their cooperation in the experiment at NSRR. Part of this work was supported by the Grant-in-Aid for Scientific Research No.61420048 from the Japanese Ministry of Education, Science and Culture (1986-1987).

NOMENCLATURE

English

a_i : interfacial area concentration
C_0 : distribution parameter
D_c : diameter of the collimator
g : gravity
j : volumetric flux of the mixture
j_g : superficial gas velocity
j_l : superficial liquid velocity
L : length of a unit slug bubble and a liquid slug
L_B : length of a slug bubble
L_c : length of the collimator
s : gap thickness of the duct
t : thickness of the duct
t_B : thickness of a slug bubble

V_{gj} : drift velocity
v_g : gas velocity
w : width of the duct
w_B : width of a slug bubble

Greek

α : overall mean void fraction
$\Delta\rho$: difference of the density between two phases
ρ_g : density of the gas
ρ_l : density of the liquid

BIBLIOGRAPHY

Akagawa, K. and Sakaguchi, T., 1966, Fluctuation of Void Ratio in Two-Phase Flow, (2nd Report: Analysis of Flow Configuration Considering the Existence of Small Bubbles in Liquid Slugs), *Bulletin of JSME*, vol. 9, no. 33, pp. 104-110.

Asanuma, T. *et al.*, 1986, *Handbook of Flow Visualization*, Japanese Society of Flow Visualization, Asakura, Tokyo, (in Japanese).

Fujine, S., Yoneda, K. and Kanda, K., 1985, Digital Processing to Improve the Image Quality in Real-Time Neutron Radiography, *Nucl. Instr. Method*, vol. 228, pp. 541-548.

Griffith, P. and Wallis, G. B., 1961, Two-Phase Slug Flow, *J. Heat Transfer*, vol. 83, pp. 307-320.

Griffith, P., 1963, The Prediction of Low Quality Boiling Void, ASME preprint 63-HT-20.

von der Hardt, P. and Rottiger, H., 1981, *Neutron Radiography Handbook*, D. Reidel, Dordrecht, Holland, p. 19.

Harris, D. H. C. and Seymour, W. A. J., 1986, Applications of Real Time Neutron Radiography at Harwell, in *Neutron Radiography*, ed. J. P. Barton *et al.*, D. Reidel, Dordrecht, Holland, pp. 595-600.

Ishii, M., 1977, One-Dimensional Drift Flux Model and Constitutive Equations for Relative Motion between Phases in Various Two-Phase Flow Regimes, ANL-77-47.

Jones, O. C., Jr. and Zuber, N., 1979, Slug-Annular Transition with Particular Reference to Narrow Rectangular Ducts, in *Two-Phase Momentum, Heat and Mass Transfer in Chemical, Process and Energy Engineering Systems*, vol. 1, ed. F. Durst, G. V. Tsiklauri and N. H. Afgan, Hemisphere, Washington, D.C., pp. 345-355.

Robinson, A. H. and Wang, S. L., 1983, High Speed Motion Neutron Radiography of Two-Phase Flow, *ibid.*, pp. 653-659.

Serizawa, A., 1983, Turbulence Structure of Bubbly Flow, *Proc. 2nd Symp. on Multiphase Flow, Tokyo*, pp. 99-118.

Stewart, P. A. E., 1983, Aero Engine Applications of Cold Neutron Fluoroscopy at Rolls-Royce Limited, in *Neutron Radiography*, ed. J. P. Barton and P. von der Hardt, D. Reidel, Dordrecht, Holland, pp. 526-633.

Tamaki, M., Ohkubo, K., Ikeda, Y. and Matsumoto, G, 1986, Analysis of Two-Phase Counter Flow in Heat Pipe, *ibid.*, pp. 609-616.

THE EFFECT OF A FLAT-PLATE-TYPE OBSTACLE ON A THIN LIQUID FILM FLOW

T. FUKANO, T. TANAKA, K. KUTARAGI, and M. KANAMORI
Department of Mechanical Engineering
Faculty of Engineering
Kyushu University, 36, Fukuoka, Japan

ABSTRACT

Detailed visual observations and measurements of local film thickness, axial distributions of static pressure and gas velocity along a flat-plate-type flow obstacle (with and without a projection used as a simulation of a grid type spacer in a nuclear reactor) were conducted with special attention to drypatch formation, or liquid film breakdown. The effect of the inclination of the test section or gravity was also investigated. The results made it clear that, the flat-plate itself did not promote drypatch formation but the projection had a strong effect, even in the case where the projection is in contact with the surface at only one point. In general, the drypatch occured just in front of the projection in a wide range of both air and water flow rates in an inclined duct because the gravitational force acts to remove water downward as does the action of the horse shoe vortex formed stationally just in front of the projection.

INTRODUCTION

In some steam generators for nuclear power stations and chemical plants a lot of trouble has been experienced, in which the gas/vapor-liquid two-phase flow phenomena had serious effects on the safety operation of these plants. Especially when liquid flow rate is low, important problems which lead to a tube failure have arisen. They probably arise from a decreased heat removal from the heated tube surface and from a local chemical concentration caused by the dryout of liquid film.

Recently the heat removal from a heating surface near the spacer that supports nuclear fuel rods attracted a designer's attention from the view point of the economic design and the safe operation of a nuclear reactor. The flow pattern in the exit section of the steam generator of a BWR is supposed to be a thin liquid film flow that is accompanied by a high speed vapor flow which is generated from cooling water.

The grid type spacer may have a strong effect on such a liquid film flow because it usually has projections which jut into the liquid film and directly contact a fuel rod surface. Shiralkar and Lahey(1973) did farsighted research on this problem. They made it clear through visual observation that the upstream critical heat flux may have been caused by the occurrence of a drypatch just in front of the flow obstruction.

Initially, one of the present authors [Fukano, et al.(1985)] investigated the liquid film flow without flow obstruction. Theoretically and experimentally, the author bore in mind the fluid dynamic conditions that caused the film breakdown. The author clarified the two different mechanisms:

(a) the generation of a viscous wave in the low gas region;

(b) the liquid supplied by disturbance waves in the high gas flow region that caused the film breakdown on the bottom of the channel.

Then the author created the effect of a flat-plate-type obstruction on liquid film flow in a horizontal rectangular duct [Fukano, et al.(1986)].

Shiina (1988) did experimental research on the effect of a flat-plate-type flow obstruction on the heat transfer near an obstacle that was under an air single phase flow in a horizontal duct. The authors made it clear that a heat transfer enhancement could be expected due to the increase in the turbulence level at the exit of the spacer.

In this research, secondarily, the flat-plate-type flow obstruction with and without a projection, which has more realistic geometry than that used by Taitel and Lahey, were inserted in the channel and the detailed visual observation and measurements of the film thickness, the static pressure, and the gas velocity distributions near the flow obstruction were made. The effects of gravity on a drypatch formation were also investigated by using an inclined test section.

EXPERIMENTAL APPARATUS AND PROCEDURE

Figure 1 shows a schematic diagram of the experimental apparatus used in this experiment. The test section was made of acrylic resin to permit observation of the flow pattern. The separate air and water streams were joined and accelerated gradually in a converging entry section which preceded the film measuring section. Air was supplied from a compressor to the air-water joining section via an orifice flow meter. Water from a constant head tank was supplied through a rotameter to the air-water joining section. Its bottom was made of spongy metal, through which water was introduced into the test section. The inclination angles of the test section were 0° and 30°.

Figure 2 shows the measuring section, A being for local film thickness, B for static pressure and C for gas velocity. The flat-plate-type flow obstruction with a projection used in this

OBSTACLE ON A THIN LIQUID FILM FLOW

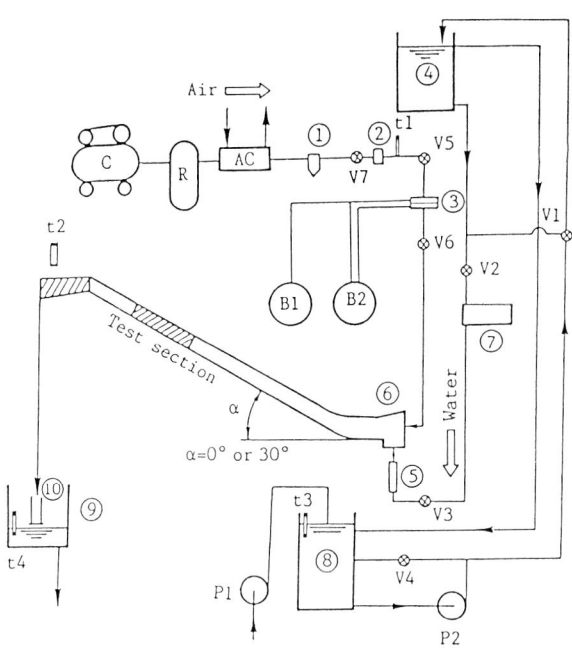

1: Air filter
2: Reducing valve
3: Orifice flow meter
4: Constant head tank
5: Rotameter
6: Air-water mixing section
7: Water filter
8,9: Water storage tank
10: Measuring cylinder

C: Compressor
R: Surge tank
AC: After cooler
B1,B2: Bourdon pressure gauge
P1,P2: Pumps
V1~V7: Valves
t1~t2: Thermometers

FIGURE 1. Experimental apparatus

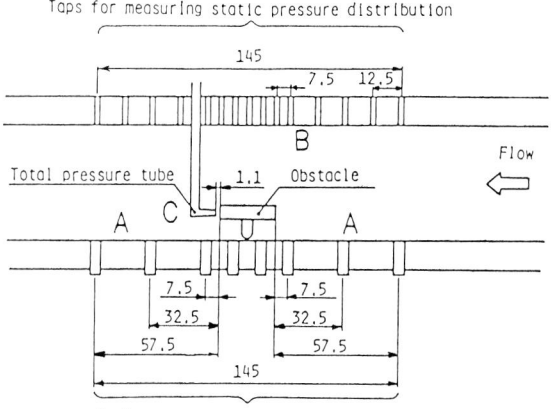

FIGURE 2. Locations of probes relative to the obstacle

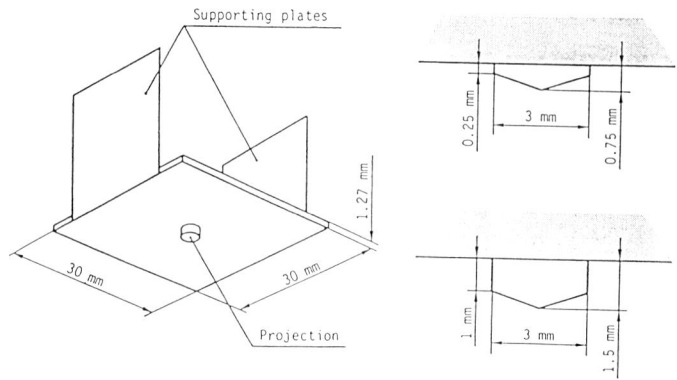

Flat-plate-type obstacle   Geometries of projections

FIGURE 3. Shapes of the flat-plate-type obstacle

experiment is shown in Fig. 3 with the detailed shape of the projection. The parts of the obstruction were made of acrylic resin and the projection was glued at the center of the plate.

The edge of the plate was not rounded so as not to complicate the effect of the edge of the plate. The thickness of the plate was determined to be 1.27 mm, taking into consideration the blockage ratio of the spacer in one of the actual reactors.

In this experiment the size of the gap, s, between the flat plate and the bottom of the duct was either 0.75 mm or 1.50 mm. The projection, if attached to the flat-plate obstruction, was always set so that its head directly contacted the duct surface. The diameter of the projections were 3 mm, and the shape was altered a bit according to the gap size as already shown in Fig. 3.

Figure 4 shows the location of the obstacle relative to the duct. The locations of the electrode for local film thickness measurement are shown in Fig. 5. A schematic diagram of the film thickness measuring device (the constant current method) is shown in Fig. 6. Each sensor probe which consisted of a pair of triple concentric rings was mounted flush with the bottom surface of the test section at the locations designated by P1∿P8 in order of the flow direction. The center and the outer electrodes were for constant current power supply and the center and the middle electrodes were for the measurement of the voltage drop due to the liquid film. The velocity distribution of the gas flow was measured by traversing a total pressure tube as shown in Fig. 7 and by subtracting the static pressure measured at the duct's upper surface. The range of the experiment is shown by mark ● in Fig. 8. Γ_c and j_{GC} in the axes stand for the volumetric flow rates of liquid per unit duct width and gas per unit area, respectively. Subscript c means that they were measured at the center of the duct width. The boundaries of the flow patterns without any obstruction are shown by a dot-dash line and by a solid line[Fukano, et al.

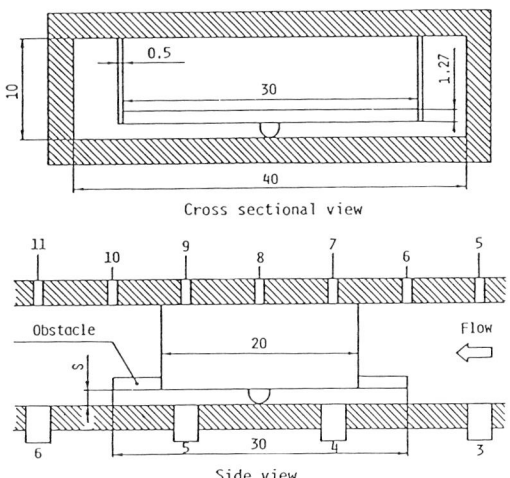

FIGURE 4. Location of the obstacle

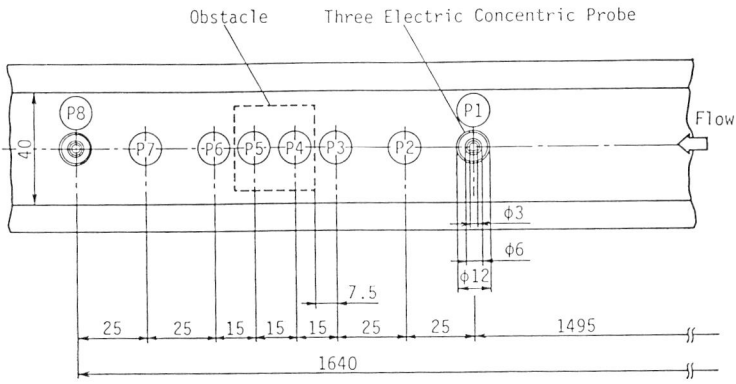

FIGURE 5. Locations of the probe for film thickness measurement

(1979)]. D is the disturbance wave region, R the ripple region and A the annular flow region. In the NW region the breakdown of the liquid film occured on the bottom surface of the test section. It must be noticed that the drypatch occurrence due to the effect of the obstruction was investigated in the region because the bottom is always covered by a liquid film. In the experiment special attention was paid to the setting of the flow rates of both phases, especially to the liquid phase, so that the quantitative comparison of the results which may differ due to the effect of gravity can be made in detail.

RESULTS AND DISCUSSION

The purpose of this investigation was to clarify the effect of the flow obstruction on a liquid film. The liquid film flow

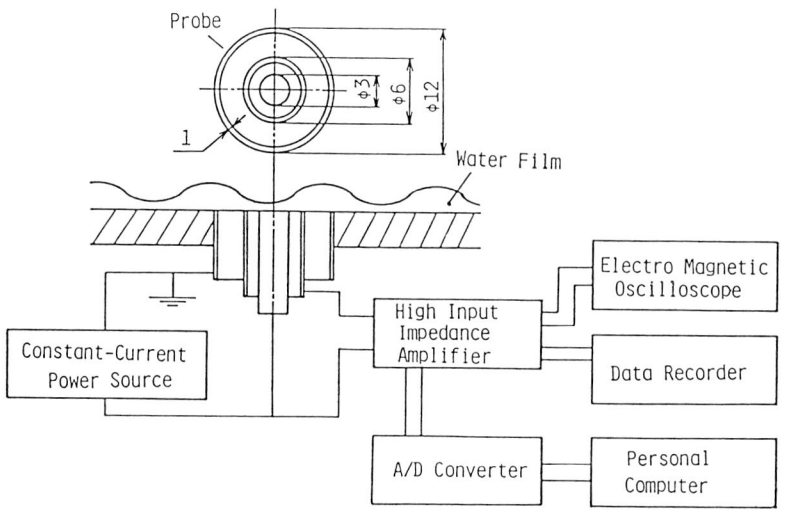

FIGURE 6. Constant current method for film thickness measurement

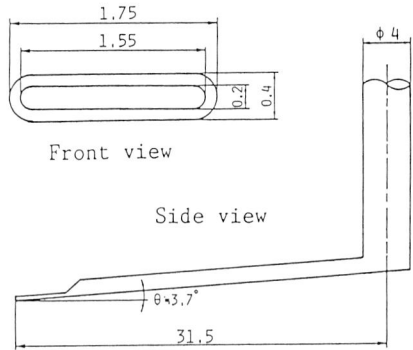

FIGURE 7. Total pressure probe

was controlled mainly by a gas flow. Therefore the effect of the obstacle on the gas flow must have been reflected indirectly in the liquid film flow. In the following, the static pressure distribution in the axial direction and the velocity distribution in the wake of the obstruction is discussed in relation to the liquid film breakdown.

The Effect of the Obstacle on the Gas Flow

Static pressure distribution. Figure 9 shows an example of the static pressure distribution in the flow direction when a gas phase alone flows with the velocity j_{GC} of 64 m/s. The one dot-dash line was drawn by assuming that the pressure gradient near the obstacle was equal to that in the region far upstream of the obstacle and by taking the blockage effect of the obstacle into consideration.

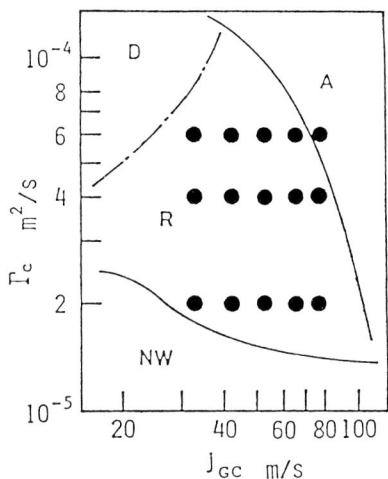

FIGURE 8. Experimental range

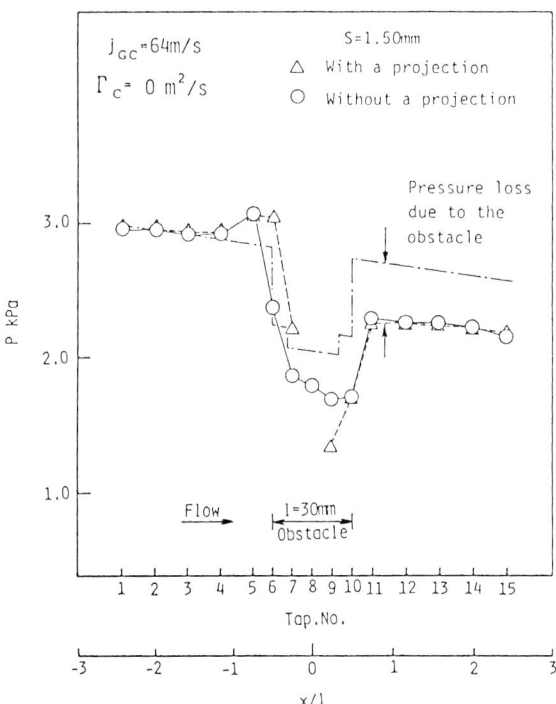

FIGURE 9. Static pressure distribution along the duct axis

It must be noticed that the static pressures increased just in front of the flat-plate-type obstacle in both cases, with and without a projection. This signified that the increase was due to the stagnation of the gas flow at the front edge of the

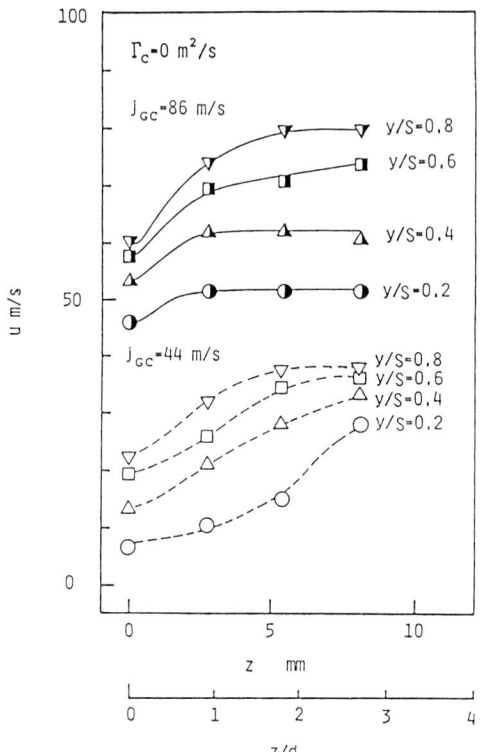

FIGURE 10. Velocity profile of the wake of the projection in the duct width direction

plate, and also just behind the obstacle due to the sudden increase in the flow area.

The effect of the projection on the pressure distribution can be seen in Fig. 9. Its effect, however, was more strongly noticed in the velocity distribution in the wake of the projection as shown in Fig. 10, that is, the wake width becomes wider as the gas velocity decreases, especially in the region closed to the bottom. H is the duct height, y the distance from the bottom of the duct, and z/d the coordinate in the direction of duct width normalized by the diameter of the projection. Finally, z=0 represents the center of the duct width, or the axis of the projection.

Velocity profile. Figures 11(a) and (b) show the velocity profiles of air in the direction of duct height measured on the line which was 1.1 mm downstream of the trailing edge of the plate. The solid lines in the figures show those in the case without projection. As shown in the figure, the velocity of the gap flow is rather large, or in some regions larger than that of the local value which the solid line takes at the corresponding location. On the other hand, however, where the projection existed, the velocity in the wake of the projection even took a negative value when the gap was 0.75 mm. Therefore

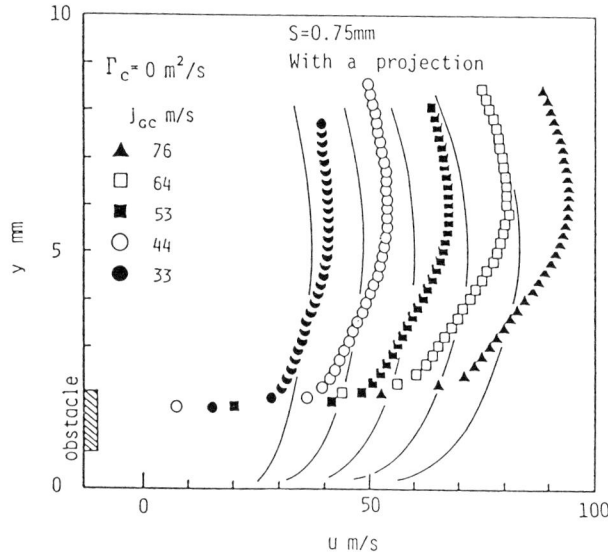

(a) with projection

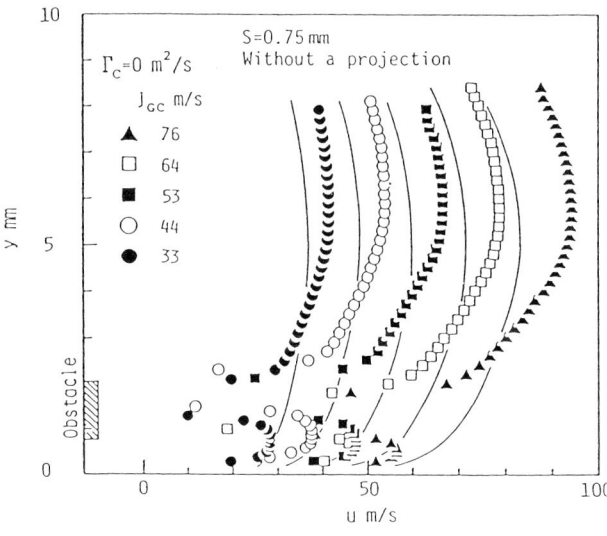

(b) without projection

FIGURE 11. Velocity profile of air flow in the direction of duct height

the data taken in those regions were omitted in Fig. 11(b).

Even when the water film existed, the velocity profile of the gas flow was not affected strongly, as shown in Fig. 12. Figure 13 shows in detail that the turbulence level of the air flow measured by a hot wire velocimeter was higher on the gas-liquid interfacial side than on the smooth solid surface side.

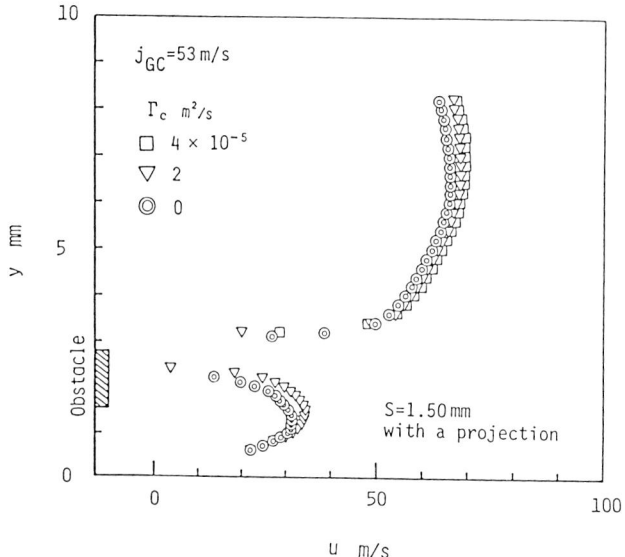

FIGURE 12. Effect of the liquid film on the air velocity profile

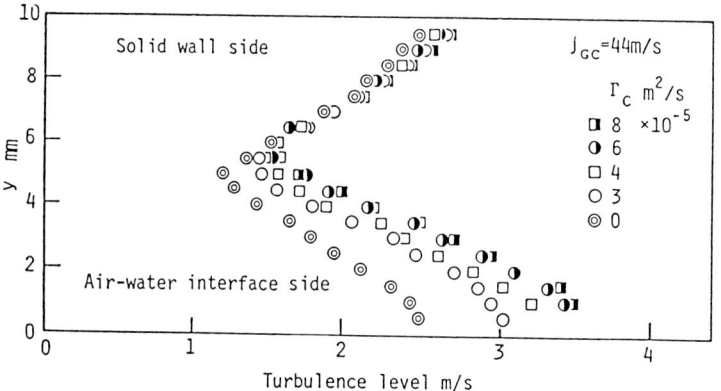

FIGURE 13. Effect of liquid film on the turbulence level of the air flow

The Effect of Inclination of the Test Section on the Liquid Film in the Case of No Obstacle

If the test section was not horizontal the gravitational force acted on the liquid film in the opposite direction to the flow of liquid because in the present case two phases flowed upwards. Accordingly, the film thickness t_{fm} must have been larger in the case of $\alpha = 30°$ than in the case of $\alpha = 0°$, as clearly shown in Fig. 14. Also the wave height, h_{fm}, became larger, as shown in Fig. 15, signifying that the increase in the interfacial shear force due to this increase in the wave height, or to the interfacial roughness, was not so large that

OBSTACLE ON A THIN LIQUID FILM FLOW

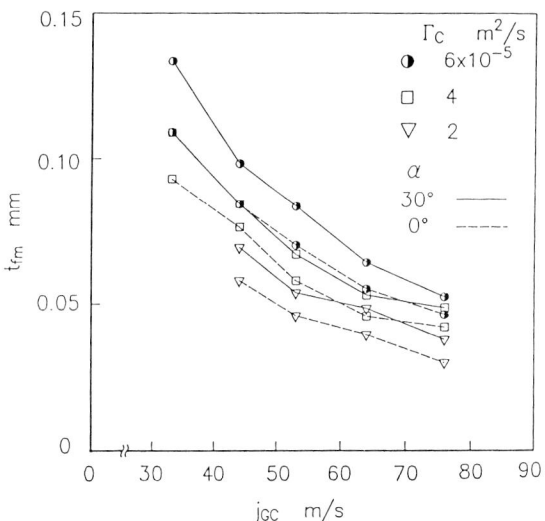

FIGURE 14. Mean film thickness

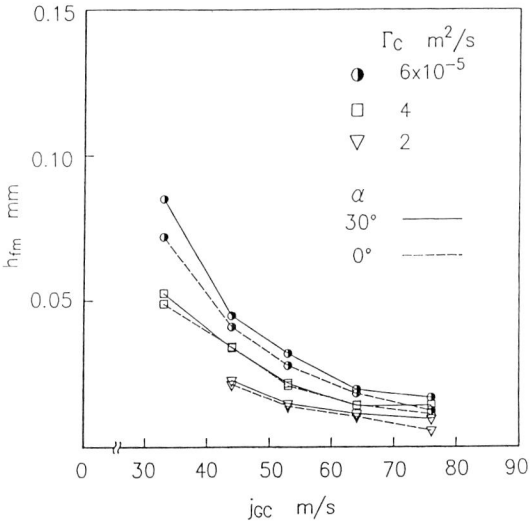

FIGURE 15. Wave height

the average film thickness did not become smaller than that in the case of horizontal flow.

The Effect of the Flat-Plate (Without a Projection) on the Liquid Flow

Figure 16 shows the time-averaged film thickness distribution along the flat-plate-type obstacle without a projection. The flow direction of two phases is shown by an arrow to be from left to right. The location of the obstacle is also shown.

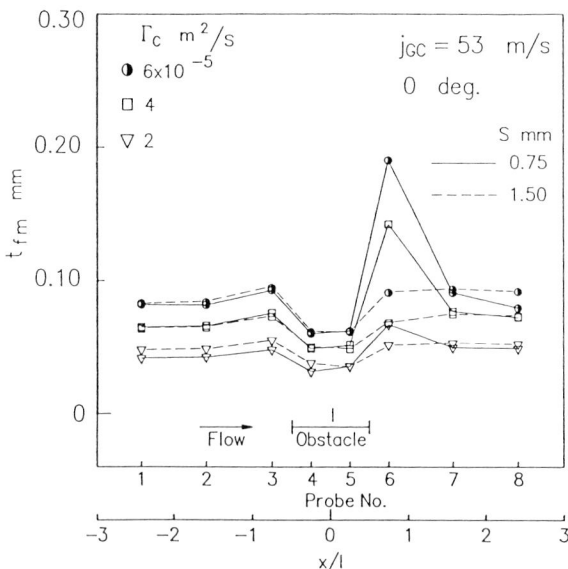

FIGURE 16. Film thickness distribution along the obstacle without a projection

In the region far upstream of the obstacle the liquid flow was not affected by the obstacle. Film thickness increased just in front of the plate due to the increase in the static pressure caused by the stagnation of the air flow (Fig. 9). Air could flow rather easily even in a small gap, as already shown in Fig. 11. The liquid flow rapidly accelerated in the gap and the film thickness decreased. At the exit the cross sectional area was considerably enlarged for a gap flow. Therefore, the liquid film decelerated there due to the sudden increase in static pressure (see Fig. 9) and its thickness increased abruptly. This increasing rate of the film thickness became larger as the gap became small because the enlarged area ratio from the gap to the exit, or the static pressure increased ratio, was much larger in the case of small gap. Additionally, the abrupt decrease in the interfacial shear stress was due to the sudden decrease in the gap flow velocity just behind the plate.

The general behavior of the film thickness described above held good in any air flow rate, as shown in Fig. 17(a) and (b). Also, the inclination of the test section had no effect on this general behavior, as shown in Fig. 18(a) and (b) if the flat plate had no projection.

The Effect of the Projection on the Liquid Film Flow

The effect of the obstacle always appeared more clearly in the case of the smaller gap. Therefore, in the following sections the experimental results (mainly the case of s=0.75 mm) are shown and discussed.

OBSTACLE ON A THIN LIQUID FILM FLOW

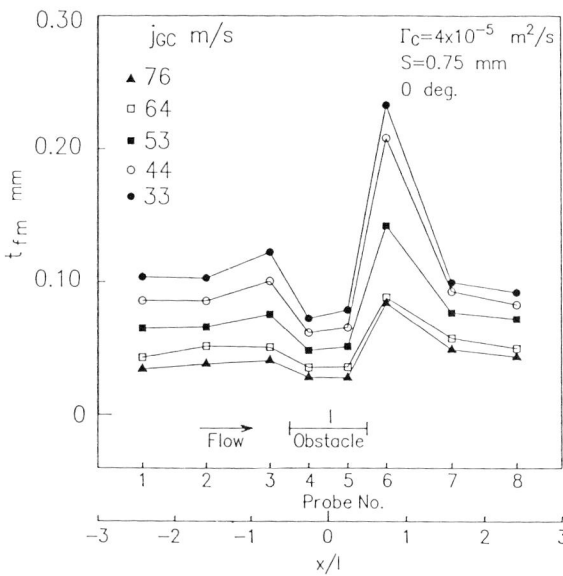

(a) Effect of j_{GC} (S=0.75mm)

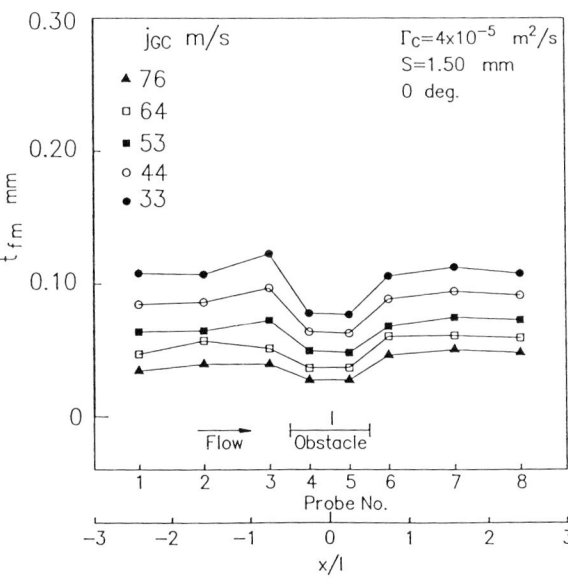

(b) Effect of j_{GC} (S=1.5mm)

FIGURE 17. Film thickness distribution along the obstacle without a projection ($\alpha = 0°$)

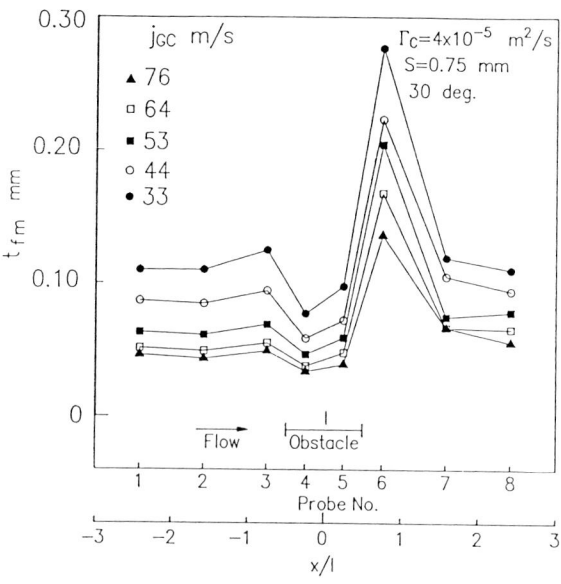

(a) Effect of j_{GC} (S=0.75mm)

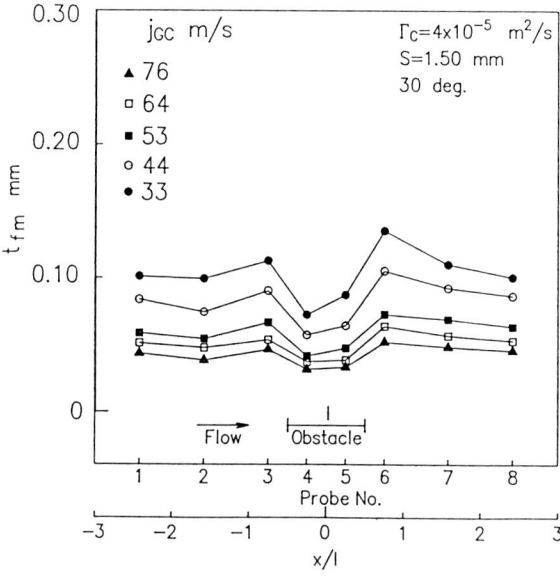

(b) Effect of j_{GC} (S=1.5mm)

FIGURE 18. Film thickness distribution along the obstacle without a projection ($\alpha = 30°$)

The liquid film thickness always changed with time because there were many waves on the surface of the liquid film. For example, this was shown by the ripple in the region far upstream of the obstacle. If the plate had no projection, then the fluctuation of the liquid film was not altered severely by the existence of the plate. In this case the change of the flow pattern or of the wave pattern was hardly noticed by visual observation.

On the other hand, if the plate had a projection, the fluctuation pattern of the film thickness with time changed drastically along the obstacle as shown in Figs. 19 and 20. Figure 19 represents the case of j_{GC}=33 m/s and Γ_c=4×10^{-5} m^2/s, and Fig. 20 for j_{GC}=76 m/s and Γ_c=4×10^{-5} m^2/s.

In the region far upstream of the obstacle, P1 and P2 in the figures, the surface waves had a fine structure, so the amplitude of the fluctuation was small and the frequency was high. This feature held in the region on the upstream side of the obstacle, but changed drastically on the downstream side, especially in the close vicinity of the projection. That is, the flow pattern changed into a disturbance wave flow with the rather regular passing of the disturbance wave. Comparisons of (a) for α=0° and (b) for α=30° revealed that the regularity was more clearly seen in the case of α=30°.

The changes of the wave structure and the film thickness along the obstacle can be seen clearly in Figs. 21(a) and (b) and 22(a) and (b), the flow conditions of which correspond to those of Figs. 19(a) and (b) and 20(a) and (b), respectively. These figures show the cumulative frequency of the film thickness fluctuation plotted on normal probability distribution paper. The occurrence of large scale waves was noticed typically at P5 and P6 in Figs. 21(a) and (b) as shown by an arrow.

The changes of film thickness along the obstacle are plotted in Figs. 23(a) and (b) and 24(a) and (b) for various flow conditions of air flow with a constant water flow rate Γ_c of 4×10^{-4} m^2/s where (a) represents the time averaged film thickness and (b) represents the wave height h_{fm}, where h_{fm} is defined as ($t_{fmax}-t_{fmin}$). These figures show the general trend of the film thickness change along the flat-plate obstacle discussed in the previous section. In detail, however, it must be noticed that the existence of the projection causes the film breakdown[Fig. 24(a)] and that this trend became effective in the intermediate range of air flow rate, that is, j_{GC}=44∼64 m/s.

The experimental results, especially those of the film thickness showed the important effects of the existence of the projection and the inclination of the test section, or the gravitational force on the liquid film breakdown . The number and the location of the film thickness measurement were too few to clarify the more detailed change of the film thickness near the projection. They were insufficient because the electrode was rather large compared to the size of the projection and the scale of the change of the flow configuration of the liquid

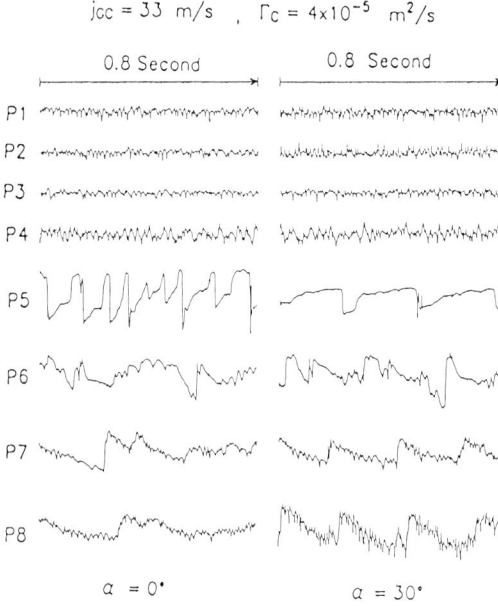

FIGURE 19. Change of the film thickness with time (j_{GC}=33 m/s, with projection)

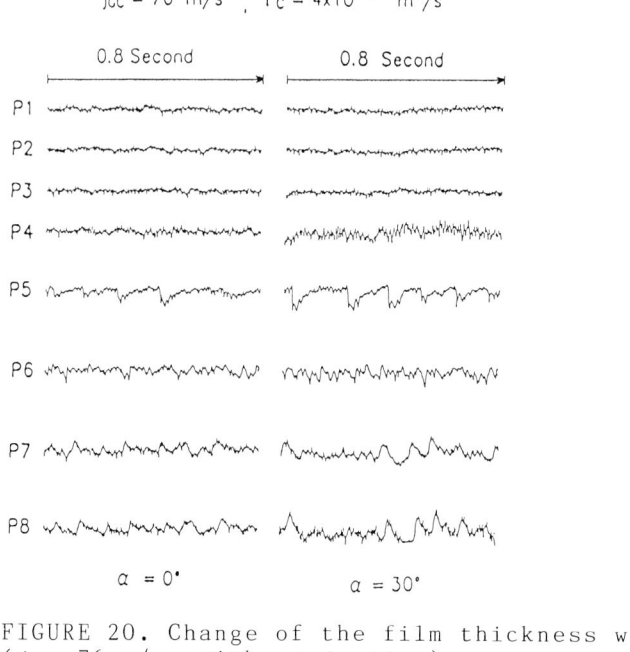

FIGURE 20. Change of the film thickness with time (j_{GC}=76 m/s, with projection)

OBSTACLE ON A THIN LIQUID FILM FLOW

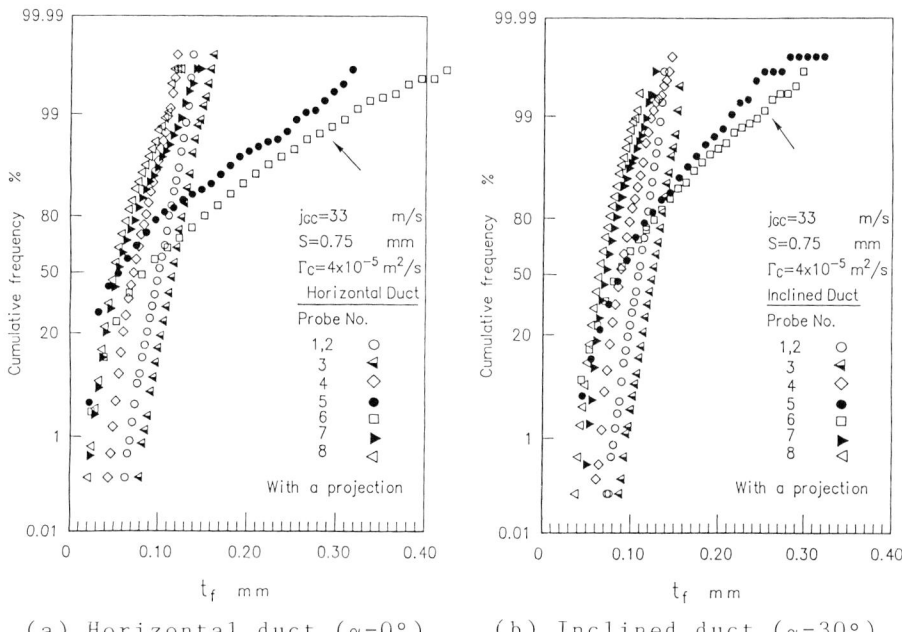

(a) Horizontal duct ($\alpha=0°$) (b) Inclined duct ($\alpha=30°$)

FIGURE 21. Cumulative frequency of the film thickness (with a projection, $j_{GC}=33$ m/s)

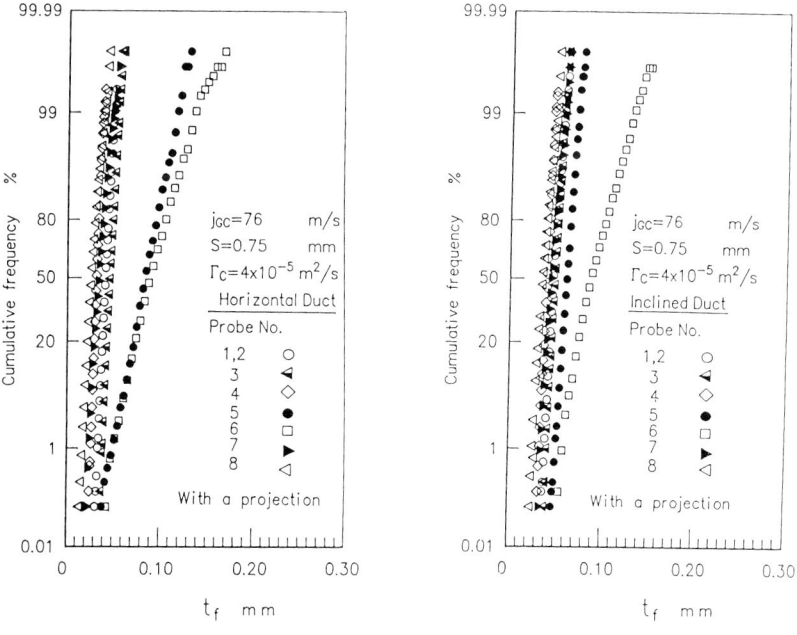

(a) Horizontal duct ($\alpha=0°$) (b) Inclined duct ($\alpha=30°$)

FIGURE 22. Cumulative frequency of the film thickness (with a projection, $j_{GC}=76$ m/s)

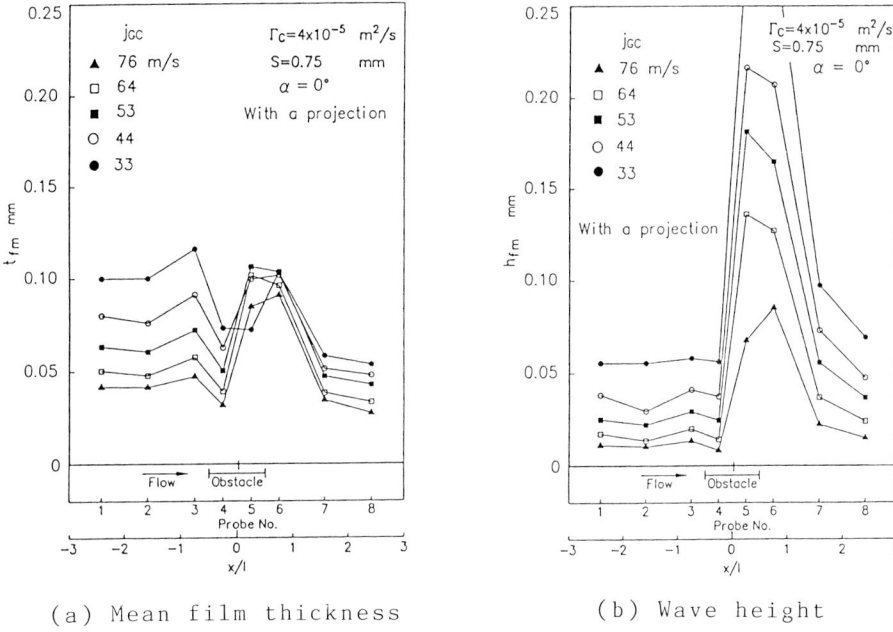

(a) Mean film thickness (b) Wave height

FIGURE 23. Film thickness distribution along the obstacle with a projection ($\alpha=0°$)

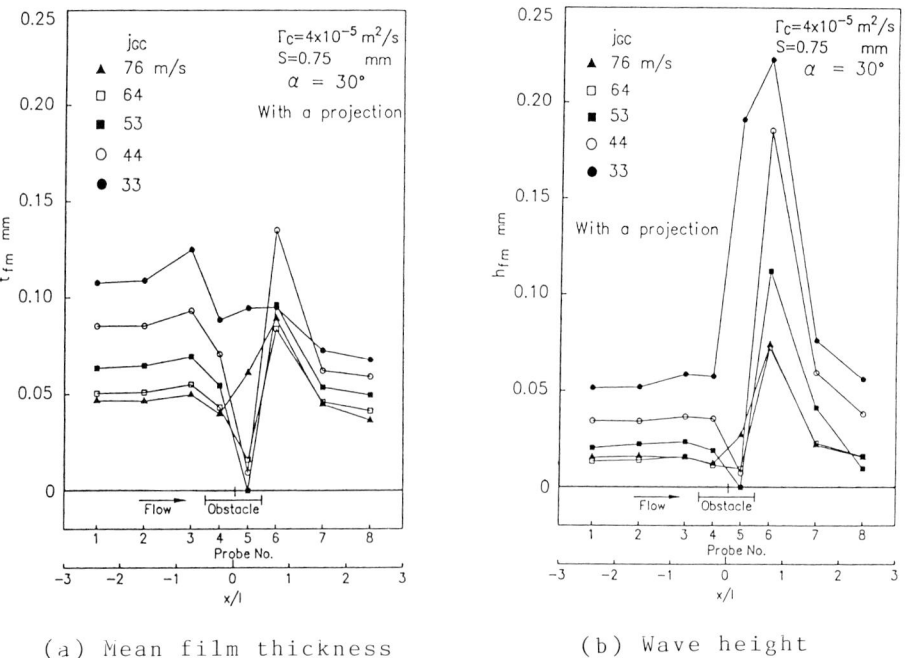

(a) Mean film thickness (b) Wave height

FIGURE 24. Film thickness distribution along the obstacle with a projection ($\alpha=30°$)

OBSTACLE ON A THIN LIQUID FILM FLOW

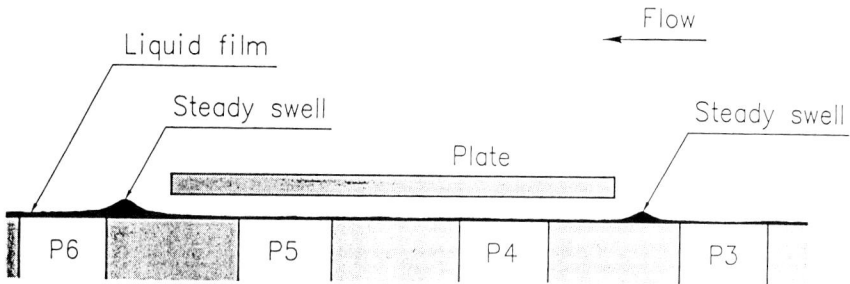

FIGURE 25. Schematic drawing of the film thickness distribution (without a projection)

film, and to the size of the drypatch. So, the detailed observation was done in these experiments by using a duct made for this special purpose. In the following section this visual observation is discussed.

Visual Observation

In the case of the flat-plate without projection. Figure 25 shows the schematic drawing of the film thickness distribution in the case of the flat-plate obstacle without projection. As already described in Fig. 16, the film thickness increases at both locations just in front of and behind the flat-plate. The axial scale of this stationary swell was small, however, as shown in comparison with the electrode, in Fig. 16. Therefore the height of the swell was actually larger than the measured film thickness. The flat-plate itself did not play an important role in the film breakdown.

In the case with the projection. Although flow configuration near the obstacle was quite complicated depending on the flow conditions of both phases and the geometric condition when the flat plate had a projection, the following features were held in common regardless of the experimental conditions. That is, (1) the liquid swell was formed at two places, just in front of and behind the flat plate, as discussed in the previous section; (2) the rear liquid-swell increased its height at the center part which corresponded to the wake of the projection; and, (3) in the wake of the projection the liquid film formed a narrow stream with a larger thickness and flowed with a velocity that was considerably lower than those of the oncoming liquid film.

These three common phenomena kept the liquid film stable. From the view point of the limit of the maximum heat flux in an actual nuclear reactor, attention must be paid to the occurrence of drypatches. The following discussion focusses on this problem.

Area maps of the occurrence of the drypatch are summarized in Figs. 26(a) through (d), with the flow rate of both phases being the two axes. The occurrence of the drypatch was always caused by the existence of the projection. The flow

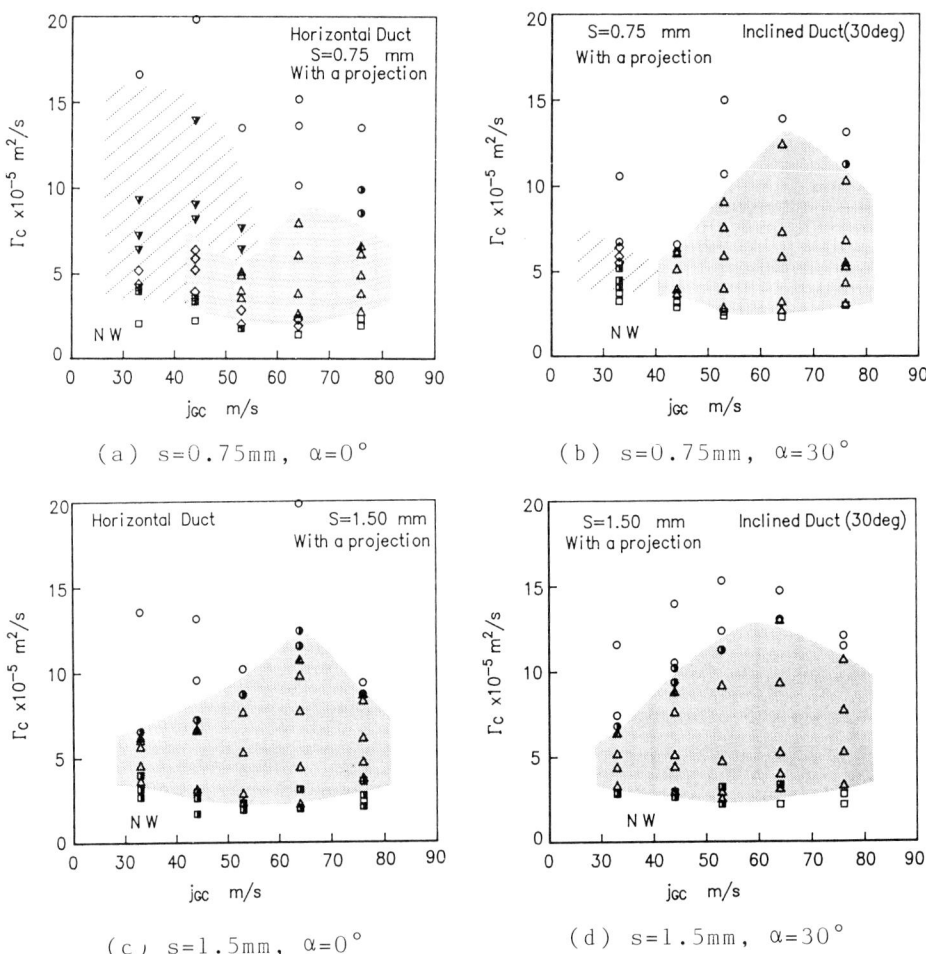

FIGURE 26. Area map of the occurrence of drypatch (with a projection)

configuration near the projection can be divided fundamentally into four regions as shown in these figures. That is, (1) no drypatch region, which is shown by the symbol ○ in Figs. 26, (2) front drypatch region △, (3) rear drypatch region ▽, and (4) NW region where the projection played no role in the drypatch occurrence[□]. As explained briefly in Table 1, even in the no drypatch region, the liquid film was very thin just in front of the projection. By reducing the liquid flow rate Γ_c, a drypatch with a narrow crescent shape appeared in close vicinity of the tip of the projection. This regime is expressed by the symbol ◐. The drypatch area became wider and wider as Γ_c was reduced until a steady large-scale-crescent-dry-area as shown by △ appears. This region is also shown by the dotted, shadowed area in the figures. In the region between ◐ and △, the dry area fluctuates with time, which is expressed by the symbol ▲.

TABLE 1. Classification of the geometry of the drypatch and the symbol used in Figs. 26(a)-(d).

Front drypatch		Rear drypatch	
Symbol	Flow configuration	Symbol	Flow configuration
○	Without drypatch / Somewhat thinner film	▼	Unsteady drypatch
◐	Thinner film / Drypatch	◇	Steady drypatch
▲	Fluctuation of Drypatch area	◧	Transition between ◇ and □
△	Stationary Drypatch area	□	NW region: Drypatch occurs independent of the projection

Although these front drypatches occured in the wide range of the flow and geometric conditions, it must be noticed that the liquid film was apt to break under the gas flow rate condition j_{GC} of 60 m/s∼70 m/s, which was estimated to be approximately equal to the vapor exit velocity of the steam generator of a BWR. And the front-drypatch region was wider in the case of the inclined duct than that of the horizontal duct, especially when the gap size was small, s=0.75 mm. This was because the liquid was rejected by the gravitational force as well as by the action of the horse-shoe vortex in the region just in front of the projection. If the gap was wide, s=1.5 mm, the effect of the projection was weakened because its scale relative to the gap became small.

This effect was also seen in the rear-drypatch occurrence as clearly shown, especially, in Figs. 26(c) and (d) of s=1.5 mm. If the gap was small, s=0.75 mm, however, the effect of the projection on the drypatch formation was quite strong. The rear-drypatch region is represented by slashed area, and also by the marks ▼ and ▽. In the regime expressed by ▽ a steady large scale dry area was observed and in almost all cases this drypatch was accompanied by the steady front-drypatch. Therefore, this regime is expressed by the combined mark of ◇.

The rear-drypatch region was wider in the horizontal duct than in the inclined duct because water could be supplied from the rear part of the flat plate to the liquid deficit wake region due to gravity in the inclined duct. Furthermore, the rear-drypatch region became wider as the gas velocity was reduced. This was attributable to the fact that the width of the wake of the projection in the gas flow became wider as the gas flow rate was reduced, as already shown in Fig. 10.

CONCLUSION

Flow configuration near the flat-plate-type flow obstruction simulating a grid type spacer in a nuclear reactor was experimentally investigated. In this research the measurements of the axial distribution of the film thickness, the static pressure, and the velocity profile as well as detailed observations were conducted with special attention paid to the drypatch formation near the obstacle. The results are summarized as follows.

1) The flat-plate itself had no effect on the drypatch formation. At both places just in front of and behind the edges of the flat plate, the liquid swells were created due to the static pressure change along the gap between the flat-plate and the bottom of the duct. Then it relieved the occurrence of drypatch.

2) The projection was responsible for the formation of drypatch.

3) Two types of drypatch formation were observed; one was the front-drypatch and the other was the rear-drypatch.

4) The front-drypatch was apt to occur under the air velocity of 60∼70 m/s, and the rear-drypatch occured under lower air velocity.

5) The front-drypatch occured in the wider flow rate ranges of both phases in the inclined duct rather than in the horizontal duct.

6) Even in the case where the drypatch did not occur, the flow pattern changed from a ripple flow to a disturbance wave flow as the liquid film passed by the projection.

NOMENCLATURE

- d : Diameter of the projection   mm
- H : Duct height   mm
- h_{fm} : Wave height   mm
- j_{GC} : Volumetric flux of air measured at the center of the duct width $m^3/m^2 s$
- ℓ : Axial length of the flat-plate   mm
- s : The size of the gap between the flat-plate and the bottom of the duct   mm
- t_{fm} : Mean film thickness   mm
- u : Absolute velocity of air flow   m/s
- x : Axial coordinate
- y : Distance upward from the bottom of the duct
- z : Coordinate in the duct width direction
- α : Angle of the duct inclination
- Γ_c : Water volumetric flow rate in a unit width measured at the center of the duct width m^2/s

Flow Pattern Symbol
- A : Annular flow

D : Disturbance wave flow
NW : Film breakdown
R : Ripple

ACKNOWLEDGMENTS

The authors are indebted to Dr. Fukuhara, Dr. Kadoguchi, Ms. Yoshimura and the research colleagues in their laboratory for their help in completing this paper.

REFERENCES

Fukano, T., Ishida, K., Morikawa, K., Nomura, H., Takamatsu. Y., and Sekoguchi, K., 1979, Liquid Films Flowing Concurrently with Air in Horizontal Duct (Part I, Flow Pattern), *Bull. of the JSME*, 22-172, p.1374.

Fukano, T., Itoh, A., and Ousaka, A., 1985, Breakdown of a Liquid Film Flowing Concurrently with Gas in Horizontal Lines, *PhysicoChemical Hydrodynamics*, 6-1/2, p.23.

Fukano, T., Morikawa, K., and Tominaga, A., 1986, The Effect of a Flow Obstacle on Liquid Film Flowing Concurrently with Air in a Horizontal Rectangular Duct, *Trans. JSME*, 52-477, p.2052.

Shiina, K., Nakamura, S., and Shimizu, N., 1988, Enhancement of Forced Convective Heat Transfer in a Rectangular Channel Using Thin Plate-Type Obstacles (1st report), *Trans. JSME*, 54-497, p.148.

Shiralkar, B. S., and Lahey, R. T. Jr., 1973, The Effect of Obstacle on a Liquid Film, *J. of Heat Transfer*, 95-4, p.528.

DISTURBANCE WAVE IN BOILING FLOW

S. NAKANISHI
Dept. of Chemical Engineering
Himeji Institute of Technology
2167, Shosha, Himeji, Japan
S. YAMAUCHI and T. SAWAI
Dept. of Mechanical Engineering
Takamatsu National College of Technology
355, Chokushi-cho, Takamatsu, Japan

ABSTRACT

A disturbance wave in boiling flow was experimentally investigated through film thickness measurements in a steam-water system as well as flow visualization in a Freon-113 system. Comparing the results with the corresponding adiabatic flow data, we established a new kind of hydrodynamic non-equilibrium, and identified it with that in liquid distribution between the base film and the bodies of the disturbance wave. The decaying process of this non-equilibrium was examined in detail in a non-heated section. Interaction of the disturbance wave with the dryout phenomenon was then clearly demonstrated. Finally, the effects of artificial hydrodynamic non-equilibrium associated with the disturbance wave were discussed.

INTRODUCTION

As is well known, in most cases of two-phase annular flow large surface waves travel on the annular liquid film adhering to the channel wall. These waves, named "disturbance waves", have a great influence on characteristics of the annular flow. These characteristics include interfacial shear stress between the liquid film and gas core, and liquid entrainment from the film with the gas flow. These two topics remain subjects of intense investigation by a number of authors(Hewitt et al., 1970; Whalley at al., 1987).
The disturbance wave itself is one of the most absorbing topics in the field of two-phase flow investigation. Though there already exists a huge accumulation of theoretical and experimental works, there are still many issues to be solved. For example: the origin of the disturbance wave, its characteristics as a stochastic process, hydrodynamic non-equilibrium under flow developing conditions involving boiling flow, and so on. Our interests have particularly been focused on the effect of boiling on the behavior of the disturbance wave. In the course of our investigation of wall temperature fluctuation of evaporating tubes in dryout regime(Nakanishi et al., 1982a), it was suggested that passage of disturbance waves induce momentary drops of wall temperature, giving a stochastic character to the temperature

fluctuation. This suggestion was confirmed, through measurement of the thickness of the annular liquid film at the exit of a test section for a CHF experiment in our heat transfer loop with medium pressure(3MPa) water(Nakanishi et al., 1985a, 1986). A simulation model of the wall temperature fluctuation was constructed(Nakanishi et al., 1984) which required detailed information on the stochastic characteristics of some parameters of the disturbance wave. In the same study, we found interesting the wave profile of the boiling flow largely differing from that of the adiabatic air-water two-phase flow with the same superficial velocities. This suggests that boiling should play an important role in the behavior of the disturbance wave in boiling flow. A possible explanation for this difference is that there is some hydrodynamic non-equilibrium due to phase change. This was experimentally studied earlier under rather limited conditions by Brown et al.(1975). Not only was more information required for developing the dryout modeling, interest in this non-equilibrium phenomenon led to investigation of the disturbance wave in a boiling channel in more detail(Nakanishi et al., 1982b; Nakanishi et al., 1985b; Nakanishi et al., 1987; Sawai et al., 1989). The experimental works were conducted on a heat transfer loop with steam-water system(Nakanishi et al., 1982b; Nakanishi et al., 1985b; Nakanishi et al., 1987) and a Freon boiling test rig for visualization. In this paper, results obtained are presented including some of experimental works with artificial non-equilibrium(Ishigai et al., 1974; Nakanishi et al., 1979).

EXPERIMENTAL APPARATUS AND METHODS

Boiling Steam-Water System

A test rig for the previous dryout study(Nakanishi et al., 1982a) was used to measure the thickness of water film in boiling flow. Figure 1 shows schematically the test section employed for this purpose. It comprised an upstream heated section(heated section I), a conductance probe block, a downstream heated section(heated section II), and was set into vertical position. The two heated sections I and II were made of AISI-304 stainless steel tube with 4 mm I.D. and 6 mm O.D., and were 1.35 m and 0.15 m long, respectively. They were connected electrically in series with a pair of flanges, between which the conductance probe block was inserted. Heating of the test section was made by directly passing an alternating current through it, so that heat flux was kept uniform over its whole length. Figure 2 shows the detail of the conductance probe block for film thickness measurement. This block was made of bakelite resin. Two pairs of conductance probes were located at a distance of 27 mm along the flow direction. Each probe consisted of a pair of electrodes made of 0.9 mm stainless steel wire which were set 3 mm apart and whose tips set flush with the surface of the hole for water-steam mixture. The 50 mm length of the block acted as an electrical and thermal insulator. This block's effect of interruption of heating on the characteristics of the disturbance wave was shown to be negligibly small, from examination of our data on

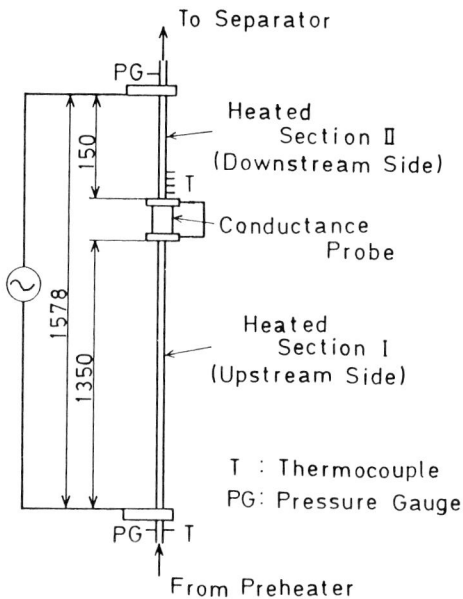

FIGURE 1. Test section for steam-water system. (Standard arrangement).

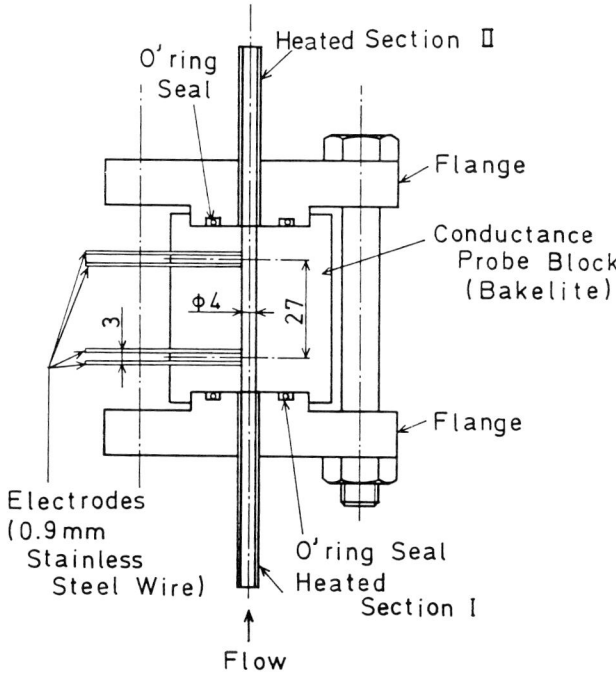

FIGURE 2. Conductance probe block.

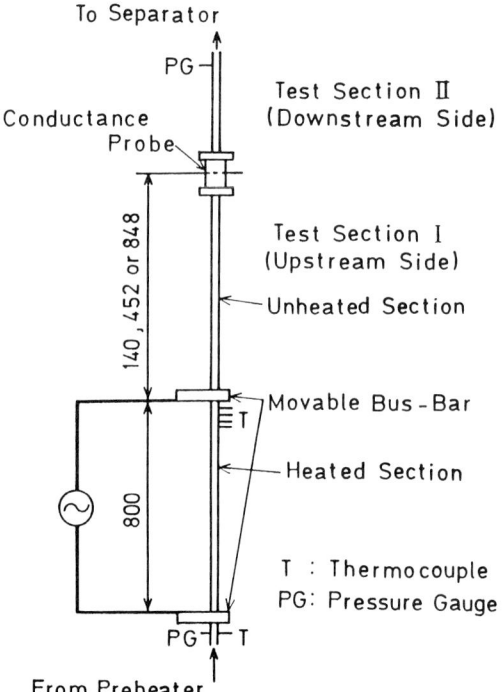

FIGURE 3. Test section for steam-water system. (Modified arrangement).

hydrodynamic non-equilibrium and that of Brown et al.(1975). To enhance electrical conductivity of liquid phase water, 1.25 mg of potassium chloride was added to each 1 kg of feed water for the test rig for high quality runs (higher than 0.8). For medium quality runs (from 0.5 to 0.8) 3.75 mg was added. To minimize effects due to polarization a 5 kHz sinusoidal current was applied to the conductance probes, whose output signals were conditioned to eliminate superfluous components from disturbance waves on the film surface, using an AD converter/personal computer system. To avoid inaccuracies caused by permanent deformation of the conductance probe block, the block was renewed every several runs.

The test conditions were as follows: pressure 2.95 MPa, mass flux 300, 500, and 700 $kg/m^2 s$, and heat flux up to 0.7 MW/m^2.

For examination of non-equilibrium effects due to boiling on the disturbance wave, a modification was applied to the test section described above, as shown in Fig. 3. The test section I was heated only over a limited length, 0.8 m. A non-heated zone, whose length was variable was located below. To achieve this arrangement a pair of movable bus-bars were employed, as can be seen from Fig. 3. Measurement of film thickness was conducted at the conductance probe block connected just downstream of the non-heated zone. The test section II was not heated.

The test conditions were as follows: pressure 2.95 MPa,

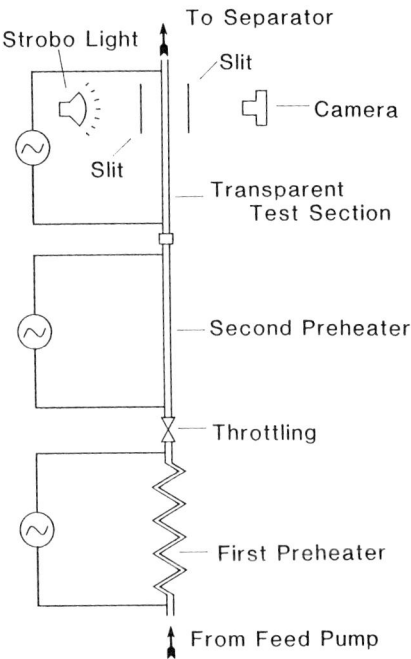

FIGURE 4. Test rig for Freon-113 system.

mass flux 300 and 500 kg/m^{2}s, heat flux up to 1.33 MW/m^2, and ratio of non-heated length to tube diameter 35, 113, and 212.

Freon Test Loop

To visualize flow boiling in the tube, a heat transfer loop for Freon-113 was constructed as shown in Fig. 4. Freon-113 was heated to a certain temperature-slightly under the saturation temperature-in the first preheater, and then heated up to a specified quality in the second preheater, before entering into a transparent vertical test section. The test section was a Pyrex glass tube, 7 mm I.D./10 mm O.D. and 1 m in length, whose outer surface was thinly coated with an electrically conductive and transparent film of tin-oxide, using sputtering. Heating of the test section was conducted directly by passing an alternating current through it, and the effective heating length was 0.9 m. The test section was connected with the second preheater without any change of cross-section, so that no disturbance was introduced at the joint.

Visual observation was made, using illumination from a stroboscopic lamp, and sequence photographs were taken with a streak camera in order to determine the characteristics of movement of the disturbance wave and dryout front.

The test conditions were as follows: pressure at the exit of the test section ranged from atmospheric to 0.4 MPa, mass flux 100-500 kg/m^{2}s, and heat flux up to 77 kW/m^2.

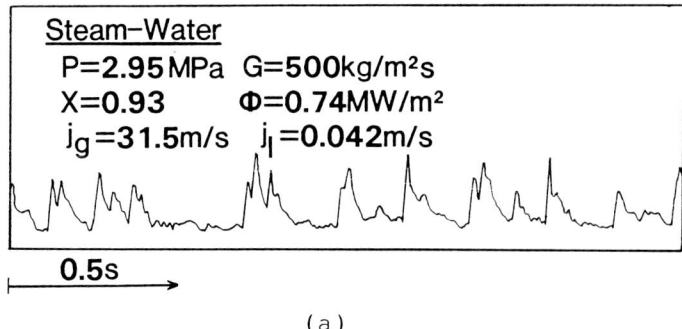

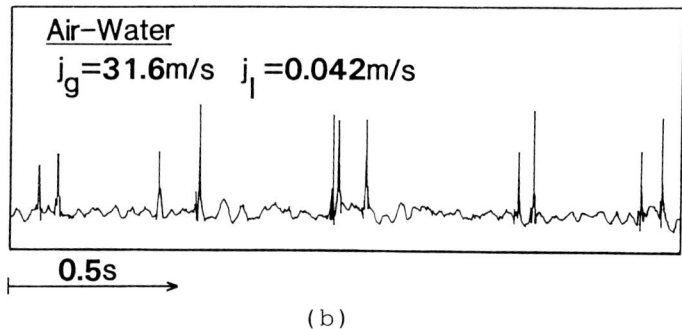

FIGURE 5. Recordings of liquid film thickness signals.
(a) Boiling steam-water; (b) adiabatic air-water

MAIN PARAMETERS OF DISTURBANCE WAVE IN BOILING FLOW

Wave Form of Disturbance Wave

Figure 5a shows a typical example of film thickness fluctuation, recorded in a boiling steam-water system. To facilitate identification of signal peaks corresponding to the disturbance wave, high frequency components were filtered out by the 21-point moving averaging method(Savitzky et al., 1964). As seen in Fig. 5a, the wave profile of the disturbance wave comprised a fairly steep front side and a long-tailed rear side, which made it difficult to clearly discriminate the body of the disturbance wave from the base film. Individual waves were rarely isolated from each other, but traveled in the form of wave packets comprising several waves. The adiabatic air-water system behaved in a different manner.

Figure 5b shows a typical example of film thickness recording in the adiabatic air-water system, with the same superficial velocities as those in the steam-water system of Fig. 5a. In this case, because of steepness of the signals corresponding to the disturbance wave, filtration of signals was successfully executed with the 21-point adaptive smoothing method(Kawata et al., 1984). As seen from the Fig 5b, the spike-like signals for the disturbance wave made

DISTURBANCE WAVE IN BOILING FLOW

discrimination of it from the base film much easier, and the signals were well separated from each other.

Propagation Velocity

Figure 6 shows the probability density distribution of the disturbance wave for the run corresponding to Fig. 5a. The solid curve in the Fig 6. represents a logarithmic Gaussian distribution whose mean value and standard deviation are equal to the corresponding values for the experimental data. Agreement is comparatively good; this observation is valid for our other data with a boiling steam-water system.

In Figure 7, the mean propagation velocity, $\overline{U_d}$, and its standard deviation, σ_{Ud}, for a boiling steam-water system are plotted against the quality, x, at a fixed mass flux(300 kg/m^2) for various values of heat flux. In the high quality region such as in Fig. 7, the mean value decreases monotonously, and the standard deviation does the same, but more slowly. This slowness is the reason why the probability density distribution takes a similar form over a wide range of quality. No significant effect of heat flux was observed on the propagation velocity for any of the conditions investigated. Figure 8 shows the propagation velocity for a medium quality region; data with air-water system are also plotted for comparison. The variations with quality are relatively small. Figure 9 represents the data with boiling Freon-113 flow, which shows the similar features as those with the steam-water.

The propagation velocities, $\overline{U_d}$, for boiling flow were successfully correlated in term of $\tilde{v}$, the ratio of shear stress, τ_i, to mass velocity, G. This is shown in Fig. 10. The use of frictional velocity succeeded in correlating the data for adiabatic air-water flow(Nakanishi et al., 1979); the shear stress was estimated with the correlation proposed by Henstock and Hanratty(1970).

Time Separation between Waves

Two types of probability density distributions of the time separation between successive waves were observed. These were the one peak and plural peak types. The former was well approximated by a logarithmic Gaussian distribution, and observed in the range of quality lower than 0.9 for the boiling steam-water system. The latter had a peak at about 100 ms, and the former at 240-300 ms for the boiling steam-water system(see Fig. 11), and was observed in the range of quality higher than 0.9 to the dryout. The plural peak distribution was reported by Hall-Taylor and Nedderman(1968), who found four peaks in the spatial wave separation data for the air-water system.

Figure 12 gives the mean time separation, $\overline{T_d}$, and its standard deviation, σ_{Td}, corresponding to the condition of Fig. 7. For the time separation, effect of heat flux was also weak. In this high quality region, the time separation increases with the quality, which means that waves gradually disappear when approaching the dryout region. The ratio of the standard deviation to the mean value remains nearly constant in this quality range. Figure 13 represents the time separation for boiling Freon-113 flow.

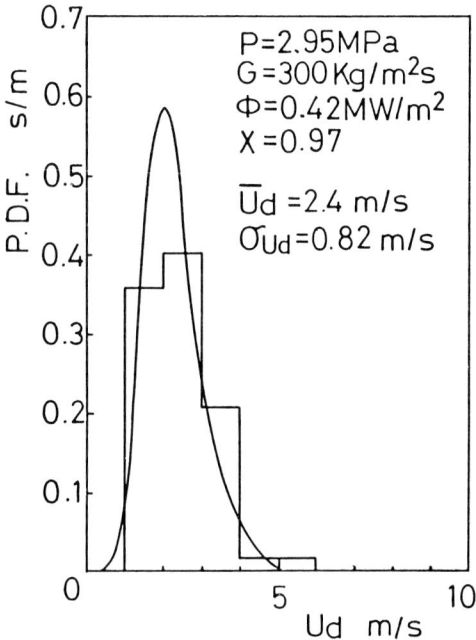

FIGURE 6. Probability density distribution of propagation velocity of disturbance wave (Steam-water).

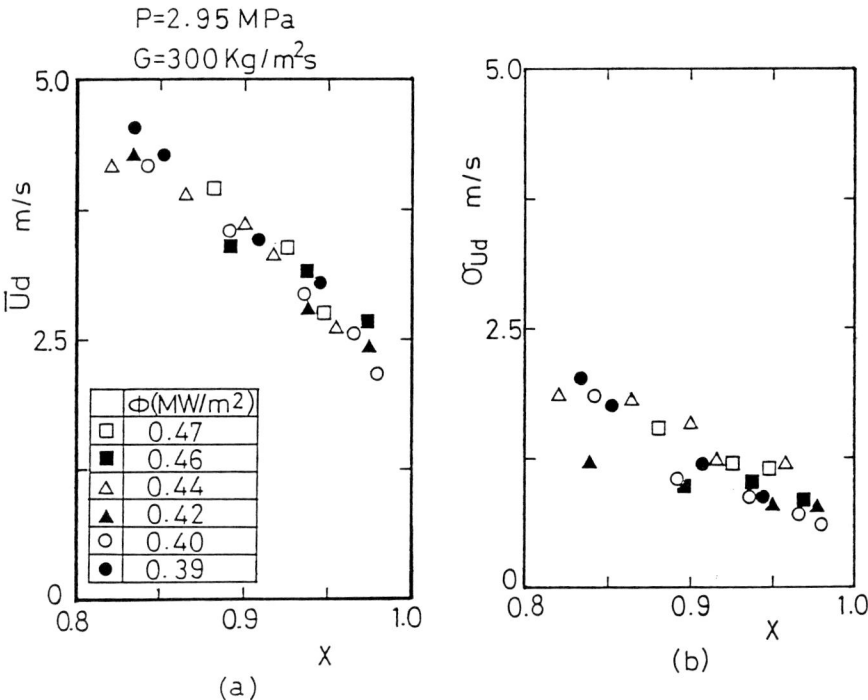

FIGURE 7. Propagation velocity of disturbance wave. (a) Mean value; (b) standard deviation.

DISTURBANCE WAVE IN BOILING FLOW

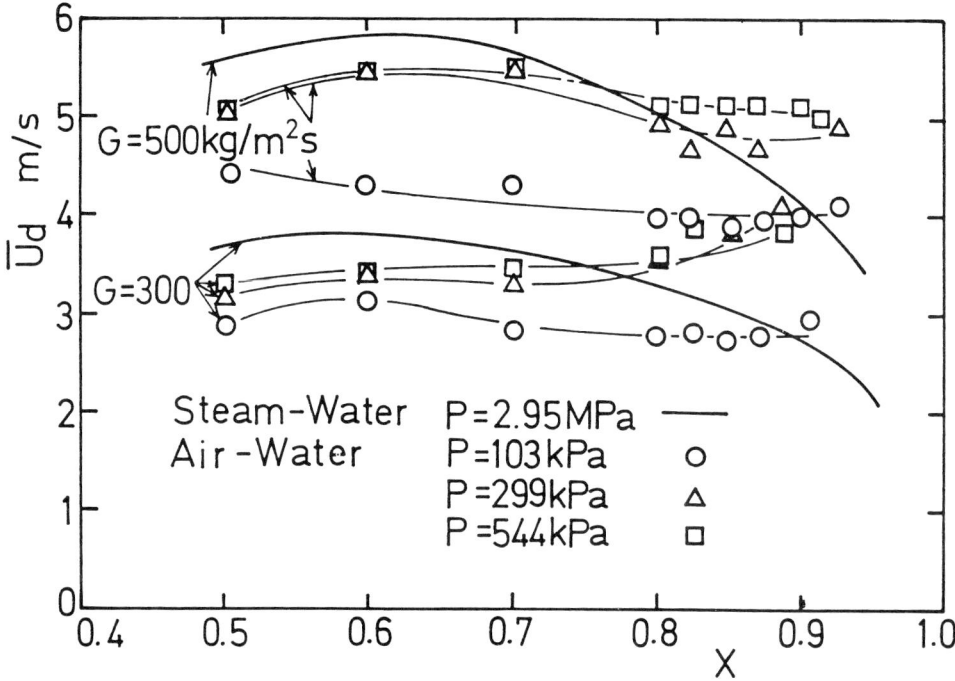

FIGURE 8. Mean propagation velocity. Comparison of boiling flow data with adiabatic ones.

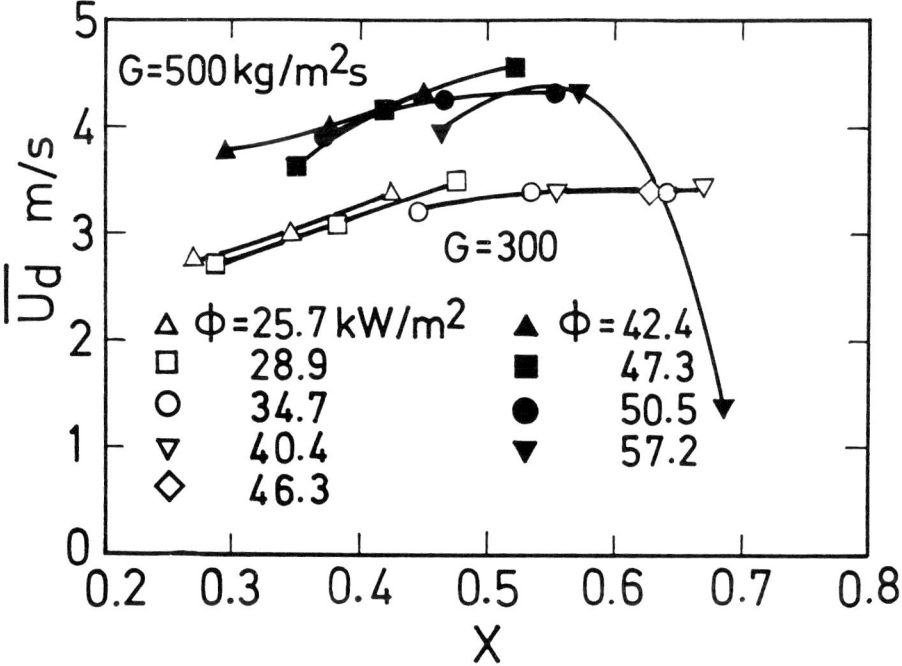

FIGURE 9. Mean propagation velocity (Freon-113).

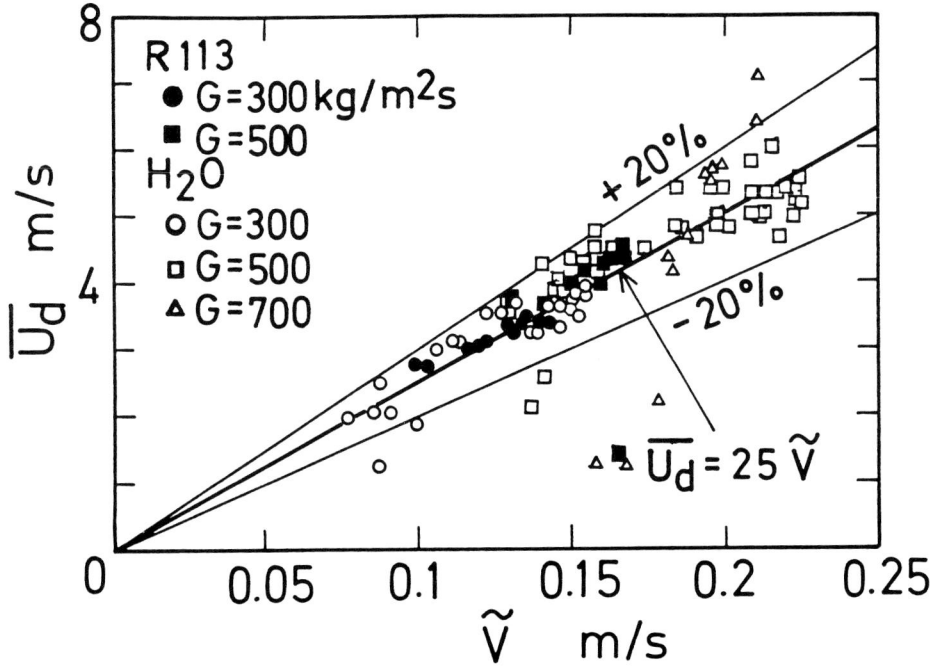

FIGURE 10. Correlation of mean propagation velocity in boiling flow.

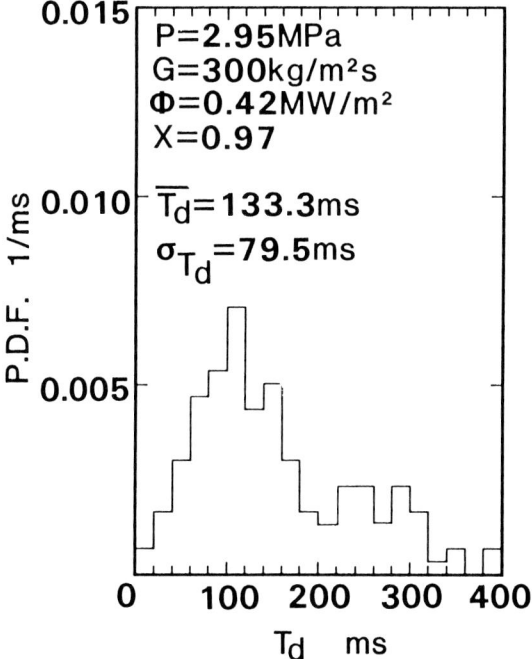

FIGURE 11. Probability density distribution of time separation of disturbance wave (Steam-water).

DISTURBANCE WAVE IN BOILING FLOW

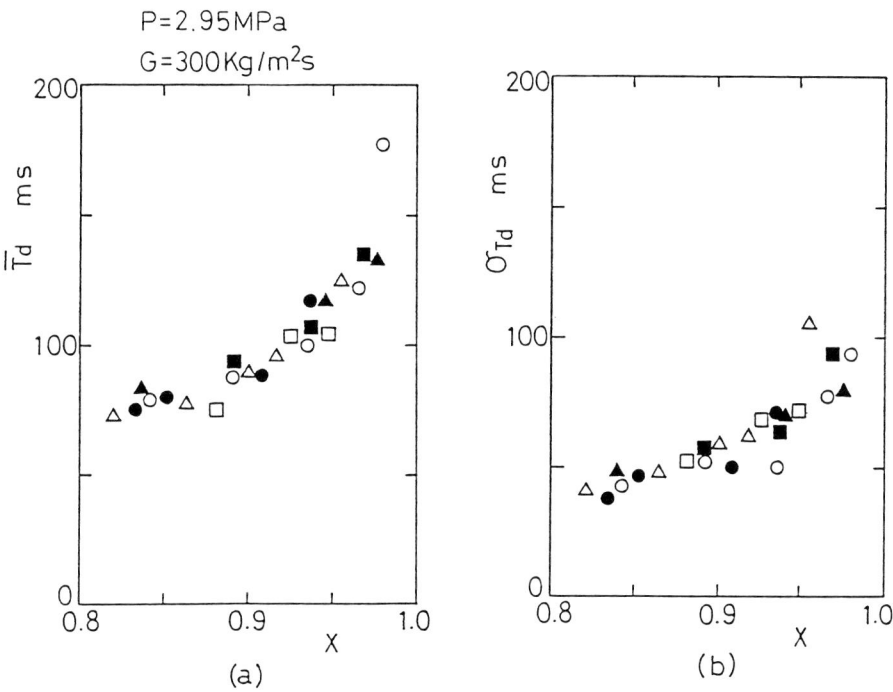

FIGURE 12. Time separation of disturbance wave (Steam-water). (a) Mean value; (b) standard deviation.

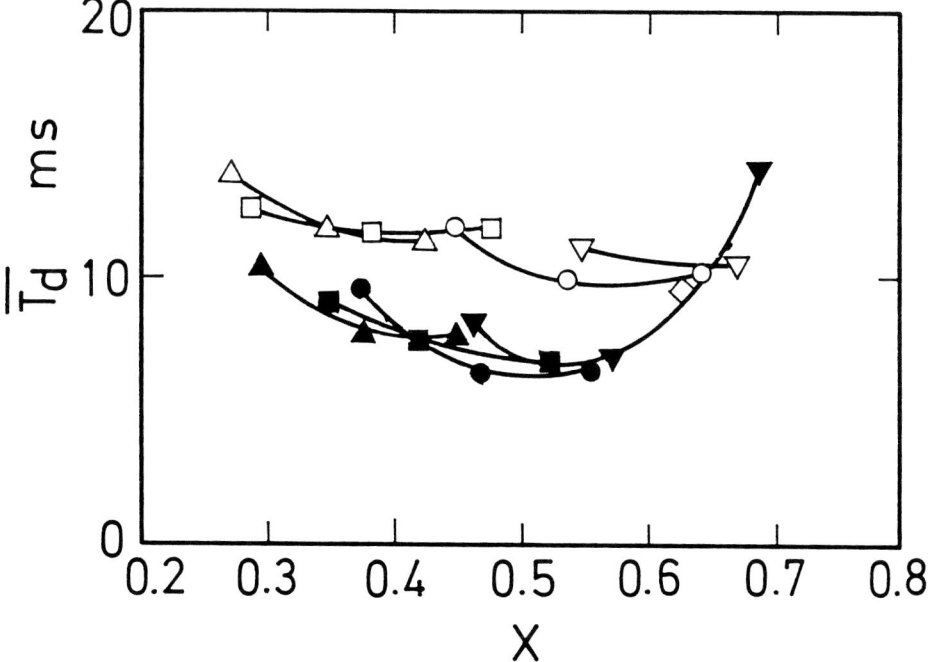

FIGURE 13. Mean time separation (Freon-113). Legends are given in Fig. 9.

Spatial Wave Separation

   Here, we define the mean spatial wave separation as the product of the mean propagation velocity and the time separation. The data for boiling water flow are plotted in Fig. 14, where those of air-water systems are also included for comparison. A difference in the behavior in the neighborhood of the dryout region is clearer. The boiling flow seems not to have suffered from the effect of decrease in liquid flow rate even at the highest quality. Figure 15 is the Freon-113 data, showing the similar tendency.

NON-EQUILIBRIUM DUE TO BOILING

Comparisons of Adiabatic and Boiling Flow

   Comparing the disturbance wave data from boiling flow with those from adiabatic air-water flow, we found significant differences between them even when the same density ratio of gas and liquid phases was realized in both the flows. This was done by adjusting the system pressure of the air-water system. For example, the mean spatial wave separation (see Fig. 14) exhibits totally different behavior at the high quality region in the neighborhood of the dryout point. In the boiling steam-water system, the wave separation takes a constant approximate value (0.3-0.4 m), which suffers weak influence of the mass flux, irrespective of the quality even beyond the dryout point. On the contrary, in the adiabatic air-water system, the value begins to increase drastically at a certain quality, corresponding to the sharp increase in the time separation. Also the effect of the mass flux was rather remarkable. A similar situation was observed in the time separation. However, with the propagation velocity, the air-water system takes a nearly constant value irrespective of the quality, while the boiling water flow shows a decrease with the quality (see Fig. 8). As for the ratio of the standard deviation to the mean value, that of the air-water flow is much smaller for the propagation velocity but significantly larger for the time separation compared to that of the boiling water flow. These differences cannot be interpreted as only the effect of the difference in physical properties. The evidence suggests existence of a new type of hydrodynamic non-equilibrium due to flow boiling, other than the imbalance of distribution of the liquid flow between the film and entrainment, which was proposed by the Harwell group(Tong et al., 1972). We are conducting experimental studies, described below, to identify this new type of non-equilibrium.

Recovery of Equilibrium in Non-Heated Channel

   The experimental works were conducted with the boiling steam-water system, and the test section illustrated in Fig. 3 was used to investigate the recovery process, in which the hydrodynamic non-equilibrium due to boiling attenuated at the non-heated section located just downstream of the heated section. The length of the non-heated section was varied between 35 and 212 diameters of the test section.

DISTURBANCE WAVE IN BOILING FLOW

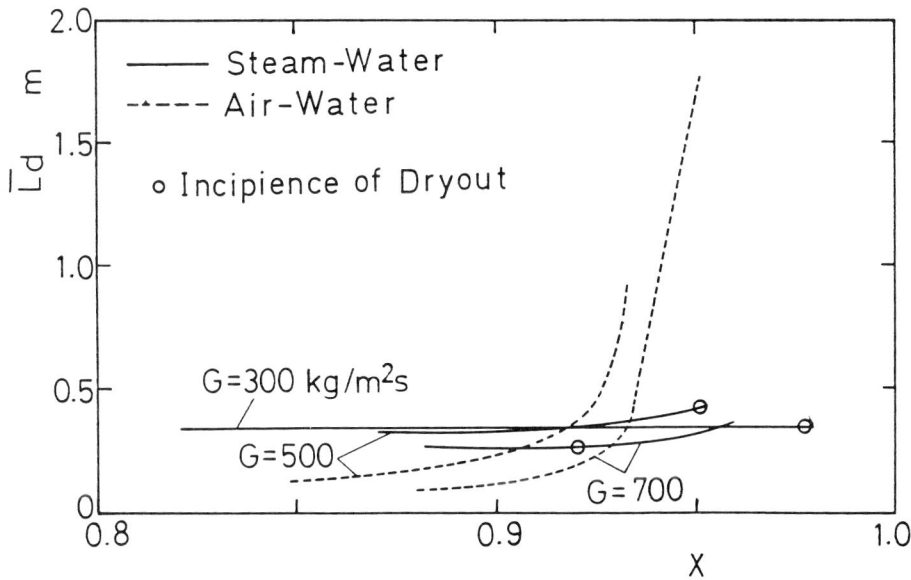

FIGURE 14. Mean spatial wave separation. Comparison of boiling flow data with adiabatic ones.

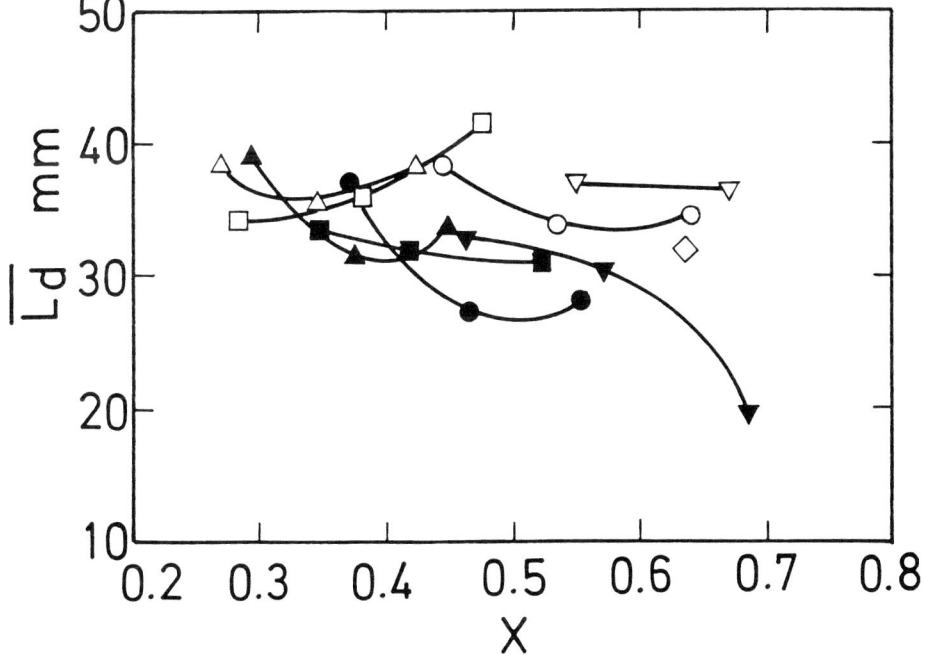

FIGURE 15. Mean wave separation (Freon-113). Legends are given in Fig. 9.

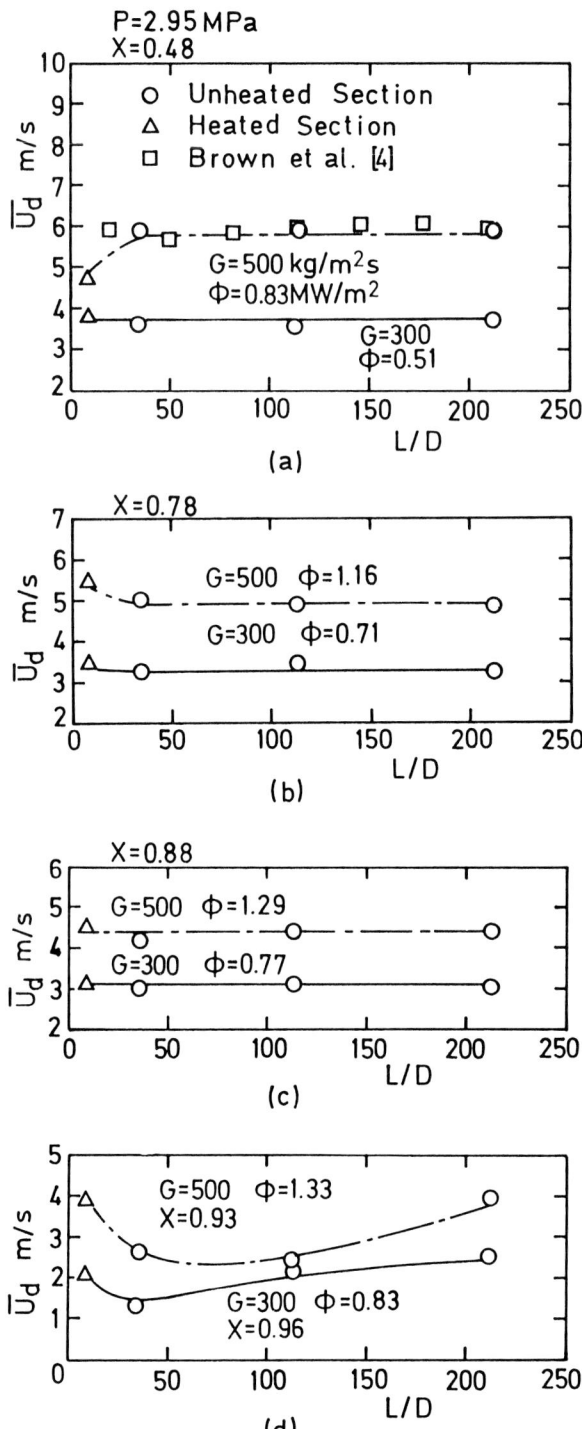

FIGURE 16. Variation of propagation velocity along non-heated channel (Steam-water).

The mean value of the propagation velocity is shown against the length of the non-heated section in Fig. 16. In Fig. 16a, the data points from the paper by Brown et al. (1975) are plotted for reference; their condition was the mass flux 297 kg/m^{2}s and the quality 0.53. The propagation velocity seems to take its final value after a very short non-heated section except for the case where the dryout occurred at the exit of the heated section (Fig. 16d). As seen from Fig. 16a, this agrees with the observation by Brown et al., who attributed its reason to their experimental condition which realized an equilibrium condition by chance. But this was not the case for the conditions corresponding to Fig. 16b and 16c. The effect of the non-equilibrium appears in the standard deviation, as shown in Fig. 17; it decreases along the non-heated section approaching that of the adiabatic air-water system.

The time separation is shown against the length of the non-heated section in Fig. 18. The effect of the non-equilibrium can be easily understood through replotting this against the quality, taking the length of the non-heated length as a parameter, as in Fig. 19. As the length increases, the form of the curve against the quality approaches that for the adiabatic air-water system, suggesting attainment of equilibrium state. This becomes more clear when the mean spatial wave separation is plotted (Fig. 20).

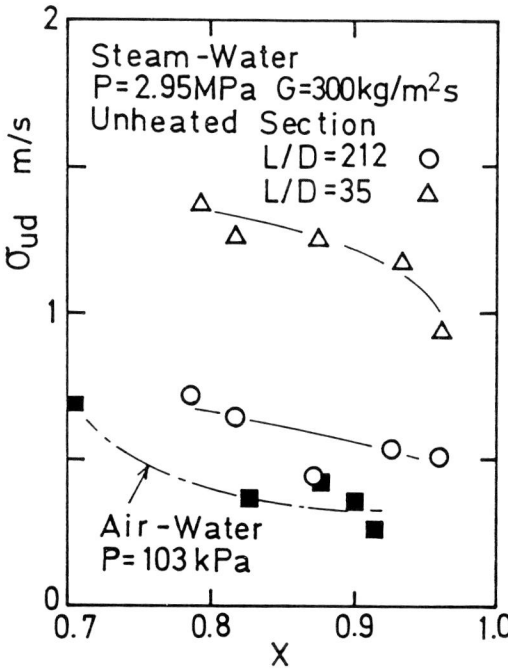

FIGURE 17. Variation of standard deviation of propagation velocity along non-heated channel (Steam-water).

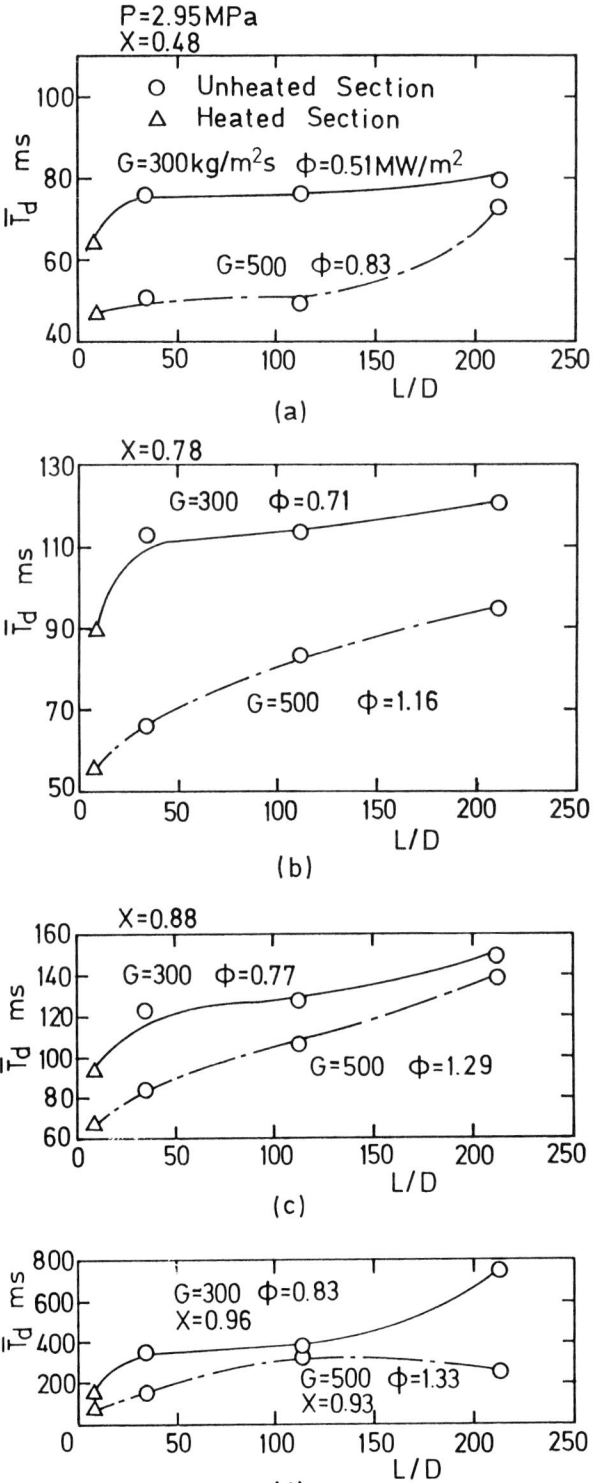

FIGURE 18. Variation of time separation along non-heated channel (Steam-water).

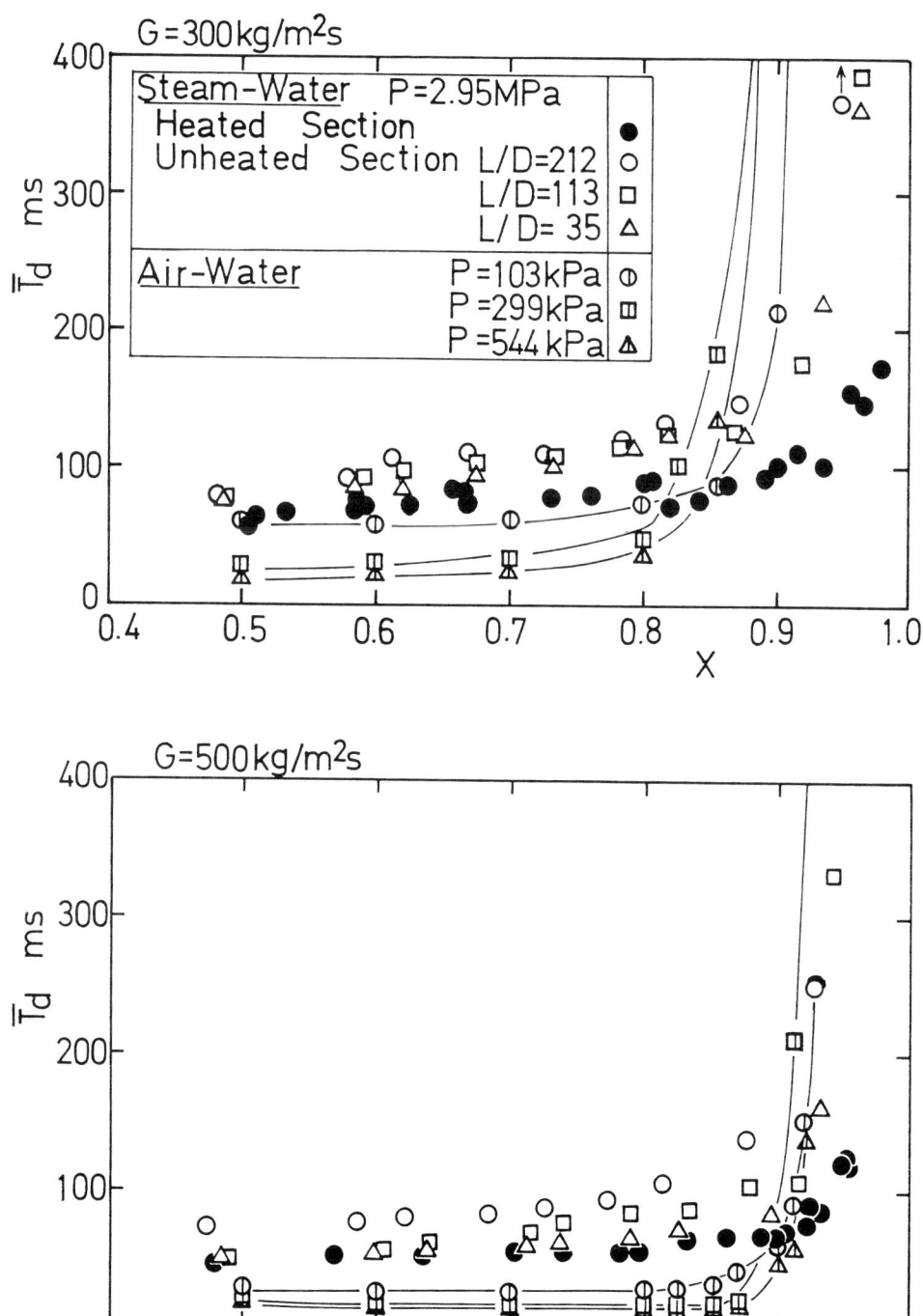

FIGURE 19. Mean time separation. Comparison of boiling flow data with adiabatic ones.

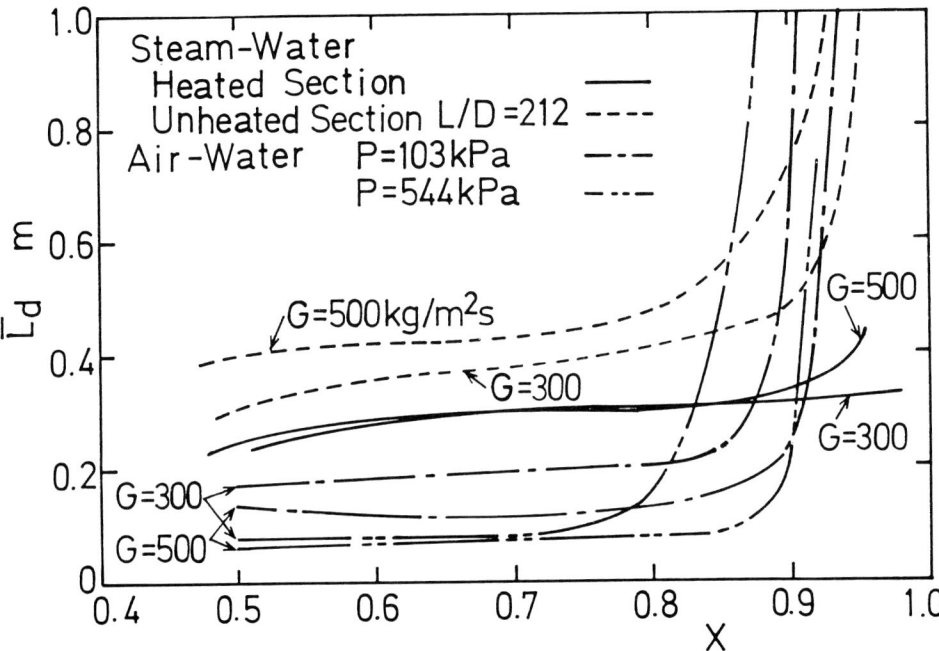

FIGURE 20. Mean wave separation. Comparison of boiling flow data with adiabatic ones.

Results from Visualization

To identify the true character of the non-equilibrium concerned, a visualization study was conducted at the test section shown in Fig. 4. Figure 21 is a typical example of sequence photographs of the behavior of the disturbance waves in the heated channel. The shadow rings indicate the disturbance waves. Each width presents a measure of each wave, so development of the disturbance waves in the heated boiling channel can be investigated by tracing this quantity in detail. Hereafter it is called "wave width" and is denoted W_d. Figure 22 shows the wave width against the quality. Naturally the former decreases with the latter, but given the quality and the mass flux, the higher the heat flux, the larger the wave width. This also suggests existence of the hydrodynamic non-equilibrium, because the heat flux must govern its intensity. The higher the heat flux, the shorter the distance corresponding to the same quality range and, in turn, the stronger the upstream effect at the same quality. In Fig. 23 the propagation velocity is plotted against the wave width; these two quantities clearly have a positive correlation.

General Discussions

From the results presented above, existence of the new type of hydrodynamic non-equilibrium was established. The notion of the entrainment-film flow non-equilibrium predicts slower propagation of the density wave in high quality

DISTURBANCE WAVE IN BOILING FLOW

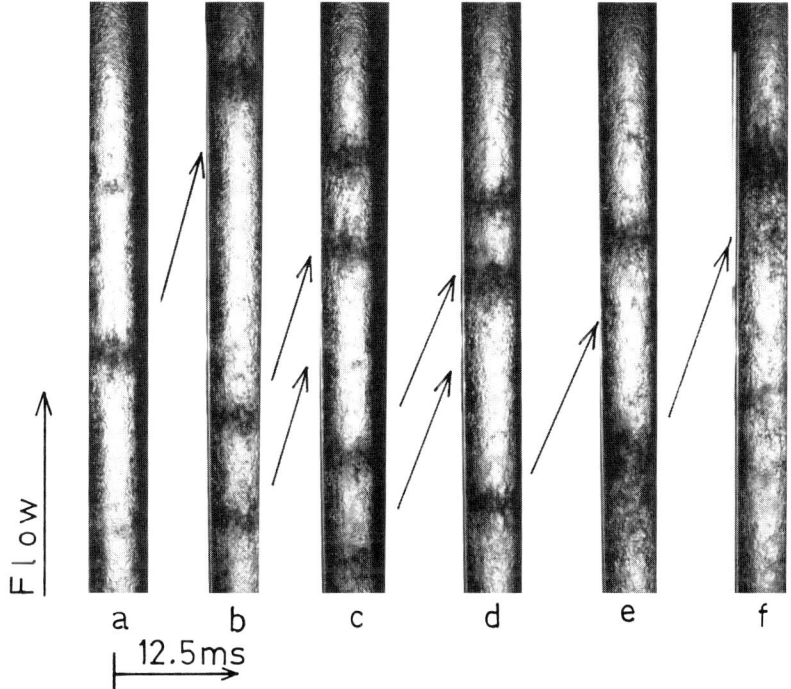

FIGURE 21. Sequence photographs of disturbance waves in boiling flow (Freon-113). $G=300$ kg/m^{2}s; $\phi=28.9$ kW/m^2; $x=0.48$; $P=162$ kPa.

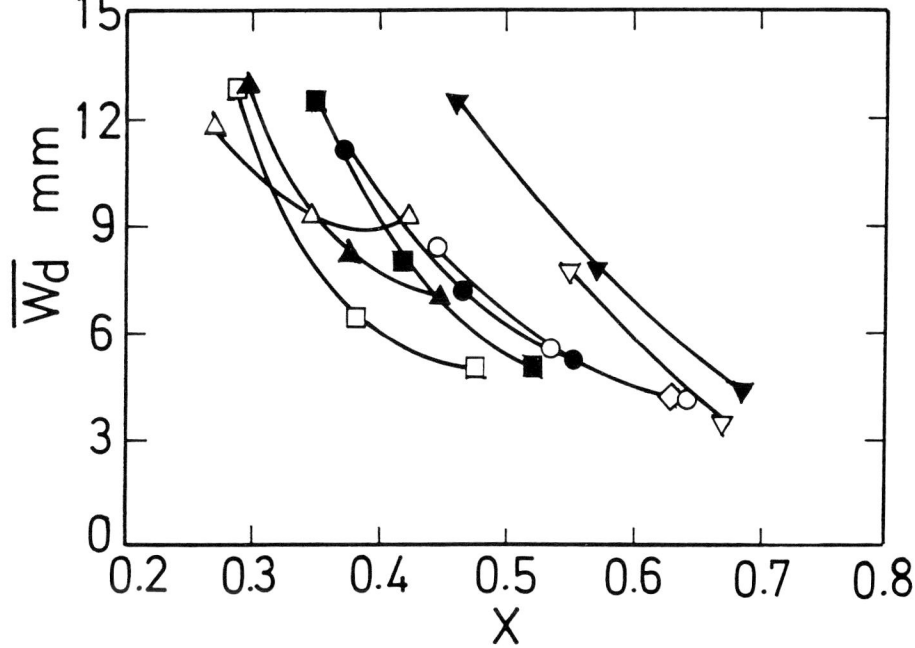

FIGURE 22. Mean width of disturbance wave (Freon-113). Legends are given in Fig. 9.

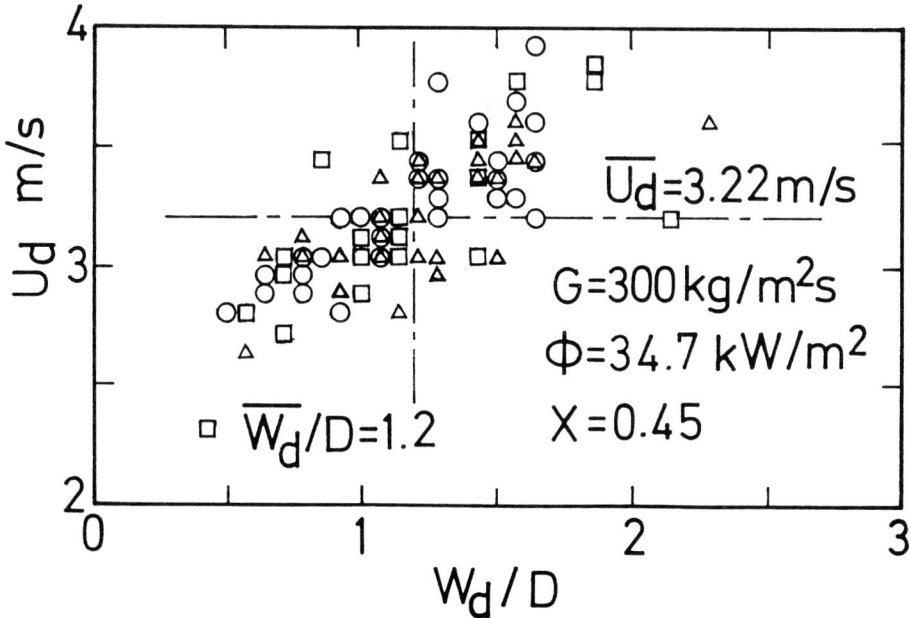

FIGURE 23. Correlation between propagation velocity and wave width (Freon-113).

boiling flow whose film flow rate must be smaller than that of the equilibrium flow. The only possible mechanism is non-equilibrium distribution of liquid between the base film and the body of the disturbance wave. In a high quality region, the mass of the disturbance wave traveling over the base film is more excessive in the boiling flow than in the equilibrium flow, because the film is made thinner along the channel by evaporation. If this non-equilibrium flow enters a non-heated channel, the disturbance wave in non-equilibrium state advances along the channel, feeding its excessive liquid to the base film and recovering the equilibrium state. In this process, the disturbance wave attenuates and the time separation of the waves increases along the channel. This picture well explains the variation of the film thickness recordings along the non-heated channel following a boiling channel(Fig. 24).

DISTURBANCE WAVE IN DRYOUT REGION

Results from Film Thickness Measurement

In order to clarify the relation between the disturbance wave and the wall temperature fluctuation in the dryout region, a number of special runs were conducted with the test section in Fig. 1. Dryout was deliberately realized in the heated section II, located downstream of the conductance probe block.

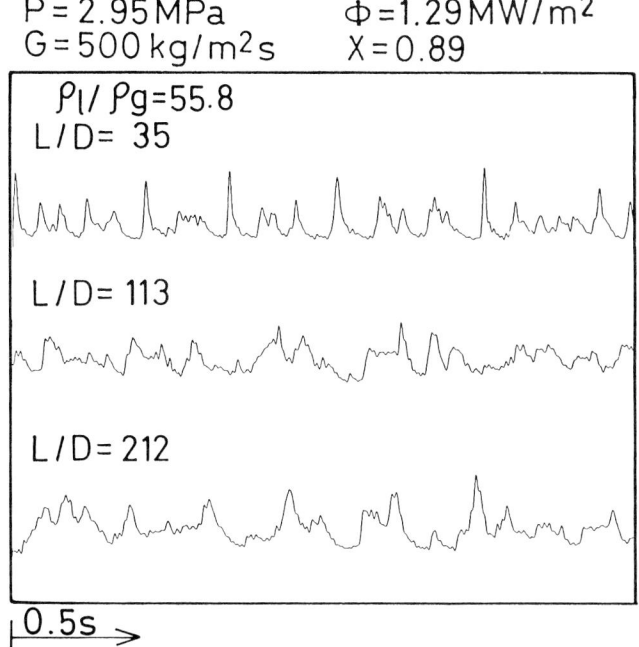

FIGURE 24. Change of wave profile along non-heated channel (Steam-water).

Figure 25a is a pair of simultaneous recordings of the liquid film thickness at the probe block and at the wall temperature at a point 20 mm downstream of the inlet of the section II. In this case the dryout point was set at just downstream of the inlet of the section II. It can be observed that when a wave packet comprising several large disturbance waves arrives, the wall temperature drops. If the wave or wave packet is small, or if the time interval between waves is long, the temperature rises. Figure 25b shows a case where the dryout point is set just upstream of the probe block. Here, arrival of disturbance waves is more sparse than in the above case, so that the correspondence of the temperature drop with the wave arrival is more clear. This strongly suggests that the wall surface in the post-dryout surface is intermittently quenched by liquid mass fed by disturbance waves, which supports our previous conjecture(Nakanishi et al., 1982a).

Results from Visualization

Behavior of disturbance waves in the dryout region is now in progress at the visualization test rig illustrated in Fig. 4. Figure 26 is a typical example of sequence photographs showing movement of disturbance waves. As seen from Fig. 26, the dryout front is washed away and is driven to advance downstream by the disturbance waves arriving there. In this period, neither any film break-up nor any dry-patch is observed on the liquid film. When the dryout front moves too far downstream to be fed a quantity of liquid

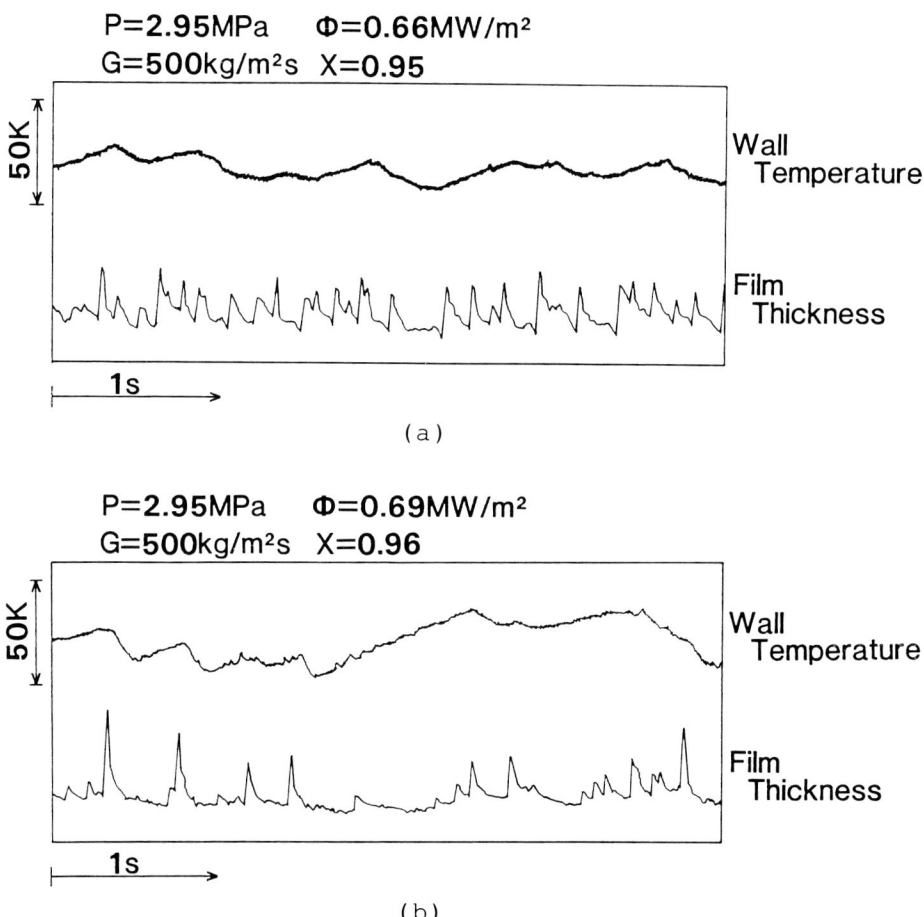

Figure 25. Correspondence of wall temperature peaks with passing of disturbance waves (Steam-water).

enough for continuing its advance, it recedes more quickly than in the advancing phase and sometimes even instantaneously. In the latter case, the liquid film disappears at nearly the same time over some length from the dryout front, which makes it difficult to define its receding velocity. In this manner, the dryout front repeats its up- and downstream movement. Our earlier preliminary investigation with a rectangular channel(Nakanishi et al., 1982b) gave the similar result. In conclusion, the picture of the dryout phenomenon obtained here supports the conclusion for the steam-water system derived in the preceding section.

Velocity of Dryout Front

The velocity, U_f, of the dryout front was determined from the visualization above described. Figure 27 shows an example of variation of the front velocity against the time lapse. Positive values correspond to the advancing phase and

DISTURBANCE WAVE IN BOILING FLOW

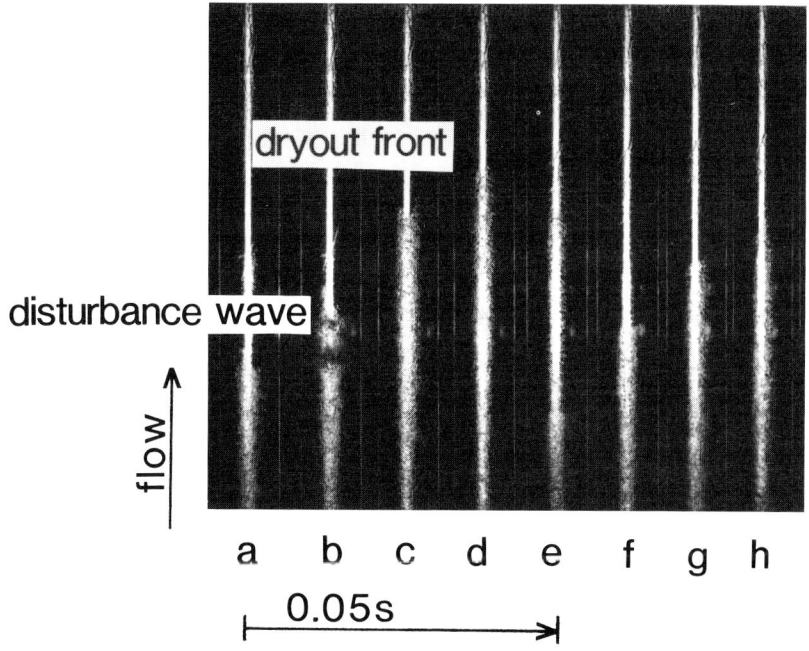

FIGURE 26. Disturbance wave in dryout region (Freon-113). $G=300$ kg/m^{2}s; $\phi=47.2$ kW/m^2; $x=0.80$; $P=176$ kPa.

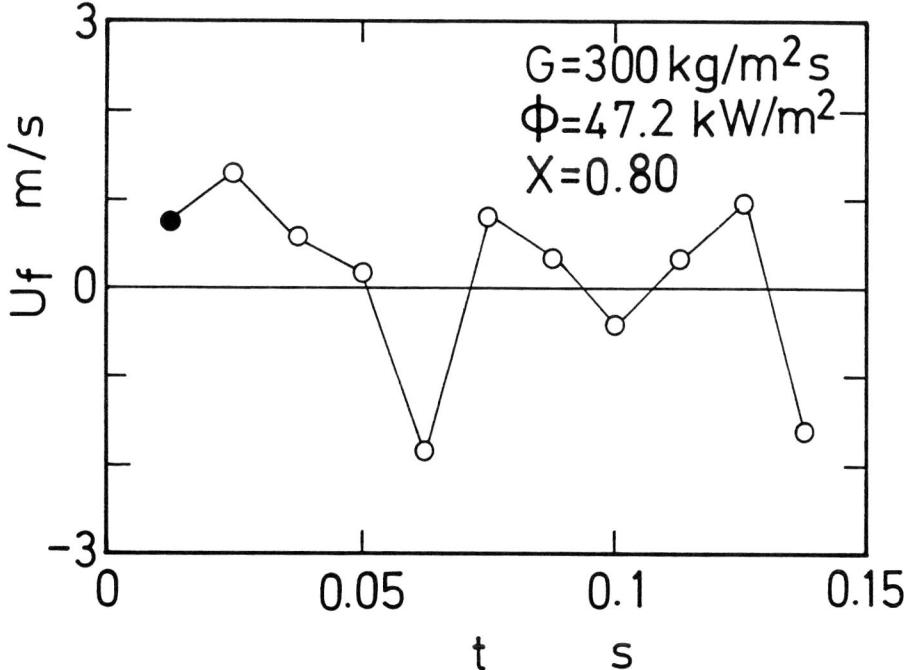

FIGURE 27. An example of variation of velocity of dryout front (freon-113).

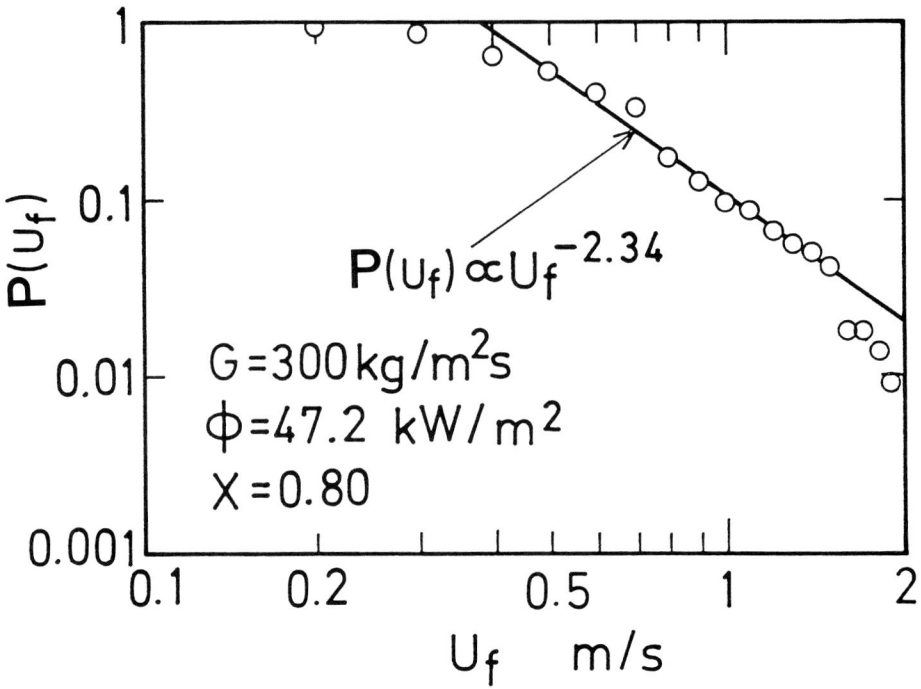

FIGURE 28. An example of probability function of advancing velocity of dryout front (Freon-113).

negative ones to the receding one. The latter was less accurate than the former, as stated above. The probability density distribution of the front velocity was obtained for the advancing phase only because of low accuracy in the data for the receding phase. As regards the data obtained in this study, probability, $P\{U_f\}$, in which the front velocity is larger than U_f is in inverse proportion to D-th power of U_f, where D is a fractal dimension(Mandelbrot 1983). Figure 28 shows a typical plotting, where D is somewhat larger than 2. The value of D needs more detailed investigation.

EXPERIENCES WITH ARTIFICIAL HYDRODYNAMIC NON-EQUILIBRIUM

Velocity Transient in Gas Phase

Stimulated by the study of the premature and temporal dryout triggered by a sudden decrease of flow rate, particularly of steam feed to an evaporating tube(Ishigai et al., 1974), we investigated transient flow in a disturbance wave regime in a parallel air-water shear flow(Nakanishi et al., 1979). Experimental works were conducted with a transparent rectangular channel, 200 mm in width, 20 mm in height, and 1.9 m in length. Flow transient was triggered by a stepwise change in gas flow rate; consequently a hydrodynamic non-equilibrium state was realized for a moment. For the case of gas flow increase, the transient process was the blow-off of the liquid hold-up which becomes excessive in the after-transient condition, while in case of gas flow decrease it was the recovery of the liquid hold-up by propagation of the after-transient thick liquid film. In the former case, the transient was much shorter and the disturbance wave did not suffer from any significant modification due to the transient. In the latter case, the disturbance wave completely died out in the transient period according to the experimental conditions. The recovering velocity was successfully analyzed with the kinematic wave theory(Lighthill et al., 1955), though the effect of the disturbance wave was significant if the surface wave on the film did not completely die out. In the both cases, recovery from the non-equilibrium state was relatively quick.

Liquid Film Sucking

Another experimental study(Sawai et al.,1989) was conducted with the same apparatus used above. To create non-equilibrium liquid distribution between the base film and the disturbance waves, the liquid film is partly sucked through a porous wall located in the middle of the test section bottom. Film thickness was measured by a number of conductance probes along the test channel, to observe the recovering process of non-equilibrium. It was confirmed that the liquid film acquired its equilibrium thickness when supplied with liquid from the rear sides of disturbance waves traveling over it. As a result, the disturbance waves attenuated or died out according to circumstances. The decaying speed is relatively low. These results fit in well with our notion of the hydrodynamic non-equilibrium associated with boiling.

CONCLUSIONS

Experimental investigation of the disturbance wave in flow boiling revealed its definite difference from that of the corresponding adiabatic flow. This was attributed to the hydrodynamic non-equilibrium other than in liquid distribution between the liquid film and the entrainment, i.e., the non-equilibrium in liquid distribution between the base film and the bodies of the disturbance wave. The growth and decay of the disturbance wave along the boiling channel and along the adiabatic channel were observed, so as to make clear a number of interesting properties of the disturbance wave. The role of the disturbance wave in the dryout region was also clarified; the wall temperature fluctuation can be predicted, provided that our knowledge about the properties of the disturbance wave, including stochastic properties, is made sufficient. Study in this direction is continuing.

NOMENCLATURE

English

D   inner diameter of channel tube [m]
D   fractal dimension [-]
G   mass flux [$kg/m^2 s$]
L   distance from inlet of non-heated section [m]

L_d spatial wave separation of disturbance wave [m]
P pressure [MPa]
$p\{.\}$ probability density distribution [-]
T_d temporal wave separation, or time separation of disturbance wave [ms]
U_d propagation velocity of disturbance wave [m/s]
U_f velocity of dryout front [m/s]
$\dot{U}_{Wd}$ rate of width change of disturbance wave [m/s]
$\tilde{v} = \tau_i/G$ equivalent velocity [m]
W_d width of disturbance wave [m]
x thermodynamic equilibrium quality [-]

Greek

σ standard deviation [-]
τ_i interfacial shear stress [Pa]
ϕ heat flux [W/m^2]

Subscripts, Superscripts, and Other
   —   mean value

REFERENCES

Brown, D. J., Jensen, A., and Whalley, P. B., 1975, Non-Equilibrium Effects in Heated and Unheated Annular Two-Phase Flow, A.S.M.E. Paper, 75-WA/HT-7.

Hall-Taylor, N. S., and Nedderman, R. M., 1968, The Coalescence of Disturbance Waves in Annular Two-Phase Flow, Chem. Eng. Sci., vol. 23, pp. 551-564.

Henstock, W. H., and Hanratty, T. J., 1970, The Interfacial Drag and the Height of the Wall Layer in Annular Flows, A. I. Ch. E. Journal, vol. 26, pp. 990-999.

Hewitt, G. F., and Hall-Taylor, 1970, Annular Two-Phase Flow, Pergamon Press, pp. 98-127.

Ishigai, S., Nakanishi, S., Yamauchi, S., and Masuda, T., 1974, Effect of Transient Flow on Premature Dryout in Tube, Proc. 5th International Heat Transfer Conf., vol.4, pp. 300-304.

Kawata, S., and Minami, S., 1984, Adaptive Smoothing of Spectropic Data by Linear Mean Square Estimation, Applied Spectroscopy, vol. 38, no. 1, pp. 49-58.

Lighthill, M. J., and Whitham, G. B., 1955, On Kinematic Waves: I. Flood Movement in Long Rivers, II. Theory of Traffic Flow on Long Crowded Roads. Proc. Roy, Soc. London, Ser. A, vol. 229, pp. 281-345.

Mandelbrot, B. B., 1983, The Fractal Geometry of Nature, originally published by W. H. Freeman and Company.

Nakanishi, S., Ishigai, S., and Yamauchi, S., 1979, Transient Behavior of Two-Phase Shear Flow, Two-Phase Momentum, Heat and Mass Transfer in Chemical, Process, and Energy

Engineering Systems, vol. 1, Hemisphere Publishing Corp., pp. 315-326.

Nakanishi, S., Yamauchi, S., Ishigai, S. and Kotani, H., 1982a, Wall Temperature Fluctuation of the Evaporating Tube at the Dryout Region, Proc. 7th International Heat Transfer Conf., vol.4, pp. 315-320.

Nakanishi, S., Ozawa, M., Ishigai, S. and Miwa, E., 1982b, Dryout Phenomenon in Two-Phase Shear Flow, Heat Transfer in Nuclear Reactor Safety, Hemisphere Publishing Corp., pp. 487-498.

Nakanishi, S., Kaji, M., Yamauchi, S., Kazuoka, Y., and Sawai, T., 1984, Wall Temperature Fluctuations of Evaporating Tube at Onset of Dryout (2nd Report, A New Model of Fluctuation Generating Mechanism), Trans. J.S.M.E., Ser. B, vol. 50, no. 452, pp. 1152-1158.

Nakanishi, S., Kaji, M., Yamauchi, S., and Sawai, T., 1986, An Experimental Study on Disturbance Waves in Boiling Two-Phase Flow, Heat Transfer:Japanese Research, vol. 15, no. 3, pp.87-101; originally published in Trans. J.S.M.E., ser. B, vol. 51, no. 463(1985a), pp. 1026-1031.

Nakanishi, S., Kaji, M., Yamauchi, S., Kazuoka, Y., and Sawai, T., 1985b, Behavior of Disturbance Waves in the Dryout Region, PhysicoChemical Hydrodynamics, vol. 6, no. 1/2, pp. 157-164.

Nakanishi, S., Kaji, M., Yamauchi, S., and Sawai, T., 1987, Experimental Studies of Disturbance Waves in Boiling Two-Phase Flow(Continued Report; Effects of Hydrodynamic Non-Equilibrium and Density Ratio), Trans. J. S. M. E., ser. B, vol. 53, no. 487, pp. 1091-1096.

Savitzky, A., and Golay, M. J. E., 1964, Smoothing and Differentiation of Data by Simplified Least Squares Procedures, Analytical Chemistry, vol. 36, no. 8, pp. 1627-1639.

Sawai, T., Yamauchi, S., and Nakanishi, S., 1989, Behavior of Disturbance Wave Under Hydrodynamic Non-Equilibrium Conditions, International Journal of Multiphase Flow, vol. 15, no. 3, pp341-356.

Tong, L. S., and Hewitt, G. F., 1972, Overall Viewpoint Flow-Boiling CHF Mechanism, A. S. M. E. Paper, 75-WA/HT-54.

Whalley, P. B., 1987, Boiling, Condensation and Gas-Liquid Flow. Oxford University Press, pp. 78-95.

TWO-EQUATION TURBULENCE MODELING OF AN UPPER-PLENUM WATER POOL ABOVE A HORIZONTAL PERFORATED PLATE WITH STEAM CONDENSATION

S.G. BANKOFF
Chemical Engineering Department
Northwestern University
Evanston, IL 60208
C.L. CHEN
Singer Corporation
Columbia, MD

ABSTRACT

The PHOENICS code is used to simulate the flow and heat transfer in an upper-plenum water pool above a horizontal perforated plate through which steam is injected. At high steam velocities no water falls downwards through the holes, but as the steam velocity is reduced, the steam region in the plenum is reduced to "flame-like" steam jets, which can be modeled as point boundary sources of momentum, mass and heat. When the length of the jets is reduced to about 2-3 hole diameters, water leakage downwards, characterized as "weeping", begins. A k-ε turbulence model was used to predict the velocities and temperatures in a narrow pool between two flat glass plates. Reasonably good agreement between calculations and experiment is shown.

INTRODUCTION

Flooding occurs in a gravity-dominated countercurrent gas-liquid flow when for a given gas flow rate the downwards liquid flow cannot be further increased, or vice versa. It depends, among other things, on the configuration of the system, inlet and outlet conditions, presence or absence of heat and mass transfer, and the physical properties of the fluids (Tien and Liu, 1979; Bankoff and Lee, 1983; Liu and Tien, 1982). The problem of particular interest in the present case is motivated by a hypothetical nuclear reactor loss-of-cooling accident in which the emergency core cooling system water is held up in the upper plenum above the upper tie plate by the steam escaping from the hot reactor core.

The upper tie plate is a horizontal perforated plate, below which are flow resistances which can be idealized as parallel channels. Several studies (Speyer and Kmetyk, 1977; Piggott and Ackerman, 1982; Wallis, et al., 1981) of air-water-flooding in parallel channels have noted the possibilities for various modes of interactions between channels. Previous work in this laboratory (Bankoff, et al., 1981; Dilber and Bankoff, 1985) and elsewhere (Merilo, et al., 1979) has dealt with steam-water flooding in and above a horizontal perforated plate. There is essentially no difference between air-water flooding and steam-water flooding if the entering water is at the saturation temperature. However, as the inlet water subcooling increases from zero, condensation effects play in an important role in determining top-dominated flooding (although not in bottom-dominated flooding, where the exiting water is always nearly saturated). The horizontal perforated plate with flow channels beneath it is an important example of a top-dominated flooding geometry. As the steam flow rate is reduced, keeping the inlet water flow rate and subcooling constant, the mass velocity of steam upwards through the pool decreases towards zero. In the process the plenum pool changes from a violently mixed two-phase region to an essentially single-phase pool, with "flame-like" steam jets issuing from the bottom holes. When the steam jet length is more than about three hole diameters, the jet outline appears to be stationary to the unaided eye. The flow of steam through the hole is then exactly matched by condensation at the jet surface. However, as the flow of steam is reduced further, the jet length decreases, and begins to oscillate. When the jet length is, on the average, one or two hole diameters, the amplitude of the oscillations increases, and water begins to leak downwards through the hole. This is termed "weeping" in the chemical engineering literature. As the steam flow is further reduced, the jet becomes unstable, and liquid water pours through the hole, condensing the up-flowing steam below the plate. At this point, termed "dumping", the entire pool is unstable, and drains into the lower plenum.

The major interest in this work was the onset of weeping, which, in a nuclear reactor accident, would signal the point at which water begins to penetrate into the overheated reactor core. Data and correlations of this type have already been reported (Bankoff, et al., 1981; Dilber and Bankoff, 1985). Additional data on weeping with flow tubes below the perforated plate have been obtained (Chen 1989). However, because of the small scale of the apparatus and the low pressure, it is difficult to determine the reliability of these correlations for other geometries at high pressure and much larger scales, as in an reactor accident. It seems possible that the plenum hydrodynamics just before weeping might be successfully modeled by a single-phase k-ϵ turbulent transport model, in which the small bottom steam jets are replaced by boundary sources of heat, mass and momentum. While no temperature traverses of the pool just prior to weeping were taken, for the above reasons, it is possible to obtain qualitative agreement with observations from the location of the pool isotherms. Calculations were performed for 48 runs, covering a range of inlet water locations, delivery directions (horizontal vs. vertical downwards), and temperature, as well as plate perforation ratio. All show the same general features. Far from the onset of weeping, the water in the near vicinity of the plate is saturated. However, as the experimental weeping point is approached, cold water begins to penetrate downwards. At the onset of weeping, subcooled water penetrates to within three hole diameters of the plate, in agreement with experiment. Only one set of typical calculations is given here.

TWO-EQUATION TURBULENCE MODELING

EXPERIMENTAL APPARATUS

A schematic diagram of the experimental apparatus is shown in Fig. 1. Cold water coming from the building supply was used either directly or fed through a heat exchanger which controlled the reserve tank water temperature. Steam from the building supply at about 0.9 MPa was passed through a separator and throttling valve to produce saturated steam for the test channel. The steam flow rate was measured by either of two venturis with nominal diameters of 31.8 and 50.8 mm. The test section consisted of a vertical aluminum channel 2.14 m long and 280 x 97 mm. in cross-section, fitted with a horizontal brass perforated test plate (Fig. 2). Cold water entered the upper chamber through a vertical tube, whose height above the test plate could be varied. Vertical downwards injection was obtained by using an open-end tube, and horizontal injection by a closed-end tube with holes drilled in the wall near the end.

Steam entered the lower chamber through two ports feeding a horizontal pipe with a number of equally-spaced downward-facing holes. Two side-wall ports near the middle of the upper chamber removed the water overflow, while the uncondensed steam, if any, was vented at the top. Temperatures were measured by six thermocouples. The perforated plate, located near the center of the test section, had drilled holes (Fig. 3) to which were attached pyrex glass tubes. Their length could be varied to provide an adjustable flow resistance.

EXPERIMENTAL PROCEDURE

The geometrical parameters which were varied included the water injection direction and height above the plate, the tube length, and the numbers and spacing of the holes. These included both horizontal and vertical directions; injection heights of 0.178 and 0.356 m; tube lengths of 0.406 m, 0.203 m, and 0.102 m; and two plate configurations, one with 12 tubes and the other six tubes in two rows.

A stable two-phase pool was initially established by setting a high steam flow rate, followed by injection of water at a fixed flow rate and temperature. At this stage there was no water within the tubes, and the condensing steam jets were stable. The steam flow rate was then gradually reduced until collapse of one or more steam jets occurred, at which time water penetration downwards began. Upon further reducing the steam flow rate, the 100% penetration point (dumping) was attained, in which the entire upper pool suddenly drained through the perforated plate. The data were replicated several times for each set of conditions.

EXPERIMENTAL RESULTS

Visual Observations

Under different experimental conditions various phenomena were observed. Violent condensation and rapid vapor bubble collapse took place at low water injection temperatures and large water flow rates. With horizontal water injection, penetration began at the tubes closest to the outer channel wall, while for vertical water injection, penetration began at the tubes closest to the channel center. This corresponded in each case to the location of greatest thinning of the saturated water layer above the plate. The steam jet length before weeping began was about one to three tube diameters. A penetrated tube might fill with steam again if the pool began to oscillate up and down. However, rapid dumping usually followed the oscillation phase upon further reduction of the steam flow rate. On the other hand, part of the pool might refill again after dumping, so that cyclic dumping and refill might develop.

Propagation of penetration from one tube to another was observed only when the pool temperature was close to saturation, and the longest tubes were used.

Results

Water injection temperatures of $40^\circ C$, $60^\circ C$, and $80^\circ C$ were used at five water flow rates in equal intervals between 0.075 and 0.379 kg/s. Weeping and dumping data were taken for each possible combination of all parameters, for a total of 96 runs. The steam and water enthalpy injection rates, $\dot{h}_s$ and $\dot{h}_w$, both taken relative to saturated liquid, were calculated for each run. In the ideal case these two quantities are equal, corresponding a condensation efficiency of unity, implying complete condensation and saturated exit water.

Fig. 4 shows a typical $\dot{h}_s$ vs. $\dot{h}_w$ plot for the onset of downwards penetration, or weeping, at three different water injection temperatures. The intersection of these curves with the vertical axis represents the steam flow rate needed to maintain the pool above the plate in the absence of condensation. The increase in slope with increased water subcooling shows that cold water is penetrating close to the perforated plate at the onset of downwards penetration, thereby reducing the effective steam velocity through the holes. If the water inlet tube is raised, still with horizontal injection, the water subcooling effect does not appear, since cold water never reaches the plate (Fig. 5).

ANALYSIS

From the experimental results it was observed that water penetration began shortly after the steam jets became unstable and started to collapse, corresponding to complete condensation near the penetrated tubes. For this reason a single-phase flow model for the upper plenum can be used to predict the onset of weeping. For complex upper head geometries, such as in a PWR, a three-dimensional calculation would be required. However, we show here a two-dimensional calculation, for a low horizontal water injection case, using PHOENICS, a flexible multidimensional two-fluid code (CHAM, Ltd.) with numerous built-in options, developed by Spalding and co-workers (Spalding, 1981; Canton, et al., 1983; Launder and Spalding, 1972). In this code all conservation equations are expressed in the form:

$$\frac{\partial}{\partial t}(\alpha\rho\phi) + \text{div}(\alpha\rho\sigma\phi - \alpha\Gamma_\phi \text{ grad } \phi) = \alpha S_\phi \qquad (1)$$

where ϕ is any dependent variable, Γ_ϕ is an exchange coefficient, and S_ϕ is a volumetric source term. The method used for solving the finite domain equations is fully implicit, using upwind differencing. The code contains an optional k-ε two-equation turbulent transport model. The equations for k and ε are expressed as follows:

TWO-EQUATION TURBULENCE MODELING

$$\frac{Dk}{Dt} = \frac{1}{\rho} \frac{\partial}{\partial x_k} [\frac{\mu_t}{\sigma_k} \frac{\partial k}{\partial x_k}] + \frac{\mu_t}{\rho} (\frac{\partial U_i}{\partial x_k} + \frac{\partial U_k}{\partial x_i}) \frac{\partial U_i}{\partial x_k} \quad (2)$$

$$\frac{D\epsilon}{Dt} = \frac{1}{\rho} \frac{\partial}{\partial x_k} [\frac{\mu_t}{\sigma_\epsilon} \frac{\partial \epsilon}{\partial x_k}] + \frac{C_1 \mu_t}{\rho} \frac{\epsilon}{k} (\frac{\partial U_i}{\partial x_k} + \frac{\partial U_k}{\partial x_i}) \frac{\partial U_i}{\partial x_k} - C_2 \frac{\epsilon^2}{k} \quad (3)$$

Following the recommendations of Launder and Spalding (1972,1974), $C_1 = 1.44$, $C_2 = 1.92$, $\sigma_k = 1.0$, $\sigma_\epsilon = 1.3$, $C_\mu = 0.09$. For the upper plenum calculation, single-phase two-dimensional steady-state flow was assumed, and the pressure-correlation equations were solved simultaneously for all cells in the domain, instead of iterating slab by slab. Similarly, the fluid pressure, p, y-direction velocity, v, z-direction velocity, w, turbulent kinetic energy, k, dissipation rate of turbulent kinetic energy, ϵ, and specific enthalpy, h, were computed over the whole domain at each iteration. The temperature effect on the viscosity was included. The turbulent exchange coefficients for velocity and enthalpy were given by

$$\Gamma_t = \frac{C_\mu \rho k^2}{\epsilon} + \mu_\ell \quad (4)$$

$$\Gamma_h = \frac{\mu_t}{\sigma_{t,h}} + \frac{\mu_\ell}{\sigma_{\ell,h}} \quad (5)$$

where $\sigma_{\ell,h} = \frac{C_p \mu_\ell}{k}$

$\sigma_{t,h} = 0.9$

Other physical properties were assumed to be constant, except that the Boussinesq buoyancy effect was included as follows:

$$\rho = \rho_{ref} (1-\beta(h-h_{ref})) \quad (6)$$

A two-dimensional 14 x 14 nonuniform grid was used (Fig. 6) with smaller cells near the test plate. Steam was injected at three locations, corresponding to the six-hole experimental plate. The corresponding mass, momentum and enthalpy sources were included in the boundary conditions.

RESULTS AND DISCUSSION

Twelve experimental runs were simulated, but space limitations allow discussion of only one case. The convergence criterion adopted was that the sum of the absolute volumetric continuity errors for all cells was to be less than 5% of the total volumetric inflow rate.

Figures 7 and 8 show some typical velocity and temperature distributions. Cold water is injected into the domain horizontally, and separates into two streams along the wall. Two large vortices are created above and below the incoming water jet. This was confirmed by experimental observation. The corresponding temperature distribution in the pool show that cold water arrives first at the outer edge of the plate. The saturated region close to the center of the channel can be explained as a two-phase flow region.

Figure 9 shows the turbulent kinetic energy and dissipation rate distributions. Both show a maximum near the water jet entrance, which is consistent with previous work (Launder and Spalding 1972). The saturated water layer at the bottom is always thinnest near the outer wall. Hence, the criterion chosen for the prediction of the onset of weeping for zero downwards injection velocity is that the temperature at the third cell above the outermost steam inflow hole becomes less than the saturation temperature. The center of the third cell is 23mm above the test plate, which is about two tube diameters. Based on our experimental observations, the steam jet length is about two to three diameters when it becomes unstable. Using this criterion, Figs. 10 and 11 show the calculated and predicted curves of weeping for several injection subcoolings and flow rates. One sees that higher steam flow rates are needed for lower water injection temperatures, as confirmed by experimental data.

A comparison of the predicted and experimental steam flow rate for onset of weeping is shown in Fig. 12 for the six-hole plate with low and horizontal water injection. The agreement is quite good, although the predicted values are lower than the experimental data at high steam flow rates, probably because a fixed steam jet length was chosen. Actually, the steam jet length should be proportioned to the steam flow rate. If this effect were included, the prediction would be better.

CONCLUSIONS

Countercurrent flow limitation of steam and water through a perforated plate is controlled by the condensation of steam in the upper plenum. Water subcooling is an important factor in determining the water penetration. When a large amount of cold water arrives at the perforated plate, the whole pool begins to oscillate, and penetration occurs for all the tubes. The tube length below the test plate has little influence for all the tubes, except possibly in the propagation to neighboring tubes when the pool is close to saturation. The k-ε turbulent model gave qualitative agreement with experiment, which is rather remarkable, considering the complexity of the geometry and the simple flows from which the model parameters were derived.

ACKNOWLEDGEMENT

This work was supported by the U.S. Nuclear Regulatory Commission under Grant No. NRC-G-04-81-020. We thank Professors M. C. Yuen and R. S. Tankin for their help in the design and construction of this equipment.

REFERENCES

1. C.L. Tien and C.P. Liu, (1979) "Survey on Vertical Two-Phase Countercurrent Flooding," EPRI NP-984, Electric Power Research Inst., Palo Alto, CA.

2. S.G. Bankoff and S.C. Lee, (1983) "A Comparison of Flooding Models for Air-Water and Steam-Water Flow," Advances in Two-Phase Flow and Heat Transfer, Vol. 1, S. Kakac and M. Ishii, eds., pp.745-780, Martinus Nijhoff, The Hague, The Netherlands.

3. C.P. Liu and C.L. Tien, (1982) "A Review on Gas-Liquid Countercurrent Flow Through Multiple Paths," Heat Transfer in Nuclear Reactor Safety, S.G. Bankoff and N. Afgan, eds., pp.421-446, Hemisphere Publ. Co., N.Y.

4. D.M. Speyer and L. Kmetyk, (1977) "Flooding in Multi-Channel Two-Phase Counterflow," Nuclear Reactor Safety Heat Transfer, A.A. Bishop, F.A. Kulacki, S.G. Bankoff, and O.C. Jones, eds., ASME, N.Y., pp.55-62.

5. B.D.G. Piggott and M.G. Ackerman, (1982) "A Study of Countercurrent Flow and Flooding in Parallel Channels", Heat Transfer in Nuclear Reactor Safety, S.G. Bankoff and N. Afgan, eds., pp.361-378, Hemisphere Publ. Co., N.Y.

6. G.B. Wallis, A.S. Karlin, C.R. Clark, III, D. Bharathan, Y. Hagi, and H.J. Richter, (1981) "Countercurrent Gas-Liquid Flow in Parallel Vertical Tubes," Int. J. Multiphase Flow, Vol. 7, pp.1-19.

7. S.G. Bankoff, R.S. Tankin, M.C. Yuen and C.L. Hsieh, (1981) "Countercurrent Flow of Air/Water and Steam/Water through a Horizontal Perforated Plate," Int. J. Heat Mass Transfer, Vol. 24, pp.1381-1395.

8. I. Dilber and S.G. Bankoff, ((1985) "Countercurrent Flow Limits for Steam and Cold Water Injection through a Horizontal Perforated Plate with Vertical Jet Injection," Int. J. Heat Mass Transfer, Vol. 28, pp.2382-2385.

9. M. Merilo, M. Colah, and R.B. Duffey, (1979) "Condensation-Induced Transition from Bubbling to Liquid Downflow in a Turbulent Two-Phase Pool," 18th National Heat Transfer Conference, San Diego.

10. D.B. Spalding, (1981) " General Purpose Computer Program for Multi-dimensional One and Two Phase Flow," J. Math and Computers in Simulation, Vol. 23, pp.267-276.

11. M.C. Canton, H.I. Rosten, D.B. Spalding, and D.G. Tatchell, (1983) Phoenics: An Instructional Manual, Concentration, Heat and Momentum, Ltd. (CHAM), Wimbledon, London, England.

12. B.E. Launder and D.B. Spalding, (1972) Mathematical Models of Turbulence, Academic Press, London, England.

13. B.E. Launder and D.B. Spalding, (1974) "The Numerical Computation of Turbulent Flows," Comp. Meths. Appl. Eng., pp.269-289.

14. C.L. Chen, PhD Dissertation, Chemical Engineering Department, Northwestern University, Evanston, IL (in preparation).

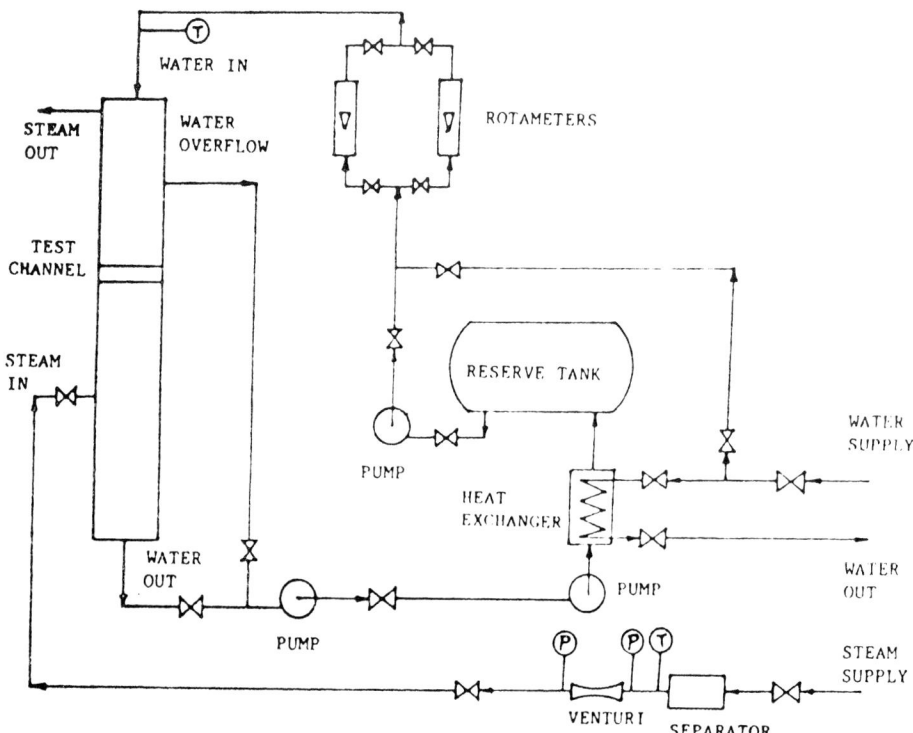

FIGURE 1. Schematic diagram of experimental system

TWO-EQUATION TURBULENCE MODELING

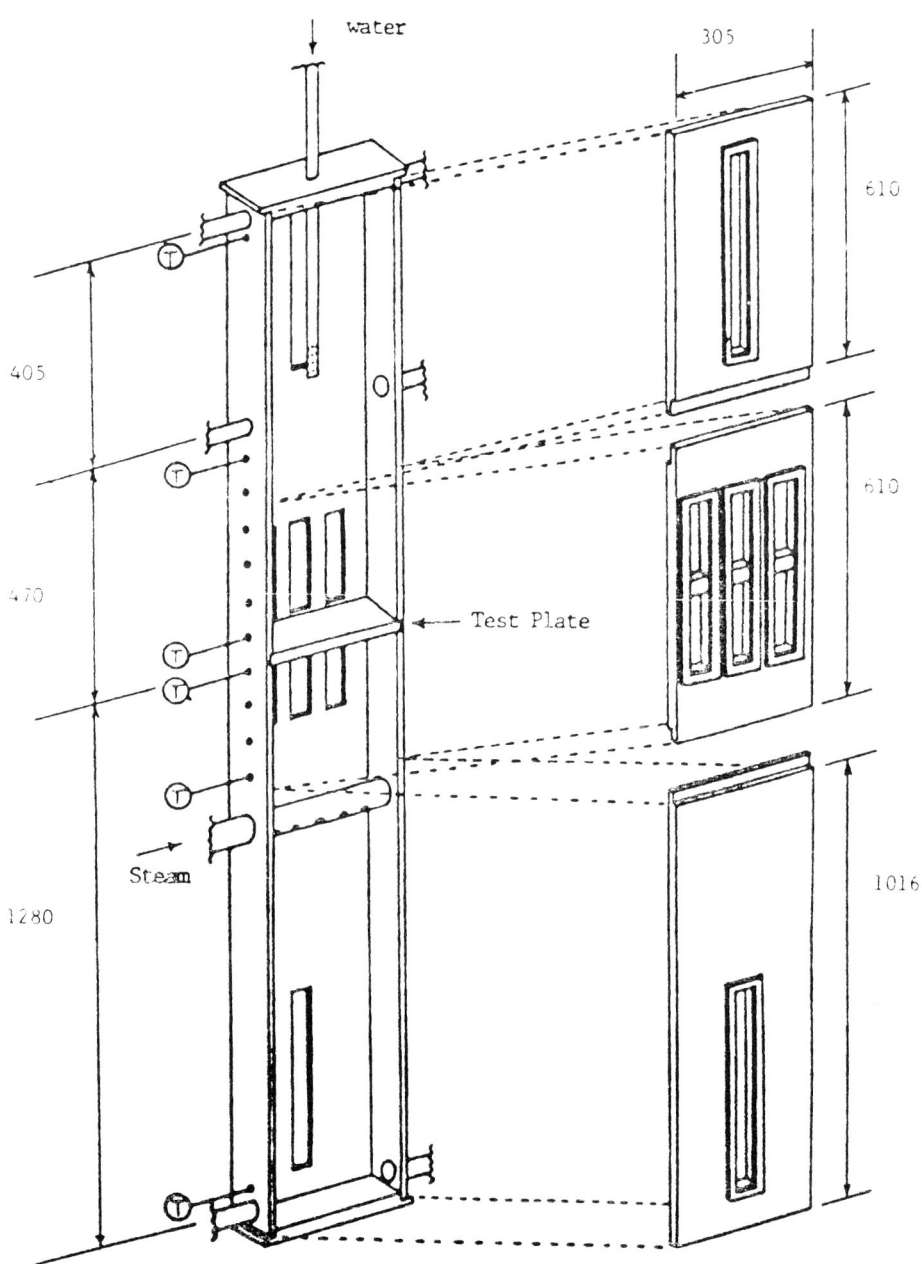

FIGURE 2. Exploded view of test section with perforated test plate (measurements in mm)

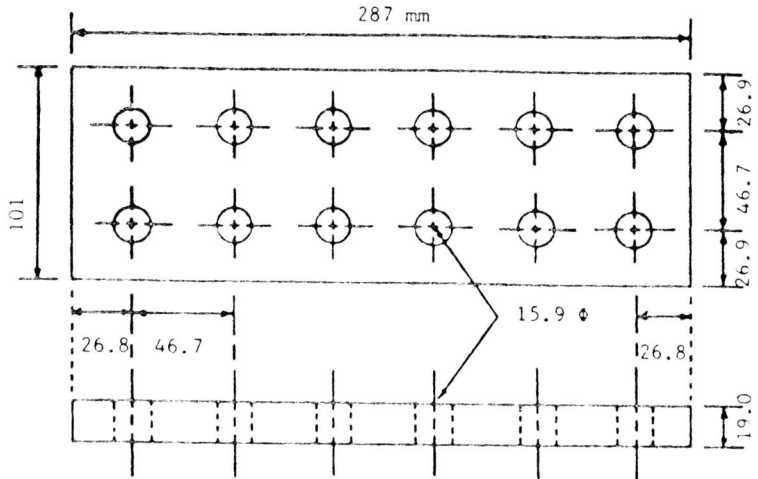

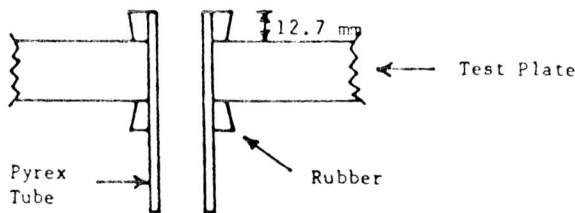

FIGURE 3. Test plate and tube

TWO-EQUATION TURBULENCE MODELING

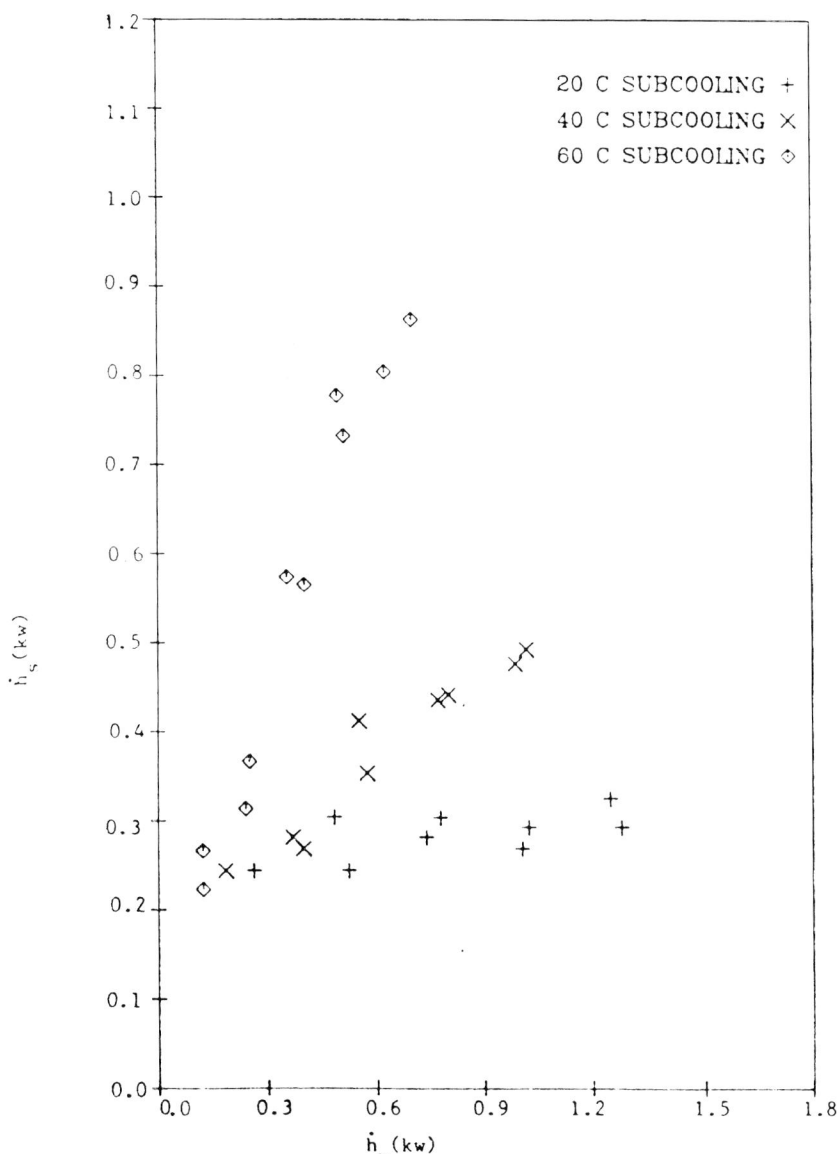

FIGURE 4. Steam and water enthalpy flow rates relative to saturated water at onset of weeping. 6 tubes, 11.1 mm diam. and 406 mm long; horizontal water injection at height of 178 mm.

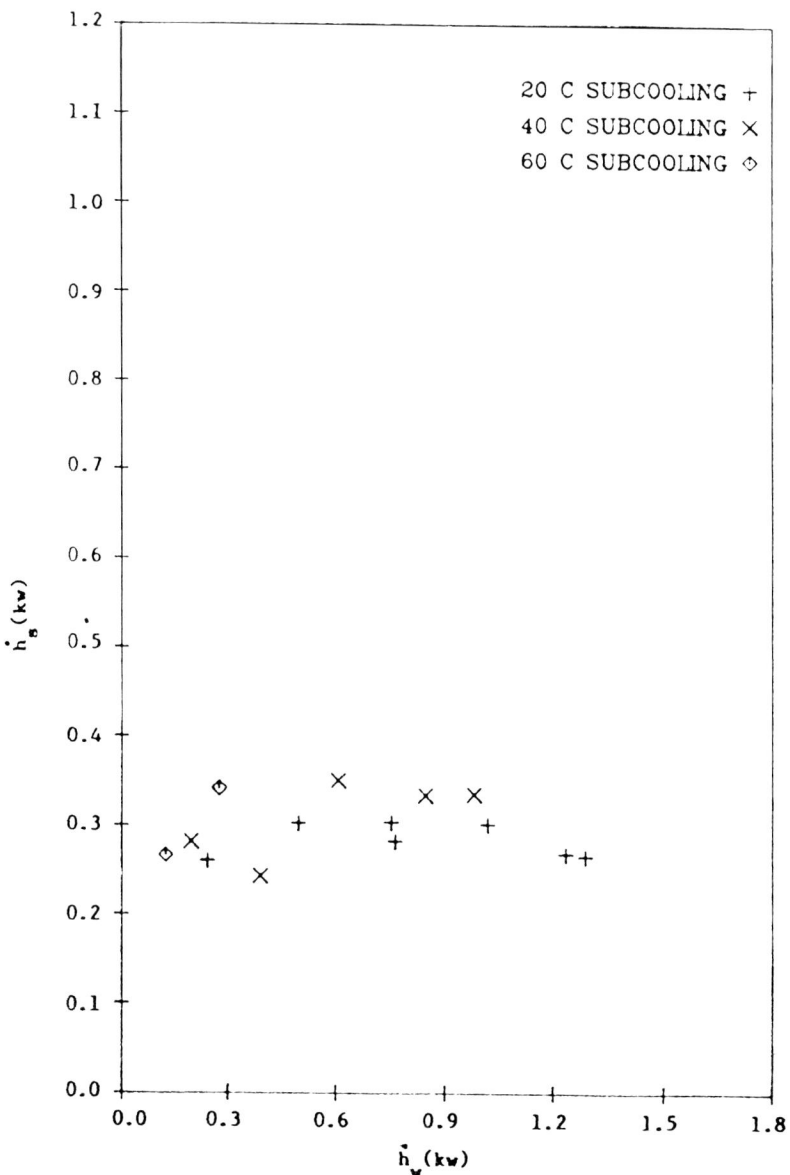

FIGURE 5. Steam and water enthalpy flow rates relative to saturated water at onset of weeping. 6 tubes, 11.1 mm diam. and 406 mm long; horizontal water injection at height of 356 mm.

TWO-EQUATION TURBULENCE MODELING 225

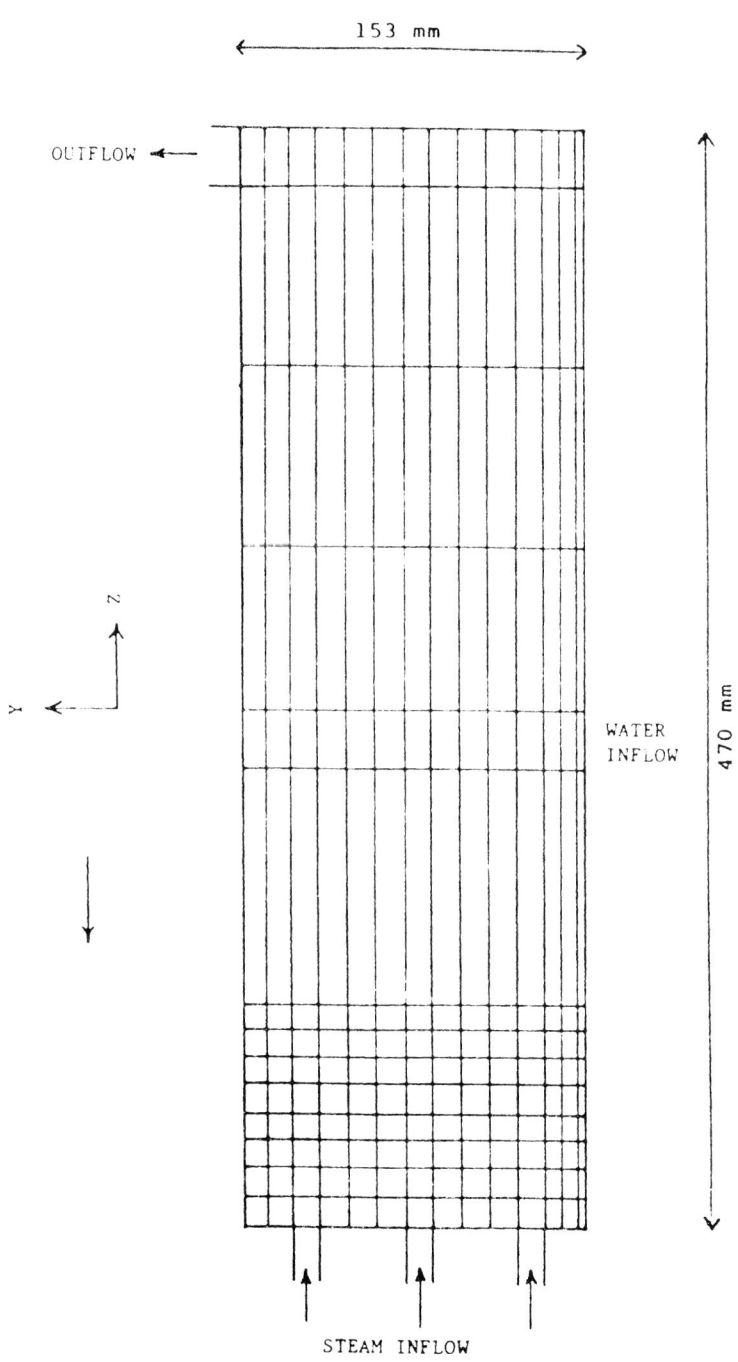

FIGURE 6. PHOENICS finite-domain grid.

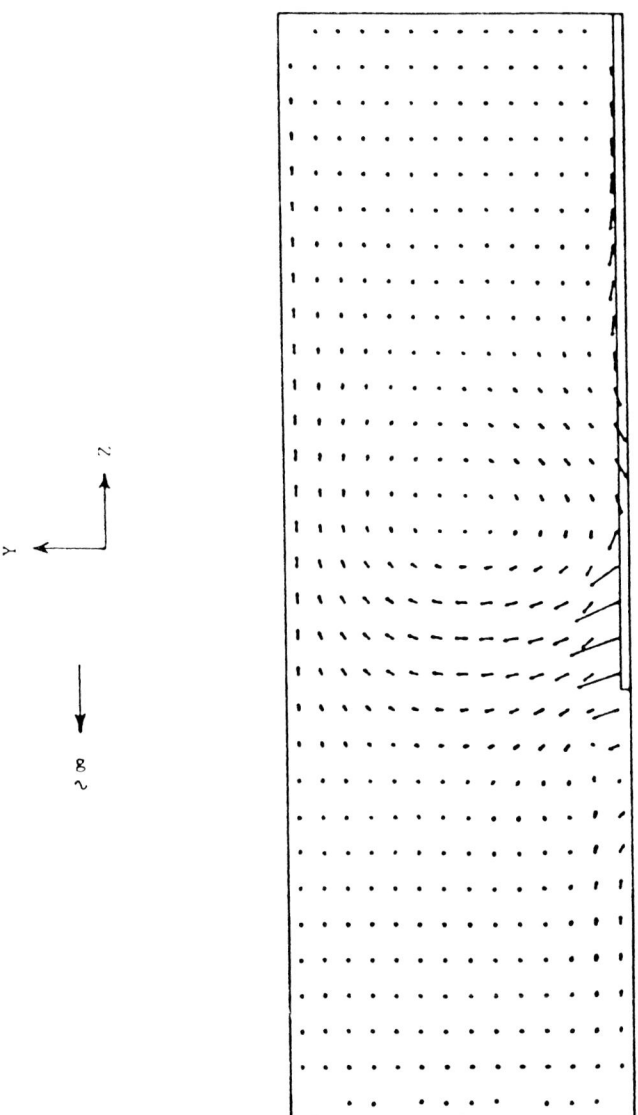

FIGURE 7. Velocity field. Steam injection velocity, 30 m/s. Water injection rate, 0.227 kg/s. Water injection temperature, 60°C.

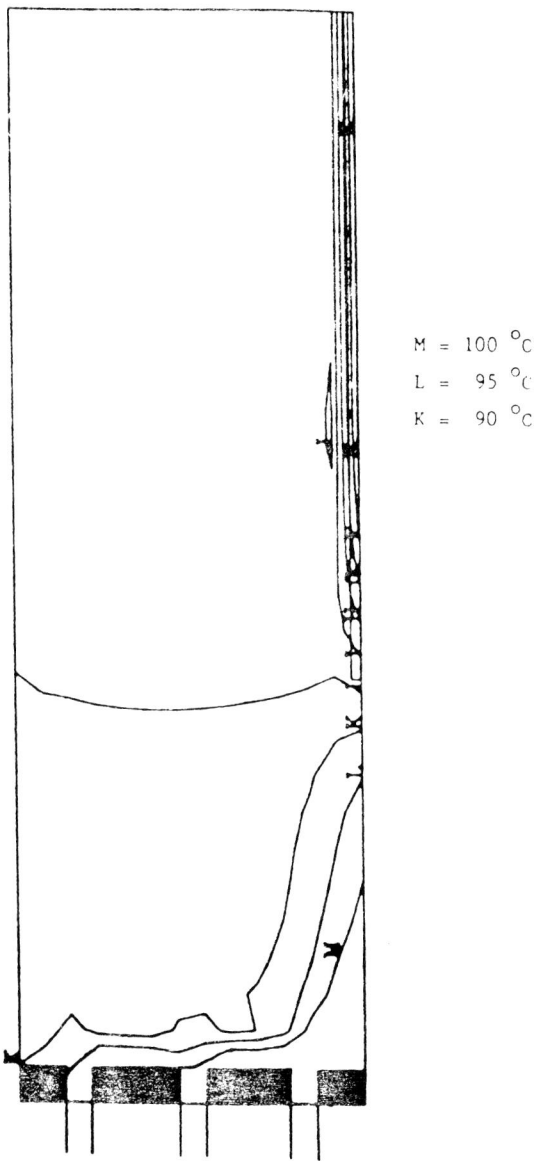

FIGURE 8. Temperature contours. Steam injection velocity, 30 m/s. Water injection rate, 0.227 kg/s. Water injection temperature, 60°C.

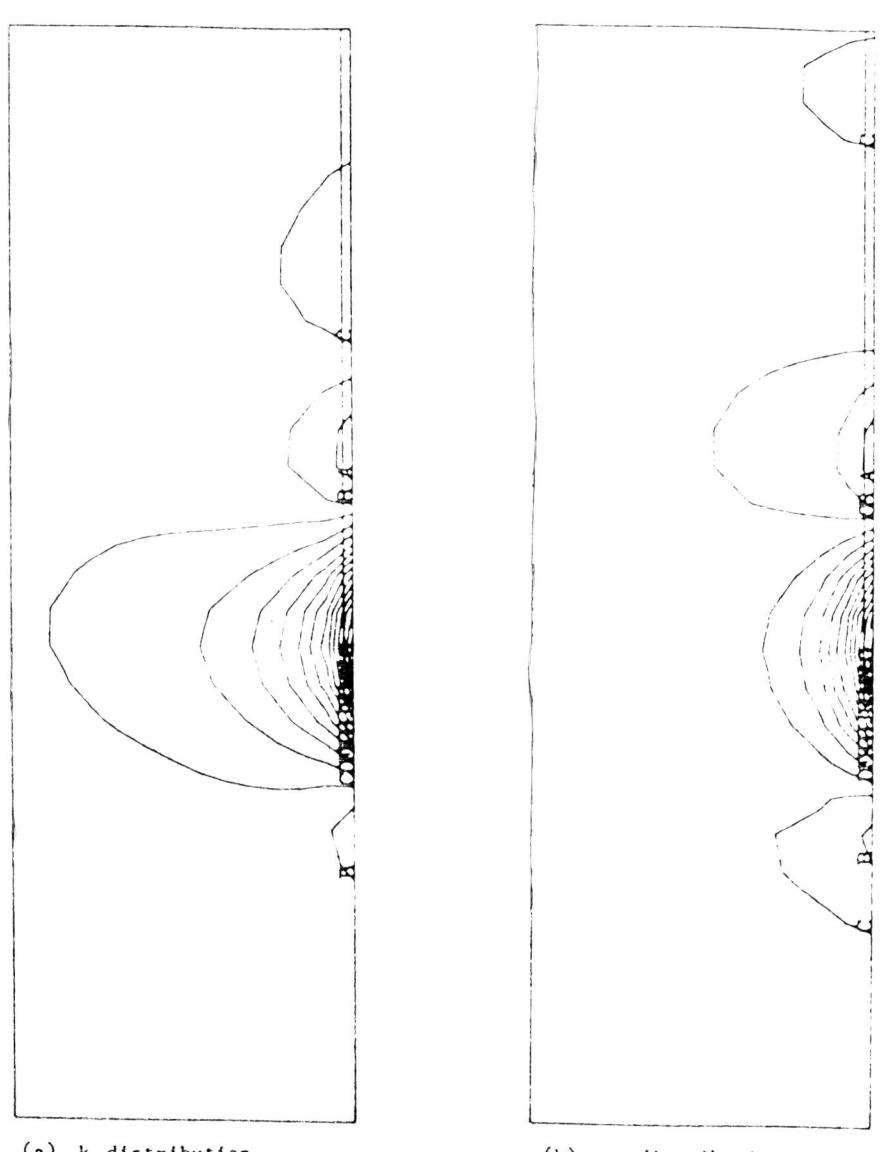

(a) k distribution   (b) ε distribution

FIGURE 9. Turbulent kinetic energy and dissipation rate contours. Steam injection velocity, 30 m/s. Water injection rate, 0.227 kg/s. Water injection temperature, 60°C.

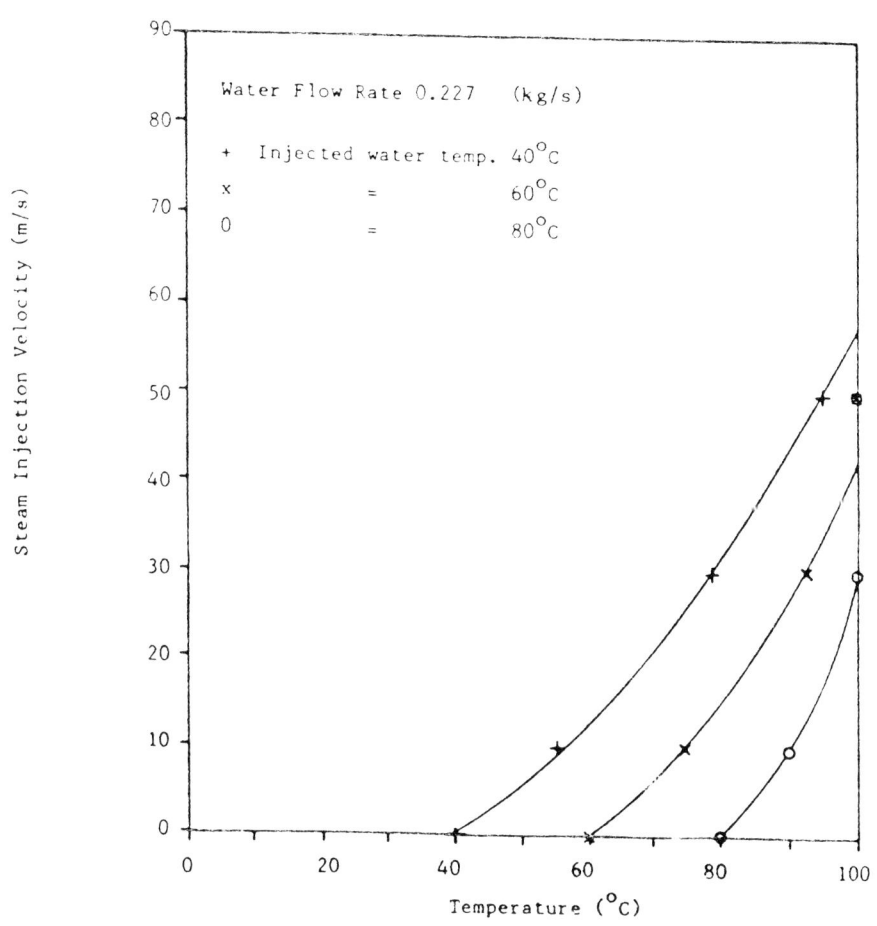

FIGURE 10. Calculated temperature at the third cell above the outermost plate perforation.

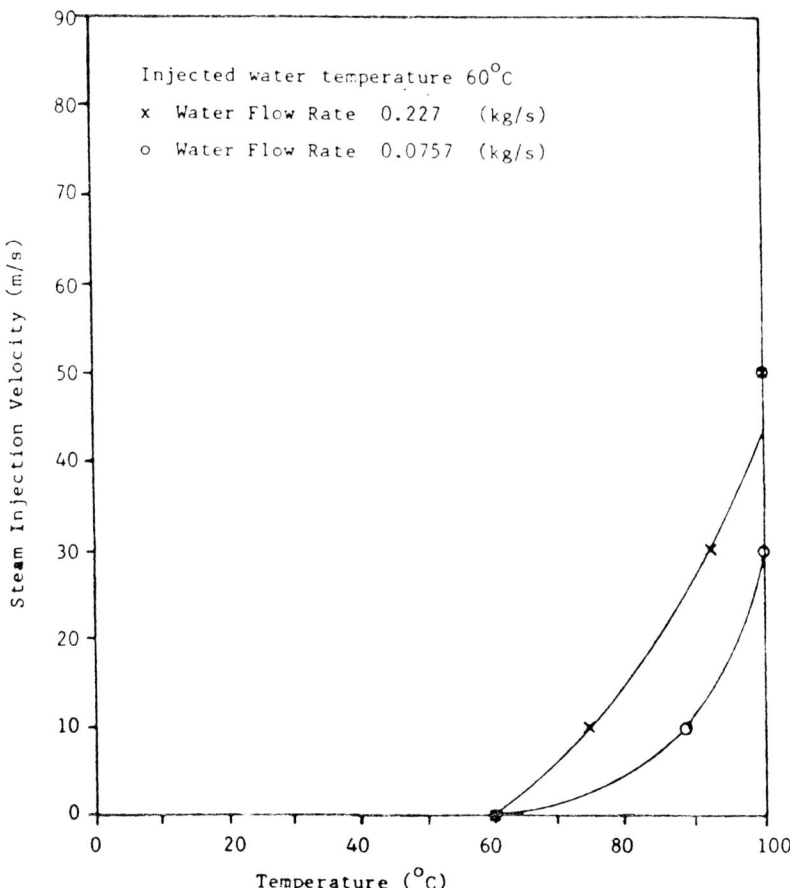

FIGURE 11. Calculated temperature at the third cell above the outermost plate perforation.

TWO-EQUATION TURBULENCE MODELING

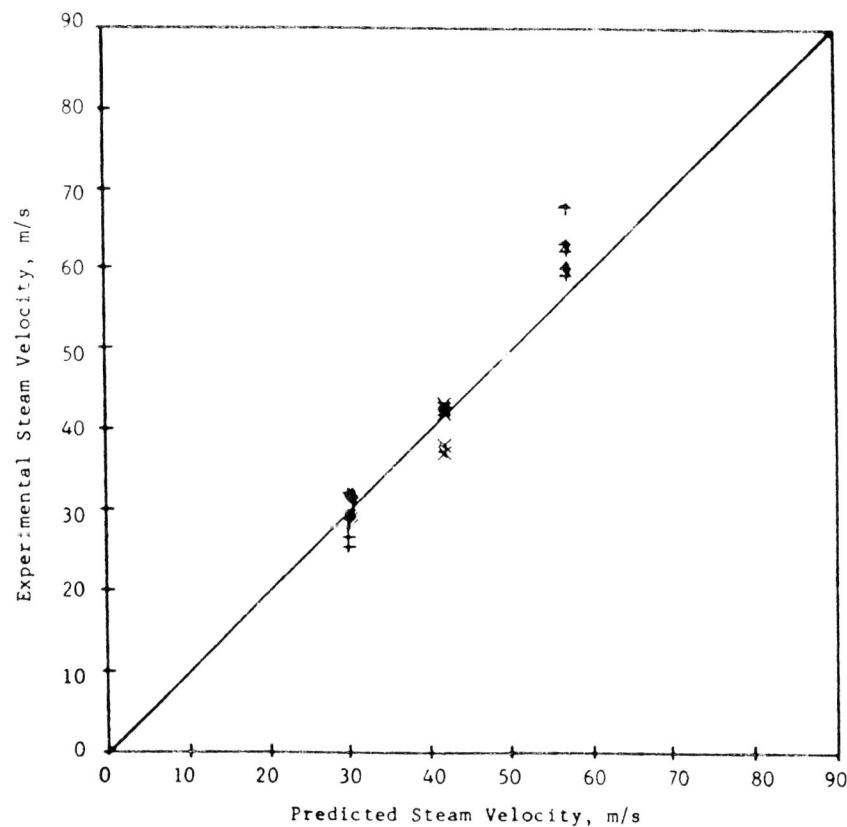

FIGURE 12. Comparison of calculated and experimental steam velocities at onset of weeping with horizontal water injection for twelve runs.

BUBBLE COALESCENCE AND TRANSITION FROM WALL VOID PEAKING TO CORE VOID PEAKING IN TURBULENT BUBBLY FLOW

I. ZUN and I. KLJENAK
Faculty of Mechanical Engineering
University of Ljubljana
Lujubljana, Yugoslavia
A. SERIZAWA
Department of Nuclear Engineering
Kyoto University
Kyoto, Japan

ABSTRACT

Two types of tube models were constructed to study the hydrodynamics of bubble-to-slug transition in vertical upward flow. First, a one-dimensional tube model was developed to compare the predictions with the existing flow regime maps. A three-dimensional model was then constructed, subdividing the whole tube into a bundle of conduits where the bubbles were moving upwards and coalesced according to the principles defined by the one-dimensional model.

The conceptual distinction between the one-and three-dimensional tube models was in the possible initial distribution of bubbles due to transverse lift and dispersion before the process of coalescence began. After the process of coalescence took place, bubble lateral migration towards the tube center was allowed whenever the boundary conditions on bubble segregation were being changed.

From the three-dimensional model, a two-dimensional version was developed to spare computing time on void fraction profile predictions. Comparison with the experimental results on convex to concave void fraction profile evolution was promising.

INTRODUCTION

The present state of knowledge about bubble to slug transition regime (BTS) is mostly empirical. It is based mainly on visual observation of interface configuration with reference to the volumetric fluxes of both phases and their static physical properties. There have been some attempts to search for the first principles which govern flow regime changes as well, but no final physical picture has yet been constructed.

To understand the physics of BTS regime, two kinds of research seem to be promising. The first one is in the stage of quantitative experimental observation of fluctuating parameters (third order parameters; Hewitt, 1978), as in the work of Jones and Zuber (1975), Lübbesmeyer and Leoni (1983), Tutu (1984), and Matsui (1984). The second aspect considers a certain geometrical picture of bubble space arrangement in order to find static parameters, such as the mean displacement of bubbles from each other, which would indicate a high probability of bubble coalescence (Radovcich, 1962; Bilicki and Kestin, 1987). This approach has led in recent years to a dynamic simulation of bubbly flow in Lagrangian frame (Stuhmiller et al., 1983).

In this work, two types of conduit models were constructed to study the hydrodynamics of BTS evolution. First, a one-dimensional tube model was developed, very similar to one recently proposed by Bilicki and Kestin (1987). The authors postulate here that an unstable bubbly vertical flow gradually develops into slug flow due to wake drift which tends to distribute bubbles to move upwards in chains. The bubbles are first distributed within a vertical array into groups, which at sufficient bubble concentration results in bubble coalescence.

The contribution of this paper is in the construction of a two-dimensional picture which allows us to take into account the principles of bubble transverse lift (Žun, 1980), bubble lateral dispersion (Žun, 1985) and bubble lateral segregation (Žun, 1988b). As distinct from Stuhmiller's work (1983), this for the first time enables prediction of the evolution from concave to convex void fraction profile. Once the bubble size space distribution is known, local intefacial area concentration evolution can be determined, which is indespensible to interfacial transfer terms calculation (Kataoka et al., 1985). Experimental support for the theory is given by Serizawa's measurements (1974).

THEORY

One-Dimensional Model

To describe BTS evolution, a one-dimensional model was constructed which considered only vertical bubble movements in variable slip conditions due to wake drift. Bubble slip in vertical direction generates wake behind the bubble which is superimposed on liquid velocity. If the bubble appears within the wake region of a preceding bubble, then it accelerates relative to bulk liquid velocity and eventually collides with its predecessor. After coalescence, the resulting larger bubble accelerates to a correspondingly higher terminal speed which increases wake drift. The process of bubble drift and coalescence thus continues.

The authors considered the movement of gas elements in z (vertical) direction as shown in Fig. 1. At z = 0 the elements are generated as spheres of diameter d_{be0}. The velocity of each element i is a sum of instantaneous local liquid velocity and bubble relative velocity.

$$v_{bi}(t) = v_l(z_i,t) + v_{bl} \tag{1}$$

TURBULENT BUBBLY FLOW

Here, v_{bl} is taken invariant to t as long as the bubble size is constant. This is possible, if one neglects the bubble expansion due to static pressure decrease and the influence of bulk liquid turbulence on slip conditions.

Given sufficient distance between the elements, liquid velocity is equal to $v_{l\infty}$. Behind the elements liquid velocity is increased due to wake perturbation as suggested by Bilicki and Kestin (1987)

$$v_1 \sim v_{bl}(C_{D\infty} A/(\beta^2 z^2))^{1/3}, \qquad (2)$$

where

$$C_{D\infty} = 4gd_{be}/(3v_{bl}^2) \qquad (3)$$

The increased liquid velocity is now defined by

$$v_1(z,t) = v_{l\infty} + k_1(d_{be}/2)(v_{bi}(t)-v_{l\infty})^{1/3}(z_i-z)^{-2\zeta/3} \qquad (4)$$

A constant drag coefficient for a single-bubble system in an infinite medium is assumed in (2). There are two principal weaknesses in such physical picture. The first is due to the influence of bubble size on the bubble intrinsic nonrectilinear motion, and second is due to the influence of multi-bubble surroundings where all bubbles are not travelling in the same line. For that reason, ζ in exponent of $(z_i - z)$ is introduced to account for "efective drift"; $\zeta > 1$. Eq.(4) is assumed to be valid during the entire arrival time. The velocity of the liquid adjacent to the bubble tail would than be

$$v_1(z_i-d_{be}/2,t) = v_{bi}(t). \qquad (5)$$

From that,

$$k_1 = (v_{bi}(t) - v_{l\infty})^{2/3}(d_{be}/2)^{2\zeta/3-1}, \qquad (6)$$

and

$$v_1(z,t) = v_{l\infty} + (d_{be}/(2(z_i-z)))^{2\zeta/3} (v_{bi}(t)-v_{l\infty}). \qquad (7)$$

Fig. 1 shows a situation where element i, outside the influence from el. (i-1), disturbs the liquid and thus accelerates element (i+1). It is expected that element (i+1) collides after sufficient time with element i.

Finally, average void fraction at chosen location is determined by

$$\bar{\alpha} = \bar{T}_r/(\bar{T}_a+\bar{T}_r) \qquad (8)$$

where $\bar{T}_r$ represents mean residence time interval during which the gas phase occupies observed location, and $\bar{T}_a$ arrival time, defined by the time interval during which the liquid occupies observed location.

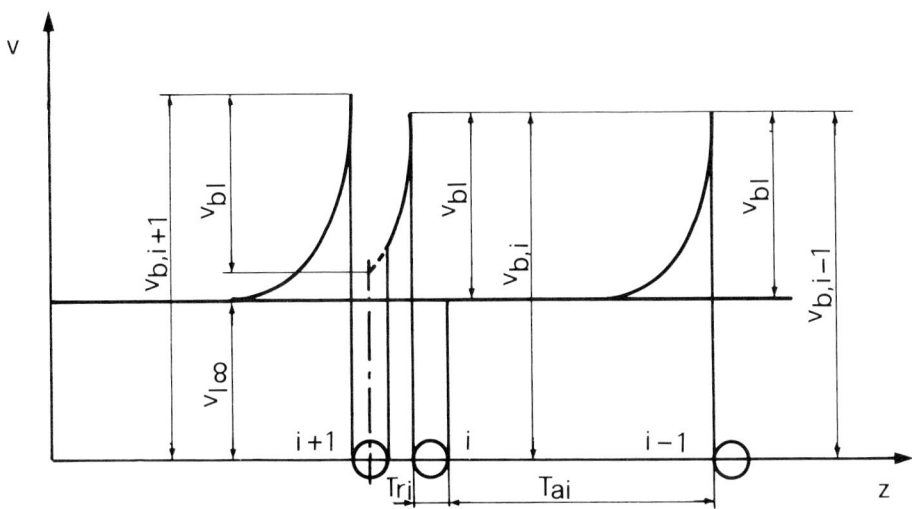

FIGURE 1. Wake drift on bubble moving in a vertical array.

Two-Dimensional Model

Nonhomogeneous void fraction radial distribution is a result of two simultaneous processes having different time scales. The first process is bubble segregation according to bubble size, which is assumed to be caused by the transverse lift force being balanced by turbulent surroundings (Žun 1988b). The second process is bubble coalescence which causes significant bubble growth. In the present model, these two simultaneous processes are separated, for the sake of computing time. Such a physical picture is possible since the time scale of bubble transverse migration is much smaller than the time scale of bubble coalescence as observed by experiments (see Results & Discussion).

To reduce the enormous computing time required by Lagrangian frame, the void fraction profile can be predicted in Eulerian frame with a dispersion model first, and the results taken as an input to bubble coalescence simulation.

Bubble Segregation Modeling

The bubble dispersion model, developed by Žun (1985; 1988a), where large scale turbulent liquid eddies are supposed to restrain bubble transverse penetration due to lift and diffusion, points to bubble size as a key parameter for void peak radial location formation in a turbulent shear flow. In this model, the total bubble flux for the cocurrent upflow is described by

TURBULENT BUBBLY FLOW

$$\underline{j}_b = (0,0,\bar{a}\bar{v}_{1,z}) + (0,0,\bar{a}\bar{v}_{bl,z})$$

$$- \frac{\bar{a}C_M \varrho_1 v_{b\infty}^2 (\partial \bar{v}_{1,z}/\partial x_1, \partial \bar{v}_{1,z}/\partial x_2, 0)}{(g\Delta\varrho)}$$

$$+ (\overline{a'v'_{bl,x1}}, \overline{a'v'_{bl,x2}}, 0)$$

$$- (\overline{a'v'_{1,x1}}, \overline{a'v'_{1,x2}}, 0) , \qquad (9)$$

where the five terms on the r.h.s. represent, respectively:

(i)   The convective bubble flux with $\bar{a}$ being time-averaged void fraction and $\bar{v}_1$ the time-averaged local liquid velocity,

(ii) The contribution by the gravitational field over the time-averaged local bubble relative velocity $\bar{v}_{bl}$,

(iii) The bubble transverse migration term whose basic principle is explained by the transverse lift force

$$\underline{F}_T = C\varrho_1 (\nabla \times \underline{v}_1) \times \underline{v}_{bl} , \qquad (10)$$

(iv) The bubble fluctuation term which is defined by the bubble lateral diffusion, and

(v)   The rate of local transport of void fraction associated with eddy diffusivity.

From (9), the void fraction radial profile was defined (Žun 1985) by

$$\bar{a}^{-1}\frac{d\bar{a}}{dx} = \frac{C_M v_{b\infty}^2 (\varrho_1/g\Delta\varrho) d\bar{v}_{1,z}/dx}{\varepsilon_M (1 + (\varsigma/\pi f_{b\infty}) d\bar{v}_{1,z}/dx + (\varsigma/2\pi f_{b\infty})^2 (d\bar{v}_{1,z}/dx)^2) - D_b} \qquad (11)$$

In the case of polydispersed bubbly flow, the bubble segregation occurs according to bubble size and its intrinsic oscillatory motion. This was recently demonstrated experimentally and theoretically for vertical upward flow (Žun 1988b). Bubbles having weak intrinsic oscillatory motion tend to remain in the core region, forming a void fraction peak at the tube centerline which is, as long as the bubble size is kept constant, a rather stable bubble flow regime called void coring. The result of the combination of void coring and wall void peaking flow regime is possible and rather stable.

To describe the bubble polydispersed system (due to the bubbles of multiple sizes), the distribution of the bubble equivalent sphere diameter should be determined in advance.

The solution of (11) then gives, in accordance with $d_{be,i}$, a set of particular solutions, and the entire void fraction profile is obtained by summation. This is based on the postulate that each particular bubble size retains its preferable radial location even when it appears in a multibubble structure (Žun, 1988b).

Bubble Coalescence Modeling

The whole tube is subdivided into a bundle of conduits through which the bubbles are moving. Axes of these conduits are uniformly distributed in concentric rings. The cross-sectional area of conduits is constant. The number of conduits in each ring is therefore in proportion to the ring area (Fig. 2). Because the bubble coalescence model does not account for lateral collisions of bubbles, the shape of the conduit is not important.

The bubbles move along each particular conduit axis at a concentration defined at input by (11). Each bubble has to be followed along its way, hence the local instant formulation replaces time averaged terms in (9). The convective nonstationary liquid velocity (eq.(7)) which appears in the first term, contributes to the bubble coalescence, as described by one-dimensional model. Once the two bubbles coalesce, the resulting larger bubble tends to segregate to the adjoining conduit in direction towards the tube center, since the bubble transverse lift force normally weakens rapidly as long as the initial bubble shape is ellipsoidal (Žun, 1988b). Such bubble penetration is restricted by the principle of bubble lateral diffusion which is in present local instant formulation defined by the space available in the conduit (Fig. 3).

A necessary condition to satisfy the principle of bubble diffusion is that bubble (element) transition from conduit C_{i+1} to C_i is possible if there is no collision, and δ_1 and δ_2 are greater than a limiting value δ_0. For bubbles with intrinsic nonrectilinear motion (that is, at terminal conditions) δ_0 is a function of bubble size

$$\delta_0 = k_d d_{be}, \qquad (12)$$

where k_d is a constant, whereas for bubbles with intrinsic rectilinear motion δ_0 is a constant which is much smaller than in case (12). This prescribes greater bubble penetrating ability. Bubbles of equivalent sphere diameter greater than the conduit diameter are considered as "slugs".

The process of bubble coalescence can be further simplified into a two-dimensional model using an array of conduits as shown in Fig. 4. During the bubble transition from the outer to inner tube ring the number of bubbles is conserved, which yields the following boundary condition:

$$c_i' - c_i = A_{i+1}/A_i (c_{i+1} - c_{i+1}') \qquad (13)$$

This requires that bubble concentration in ring i increases by factor A_{i+1}/A_i more than bubble concentration in ring i+1

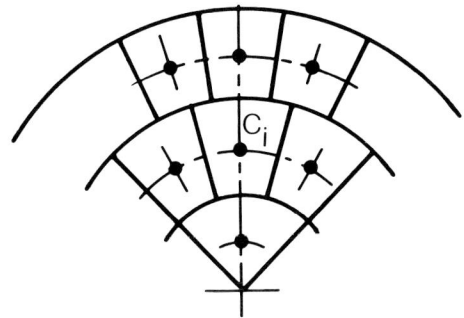

FIGURE 2. The bundle of conduits

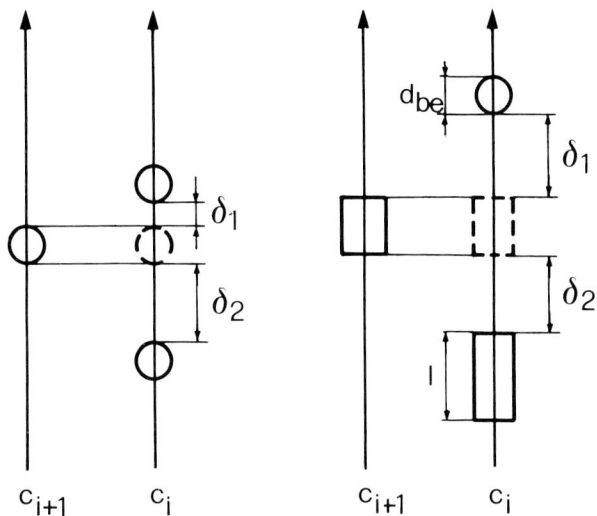

FIGURE 3. Bubble transverse penetration limitation

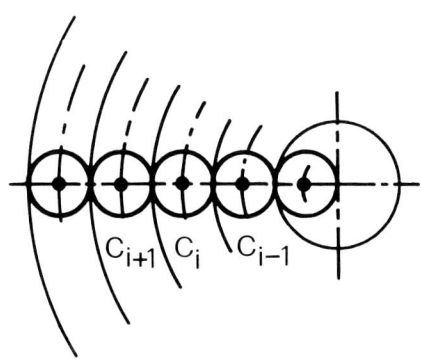

FIGURE 4. Two-dimensional arrangement of conduits

decreases. Since in a two-dimensional model each conduit represents a corresponding ring, for every bubble which leaves from conduit C_{i+1}, A_{i+1}/A_i bubbles arrive to conduit C_i. Thus each bubble arrival in C_i from C_{i+1} should trigger a new bubble generation with probability of $(A_{i+1}/A_i - 1)$. There are the same limitations of such bubble generation as by bubble transition, that is the bubble generation should be generated within the space limits δ_1, δ_2 apart from existing bubbles. The transfer of bubbles is not allowed until generation of bubbles is successfully accomplished.

RESULTS AND DISCUSSION

One-Dimensional Model

A one-dimensional flow of gas elements (bubbles) was numerically constructed in Lagrangian frame for the tube distance L. The solution was obtained by discrete simulation with variable time step. The velocity of gas element (bubble) is in each discrete time defined by (1) and (7). The time step is defined by

$$\Delta t = \Delta x / v_{b,max} \qquad (14)$$

with $v_{b,max}$ being the maximum element velocity in time t. Δx is choosen in advance and should be $\leq d_{be}/2$. Coordinates of element centers are determined by

$$z_i(t_i + \Delta t) = z_i(t_i) + v_{bi}(t_i)\Delta t. \qquad (15)$$

If l_i denotes the length of an element i (initially, for the element before the first coalescence takes place $l_i = d_{be,i}$), then the condition for coalescence is defined by

$$|z_i - z_{i+1}| \leq (l_i + l_{i+1})/2 . \qquad (16)$$

Boundary conditions: At $z = L$ tube ends to the atmosphere. Velocity of the last element before $z = L$ is determined by $v_{l\infty} + v_{bl}$ since there is no wake drift. Initial velocity at $z = -d_{be}/2$ also equals $v_{l\infty} + v_{bl}$. Velocity of an element at each time t is otherwise defined by (1) and (7), if d_{be} is replaced by l_i.

Bubble generation at $z = 0$ was determined by Erlang distribution of arrival times (Lackmé, 1965; Žun et al. 1982).

$$F(t) = 1 - e^{-\lambda t}(1 + \lambda t + (\lambda t)^2/2 + (\lambda t)^3/6), \qquad (17)$$

$$\lambda = n/\bar{T}_a . \qquad (18)$$

An example of bubble growth evolution due to coalescence is shown in Fig. 5. Fig. 6 shows the effect of initial conditions chosen, if the arrival times are defined by first and fourth order Erlangian functions.

At the fixed locations downstream from the tube, element residence times, arrival times, element lengths and void

TURBULENT BUBBLY FLOW

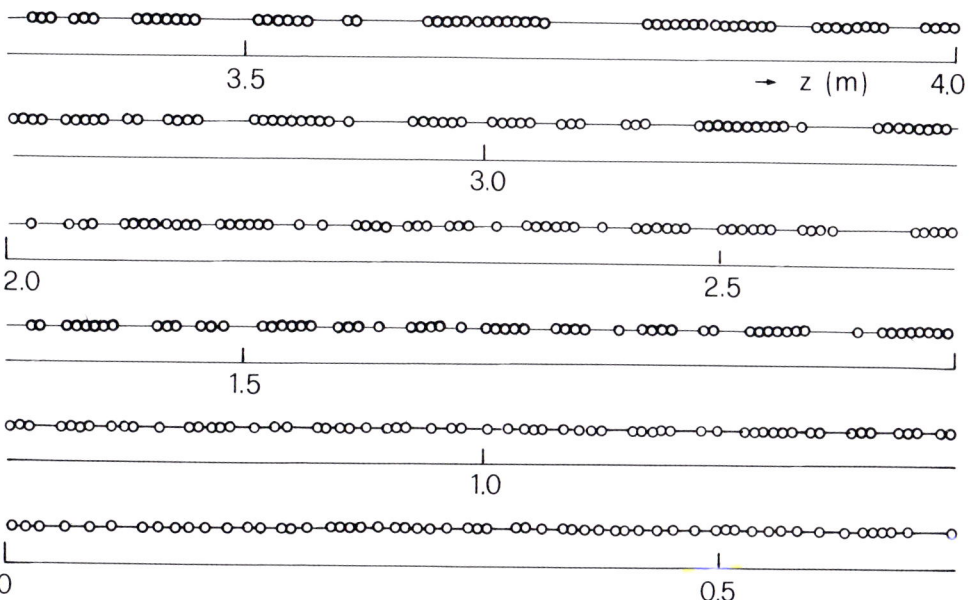

FIGURE 5. The process of BTS evolution (j_l = 1.0 m/s, j_g = 1.4 m/s)

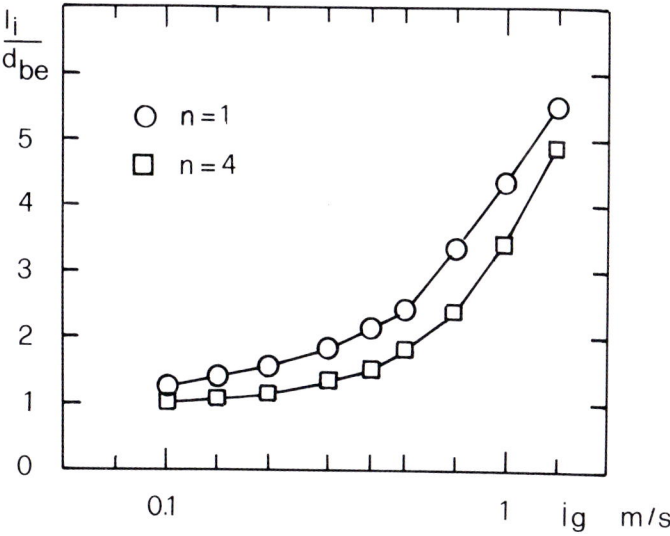

FIGURE 6. Gas element evolution from Erlang n^{th} order distribution function

fraction were measured. The void fraction was calculated from (8). The one-dimensional model was successfully applied to predict the BTS evolution in comparison to empirical flow regime maps available at present (Kljenak, 1988).

Two-Dimensional Model

Measurements of developing bubbly flow by Serizawa (1974) have shown that the void peak near the wall is formed in a rather short distance downstream from the tube entrance, less than 10D. This accords with experimental results by Herringe and Davies (1976) who observed for various air/water mixers void peaks near the wall as early as at 8D. The void coring profile evolution on the other hand requires more then 20D, as observed by Serizawa (Fig. 7). This observation supports the theoretical assumption which proposes two different time scales for the process of bubble segregation and coalescence and thus enables the authors to treat them separately.

Input parameters

The bubble dispersion model, Eq. (11), was utilized first. This comprises the liquid volumetric flux, the area averaged void fraction and the mean bubble size required as input parameters. The uncertainty principle also requires a bubble size distribution function and the portion of unsegregated bubbles due to non-developed bubbly flow

$$\bar{a}_c = k_c \langle \bar{a} \rangle . \qquad (19)$$

The data on bubble free rise properties were taken from Žun & Malahovsky (1982).

The predictions of void fraction profile by the dispersion model served as an input at z = 8D to the simulation process of bubble coalescence. Here, the bubble concentration in each conduit was calculated from void fraction profile based on the assumed first order Erlang distribution of arrival times, Eq. (14).

The liquid velocity and bubble relative velocity profiles, which are requested as input parameters in (7), were taken from Serizawa's measurements (1974), Fig. 8.

The weakness of such an assumption is in the replacement of instantaneous values of bubble relative velocity with a time-averaged value. This is supposed to be compensated by v_l values taken for bubbly mixture instead for the bulk liquid and by appropriate ζ in (7). The question is what are the best parameters to put in. A more appropriate way would be to consider time averaged bulk liquid velocity profile for $v_{l\infty}$ and individual bubble rise velocity based on corresponding drag coefficient (instead of (3)). There is a possibility to predict v_{bl} radial profiles, too, based on dispersion model (Žun et al., 1987), but this is not yet tested enough to be applied in present analysis.

TURBULENT BUBBLY FLOW

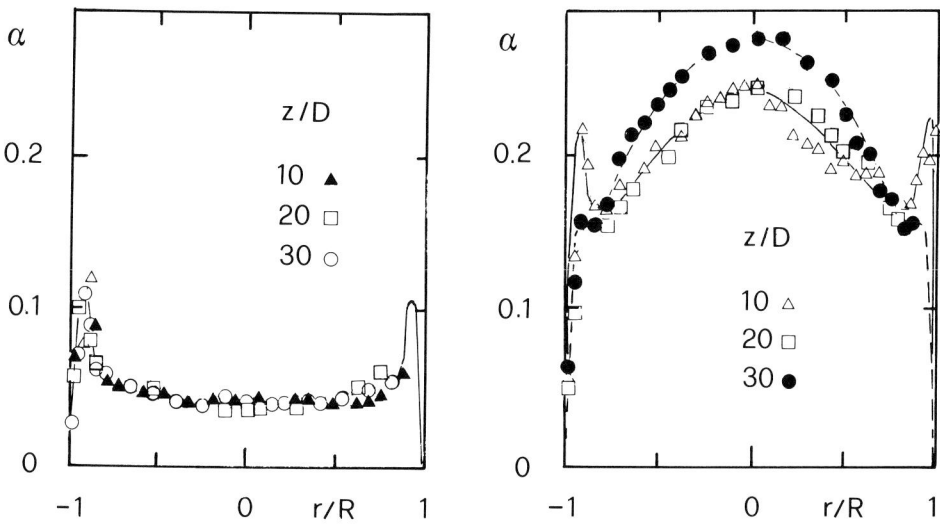

FIGURE 7. Entrance effect upon flow arrangement (j_1 = 0.59 m/s)

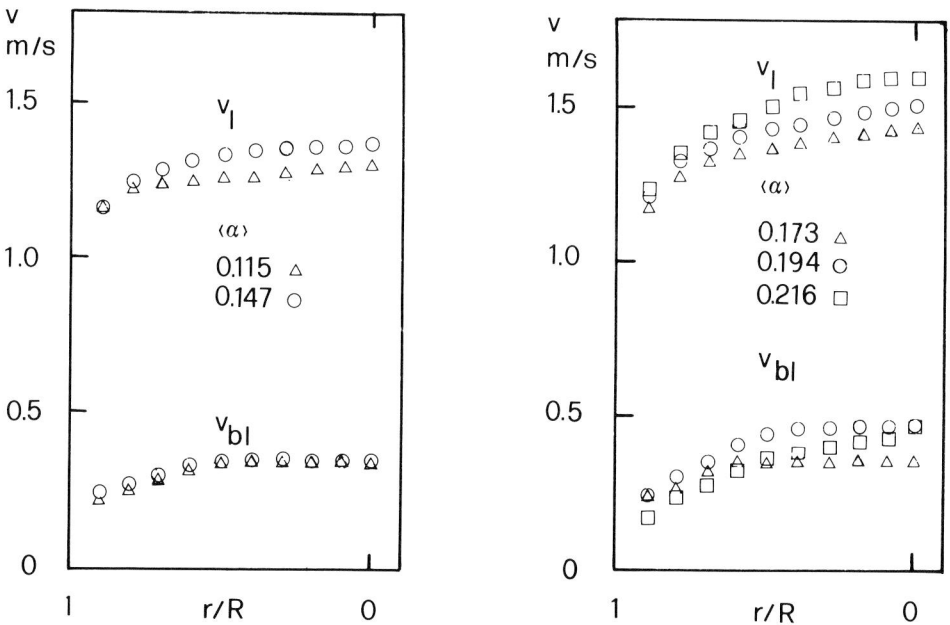

FIGURE 8. Liquid velocity and bubble relative velocity profiles (Serizawa, 1974)

Predictions

Two-dimensional predictions of void fraction profiles as measured by Serizawa were studied at j_l = 1.03 m/s and average void fractions: 0.046, 0.115, 0.147, 0.173, 0.194, and 0.216. Initial mean bubble equivalent sphere diameter was taken as 3.25 mm. Since the bubble size distribution of injected bubbles was not known, only 5% dispersion of normal distribution was taken. The limiting bubble size above which the bubbles tend to segregate towards the tube center was calculated to be 5 mm in all cases. Below this value k_d = 20 was taken in (12) which rather strongly restricted bubble migration towards the tube center. This enabled the authors to keep the wall void peak at the same location as being predicted by (11) as long as the bubbles do not coalesce. For the bubbles larger than 5 mm, δ_0 = 2 mm proved to be the best fit.

Bilicki & Kestin's (1987) model predicts the BTS transition based on ratio a/d_{be}, where a measures the limiting distance between the two adjacent bubbles which are in a position to coalesce:

$$a/d_{be} = (1.6/\bar{\alpha}^{1/3}) - 1. \qquad (20)$$

According to visual observation (Serizawa, 1974), the onset of BTS regime appeared at $\alpha_c \approx 24.6\%$. For d_{be} = 5mm this gave a $\approx$ 8mm. Such limiting distance accords with the present analysis which points out two non-equilibrium processes. The first one is bubble transverse wall migration which forms a rather stable bubble layer near the wall before the bubbles start to coalesce. For these bubbles k_d = 20 or δ_0 = 65 mm. The second process consists of bubble coalescence and consequently, segregation of coalesced bubbles in the core region. For these bubbles δ_0 < a is required.

ζ in (7) was taken as 3.0 for the choosen input velocity profiles. k_c in (19) had to be increased a little bit when increasing $\langle \alpha \rangle$ which is logical if the input bubble size distribution is not known (k_c = 0.4-0.6).

The present model gives, at rather high void fraction in some conduits, gas slugs of length to a couple of conduit diameters. These slugs stand for larger Taylor bubbles which in reality do not yet occupy the whole tube, but are large enough to occupy several conduit diameters used in the model. The largest discrepancy is developed in such a case at the tube center. For this reason, a corresponding correction was used (Kljenak, 1988).

The final predictions correspond to axial distance 30D from the tube entrance and are shown in Fig. 9. Here, continuous curves represent the predictions of the bubble dispersion model (11) and squares represent the predictions when bubble coalescence is included. From these figures it can be concluded that BTS transition is a gradual and systematic evolution of the interfacial configuration, as already discussed by Stuhmiller et al. (1983). Within the range of

TURBULENT BUBBLY FLOW

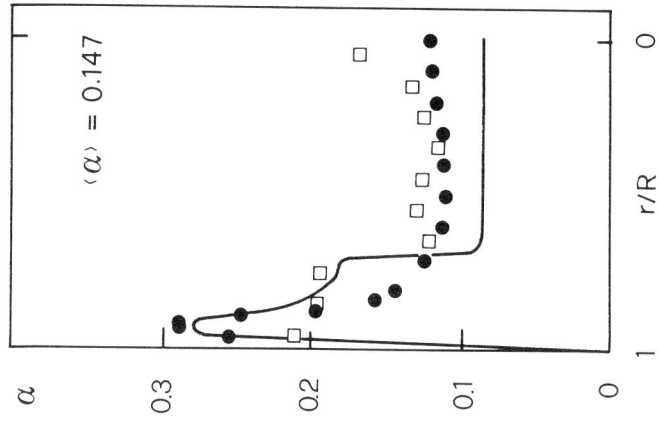

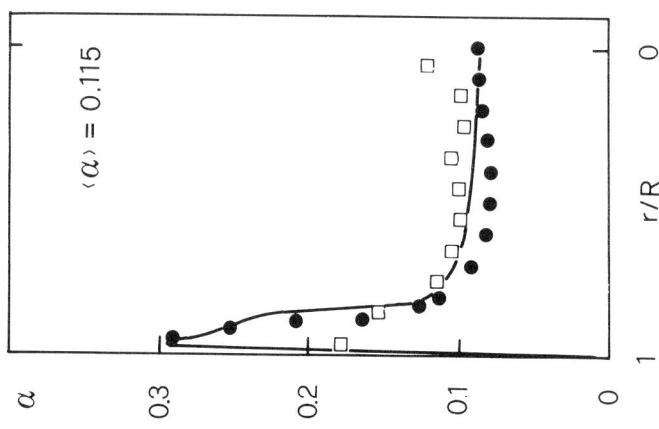

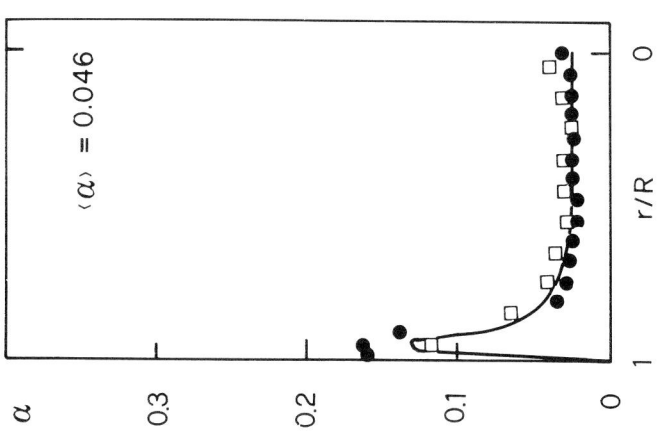

FIGURE 9a. Void fraction profiles ($j_l = 1.03$ m/s, $z/D = 30$);
● experiment (Serizawa, 1974);
— theory, dispersion model only, Eq. (11), input at $z/D = 8$;
□ final predictions after coalescence simulation

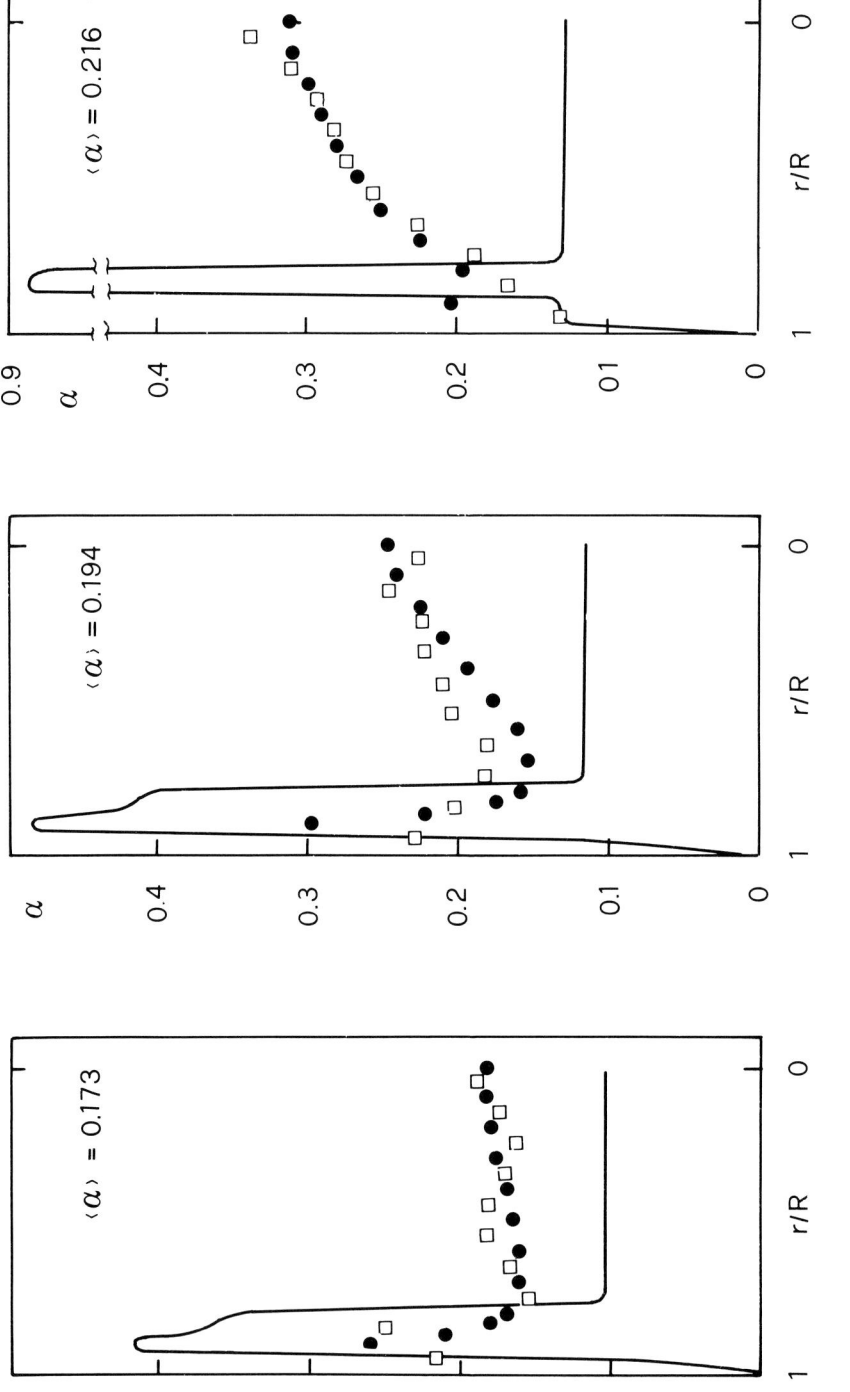

FIGURE 9b. Void fraction profiles (j_l = 1.03 m/s, z/D = 30);
● experiment (Serizawa, 1974);
— theory, dispersion model only, Eq. (11), input at z/D = 8;
☐ final predictions after coalescence simulation

TURBULENT BUBBLY FLOW

experimental data, the bubble dispersion model should be implemented by the bubble coalescence model when $\langle\alpha\rangle$ exceeds approx. 15%.

CONCLUSIONS

To study the hydrodynamics of bubble-to-slug transition in vertical upward flow, a two-dimensional model was constructed. The studies of concave to convex void fraction profile evolution have shown that there is no abrupt transition from bubble to slug flow. Bubble transverse lift, bubble lateral dispersion, bubble lateral segregation and wake drift are probably the principle mechanisms which govern the process of coalescence.

ACKNOWLEDGEMENTS

The computation was done at the LAKOS computer centre, Ljubljana. Financial support was given by the Slovene Research Community.

NOMENCLATURE

A cross-section
a distance between two adjacent bubbles
C conduit
C_M bubble transverse migration coefficient
C_D drag coefficient
c bubble concentration
D_b bubble dispersion coefficient
d_{be} bubble equivalent sphere diameter
f_b bubble intrinsic oscillation frequency
g gravity constant
j volumetric flux
k factor
n factor
T_a arrival time
T_r residence time
t time
v velocity
x coordinate
z coordinate
α void fraction
β proportionality constant
δ distance
ε_M eddy diffusivity momentum of liquid phase
ζ attenuation factor
λ parameter
ϱ density
$\Delta\varrho$ $\varrho_l - \varrho_d$

Subscripts

b   bubble
bl   relative
x   coordinate
z   coordinate
b∞   bubble free rise condition

BIBLIOGRAPHY

Bilicki, Z., and Kestin, J., 1987, Transition Criteria for Two-Phase Flow Patterns in Vertical Upward Flow, Int.J. Multiphase Flow, vol.13, pp. 283-294.

Herringe, R. A., and Davis, M. R., 1976, Structural Development of Gas-Liquid Mixture Flows, J.Fluid Mech., vol. 73, pp.97-123.

Hewitt, G. F., 1978, Measurement of Two Phase Flow Parameters, Academic Press, London.

Jones, O. C. Jr., and Zuber, N., 1975, The Interrelation Between Void Fraction Fluctuations and Flow Patterns in Two-Phase Flow, Int.J.Multiphase Flow, vol.2, pp. 273-306.

Kataoka, I., Ishii, M., and Serizawa, A., 1985, Interfacial Area in Two-Phase Flow: Formulation and Measurement, Multiphase Flow and Heat Transfer - HTD, eds. V. K. Dhir, J. C. Chen, O. C. Jones, vol.47, pp. 131-140, ASME.

Kljenak, I., 1988, The Structure of Bubbly Flow, MSc Thesis, Univ.of Ljubljana, Fac. Mech. Eng.

Lackmé, C., 1965, Some Statistical Properties of Two-Phase Flow in Vertical Tubes, Symp. Two-Phase Flow, vol.2, paper D2, Exeter.

Lübbesmeyer, D., and Leoni, B., 1983, Fluid-Velocity Measurements and Flow-Pattern Identification by Noise-Analysis of Light-Beam Signals, Int.J.Multiphase Flow, vol.9, pp. 665-679.

Matsui, G., 1984, Identification of Flow Regimes in Vertical Gas-Liquid Two-Phase Flow Using Differential Pressure Fluctuations, Int.J.Multiphase Flow, vol.10, pp. 711-720.

Radovcich, N. A., 1962, The Transition From Two-Phase Bubble Flow to Slug Flow, MSc Thesis, MIT.

Serizawa, A., 1974, Fluid-Dynamic Characteristics of Two-Phase Flow, Doctoral Thesis, Kyoto University.

Stuhmiller, J. H., Ferguson, R. E., Wang, S. S., and Agee, L. J., 1983, Two-Phase Flow Regime Modeling, in Transient Two-phase Flow, Proc. 3rd CSNI Specialist Meeting, eds. M. S. Plesset, N. Zuber, I. Cattom, pp. 353-368, Hemisphere, Washington.

Tutu, N. K., 1984, Pressure Drop Fluctuations and Bubble-Slug Transition in a Vertical Two-Phase Air-Water Flow, Int.J. Multiphase Flow, vol.10, pp. 211-216.

Žun, I., 1980, The Transverse Migration of Bubbles Influenced by Walls in Vertical Bubbly Flow, Int.J.Multiphase Flow, vol. 6, 583-588.

Žun, I., and Saje, F., 1982, Statistical Characteristics of Bubble Flow, Proc, 3 CEC, vol.2, pp. 112-119, Graz.

Žun, I., and Malahovsky, A., 1982, The Terminal Properties of Single Gas Bubbles, Proc. 3CEC, vol.2, pp. 120-127, Graz.

Žun, I., 1985, The Role of Void Peaking in Vertical Two-Phase Bubbly Flow, in Papers presented at the 2nd Int. Conf. on Multi-Phase Flow, London 1985, pp. 127-140, ed. BHRA The Fluid Engineering Centre, Cranfield.

Žun, I., Serizawa, A., Kataoka, I., and Kljenak, I., 1987, Bubble Relative Velocity in Two-Phase Shear Flow, European Two-Phase Flow Group Meeting, Paper 3-I-3, Proc. 6th Yugoslav Congr. Chem. Eng., vol.2, pp. 6-8, Dubrovnik.

Žun, I., 1988a, Transition from Wall Void Peaking to Core Void Peaking in Turbulent Bubbly Flow, Transient Phenomena in Multiphase Flow, ed. N. H. Afgan, pp. 225-245, Hemisphere, Washington.

Žun, I., 1988b, The Mechanism of Bubble Non-Homogeneous Distribution in Two-Phase Shear Flow, ANS Proc. 1988 Nat. Heat Transfer Conf., vol.3, pp. 106-113, ANS, La Grange Park.

PHASE SEPARATION AND DISTRIBUTION PHENOMENA

EFFECTS OF DROPLET-SIZE DISTRIBUTION AND FLOW-BLOCKAGE UPON INERTIA COLLECTION OF DROPLETS BY HORIZONTAL CYLINDERS IN DOWNWARD FLOW OF GAS-LIQUID MIST

T. AIHARA
Institute of Fluid Science
Tohoku University, Sendai 980, Japan
W.-S. FU
Institute of Mechanical Engineering
National Chiao Tung University, Hsinchu, Taiwan, R.O.C.
Y. SUZUKI
Mitsui and Company Ltd., Tokyo, Japan

ABSTRACT

This paper describes a trajectory analysis of the inertia collection of monodisperse- and polydisperse-droplets by horizontal circular cylinders in a downward flow of gas-liquid mist, in order to obtain fundamental data on heat-exchanger design. The effect of far-upstream droplet-size distribution and the blockage effect of gas-phase flow on the local collection efficiency and on the velocity and size-distribution of droplets on impingement are examined. Furthermore, general correlating equations for these results and a new equivalent diameter of polydisperse-droplets are proposed.

INTRODUCTION

Cooling of heated bodies by suspending water droplets in a gas stream has a remarkably improved performance of heat transfer in comparison with single-phase gas cooling. Accordingly, gas-water mist cooling is considered to be available for a significant reduction in the size and weight of heat exchangers and for emergency cooling at peak loads or in accidents of normally gas-cooled equipment.

The heat transfer of gas-water mist flow is an extremely complicated phenomenon which depends on the size distribution of water droplets in free-stream, the droplet trajectories, and the flow behavior and evaporation of the water film formed on a heated body. Consequently, although many theoretical and experimental studies have been made, the majority of existing theoretical studies are focused on either the heat transfer from a dry surface or the flow behavior and evaporation of a water film, without consideration for the droplet trajectories, as in Hodgson and Sunderland(1968), Thomas and Sunderland(1970), Finlay(1971), Aihara et al.(1979), Nishikawa and Takase(1979), and Aihara and Fu(1982). The droplet trajectories or the local collection efficiency were taken into consideration only by Goldstein et al.(1967) and Lu and Heyt(1980).

With regard to the droplet trajectory and local collection efficiency in the case of no heat transfer, many analyses have hitherto been made by using various models. As for gas-phase streamlines around a circular cylinder or a sphere, unseparated potential flow models have been adopted, as in Healy(1970), and Morsi and Alexander(1972). As for a normal flat plate, Hess' free-streamline theory(1973) has been used to calculate the gas-phase streamlines, as in Ushiki et al.(1977).

Very recently, Aihara and Fu(1986) have used the inviscid bluff-body wake model by Kiya and Arie(1977), including far-wake displacement effect, and made a trajectory analysis of the local impinging velocities and partial collection efficiency of monodisperse- and polydisperse-droplets by a normal finite flat plate and a circular cylinder that were immersed in a uniform gas-liquid mist flow in the nongravitational field. In its practical application, however, the effect of the wind-tunnel blockage or flow-blockage in tube banks and the gravity effect cannot be neglected.

From the above viewpoint, a trajectory analysis of the inertia collection of droplets by horizontal circular cylinders in a downward flow of gas-liquid mist was carried out in this study, taking into account the far-upstream droplet-size distribution, gravity, and flow blockage effect. First, the droplet trajectories, the local impinging velocities onto the cylinders, and the local collection efficiency were numerically analyzed for monodisperse-droplets. In the case of polydisperse-droplets having a Rosin-Rammler distribution, the effects of the far-upstream droplet-size distribution and other pertinent parameters were examined upon the size-distribution on impingement and the total collection efficiency; furthermore, general correlating equations for these results and a new equivalent diameter of polydisperse-droplets were proposed.

CASE OF MONODISPERSE-DROPLETS

Physical Model and Governing Equations

The heat-transfer rate from the front half surface of a cylinder in air-water mist flow was much greater than that from the rear half surface, as observed in the experiments by Hodgson et al. (1968) and Basilico et al.(1981). Accordingly, in this study, the authors considered only the trajectories of droplets impinging onto the front surfaces of cylinders.

In order to make the analysis easy, the flow field was divided into two regions: one was the far-upstream region where the existence of the cylinders did not affect the gas flow, that is $|v_g/u_g|<10^{-15}$; the other was the near-upstream region where their existence was not negligible. A full description is given here only of the near-upstream region; and the motion of droplets in the far-upstream region is described in the Appendix.

Figure 1 shows the physical model of the near-upstream region in the case of a horizontal cylinder placed in the middle of a channel of width b or one of the cylinders in a cross-flow array

HORIZONTAL CYLINDERS

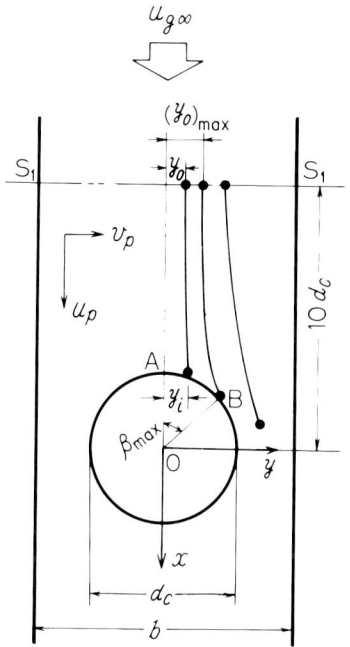

FIGURE 1. Physical model of the near-upstream region.

with a pitch b. The starting line S_1 for calculation of droplet trajectories was located at a distance $10d_c$ upstream the cylinder.

The authors considered a droplet of diameter d_p, which passed through a point y_o on the line S_1, and then impinged onto the cylinder at the position y_i. The interception effect was taken into account for judgement as to the droplet impingement. The authors let $(y_o)_{max}$ be the point on the line S_1 through which the droplet just grazing the cylinder at point B passed. If y_o was greater than this critical value $(y_o)_{max}$, the droplet flew away downstream without impinging onto the cylinder.

For simplification, the following were assumed:
(1) The gas flow is steady, two-dimensional and laminar.
(2) The thickness of the gas-phase boundary layer is very small, compared with the cylinder diameter d_c.
(3) The liquid droplets are spherical, rigid particles and are uniformly distributed in the gas-phase far-upstream.
(4) The cloud of droplets at a low concentration does not affect the gas-phase flow pattern; and the droplet-droplet interaction is negligible.
(5) The gas-droplet interaction follows the drag law for a single rigid particle.
(6) All physical properties remain constant; liquid density is much greater than gas density.
(7) The influence of electrostatic force, free-stream turbulence and the Saffman effect are negligible.

The validity of these was discussed in the authors' previous report(Aihara and Fu, 1986). Subject to the above assumptions, the equations of the motion of a small droplet were derived as follows:

$$\frac{dU_p}{d\tau} = \frac{C_D Re_p}{24K} |W_r| (U_g - U_p) + \frac{1}{Fr^2}, \qquad (1)$$

$$\frac{dV_p}{d\tau} = \frac{C_D Re_p}{24K} |W_r| (V_g - V_p), \qquad (2)$$

where the dimensionless variables are defined as:

$$X = x/d_c, \quad Y = y/d_c, \quad U_p = u_p/u_{g\infty}, \quad V_p = v_p/u_{g\infty},$$
$$U_g = u_g/u_{g\infty}, \quad V_g = v_g/u_{g\infty}, \quad W_r = w_r/u_{g\infty}, \qquad (3)$$
$$\tau = tu_{g\infty}/d_c.$$

The droplet freestream Reynolds number Re_p, inertia parameter K, and Froude number Fr are defined as:

$$Re_p = d_p u_{g\infty}/\nu_g, \qquad (4)$$

$$K = \frac{1}{18}\left(\frac{d_p}{d_c}\right)\left(\frac{\rho_p}{\rho_g}\right) Re_p, \qquad (5)$$

$$Fr = u_{g\infty}/\sqrt{gd_c}. \qquad (6)$$

With regard to the droplet drag coefficient C_D, Morsi and Alexander's polynomial approximation(1972) was used; and the following initial condition was adopted on the basis of the verification in the Appendix:

$$\text{at } X = -10, \quad U_p = U_g \text{ and } V_p = V_g. \qquad (7)$$

The components of gas velocity u_g and v_g were evaluated using БЛОХ's model(Stepanov, 1962) of an inviscid flow with flow blockage but no separation, because it was found by Aihara and Fu(1986) that there was little influence of gas flow separation upon the local partial collection efficiency in the case of a single cylinder in a uniform flow, suspending droplets of $d_p/d_c \geq 10^{-3}$.

БЛОХ's complex potential $F(z)$ and complex velocity $w(z)$ are represented by:

$$F(z) = z - \frac{\gamma}{2\pi i} \ln \frac{\sinh \bar{\chi}(z - ia)}{\sinh \bar{\chi}(z + ia)} + \frac{e\bar{\chi}}{2\pi} \coth \bar{\chi} z, \qquad (8)$$

$$w(z) = 1 - \frac{\gamma \bar{\chi} \sin 2\bar{\chi} a}{\pi \cosh 2\bar{\chi} z - \cos 2\bar{\chi} a} - \frac{e\bar{\chi}^2}{2\pi \sinh^2 \bar{\chi} z}, \qquad (9)$$

with

$$z = x + iy \quad \text{and} \quad \bar{\chi} = \pi d_c/b , \tag{10}$$

where e is the strength of a doublet at the center O of a cylinder and γ is the circulation at a distance $\pm a$ from the center, as shown in Fig. 2. The stream function corresponding to the cylinder surface was obtained from Eq.(8), as follows:

$$y + \frac{\gamma}{4\pi} \ln \frac{\cosh 2\bar{\chi}x - \cos 2\bar{\chi}(y-a)}{\cosh 2\bar{\chi}x - \cos 2\bar{\chi}(y+a)}$$
$$- \frac{e\bar{\chi} \sin 2\bar{\chi}y}{2\pi \cosh 2\bar{\chi}x - \cos 2\bar{\chi}y} = 0 . \tag{11}$$

The values of e, γ, and a were determined so as to satisfy the following conditions:

at $y = 0$, $x = \pm d_c/2$,

at $x = 0$, $y = \pm d_c/2$ and $d^2y/dx^2 = \mp 2/d_c$. (12)

By substituting the obtained e, γ, and a into Eq.(9), the velocity components u_g and v_g were determined.

БЛОХ's model(Stepanov, 1962) gives relatively good stream functions up to the far-upstream region; however, the fidelity in conformal mapping of the outline of a cylinder tends to deteriorate with an increase in the blockage ratio d_c/b. Accordingly, the present numerical analysis was carried out only for $d_c/b \le 0.4$. In the case of $d_c/b = 0$, the usual complex potential for a single cylinder in a uniform flow was used.

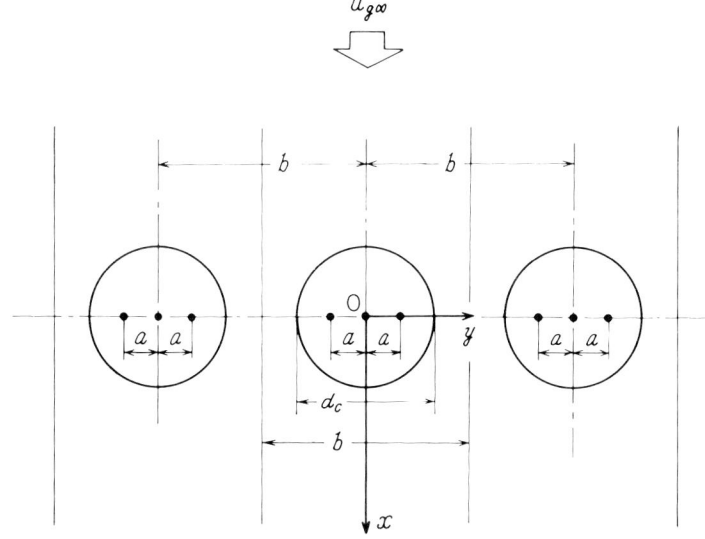

FIGURE 2. БЛОХ's model for an inviscid flow past a transverse array of cylinders.

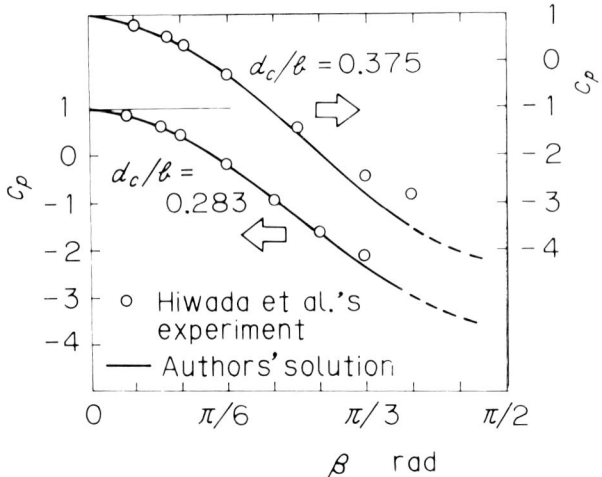

FIGURE 3. Comparison of the present solution with Hiwada and Mabuchi's experimental results(1980) in respect of pressure coefficient C_p on cylinder surfaces; $Re_g = 1.6 \times 10^4$.

Since the method of numerical calculation was the same as in the authors' previous study(Aihara and Fu, 1986) except for the time increment and a consideration of the interception effect, its description is omitted here.

Numerical Results and Discussion

The numerical calculations were made for the range of $Re_p = 4.7-200$, $K = 0.14-57$, $Re_g = 6.7 \times 10^3 - 3.3 \times 10^4$, $Fr = 2.8-14.3$ and ∞, and $d_c/b = 0-0.4$. The numerical results beyond those given here are available in Fu and Aihara(1985).

Figure 3 is a typical plot of the pressure coefficient C_p on the cylinder surface. These theoretical values agree well with the experimental results by Hiwada and Mabuchi(1980) for $\beta < \pi/3$ rad.

Figure 4 shows the gas streamlines and droplet trajectories in the vicinity of a cylinder; the smaller the droplet size d_p/d_c, the more the droplet is liable to bypass the cylinder.

Figure 5 presents the relations between the starting point y_o of trajectory-calculation and the impinging point y_i of the droplet onto the cylinder. The grazing point(point B in Fig.1) shifts downstream with an increase in the droplet size; and the y_i/d_c curves end at the respective grazing points, which are connected by a dashed-line in Fig. 5.

The velocities of droplets on impinging onto the cylinder are plotted in Fig. 6. With an increase in droplet diameter, the impinging velocity in x-direction u_{pi} increases, but that in y-direction v_{pi} decreases. Generally, both of the velocity components increase on being apart from the stagnation point.

HORIZONTAL CYLINDERS

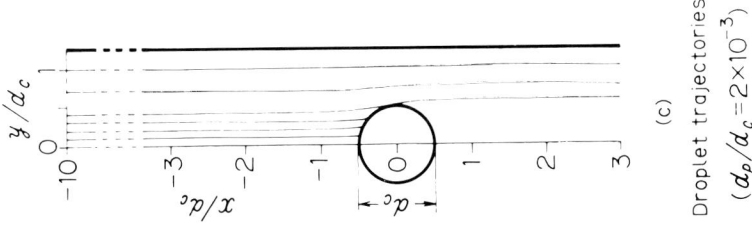
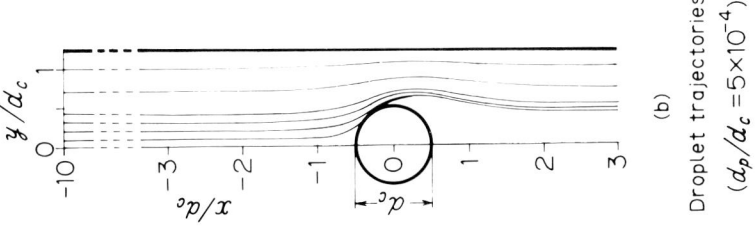
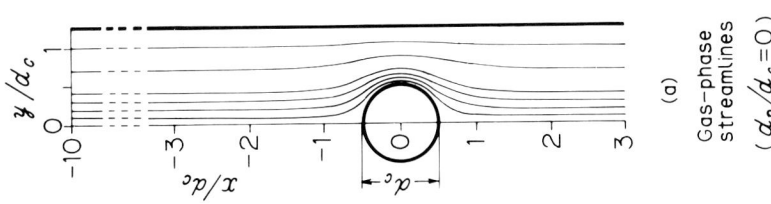

FIGURE 4. Gas streamlines and droplet trajectories near a cylinder with flow blockage; $Re_g = 1.6 \times 10^4$, $Fr = 7.14$, $K(d_p/d_c)^{-2} = 7.5 \times 10^5$, $d_c/b = 0.4$.

260 DISTRIBUTION PHENOMENA

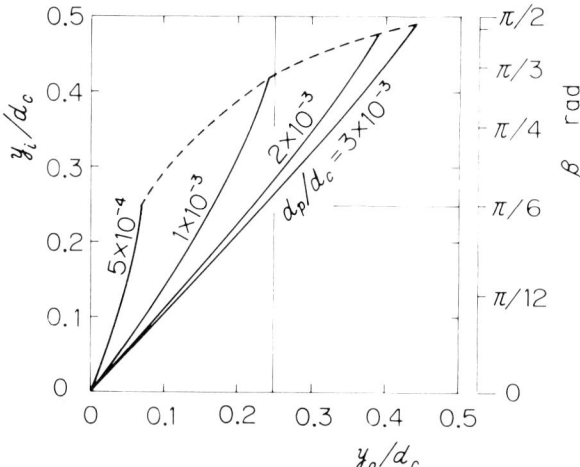

FIGURE 5. Relations between starting point of calculation y_o and impinging point of droplet y_i; the values of parameters are the same as in Fig. 4.

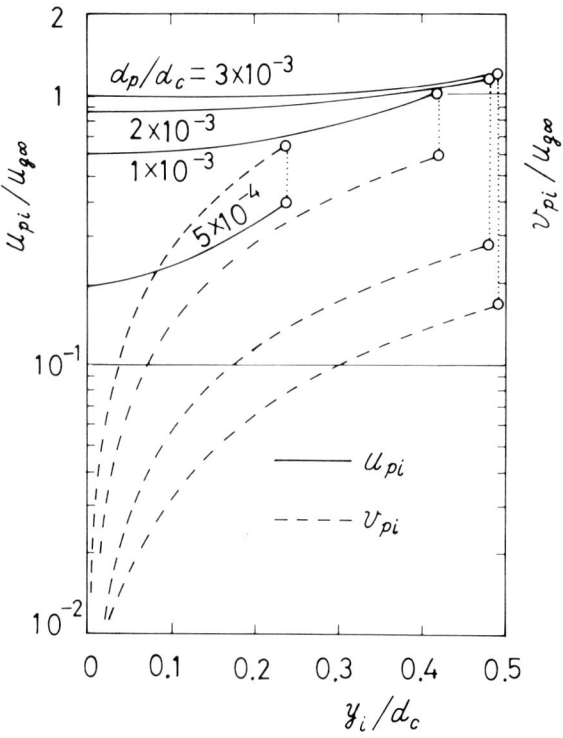

FIGURE 6. Velocity components u_{pi} and v_{pi} of droplets impinging onto a circular cylinder; $Re_g = 1.6 \times 10^4$, $Fr = 7.14$, $K(d_p/d_c)^{-2} = 7.5 \times 10^5$, $d_c/b = 0.4$; symbol O refers to the grazing point.

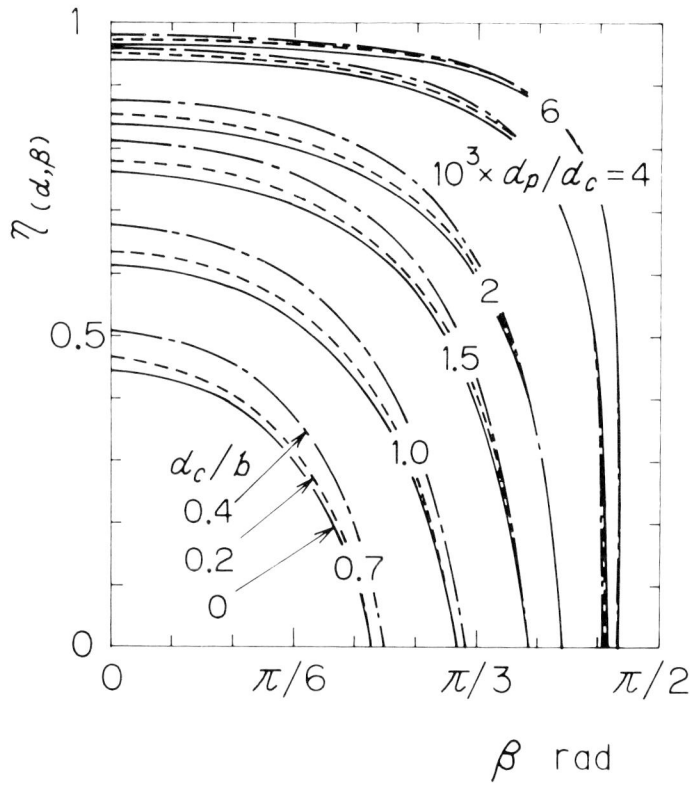

FIGURE 7. Effect of the droplet diameter d_p and blockage ratio d_c/b on the distributions of local partial collection efficiency $\eta_{(\alpha,\beta)}$ in the nongravitational field(Fr = ∞); Re_g = 1.66 $\times 10^4$, $K(d_p/d_c)^{-2}$ = 7.7$\times 10^5$.

Figure 7 presents the distributions of the local partial collection efficiency $\eta_{(\alpha,\beta)}$ which was calculated with Eq.(13).

$$\eta_{(\alpha,\beta)} = dy_0/dy_i . \qquad (13)$$

The derivation of Eq.(13) is given in the previous report(Aihara and Fu, 1986). The value of $\eta_{(\alpha,\beta)}$ increases with an increase in droplet size d_p and blockage ratio d_c/b. The effect of flow blockage is remarkable for the smaller droplets whose inertia force is weak.

Figures 8 and 9 show the effect of gravity on the local partial collection efficiency $\eta_{(\alpha,\beta)}$. The effect of gravity becomes greater with a decrease in the gas freestream velocity $u_{g\infty}$ and on approaching the grazing point. In the case of Re_g = 3.2$\times 10^4$, there was almost no recognizable effect of the gravity over the range of Fr $\geq$ 14.3 and K = 0.73-57.

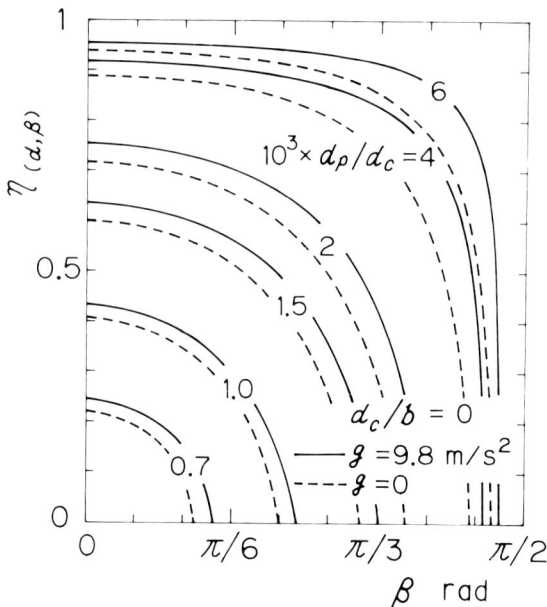

FIGURE 8. Effect of gravity on local partial collection efficiency $\eta_{(d,\beta)}$; $Re_g = 0.67 \times 10^4$, $Fr = 2.86$ and ∞, $K(d_p/d_c)^{-2} = 3.1 \times 10^5$.

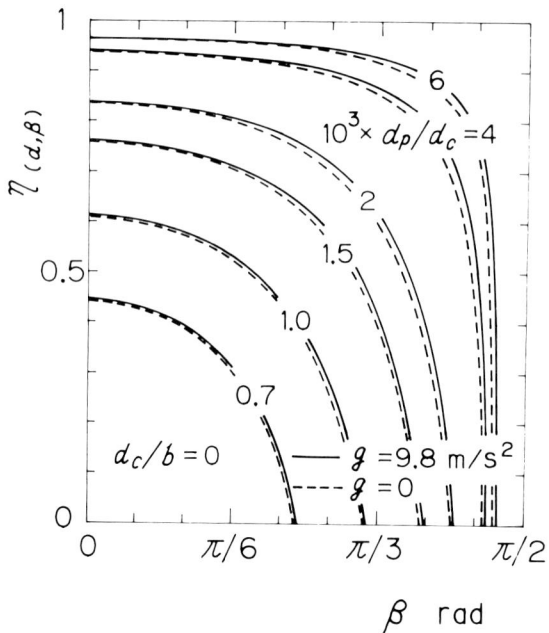

FIGURE 9. Effect of gravity on local partial collection efficiency $\eta_{(d,\beta)}$; $Re_g = 1.66 \times 10^4$, $Fr = 7.14$ and ∞, $K(d_p/d_c)^{-2} = 7.7 \times 10^5$.

CASE OF POLYDISPERSE-DROPLETS

Droplet-Size Distributions Before and After Impingement onto Cylinders

In considering a gas stream containing polydisperse-droplets, the authors first assumed that the droplet size distribution could be approximated by a mass-basis Rosin-Rammler distribution, expressed by Eq.(14). Its applicability has been experimentally verified by Aihara et al.(1979), Hishida et al.(1981) and Aihara et al.(1985).

$$f_{w(d)} = \frac{n}{d_o} \left(\frac{d_p}{d_o} \right)^{n-1} \exp\left[-\left(\frac{d_p}{d_o} \right)^n \right], \tag{14}$$

where $f_{w(d)}$ is the mass-basis size distribution function, d_o the size parameter and n the dispersion parameter.

Then, the mass-basis cumulative oversize fractions were calculated, utilizing the numerical results in the preceding section, after the same method as in the authors' previous report (Aihara and Fu, 1986), over the ranges of $d_o/d_c = 8\times10^{-4}$ to 2×10^{-3} and $n = 1$ to 4.

Some of the results obtained are shown in Figs. 10(a) and 10(b), where the size distributions of droplets in freestream are plotted by dashed lines and those of droplets impinging onto the cylinder, by solid curves. The smaller the size parameter d_o, the greater the differences become between the droplet-size distributions before and after the impingement. This tendency is more remarkable in a region of smaller droplet size. There was little influence of the blockage ratio on the differences between both the size distributions.

Log-normal distribution has been frequently used for the size-distribution analysis of drops captured by a body (Adler and Marshall, 1951; Ôyama et al., 1953). Figures 11(a) and 11(b) are replots of the cumulative oversize fractions on the log-probability papers. In this case, the size distribution of the droplets on impinging onto the circular cylinder represents a nearly straight line. This suggests the possibility of the following mistake: if a blunt body of large size is used for the inertia collection of liquid droplets originally having a Rosin-Rammler distribution, then those polydisperse droplets may be taken to be droplets having a log-normal distribution.

GENERAL CORRELATING EQUATIONS

Local Partial Collection Efficiency

As may be seen from Fig. 12, the local partial collection efficiency is well correlated with Eqs.(15)-(17).

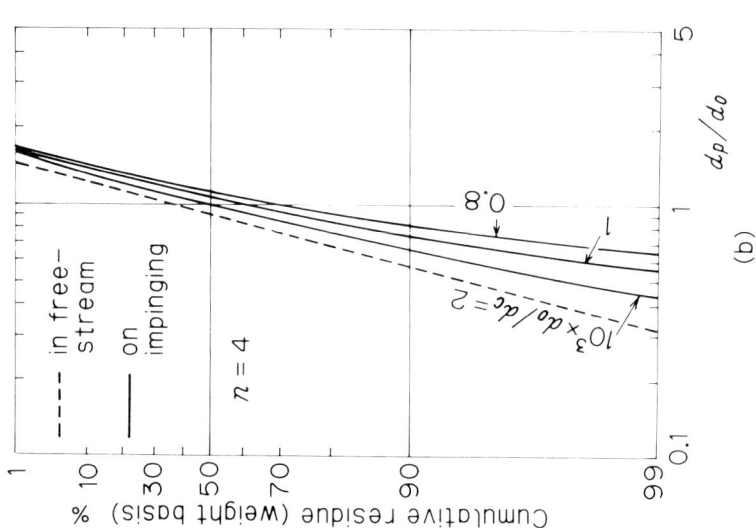
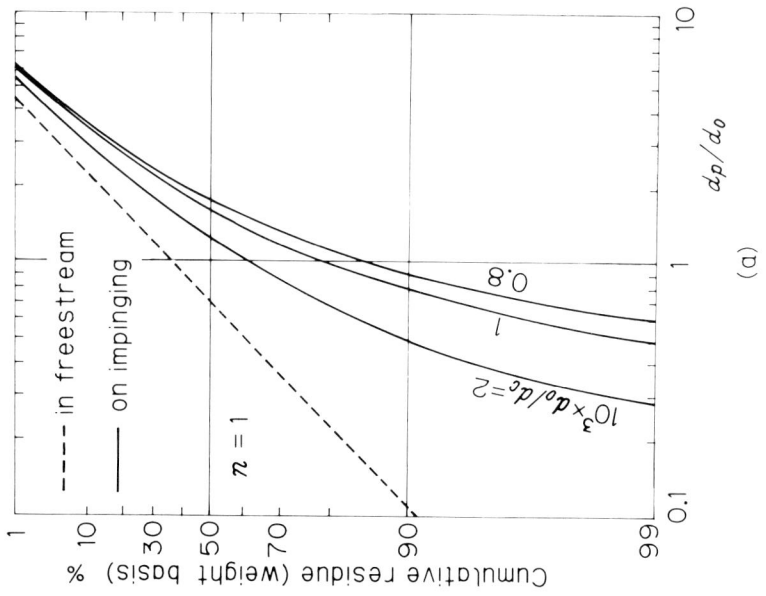

FIGURE 10. Size distributions of droplets in freestream and on impinging onto circular cylinders, on Rosin-Rammler papers; $Re_g = 1.6 \times 10^4$, $Fr = 7.14$, $K(d_p/d_c)^{-2} = 7.5 \times 10^5$, $d_c/b = 0.4$.

HORIZONTAL CYLINDERS

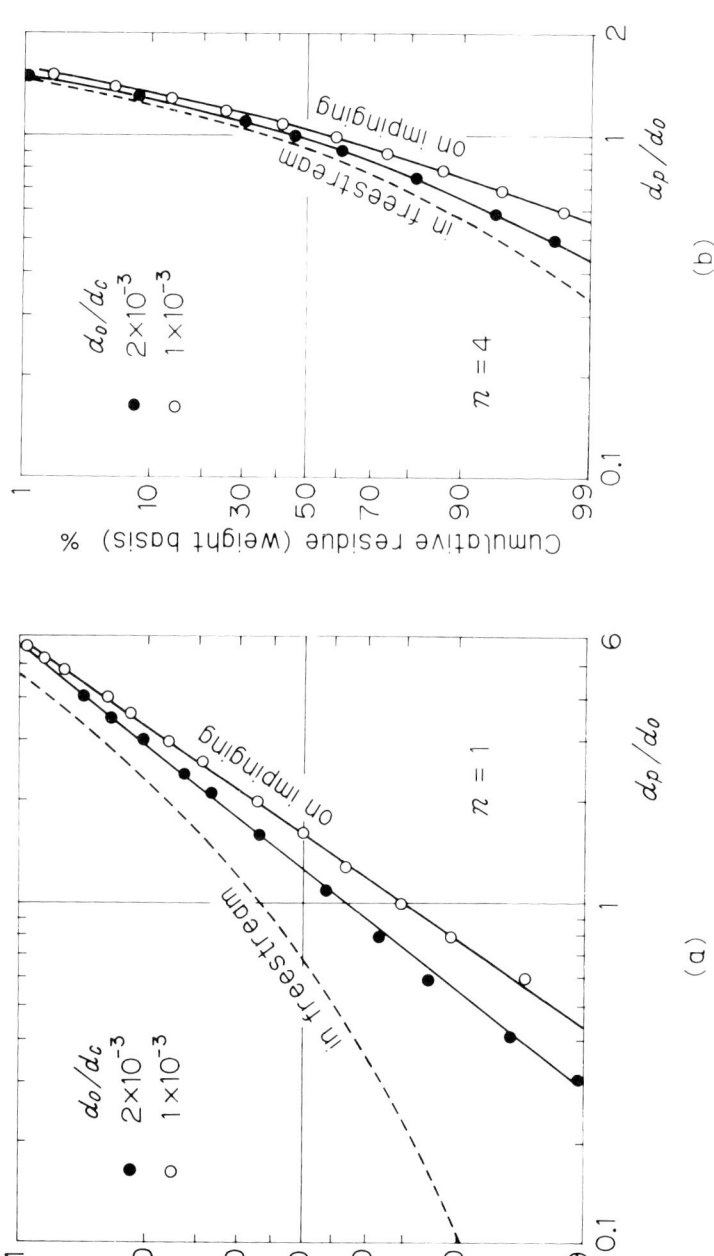

FIGURE 11. Size distributions of droplets in freestream and on impinging onto circular cylinders, on log-probability papers; the values of parameters are the same as in Fig. 10.

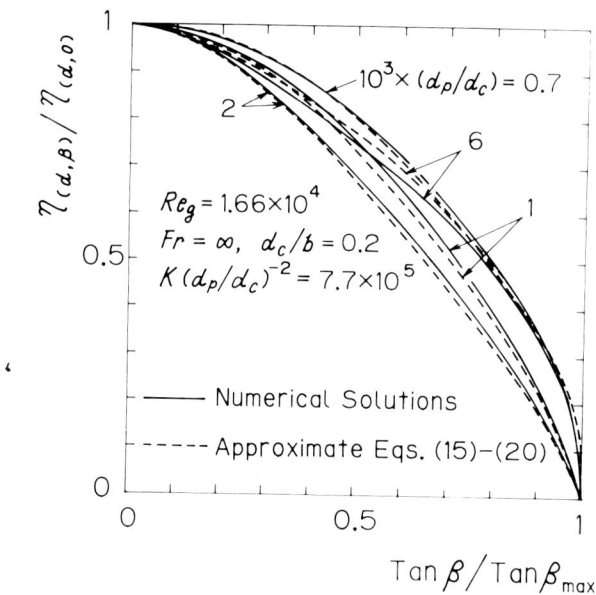

FIGURE 12. General correlation of local partial collection efficiency $\eta_{(d,\beta)}$; $Re_g = 1.66 \times 10^4$, $Fr = \infty$, $K(d_p/d_c)^{-2} = 7.7 \times 10^5$, and $d_c/b = 0.2$.

$$\frac{\eta_{(d,\beta)}}{\eta_{(d,0)}} = \left[2\left\{\left(\frac{\tan\beta}{\tan\beta_{max}}\right)^{m_1} + 1\right\}^{-1} - 1 \right]^{m_2} \quad (15)$$

with

$$m_1 = 2\left[1 - 0.12 \ln K + 0.033 K^{0.5}\right] \quad (16)$$

and

$$m_2 = 0.233\left[1 + 0.024\left(\ln\frac{K}{1.4}\right)^2 - 0.065 \ln Re_g - \frac{0.093}{Fr}\right]^{-1}. \quad (17)$$

Equations (15)-(17) are applicable for the range of the present calculation within the maximum error of ±0.067 in $\eta_{(d,\beta)}$ (within ±3 percent for the majority of the results). Here β_{max} is the azimuth angle of the grazing point and was approximated by Eq. (18) within the maximum error of ±0.045 rad (within ±3 percent for the majority of the present results):

$$\beta_{max} = \frac{\pi}{2}\eta_{(d,0)}^{0.86} - 0.047\left[1 + 2.5\left(\frac{d_c}{b}\right) - \frac{2}{Fr}\right]. \quad (18)$$

The partial collection efficiency at the stagnation point $\eta_{(d,0)}$ was also correlated with the following equations within the maximum error of 0.02 for the range of the present calculation:

HORIZONTAL CYLINDERS

$$\eta_{(d,0)} = \left(1 - \frac{1}{K^*}\right)\left[1 + \frac{3.8}{Re_p^{0.88}}\left(\frac{d_c}{b}\right)^{1.5}\right]\left[1 + \frac{0.72}{(Re_p\,Fr)^{0.78}}\right], \quad (19)$$

with

$$K^* = 2.99 \exp\{0.054(\ln K + 10.2)\ln K\}. \quad (20)$$

Local Total Collection Efficiency and Equivalent Diameter

The local total collection efficiency $\eta_{(\beta)}$, expressed by Eq. (21), is required for the analysis of heat- and mass-transfer of gas-liquid mist flow.

$$\eta_{(\beta)} = \int_{(d_p)_{min}}^{(d_p)_{max}} f_{w(d)}\,\eta_{(d,\beta)}\,d\,d_p. \quad (21)$$

It is, however, somewhat troublesome to numerically integrate Eq.(21) with Eqs.(14) through (20), on each occasion.

Hence the authors propose a new equivalent diameter d_e for practical convenience. This is defined by Eq.(22), as the diameter of imaginary monodisperse-droplets which would bring about the same rate of total impingement onto the cylinder as the polydisperse-droplets having the size distribution of Eq.(14).

$$\int_0^{\beta_{max}} \eta_{(d_e,\beta)}\,d\beta = \int_0^{\beta_{max}} \eta_{(\beta)}\,d\beta. \quad (22)$$

The values of equivalent diameter d_e were evaluated for the pertinent parameters of various values, by solving Eq.(22) with successive approximation method with the 1-%-diameter as $(d_p)_{max}$ and the 99.9-%-diameter as $(d_p)_{min}$; then, a general correlating equation for d_e was derived as

$$\frac{d_e}{d_c} = 0.196\left(\frac{d_o}{d_c}\right)^P n^{1.17} Re_g^{-0.057}\left[1 - 0.04\left(\frac{d_c}{b} + \frac{1}{Fr}\right)\right] \quad (23)$$

with

$$P = 0.73\,n^{0.18}, \quad (24)$$

which is applicable for n = 2-4 with the maximum error of +2 to -4 percent (within ±1 percent for the majority of the present results).

Figure 13 shows a comparison between the exact value $\eta_{(\beta)}$, according to Eq.(21), and the approximate value $\eta_{(d_e,\beta)}$, obtained with the equivalent diameter d_e from Eq.(23). The approximate collection efficiency $\eta_{(d_e,\beta)}$ is close to the true total collection efficiency $\eta_{(\beta)}$ in the region of $\beta \leq \pi/3$ which is important for the heat transfer; however, the angle of the grazing point β_{max} is underestimated by the use of d_e, because the polydisperse-droplets contain the droplets of diameters greater than d_e.

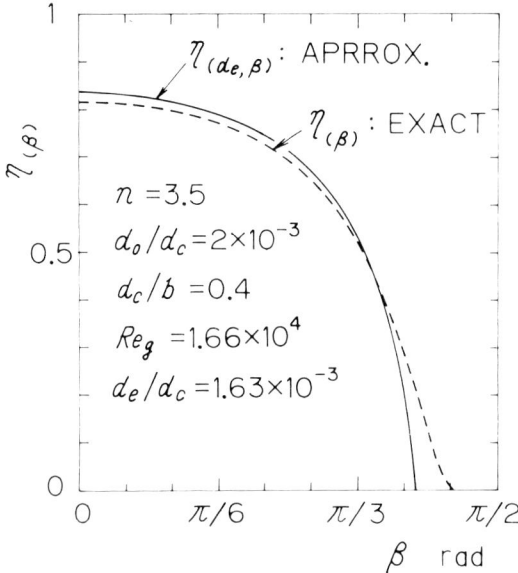

FIGURE 13. Comparison between exact value $\eta_{(\beta)}$ and approximate value $\eta_{(d_e,\beta)}$.

Impinging Velocity of Droplet

The tangential component of the impinging droplet velocity w_i is necessary for the analysis of the formation and flow behavior of the liquid film on the cylinder surface. The authors let α be the angle between the impingement direction of a droplet of diameter d_ρ at the azimuth angle β, and the tangent to the cylinder surface or liquid film.

The values of tangential velocity $|w_i|\cos\alpha$ were numerically calculated for the pertinent parameters of various values. Those numerical results were correlated with the following equations within the maximum error of +0.085 to -0.011 (within ±5 percent for the majority of the present results):

$$\frac{|w_i|}{u_{g\infty}}\cos\alpha = \left[1 + 0.012\,\frac{Re_p^2}{Fr^{3.4}} + \frac{0.21}{K^{0.52}}\left\{1 + \frac{\pi}{2}\left(\frac{d_c}{b}\right)^2\left(3 - \frac{1}{K^{0.52}}\right)\right\}\right]\sin\beta. \quad (25)$$

CONCLUSIONS

A trajectory analysis of the inertia collection of monodisperse- and polydisperse-droplets by horizontal circular cylinders in a downward flow of gas-liquid mist was made, taking into account the far-upstream droplet-size distribution, gravity, and flow blockage effect. The results obtained are summarized as follows:

HORIZONTAL CYLINDERS

(1) The smaller the droplet size, the more the droplet is liable to bypass the cylinder; and the grazing point of the droplet on the cylinder shifts downstream with an increase in the droplet size.

(2) With an increase in droplet diameter, the impinging droplet velocity increases in x-direction, but decreases in y-direction. Generally, both of the velocity components increase on being apart from the stagnation point.

(3) The local partial collection efficiency increases with an increase in the droplet size and blockage ratio; and the effect of flow blockage is remarkable for the smaller droplets.

(4) The effect of the gravity on the local partial collection efficiency becomes greater with a decrease in the gas freestream velocity and on approaching the grazing point. In the case of the gas Reynolds number $Re_g = 3.2 \times 10^4$, there was almost no recognizable effect of the gravity.

(5) The smaller the droplet size, the greater the differences become between the droplet-size distributions before and after the impingement. If a blunt body of large size is used for the inertia collection of liquid droplets originally having a Rosin-Rammler distribution, then those polydisperse-droplets may be mistaken to be droplets having a log-normal distribution.

(6) A new equivalent diameter was proposed as the diameter of imaginary monodisperse-droplets which would bring about the same rate of total impingement onto the cylinder as the polydisperse droplets.

(7) General correlating equations were derived for the local partial collection efficiency, the azimuth angle of the grazing point, the equivalent diameter, and the tangential component of the impinging droplet velocity.

ACKNOWLEDGEMENTS

The authors wish to express their sincere gratitude to Professor H. Daiguji, Tohoku University for his advice on the application of the БЛОХ's model, to Dr. J.K. Kim for his assistance in making numerical calculation, and to Research Associate, Mr. M. Hongoh, Institute of Fluid Science, Tohoku University for his assistance in drawing the figures. The research was carried out with the Grants-in-Aid for Scientific Research B-58460105 of the Ministry of Education, Science and Culture of Japan.

APPENDIX

Motion of a Droplet in Far-Upstream Region

The velocity u_p of a droplet varies with the travelling distance x from the location at which the droplet is introduced into a vertical, downward, gas stream with a velocity $u_{g\infty}$. Figures 14 and 15 are typical plots of the numerical solutions obtained

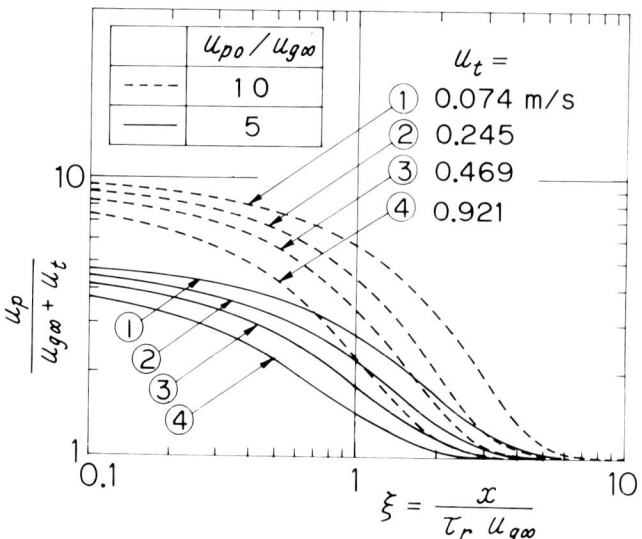

FIGURE 14. Effect of initial droplet velocity u_{po} upon a decrease in droplet velocity u_p in vertical, downward gas flow of $u_{g\infty}$ = 5 m/s.

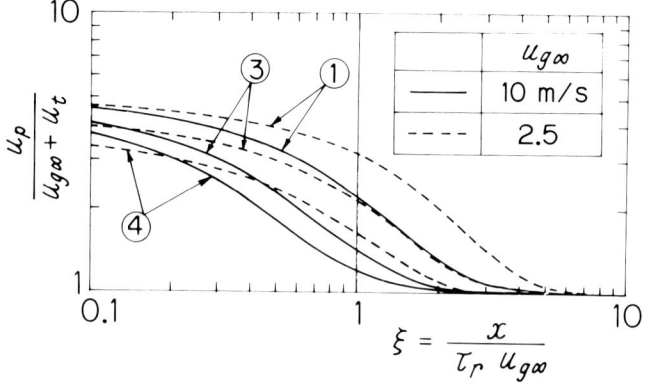

FIGURE 15. Effect of gas velocity $u_{g\infty}$ upon a decrease in the velocity u_p of a droplet with initial velocity of $u_{po} = 5u_{g\infty}$; items without mention here are the same as in Fig. 14.

with Eq.(1) for ρ_p/ρ_g = 840, corresponding to water drops in an air stream. In the figures, u_{po} and u_t are the initial velocity of the droplet at x = 0 and its terminal settling velocity in still gas, respectively; τ_r is the relaxation time defined as

$$\tau_r = \frac{1}{18}\left(\frac{\rho_p}{\rho_g}\right)\frac{d_p^2}{\nu_g} \quad . \tag{26}$$

HORIZONTAL CYLINDERS

As may be seen from the figures, the droplet velocity u_p decreases to its terminal value $(u_{g\infty} + u_t)$ within a short travelling time; for example, the value of u_p decreases to $2u_{g\infty}$ within about 0.1 sec in the case of $u_t = 0.469$ m/s, $u_{g\infty} = 5$ m/s, and $u_{po} = 50$ m/s.

Effect of Initial Condition upon Grazing Trajectories

According to the authors' numerical experiment (Fu and Aihara, 1985), when the starting line S_1 for numerical integration of Eqs.(1) and (2) was selected to be at $X = -10$, the initial condition of droplet motion exerted little influence upon the droplet trajectory in the near-upstream region. For example, the value of $(y_0)_{max}$, computed using the initial condition of $U_p = 0$ or 2, agreed with that computed using the initial condition of $U_p = 1$ within a deviation of ± 0.2 percent for $d_p/d_c < 1.5 \times 10^{-3}$ and within ± 2 percent for $d_p/d_c < 5 \times 10^{-3}$.

Consequently, it may be concluded from the above discussions that the initial condition of Eq.(7) enables one to predict droplet trajectories within an error of ± 2 percent in the liquid spray systems which one encounters in practical applications.

NOMENCLATURE

b = channel width or transverse pitch of cylinders
C_D = drag coefficient of a droplet
C_p = pressure coefficient, $(p - p_\infty)/(\rho_g u_{g\infty}^2/2)$
d_c = diameter of circular cylinder
d_e = equivalent diameter, Eq.(22)
d_p = diameter of droplet
d_0 = size parameter in Rosin-Rammler equation (14)
Fr = Froude number, Eq.(6)
$f_{w(d)}$ = mass-basis size distribution function, Eq.(14)
g = gravitational acceleration
K = inertia parameter, Eq.(5)
n = dispersion parameter in Rosin-Rammler equation (14)
Re_g = gas Reynolds number, $d_c u_{g\infty}/\nu_g$
Re_p = droplet freestream Reynolds number, Eq.(4)
t = time
u, v = velocities in x- and y-directions
$\boldsymbol{w_i}$ = impinging velocity vector of droplet onto a cylinder
$\boldsymbol{w_r}$ = relative velocity vector between gas-phase and droplet
x, y = Cartesian coordinates as shown in Fig.1
y_i = position of droplet impinging onto a cylinder, Fig.1

y_o = position of droplet crossing the line S_1, Fig.1

$(y_o)_{max}$ = y_o of the grazing droplet trajectory, Fig.1

ν = kinematic viscosity of fluid
ρ = fluid density
β = azimuth angle from the forward stagnation point
$\eta_{(d,\beta)}$ = local partial collection efficiency at a position of angle β, for a droplet of diameter d_p, Eq.(13)
$\eta_{(\beta)}$ = local total collection efficiency, Eq.(21)

Subscript

d = droplet of diameter d_p
g = gas phase
i = on impinging
p = droplet
∞ = far-upstream or freestream

REFERENCES

Adler, C. R., and Marshall, W. R.,Jr., 1951, Performance of Spinning Disk Atomizers-Part II, Chem. Engng. Progr., Vol. 47, No. 12, pp. 601-608.

Aihara, T., and Fu, W.-S., 1982, Heat Transfer from a Wedge in Air-Water Mist Flow (Theoretical Study of a Vertical Infinite Wedge of Uniform Wall Temperature), Transactions of Japan Society of Mechanical Engineers (Ser.B), Vol. 48, No. 436, pp. 2536-2546. (In Japanese)

Aihara, T., and Fu, W.-S., 1986, Effects of Droplet-Size Distribution and Gas-Phase Flow Separation Upon Inertia Collection of Droplets by Bluff-Bodies in Gas-Liquid Mist Flow, International Journal of Multiphase Flow, Vol. 12, No. 3, pp. 389-403.

Aihara, T., Shimoyama, T., Hongoh, M., and Fujinawa, K., 1985, Instrumentation and Error Sources for the Measurement of the Local Drop-Size Distribution by an Immersion-Sampling Cell, Proceedings, 3rd International Conference on Liquid Atomisation and Spray Systems, P. Eisenklam and A. Yule, ed., Institute of Energy, London, Vol. 2, pp. VC/5/1-VC/5/11.

Aihara, T., Taga, M., and Haraguchi, T., 1979, Heat Transfer from a Uniform Heat Flux Wedge in Air-Water Mist Flows, International Journal of Heat and Mass Transfer, Vol. 22, No. 1, pp. 51-60.

Basilico, C., Jung, G., and Martin, M., 1981, Etude du Transfert Convectif Entre un Cylindre Chauffe et un Ecoulement d'Air Charge de Gouttelettes d'Eau, International Journal of Heat and Mass Transfer, Vol. 24, No. 3, pp. 371-385.

Finlay, I. C., 1971, An Analysis of Heat Transfer During Flow of an Air/Water Mist Across a Heated Cylinder, Canadian Journal of Chemical Engineering, Vol. 49, June, pp. 333-339.

Fu, W.-S. and Aihara, T., 1985, Heat Transfer from a Heated Body in Air-Water Mist flow(2nd Report), Transactions of Japan Society of Mechanical Engineers(Ser.B), Vol. 51, No. 463, pp. 874-881. (In Japanese)

Goldstein, M. E., Yang, W.-J., and Clark, J. A., 1967, Momentum and Heat Transfer in Laminar Flow of Gas with Liquid-Droplet Suspension over a Circular Cylinder, ASME Journal of Heat Transfer, Vol. 89, No. 2, pp. 185-194.

Healy, J. V., 1970, Perturbed Two-Phase Cylindrical Type Flows, Physics of Fluids, Vol. 13, No. 3, pp. 551-557.

Hess, J. L., 1973, Analytical Solutions for Potential Flow over a Class of Semi-Infinite Two-Dimensional Bodies Having Circular-Arc Noses, Journal of Fluid Mechanics, Vol. 60, No. 2, pp. 225-239.

Hishida, K., Maeda, M., and Ikai, S., 1981, Study on Heat Transfer in a Binary Mist Flow, Transactions of Japan Society of Mechanical Engineers(Ser.B), Vol. 47, No. 419, pp. 1279-1286. (In Japnanes)

Hiwada, M., and Mabuchi, I., 1980, Flow Around and Heat Transfer from a Circular Cylinder with Large Blockage Ratios, Transactions of Japan Society of Mechanical Engineers (Ser.B), Vol. 46, No. 409, pp. 1750-1759. (In Japanese)

Hodgson, J. W., Saterbak, R. T., and Sunderland, J. E., 1968, An Experimental Investigation of Heat Transfer from a Spray Cooled Isothermal Cylinder, ASME Journal of Heat Transfer, Vol. 90, No. 3, pp. 457-463.

Hodgson, J. W., and Sunderland, J. E., 1968, Heat Transfer from a Spray-Cooled Isothermal Cylinder, Industrial and Engineering Chemistry Fundamentals, Vol. 7, No. 4, pp. 567-572.

Kiya, M., and Arie, M., 1977, An Inviscid Bluff-Body Wake Model Which Includes the Far-Wake Displacement Effect, Journal of Fluid Mechanics, Vol. 81, No. 3, pp. 593-607.

Lu, C. C., and Heyt, J. W., 1980, Heat Transfer from Two-Phase Boundary Layers on Isothermal Cylinder: Influence of Drop Trajectory, AIChE Journal, Vol. 26, No. 5, pp. 762-769.

Morsi, S. A., and Alexander, A. J., 1972, An Investigation of Particle Trajectories in Two-Phase Flow Systems, Journal of Fluid Mechanics, Vol. 55, No. 2, pp. 193-208.

Nishikawa, N., and Takase, H., 1979, Effects of Particle-Size and Temperature Difference on Mist Flow over a Heated Circular Cylinder, ASME Journal of Heat Transfer, Vol. 101, No. 4, pp. 705-711.

Oyama, Y., Eguti, M., and Endou, K., 1953, Studies on the Atomization of Water Droplets, Kagaku Kogaku, Society of Chemical Engineers, Japan, Vol. 17, No. 7, pp. 269-275. (In Japanese)

Stepanov, G. Ju., 1962, <u>Hydrodynamics of Cascade Turbomachinery</u>, p. 58, National Publisher, Physicomathematical Literature, Moscow. (In Russian)

Thomas, W. C., and Sunderland, J. E., 1970, Heat Transfer between a Plane Surface and Air Containing Suspended Water Droplets, <u>Industrial and Engineering Chemistry Fundamentals</u>, Vol. 9, No. 3, pp. 368-374.

Ushiki, K., Kubo, K., and Iinoya, K., 1977, Inertial Separation of Particle by Ribbon (Effects of Incidence Angle and Deviation from Stokes Drag), <u>Kagaku Kogaku Ronbunshu, Society of Chemical Engineers, Japan</u>, Vol. 3, No. 2, pp. 172-178. (In Japanese)

VOID FRACTION DISTRIBUTION IN TWO-PHASE SINGLE-COMPONENT CONDENSING FLOW

S. TODA and Y. HORI
Department of Nuclear Engineering
Tohoku University
Aramaki-Aoba, Sendai, Japan 980

ABSTRACT

The behavior of condensing vapor bubbles in subcooled bulk flow is a complicated phenomenon influenced by both heat and mass transfer. To clarify the characteristics of such a thermal-nonequilibrium two-phase flow, an experimental research has been done using a boiling water loop. The movement of vapor bubbles flowing in the transparent test section was recorded by a video camera. An image-processing technique was newly introduced to measure the distributions of void fraction along the test section. The average density of two-phase fluid was also measured by using a density meter. These void fraction data plotted against inlet liquid subcooling were located midway between two curves which were predicted respectively from the Saha-Zuber subcooled boiling theory and an assumption considering the thermal-equilibrium condition. From these visual observations, two different types of bubbles were found in their condensing process during flowing in the same subcooled bulk flow, and they were largely influenced by subcooling distributed in their surrounding liquid.

INTRODUCTION

In the early stage of loss-of-coolant accident (LOCA) in a pressurized water reactor (PWR), sudden coolant boiling occurs on the surface of fuel rods. In the next stage, it changes into subcooled boiling because cool water is immediately poured from the emergency core cooling system (ECCS). Flowing up in the subcooled water, the generated bubbles are abruptly reduced in volume and disappear by condensation. This flow can be regarded as a transient two-phase flow under a thermal-equilibrium condition. These behaviors of the bubbles greatly affect the thermo-hydrodynamic and nuclear-kinetic characteristics of nuclear reactor power plants. On the analysis of the subcooled boiling stage in an uniformly heated tube, several works (1),(3)-(6) have been reported. These studies concentrate upon the problem of void fraction distribution along the tube channel which is uniformly heated at a constant heat flux. When the two-phase mixture of saturated or superheated vapor and subcooled liquid flows into an unheated tube, void fraction decreases with condensation of

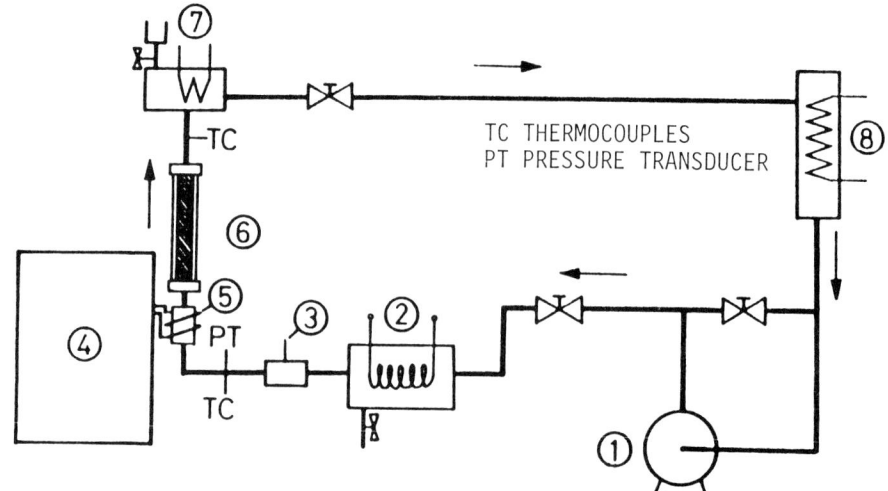

1. MAGNET PUMP
2. PREHEATING TANK
3. DRAG-DISC TYPE FLOWMETER
4. HIGH FREQUENCY INDUCTION HEATING MACHINE
5. WORK COIL
6. TEST SECTION
7. PRIMARY COOLING TANK
8. SECONDARY COOLING TANK

FIGURE 1. Schematic diagram of experimental apparatus

vapor phase along the flow direction until the flow reaches the thermal-equilibrium condition. Regarding this complicated phenomenon, Hori et al. (2) measured temperature fluctuations in the two-phase mixture using microthermocouples. And they estimated the decrease of liquid bulk subcooling caused by the condensation of vapor phase.

The purpose of this investigation was to know the behavior of these condensing vapor bubbles in subcooled bulk flow and to clarify the influences of experimental parameters such as inlet liquid subcooling and flow velocity. In this study, a single-component two-phase mixture flowed up in a vertical circular tube at atmospheric pressure. The movement of liquid-vapor interface at the transparent test section was visually recorded on video tapes. The void fraction distributions and the condensing behaviors of vapor bubbles were measured by using an image processing technique under various inlet velocities and subcoolings. The void fraction data were compared with those estimated from average density data of two-phase mixture obtained by using a vibration type density meter. Finally, the obtained void fraction characteristics were compared with the previous theories of subcooled boiling, and the condensing behaviors of individual bubbles were discussed.

EXPERIMENTAL APPARATUS AND PROCEDURE

A schematic diagram of the experimental apparatus is shown in Fig. 1. A small-scale boiling loop was filled up with dis tilled water. Experiments were carried out at nearly atmospheric pressure. The working fluid was circulated through the

VOID FRACTION DISTRIBUTION

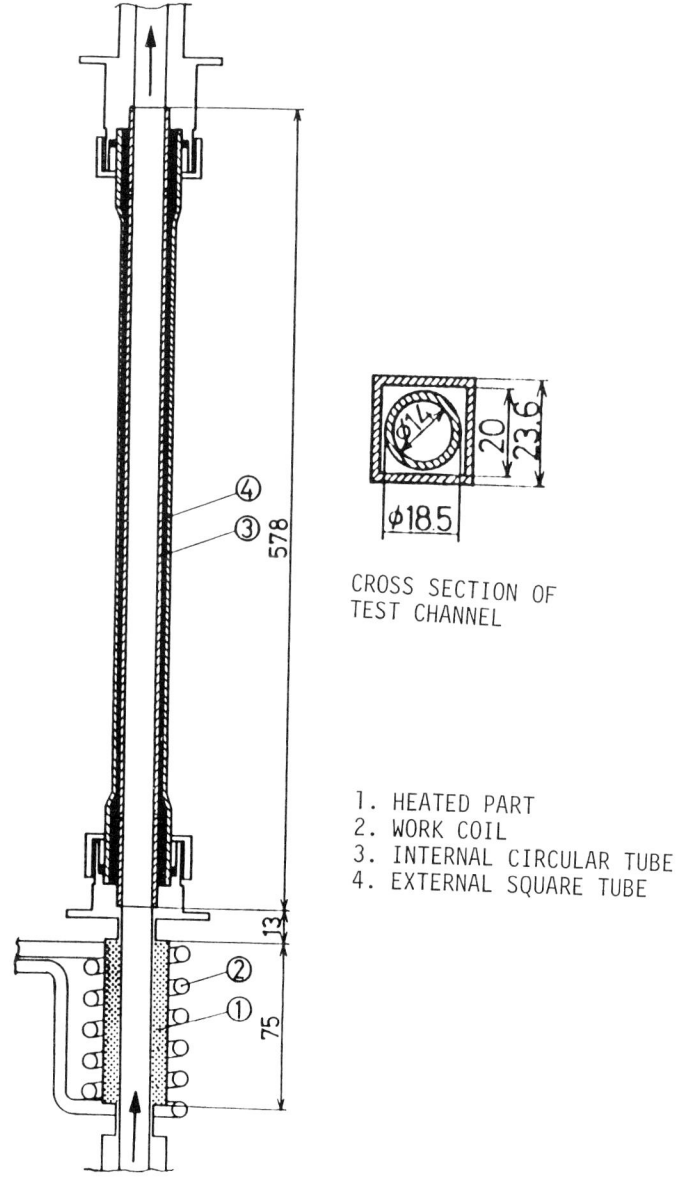

FIGURE 2. Test section for visualized experiment

loop by magnet pump 1, and its flow rate was measured by a drag-disc type flowmeter 3. The inlet liquid temperature was controlled by regulating electrical heat input in preheating tank 2. A part of the test section 6 was heated by a high-frequency induction heating machine, 4. The vapor bubbles generated at its internal heated wall flowed up in the test section, and were mixed with subcooled bulk water. They diminished by condensation. Thereafter, the flow reached the thermal-equilibrium condition. After passing the test section 6, the remaining uncondensed vapor bubbles were separated from liquid phase, and condensed completely in primary and

278   DISTRIBUTION PHENOMENA

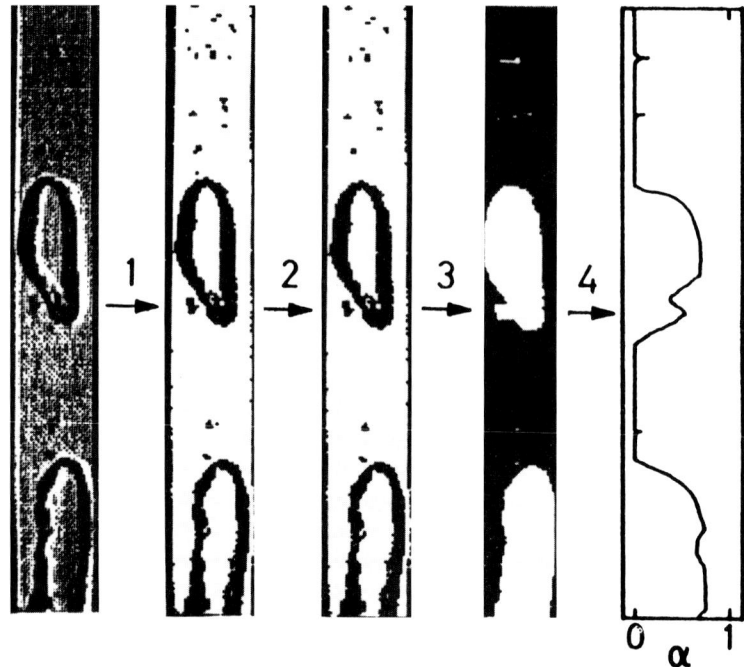

1. DETECTING INTERFACES (TWO-VALUING)
2. ERASING INTERRUPTIVE DOTS (ERASING SINGLE POINTS)
3. SEPARATING TWO PHASES
4. CALCULATING TWO-PHASE FLOW PARAMETERS

FIGURE 3. Process of measuring two-phase flow using the video image processor

secondary cooling tanks 7 and 8.

The details of the test section are illustrated in Fig. 2. The section was a vertical circular tube of 14 mm in internal diameter, in which the working fluid flowed up. The lower part of the test section, a thick walled cylinder, was made of stainless steel (SUS 304) and surrounded by the work coil of the high frequency induction heating machine. This part was heated by electrical induction heating, and then subcooled boiling occurred at its internal wall surface. The upper visualized part of the test section was transparent, and was composed of double tubes of Pyrex glass. The inner tube was circular and the outer was square. The gap between them was filled up with distilled water, which was designed to be free from the optical distortion caused by refraction.

The pictures of vapor bubbles moving at the transparent part were recorded at time intervals of 1/60 second on video tapes by a video camera. The camera had a 1/1000 second electrical shutter and, within the range of the experimental conditions, the recorded pictures were sufficiently clear without shading. A 300-watt electric lump was placed on the opposite side of the test section to the camera. Lighted backward by the beam passing through the transparent test section, the liq-

VOID FRACTION DISTRIBUTION

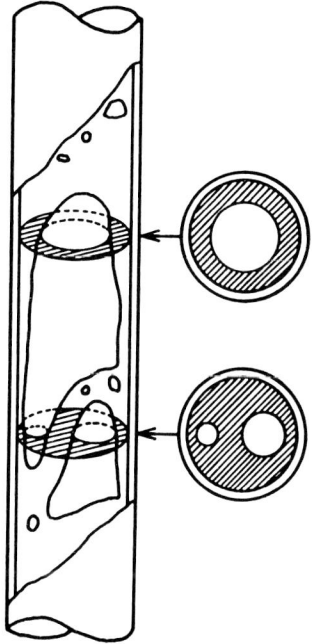

FIGURE 4. Three-dimensional shape model of vapor bubble

uid-vapor interface was observed as a black line because of the different refractive indexes between two phases. Then an image processing technique was newly introduced to measure the time-varying location of liquid-vapor interface.

The detecting of the interface proceeded as demonstrated in Fig. 3: First, the recorded data of video pictures were transmitted to a video image processor which was connected to a personal computer. The processor had a function to digitize video image pictures into a 256×256 dot matrix with a resolution of 256 levels of black-and-white contrast. Second, a threshold level of contrast was properly fixed. Dots whose contrast levels were lower than this threshold level were transformed into black dots, and the others into white dots, in what is generally called the "two-valuing" process. The liquid-vapor interface was brought into high-contrast relief, as shown in Fig. 3.1. Next, in this two-valuing process, isolated dots which might cause small errors in the analysis were often generated due to the uneven lighting. To avoid such errors, these isolated dots were eliminated by a filtering program, Fig. 3.2. Then, the graphic drawing of the two-phase interface was carried out by repeated horizontal scanning, Fig. 3.3. Finally, the distributions of void fraction were calculated by assuming a three-dimensional shape model of vapor bubbles, Fig. 3.4. This model is illustrated in Fig. 4. The horizontal cross-sectional shapes of flowing bubbles were supposed to be circular at any measuring location of the test section. From the results of preliminary experiments, the application of this model was found to be acceptable. The details are discussed later.

The average density of the two-phase mixture was also meas-

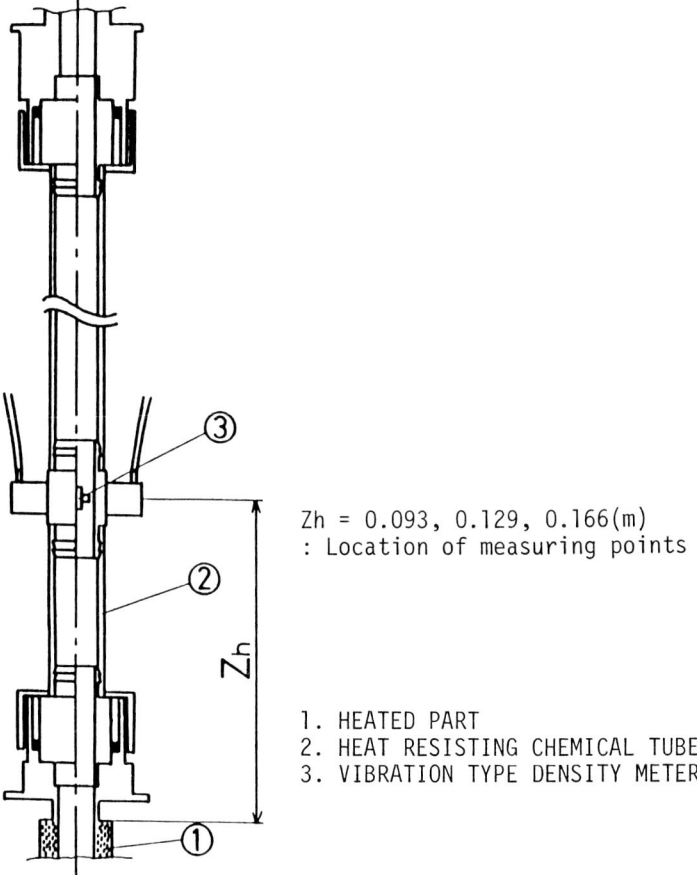

FIGURE 5. Test section for measuring the average density of two-phase mixture

ured by using a vibration type density meter. Figure 5 shows the test section of this measuring experiment. The measurements were carried out at three measuring locations which, in the figure, are represented by the distance Z_h. The density meter had a high-frequency vibrating plate located parallel to the flow direction and at the center of the flowing channel. The fluid density was detected from the variation of its vibrating frequency. Void fraction was calculated from the next equation assuming the homogeneous two-phase flow model.

$$\rho_m = \alpha \rho_g + (1-\alpha)\rho_\ell \tag{1}$$

The void fraction data obtained from the visualized experiments were always compared with these data.

The actual experiments were performed as follows: The inlet water temperature and velocity were set at their initial values. The inlet liquid subcooling was defined against the inlet pressure of the test section. The induction heating started after the whole system of the experimental apparatus became steady at the initially set conditions. Several tens of seconds later, the subcooled boiling initiated at the

VOID FRACTION DISTRIBUTION

heated part reached a steady state, and the vapor generation rate became constant. Then, the measurement using the video camera or the density meter started to run for several seconds. Finally, the data obtained were analyzed to calculate the variation of void fraction, bubble volume, bubble velocity and so on. In this experiment, the liquid inlet velocity was ranged from 0.43 to 1.06 m/s, and the inlet subcooling from 6.6 to 13.8 K. Heat input was set at a constant value of 2.47 kW.

RESULTS AND DISCUSSION

The Accuracy of the Measuring Method Using an Image Processing Technique

In some cases, the three-dimensional shape model (see Fig. 4) supposed in this study might be different from the actual shape of vapor bubbles. Then, in the flow including the bubbles complicated in shape such as churn flow, it might be difficult to maintain accuracy of measurement. In this experiment, preliminary experiments were carried out to confirm the accuracy of this measuring method for practical usage. In these experiments, the inlet liquid temperature maintained the saturated condition. Therefore, the bubbles generated at the heated part flowed up without condensing, and the void fraction distribution along the visualized section did not change in the saturated bulk flow. These experiments were made at atmospheric pressure, inlet liquid velocity in the range of 0.43 to 0.85 m/s and total heat input by induction heating in the range of 1.48 to 1.96 kW.

The results obtained were compared with the theoretical values as follows: The relation between void fraction and quality is given by

$$\alpha = \frac{Gx}{\rho_g u_g} \qquad (2)$$

The mean vapor drift velocity in the saturated water-vapor upward flow was calculated from the next equation presented by Kroeger and Zuber (3).

$$u_g = 1.13(U_g+U_\ell) + 1.18\left[\frac{\sigma(\rho_\ell-\rho_g)g}{\rho_\ell^2}\right]^{0.25} \qquad (3)$$

Substituting Eq. (2) into Eq. (3), one obtains

$$\alpha = x/\left\{1.13\left(x\frac{\rho_\ell-\rho_g}{\rho_\ell}+\frac{\rho_g}{\rho_\ell}\right) + 1.18\frac{\rho_g}{G}\left[\frac{\sigma(\rho_\ell-\rho_g)g}{\rho_\ell^2}\right]^{0.25}\right\} \qquad (4)$$

The thermal-equilibrium quality is given by

$$x_{eq} = \frac{Q - GAC_{p\ell}\Delta T_{in}}{GAH_{\ell g}} \qquad (5)$$

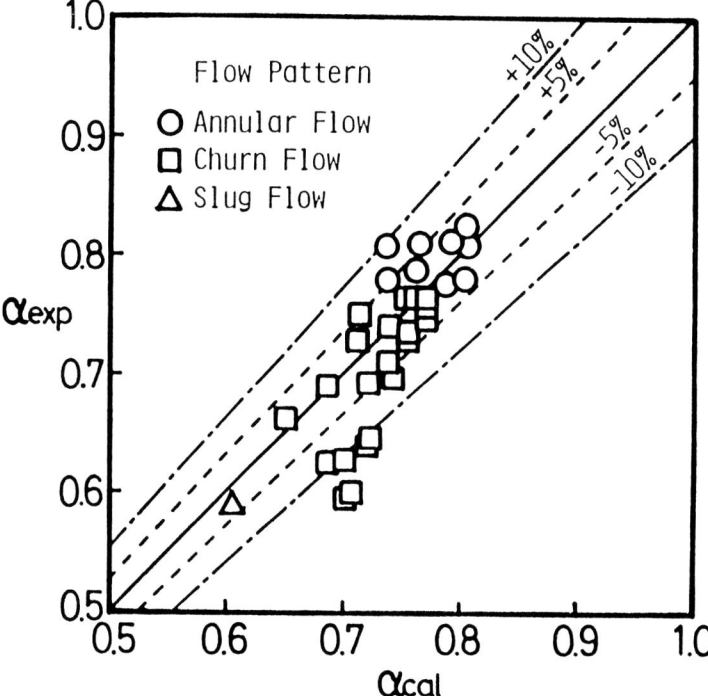

FIGURE 6. Comparison of void fraction between experimental data and theoretical values under saturated condition

In saturated condition, all heat input was consumed on vaporizing as latent heat. Then, the quality is given by the next simple equation.

$$x = \frac{Q}{GAH\ell_g} \qquad (6)$$

Substituting Eq. (6) into Eq. (4), one obtains the relation between void fraction and quality under saturated condition. The comparison of void fraction between the present experimental data and theoretical values calculated by the above-mentioned method is shown in Fig. 6, where flow patterns are also presented. The data obtained were plotted within the scatter of 10 percent of the theoretical values, so this measuring method was acceptably accurate. And, the differences between flow patterns did not affect the accuracy of the measured results. Especially, even in the case of churn flow, the measurement errors caused by the complexity of bubble shape may have been counterbalanced by time-averaging process assuming the uniformity of cross-sectional void distribution, although the shape of flowing bubbles was very complicated and unsymmetrical.

<u>Local Distributions of Time-Averaged Detecting Ratio of Vapor Phase</u>

The contour lines in Fig. 7 represent the local distributions

VOID FRACTION DISTRIBUTION

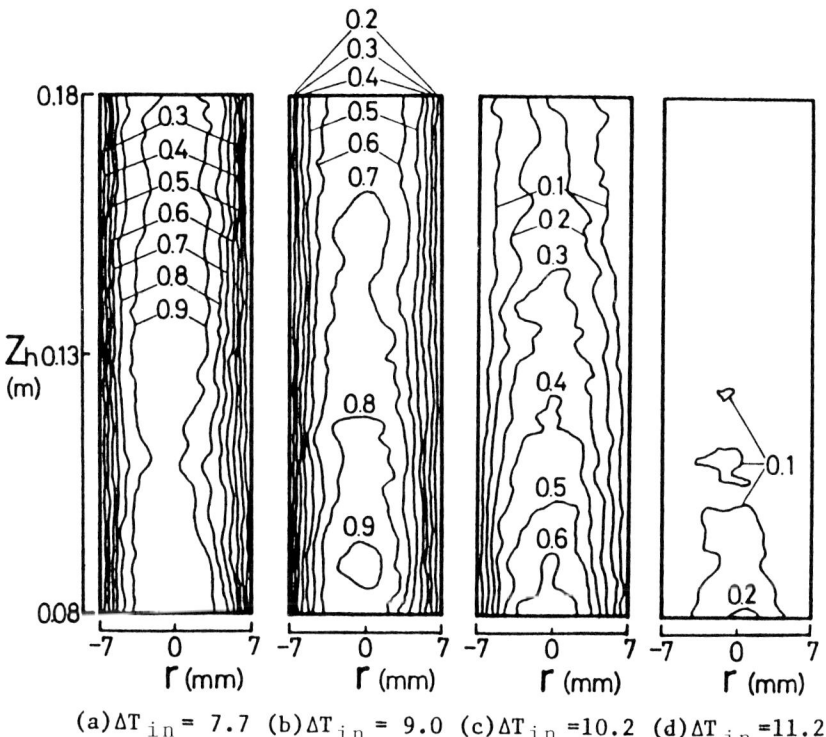

(a) $\Delta T_{in} = 7.7$ (b) $\Delta T_{in} = 9.0$ (c) $\Delta T_{in} = 10.2$ (d) $\Delta T_{in} = 11.2$

FIGURE 7. Local distribution of time-averaged detecting ratio of vapor phase

of the time-averaged void detecting ratio of vapor phase at a inlet liquid velocity of 0.64 m/s and different inlet sub-coolings. This ratio indicates the proportion of the detecting time length of vapor bubbles to the whole measuring time. These averaged distributions were obtained from 126 image frames. The averaging time length was about 2 seconds. Z_h is the distance from the end of the heated part, and ΔT_{in} is the inlet liquid subcooling. The test section below the point of $Z_h=0.08$(m) was made of steel, so it was not measured. The figure shows that the lines tend to be almost symmetrical with respect to the center axis of the cylindrical measuring section. Therefore, the distribution of vapor bubbles was uniform over the horizontal cross section.

In the case of Fig. 7(a), the observed flow pattern was churn flow with extremely disturbed liquid-vapor interfaces. At the lower section, the distance between the inner contour lines was decreased locally. From this result, the following phenomenon was expected to occur: under low inlet subcooling conditions, the flow at the exit of the heated part was in thermal-nonequilibrium, including separately both superheated vapor and subcooled liquid. At the initial stage, the vapor phase distributed near to the internal wall was condensed into the surrounding subcooled liquid. And at the next stage, this liquid phase increased by condensation was vaporized again due to contacting with and heating by the remaining superheated vapor in the internal bulk flow. Then, downstream

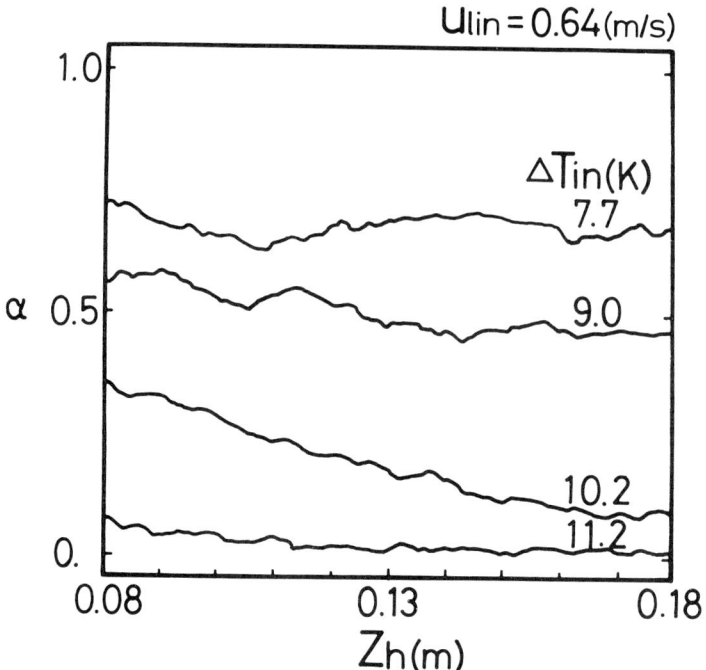

FIGURE 8. Void fraction distribution along flow direction

from this point, the tendency of void fraction changed from "decreasing" to "increasing". That phenomenon will be explained in the next.

As seen from these figures, the small increase of inlet liquid subcooling greatly influenced the distributions of vapor void. For instance, in the case of Fig. 7(c), the vapor void along the flow direction was remarkably decreased. In this case, the observed flow pattern was slug flow accompanied with fine vapor bubbles collapsing rapidly.

Void Fraction Distributions along the Test Section

Figure 8 shows void fraction distributions along the flow direction under the same experimental conditions as shown in Fig. 7. These data were obtained by assuming the three-dimensional shape model of vapor bubbles. As has been stated above, the tendency of void fraction distribution at $\Delta T_{in}=7.7(K)$ changed from decreasing to increasing after the point of $Z_h=0.11(m)$. Since the thermal-equilibrium quality x_{eq} at $\Delta T_{in}=7.7(K)$ was calculated to be negative, the void fraction gradually decreased after the measuring section and finally reached zero. Under a comparatively low inlet subcooling condition, the variation of void fraction was often fluctuated along the flow direction. It was expected that these fluctuations indicate the highly thermal-nonequilibrium transients caused by the coexistence of condensation and vaporization in superheated vapor and subcooled liquid. However, under a high inlet subcooling condition, such kinds of fluctuations could not be observed. These phenomena will be analyzed in detail by our future study.

VOID FRACTION DISTRIBUTION

As seen in Fig. 8, the variation of void fraction in the case of $\Delta T_{in}=10.2(K)$ showed a uniform reduction. This data showed that each of the flowing vapor bubbles condensed at a constant rate. But, from the observational results, these bubbles showed the complicated behaviors to be quite different from each other. The individual behaviors of bubbles are discussed later. In the case of $\Delta T_{in}=11.2(K)$, the distribution of void fraction showed the tendency to gradually decrease and approach zero. Thus, most of the vapor bubbles generated in the heated part appeared to have been condensed before they flow into the measuring section.

Effect of Inlet Liquid Subcooling

As stated in the introduction, the theoretical analyses of the void fraction distribution in a uniformly heated vertical tube where subcooled surface boiling occurred all over the testing tube have been presented by many researchers. Thermal boundary conditions in their experimental systems were fundamentally different from the present case. In the authors' system, an unheated test section on the latter part of the test tube followed close behind the heated boiling section. However, during flowing at the heated part where vapor bubbles were generated, the flow was assumed to be similar to those in the above-mentioned previous studies. Consequently, the void fraction at the exit of the heated part was estimated by introducing the Saha and Zuber subcooled boiling theory.

Saha and Zuber (7) developed an empirical formula for determining the thermal-equilibrium quality at the point of net vapor generation as

$$x_{eqB} = \begin{cases} -0.0022 \dfrac{q_w}{\rho_\ell H_{\ell g}} \dfrac{D_e}{a_\ell} & (Pe \leq 70000) \\ -154 \dfrac{q_w}{\rho_\ell H_{\ell g}} \dfrac{1}{u_{\ell in}} & (Pe > 70000) \end{cases} \quad (7)$$

where Pe is the Peclet number defined by

$$Pe = \frac{G D_e C_{p\ell}}{k_\ell} \quad (8)$$

Ahmad (1) analyzed the liquid subcooling distribution along the flow direction and proposed a simple relation between the true local quality and thermal-equilibrium quality as

$$x = \frac{x_{eq} - x_{eqB}\exp\left(\dfrac{x_{eq}}{x_{eqB}} - 1\right)}{1 - x_{eqB}\exp\left(\dfrac{x_{eq}}{x_{eqB}} - 1\right)} \quad (9)$$

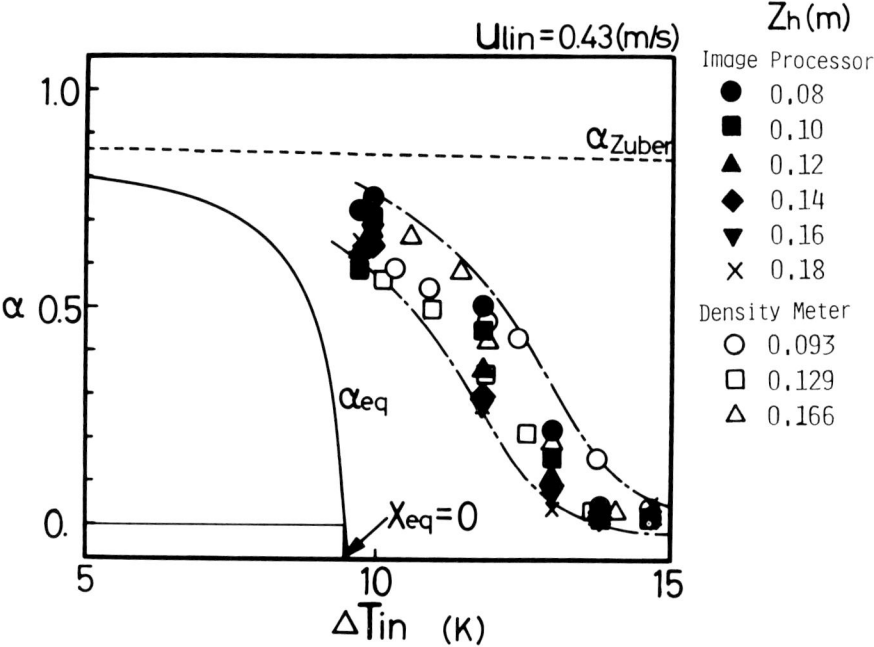

FIGURE 9(a). Variations of void fraction versus inlet liquid subcooling [u_{lin}=0.43(m/s)]

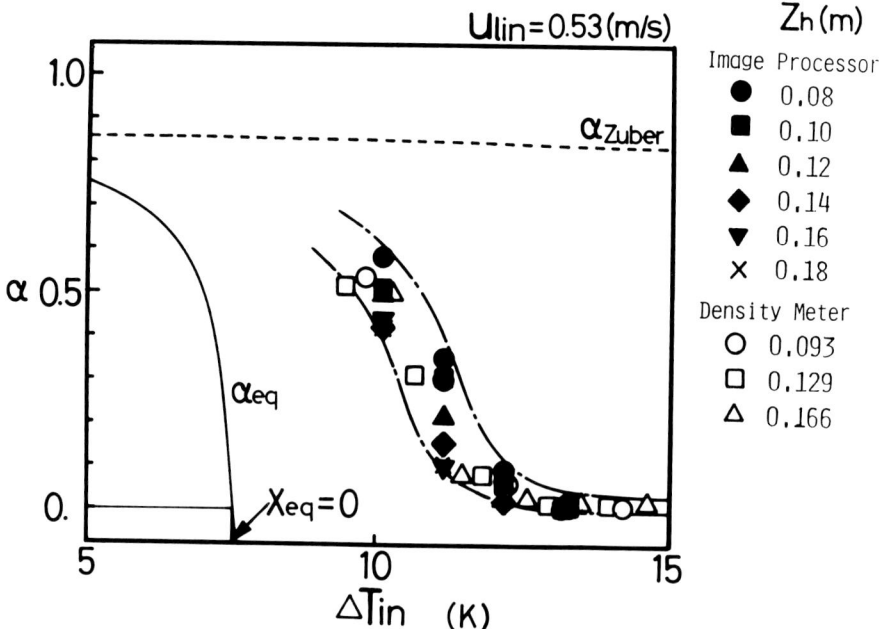

FIGURE 9(b). Variations of void fraction versus inlet liquid subcooling [u_{lin}=0.53(m/s)]

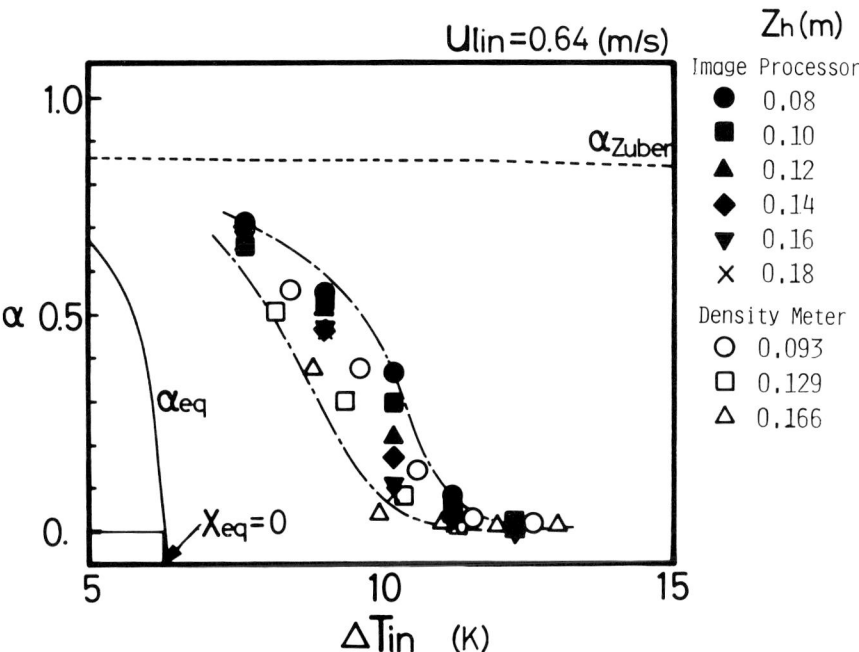

FIGURE 9(c). Variations of void fraction versus inlet liquid subcooling [$u\ell_{in}$=0.64(m/s)]

The true local quality was obtained by substituting both Eq. (5) and Eq. (7) into Eq. (9). The vapor void fraction was calculated by substituting Eq. (9) into Eq. (4).

Figures 9(a)-(c) represent the variation of void fraction plotted versus inlet liquid subcooling at different inlet velocities. The experimental data measured by the image processor and the density meter were plotted in these figures respectively. Dotted lines show the void fractions calculated from the above-mentioned Saha and Zuber's equation, and solid lines show those calculated by assuming the thermal-equilibrium condition. The value of Z_h represents the vertical locations of measuring points. As seen in these figures, all present data were distributed midway between the above-mentioned two curves. And the data obtained from the density meter showed almost the same tendency as those obtained from the image processor. The void fraction calculated from the Saha and Zuber's theory hardly decreased against the increase of inlet liquid subcooling within the present experimental condition. On the contrary, the void fraction calculated from the thermal-equilibrium assumption showed the sudden decrease as inlet subcooling increased. In the actual experiment, the remarkable decrease of void fraction was observed at the range of subcooling higher than the decreasing curve of void fraction estimated from the thermal-equilibrium assumption.

From the results seen in Figs. 9(a)-(c), it was reconfirmed that the flow pattern at the outlet point of the heated part was almost similar to that of the uniformly heated subcooled boiling flow. And, the flow approached the thermal-equilibri-

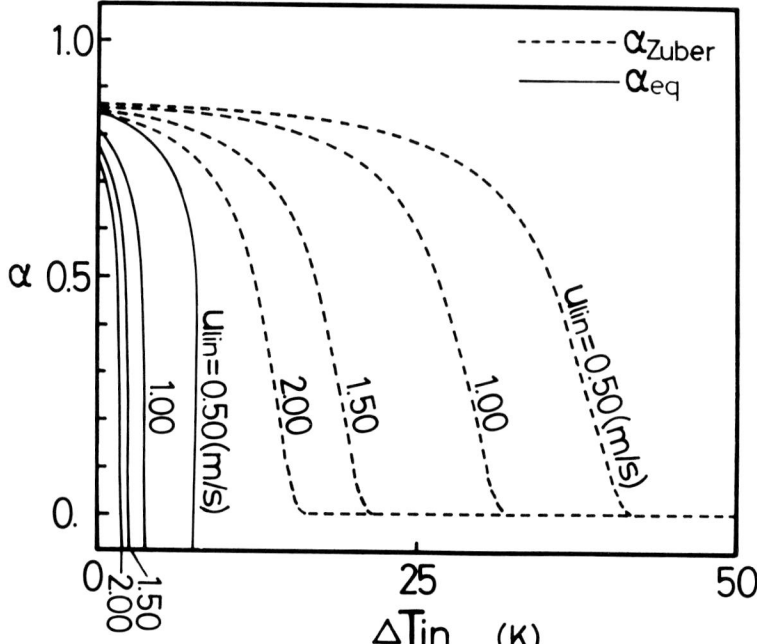

FIGURE 10. Variations of void fraction versus inlet liquid subcooling calculated from the Saha and Zuber's theory(6) and the thermal-equilibrium assumption

um condition during flowing up in the unheated section. This approaching process was greatly affected by the inlet liquid subcooling. As the inlet subcooling increased, the condensation of vapor bubbles occurred at the upstream side of the flowing channel, and the condensation was remarkably enhanced.

Figure 10 shows the variation of void fraction calculated from Saha and Zuber's theory versus high inlet subcooling compared with the cases of Fig 9. Variation tendencies in these curves were very similar to the obtained experimental data. Therefore, it may be proposed that the void fraction profiles obtained from the present experiment can be predicted from the modified theory by introducing the condensing characteristics of individual vapor bubbles into previous theories of subcooled boiling in the uniformly heated channel, for instance, as proposed by Saha and Zuber.

Condensing Behaviors of Vapor Bubbles

As stated above, remarkable condensation occurred under high inlet subcooling condition, and its decreasing curve of time-averaged void fraction along the test tube was comparatively smooth. From these experimental results, it was postulated that the behaviors of individual bubbles were hardly different, and they collapsed at a constant rate. But, as a matter of fact, it was found that there existed many different sizes of vapor bubbles flowing up in the test section, and they showed different condensing behaviors. A few examples are given to clarify this explanation.

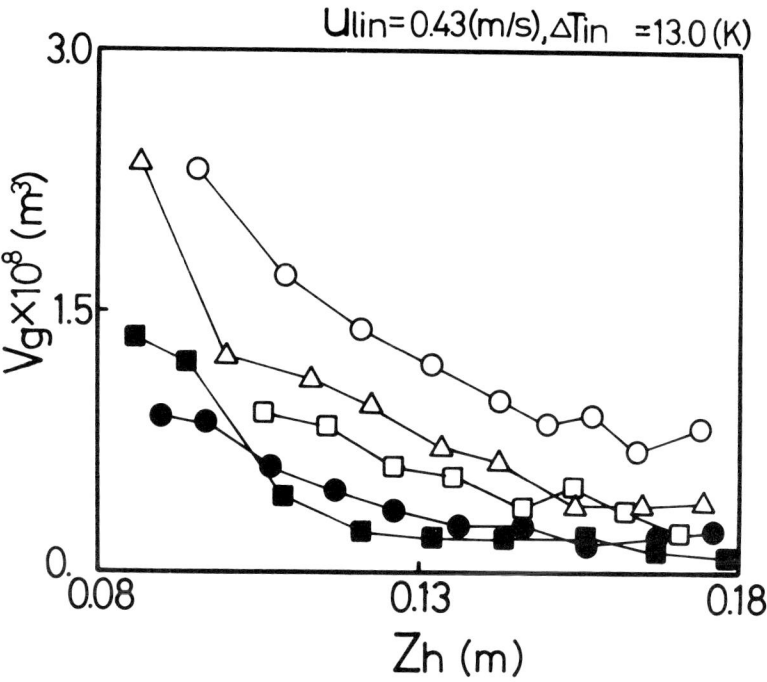

FIGURE 11. Collapsing behaviors of a single bubble

Figure 11 describes the collapsing behaviors of a single bubble under the condition of $u\ell_{in}$=0.43(m/s) and ΔT_{in}=13.0(K), in which the conspicuous decrease of void fraction along the flow direction was observed. V_g is the volume of condensing vapor bubbles obtained by supposing the three-dimensional shape model (see Fig. 4). Z_h indicates the distance from the upper end of the heated part to the gravity center of a vapor bubble. As seen in this figure, the vapor bubbles flowing into the visualized test section had different volumes, and collapsed showing different condensing behaviors (as shown later in Figs. 12(a) and (b)). As is well known, the departure of vapor void from the inner wall of the heated part was intermittent. And as the generated bubbles flowed up in the test section, they repeated coalescing or separating mutually with condensation. These phenomena brought the nonuniform temperature in liquid phase surrounding bubbles at the unheated section. This nonuniformity of liquid temperature was described in detail in reference (2). Each bubble showed different collapsing behaviors due to different subcoolings of its surrounding liquids.

Figures 12(a) and (b) present two examples of actual collapsing behaviors of bubbles under the same experimental condition of $u\ell_{in}$=0.64(m/s) and ΔT_{in}=10.2(K). These collapsing behaviors were found to be considerably different. One type is represented by a bubble becoming smaller slowly as shown in Fig. 12(a), and the other collapsing rapidly with the disturbance of liquid-vapor interface as shown in Fig. 12(b). From these figures, it was confirmed that the subcoolings of their surrounding liquid were quite different. When the bubble interface was disturbed, the condensation of vapor

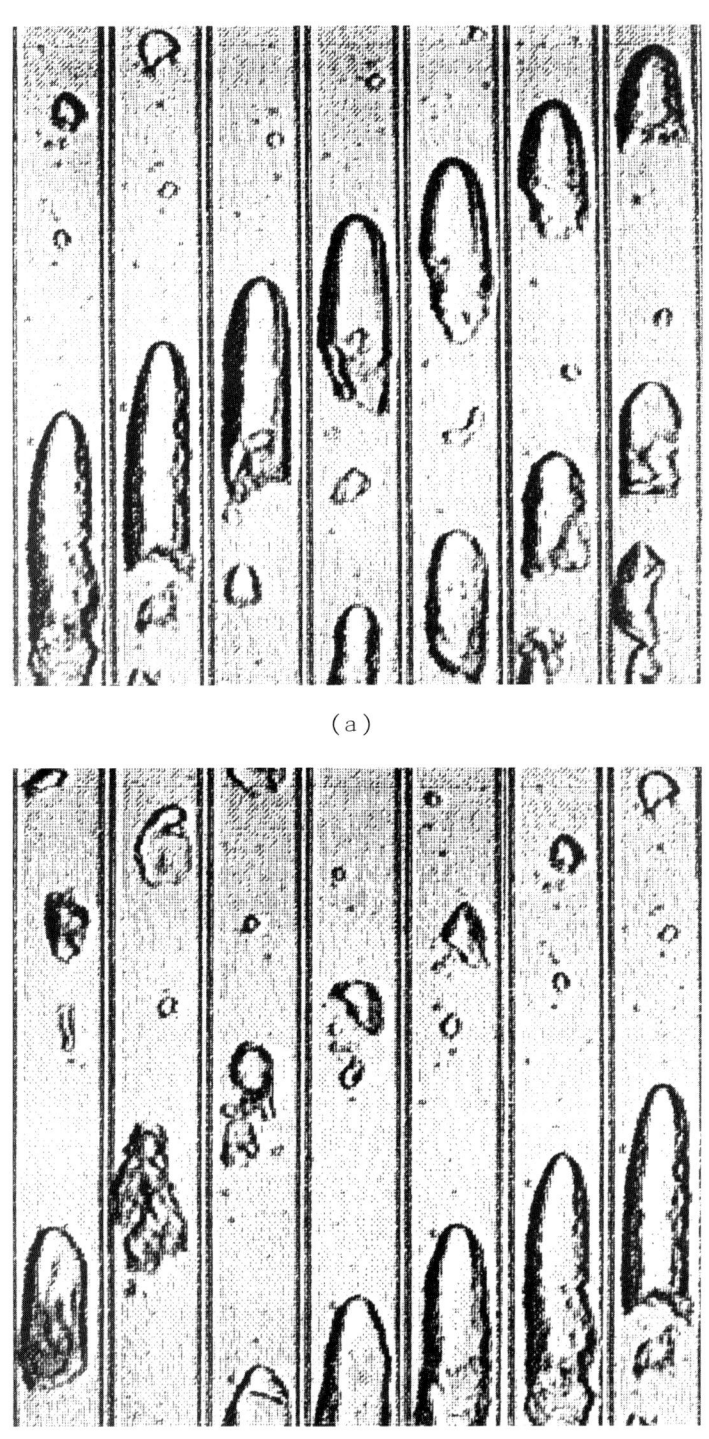

(a)

(b)

FIGURE 12. Typical behaviors of condensing bubble (1/60 second per one frame)

bubbles was suddenly enhanced due to increased interfacial areas.

CONCLUSION

To investigate the two-phase flow phenomena with condensing vapor phase in the subcooled bulk flow, research was undertaken using a boiling water loop. The movement of vapor bubbles at the transparent test section was recorded on video tapes. Video image processing technique was newly introduced to measure the location of liquid-vapor interface precisely. The distributions of void fraction and the volumetric variations of a single bubble along the test section were obtained by assuming a three-dimensional shape model of vapor bubbles. On the other hand, the measurement of the average density of two-phase flow was also carried out by using a vibration type density meter. The following conclusions were reached.

1. The small increase of inlet liquid subcooling extremely affected the distributions of void fraction along the test section.
2. Under low inlet subcooling conditions, the fluctuation of void fraction distribution was observed. Both superheated vapor and subcooled liquid existed together at the outlet of the heated part, and heat and mass transfers between these two phases caused this fluctuation.
3. The obtained void fraction data were distributed midway between two curves calculated from Saha and Zuber's theory and the thermal-equilibrium assumption. And it was expected that the void fraction profiles could be accurately predicted by introducing the condensing characteristics of individual bubbles into previous theories of subcooled boiling two-phase flow.
4. The vapor bubbles with many sizes flowing up in the test section showed different patterns of collapsing due to the nonuniformity of surrounding liquid temperature.
5. The newly developed method using a video image processing system was sufficiently satisfactory for measuring the flow parameters in the two-phase flow.

ACKNOWLEDGMENTS

The authors are indebted to Mr. K. Kurokawa for his helpful discussion, and Mr. H. Shinbo and Mr. K. Agawa for their assistance in this experiment. This work was supported by a Grant-in-Aid for Development of Scientific Research (No.62460229) from the Ministry of Education in Japan.

NOMENCLATURE

A	cross-sectional area of the test tube (m^2)
a	thermal diffusivity (m^2/s)
C_p	specific heat at constant pressure (J/kg K)
D_e	hydraulic diameter (m)
G	mass velocity ($kg/m^2\ s$)
g	acceleration due to gravity (m/s^2)

$H_{\ell g}$	latent heat of vaporization (J/kg)
k	thermal conductivity (W/m K)
Pe	Peclet number ($=GD_e C_{p\ell}/k$)
Q	heat input power (W)
q_w	heat flux on the heated wall of the test section (W/m^2)
T	temperature (K)
ΔT_{in}	liquid inlet subcooling temperature (K)
t	time (s)
U	superficial flow velocity (m/s)
u	flow velocity (m/s)
x	quality
x_{eq}	thermal-equilibrium quality
x_{eqB}	thermal-equilibrium quality at the point of net vapor generation in the heated part
Z_h	distance along the channel measured from the end of the heated part (m)
α	void fraction
ρ	density (kg/m^3)
ρ_m	average density of two-phase mixture (kg/m^3)
σ	surface tension (N/m)

Subscript

ℓ	liquid phase
g	vapor phase
s	saturation value
in	inlet value

REFERENCES

1. Ahmad, S. Y.(1970), Axial Distribution of Bulk Temperature and Void Fraction in a Heated Channel with Inlet Subcooling, Trans. ASME, Ser. C, 92-4, pp. 595-609.

2. Hori, Y., Toda, S. and Kurokawa, M.(1988), Two-Phase Flow Phenomena with Condensed Vapor Phase in Subcooled Bulk Flow in a Tube, Proc. 1st KSME-JSME Thermal and Fluids Engineering Conf., Seoul, 2, pp. 218-223.

3. Kroeger, P. G. and Zuber, N.(1968), An Analysis of the Effects of Various Parameters on the Average Void Fractions in Subcooled Boiling, Int. J. Heat Mass Transfer, 11, pp. 211-233.

4. Larson, P. S. and Tong, L. S.(1969), Void Fractions in Subcooled Flow Boiling, Trans. ASME, Ser. C, 91, pp. 471-476.

5. Levy, S.(1967), Forced Convection Subcooled Boiling-Prediction of Vapor Volumetric Fraction, Int. J. Heat Mass Transfer, 10, pp. 951-965.

6. Rouhani, S. Z. and Axelsson, E.(1970), Calculation of Void Volume Fraction in the Subcooled and Quality Boiling Region, Int. J. Heat Mass Transfer, 13, pp. 383-393.

7. Saha, P. and Zuber, N.(1974), Point of Net Vapor Genera-

tion and Vapor Void Fraction in Subcooled Boiling, Proc. 5th Int. Heat Transfer Conf. Tokyo, 4, 175-179.

COUNTERCURRENT GAS-LIQUID FLOW IN A BOILING WATER REACTOR CORE DURING POSTULATED LOSS-OF-COOLANT ACCIDENTS

M. MURASE, H. SUZUKI, Y. KATOAOKA, and S. HATAMIYA
Energy Research Laboratory
Hitachi, Ltd.
1168 Moriyama, Hitachi, Ibaraki, 316 Japan

ABSTRACT

Countercurrent gas-liquid flow in parallel vertical channels is theoretically and experimentally evaluated for a boiling system simulating a BWR core. The parallel channel models are applied to the LOCA analysis program, SAFER, and verified by Two Bundle Loop test analysis. Applying the parallel channel effects to the fuel bundle, water drain tubes with the restricted bottom ends are developed in order to mitigate CCFL at the core outlet.

INTRODUCTION

During postulated loss-of-coolant accidents (LOCAs) in BWRs, emergency core spray water entering the core is opposed by an upward steam flow, which is produced by heat transfer from the fuel rods and by flashing due to depressurization. It has been clarified by experimental results of the SSTF [1] and small-scaled gas-liquid systems [2-5] simulating a BWR core that three flow patterns, namely, liquid down-flow, countercurrent flow and cocurrent up-flow appear simultaneously in parallel vertical channels under such countercurrent flow conditions. There are two effects due to the occurrence of these flow patterns under BWR LOCA conditions. First, a high heat removal rate is expected in the cocurrent up-flow bundles due to the high steam flow rate as demonstrated by the Two Bundle Loop [6]. Second, core reflooding comes earlier due to the high water flow rate in the liquid down-flow bundles [1].

This paper summarizes the authors' studies on thermal-hydraulics related to reactor safety after the former Japan-U.S. Seminar on Two-Phase Flow Dynamics, 1984. First,

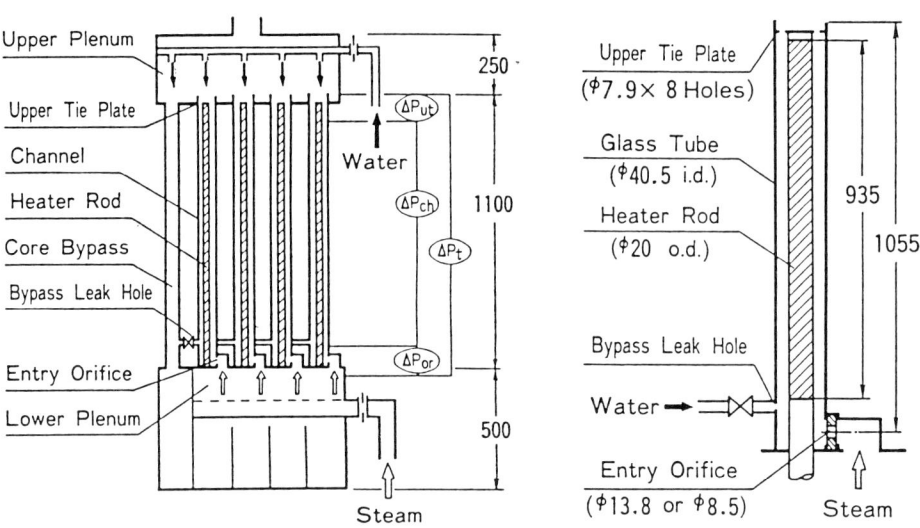

Fig.1 Schematic diagram of test apparatus (units:mm)

Fig.2 Detailed description of channel (units:mm)

countercurrent gas-liquid flow is theoretically and experimentally evaluated for a boiling system simulating a BWR core (7). Then, parallel channel models to evaluate thermal-hydraulic behavior in a BWR core are applied to the LOCA analysis program, SAFER (8), and verified by Two Bundle Loop test analysis (9). Finally, applying the parallel channel effects to the fuel bundle, water drain tubes with restricted bottom ends are developed in order to mitigate countercurrent flow limiting (CCFL) and increase the falling water flow rate at the upper tie plate (10).

COUNTERCURRENT GAS-LIQUID FLOW IN PARALLEL CHANNELS

Test Apparatus

Figure 1 shows a schematic diagram of the test apparatus. The apparatus simulated the inside of a shroud in a BWR core and consisted of four circular channels with a single heated rod. Water was injected into the upper plenum to simulate a core spray system and steam was injected into the lower plenum to simulate steam generation due to depressurization boiling. Leak flows between the core bypass and channels were simulated by a bypass leak hole at the bottom of each channel. The bypass leak flow rate was controlled by a flow control valve.

Figure 2 shows a detailed description of the channel and Table 1 compares main specifications of the test apparatus

Table 1 Specifications of test apparatus

Item	Units	BWR	Apparatus	Ratio
Channel flow area: A_{ch}	cm²	98	9.74	0.10
Upper tie plate flow area: A_{ut}	cm²	79	3.92	0.05
Entry orifice flow area: A_{or}	cm²	30	1.50	0.05
		11.3	0.57	0.05
Channel hydraulic diameter: D_{ch}	mm	12.8	20.5	1.6
Heated rod diameter: D_H	mm	12.5	20.0	1.6
Heated length: H_H	m	3.71	0.935	0.25
Volumetric averaged power density	W/cm³	166	4.4	0.038

with a BWR core. Each channel was made of a glass tube with an inner diameter of 40.5 mm and a single electrically heated rod with an outer diameter of 20 mm was installed in the channel. Each channel had an upper tie plate and side entry orifice at the top and bottom of the channel. The orifice diameter was 13.8 or 8.5 mm, which simulated the orifice in the central or peripheral bundle in a BWR core. The orifice plate in each channel could be changed in accordance with the test conditions. The axial power distribution of the heater rod was uniform, its maximum power was 4 kW, and the power gave a volumetric averaged power density of 4.4 W/cm³, which was 3.8 % of a BWR bundle under normal operating conditions. The tests were performed at atmospheric pressure and saturated temperature.

Calculation Methods

In the calculations of this study, flow patterns in a single channel were determined from mass balance equations based on the CCFL correlations at the upper tie plate and entry orifice, and on the steam generation rate in the channel. Then the pressure drop characteristic was calculated for each flow pattern. Finally, countercurrent behavior in boiling channels under slow transient conditions were calculated. The calculations were performed for the conditions of positive gas velocity (gas up-flow) and saturated temperature.

Transient mass balance equations for liquid and gas in the channel are:

$$(dM_f/dt) = G_{f,or} + G_{f,bp} + G_{f,E} - G_{f,ut} , \qquad (1)$$

$$(dM_g/dt) = G_{g,or} + G_{g,E} - G_{g,ut} , \qquad (2)$$

where G is mass flow rate. The liquid mass flow rates at the entry orifice and upper tie plate, $G_{f,or}$ and $G_{f,ut}$, were given by the CCFL correlations as functions of gas flow rates, $G_{g,or}$ and $G_{g,ut}$, respectively. The bypass leak flow rate, $G_{f,bp}$, was given as a function of differential pressure across the bypass leak hole. The steam generation rate in the channel, $G_{g,E} = -G_{f,E}$, was given by a heat balance equation.

When (dM_f/dt) in Eq. (1) was positive, liquid mass and mixture level in the channel increased and a steady state was reached after the mixture level met the top of the channel. Under the conditions, liquid flow rate at the upper tie plate did not follow the CCFL correlation and was given by the steady state condition, $(dM_f/dt) = 0$ in Eq. (1). When (dM_f/dt) in Eq. (1) was negative, the liquid mass and mixture level in the channel decreased, then the steam generation rate in the channel decreased and the falling liquid flow rate at the upper tie plate increased. A steady state was reached at the condition, $(dM_f/dt) = 0$.

Total pressure drop in the channel (i.e. differential pressure between the upper and lower plenums) was expressed by a summation of the pressure drops at the entry orifice, in the channel and at the upper tie plate. In the channel, acceleration loss, static head and friction loss were considered.

$$\Delta P_t = \Delta P_{or} + \Delta P_{ch} + \Delta P_{ut} ,$$

$$\Delta P_{ch} = \Sigma \left\{ \Delta P_{AC}(k) + \Delta P_{SH}(k) + \Delta P_F(k) \right\} , \qquad (3)$$

where the channel is divided into multiple nodes. Friction loss at the entry orifice ΔP_{or} was divided into the friction losses for a single-phase gas flow and liquid flow because falling liquid flow rate was low at a high gas flow rate and gas flow rate was low at a high falling liquid flow rate as expected from the CCFL correlation. Pressure drop calculations for cocurrent up-flow were extended for countercurrent flow in the calculations of ΔP_{ch} and ΔP_{ut}. In the calculations, the evaporation rate γ above the mixture level, $\gamma = 0$, was used.

Figure 3 shows the calculated pressure drop in a single channel and flow region transitions in parallel channels. In the channel of the apparatus, the total pressure drop depended mainly on the friction loss at the entry orifice, the static head in the channel, and friction loss at the upper tie plate.

Flow region transitions in parallel channels are also shown in Fig. 3. Each channel was connected with the upper and lower plenums at the upper and lower entries, and the pressure drop had to be identical for all channels. Three

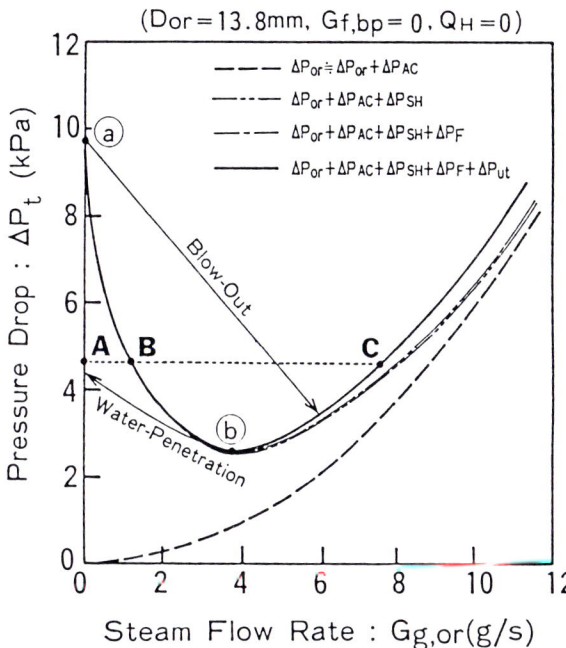

Fig.3 Calculated pressure drop in single channel and flow region transition in parallel channels

values of gas flow rate were basically possible at a given pressure drop, where flow patterns were liquid down-flow ($G_{g,or}=0$) or counter-current flow ($G_{g,or}>0$) in region A, counter-current flow ($G_{f,or}<0$) or gas up-flow ($G_{f,or}=0$) in region B, and gas up-flow in region C. The gas up-flow in regions B and C became cocurrent up-flow with a positive bypass leak flow. Figure 3 shows that multiple combinations of regions A, B and C possibly appeared for given boundary conditions. However, a unique combination could be obtained according to the following characteristics in parallel channels.

(1) Region B, with its negative pressure drop gradient, could appear in only one channel because its state was not as stable as those in the other regions (5).

(2) On increasing gas flow rate, the flow pattern changed from regions A to C in one of the liquid down-flow channels at the peak pressure drop point ⓐ (blow-out transition) because the pressure drop could not increase in the channels (11).

(3) On decreasing the gas flow rate, the flow pattern changed from regions C to A in one of the gas up-flow channels at the local minimum pressure drop point ⓑ (water-penetration transition) because the pressure drop could not decrease in the channels (11).

(4) In a parallel channel system with different channel characteristics, such as heater power and entry orifice size, blow-out and water-penetration transitions occured easily in channels with lower static head or pressure drop and in channels with higher static head or pressure drop, respectively (5).

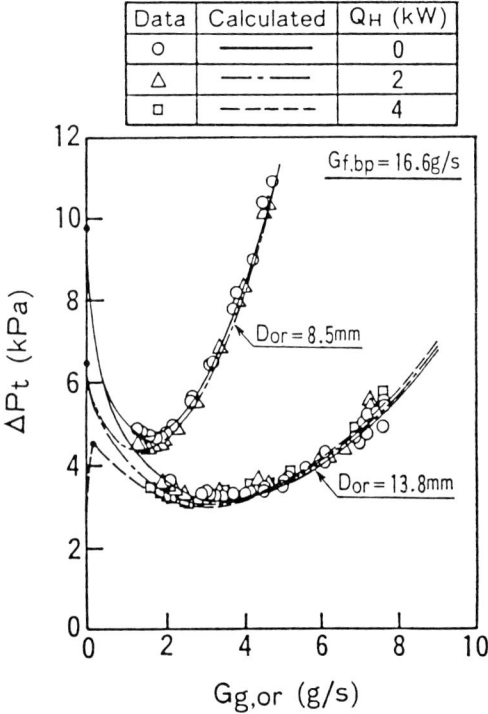

Fig.4 Comparison of calculated and measured pressure drop in single channel

Using the above mentioned characteristics and the pressure drop curve shown in Fig. 3, a unique combination of flow patterns could be calculated following a change of the gas flow rate.

Pressure Drop in a Single Channel

The calculated pressure drop curves for the steam flow rate at the entry orifice are compared with the measured values in Fig. 4. The effect of the heater power on the pressure drop data was not clearly indicated. The calculated pressure drop decreased, with increasing heater power, due to the static head decrease at a low steam flow rate and then slightly increased mainly due to the friction loss increase in the channel and at the upper tie plate in the region of high steam flow rates. The calculated pressure drop for the channel with a 13.8 mm orifice and 4 kW heater power was low at a low steam flow rate because (dM_f/dt) in Eq. (1) was negative and a mixture level was formed in the channel. In the channel, the flow pattern was countercurrent flow in region A. The pressure drop for the channel with an 8.5 mm orifice became high, especially at a high steam flow rate because of the high friction loss at the entry orifice. As shown in Fig. 4, the calculated pressure drops agreed with the data within ±0.3 kPa or about ±10 %, and showed similar trends to the data.

Behavior in Parallel Channels

The calculated and measured differential pressures for steam flow rates into the lower plenum are compared in Figs. 5 and 6. Three of four channels were used for the tests,

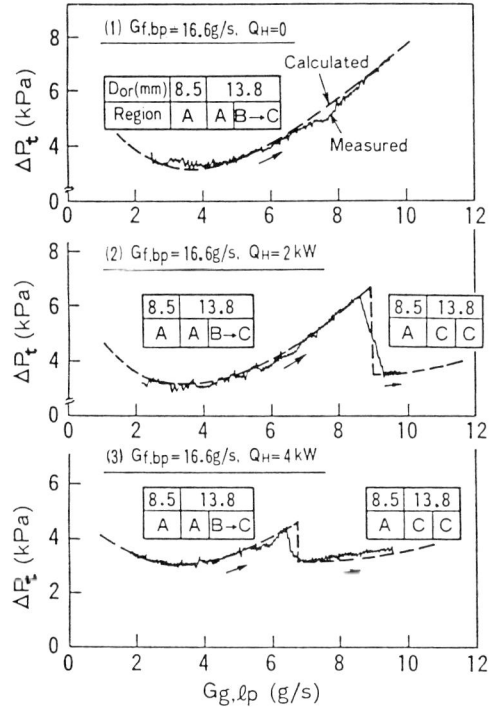
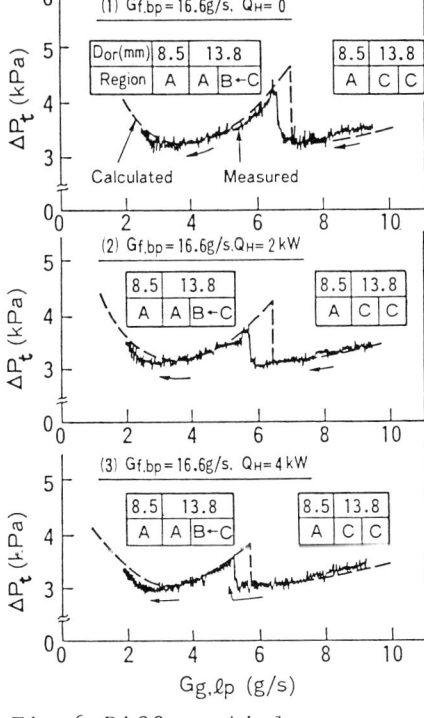

Fig.5 Differential pressure vs. steam flow rate in three channels on increasing steam flow rate

Fig.6 Differential pressure vs. steam floe rate in three channels on increasing steam flow rate

and the other channel was closed by a plate at the orifice position. Two of the three channels had a 13.8 mm orifice and one channel had an 8.5 mm orifice. Slightly subcooled spray water was injected into the upper plenum at the flow rate of $G_{f,s}=150$ g/s. Spray injection was stopped when the mixture level in the upper plenum became higher than about 100 mm and the mixture level was controlled within 50-100 mm. Steam from the boiler was injected into the lower plenum and its flow rate was increased from zero to 10 g/s or decreased from 10 g/s to zero. The initial bypass leak flow rate, which was measured under the conditions without spray and steam injections, was $G_{f,bp}=16.6$ g/s for each channel.

Figure 5 compares the calculated and measured differential pressures on increasing the steam flow rate. At a low steam flow rate, flow patterns were liquid downflow (region A) in two channels and countercurrent flow filled with two-phase mixture (region B) in one channel.

As increasing the steam flow rate, region B became region C following the pressure drop curve in Fig. 4, and the blow-out transition from regions A to C occured at the peak pressure drop point as shown in Fig. 3. The gas flow rate at the transition, $G_{g,tr}$, is expressed by the equation:

$$G_{g,tr} = G_{g,a} N_A + G_{g,c} N_C , \qquad (4)$$

where $G_{g,a}$ and $G_{g,c}$ are gas flow rate at the peak pressure drop in the down-flow (region A) and up-flow (region C) channels, N_A and N_C are the numbers of the down-flow and up-flow channels, respectively. In the case of $Q_H=0$, the peak pressure drop was high and the blow-out transition occured outside the test condition region. At higher heater power, the value of the peak pressure drop became lower and the blow-out transition occured at the lower steam flow rate. Steam started to flow into two channels with the transition and the differential pressure became lower after the transition. Figure 5 shows that the up-flow (region C) occured easily in the channels with a large orifice because the pressure drop was low for a given steam flow rate at the entry orifice (see Fig. 4). The calculated transition point agreed well with the measured.

Figure 6 shows the differential pressure versus steam flow rate on decreasing the steam flow rate. In the region of a high steam flow rate, flow patterns were liquid down-flow (region A) in the channel with a small orifice and cocurrent up-flow (region C) in the channels with a large orifice. On decreasing the gas flow rate, the water-penetration transition occured in one of the two up-flow channels at the local minimum pressure drop point as shown in Fig. 3. The gas flow rate at the transition point, $G_{g,tr}$, is expressed by the equation:

$$G_{g,tr} = G_{g,a} N_A + G_{g,b} N_C , \qquad (5)$$

where $G_{g,a}$ and $G_{g,b}$ are gas flow rates at the local minimum pressure drop in the down-flow (region A) and up-flow (region C) channels, respectively. The differential pressure increased due to the transition because steam flowed into only one channel after the transition. At higher heater power, the transition occured at the lower steam flow rate because the steam flow rate at the local minimum pressure drop, $G_{g,b}$, became lower. The calculated transition points gave about a 10 % higher steam flow rate than the measured because the pressure drop gradient was low around the local minimum pressure drop point.

LOSS-OF-COOLANT ACCIDENTS

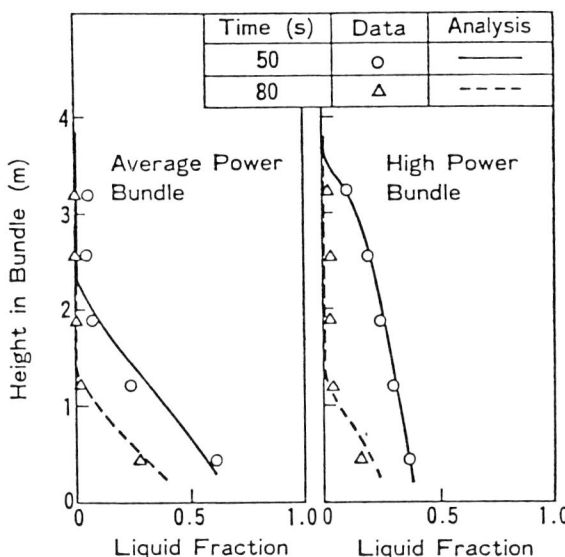

Fig.7 Axial liquid fraction distribution (TBL large break test analysis)

Application to Transient Analysis

Parallel channel models, which evaluated flow patterns in a BWR core, were developed and applied to the LOCA analysis program, SAFER. Then, a TBL test analysis was performed as one of the verifications for the models.

The Two Bundle Loop (TBL) had two full-size, electrically heated bundles with the powers of 4 and 6 MW and it simulated the length from the bottom of the jet pumps to the upper plenum with the same length as the BWR/5 plant. Figure 7 compares axial liquid fraction distributions between the TBL data and SAFER analysis in the case of a recirculation line 100% break. Since the flow pattern was countercurrent in the average power bundle, liquid remained in the bottom part of the bundle due to the low steam velocity. On the other hand, the liquid distributed rather uniformly in the high power bundle after 50 seconds due to the high steam velocity. Therefore, high heat removal rates were expected for the high power bundle. At 80 seconds, both bundles became countercurrent flow due to the decrease of the steam flow rate from the lower plenum to the core. Agreement between the calculation and measured liquid distribution showed the correctness of SAFER with its parallel channel models.

MITIGATION OF COUNTERCURRENT FLOW LIMITING

Concepts of Water Drain Tube

Figure 8 outlines the concept of the water drain tubes. The increase of the falling water flow rate into the core was achieved by using tubes with a contracted flow area at a bottom end. Because the ascending steam flow rate into the

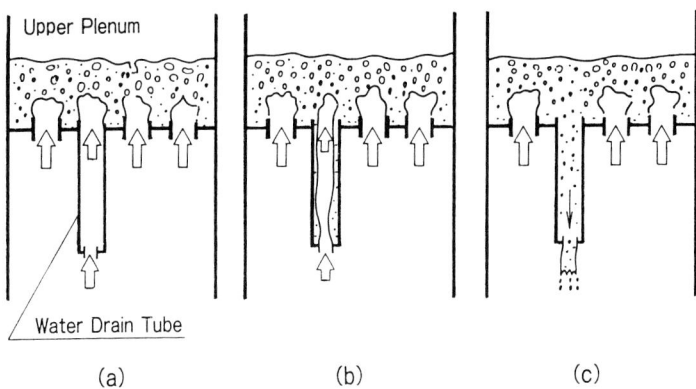

Fig.8 Flow pattern change in water drain tube

tube was restricted at the lower end, the steam velocity at the upper end of the tube was lower than that in the holes of the perforated plate (Fig. 8-a). Therefore, water could penetrate into the tube even when it could not fall through the perforated plate holes due to the high steam velocity. The static head in the tube increased because of water accumulated inside the tube owing to the CCFL at the lower end, and the ascending steam flow rate into the tube decreased due to high flow resistance (Fig. 8-b). Finally, the steam flow rate became zero and only water fell through the tube when the static head inside the tube became higher than the pressure drop of ascending steam flow across the perforated plate (Fig. 8-c).

When water in the upper plenum penetrated into the tube, part of the steam in the upper plenum was carried down with falling water. Steam carry-under into the tubes reduced the falling water flow rate because the static head in the tube decreased. In order to decrease the steam carry-under through the tubes, steam vent tubes attached at the outlets of the perforated plate holes were effective.

Test Apparatus

Figure 9 shows a schematic diagram of the experimental apparatus. The apparatus simulated a BWR fuel channel and had a 134x134 mm square channel. Polycarbonate viewing windows were used in the channel and upper plenum to facilitate visual observation. Two typical plates simulating BWR upper tie plates were used in the test. Test plate dimensions were 134x134 mm by 15 mm thick. One test plate had 25 holes, but no water drain tubes. The diameter of the

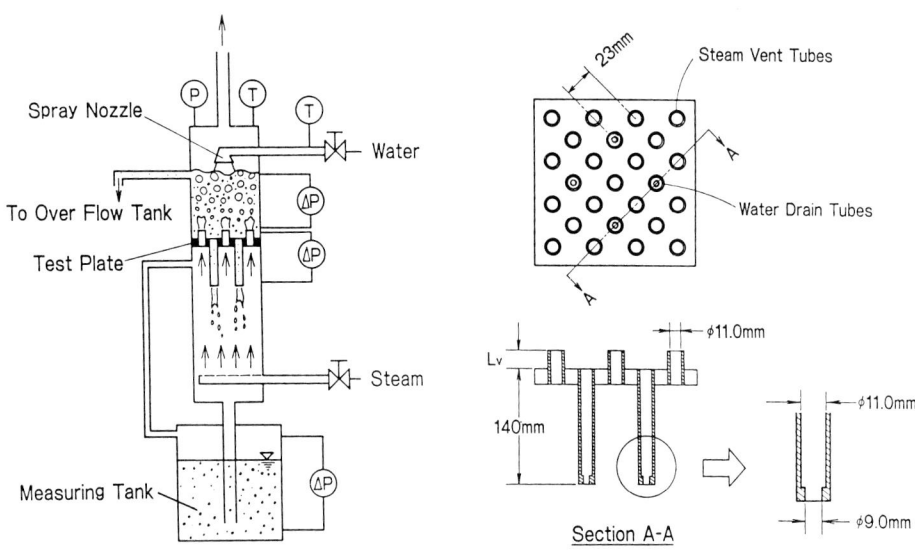

Fig.9 Schematic diagram of experimental apparatus

Fig.10 Configuration of test plates

holes was 12.5 mm. The other test plate shown in Fig. 10 had 4 water drain tubes and 21 steam vent tubes surrounding each water drain tube. The water drain tube was 140 mm long with an 11 mm inner diameter pipe having an orifice at the bottom end for which the reduced flow area was 70 %. Six test plates with having different length of vent tubes (0, 10, 20, 30, 40 and 60 mm) were provided to prevent steam carry-under into the water drain tubes.

In the tests, steam was supplied to the lower part of the test channel and slightly subcooled water was sprayed into the upper plenum. Water in the upper plenum, however, became saturated because subcooled water mixed well with steam in the upper plenum before falling. The two-phase mixture level inside the upper plenum was kept constant by drain off through an overflow pipe. The falling water flow rate through the perforated plate was measured from the increasing rate of the water level in the measuring tank.

Falling Water Flow Rate

The total water flow rate through the perforated plate with the water drain tubes was obtained by summing up the water flow rate through the perforated plate holes and the water flow rate through the tubes.

The former water flow rate was calculated from a CCFL characteristics for a perforated plate holes, which are generally represented by Wallis-type correlation (12) using

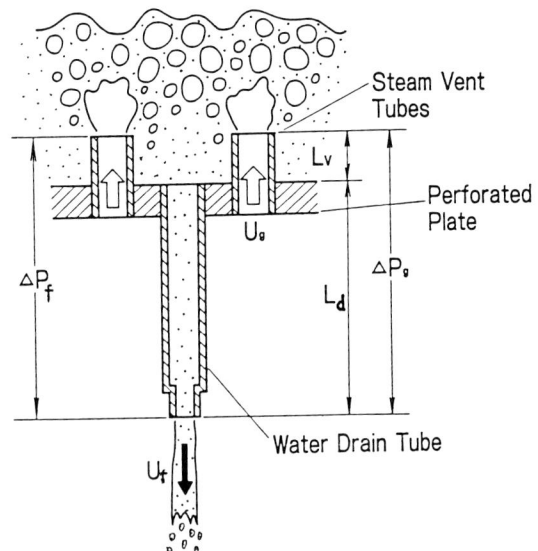

Fig.11 Concept of evaluation model

Kutateladze number.

$$(K_g)^{1/2} + m_K (K_f)^{1/2} = C_K, \quad (6)$$

$$K_i = j_i^2 (\rho_i^2 / (\rho_f - \rho_g) \sigma g)^{1/4},$$
$$(i = g \text{ or } f), \quad (7)$$

where K_i is the Kutateladze number and j_i is the volumetric flux of the gas (g) or liquid (f), respectively. σ is the surface tension of the liquid. The falling water flow rate through the plate holes for a given ascending steam flow rate was calculated using Eq. (6).

The latter water flow rate through the water drain tubes was calculated from a model based on pressure balance of fluids. In the model, water was assumed to fall down through the tubes and steam was assumed to ascend through the perforated plate holes as shown in Fig. 11. Water falling through the plate was formed into minute droplets, and some of them were entrained in the ascending steam flow. After passing through the plate, part of the ascending steam was carried down with falling water. Then, pressure drops for each phase are expressed as follows:

$$\Delta P_g = \rho_g^* g (L_d + L_v) + \frac{1}{2} \rho_g^* u_g^2 (1 + \lambda_v \frac{L_v}{D_v} + \zeta), \quad (8)$$

$$\Delta P_f = \rho_f^* g (L_d + L_v) - \frac{1}{2} \rho_f^* (\beta u_f)^2 (\frac{1}{\beta^2} + \lambda_d \frac{L_d}{D_d} + \zeta_d), \quad (9)$$

where
$$\frac{1}{\rho_i^*} = \frac{X_i}{\rho_g} + \frac{(1-X_i)}{\rho_f}, \quad (i = g \text{ or } f). \quad (10)$$

Terms of the right-hand side of Eqs. (8) and (9) indicate static head, acceleration loss, friction loss and pressure drop due to the flow area change, respectively. In Eq. (9), β is a contraction ratio and u_f is the velocity of falling fluid at the tube lower end. In Eq. (10), ρ_i^* is the mixture mean density (13) of ascending fluid or falling fluid. X_i

LOSS-OF-COOLANT ACCIDENTS

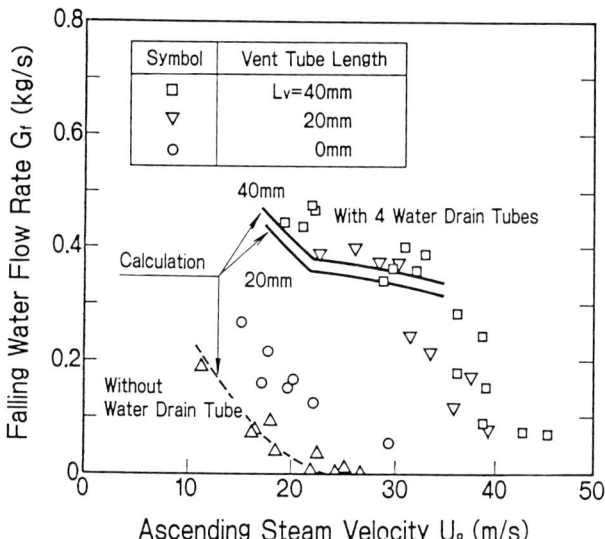

Fig.12 Comparison between calculation and measurement

is steam quality for ascending fluid or falling fluid. Since the flow was turbulent, the friction loss coefficient λ was calculated using the Blasius equation. The pressure drop coefficient ζ due to contraction of flow area was determined experimentally. Assuming ΔP_g equals ΔP_f, a correlation between u_g and u_f is derived from Eqs. (8) and (9), and the water flow rate through the water drain tubes is calculated.

$$\Delta P_g = \Delta P_f, \qquad (11)$$

$$(G_f)_d = N_d \rho_f^* u_f A_d, \qquad (12)$$

where N_d is the number of the water drain tubes and A_d is the flow area at the tube lower end.

Test Results and Discussion

Figure 12 shows the relationship between steam velocity at the plate holes and falling water flow rate. First, the CCFL characteristics of the test plate without water drain tubes were measured and calculated using the CCFL correlation, Eq. (6), where numerical constants $m_K = 1.0$ and $C_K = 2.0$ were used. The broken line shows the calculated results. The falling water flow rate became almost zero in the region of $u_g > 22$ m/s. On the other hand, the test plate

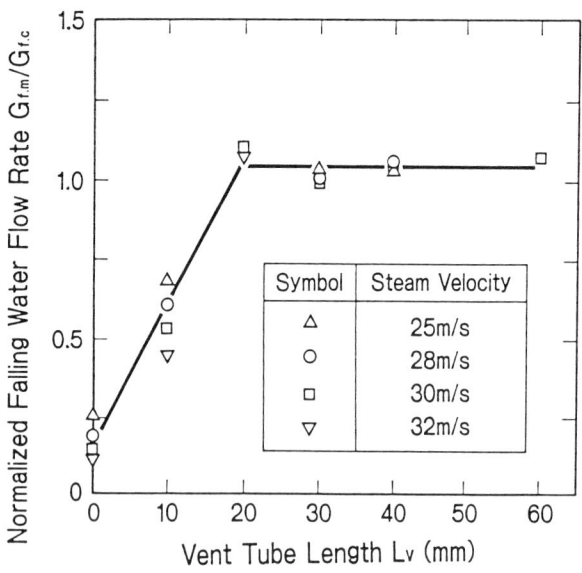

Fig.13 Effect of steam vent tubes

with water drain tubes provided a much higher falling flow rate, especially in the case with steam vent tubes longer than 20 mm. The solid lines show the value calculated from Eqs. (8)-(12) and (6), where mixture mean densities defined by Eq. (10), $\rho_g^* = \rho_g$ and $\rho_f^* = \rho_f$, were used (quality of ascending fluid X_g =1 and quality of falling fluid X_f =0). The calculated values agreed with the data within 10 % at steam velocities below 30 m/s. At higher steam velocities than 30 m/s, the falling water flow rate decreased due to the flow pattern transition in the water drain tubes from liquid down flow to steam up-flow. The falling water flow rate of the plate with water drain tubes but no steam vent tubes (L_v=0) was much lower than the plate with steam vent tubes longer than 20 mm, because of steam carry-under into the water drain tubes that was observed through the poly-carbonate viewing tubes.

Figure 13 shows the effect of the length of steam vent tubes on the falling water flow rate. The measured falling water flow rates are normalized by the calculated ones under the condition of no steam carry-under. In the region of L_v<20 mm, the falling water flow rates became low, which indicates that steam carry-under occured. In the region of L_v>20 mm, however, the measured falling water flow rates agreed with the calculated ones under the assumption of no steam carry-under. Therefore, steam vent tubes longer than 20 mm were judged effective to prevent steam carry-under into the water drain tubes.

In this study, experiments were performed for perforated plates simulating a BWR upper tie plate. However, the technique to mitigate the CCFL at the core outlet can be applied to other types of reactors such as PWRs. The required falling water flow rate from the tubes was obtained by using Eqs. (8)-(12) and selecting suitable values for the parameters such as N_d and A_d in Eq. (12). The region of ascending steam velocity, where the water drain tubes work effectively, can be changed by the contracted ratio β of the flow area at the lower end of the tubes.

CONCLUSIONS

Countercurrent gas-liquid flow has been theoretically and experimentally evaluated for a boiling system simulating a BWR core. In a single channel, the calculated pressure drop for steady state conditions agreed with the measured values within ±10 % and gave similar trends to data for the effects of the bypass leak flow rate and heater power. In parallel channels, flow pattern transitions occurred at the peak pressure drop point on increasing the steam flow rate, but at the local minimum pressure drop point on decreasing the steam flow rate. The calculated behavior in parallel channels exhibited similar trends as the measured values. The calculated steam flow rates at the transitions agreed well with data on increasing the steam flow rate but were about 10 % higher than the measured on decreasing the steam flow rate. Steam up-flow or cocurrent up-flow occurred easily in a channel with low pressure drop for the steam flow rate at the entry orifice, namely, a large entry orifice, high power, or low bypass leak flow rate. The parallel channel models were applied to the LOCA analysis program, SAFER, and verified by TBL test analysis. The calculated axial liquid fraction distributions agreed well with the measured.

Applying the parallel channel effects, a method to mitigate CCFL at the core outlet has been developed. Water drain tubes with a contracted flow area at a bottom end and steam vent tubes were attached to perforated plates at the core outlet. It was experimentally confirmed that a large amount of water could fall through the water drain tubes, even when water could not fall through the perforated plates holes due to CCFL. The calculated falling water flow rates agreed with the measured within 10 % when the perforated plate had water drain tubes and steam vent tubes longer than 20 mm. When the steam vent tubes were shorter than 20 mm, the falling water flow rate through the water drain tubes decreased because of steam carry-under into them. At higher steam velocities than 30 m/s, the falling water flow rate also decreased due to the flow pattern transition in the water drain tubes from liquid down flow to steam up-flow.

NOMENCLATURE

A = flow area, m^2
D = diameter, m
G = mass flow rate, kg/s
g = gravitational acceleration, m/s^2
H = height, m
j = volumetric flux, m/s
K = Kutateladze number
L = length, m
M = fluid mass, kg
N = number of channels
N_d = number of water drain tubes
ΔP = pressure drop, N/m^2=Pa
Q_H = heater power, kW
t = time, s
u = velocity, m/s
X = quality

Greek Symbols

β = contraction ratio
γ = evaporation ratio
ζ = pressure drop coefficient
λ = friction coefficient
ρ = density, kg/m^3
ρ^* = mean density, kg/m^3
σ = surface tension, N/m

Subscripts

d = water drain tube
f = liquid
g = gas
v = steam vent tube

REFERENCES

(1) J. A. Findlay, "BWR Refill-Reflood Program Task 4.4-Evaluation of Parallel Channel Phenomena," NUREG/CR-2566 (1982).

(2) W. M. Conlon and R. T. Lahey, Jr., "An Experimental Investigation of Boiling Water Nuclear Reactor Parallel Channel Effects During a Postulated Loss-of-Coolant Accident," NUREG/CR-2971 (1982).

(3) M. R. Fakory and R. T. Lahey, Jr., "An Investigation of BWR/4 Parallel Channel Effects During a Hypothetical Loss-of-Coolant Accident for Both Intact and Broken Jet Pumps," Nucl. Technol. 65 (1984) 250-265.

(4) M. Murase and H. Suzuki, "Evaluation of Countercurrent Gas/Liquid Flow in Parallel Channels with Restricted End," Nucl. Technol. 68 (1985) 408-417.

(5) M. Murase and H. Suzuki, "Countercurrent Gas/Liquid Flow in Parallel Channels Simulating a Boiling Water Reactor Core," Nuclear Engineering and Design 95 (1986) 79-89.

(6) M. Murase and M. Naitoh, "BWR Loss of Coolant Integral Tests with Two Bundle Loop, (II) Effect of ECCS Activation Modes on Thermal-Hydraulic Characteristics," J. Nucl. Sci. Technol. 22(4) (1985) 301-312.

(7) M. Murase, et al., "Countercurrent Gas-Liquid Flow in Boiling Channels," J. Nucl. Sci. Technol. 23(6) (1986) 487-502.

(8) B. S. Shiralkar, et al., "Evolution of LOCA Analysis at General Electric," Proc. Int. Nucl. Power Plant Thermal Hydraulics and Operations Topical Meeting, No.E4 (1984).

(9) H. Suzuki, et al., "Development of Parallel Channel Models and Their Application to BWR LOCA Analysis," Proc. Second Int. Nucl. Power Plant Thermal Hydraulics and Operations Topical Meeting, No.3B-4 (1986).

(10) S. Hatamiya, et al., "Mitigation of Countercurrent Flow Limiting at the Core Outlet During a LOCA," ANS Transactions, 55 (1987) 744-745.

(11) G. B. Wallis, et al., "Countercurrent Gas-Liquid Flow in Parallel Vertical Tubes," Int. J. Multiphase Flow 7 (1981) 1-19.

(12) G. B. Wallis, One Dimensional Two-Phase Flow, McGraw-Hill Book Co., New York (1969).

(13) J. G. Collier, Convective Boiling and Condensation, McGraw-Hill Book Co., New York (1972).

Fig. 6. Dual beam X-ray scanner. (See page 429 for original placement.)

Fig. 7. High-speed analog processing circuit. (See page 429 for original placement.)

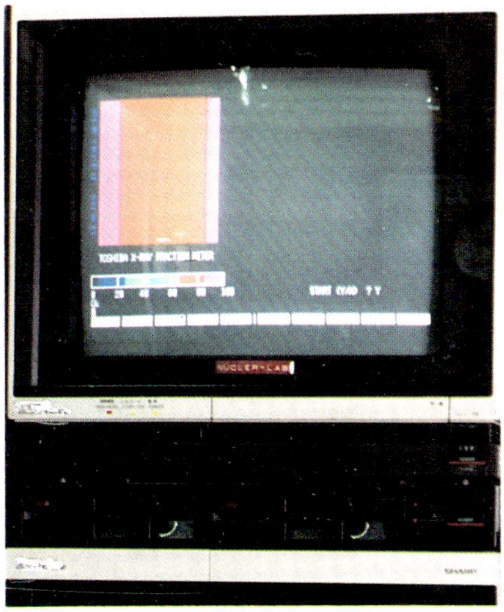

Fig. 8. Color display. (See page 429 for original placement.)

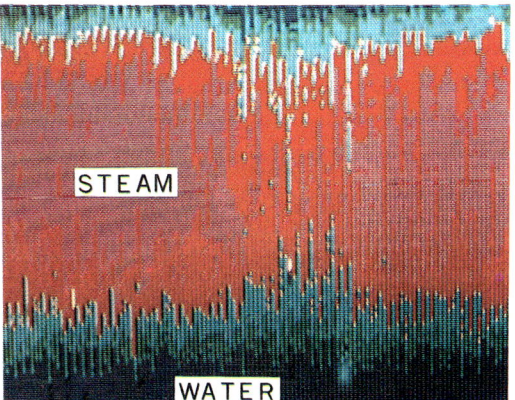

Fig. 9. A measured flow pattern (annular flow). (See page 429 for original placement.)

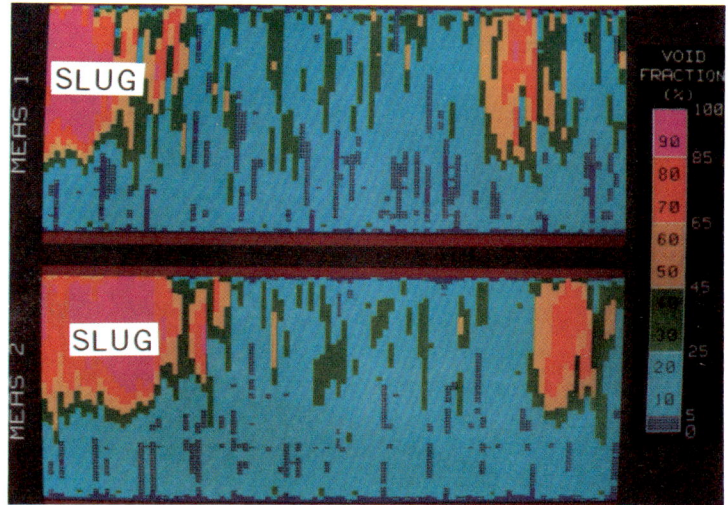

Fig. 10. Measured flow patterns (slug flow showing time difference between cross-sections 1 and 2). (See page 430 for original placement.)

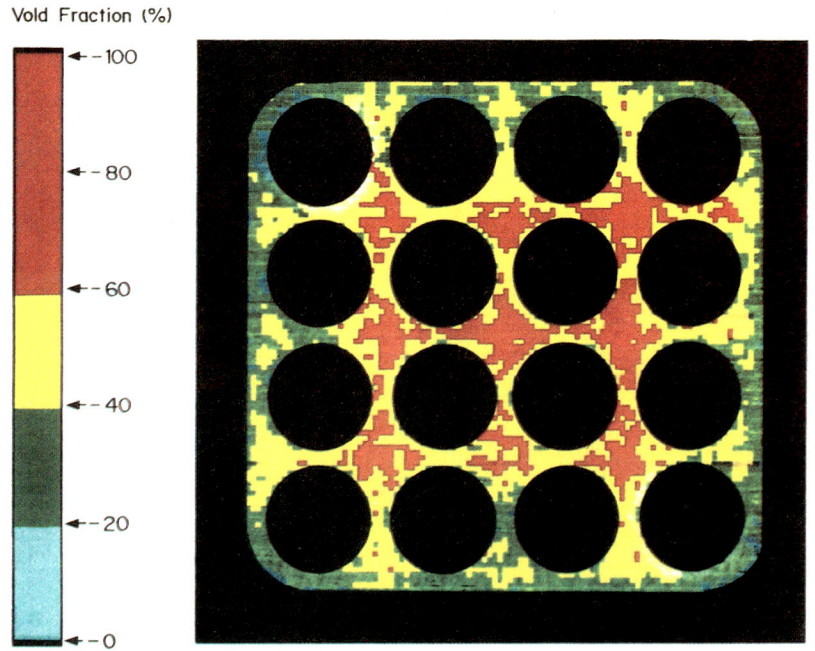

Fig. 17. Typical CT picture. PRESS = 0.98 (MPa), Tin = 447 (K), QUALITY = 0.800 (%), MASS FLUX = 3.00×10^6 (kg/m$^2 \cdot$ h). (See page 435 for original placement.)

(a) Original image (b) Pseudo-color image

PHOTO 1. Video image of the calibration sample taken by the optical method. (See page 150 for original placement.)

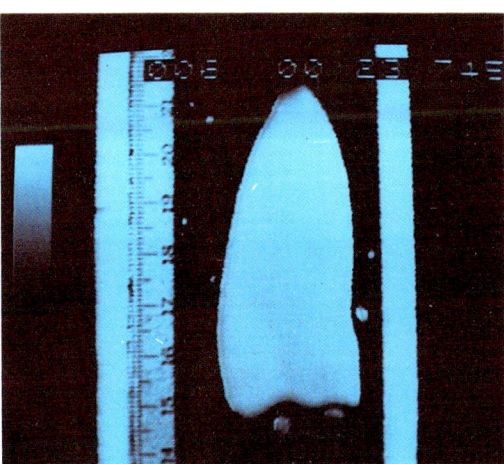 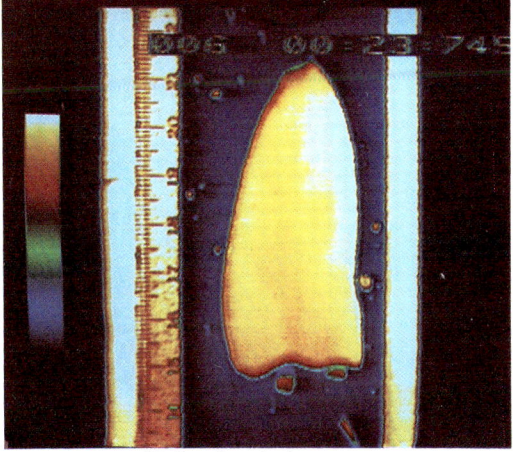

(a) Original image (b) Pseudo-color image

PHOTO 2. Slug flow image taken by the optical method. (See page 150 for original placement.)

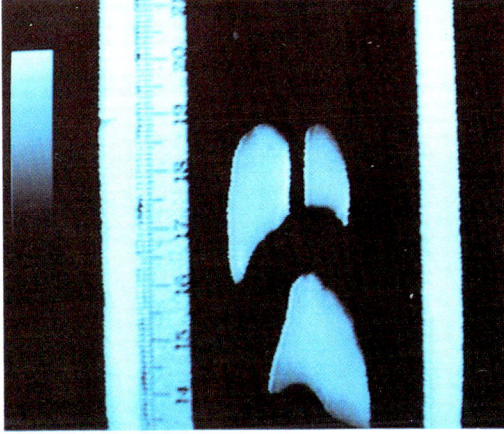 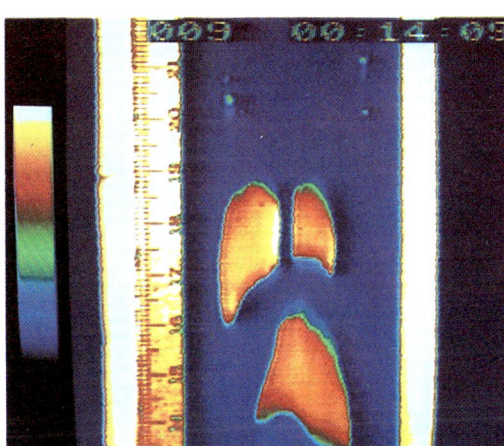

(a) Original image (b) Pseudo-color image

PHOTO 3. Slug flow image taken by the optical method. (See page 150 for original placement.)

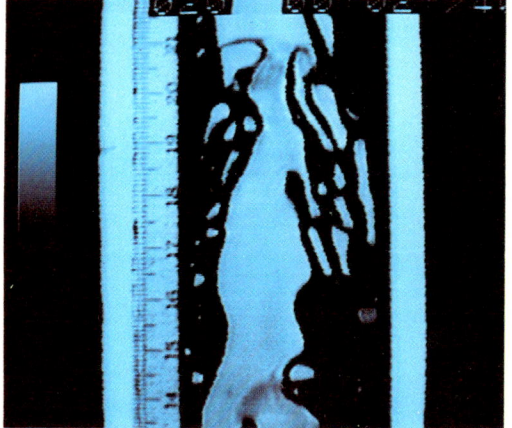
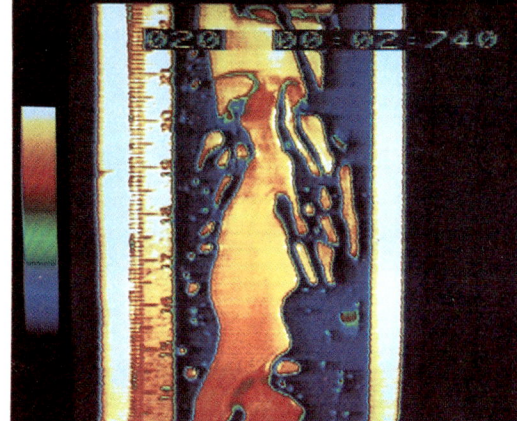

(a) Original image   (b) Pseudo-color image

PHOTO 4. Froth flow image taken by the optical method. (See page 151 for original placement.)

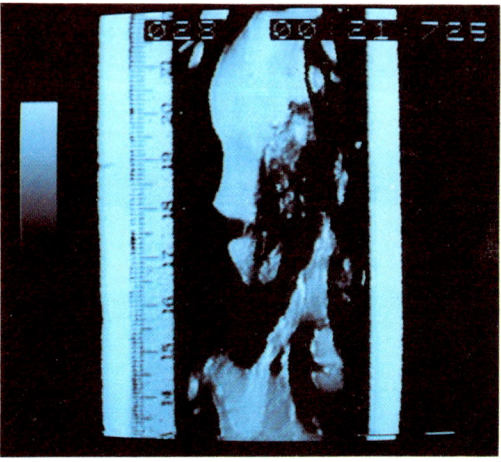
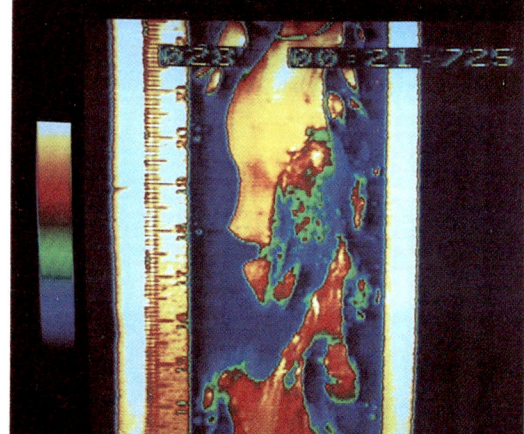

(a) Original image   (b) Pseudo-color image

PHOTO 5. Froth flow image taken by the optical method. (See page 151 for original placement.)

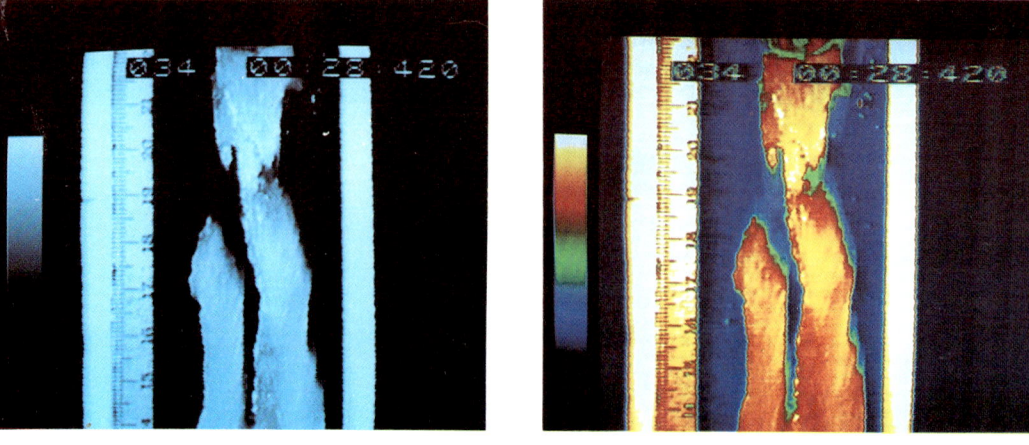

(a) Original image   (b) Pseudo-color image

PHOTO 6. Annular flow image taken by the optical method. (See page 151 for original placement.)

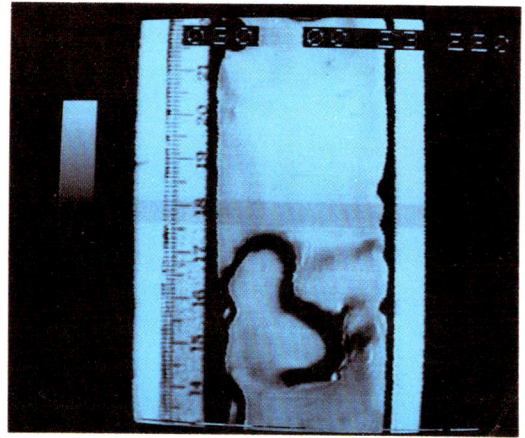 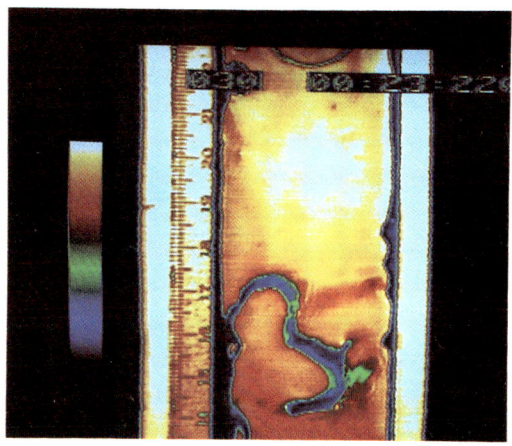

(a) Original image   (b) Pseudo-color image

PHOTO 7. Annular flow image taken by the optical method. (See page 152 for original placement.)

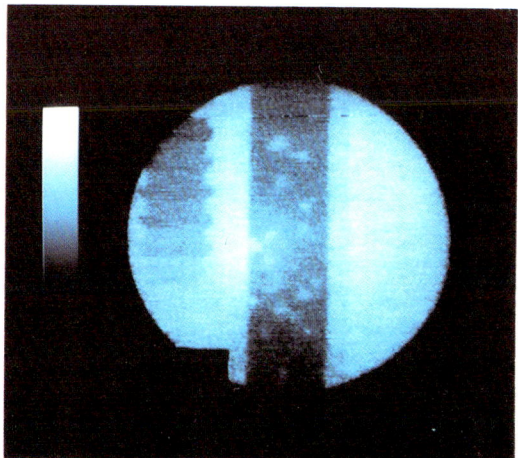 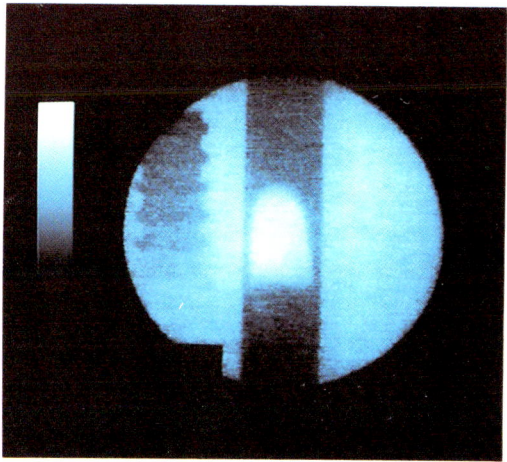

PHOTO 8. Bubbly flow image taken by NRG. (See page 152 for original placement.)   PHOTO 9. Slug flow image taken by NRG. (See page 152 for original placement.)

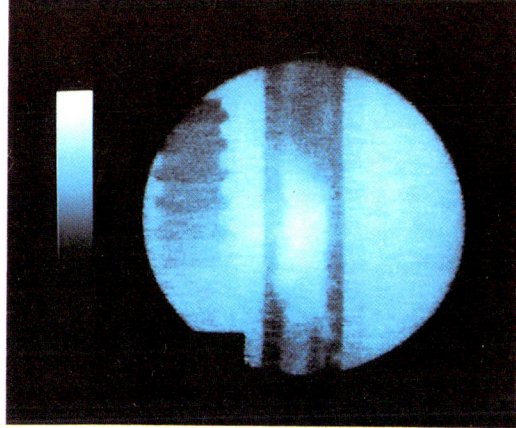 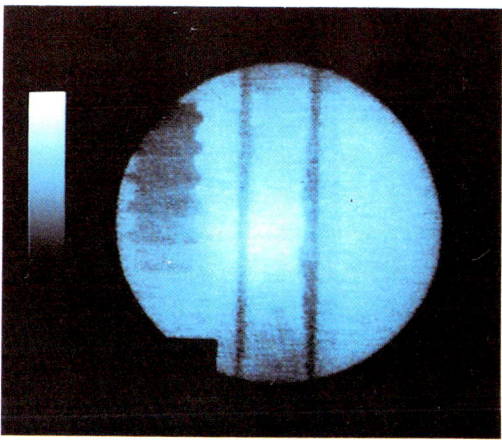

PHOTO 10. Froth flow image taken by NRG. (See page 152 for original placement.)   PHOTO 11. Annular flow image taken by NRG. (See page 152 for original placement.)

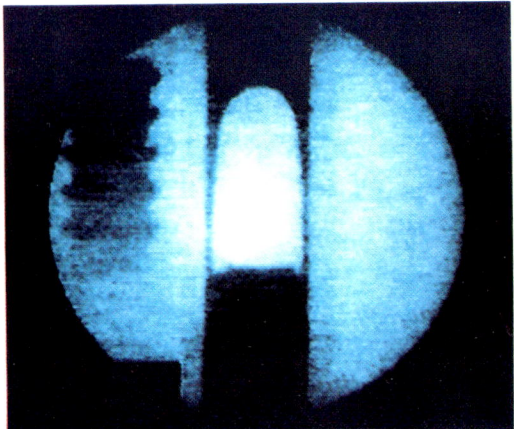

PHOTO 12. Original image of a stagnant slug bubble taken by NRG. (See page 153 for original placement.)

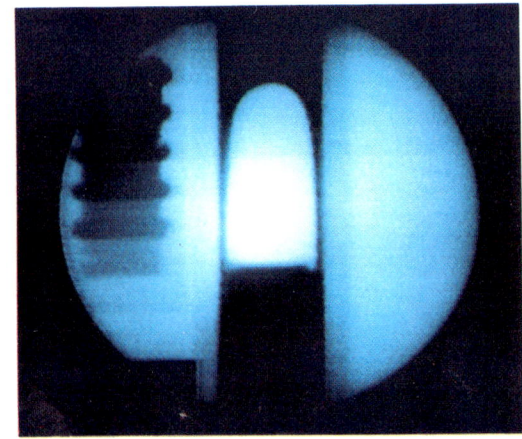

PHOTO 13. Integrated image of a stagnant slug bubble taken by NRG. (See page 153 for original placement.)

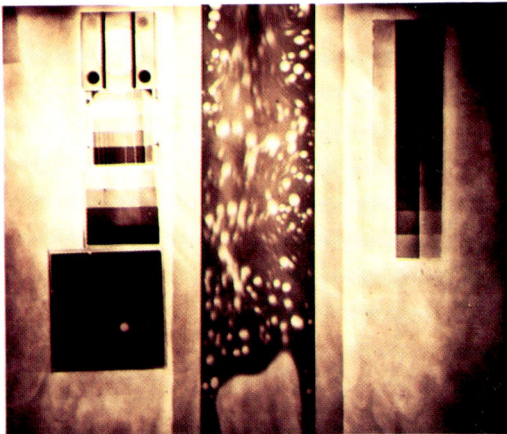

PHOTO 14. Deformed-bubble image taken by pulsed NRG. (See page 153 for original placement.)

PHOTO 15. Slug flow image taken by pulsed NRG. (See page 153 for original placement.)

PHOTO 16. Froth flow image taken by pulsed NRG. (See page 153 for original placement.)

PHOTO 17. Annular flow image taken by pulsed NRG. (See page 153 for original placement.)

THE ANALYSIS OF PHASE SEPARATION PHENOMENA IN BRANCHING CONDUITS

R.T. LAHEY, JR.
Rensselaer Polytechnic Institute
Troy, New York 12180-3590

INTRODUCTION

Two-phase flow in branching conduits occurs in many industrial applications such as nuclear reactors, chemical process plants and in the petroleum industry. Phase separation in branching conduits can have a profound effect on system performance. Examples include the effect of phase separation in the main coolant piping on the effectiveness of emergency core cooling for a light water nuclear reactor (LWR) during a hypothetical loss-of-coolant accident (LOCA); the effectiveness of wet steam injection systems for enhanced oil recovery; and the design of (liquid) slug catchers and phase separators for off-shore oil well platforms.

Let us first review the lessons learned to date from the experimental data and the analysis of the observed phenomena. First of all we consider data taken for dividing two-phase flows. Collier [1975] presented the data of St. Pierre, which were taken in a horizontal tee in which the diameter of the inlet pipe and run was, $D_1 = D_2 = 0.038$ m, while the side branch was, $D_3 = 0.025$ m. These data showed that over a wide range of mass extraction ratios (w_3/w_1) almost complete phase separation occurred (i.e., $w_3 x_3 = w_1 x_1$). These trends are reasonable since the vapor phase normally has far less axial inertia than the liquid phase (i.e., $\rho_G u_G^2 \ll \rho_L u_L^2$), thus the vapor can be expected to more easily turn the corner into the side branch.

Honan & Lahey [1981] ran air/water phase separation experiments in vertical 0.038 m diameter ($D_1 = D_2 = D_3$) plexiglas Wye and Tee test sections for bubbly and churn-turbulent flows. Their data also showed that almost complete phase separation occurred and that the phase separation ratio, (x_3/x_1), was strongly affected by the flow split (w_2/w_3).

Zetzmann [1982] presented data for low pressure (p < 0.3 MPa) air/water two-phase flows using Tee and 45° Wye test sections in vertical and horizontal configurations. He only performed experiments for a narrow range of mass extraction ratios for the Tee test section, however, data were taken over a wider range for the Wye test section.

Azzopardi & Whalley [1982] took annular flow data in Tees for horizontal and vertical orientations. They also found that the phase separation phenomena was strongly effected by the mass extraction ratio (w_3/w_1), flow regime, operating conditions and test section geometry and orientation. Reimann & Seeger [1983] took data for stratified horizontal flows. Again, they found that phase separation depended strongly on the mass extraction ratio, flow regime and test section geometry (eg, the orientation of the side branch).

To date, several researchers have tried to develop empirical models for the prediction of phase separation in branching conduits. Zetzmann [1982] used the fact that the parameters which most strongly effected phase separation were the mass extraction ratio (w_3/w_1), the diameter ratio (D_3/D_1), the inlet mass flux (G_1) and the angle (θ) between the branch and run. He developed a correlation based on his data. Unfortunately, his data base was limited and thus his correlation is of limited practical value. Henry [1981], who took phase separation data in a horizontal Tee, also proposed a correlation. Again, this correlation is only valid for a limited range of conditions.

Seeger [1985] took phase separation data in a Tee for both air/water and steam/water flows. His steam/water data deserve special attention, since they represent high pressure conditions. Significantly, the observed phase separation was similar to that in air/water flows. Thus, flashing does not appear to have a significant influence on phase separation. Unfortunately Seeger did not propose a mechanistic phase separation model. Rather, he also proposed a correlation which is valid over a limited range of conditions.

Saba & Lahey [1984] took air/water data in horizontal Wyes and Tees of 0.038 m diameter ($D_1 = D_2 = D_3$). These data were for relatively high mass extraction ratios (ie, $w_3/w_1 \geq 0.3$). They proposed a mechanistic model which was based on the concept of dividing "stream-lines". Unfortunately their model is only valid for the higher mass extraction ratios (ie, $w_3/w_1 \geq 0.3$), where a fairly large portion of the gas arriving at the junction is extracted through the side branch.

Azzopardi & Whalley [1982] and Azzopardi & Baker [1981] developed phenomenologically-based models for low mass extraction ratios (i.e., $w_3/w_1 < 0.1$) for annular and bubbly/churn flows, respectively. They proposed a "zone of influence" from which each phase was extracted through the branch. As previously noted, their models are only applicable for low mass extraction ratios.

Let us now consider phase separation for impacting two-phase flows. Hong [1978] conducted phase separation experiments in relatively small impacting Tees ($D_1 = D_2 = D_3 = 0.525$ mm). He reported equal phase distribution over a wide range of extraction ratios ($0.15 \leq w_3/w_1 \leq 0.85$) and system operating conditions.

More recently, Azzopardi et al. [1986] reported annular flow phase separation data from an experiment which used an impacting Tee ($D_1 = D_2 = D_3 = 32$ mm). The observed data trends were quite different from those of Hong, but were very similar to those of Hwang [1986]. The reason for this discrepancy is not clear, however, the Weber number, which is a measure of the importance of the surface tension, was several orders of magnitude smaller in Hong's experiment than in Hwang's [1986] and Azzopardi's [1986] experiments. Moreover, the axial inertia of the two-phase flow streams exiting the various branches was quite small in Hong's experiment. It is suspected that Hong's data may have been influenced by the hydrostatic head of the fluid in the lines leading to the phase separators and by surface tension effects. If these suspicions are correct, then Hong's data should be considered inapplicable for most cases of practical concern.

It the study of Hwang [1986] experimental data on phase separation in dividing and impacting horizontal Wye and Tee junctions were reported. Special emphasis was given to dividing flows having low mass extraction rates (i.e., $w_3/w_1 < 0.3$) since very little data existed for these conditions. It should be noted that Honan & Lahey [1981] and Zetzmann [1982] investigated dividing two-phase flows in test sections having a vertical orientation, while Saba & Lahey [1984], and the data of Hwang [1986], were taken for a horizontal configuration. While test section orientation is not significant in some cases of interest, the location of the side branch can have a significant effect for horizontal conduits [Smoglie et al, 1987] in which stratification occurs.

DISCUSSION OF EXPERIMENTAL PROGRAM

From the data and insight gained from previous research we know that pronounced phase separation can occur in branching conduits and that the degree of phase separation is strongly dependent on the flow split (w_2/w_3), conduit geometry (eg, D_3/D_1), fluid properties and the flow regime at the inlet of the junction. Based on these insights Hwang [1986] conducted an extensive experimental program in which special attention was given to dividing flows having low mass extraction rates. We shall summarize these data herein.

The air/water test loop used in the experimental program of Hwang [1986] is shown in Fig.-1. It consisted of a branching test section, open air and closed water loops, and the related instrumentation and computer-based data acquisition system.

The test sections (ie, the Tees and Wyes) were made from plexiglas and each branch had the same dimensions (ie, $D_1 = D_2 = D_3 = 0.038$ m I.D.). There was numerous static pressure taps along the test section which provided detailed information on the pressure distribution, especially in the region of the junction. Also, to assure fully developed flow, additional copper piping having the same I.D. as the test section was installed on all branches. To determine the flow quality at the exit

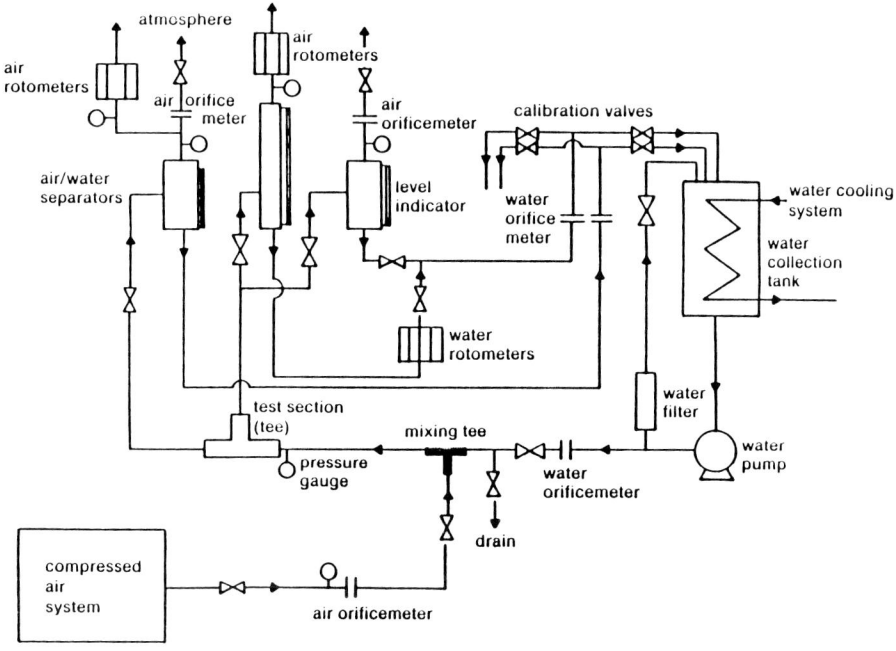

FIGURE 1. Air/Water Loop.

from the branches, the mass flow rate of each phase was separately measured using orifices or rotameters after the air/water mixture was separated in special separator tanks. Having these flow rates, the flow quality at section-i is given by,

$$x_i = \frac{w_{G_i}}{w_{G_i} + w_{L_i}} \qquad (i = 1, 2, 3)$$

To assess the accuracy of these data the continuity equation for both phases was applied. Any data which did not satisfy continuity within ±5% was rejected.

Phase separation experiments for dividing and impacting flows were performed for the horizontal test section configurations shown in Figs. 2&3 and the following operating conditions:

Mass fluxes

(i) Low mass flux:

$G_1 \cong 1,350$ (kg/m²-s)

(ii) Medium mass flux:

$G_1 \cong 2,050$ (kg/m²-s)

(iii) High mass flux:

$G_1 \cong 2,700$ (kg/m²-s)

Inlet Qualities

x_1 = 0.2%, 0.3% and 0.4%

Mass Extraction Ratios (w_3/w_1)

~0.02 ≤ w_3/w_1 ≤ ~0.95

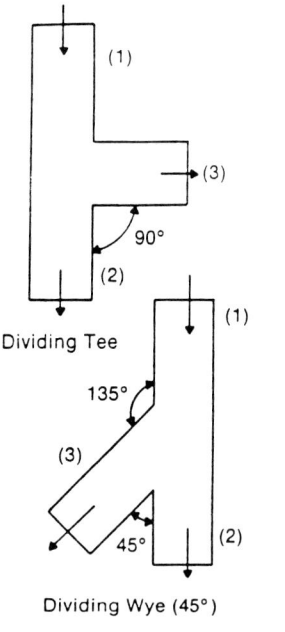
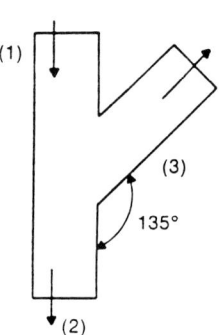

FIGURE 2. Test Sections & Flow Configurations.

BRANCHING CONDUITS 317

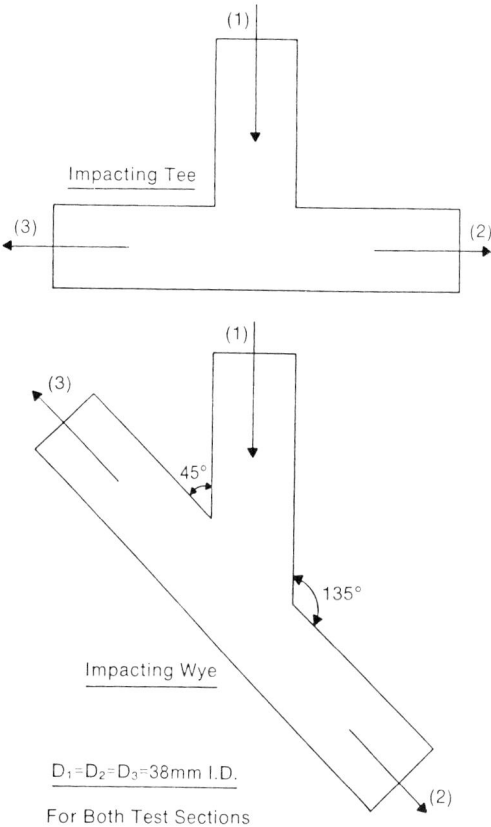

FIGURE 3. Test Section Configurations.

High speed (0.00025 sec shutter speed) still photography was also performed to provide enhanced visualization of the flow at and around the junction.

A summary of the observed trends in the phase separation data taken in this study are shown in Figs. 4&5. These data trends are typical ones for horizontal branching conduits. We note in Fig. 4, that for large extraction ratios (w_3/w_1) in dividing flows, essentially complete phase separation (ie, $x_3/x_1 = [w_3/w_1]^{-1}$) occurs no matter what the angle (θ) of the side branch is. In contrast, for intermediate extraction ratios there is a peak in x_3/x_1 and the degree of phase separation depends on θ. Moreover, it can be noted that for low extraction ratios there are values of x_3/x_1 less than unity and the data trends indicate $x_3/x_1 \to 0.0$ as $w_3/w_1 \to 0.0$.

These data trends make sense if we recall that the liquid phase normally has more axial inertia than the vapor phase and thus it is much harder for it to change direction and exit through the side branch than the gas phase. This is particularly true for large θ. Moreover, the data trends for low extraction ratios are consistent with the notion of a "zone of influence" which is close to the side branch opening. In this region the axial inertia of the liquid phase may be relatively low and the concentration of vapor may also be low. Thus we expect $x_3/x_1 < 1.0$.

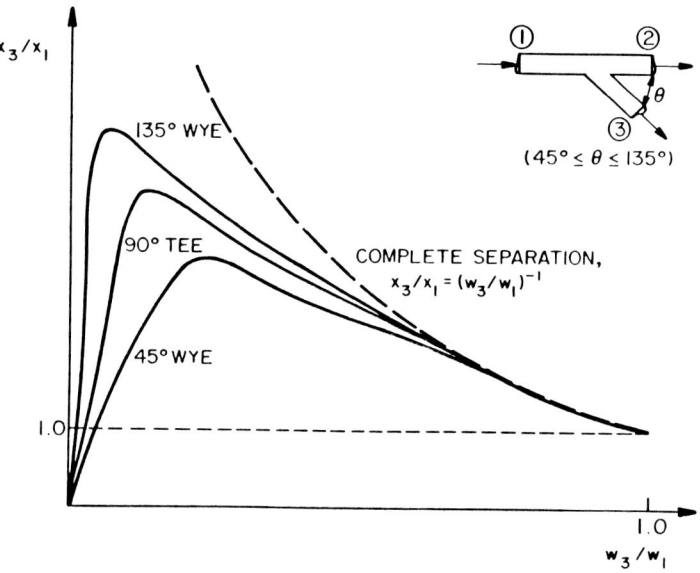

FIGURE 4. Phase Separation Trends in Dividing Wyes and Tees.

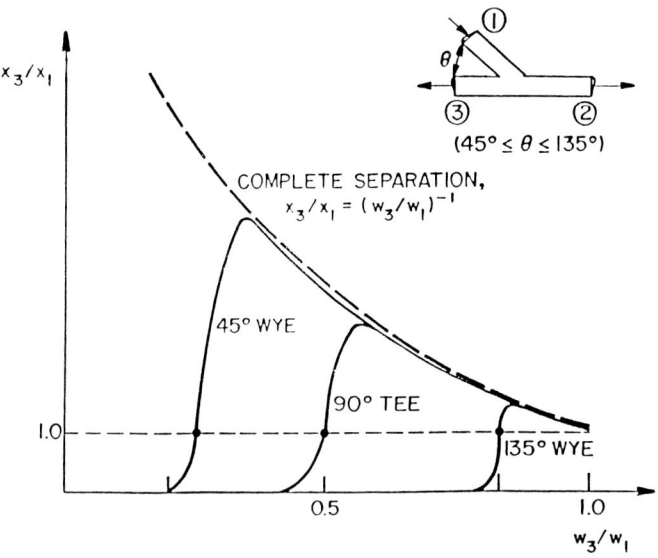

FIGURE 5. Phase Separation for Impacting Flows.

Figure-5 summarizes the data trends seen for impacting stratified two-phase flows in horizontal branching conduits. It can be seen there is an abrupt transition from no gas flow through branch-3 to having all the gas flow through branch-3 as w_3/w_1 is increased. For example, we note that for an impacting Tee ($\theta=90°$) that, as expected, $x_3/x_1 = 0.5$ when $w_3/w_1 = 0.5$). However, this branching conduit acts like a fluidic switch as the mass extraction ratio, w_3/w_1, is either increased or decreased.

BRANCHING CONDUITS

As for dividing two-phase flows, the effect of the branch angle (θ) on the phase separation ratio (x_3/x_1) for impacting flows is explainable in terms of the large differences in the axial inertia of the liquid and gas phase.

DISCUSSION OF ANALYSIS

In the analysis of phase separation in branching conduits there are a number of possible state variables, however, for steady-state conditions they can be categorized as: the three flow rates, w_i, and flow qualities, x_i, and the junction pressure drops, Δp_{13_J} and Δp_{12_J}. This gives a total of eight variables. Of these, we normally specify three variables (eg; w_1, x_1 and w_3, or, x_1, Δp_{13_J} and Δp_{12_J}, etc.). We must then have relationships for the remaining five variables. Four of the equations which are needed to calculate phase separation in branching conduits are:

The steady-state continuity equations:

$$w_1 = w_2 + w_3 \qquad \text{(mixture)} \tag{1}$$

$$w_1 x_1 = w_2 x_2 + w_3 x_3 \qquad \text{(vapor phase)} \tag{2}$$

and the steady-state mixture momentum equations:

$$\Delta p_{13} = \Delta p_{1-1_J} + \left(\Delta p_{13}\right)_J + \Delta p_{3_J-3} \qquad \text{(branch)} \tag{3}$$

$$\Delta p_{12} = \Delta p_{1-1_J} + \left(\Delta p_{12}\right)_J + \Delta p_{2_J-2} \qquad \text{(run)} \tag{4}$$

where the static pressure change in the inlet and outlet branches is given by:

$$\Delta p_{i-i_J} = \Delta p_{i_J-i} = K_i \frac{G_i^2}{2\rho_L} \phi_{Lo_i}^2 + \bar{\rho}_i \, g \, L_i \sin\theta_i \qquad (i = 1, 2, 3) \tag{5}$$

and,

$$K_i = f_i \frac{L_i}{D_H}$$

The static pressure change in the junction is given by:

$$\left(\Delta p_{13}\right)_J = \frac{\langle \rho_{H_3} \rangle}{2} \left[\frac{G_3^2}{\langle \rho_3''' \rangle^2} - \frac{G_1^2}{\langle \rho_1''' \rangle^2} \right] + \frac{G_1^2}{2\rho_L} K_{13} \Phi_{13} \tag{6}$$

$$\left(\Delta p_{12}\right)_J = \frac{K_{1-2}}{2} \left[\frac{G_2^2}{\langle \rho_2' \rangle} - \frac{G_1^2}{\langle \rho_1' \rangle} \right] \tag{7}$$

The details of these pressure drop equations have been reported elsewhere [Lahey, 1986; Hwang & Lahey, 1988] and thus will not be repeated herein.

Let us now consider the specification of the remaining (fifth) equation for phase split in a Tee.

The Analysis of Phase Separation in a Dividing Tee

Let us postulate the existence of mean "stream lines" for each phase, such as that shown in Fig.-6. We can employ a modified version of Euler's equations of motion for a fluid particle traveling on a curved streamline. In the s-direction (ie, tangent to the streamline), we have,

$$\frac{\partial p}{\partial s} = -\rho_k u_k \frac{\partial u_k}{\partial s} + F_{D_{k,s}} \tag{8}$$

where u_k is the velocity of phase-k along the streamline and $F_{D_{k,s}}$ is the component of the interfacial drag force per unit volume in the s-direction.

In the normal (n) direction, we have,

$$\frac{\partial p}{\partial n} = \rho_k \frac{u_k^2}{R_k} + F_{D_{k,n}} \tag{9}$$

where R_k is the radius of curvature of the streamline for phase-k and $F_{D_{k,n}}$ is the component of the interfacial drag force per unit volume in the direction normal to the streamline

In the region of the junction the flow is very complicated and exact analysis is difficult if not impossible. Nevertheless, a simplified phenomenological approach can be used. As shown in Fig.-7 this approach is based on the idea that the gas and liquid flows going into the side branch are coming from regions (ie, "Zones of Influence") bounded by dividing streamlines. The initial positions of these streamlines in the inlet branch are given by δ_L and δ_G, respectively, for the liquid and gas phase. For a given geometry (D_1 and D_3), fluid properties and inlet flow conditions (w_1, x_1, ...) a method for calculating δ_G and δ_L was developed by Hwang et al [1988].

Whether the gas and liquid flow into the side branch or not, depends on the balance of forces on each phase. Figure-8 shows the relevant forces acting on the gas and liquid at a typical streamline crossing. This figure indicates that particles on the two streamlines have velocities $\underline{u}_G$

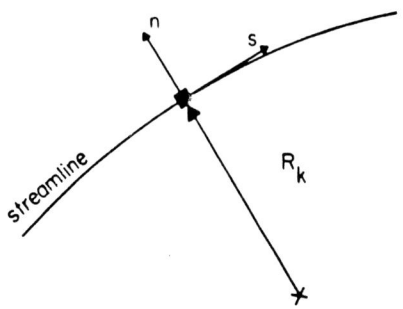

FIGURE 6. A Typical Streamline.

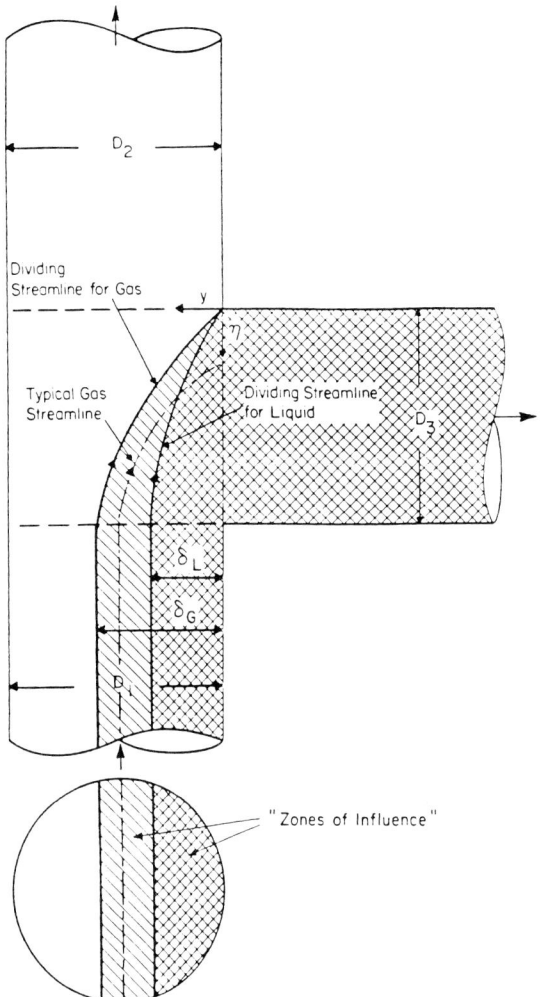

FIGURE 7. Phase Separation Model Based on Dividing Streamlines.

and $\underline{u}_L$, respectively. Due to phasic slip, a volumetric drag force, F_{D_G}, acts on the gas and an equal and opposite drag force, $F_{D_L} = -F_{D_G}$, acts on the liquid. Both drag forces act in a direction parallel to the relative velocity, $(\underline{u}_G - \underline{u}_L)$, vector. Due to the motion along curved streamlines, the centrifugal forces per unit volume, $\rho_G u_G^2/R_G$ and $\rho_L u_L^2/R_L$, act on the gas and liquid phase, respectively, in directions normal to their respective streamlines.

Applying the modified Euler s & n equations, Eqs. (8) & (9), to the gas phase yields,

$$\frac{\partial p}{\partial s_G} = -F_{D_G} \sin\gamma - \rho_G u_G \frac{\partial u_G}{\partial s_G} \qquad (10)$$

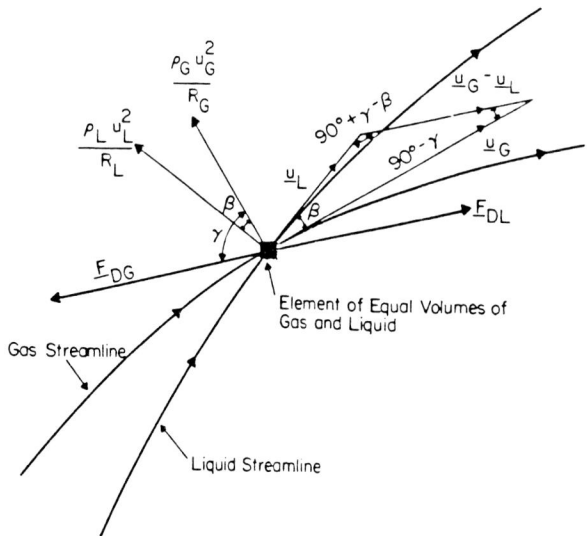

FIGURE 8. Balance of Force.

$$\frac{\partial p}{\partial n_G} = F_{D_G} \cos\gamma + \rho_G \frac{u_G^2}{R_G} \qquad (11)$$

Similarly for the liquid phase,

$$\frac{\partial p}{\partial n_L} = -F_{D_L} \sin(\gamma - \beta) - \rho_L u_L \frac{\partial u_L}{\partial s_L} \qquad (12)$$

$$\frac{\partial p}{\partial n_L} = -F_{D_L} \cos(\gamma - \beta) + \rho_L \frac{u_L^2}{R_L} \qquad (13)$$

Dynamic equilibrium exists when the resultant volumetric forces acting on the gas and the liquid are equal in magnitude and direction. This condition is shown graphically in Fig.-9 in which, for simplicity, we have neglected spatial acceleration, but the other forces in Eqs. (10)-(13) are identified.

If we denote $F_D \stackrel{\Delta}{=} |F_{D_L}| = |F_{D_G}|$, then the equilibrium condition is easily deduced from Fig.-9 as:

$$2F_D \sin\gamma = \frac{\rho_L u_L^2}{R_L} \sin\beta \qquad (14)$$

$$2F_D \cos\gamma = \frac{\rho_L u_L^2}{R_L} \cos\beta - \frac{\rho_G u_G^2}{R_G} \qquad (15)$$

Dividing Eq. (14) by Eq. (15) we obtain:

BRANCHING CONDUITS

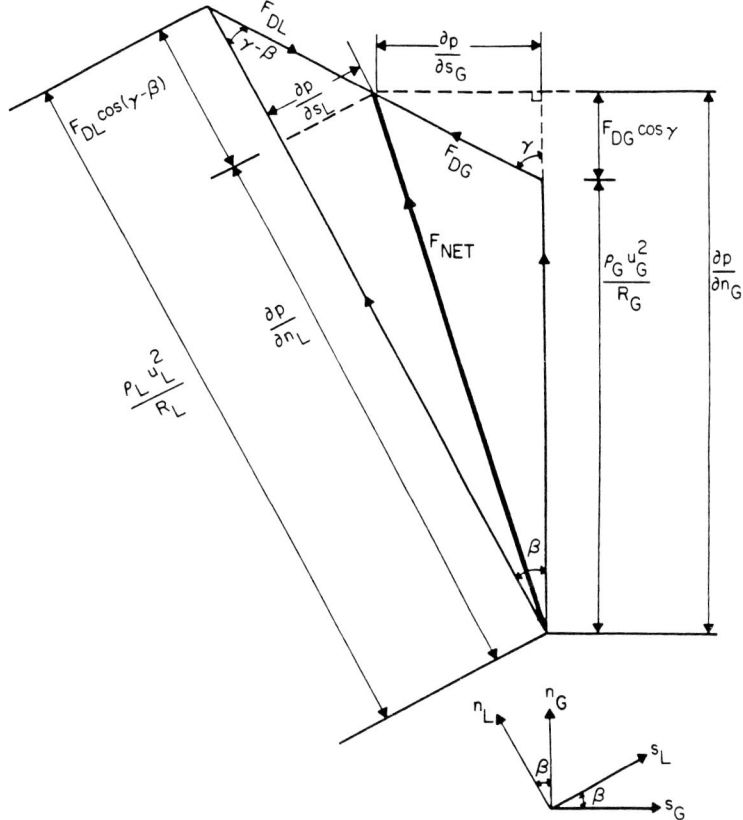

FIGURE 9. Force Vector Diagram.

$$\tan\gamma = \frac{\sin\beta}{\left[\cos\beta - \left(\frac{\rho_G}{\rho_L}\right)\left(\frac{u_G}{u_L}\right)^2\left(\frac{R_L}{R_G}\right)\right]} \tag{16}$$

Another important relation can be derived by applying the Sine Rule to the velocity triangle in Fig.-8:

$$\frac{u_G}{\sin(90°+\gamma-\beta)} = \frac{u_L}{\sin(90°-\gamma)}$$

Thus,

$$S = \frac{\Delta u_G}{u_L} = \frac{\cos(\gamma-\beta)}{\cos\gamma} \tag{17}$$

Equations (16) and (17) are very simple and convenient algebraic relations. However, it should be noted that in order to make computations with Eqs.(16) and (17) we must know the shape of the streamlines. Unfortunately, the actual shapes can only be determined numerically. However, as an approximation we can assume that the following expression is valid for the gas and liquid dividing streamlines in a dividing Tee:

$$\frac{y}{\delta_k} = 1 - \left(1 - \frac{\eta}{D_3}\right)^{m_k}, \quad \text{where } (k = G \text{ or } L) \tag{18}$$

Equation (18) satisfies the following boundary conditions:

$y = 0$, at $\eta = 0$

$y = \delta_k$, at $\eta = D_3$

$\frac{dy}{d\eta} = 0$, at $\eta = D_3$ $(m_k > 1.0)$

It can also be noted that the streamlines approach straight lines with large radius of curvature, R_k, as m_k approaches unity. This is expected to be the case for low extraction rates through the side branch (ie, small δ_k). On the other hand, as m_k approaches a value of two, the streamlines become parabolic, with relatively small radius of curvature. Numerical evaluations of the modifies Euler s & n equations have indicated that this corresponds to high extraction rates where $\delta_k \to D_1$. Hence the empirical parameter m_k is assumed to satisfy:

$m_k = 1.0$, at $\delta_k = 0$

and,

$m_k = 2.0$, at $\delta_k = D_1$

From Eq. (18) we have,

$$\left.\frac{dy}{d\eta}\right|_{\eta=0} = m_k \left(\frac{\delta_k}{D_1}\right)\frac{D_1}{D_3}$$

hence, the angle β between the dividing streamlines at the impact point ($\eta = 0$) is given by,

$$\beta = \text{Tan}^{-1}\left(m_G \frac{\delta_G}{D_1}\frac{D_1}{D_3}\right) - \text{Tan}^{-1}\left(m_L \frac{\delta_L}{D_1}\frac{D_1}{D_3}\right) \tag{19}$$

Moreover, from simple analytical geometry considerations, Eq. (18) implies the following formulation for the radius of curvature of the dividing streamline of phase-k at the impact point ($\eta=0$):

$$\frac{R_k}{D_3} = \frac{\left[1 + \left(\left.\frac{dy}{d\eta}\right|_{\eta=0}\right)^2\right]^{3/2}}{\left|\frac{d^2 y}{d\eta^2}\right|_{\eta=0}} = \frac{\left[1 + \left(m_k \frac{\delta_k}{D_1}\frac{D_1}{D_3}\right)^2\right]^{3/2}}{m_k(m_k - 1)\frac{\delta_k}{D_1}\frac{D_1}{D_3}} \tag{20}$$

We note that the minimum radius of curvature occurs when $m_k = 2.0$ and $\delta_k/D_1 = 1.0$. For this case Eq. (20) yields,

BRANCHING CONDUITS

$$\left.\frac{R_k}{D_3}\right|_{min} = \left[1 + \left(2D_1/D_3\right)^2\right]^{3/2} / 2\left(D_1/D_3\right) \tag{21a}$$

when $D_1 = D_3$, Eq. (21a) becomes,

$$R_k/D_3\big|_{min} = 5.5902 \tag{21b}$$

Rather than using Eq. (20) and correlating m_k from data, it was found to be more convenient to correlate $n_k = n_k \, (\delta_k/D_1)$, where,

$$R_k/D_3 = \frac{R_k/D_3\big|_{min}}{\left(\delta_k/D_1\right)^{n_k}} \tag{22}$$

It should be noted that this phase separation model is valid for any inlet flow regime, provided that we know the lateral distribution of the phases.

For the special case of separated flow regimes, such as stratified and annular flows, it has been found [Hwang et al, 1988] that the influence of the interfacial drag force, F_D, is relatively small and can be neglected. Moreover, the vapor and liquid may only interact at or near the stagnation point and thus the use of F_D is inappropriate. Hence, if we neglect this term we obtain from Fig.-9 that the centrifugal force on the gas is equal to that on the liquid:

$$\frac{\rho_G u_G^2}{R_G} = \frac{\rho_L u_L^2}{R_L} \tag{23}$$

Thus, Eqs. (22) and (23) yield,

$$\frac{R_G}{R_L} = \left(\frac{\rho_G}{\rho_L}\right)\left(\frac{u_G}{u_L}\right)^2 = \frac{\left(\delta_L/D_1\right)^{n_L}}{\left(\delta_G/D_1\right)^{n_G}} \tag{24a}$$

also, we can write the identity:

$$\frac{R_G}{D_3} = \left(\frac{R_L}{D_3}\right)\left(\frac{R_G}{R_L}\right) = \left(R_L/D_3\right)\left(\delta_L/D_1\right)^{n_L}/\left(\delta_G/D_1\right)^{n_G} \tag{24b}$$

Equations (16) and (17) are thus not needed for the analysis of separated flows. That is, once we get the value of R_L/D_3, R_G/D_3 can be calculated using Eq. (24b).

Extension of the Phase Separation Model for a Tee to Other Branching Conduit Configurations

The objective here is to extend the ideas just discussed for dividing flows in Tee junctions to dividing flows in Wye junctions. The physical basis of our previous model, in which the centrifugal and drag forces were identified as the dominant forces controlling phase distribution in the junction, is also appropriate for dividing Wyes of various angles (θ). While the force balance equations developed earlier for dividing flows in a Tee junction are applicable without modification for dividing flow in a Wye, the degree of phase separation will vary due to the influence of angle θ on the curvature of the streamlines.

Before proceeding with the analysis it is interesting to consider the expected influence of the side branch angle, θ, on phase distribution. For $D_1 = D_3$ (Hwang's case) and small values of θ, where θ=0 is a straight conduit, Δp_{13_J} is expected to be small. Consequently, the deviation of the dividing streamlines from straight lines should also be small, and the deviation from an equal phase split should be small. In the limit, as θ approaches zero we expect, $\Delta p_{13_J} = 0$ and $x_1 = x_2 = x_3$.

On the other hand, as shown in Fig.-10, the data indicates that as θ increases Δp_{13_J} also increases which implies a decrease in the radius of curvature of the dividing streamlines and a significant departure from an equal phase split. As we approach θ = 180°, the phase separation should approach the condition $x_3/x_1 = 1/x_1$ for $w_3/w_1 < x_1$, and $x_3/x_1 = w_1/w_3$ for, $w_3/w_1 \geq x_1$. Based on the above discussion, as well as actual measurements of the pressure distribution and visual observations of the flow in the junction [Hwang, 1986], we propose that the form of the dividing streamlines are as shown in Fig.-11. Due to the measured influence of θ on Δp_{13_J} and observations of the flow, it is reasonable to expect that the dividing streamlines begin turning towards the branch sooner as θ increases. Also for θ > 90°, visual observations indicate a stagnant zone with eddies, as shown in Fig.-11.

For cases where the angle of divergence is small (0 < θ ≤ 90°), the following expression for dividing streamlines is used:

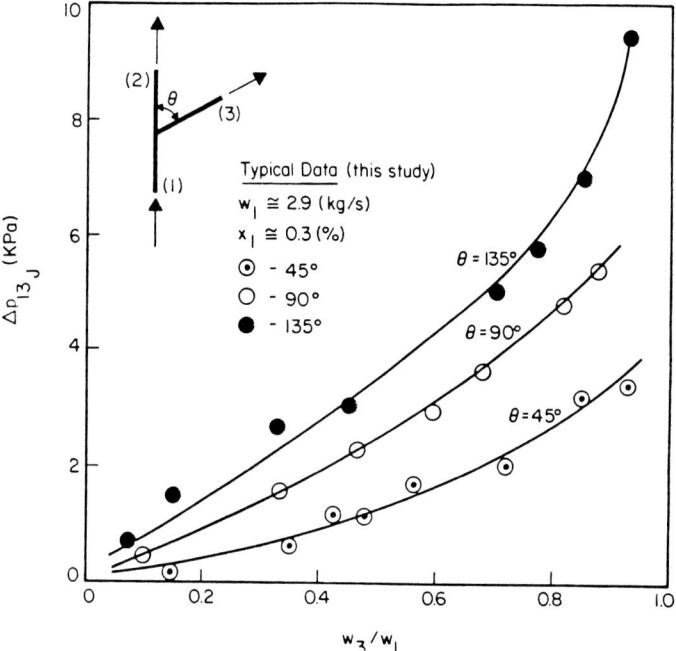

FIGURE 10. Δp_{13_J} vs. w_3/w_1 for Dividing Flows.

BRANCHING CONDUITS

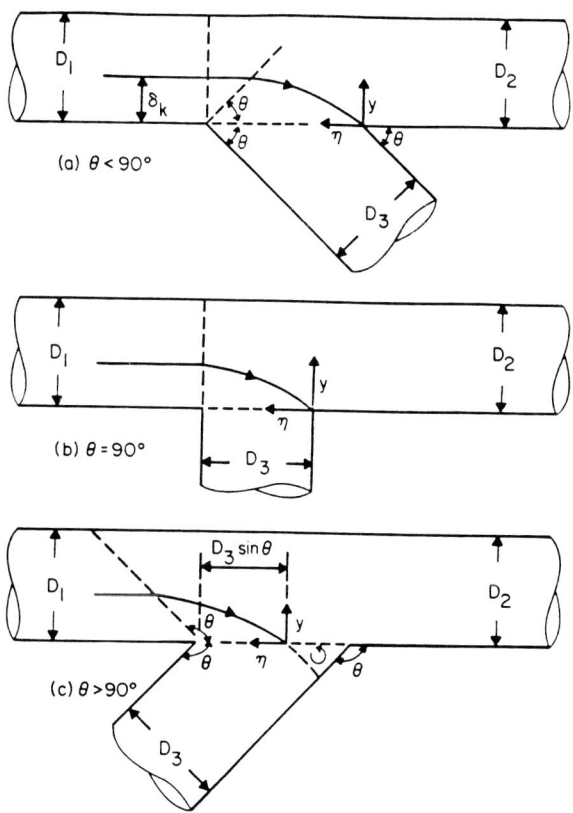

FIGURE 11. Dividing Streamline in Each Configuration for Dividing Flows.

$$\frac{y}{\delta_k} = 1 - \left(1 - \frac{\eta \sin\theta}{D_3 - \delta_k \cos\theta}\right)^{m_k} \qquad (25)$$

Equation (25) satisfies the following necessary requirements:

$y = 0$ at $\eta = 0$ (26a)

$y = \delta_k$ at $\eta = \dfrac{D_3}{\sin\theta} - \dfrac{\delta_k}{\tan\theta}$ (26b)

$\dfrac{dy}{d\eta} = 0$ at $\eta = \dfrac{D_3}{\sin\theta} - \dfrac{\delta_k}{\tan\theta}$ (26c)

From Eq. (25), we can obtain:

$$\left.\frac{dy}{d\eta}\right|_{\eta=0} = m_k \frac{\sin\theta}{\left[\dfrac{D_3/D_1}{\delta_k/D_1} - \cos\theta\right]}$$

and,

$$D_3 \frac{d^2y}{d\eta^2}\bigg|_{\eta=0} = -m_k(m_k-1)\frac{D_3/D_1}{\delta_k/D_1}\left[\frac{\sin\theta}{\frac{D_3/D_1}{\delta_k/D_1}-\cos\theta}\right]^2$$

Consequently, the radius of curvature is given by,

$$\frac{R_k}{D_3} = \frac{\left[1+\left[\frac{m_k\sin\theta}{\frac{D_3/D_1}{\delta_k/D_1}-\cos\theta}\right]^2\right]^{3/2}}{\frac{D_3/D_1}{\delta_k/D_1}\left[\frac{\sin\theta}{\frac{D_3/D_1}{\delta_k/D_1}-\cos\theta}\right]^2 m_k(m_k-1)} \qquad (27)$$

and, the angle of intersection of the streamlines by,

$$\beta = \mathrm{Tan}^{-1}\left[\frac{m_G\sin\theta}{\frac{D_3/D_1}{\delta_G/D_1}-\cos\theta}\right] - \mathrm{Tan}^{-1}\left[\frac{m_L\sin\theta}{\frac{D_3/D_1}{\delta_L/D_1}-\cos\theta}\right] \qquad (28)$$

Notice that, in the limit as θ approaches zero, R_k/D_3 approaches infinity for all values of δ_k/D_1. For $0 < \theta° \leq 90°$, we will follow the assumption adopted earlier for the case of a Tee junction that m_k ranges from 1.0 at $\delta_k/D_1 = 0$ to 2.0 at $\delta_k/D_1 = 1.0$. Hence, the minimum value of R_k/D_3 (corresponding to $\delta_k/D_1 = 1.0$ and $m_k = 2$) can be written as,

$$\frac{R_k}{D_3}\bigg|_{min} = \frac{\left[1+\left(\frac{2\sin\theta}{D_3/D_1-\cos\theta}\right)^2\right]^{3/2}}{\frac{2D_3}{D_1}\left[\frac{\sin\theta}{D_3/D_1-\cos\theta}\right]^2} \qquad (29)$$

Note that for $\theta = 90°$ and $D_3/D_1 = 1$, $R_k/D_3|_{min} = 5.5902$, as before.

Over a range, $0 < \delta_k/D_1 \leq 1$, the radius of curvature can be calculated from the same empirical relation that was used for the Tee, namely:

$$\frac{R_k}{D_3} = \frac{R_k/D_3|_{min}}{\left(\frac{\delta_k}{D_1}\right)^{n_k}} \qquad (30)$$

where n_k may be an empirical function of both $\frac{\delta_k}{D_1}$ and θ.

BRANCHING CONDUITS

For the case where $90° \leq \theta < 180°$, a similar set of relations can be written as before. The expression assumed for the dividing streamlines is given by,

$$\frac{y}{\delta_k} = 1 - \left(1 - \frac{\eta \sin\theta}{D_3 \sin^2\theta - \delta_k \cos\theta}\right)^{m_k} \tag{31}$$

which results in the following relationship for the radius of curvature, R_k, and angle, β, between the two streamlines:

$$\frac{R_k}{D_3} = \frac{\left[1 + \left[\frac{m_k \sin\theta}{\frac{D_3/D_1}{\delta_k/D_1}\sin^2\theta - \cos\theta}\right]^2\right]^{3/2}}{\frac{D_3/D_1}{\delta_k/D_1}\left[\frac{\sin\theta}{\frac{D_3/D_1}{\delta_k/D_1}\sin^2\theta - \cos\theta}\right]^2 m_k(m_k - 1)} \tag{32}$$

$$\beta = \mathrm{Tan}^{-1}\left[\frac{m_G \sin\theta}{\frac{D_3/D_1}{\delta_G/D_1}\sin^2\theta - \cos\theta}\right] - \mathrm{Tan}^{-1}\left[\frac{m_L \sin\theta}{\frac{D_3/D_1}{\delta_L/D_1}\sin^2\theta - \cos\theta}\right] \tag{33}$$

Naturally, Eq. (30) is still valid for this case and the following expression is valid for the minimum radius of curvature,

$$\left.\frac{R_k}{D_3}\right|_{min} = \frac{\left[1 + \left[\frac{2 \sin\theta}{D_3/D_1 \sin^2\theta - \cos\theta}\right]^2\right]^{3/2}}{\frac{2D_3}{D_1}\left(\frac{\sin\theta}{D_3/D_1 \sin^2\theta - \cos\theta}\right)^2} \tag{34}$$

An empirical function for $n_k(\delta_k/D_1, \theta)$ which was found [Hwang et al, 1988] to give good agreement between the analytical predictions and the data for dividing Tees is:

$$n_k(\theta = 90°, \delta_k/D_1) = 5.0 + 20.0 \, \mathrm{Exp}\left\{-53.0\left(\frac{\delta_k}{D_1}\right)\right\} \tag{35}$$

It is interesting to note that for larger extraction rates where $\delta_k \to D_1$, $n_k(\theta = 90°, \delta_k/D_1 = 1.0) \cong 5.0$. In order to account for the observed increase in the radius of the curvature for $\theta < 90°$, and the decrease in the radius of curvature for $\theta > 90°$, it was found that for dividing Wyes one can approximate the branch angle dependence by:

$$n_k(\theta, \delta_k/D_1) = n_k(\theta = 90°, \delta_k/D_1)/\sqrt{\sin\theta} \quad , \text{ for } 0° < \theta \leq 90° \tag{36a}$$

$$n_k(\theta, \delta_k/D_1) = n_k(\theta = 90°, \delta_k/D_1)\sin\theta \quad , \text{ for } 180° > \theta \geq 90° \tag{36b}$$

In addition to the above relations, the two expressions developed earlier from the force balance, Eqs. (16) and (17), remain unchanged and valid.

Finally, it should also be noted that the phase separation model for dividing flows in both Wyes and Tees is valid for all flow regimes, however, one must know what the flow regime is and how the phase velocities and void fraction are distributed in the inlet branch to use the model. Once the nonlinear algebraic equations comprising the model are solved (iteratively) to yield δ_G and δ_L, and thus, as can be seen in Fig.-7, the "Zone of Influence" is defined, and the inlet distribution of each phase is known, the flow rates w_{L_3} and w_{G_3} can be easily calculated.

Hence the dividing streamline model yields the "Fifth Equation" that we need for closure of the phase separation model for dividing two-phase flows.

The Analysis of Phase Separation in Impacting Wyes and Tees

Let us now turn our attention to the analysis of impacting two-phase flows in branching conduits.

Measurements of the pressure distribution in the region of the junction performed by Hwang [1986] suggest that the condition of equal phase split coincides with equal pressure drops, $\Delta p_{12_J} = \Delta p_{13_J}$ (ie, $p_{2_J} = p_{3_J}$). For this condition there is no net pressure gradient in the z-direction at the junction. The existence of a pressure difference, $(p_{2_J} - p_{3_J})$, will influence the two phases differently due to the differences in the inertia of each phase. Indeed, Hwang's [1986] data indicate that $x_3/x_1 < 1$ for $p_{2_J} < p_{3_J}$, and $x_3/x_1 > 1$ for $p_{2_J} > p_{3_J}$.

Based on the above discussion, the momentum balance for the control volume in Fig.-12 at the point of equal phase distribution can be written as:

$$\begin{bmatrix} \text{Rate of momentum} \\ \text{outflow from the} \\ \text{control volume in} \\ \text{the z-direction} \end{bmatrix} = \begin{bmatrix} \text{Rate of momentum} \\ \text{inflow into the} \\ \text{control volume in} \\ \text{the z-direction} \end{bmatrix}$$

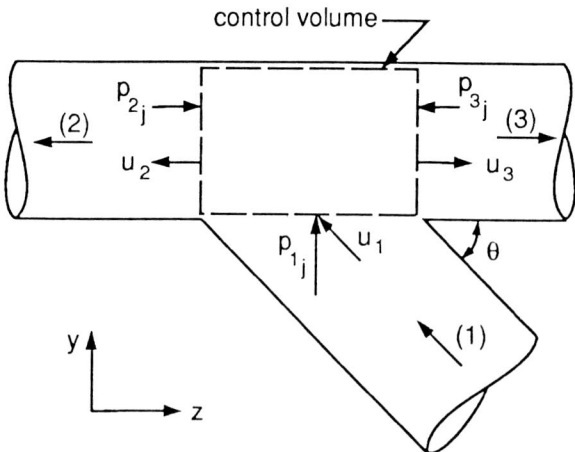

FIGURE 12. Flow Parameters for an Impacting Wye.

BRANCHING CONDUITS

Thus,

$$w_3 u_3 - w_2 u_2 = - w_1 u_1 \cos\theta \tag{37}$$

or,

$$\frac{w_3^2}{\rho_3 A_3} - \frac{w_2^2}{\rho_2 A_2} = \frac{-w_1^2 \cos\theta}{\rho_1 A_1} \tag{38}$$

Given that $x_1 = x_2 = x_3$, and assuming that the flow in the region of the junction is locally homogeneous, (ie, $\rho_1 = \rho_2 = \rho_3$), Eq. (38) yields,

$$\frac{A_1}{A_3}\left(\frac{w_3}{w_1}\right)^2 - \frac{A_1}{A_2}\left(1 - \frac{w_3}{w_1}\right)^2 + \cos\theta = 0 \tag{39}$$

If $A_2 = A_3$, Eq. (39) reduces to,

$$\frac{w_3}{w_1} = \left(1 - \frac{A_3}{A_1}\cos\theta\right)/2 \tag{40}$$

Equation (40) is presumably valid for $0 < \theta < 180°$, and is based on the assumption that $p_{2_J} = p_{3_J}$. In Hwang's experiment [1986], $A_1 = A_2 = A_3$, thus Eq. (40) yields for the flow split at equal phase distribution, $w_3/w_1 = 0.146$ for $\theta = 45°$, $w_3/w_1 = 0.5$ for $\theta = 90°$ and $w_3/w_1 = 0.845$ for $\theta = 135°$. These predictions (for $\theta = 45°$ and $90°$) are in close agreement with the experimental data trends summarized in Fig.-5.

The impacting flow in the junction is very complicated, however, the same dividing "streamline" approach as was used for dividing flows can be used for the analysis of unequal phase separation in impacting two-phase flows in Wyes and Tees.

The approach is based on the proposition that a "zone of influence" exists for each phase and that each "zone of influence" is bounded by a dividing streamline, as shown in Fig.-13. All liquid entering the junction on the right hand side of the liquid phase's dividing streamline (line-b in Fig.-13) will exit through branch 3 of the junction and the remaining liquid will exit through branch 2. The gas phase behaves in a similar fashion with its split between branches 2 and 3 defined by the gas dividing streamline (line-a in Fig.-13). The two dividing streamlines intersect at the point of impact with an angle β. Within the inlet branch, the gas and liquid dividing streamlines are assumed to be straight and separated by distances δ_G and δ_L from the conduit wall, respectively. However, as the junction region is approaches, the dividing streamlines follow curved paths due to the pressure difference, $(p_{2_J} - p_{3_J})$. For the condition of even phase distribution, as characterized by $p_{2_J} = p_{3_J}$, the dividing streamlines are assumed coincident and they remain straight until the point of impact.

The shape of the dividing streamlines is not easy to determine in an exact manner. Nevertheless, an approximation can again be made which satisfies all essential features. The model is developed for the general case of an impacting Wye with arbitrary angle θ, keeping in mind that impacting Tees correspond to the special case, $\theta = 90°$. Using the y-η coordinate system shown in Fig.-13, in which the origin is located at the point of impact, we propose, by analogy with Eqs. (18), (25), and (31), the following form for the dividing streamlines:

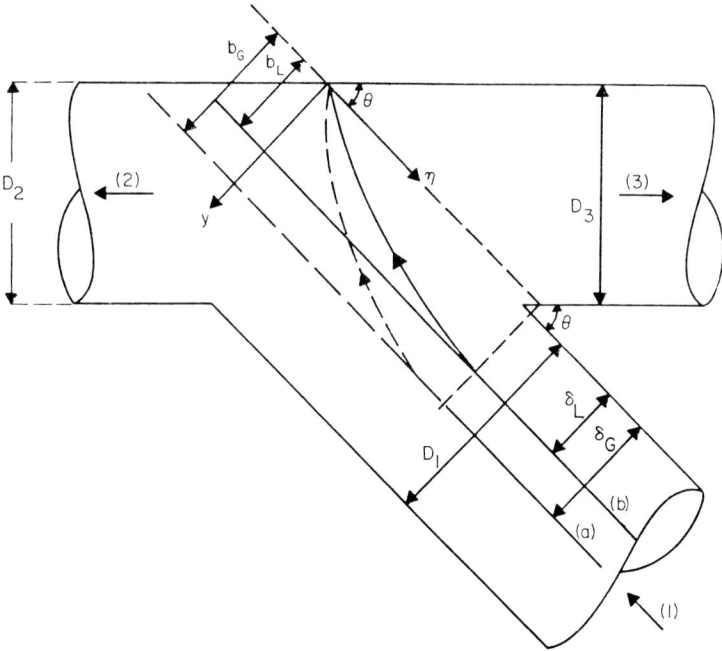

FIGURE 13. Dividing Streamlines in an Impacting Wye.

$$\frac{y}{b_k} = 1 - \left(1 - \frac{\eta \sin\theta}{D_3}\right)^{m_k} , \text{ where } k = G \text{ or } L \tag{41}$$

The distance b_k (k = G or L) is a new parameter and is defined in Fig.-13. For even phase splits, $m_k = 1$ and $b_k = 0$, the dividing streamlines are given by the straight line $y = 0$. For an uneven phase distribution ($b_k > 0$), the data [Hwang, 1986] indicates that, $1 \leq m_k \leq 2$. Equation (41) satisfies the following necessary constraints:

$$y = 0, \quad \text{at} \quad \eta = 0 \tag{42}$$

$$y = b_k, \quad \text{at} \quad \eta = D_3/\sin\theta \tag{43}$$

$$\frac{dy}{d\eta} = \frac{b_k}{D_1} \frac{D_1}{D_3} m_k \sin\theta, \quad \text{at} \quad \eta = 0 \tag{44}$$

$$\frac{dy}{d\eta} = 0 \quad \text{at} \quad \eta = D_3/\sin\theta \tag{45}$$

$$\frac{d^2y}{d\eta^2} = -m_k(m_k - 1)\frac{b_k}{D_1}\frac{D_1}{D_3}\frac{1}{D_3}\sin^2\theta, \quad \text{at} \quad \eta = 0 \tag{46}$$

Hence, from Eqs. (44) and (46) we can formulate the radius of curvature of the dividing streamline of phase-k at the point of impact ($\eta=0$) as:

BRANCHING CONDUITS

$$\frac{R_k}{D_3} = \frac{\left[1 + \left(m_k \frac{D_1}{D_3} \frac{b_k}{D_1} \sin\theta\right)^2\right]^{3/2}}{m_k(m_k - 1)\frac{D_1}{D_3} \frac{b_k}{D_1} \sin^2\theta} \tag{47}$$

and from Eq. (44), the angle β between the gas and liquid streamlines at the point of impact is given by:

$$\beta = \text{Tan}^{-1}\left(m_G \frac{D_1}{D_3} \frac{b_G}{D_1} \sin\theta\right) - \text{Tan}^{-1}\left(m_L \frac{D_1}{D_3} \frac{b_L}{D_1} \sin\theta\right) \tag{48}$$

It should be noted that for a given junction geometry (ie, a specific D_3/D_1 and θ) and given values of the density ratio (ρ_G/ρ_L), slip ratio (S) and b_G/D_1, Eqs. (16), (17), (47) and (48) provide the necessary relations for evaluating the corresponding b_L/D_1 if m_k is known. However, an iterative procedure is necessary due to the nonlinearity of these algebraic equations. In order to complete the development of the phase separation model the relationship between b_k/D_1 and w_{3k}/w_{1k} needs to be determined, as well as the functional relationship between m_k and the other parameters.

For the evaluation of w_{3L}/w_{1L} and w_{3G}/w_{1G}, we need to relate b_k and δ_k. We can use the properties that b_L/D_1 should go to unity when δ_L/D_1 becomes unity and b_L/D_1 must go to zero when δ_L/D_1 is equal to δ_E/D_1, the position of the dividing streamline for an equal phase separation. A convenient function which has these properties is:

$$\frac{b_L}{D_1} = \frac{\left(\frac{\delta_L}{D_1} - \frac{\delta_E}{D_1}\right)}{\left(1 - \frac{\delta_E}{D_1}\right)} \tag{49}$$

From Fig.-13, we note that b_G can be related to b_L, δ_L and δ_G by:

$$b_G = b_L + (\delta_G - \delta_L)$$

or,

$$\frac{b_G}{D_1} = \frac{b_L}{D_1} + \left(\frac{\delta_G}{D_1} - \frac{\delta_L}{D_1}\right) \tag{50}$$

For an even phase distribution the ratios (w_{3k}/w_{1k}) and (w_3/w_1) are equal and they are both determined by Eq. (40) for given A_3/A_1 and θ. Knowing the extraction ratio at an even phase split, $(w_3/w_1)_E$, and assuming a uniform mass flux distribution for both phases, the angle ϕ in Fig.-14 can be easily determined from geometrical consideration to be:

$$(w_3/w_1)_E = [\phi - \frac{1}{2}\sin(2\phi)]/\pi \tag{51}$$

and the depth of the "Zone of Influence" by:

$$\delta_E/D_1 = (1 - \cos\phi)/2 \tag{52}$$

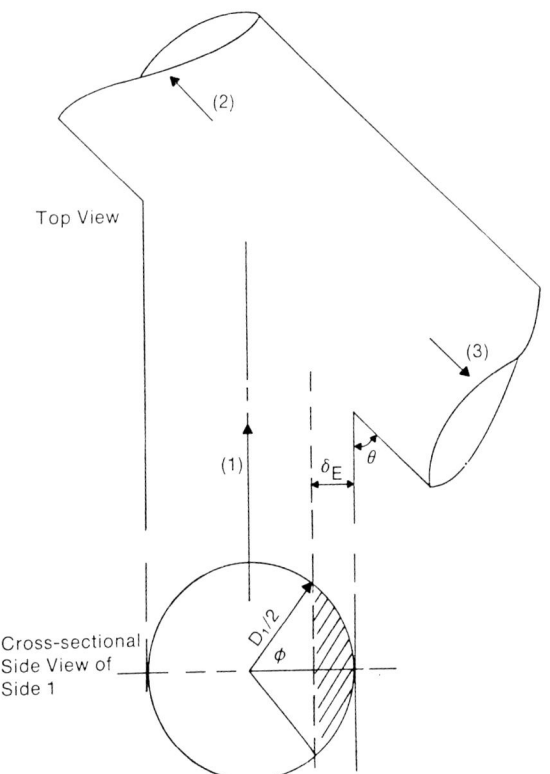

FIGURE 14. The "Zone of Influence" in an Impacting Wye.

For example, for impacting Tees ($\theta = 90°$), we get $(w_3/w_1)_E = 0.5$, $\phi = 90°$ and $\delta_E/D_1 = 0.5$.

It was proposed earlier [Hwang, 1986] that for an even phase split ($b_k = 0$) the dividing streamlines are straight ($m_k = 1$) and that their shape approaches that of a parabola ($m_k = 2$) as b_k/D_1 approaches unity. Hence, from Eq. (47) we get the following formulation for the minimum radius of curvature at $b_k/D_1 = 1$,

$$\left(\frac{R_k}{D_3}\right)_{min} = \frac{\left[1 + \left(2\frac{D_1}{D_3}\sin\theta\right)^2\right]^{3/2}}{2\left(\frac{D_1}{D_3}\right)\sin^2\theta} \qquad (53)$$

The radius of curvature varies from infinity at $b_k = 0$ down to the minimum value given by Eq. (53) at $b_k = D_1$. As before, it was assumed for simplicity that between these two extremes, the radius of curvature can be approximated by:

$$\frac{R_k}{D_3} = \frac{(R_k/D_3)_{min}}{(b_k/D_1)^{n_k}} \qquad (54)$$

BRANCHING CONDUITS

The exponent n_k was determined empirically by comparing the analytical predictions of our model with the phase separation data for impacting junctions. As before, the best agreement was obtained when Eq. (35) was used.

DATA COMPARISON

The phase separation models which have been presented herein were compared to the existing data base to appraise their predictive capability. These data were taken in Wyes and Tees of various size and orientation and for various conditions (including the high pressure steam/water data of Seeger [1985]) and inlet flow regimes (ie, stratified, bubbly and annular flows).

As can be seen in Fig.-15, more than 97% of the published dividing flow data was predicted to within ±25%. For these predictions the mean of the ratio $(x_3/x_1)_{measured}/(x_3/x_1)_{predicted}$ was 1.02 with a standard deviation of 0.103. It can be noted in Fig.-15 that some of the data taken by Hwang [1986] exhibits the most deviation from the model predictions. These are data taken for low extraction rates (w_3/w_1), where, as can be seen in Fig.-4, the slope is very steep. As a consequence any deviation between prediction and measurement is greatly amplified. The large apparent scatter seen in Fig.-15 is due to this amplification of error.

It can be seen in Fig.-16 that the dividing "streamline" model also predicts more than 95% of the available phase separation data for impacting two-phase flows to within ±25%. The mean value of $(x_3/x_1)_{measured}/(x_3/x_1)_{predicted}$ was 0.99 with a standard deviation of 0.09.

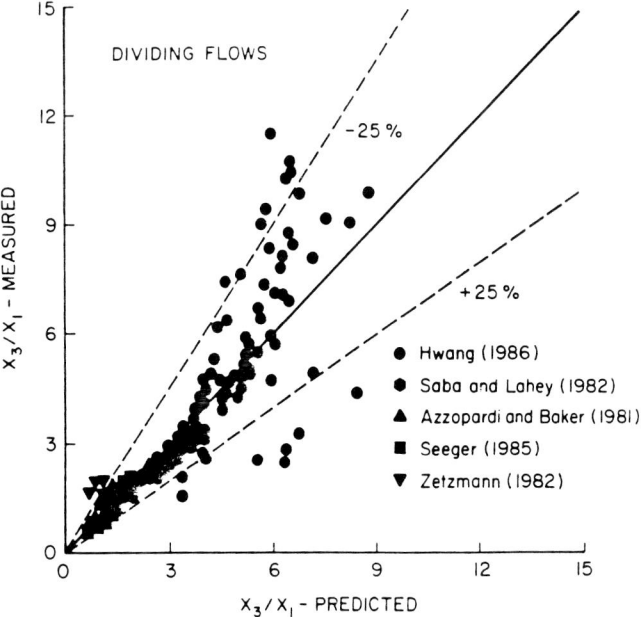

FIGURE 15. x_3/x_1 Measurement vs. Predicted (Dividing Flows)

336 DISTRIBUTION PHENOMENA

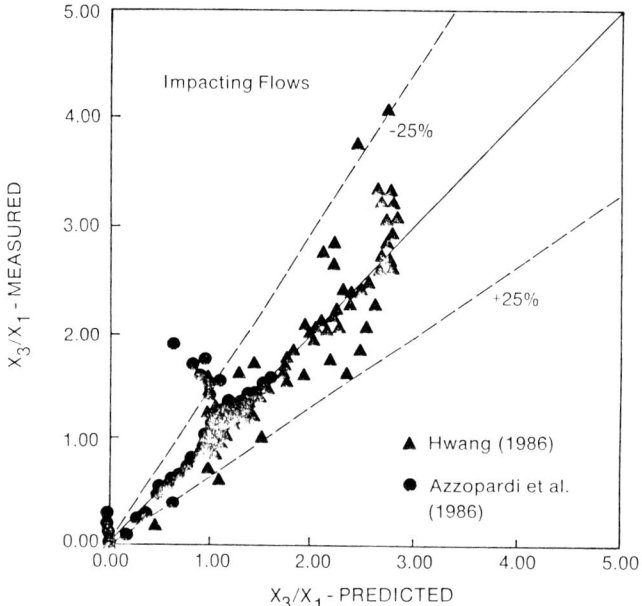

FIGURE 16. x_3/x_1 Measurement vs. Predicted (Impacting Flows)

Thus it appears that a dividing "streamline" model can predict phase separation phenomena over a very wide range of conditions. Nevertheless, to improve accuracy more detailed models may be necessary.

THE USE OF TWO-FLUID MODELS

It is hopefully clear from the previous discussion that phase separation phenomena in branching conduits is inherently multidimensional in nature. Thus it is natural to attempt to use three-dimensional two-fluid models for the prediction of phase separation.

A typical two-fluid model which is appropriate for air/water flows is given by:

<u>Continuity (Liquid)</u>

$$\frac{D_L(1-\alpha)}{Dt} + \nabla \cdot \left[(1-\alpha)u_L\right] = 0 \tag{55}$$

<u>Continuity (Gas)</u>

$$\frac{D_G \alpha}{Dt} + \nabla \cdot \left[\alpha u_G\right] = 0 \tag{56}$$

<u>Momentum (Liquid)</u>

$$(1-\alpha)\,\rho_L\,\frac{D_L \underline{u}_L}{Dt} = -\,(1-\alpha)\,\nabla p + \nabla \cdot \left[(\mu_L + \mu_L^T)\,\nabla \underline{u}_L\right] + (1-\alpha)\,\rho_L g + \underline{M}_i + \underline{M}_w \tag{57}$$

BRANCHING CONDUITS

Momentum (Gas)

$$\alpha \rho_L \frac{D_G u_G}{Dt} = -\alpha \nabla p - \underline{M}_i \tag{58}$$

where,

$$\mu_L^T = C_\mu \rho_L K_L^2 / \varepsilon_L \tag{59}$$

For simplicity, a single-phase K-ε model was used to quantify the turbulence in the continuous (liquid) phase. This transport model is given by:

Turbulent Kinetic Energy in the Liquid Phase

$$(1-\alpha) \rho_L \frac{D_L K_L}{Dt} = \nabla \cdot [(1-\alpha) \frac{\mu_L^T}{\sigma_K} \nabla K_L] + (1-\alpha) \mu_L^T (\nabla \underline{u}_L + \underline{u}_L \nabla) \cdot \nabla \underline{u}_L$$

$$- (1-\alpha) \rho_L \varepsilon_L \tag{60}$$

Dissipation Rate in the Liquid Phase

$$(1-\alpha) \rho_L \frac{D_L \varepsilon_L}{Dt} = \nabla \cdot \left[(1-\alpha) \frac{\mu_L^T}{\sigma_\varepsilon} \nabla \varepsilon_L \right] + C_1 (1-\alpha) \mu_L^T \frac{\varepsilon_L}{K_L} (\nabla \underline{u}_L + \underline{u}_L \nabla) \cdot \nabla \underline{u}_L$$

$$- C_2 (1-\alpha) \rho_L \frac{\varepsilon_L^2}{K_L} \tag{61}$$

where σ_K and σ_ε are the so-called effective Prandtl numbers for the diffusion of turbulent kinetic energy and its dissipation rate, respectively. The numerical values of the constants in the K-ε model are given by Rodi (1984) as,

$C_\mu = 0.09$, $C_1 = 1.44$, $C_2 = 1.92$, $\sigma_K = 1.0$, and, $\sigma_\varepsilon = 1.3$.

Single-phase boundary conditions at the so-called "wall point", a point near the solid boundary, were used in accordance with the findings of Marié (1987). That is, at the wall point, y_p, values of the conserved quantities are given by the "law of the wall" as:

$$u_p^+ = 2.5 \ln y_p^+ + 5.5 \tag{62a}$$

$$K_p = \frac{u^{*2}}{\sqrt{C_\mu}} \tag{62b}$$

$$\varepsilon_p = \frac{2.5 \, u^{*3}}{y_p} \tag{62c}$$

where,

$$u^+ = \frac{u_L}{u^*},$$

$$y^+ = \frac{yu^*}{\upsilon_L},$$

and,

$$u^* = \sqrt{\tau_w/\rho_L}$$

The interfacial momentum transfer terms considered in the analysis included the interfacial drag force and the virtual mass force. That is,

$$\underline{M}_i = \underline{M}_D + \underline{M}_{vm} \tag{63}$$

The drag force is normally expressed in terms of the drag coefficient, C_D, and the relative velocity, $(\underline{u}_G - \underline{u}_L)$, as:

$$\underline{M}_D = \frac{1}{8} \rho_L C_D |\underline{u}_G - \underline{u}_L| (\underline{u}_G - \underline{u}_L) A_i''' \tag{64}$$

In the branch pipes, the two-phase flow which was analyzed with this model was stratified for the flow conditions considered in this study. Therefore, a wavy-stratified drag coefficient [Andritsos & Hanratty, 1987] was used. That is,

$$C_D = \begin{cases} 0.0112 \ Re_G^{-0.2} & , \ |\underline{u}_G| \leq 5\left(\dfrac{\rho_{Go}}{\rho_G}\right)^{1/2} , \ m/s \\[2ex] \dfrac{1}{4}\left\{1+15\left(\dfrac{h_L}{D}\right)^{0.5}\left(\dfrac{|\underline{u}_G|}{5.0}\left(\dfrac{\rho_{Go}}{\rho_G}\right)^{1/2} - 1\right)\right\} & , \ |\underline{u}_G| > 5\left(\dfrac{\rho_{Go}}{\rho_G}\right)^{1/2} , \ m/s \end{cases} \tag{65}$$

where h_L is the depth of the liquid phase and ρ_{Go} is the gas phase density at STP.

In contrast, in the junction, the flow was quite mixed and a bubbly flow regime was assumed to be appropriate. The appropriate bubbly drag law was taken to be the one for distorted particles (Ishii & Zuber, 1979):

$$C_D = \frac{4}{3}R_b\left[\frac{g(\rho_L - \rho_G)}{\sigma}\left\{\frac{1+17.67 \ F(\alpha)^{6/7}}{18.67 \ F(\alpha)}\right\}^2\right]^{1/2} \tag{66}$$

where,

$$F(\alpha) = (1-\alpha)^{1.5}$$

To evaluate the interfacial area density (A_i'''), two limiting flow regimes which may occur in the branches and the junction were considered. In the bubbly flow regime, the local interfacial area density can be directly related to the local void fraction. That is,

$$A_{i_B}''' = \frac{3\alpha}{R_B} \tag{67}$$

For a stratified flow regime the global interfacial area density is given by,

$$A_{i_S}''' = 2 R \sin(\phi/2) \tag{68}$$

where R is the pipe radius and the angle ϕ, between the vertical and the point where the interface touches the conduit wall, is related to the global void fraction through,

$$\langle\alpha\rangle = 1 - \frac{1}{2\pi} (\phi - \sin\phi) \tag{69}$$

It should be noted that the interfacial area densities determined using Eqs. (68) and (69) is the values applicable to 1-D analysis. In 3-D analyses, the structure in the grid cells which span the two-phase interface must be considered. For example, if the flow is stratified, the interfacial area in each interfacial cell can be easily determined, and divided by the cell volumes to give the local interfacial area density in each cell. Cells which are in either the liquid or gas phase only have no interfacial area and thus are not affected by interfacial drag force.

In all branches of the Tee analyzed in this study, the flow regime was stratified [Lahey, 1987]. Therefore, the local 3-D interfacial area density corresponding to a stratified flow regime was used. However, in the junction, the flow regime was not well characterized, thus a weighted average of the local bubbly and the global stratified flow interfacial area density was used. The weighting factor was chosen to give the best agreement with the experimental data.

The virtual mass force may be important in the junction of the branching conduit since appreciable spatial acceleration can occur. The gas phase, with its lower inertia, accelerates more easily toward the side branch whereas the heavier liquid phase tends to continue to move in a straight line. The virtual mass force will act to retard phase separation. The virtual mass force can be written as:

$$\underline{M}_{vm} = \alpha \; \rho_L \; C_{vm} \left[\left(\frac{\partial \underline{u}_G}{\partial t} + \underline{u}_G \cdot \nabla \underline{u}_G \right) - \left(\frac{\partial \underline{u}_L}{\partial t} + \underline{u}_L \cdot \nabla \underline{u}_L \right) \right] \tag{70}$$

where C_{vm} is the virtual volume coefficient. In this analysis, a value corresponding to the bubbly flow regime, $C_{VM} = 0.5$, was used.

The set of conservation equations and closure conditions just described were solved using the general purpose flow simulation code PHOENICS. The Tee junction for which the simulation was performed was broken up into four regions: the inlet pipe, the run, the side branch and the junction. As can be noted in Fig.-17, in order to efficiently compute the 3-D flow, the two-phase flow in the inlet pipe was simulated using non-symmetric cylindrical coordinates. These results were used as inputs for the junction where a body-fitted cylindrical coordinate system was used to discretize the irregular flow domain. Lastly, the side branch and the run were simulated using body-fitted and cylindrical coordinates, respectively, and employed the results from the junction calculations as inputs.

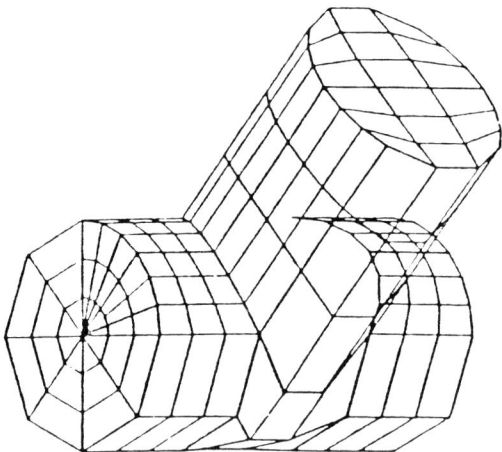

FIGURE 17. 3-D View of the Junction Grids.

In the junction, the grid was distorted in a prescribed manner such that the same topological character of the cells in the branches was retained (i.e., each cell had six faces and six adjacent neighbors). This distortion of the grid was done by specifying the coordinates of the cell corners. The body-fitted coordinate system for each of these cells are local and non-orthogonal. Figure-17 shows a 3-D view of the distorted grid cells in the junction.

The inlet and the run were each divided into 100 axial nodes, 3 radial nodes, and 12 non-uniform azimuthal nodes. The junction also had 100 axial nodes but was divided, using body-fitted coordinates, to match the grid structure of the interfacing inlet, run and side branch pipes.

A stratified two-phase mixture of a given quality and flow rate was introduced into the inlet pipe. The inlet flow rate and quality (w_1 and x_1) were in accordance with the KfK data to be predicted [Lahey, 1987]. The two-phase mixture which entered the inlet pipe was routed into the junction and the two exit pressures from the junction (p_2 and p_3) were controlled to obtain the measured mass extraction ratio (w_3/w_1). The simulation, in turn, gave the phase separation ratio (x_3/x_1). The outlet conditions from the junction were used as the inlet conditions for the run and side branch, and the static pressure profiles were also evaluated.

COMPARISONS WITH MEASUREMENTS

The simulation results obtained in this study are compared with experimental data [Lahey, 1987] in Figs.-18, 19 & 21.

Figure-18 shows the predicted phase separation ratio (x_3/x_1) for a horizontal side branch.

In all the runs evaluated, the optimum weighted-average local interfacial area density for each cell in the junction was found to be close to the global stratified flow interfacial area density. That is,

BRANCHING CONDUITS

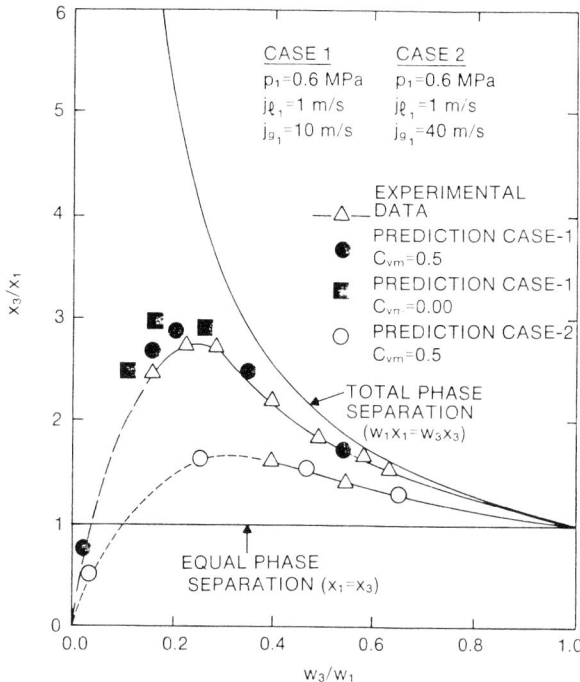

FIGURE 18. Phase Separation for Horizontal Branch.

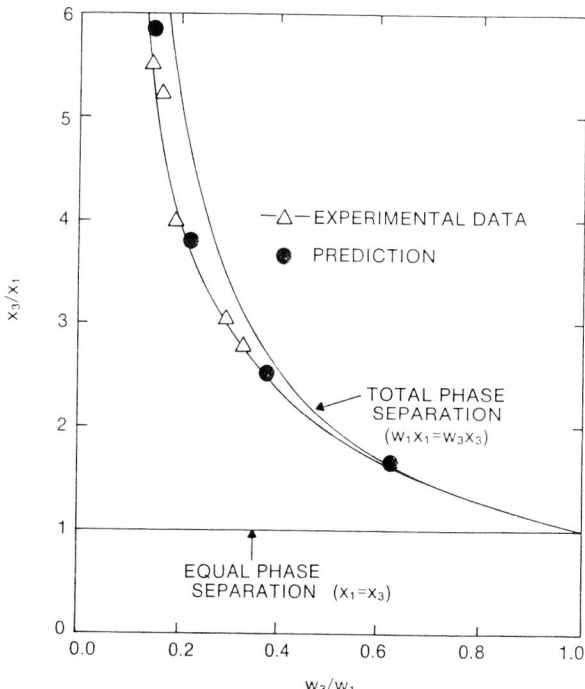

FIGURE 19. Phase Separation for Vertical Branch Above Junction.

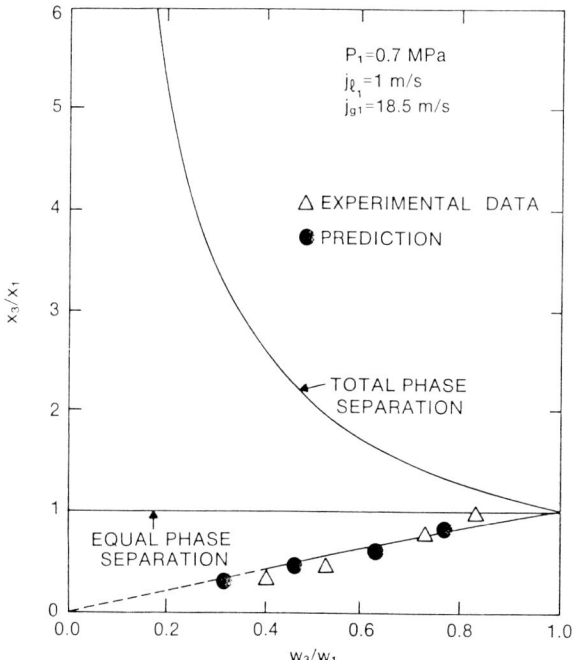

FIGURE 20. Phase Separation for Vertical Branch Below Junction.

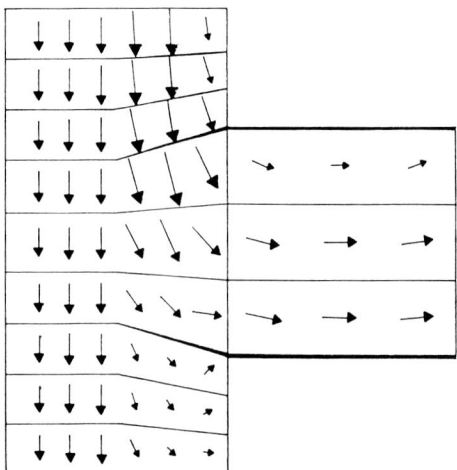

FIGURE 21. Liquid Phase Velocity Vectors for Vertical Branch Above Junction.

$$A_i''' = 0.88\, A_{i_S}''' + 0.12\, A_{i_B}''' \tag{71}$$

There is no physical basis for the use of the weighted-average given in Eq. (71), nevertheless, the flow in the junction is fairly well mixed and

BRANCHING CONDUITS

thus it is not too surprising that the local interfacial area density of the cells in the junction is close to the global value.

As can be seen in Fig.-18, eliminating the virtual mass force (ie, setting $C_{vm} = 0$) slightly increased phase separation and moved the predictions closer to the total phase separation line, and farther away from the data. Nevertheless, since the inlet liquid phase superficial velocities (j_{L_1}) were relatively low, the spatial acceleration in these runs was fairly small and thus the virtual mass force had little effect on the phase separation results.

It should be noted that this 3-D simulation accurately predicts the low mass withdrawal ratio results for a horizontal branch. This is apparently a consequence of the 3-D model adequately predicting the "Zone of Influence" in the junction. Such predictions are not possible with 2-D simulations.

Using the same drag law, virtual volume coefficient, and weighted-average interfacial area density model in the junction, Fig.-19 shows good prediction of the data from experiments having a vertical side branch above junction branch. Figure-20 shows the liquid phase velocity vectors in this case. Liquid entrainment and carry-over can be noted. The position of the dividing streamline is easy to identify. Moreover, the recirculating flow at the upper right corner of the junction indicates that the "Zone of Influence" extends beyond the junction.

Figure-21 shows that the two-fluid model adequately predicts the observed vapor pull-through for the case of a vertical side branch below the junction.

It appears that three-dimensional two-fluid models are inherently capable of predicting the observed phase separation phenomena. However, the current state-of-the-art in two-fluid modeling is such that no completely adequate models exist for the inlet flow regime, drag law and the corresponding interfacial area density in the junction of a branching conduit. This places a significant limitation on the reliability of numerical simulations of phase separation in branching conduits using two-fluid models.

SUMMARY AND CONCLUSIONS

Phase separation phenomena in branching conduits was successfully analyzed using phenomenological dividing "streamline" models and multidimensional two-fluid models. Nevertheless, significantly more research is needed before reliable predictive techniques are available. In particular, both analytical approaches require a specification of the inlet flow regime. In addition, mechanistically-based two-fluid closure models for interfacial momentum transfer (eg, interfacial drag) and the interfacial area density are needed for the junction if reliable two-fluid model predictions are to be made.

It is hoped that this paper will help stimulate the further experimental and analytical studies that are needed to produce reliable phase separation models.

ACKNOWLEDGMENT

The support given this work by the NSF is appreciated and acknowledged.

NOMENCLATURE

Latin

A_i'''	interfacial area density, 1/m
C_D	drag coefficient
C_{vm}	virtual mass coefficient
D	diameter, m
f	friction factor
F	force, N
h_L	depth of liquid in stratified flow, m
K_i	loss coefficient at location-i
K_L	liquid phase turbulent kinetic energy per unit mass, J/kg
g	gravity, m/s^2
G_i	mass flux at position-i, kg/m^2-s
L_i	length of branch-i
M_i	interfacial momentum source per unit mixture volume, N/m^3
M_w	wall force per unit mixture volume, N/m^3
p	pressure, N/m^2
R	radius of curvature, m
Re	Reynolds number ($D_H u/\upsilon$)
S	slip ratio (u_G/u_L)
$\underline{u}$	mean velocity vector, m/s
u_k	velocity of phase-k, m/s
u^+	nondimensional velocity (u/u^*)
u^*	friction velocity ($\sqrt{\tau_w/\rho_L}$), m/s
w_i	mass flow rate at position-i, kg/s
x_i	flow quality at position-i
y	distance from wall, m
y^+	nondimensional distance from the wall $\left(\dfrac{yu^*}{\upsilon}\right)$

Greek

α	local void fraction
$\langle\alpha\rangle$	global void fraction
β, γ	angles
δ_k	position of dividing streamline of phase-k, m
Δp_{ij}	static pressure change from position-i to position-j, N/m^2
ε_L	liquid phase dissipation rate per unit mass, J/kg-s
η	spatial coordinate, m
σ	effective Prandtl number
σ	surface tension, N/m
ρ	density, kg/m^3
ρ'	momentum density, kg/m^3 [Lahey & Moody, 1977]
ρ'''	energy density, kg/m^3 [Lahey & Moody, 1977]
ρ_H	homogeneous density, kg/m^3 [Lahey & Moody, 1977]
τ_w	wall shear stress, N/m^2
ϕ_{Lo}^2	two-phase friction multiplier
Φ	two-phase local loss multiplier
θ	angle between inlet and side branch
μ	viscosity, Pa-s
μ^T	turbulent viscosity, Pa-s
υ	kinematic viscosity, m^2/s

Subscripts

1	inlet
2	run
3	branch
i	interface
d	drag
vm	virtual mass force
B	bubble; bubbly flow
S	stratified flow
K	turbulent kinetic energy
ε	dissipation rate
o	at 1 atm and 20°C (STP)
J	junction
G	gas phase
L	liquid phase
H	hydraulics

REFERENCES

Anderson, J.L. and Owca, W.A., "Data Report for the TPFL Tee/Critical Flow Experiments", NUREG/CR-4164, EGG-2377, 1985.

Andritsos, N. and Hanratty, T.J., "Influence of Interfacial Waves in Stratified Gas-Liquid Flows", *AIChE Journal*, Vol. 33, No. 3, pp. 444-453, 1987.

Azzopardi, B.J., Purcis, A. & Covan, A.H., "Two-Phase Flow Split at an Impacting T", AERE-R-12179, 1986.

Azzopardi, B.J. & Whalley, P.B., "The Effect of Flow Patterns on Two-Phase Flow in a 'T' Junction," *Int. J. Multiphase Flow*, Vol. 8, No. 5, pp. 491-507, 1982.

Collier, J., "Single-Phase and Two-Phase Flow Behavior in Primary Circuit Components", Symposium on Two-Phase Flow and Heat Transfer in Water Cooled Nuclear Reactors, Dartmouth University, 1975.

Crowley, C.J. & Rothe, P.H., "Flow Visualization and Break Mass Flow Measurements in Small Break Separate Effects Experiments", Proceedings of ANS Specialist Meeting on Small Break Loss of Coolant Accident Analysis in LWRs, Monterey, Cal., 1981.

Henry, J.A.R., "Dividing Annular Flow in a Horizontal Tee", *Int. J. Multiphase Flow*, Vol. 7, pp. 343-355, 1981.

Honan, T.J. & Lahey, R.T., Jr., "The Measurement of Phase Separation in Wyes and Tees", *Nuclear Eng. & Design.*, Vol. 54, No. 1, pp. 93-102, 1981.

Hong, K.C., "Two-Phase Flow Splitting at a Pipe Tee", *J. Petroleum Technology*, 290 (1978).

Hwang, S.T., "A Study on Phase Separation Phenomena in Branching Conduits", Ph.D. Thesis, Rensselaer Polytechnic Institute, Troy, NY, 1986.

Hwang, S.T. & Lahey, R.T., Jr., "A Study on Single and Two-Phase Pressure Drop in Branching Conduits", *Experimental Thermal and Fluid Science*, Vol. 1, No. 2, 1988.

Hwang, S.T., Soliman, H.M. and Lahey, R.T., Jr., "Phase Separation in Dividing Two-Phase Flows", *Int. J. of Multiphase Flow*, Vol. 14, No. 4, 1988.

Ishii, M. and Zuber, N., "Drag Coefficient and Relative Velocity in Bubbly Droplet or Particulate Flows", *AIChE Journal*, Vol. 25, 843, 1979.

Lahey, R.T., Jr., "Current Understanding of Phase Separation Mechanisms in Branching Conduits", *Nuclear Eng. & Design*, Vol. 95, pp. 145-161, 1986.

Lahey, R.T., Jr., "Data Set No. 9, Dividing Flow in a Tee Junction", *Multiphase Science and Technology*, Vol. 3, Hemisphere Press, (Hewitt et al, editors) 1987.

Lahey, R.T., Jr. & Moody, F.J., "The Thermal-Hydraulics of a Boiling Water Nuclear Reactor", an ANS Monograph, American Nuclear Society, 1977.

Marié, J.L., "Modeling of Skin Friction and Heat Transfer in Turbulent Two-Component Bubbly Flow in Pipes", *Int. J. Multiphase Flow*, Vol. 13, No. 3, 1987.

Reimann, J. and Kahn, M., "Flow Through a Small Pipe at the Bottom of a Large Pipe with Stratified Flow", *Nuclear Science & Engineering*, Vol. 88, pp. 297-310, 1984.

Reimann, J. & Seeger, W., "Two-Phase Pressure Drop in a Dividing T-Junction, the Mechanics and Gas Liquid Flow Systems", Euromech Colloquium 176, Villard de Lans, France, Sept. 21-23, 1983.

Rodi, W., "Turbulence Models and Their Application in Hydraulics", AIHR Monograph, 1984.

Saba, N. & Lahey, R.T., Jr., "Phase Separation in Branching Conduits", *Int. J. Multiphase Flow*, Vol. 10, No. 1, pp. 1-120, 1984.

Seeger, W., "Untersuchungen zum Druckabfall und zur Massenstromumvertielung von Zweiphasenstromungen in Rechtwinkigen Rohrversweigungen", KfK 3876, Karlsruhe, FRG, 1985.

Smoglie, C., Reimann, J. and Muller, U., "Two-Phase Flow Through Small Breaks in a Horizontal Pipe with Stratified Flow", *Nuclear Eng. & Design*, Vol. 99, pp. 117-130, 1987.

Zetzmann, K., "Phasenseparation und Druckabfall in Zweiphasig Durchstroemten Vertikalen Rohrabzweigungen", Doctorate Thesis, University of Hanover, FRG, 1982.

Zuber, H., "Problems in Modeling of Small Break LOCA", NUREG/CR-0724, 1981.

ROLE OF BUBBLE BEHAVIOR IN TURBULENCE STRUCTURE OF VERTICAL BUBBLY FLOW- SIMULTANEOUS MEASUREMENT OF BUBBLE SIZE, BUBBLE VELOCITY AND LIQUID VELOCITY USING PHASE DOPPLER METHOD

K. OHBA and J. ISODA
Department of Mechanical Engineering
Kansai University
Suita, Osaka 564, Japan

ABSTRACT

The microscopic flow structure in air-water bubbly mixtures flowing through a vertical square duct was experimentally investigated by laser Doppler measurements. Local velocities of liquid and bubbles as well as bubble size were simultaneously measured using a phase Doppler velocimeter system. It was found as a result that a zig-zag motion of the bubble plays an important role in the momentum transfer of liquid flow in the direction perpendicular to the tube axis.

INTRODUCTION

The macroscopic characteristics of the flow as well as heat and mass transfer in vertical bubbly flow depends on the internal microscopic structure of the flow and heat and mass transfer, i.e. the spatial distributions and the temporal changes of the velocities of bubble and liquid, void fraction, bubble size etc. These quantities are, however, difficult to measure and current knowledge regarding the behavior of these flows is incomplete in spite of several previous investigations [Sato & Sekoguchi (1975), Serizawa et al. (1975)].

A laser Doppler velocimeter equipped with two or three photodetectors was developed [Ohba et al. (1986a)] and used to measure bubble and liquid velocities and bubble size simultaneously in air-water bubbly upflow in a vertical tube. Detailed profiles across the tube were obtained and the effect of bubble behavior on these profiles is discussed.

APPARATUS AND PROCEDURE OF EXPERIMENT

Figure 1(a) shows a schematic diagram of the present measurement system. A monochromatic light beam from a He-Ne gas laser was divided into two parallel beams by a beam splitter, and the two beams intersect each other at a measuring volume through a lens. Two photodetectors PMT1 and PMT2 were located in the lateral direction to receive light scattered or reflected laterally from bubbles.

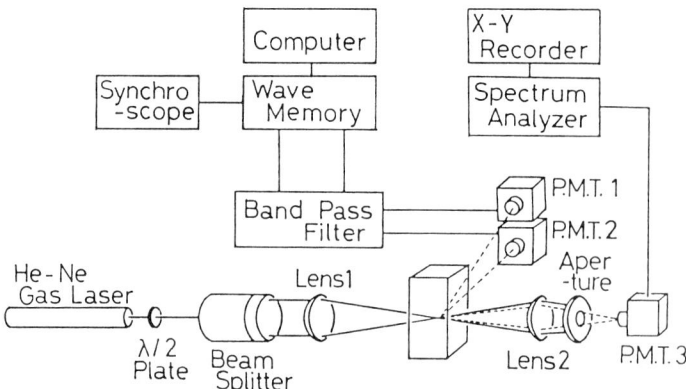

FIGURE 1(a). Schematic diagram of the measurement system in the present experiment.

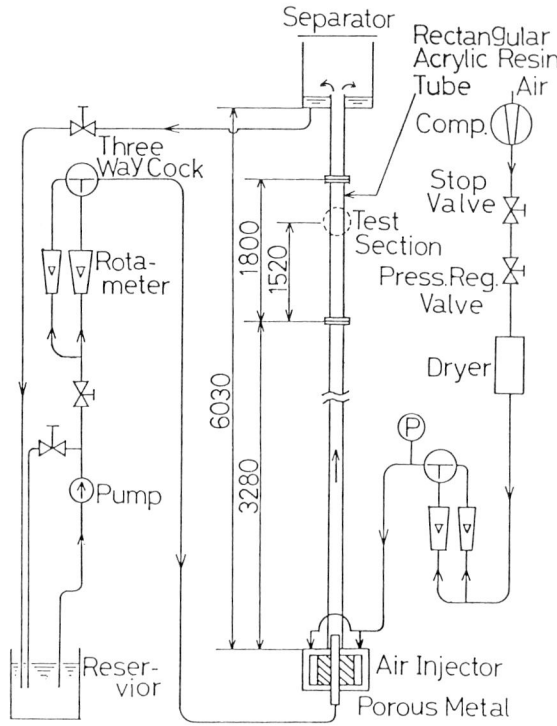

FIGURE 1(b). Schematic diagram of the flow system in the present experiment.

This optical arrangement was adopted on the basis of the following idea: two laser beams are reflected at the surface of a large particle passing through the measuring volume of a laser Doppler velocimeter (LDV). These two reflected beams intersect and interfere with each other at a definite position (an intersection point), and create a spatial fringe pattern.

PHASE DOPPLER METHOD

This fringe pattern moves as the particle moves. Hence, a photodetector located at the intersection point receives an oscillatory light signal of which frequency f_D corresponds to the Doppler frequency. One can therefore measure the particle velocity from the frequency f_D.

Furthermore, under appropriate optical conditions another photodetector looking at the same particle, but located at a different position from the first detector receives a delayed oscillatory light signal. From the time lag between the two signals from the first and second detectors as well as the Doppler frequency of the signal from the first detector one can obtain the fringe spacing created by the moving particle at which the two detectors are looking. Hence, one can obtain the size of spherical particles from the radius-fringe spacing relationship developed by Ohba & Yuhara (1986a) and Ohba & Matsuyama (1986b).

The outputs from PMT1 and PMT2 were simultaneously monitored by a synchroscope and stored in a digital wavememory for subsequent processing by a personal computer to obtain bubble size and velocity.

A very small quantity of cow milk was mixed with the water to provide scattering particles in the liquid. The third detector, PMT3, received forward-scattered light from these particles. The resultant Doppler signal was used in the normal manner to provide the local instantaneous velocity of the liquid through the use of a spectrum analyzer.

Figure 1(b) shows a schematic diagram of the flow system used in this experiment. The 17.7 mm square acrylic resin test section was oriented for vertical upflow of an air-water mixture. Measurements were made under conditions of constant ratio of gas-flow to total flow, β.

RESULTS AND DISCUSSIONS

Figures 2, 3 and 4 show slip velocities u_s ($=u_b - u_\ell$) and mean bubble diameters d_b as a function of the liquid volumetric flowrate per unit area, j_ℓ, in the left and the middle figures respectively. Photographs of bubble motion are shown in the right side of Figs. 2, 3 and 4. The circle mark represents the value on or near the center of the cross section of the duct. The rectangle mark represents the value near the tube wall. The triangle mark does the value near the middle between the center and the wall.

As shown in Fig. 2(a), u_s on the center decreases with the increase in j_ℓ, while u_s near the wall increases with the increase in j_ℓ. This is considered to be due to the fact that bubbles move up in a zig-zag way as shown in Fig. 2(c). When a bubble enters the wall region from the center region, it encounters a slower liquid velocity and in consequence the slip velocity increases because the relaxation distance of the bubble is longer than the distance travelled by the bubble during the transfer from the center to the wall region, and vice versa.

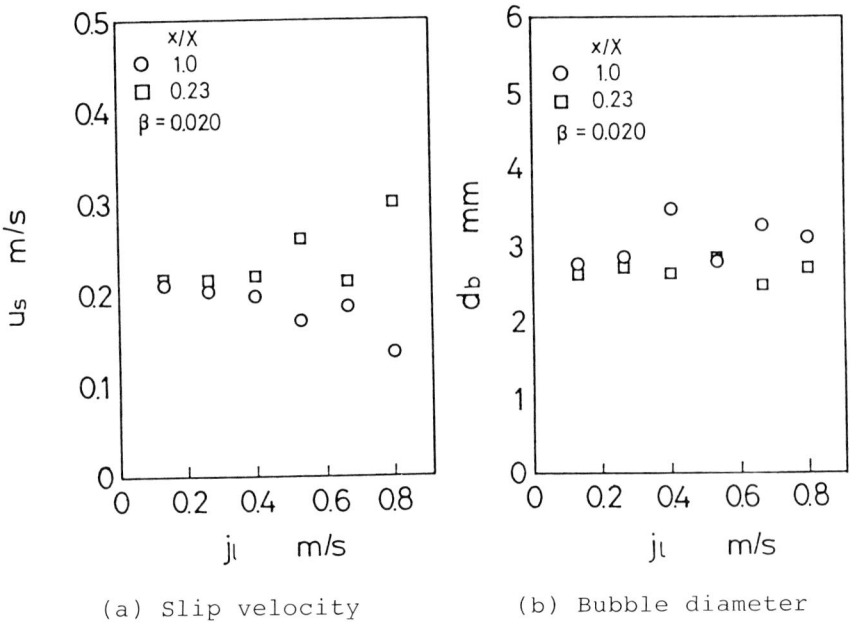

(a) Slip velocity

(b) Bubble diameter

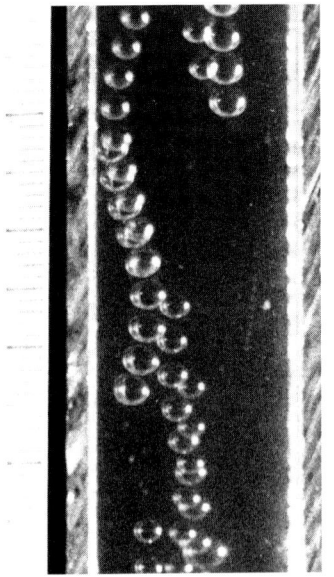

FIGURE 2. Slip velocity and bubble diameter as a function of liquid volumetric flux ($\beta = 0.02$).

(c) Behavior of bubbles ($j_\ell = 0.05$ m/s)

On the other hand, in Fig. 3(a) it is seen that u_s always increases with the increase in j_ℓ both in the center and the wall regions, although β is the same as in Fig. 2(a). This is considered to be because bubbles do not take the zig-zag motion as shown in Fig. 3(c).

PHASE DOPPLER METHOD

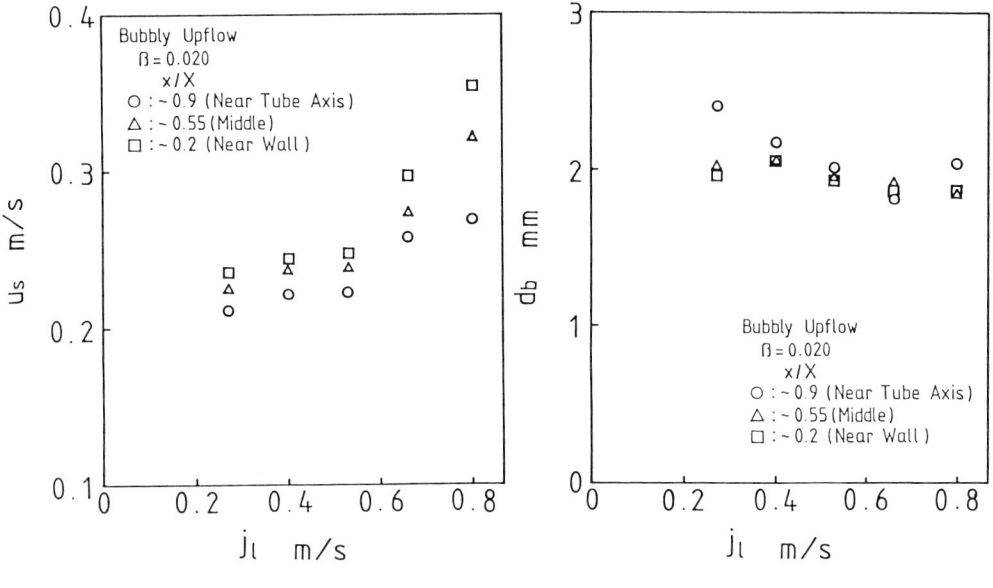

(a) Slip velocity

(b) Bubble diameter

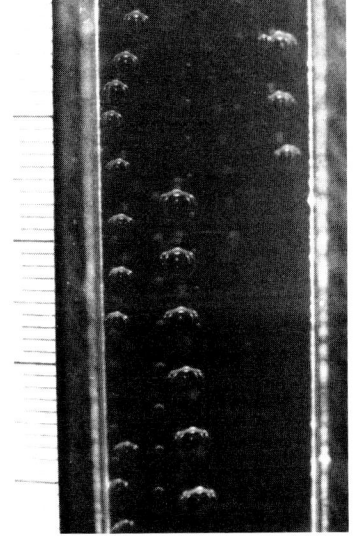

(c) Behavior of bubbles (j_ℓ = 0.13 m/s)

FIGURE 3. Slip velocity and bubble diameter as a function of liquid volumetric flux (β = 0.02).

Whether bubbles take zig-zag motion or not depends mainly on its size and shape which are dependent upon a bubble generator (an air injector) used. On the other hand, the characteristics of an air injector is strongly influenced by water quality. Hence, it is not necessarily clear why there is such a difference in bubble size as shown in Figs. 2 and 3, where bubble generation was made using tap water on different date.

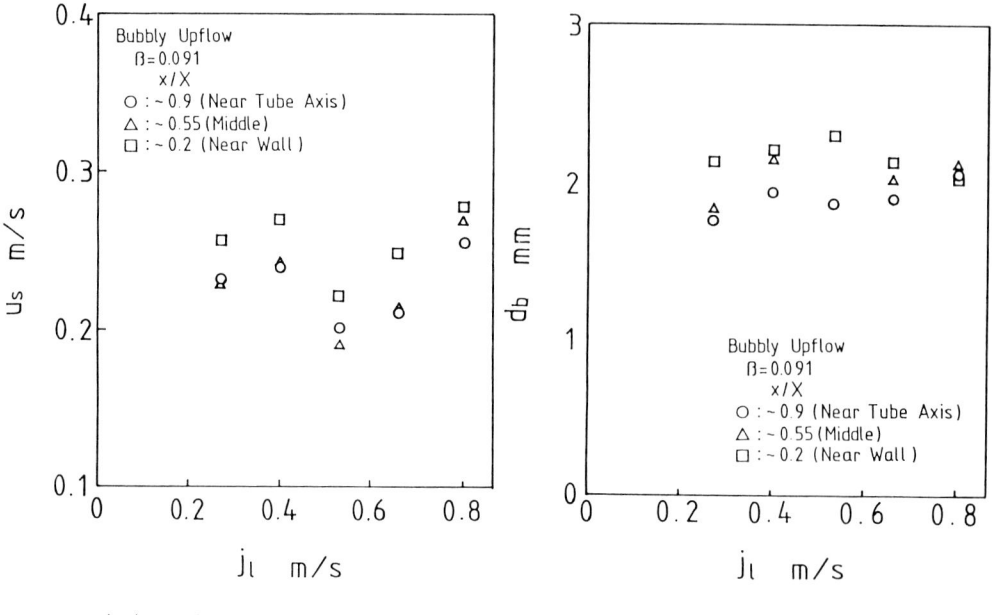

FIGURE 4. Slip velocity and bubble diameter as a function of liquid volumetric flux ($\beta = 0.091$).

It is seen from Figs. 2, 3 and 4 that the slip velocity u_s near the wall is always larger than u_s near the center. It is not clear whether this difference is real or due to the averaging of measurements taken in a zone of large liquid velocity gradient. The latter effect causes a significant difference

PHASE DOPPLER METHOD

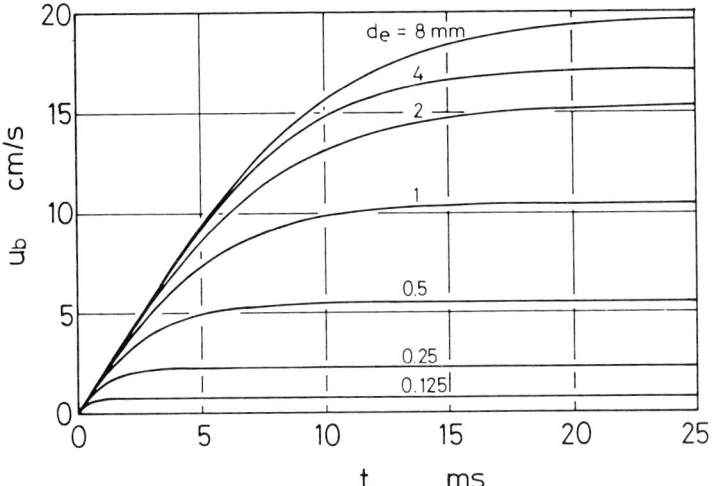

FIGURE 5. Variation of velocity with time for bubbles accelerating freely from rest (calculated).

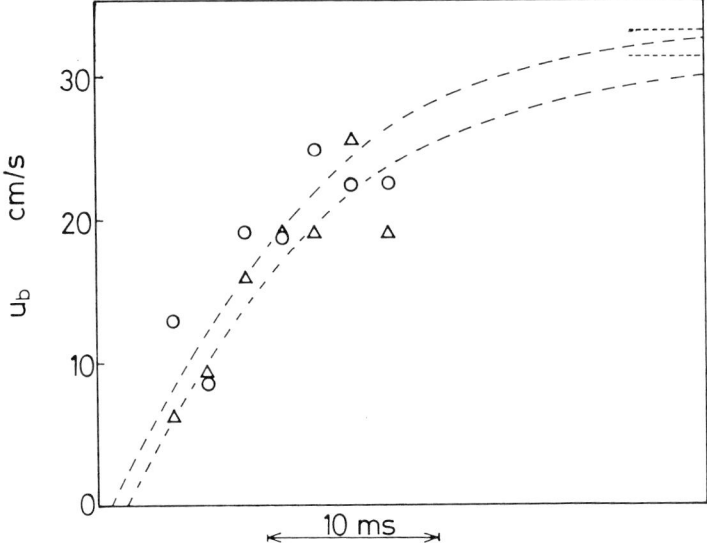

FIGURE 6. Variation of velocity with time for bubbles accelerating freely from rest (measured).

between the slip determined on different sides of a 2-mm bubble near the wall.

The slip velocity u_s is seen to have a local minimum near the region of $j_\ell = 0.5$ m/s in Figs. 2, 3 and 4. But the cause of this phenomenon is not yet elucidated. Bubble size is almost constant with no regard to the position across the duct over the wide range of j_ℓ.

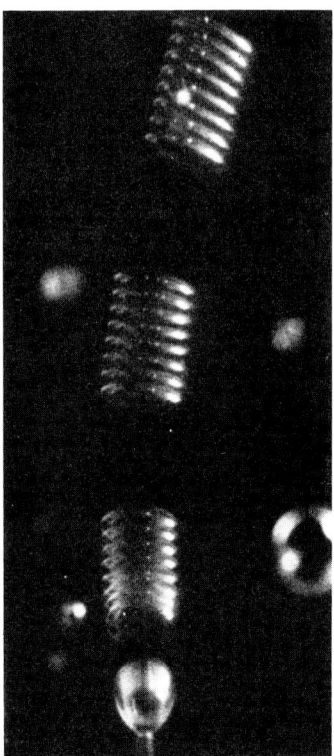

FIGURE 7. Stroboscopic photograph of bubbles leaving a circular nozzle into still water.

DISCUSSIONS ON RELAXATION TIME OF BUBBLE

We suggest in the preceding section that zig-zag motion of the bubble plays an important role in the microstructure of vertical bubbly flow. Therefore, it is important to estimate how fast a bubble can follow a variation or fluctuation of liquid phase velocity. An equation of motion of a single bubble taking account of the virtual mass of the bubble is

$$(\rho_g + \frac{1}{2}\rho_\ell)V_b \cdot \frac{du_b}{dt} = -\frac{1}{2}\rho_\ell C_D A_b (u_b - u_\ell)|u_b - u_\ell|$$
$$+ (\rho_\ell - \rho_g)gV_b \qquad (1)$$

where ρ_g, u_b, V_b and A_b are the density of the gas phase, the velocity, the volume and the cross sectional area of a bubble, respectively. The quantities ρ_ℓ, u_ℓ are the density and the velocity of the liquid phase. Of course t is time and g is the acceleration of gravity. We used as the drag coefficient

$$C_D = \frac{24}{Re}(1 + 0.15 Re^{0.687}) \qquad (2)$$

for Re < 50, and we used the value read from the standard drag curve by Clift et al. (1978) for Re > 50.

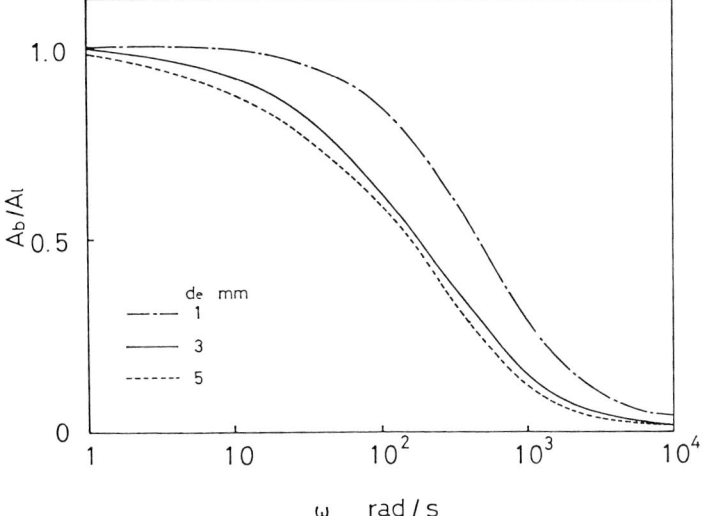

(a) Amplitude ratio

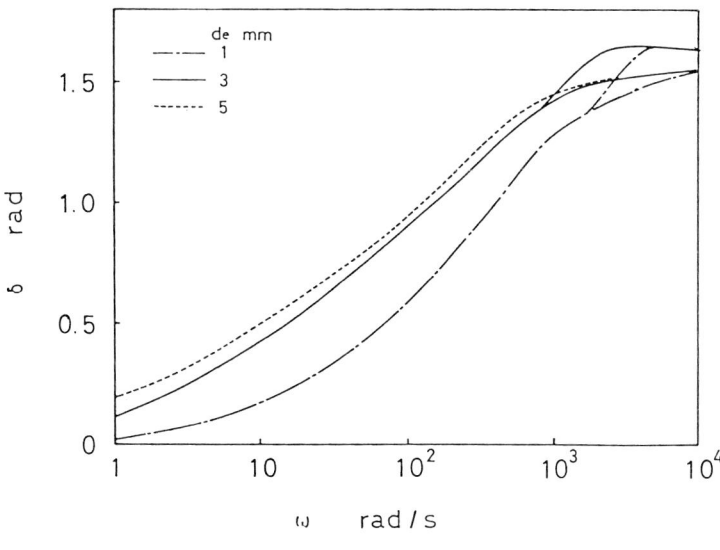

(b) Phase shift

FIGURE 8. Amplitude ratio and phase shift for bubbles entrained in oscillating fluid.

Figure 5 shows the variation of velocity with time for bubbles accelerating freely from rest, which was numerically calculated from Eq. (1) using the Runge-Kutta method. It is seen from Fig. 5 that the relaxation time for a bubble of several mm in diameter is several tens of milliseconds.

Figure 6 shows velocity variations for several bubbles following departure from the injection nozzle as determined by analysis of stroboscopic photographs such as that shown in Fig. 7. The use of virtual mass in the calculation of bubble acceleration appears appropriate based on the comparison between the data and calculations shown in Fig. 6.

In order to simulate the condition that a bubble encounters a fast liquid velocity near the center and a slow one near the wall alternatively due to its zig-zag motion, we calculated the response of a bubble when u_ℓ in Eq. (1) changes sinusoidally as shown in the following equation,

$$u_\ell = 0.25(1 + \sin\omega t). \tag{3}$$

Figure 8 shows the result of this calculation without the effects of buoyancy included. Figure 8(a) shows the amplitude ratio of the bubble velocity to the liquid one, and Figure 8(b) shows the deviation of the phase between the velocities of bubble and liquid. It is seen that the amplitude ratio begins to decrease at about 10 rad/s and the phase deviation also begins to increase at that frequency.

When the buoyancy term is included in the calculation of amplitude and phase using Eq. (1), the critical frequency was found to increase to a value near 100 rad/s.

Actual bubbles probably have an intermediate value of critical frequency between these two values of 10 and 100 rad/s. For instance, the bubbles shown in Fig. 2(c) have a lateral, zig-zag motion frequency of approximately 50 rad/s. Since gravity does not act in the transverse direction, this frequency is significantly beyond the critical value expected from these calculations.

CONCLUSIONS

Local velocities of liquid and bubbles as well as bubble size were simultaneously measured using a phase Doppler velocimetry. It was found as a result that zig-zag motion of the bubble plays an important role in the momentum transfer of liquid flow across the duct.

REFERENCES

Clift, R., Grace, J.R. and Weber, M.E., (1978), *Bubbles, Drops, and Particles*, p.171, Academic Press.

Ohba, K., Yuhara, T. and Matsuyama, H., (1986a), Simultaneous measurements of Bubble and Liquid Velocities in Two-Phase Bubbly Flow Using Laser Doppler Velocimeter, Bulletin of JSME, vol.29, no.254, p.2487.

Ohba, K. and Matsuyama, H., (1986b), Simultaneous Measurement of Size and Velocity of Large Particle, Proc. 3rd Intern. Symp. on Appl. of LA to Fluid Mech., 18.6, Lisbon.

Serizawa, A., Kataoka, I. and Michiyoshi, I., (1975), Turbulence Structure of Air-Water Bubbly Flow, Intern. J. Multiphase Flow, vol.2, p.221.

Sato, Y. and Sekoguchi, K., (1975), Liquid Velocity Distribution in Two Phase Bubble Flow, Intern. J. Multiphase Flow, vol.2, p.79.

CHARACTERISTIC STRUCTURE OF UPWARD BUBBLE FLOW

G. MATSUI
Institute of Engineering Mechanics
University of Tsukuba
Tsukuba, Ibaraki, Japan

ABSTRACT

Structural characteristics of bubble flow were investigated by experimenting with nitrogen gas-water bubble flow in a square duct and a circular pipe using a laser Doppler anemometer and a double-sensor conductance probe system. Bubble size at constant gas and liquid flows was reduced by changing the gas injection method and by adding a surfactant to water. Coring, sliding, or uniform bubble flow occurred according to the bubble size. Results of the experiment show that each bubble flow exhibits characteristic behavior in the main flow; its turbulence and bubble motion depend on the size and distribution of bubbles. Moreover, it is found that coring bubble flow has a peculiar tendency for turbulence and that sliding bubble flow, with a specific saddle-type phase distribution, has turbulent suppression. Secondary flow in bubble flow exhibits essentially similar properties to that in single-phase flow, but its velocity tends to decrease in the region where bubble concentration is highest.

INTRODUCTION

Bubble flow is one of representative flow regimes in two-phase flow which takes place in a wide range of energy conversion devices such as nuclear reactors, petroleum plants, and chemical reaction apparatus. In order to improve the efficiency of such systems, or to analyze the safety of nuclear reactors, it is essential to know the detailed structure of bubble flow which takes place in the systems. Bubble flow exhibits various patterns of internal flow structure depending on flow rate conditions, mixing methods, and/or fluid properties. However, the related flow mechanism is not perfectly clear because of complex interactions between bubbles and liquid or channel walls.

On the basis of the peculiar profile configuration of a lateral void fraction or gas-phase distribution in a cross-section of a channel, Sekoguchi et.al.(1980) have classified bubble flow into two fundamental categories, namely coring and sliding bubble flows, and their combination, intermediate bubble flows. They have indicated that the gas-phase profiles of bubble flow depend on the mixing condition and the flow rate.

In most cases, turbulence is increased if gas is injected into liquid flow (Serizawa et.al., 1975; Inoue et.al., 1976; Theofanous & Sullivan, 1982; Lance & Bataille, 1983), but Serizawa et.al. (1975) have found a turbulence suppression phenomenon; that turbulence in bubble flow decreases compared with that in all liquid flow.

In addition, some features for the behavior of bubbles have been obtained; bubbles greater than 3 mm do not exist near the wall of channel (Sekoguchi et.al., 1981), bubbles collect at the corner of the channel with a square cross-section (Sadatomi & Sato, 1982), and bubbles about 2 mm rise straight like a rigid sphere (Zun, 1980).

This work represents the further development of lateral water velocity fluctuation intensity and contour maps of water velocity from those presented in Refs. (Matsui et.al., 1986; Matsui et.al., 1987). It is also concerned with the effect of bubble size on internal characteristics of bubble flow, particularly the cross-sectional effect of bubble flow. The size of bubbles was controlled by changing the mixing conditions and by adding a surfactant to liquid. By combining the two ways, four kinds of experimental cases were set up under a fixed flow rate condition.

Experiments were carried out for upward bubble flow in a channel with a square cross-section or a circular cross-section using nitrogen gas and water as working fluids. Under fixed flow rate conditions, bubble flows classified into two fundamental categories and their combinations were encountered. Typical cases of experimental results are discussed here; the effect of bubble size and shape on flow structure, turbulence suppression, and secondary flow. From these experimental results, internal flow characteristics peculiar to bubble flow were found.

EXPERIMENTAL APPARATUS AND INSTRUMENTATION

A schematic diagram of the experimental apparatus is shown in Fig. 1. The channel system was a single closed loop and consisted of a downcomer, horizontal section, a pump, a turbine-type flowmeter, a gas-liquid mixer, a riser section, and a gas-liquid separator along the flow direction. The riser section includes the vertical test section with quick-closing valves at both ends. A channel for the test section downstream of mixer was exchangeable. The channel had a square cross-section 30 x 30 mm^2 or a circular cross-section 40 mm i.d. and was made of acrylic resin to allow visual observation and the measurement using laser beams. In the gas-liquid mixer, a sintered metal ring with 100 μm-mesh was used to make bubbles in flowing liquid by injecting gas into the liquid through the ring.

Nitrogen gas and water were used as working fluids. The nitrogen gas was injected into the flowing water through the

UPWARD BUBBLE FLOW

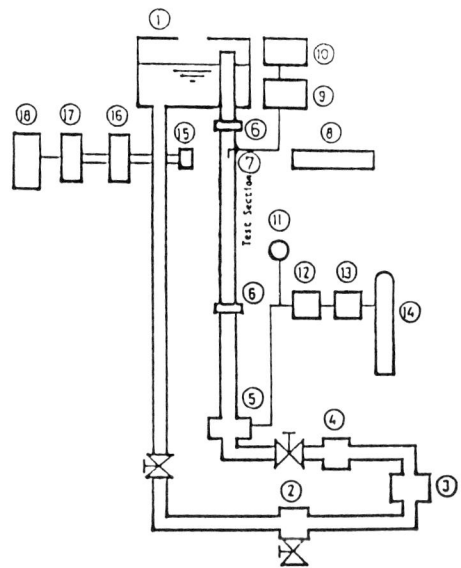

(1) Gas-liquid separator
(2) Drain
(3) Pump
(4) Turbine-type flowmeter
(5) Mixer
(6) Quick-closing valve
(7) Conductance probe
(8) Laser Doppler anemometer
(9) Counter processor
(10) Microcomputer
(11) Pressure gauge
(12) Float-type flowmeter
(13) Pressure regulator
(14) Bomb
(15) Photomultiplier
(16) Counter processor
(17) Interface
(18) Microcomputer
(19) Amplifier
(20) Low-pass filter
(21) A/D converter and Minicomputer

FIGURE 1. Schematic diagram of experimental apparatus.

mixer at the bottom part of the riser section, producing upward bubble flow. Tne gas was separated in the gas-liquid separator and was released into the atmosphere.

The water flow rate was measured by a turbine-type flowmeter, and the gas flow rate measured by a float-type flowmeter. The gas flow rate was corrected for the exit pressure of the flowmeter.

The average void fraction in the measurment section was obtained from the gas volume trapped between two quick-closing valves.

A laser Doppler anemometer (LDA) and a double-sensor conductance probe system were used to measure the structure of bubble flow on the cross-section about 1.5 m downstream of the mixer. Water velocity was measured by the former, and bubble velocity, bubble length, and local void fraction by the latter, respectively.

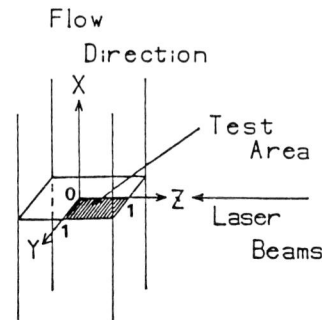 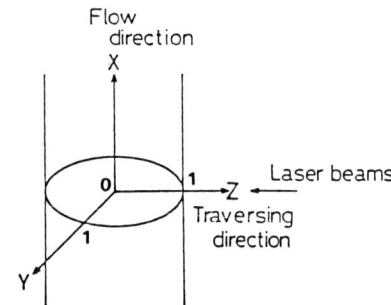

(a) Square cross-section (30x30 mm²)   (b) Circular cross-section (40 mm i.d.)

FIGURE 2. Coordinate system on the measurement cross-section.

The coordinate system on the measurement cross-section is shown in Fig.2. The measuring points by the LDA and the probe system were traversed every 1 mm from the center of the channel to the wall in the Y- or Z-direction.

The LDA-system employed has 3-beam two-component transmitting optics used in connection with an Ar-ion laser and a receiving optics located in direct forward scattering. Two 100 MHz counters were used to process the signals received from the two photo-multipliers for two velocity components. Digital data from detected signals were stored in a micro-computer through interfaces from the counters. Only signals, which simultaneously appeared on both counters, were used for further processing to insure that they are from the source particle. The average values and the fluctuations of water velocities were obtained by processing the stored data.

The double-sensor conductance probe employed has two point-electrodes with spacing of about 3 mm to detect the arrival and residence time of a bubble. The detected signals went to a counter processor and the digital data was stored in a micro-computer. The average values and fluctuations of bubble velocity, bubble length, and local void fraction were obtained by processing the stored data. The spacing between two electrodes of the probe was measured using a microscope.

The size of bubbles was controlled by changing the mixing condition and by adding a surfactant to water. The Tween 20 (Polyoxyethylene Sorbitan Monolaurate) was used as the surfactant. By adding the Tween 20 to water, the resultant value of surface tension of water was reduced by about half. Thus, by using two kinds of mixing methods and a surfactant, four kinds of experimental cases were investigated as follows:

Case (1) a 100 μm-mesh sintered metal ring was not used effectively; it was equivalent to a perforated wall.
Case (2) a surfactant was added to water under the case (1).
Case (3) a 100 μm-mesh sintered metal ring was used effectively.
Case (4) a surfactant was added to water under the case (3).

Experiments were carried out under the flow rate conditions of j_L= 0.37 to 0.93 m/s in the water superficial velocity and j_G= 0 to 0.023 in the gas superficial velocity.

EXPERIMENTAL RESULTS AND DISCUSSION

The internal behavior of bubble flow was investigated by reducing the bubble size under each constant flow rate condition for two test channels; a square cross-section, and a circular cross-section. Typical results for the four experimental cases, (1) to (4), are shown in Table 1 and Figs. 3 to 13, together with results of the experimental case (5) for water single-phase flow. The number in the figure captions for each figure corresponds to the numerical order of the experimental cases. The flow rate conditions are j_L=0.93 m/s in water superficial velocity, j_G=0.023 m/s in gas superficial velocity, $<\alpha>$=0.017 in average void fraction in the measurement section in the case of the channel with a square cross-section, and j_L=0.66 m/s, j_G=0.012 m/s, and $<\alpha>$=0.009 in the case of the channel with a circular cross-section.

Bubble Size, Shape, and Distribution

Figure 3 shows the pictures of flow in the square cross-sectional channel and frequency distribution of bubble length at two laser beam locations (the center and the wall vicinity) of the channel for each of the four experimental cases.
A horizontal white line in the pictures is a laser beam which is scattered owing to the existence of bubbles. The lateral void fraction (or phase) distribution and the lateral bubble length distribution for each bubble flow are shown in Figs. 4 and 5, respectively. Figures 3 and 4 show that the size of the bubbles is reduced in numerical order of the experimental cases and that the bubble shape changes from ellipsoidal to spherical. In the same order, the number of bubbles increases and the interfacial area also increases. The classification of bubble shape by Ishii & Mishima (1980) suggests that bubble shape in the flows of the cases (1) to (4) is wobbly, ellipsoidal, ellipsoidal, and spherical in numerical order of the cases. The lateral void fraction distributions in Fig. 5 suggest that the flow of the case (1) or (2) is classified into the category of coring bubble flow, the flow of the case (3) is classified into the category of sliding bubble flow, and the flow of the case (4) is classified into the category of intermediate bubble flow. Here, the flow of the case (4) is conveniently called uniform bubble flow, because the flow has a uniform lateral distribution of bubble length as well as a lateral void fraction distribution.

The frequency distributions of bubble length indicate that the coring bubble flow of case (1) has a tendency for devia-

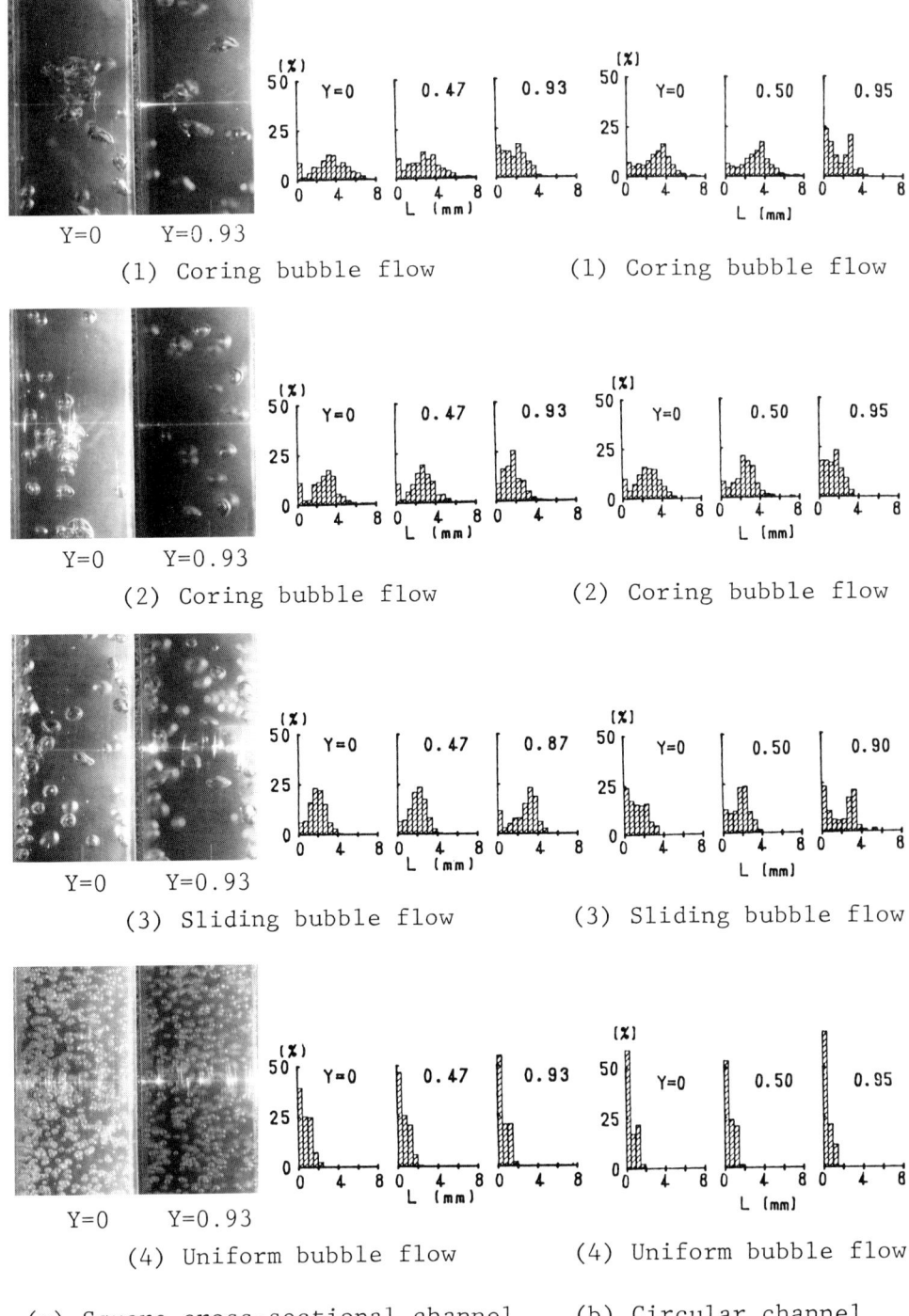

(a) Square cross-sectional channel   (b) Circular channel

FIGURE 3. Photographs and frequency distributions of bubble length.

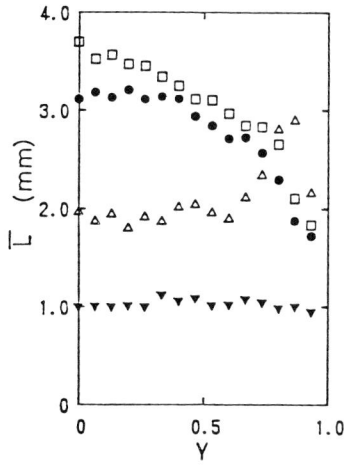
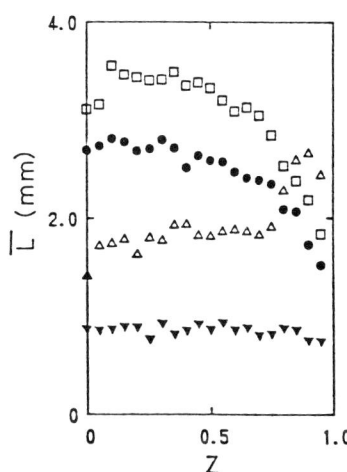

(a) Square cross-sectional channel (b) Circular channel

FIGURE 4. Lateral bubble length distributions.

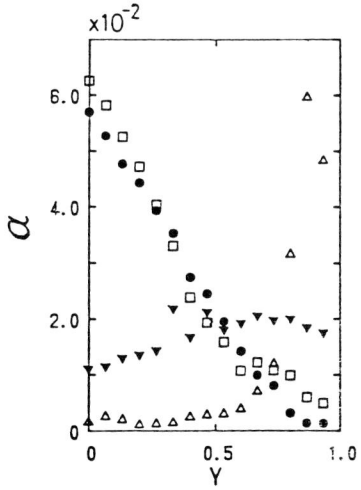
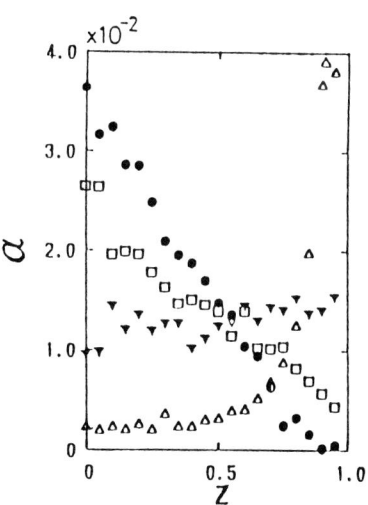

(a) Square cross-sectional channel (b) Circular channel

FIGURE 5. Lateral void fraction distributions.

TABLE 1. Flow characteristics for experimental cases.

Experimental case No.	Gas-Injection Method	Surfactant	Major Bubble Size (mm)	Bubble Shape	Lateral Bubble Distribution	Categories of Bubble Flow	Symbols in Figures
(1)	Equivalent to a perforated wall	No added	3~6	Ellipsoidal-like	Convex	Coring	□
(2)	Equivalent to a perforated wall	Added	2~5	Ellipsoidal-like or Spherical-like	Convex	Coring	●
(3)	100 μm-mesh sintered metal ring	No added	2~4	Spherical-like or Spherical	Saddle	Sliding	△
(4)	100 μm-mesh sintered metal ring	Added	1~1.5	Spherical	Uniform	Uniform	▶
(5)	Water single-phase flow						×

tion of bubble length to decrease from the center toward the channel wall and for bubbles greater than 4 mm not to be encountered near the wall. The pictures of the flow also show these trends. The coring bubble flow of case (2) has similar trends, but the deviation of bubble length decreases and the bubble shape changes to a more spherical shape compared with the flow of case (1). The sliding bubble flow of case (3) has special properties for bubble size. A few bubbles about 2 mm exist near the wall. Bubbles about 3 mm collect near the wall, exactly at the location (Y=0.87 or Z=0.9: 2 mm away from the wall) of the peak point of lateral void fraction distribution. The distance between the peak point and the wall corresponds to the radius of bubbles near the wall. Moreover, the flow has an extremely low void fraction in the core region of channel. Accordingly, this suggests that the bulk of the bubbles flow upward along the wall. The uniform bubble flow of case (4) has properties for bubble size such that the deviation of void fraction is small compared with the other bubble flows, and that the bubble length is about 1 mm. Moreover, the flow has a tendency for bubbles to distribute uniformly over the cross-section, even if the number of bubbles or gas flow rate is decreased. The flow in a channel with a circular cross-section have similar trends and properties to those in a channel with a square cross-section. The characteristics of bubble size, shape and distribution for each bubble flow are summarized in Table 1.

Main Flow Characteristics

Main mean flow and its turbulent flutuations. Mean flow or water velocity distributions in a quarter plane of the square cross-section are shown in Fig. 6(a) and their fluctuation intensity distributions are shown in Fig. 6(b), respectively. For the sake of easy comparison, lateral mean flow velocity and fluctuation intensity distributions along the Y-axis and a diagonal line are shown in Fig. 7, together with results along the Z-axis for the flows in a circular channel.

The coring bubble flows of cases (1) and (2) have a tendency for turbulence to increase toward the wall from the center along the Y-axis and for main flow velocity to be higher than that of water single-phase flow. However, the sliding bubble flow of case (3) has a peculiar tendency for turbulence to become weaker than that of water flow in the core region of the channel, (turbulence suppression), but to become stronger than that of water flow near the wall. The mean flow velocity has a peculiar tendency to be lower in the core region but higher near the wall than that of water flow. The uniform bubble flow of case (4) has a similar turbulence distribution to water flow, but a tendency for flow velocity to be higher than that of water flow. These trends are observed over the cross-section except fluctuation intensity distribution which is observed along the diagonal line for coring bubble flows. Contrary to the other bubble flows, coring bubble flows in a square cross-sectional channel have a peculiar trend that the turbulence along the diagonal line decreases toward the corner from the center.

The mean flow velocity of bubble flow over the cross-section

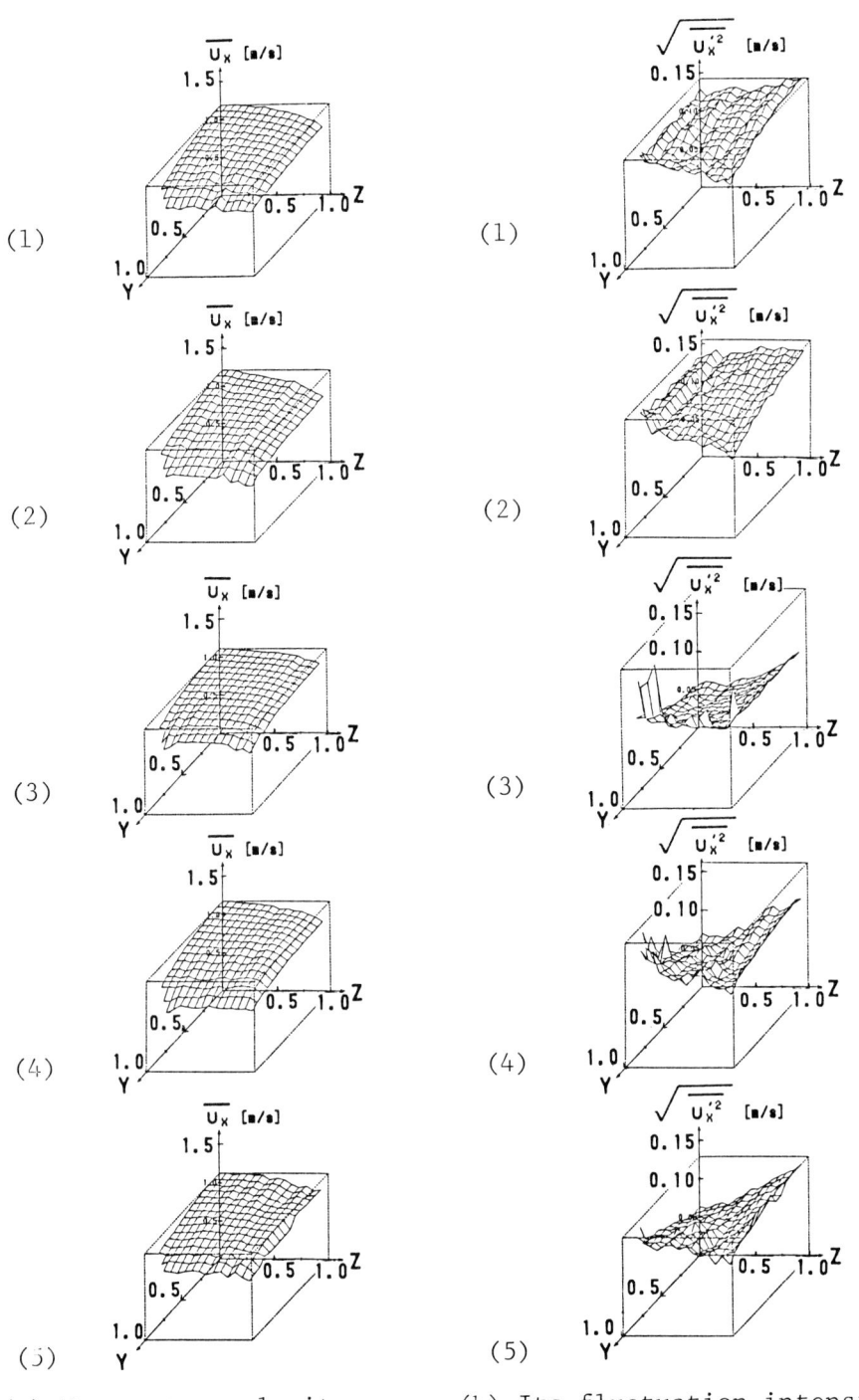

(a) Mean water velocity (b) Its fluctuation intensity

FIGURE 6. Main flow characteristics on 1/4 plane in the square cross-section.

UPWARD BUBBLE FLOW 369

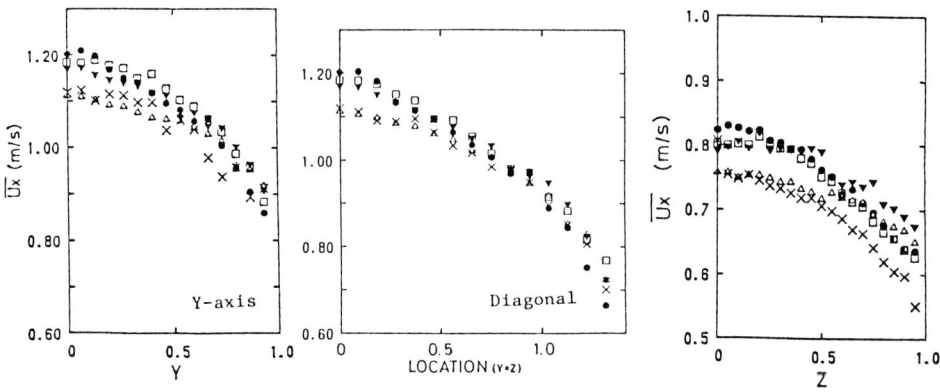

(i) Square cross-sectional channel   (ii) Circular channel

(a) Mean water velocity

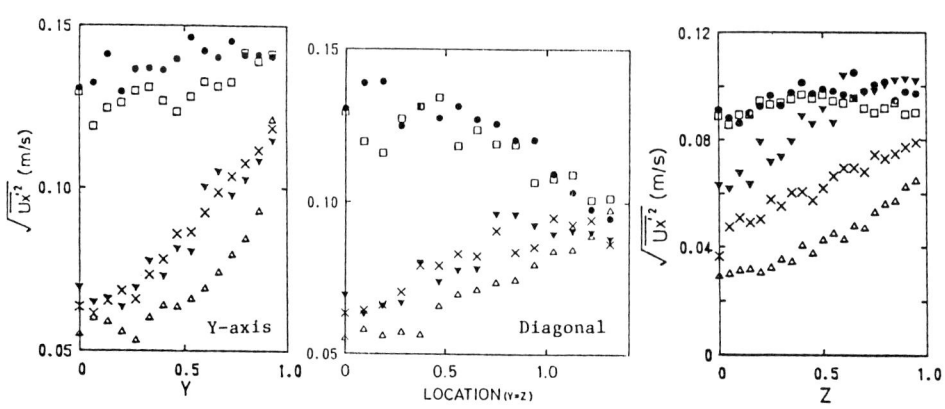

(i) Square cross-sectional channel   (ii) Circular channel

(b) Water velocity fluctuation intensity

FIGURE 7. Lateral distributions of main flow characteristics.

is high in numerical order of the cases. The tendency for velocity to increase in the order of the cases (1), (2) and (4) is interpreted to mean that because bubbles are small and spherical and distributed widely, the interaction between the bubbles and the water increases. The intensity of turbulent fluctuations of bubble flow averaged over the cross-section decreases in the order of the cases (1) (or (2)), (4), and (3). Particularly, different from the order for the mean flow velocity distribution, turbulent fluctuations have a remarkable feature that the intensity in the case (3) is lower than that in the case (4). When gas flow rate is small, turbulent fluctuations in the case (1) are lower than those in the case (2).

The flows in a circular channel have similar characteristics and trends to those in a square cross-sectional channel.

Distribution of equal mean flow velocity in a square cross-sectional channel. Figure 8 shows distributions of equal mean flow velocity or contour maps of the main flow velocity. The darker the color, the higher the flow velocity, because the velocity rises when the lattice interval becomes narrower and the velocity falls when the lattice interval becomes wider.

Compared with the equal velocity distribution of the single-phase flow (case (5)), the coring bubble flow (case (1)) has high velocity near the center and in the case (2) with added surfactant, higher velocity is widely distributed. The sliding bubble flow (case (3)) has a high velocity over a wide range, except in the center part, compared with the single-phase flow. In the case (4) with added surfactant, the flow shows a tendency for the velocity near the center to become higher and to be high over the entire range of cross-section compared with the single phase flow.

In the single phase flow, the equal velocity distribution is symmetrical to the diagonal of the channel cross-section. This is similar to the experimental results by Ohba & Yuhara (1982). Similar trends are observed in the bubble flows. This suggests that the secondary flow in bubble flow is similar to that in water single phase flow.

Turbulence induced by bubbles. Suppose that the turbulence of the liquid phase velocity of bubble flow, U_{tp}', is expressed with the sum of the component of velocity fluctuations, U_s', independent of the existence of bubbles or inherent in single phase flow and the component of liquid phase velocity fluctuations brought about only owing to disturbance of the bubbles. Also suppose that the two components are independent each other (Sato et.al., 1980). Then, the turbulence induced by the bubbles is given by the quantity ($\overline{U_{tp}'^2} - \overline{U_s'^2}$). In addition, suppose that the negative turbulence induced by the bubbles means that the bubbles would be suppressing turbulence. Figure 9 shows distributions of the turbulence induced by bubbles, which were based on this concept. Circle symbols in the figure indicate its positive value and lattice marks, its negative value. The absolute value is larger with darker color and

UPWARD BUBBLE FLOW

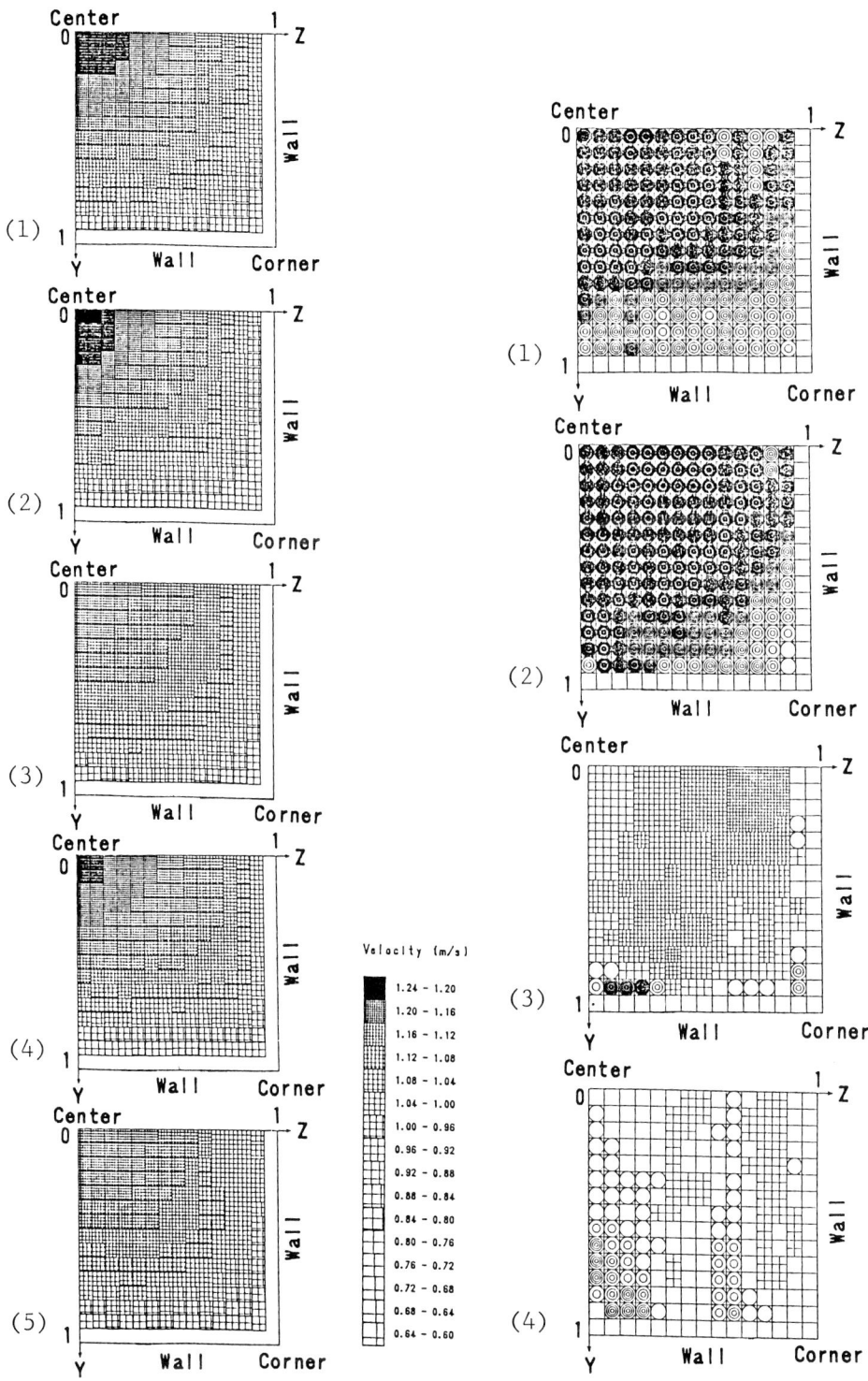

FIGURE 8. Contour maps of water velocity.

FIGURE 9. Turbulence induced by bubbles.

the color becomes lighter as it reaches zero.
In coring bubble flows, as shown in Figs. 9(1) and 9(2), it is clear that the turbulence due to the disturbance of bubbles is symmetrical to the diagonal and also has a high positive value near the center. In sliding bubble flow (case (3)), the turbulence is strong near the wall and the corner. In uniform bubble flow (case (4)), the turbulence shows a slightly higher value in some places in the quarter plane of the cross-section. From these results and the results for the phase distribution, it is considered that bubbles collect at and around the places indicating positive and strong turbulence induced by bubbles.

Turbulence suppression. Turbulence suppretion phenomena can be observed in sliding bubble flows with a saddle-type phase distribution both in square cross-sectional and circular channels. However, sliding bubble flow with a saddle-type phase distribution does not always have turbulence suppression. When turbulent suppression is occurring, the flow has a saddle-type phase distribution with extremely small void fraction in the core region of $Y<0.5$ as shown in Fig 5, and bubbles about 3 mm near the wall and bubbles about 1~2 mm in the core region in Fig. 4. Moreover, a series of experimental results show that even if bubble flows have a saddle-type phase distribution, the flows do not bring about turbulence suppression when the flows have a saddle-type phase distribution with higher void fraction or bubbles greater than 2 mm in the core region, and/or bubbles less than 3 mm near the wall.

Turbulence energy. Coring bubble flows in the cases (1) and (2) have large fluctuations of velocity in the Y-direction, compared with those of the other flows. Sliding bubble flow in the case (3) has low values of the velocity fluctuations in the Y-direction on the whole cross-section. Uniform bubble flow has small velocity fluctuations in the Y-direction and a similar trend to the single-phase flow in the case (5). Suppose that turbulence characteristics on the cross-section are symmetrical to the diagonal, that is, turbulent fluctuations in the Z-direction are equal to those in the Y-direction. Then, the turbulent energy averaged over the square cross-section takes the values 6.74, 7.39, 3.12, 3.98, and 3.84 J/m^2 in numerical order of the experimental cases. Sliding bubble flow has the lowest value in turbulent energy. Accordingly, the occurrence of turbulence suppression can be comfirmed.

Reynolds stress. Figure 10 shows distributions of Reynolds stress on the quarter plane of the square cross-section. Reynolds stress takes a small value at the location where velocity gradient is small. The results show that because the Reynolds stress takes small values when $Y \leq Z$, the velocity gradient is symmetrical to the diagonal. This trend agrees with that of the equal mean flow velocity distribution.

In coring bubble flows of the cases (1) and (2), the Reynolds stress is larger than that of single-phase flow. This suggests that turbulence intensity is large and the velocity gradient is also large. In the sliding bubble flow in case

UPWARD BUBBLE FLOW

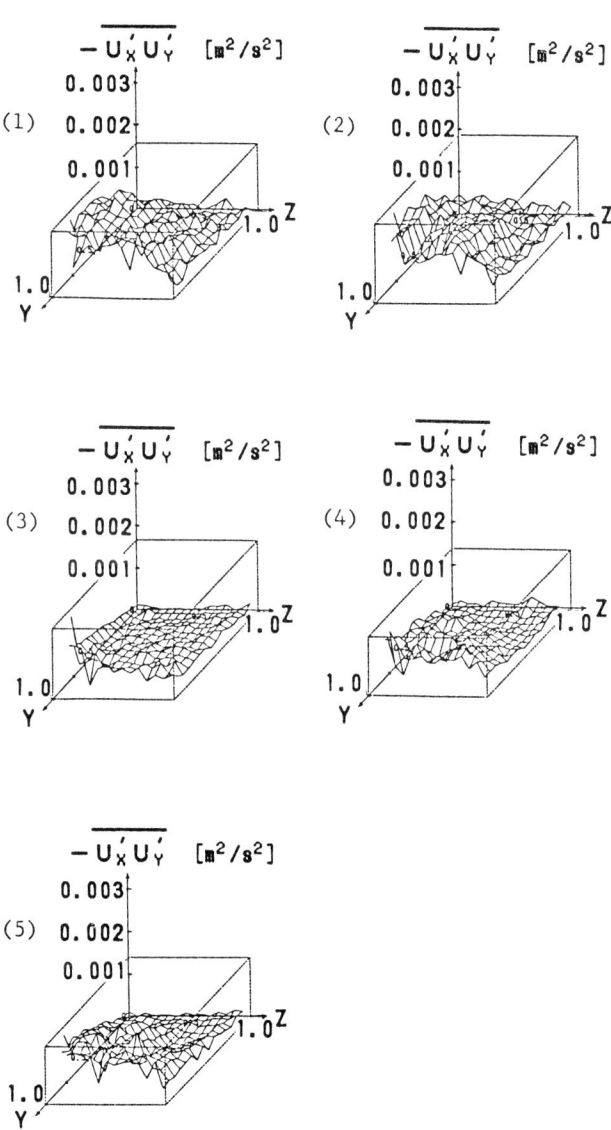

FIGURE 10. Reynolds stress.

(3), the Reynolds stress takes smaller values than that of single-phase flow. Therefore, this suggests that turbulence is suppressed and the correlation between the velocities in the X- and Y- directions is also small. Uniform bubble flow in the case (4) has a similar trend to single-phase flow.

Secondary Flow

Because flow velocities in two directions on the cross-section can not be measured by the LDA employed, concrete information on secondary flow can not be obtained. However, secondary flow generally has a symmetrical tendency to the

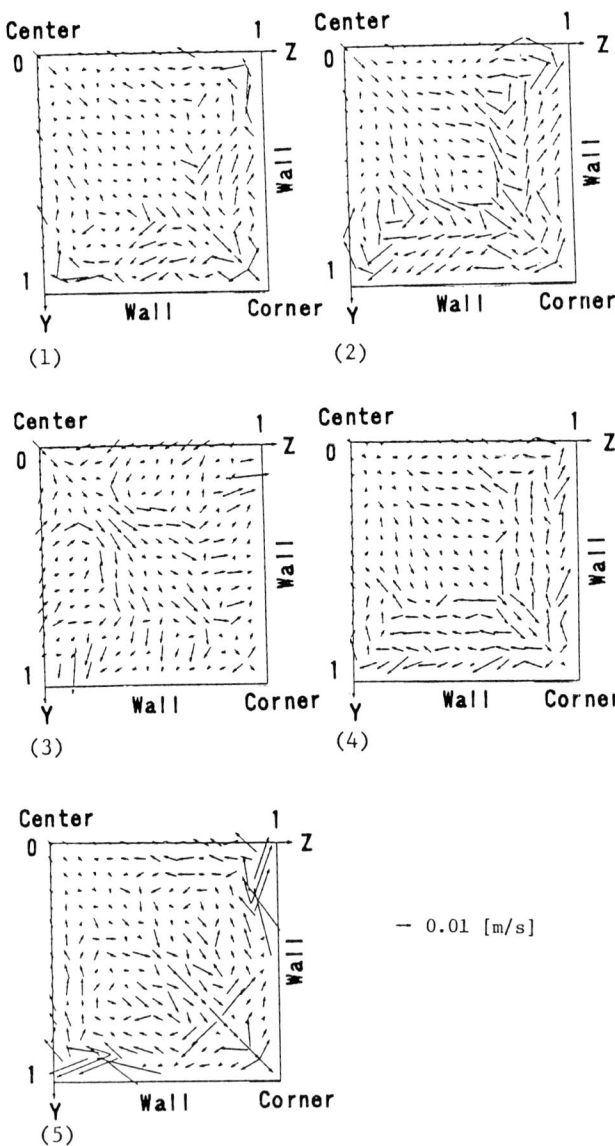

FIGURE 11. Secondary flow.

diagonal on the square cross-section. Moreover, as mentioned above, the main flow characteristics show to be almost symmetrical to the diagonal on the square cross-section. Therefore, assuming that flow on square cross-section has symmetrical properties to the diagonal, a set of flow velocities in the Y- and Z- directions can be obtained to make a diagram of secondary flow. Figure 11 shows the secondary flow on the square cross-section. Water single-phase flow has almost the same properties as the results of previous work (Melling & Whitelaw, 1976). The results of secondary flow in bubble flow show that the coring bubble flow of case (1) has slower secondary flow in the region of the channel center than that of water single-phase flow, and that the sliding bubble flow of case (3) has slower secondary flow near the wall. This is interpreted to mean that secondary flow becomes smaller owing to existence of many bubbles. In sliding bubble flow, it is supposed that bubbles collect at the corner and near the wall center because these places are the stagnation points in secondary flow.

CONCLUSIONS

Internal characteristics of bubble flow were investigated experimentally with nitrogen gas-water upward flow in a square cross-sectional channel and a circular channel under the same flow rate conditions using a laser Doppler anemometer and a double-sensor conducance probe system. The size of bubbles was controlled by changing the mixing condition and by adding a surfactant to water under the fixed flow rate or average void fraction condition. The resultant flow was coring, sliding, or uniform bubble flow.

Experimental results show that the bubble flows have characteristic structures. The main results are as follws:

1. Bubbles greater than 4 mm rise through the core region of channel, bubbles about 3 mm flow along the wall, and bubbles less than 2 mm rise straightly at each location on the cross-section of the channel.

2. The reduction in bubble size under the same flow rate conditions flattens the lateral gas-phase distribution and increases the number density of bubbles, or the gas-liquid interfacial area in the order of the cases (1), (2), and (4). In addition, the reduction in size leads to an increase in main flow velocity, but the extreme reduction does not lead to a further increase in the velocity. Infact, it brings about a decrease in turbulence.

3. The characteristics of flow in a square cross-sectional channel show a symmetrical trend to diagonal on the cross-section.

4. In sliding bubble flow, turbulence suppression can be observed. The turbulence intensity and energy averaged over the cross-section are the lowest of those of the other flows.

5. Bubbles will deviate to collect in the region where turbulence is strong.

6. Secondary flow of upward bubble flow in a square cross-sectional channel has a similar trend to that of water single-phase flow. The secondary flow tends to be slow in the regions where bubbles will collect.

7. Bubble flow in a circular channel has similar structure to bubble flow in a square cross-sectional channel.

NOMENCLATURE

    j_G : gas superficial velocity
    j_L : liquid (or water) superficial velocity
    L : bubble length
    U : water velocity in main flow (upward) direction
    X : coordinate axis in main flow direction
    Y,Z: coordinate axis on the cross-section of channel
    α : local void fraction
    $<\alpha>$: average void fraction in a measurement section

Subscripts
    s : single-phase flow
    tp : bubble flow
    X : main flow direction
    Y : Y-direction

Superscripts
    ' : turbulent fluctuation

REFERENCES

1. Inoue, A., Aoki, S., Koga, T. and Yaegashi, H., 1976, Void Fraction and Bubble and Liquid Velocity Profiles in a Vertical Pipe, Trans. JSME, vol. 42, no. 360, pp. 2521-2531.

2. Ishii, M. and Mishima, K., 1980, Study of Two-Phase Model and Interfacial Area, ANL Rept. 80-111.

3. Lance, M. and Bataille, J., 1983, Turbulence in the Liquid Phase of a Bubbly Air-Water Flow, in Advances in Two-Phase Flow and Heat Transfer, vol. I, ed. S. Kakac and M. Ishii, pp. 403-427, Matinus Nijhoff Publishers.

4. Matsui, G., Yamashita, Y. and Kumazawa, T., 1986, Effect of Bubble Size on Internal Structure of Bubbly-Liquid Flow in Vertical Square Channel, Proc., 3rd Int. Symp. on Application of Laser Anemometry to Fluid Mechanics, 4.2.

5. Matsui, G., Yamashita, Y. and Kumazawa, T., 1937, Effect of Bubble Size on Internal Characteristics of Upward Bubble Flow, Trans. JSME(B), vol. 53, no. 486, pp. 459-463.

6. Melling, A. and Whitelaw, J. H., 1976, Turbulent Flow in a Rectangular Duct, J. Fluid Mech., vol. 78, part. 2, pp. 289-315.

7. Ohba, K. and Yuhara, T., 1932, Study on Vertical Bubbly Flow Using Laser Doppler Measurement (2nd Report), Trans. JSME(B), vol. 48, no. 425, pp. 78-87.

8. Sadatomi, M. and Sato, Y., 1982, Two-Phase Flow in Vertical Noncircular Channels, Int. J. Multiphase Flow, vol. 8, no. 6, pp. 641-655.

interfacial waves were discussed. Chaotic behaviours of ripple waves and the possibility of the prediction of the onset of interfacial waves were suggested.

NOMENCLATURE

English

- C = Numerical constant
- Cp = Specific heat at constant pressure
- D = Tube inside diameter
- F = Wave frequency
- g = Gravitational acceleration
- H = Latent heat of evaporation
- h = Heat transfer coefficient
- h^+ = Dimensionless heat transfer coefficient
- L = Wave height
- L_c = Length of short condensation test section
- M = Molecular weight
- $\dot{m}$ = Local vapour condensation rate
- m_c = Homogeneous condensation rate
- P = Pressure
- Pr = Prandtl number
- q = Heat flux
- R = Tube radius
- r = Radial coordinate
- Re = Reynolds number
- Re_f = Film Reynolds number $(=\Gamma/\mu)$
- Sc = Schmidt number
- T = Temperature
- t = Time interval
- U = Average axial velocity
- u' = Velocity scale
- V = Average radial velocity
- v = Specific volume
- W = Mass flow rate
- x = Axial coordinate
- y = Distance from the tube inner surface

Greek

- β = A non-dimensional parameter expressing degree of the mass transfer effect on heat transfer
- Γ = Mass flow rate per unit peripheral length
- Δx = Axial step for the computation
- δ = Liquid film thickness
- δ_{max} = Maximum wave height
- Λ = Length scale
- λ = Thermal conductivity
- μ = Viscosity
- $\tilde{\mu}$ = Wave-effect viscosity
- ν = Kinematic viscosity
- ρ = Density
- τ = Shear stress
- ω_v = Vapour mass fraction

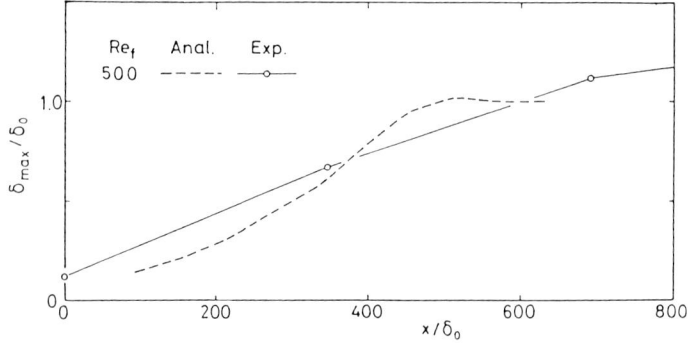

FIGURE 22. Streamwise change of maximum wave height

Navier-Stokes equations. Falling water film was dealt with as a test flow (Hagiwara, 1989a). Figure 20 demonstrates that the analysis predicts the growth and deformation of a quasi-two-dimensional wave as it proceeds downstream. The streamwise changes of the predominant wave frequency and the maximum wave height determined from the time trace of the calculated liquid film thickness, agree fairly well with experimental results (Fujita et al., 1985; Takahama et al., 1983) (see Figs 21 and 22). It indicates that the present type of numerical analysis is effective for the prediction of the flow characteristics of a wavy falling film. The present numerical method may be applicable to the prediction of the onset of disturbance waves, if the computation is carried on further towards downstream.

CONCLUSIONS

The time traces of liquid film thickness were obtained at six different axial locations, by making use of the conductance probe technique for the downward air-water annular two-phase flow. It was found that the wave height increases with axial distance, but that the wave frequency does not vary remarkably.

A simple model was introduced, in order to express the augmentation of heat and momentum transfer due to the interfacial waves. The wave effect model gives the values of the effective viscosity, expressed by the wave frequency and wave height.

Numerical analysis has been carried out for the two types of annular two-phase flow heat transfer, by making use of the wave effect model. The governing equations for each phase flow were solved numerically, with a finite difference scheme. The computed results agreed fairly well with the available experimental data. This suggests that the numerical analysis and the wave effect model are useful for predicting the enhancement of heat and momentum transfer by the interfacial waves.

Characteristics of instability of the interface and of

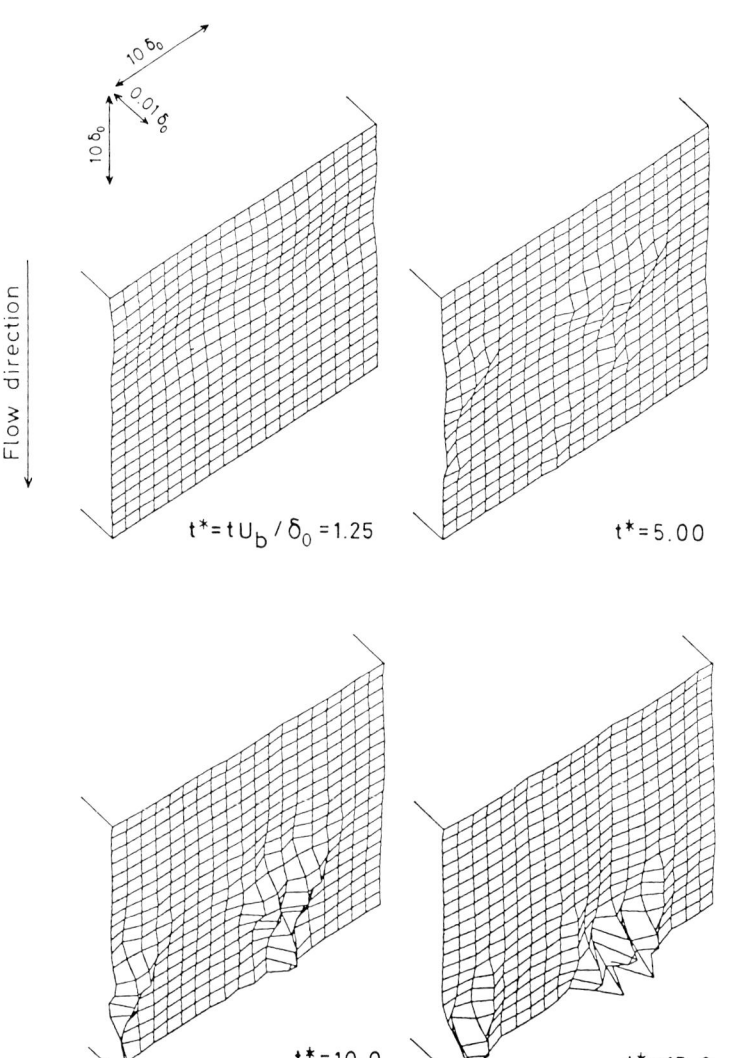

FIGURE 20. Deformation of quasi-two-dimensional wave

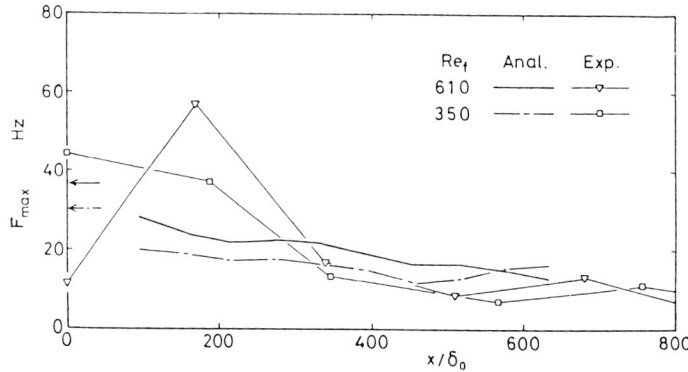

FIGURE 21. Streamwise change of predominant wave frequency

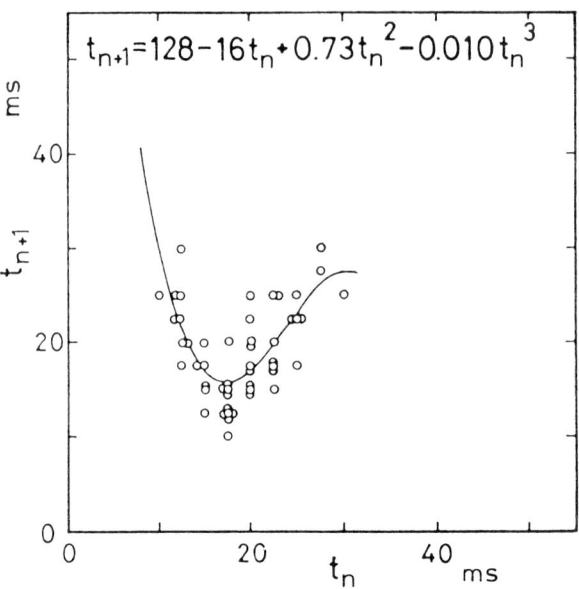

FIGURE 18. Lorenz plot of time intervals due to ripple waves of non-disturbance wave portions

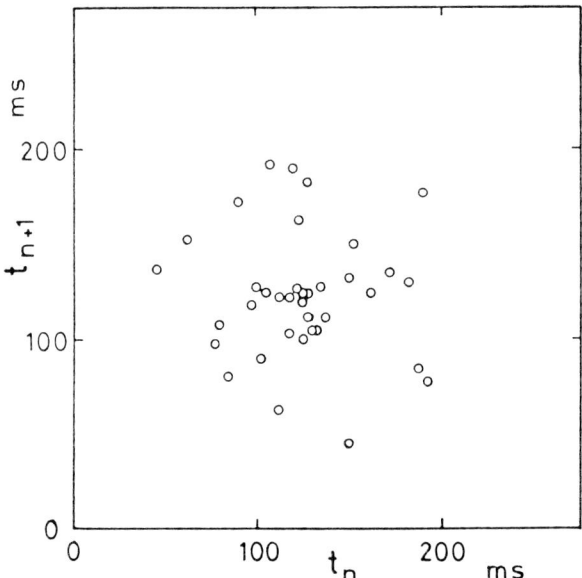

FIGURE 19. Lorenz plot of time intervals of disturbance waves

a chance to acquire a prediction method of the period or frequency of ripple waves, for some kinds of annular two-phase flows.

A numerical analysis has been initiated as a first step to predict the axial change of wave frequency and wave height. An example of the results of such analysis isshown in Figs 20-22. In the analysis are solved the time-dependent full

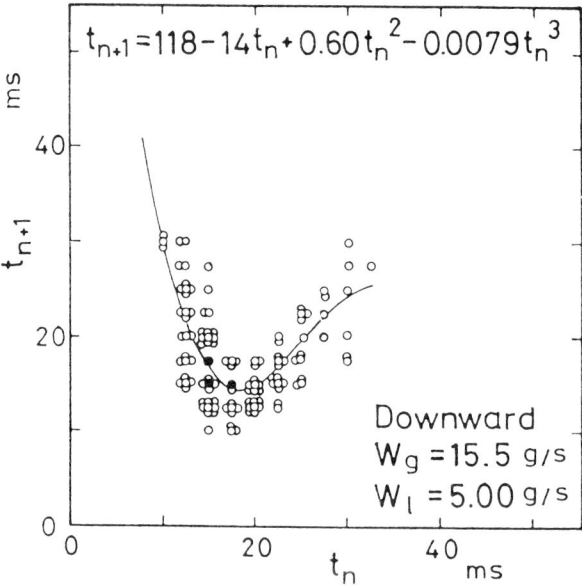

FIGURE 16. Lorenz plot of time intervals (solid circle means 10 open circles)

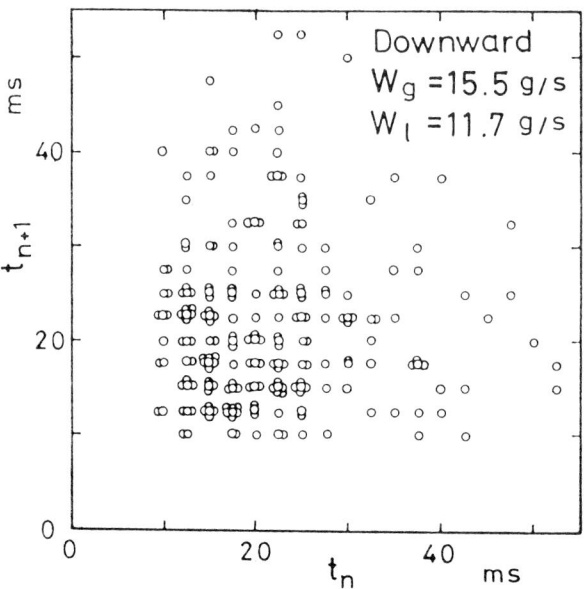

FIGURE 17. Lorenz plot of time intervals (higher liquid flow rate)

ripple waves under the influence of disturbance waves occurring with random time intervals.

A similar tendency was obtained for the time series of wall shear stress fluctuation, and also for the time traces of liquid film thickness in horizontal annular/wavy two-phase flow. The present kind of approach may suggest that there is

Particular attention was given to the set of intervals, t, between two successive maximum values of liquid film thickness and wall shear stress, appearing in their time series. The time series was obtained experimentally in a downward annular two-phase flow, at a position where the flow attained a fully developed state. The gas-phase flow was turbulent, while the liquid-phase flow was essentially laminar but with small three-dimensional ripple waves covering the interface. The Lorenz plot shows the relationship between t_{n+1} and t_n for n=1,2,3... Figure 16 shows a typical example of the Lorenz plot of the time intervals of the maxima for the liquid film thickness. Almost all the points are located in a specific region. The solid line shows the least square curve fit of the data. The plotted data can be approximated rather well by the following third order polynomial line, described by the equation

$$t_{n+1} = 118 - 14t_n + 0.60t_n^2 - 0.0079t_n^3. \qquad (28)$$

This equation indicates that a simple low-dimensional relationship exists between any two successive intervals of maxima caused by ripple waves. This suggests that the change in liquid film thickness with time by the ripple waves is chaotic, and can be explained by a simple deterministic law.

Figure 17 shows the Lorenz plot of the time intervals of the maximum values of film thickness for a higher flow rate of the liquid phase. Not only the ripple waves, but also disturbance waves were observed in this case. A simple low-dimensional function similar to Eq. (28) cannot be found between the two successive intervals.

In order to explain the difference between Figs 16 and 17, the time traces in the case of the higher liquid flow rate were divided into two parts using a conditional sampling technique: non-disturbance wave portions and disturbance wave portions. A set of intervals of the maxima $\langle t_i \rangle_{ND}$ caused by the ripple waves, free from the influence of the disturbance waves, was defined from the non-disturbance wave portions. Figure 18 represents the Lorenz plot of the set $\langle t_i \rangle_{ND}$. The tendency that almost all the points remain in a small region is the same as that for the small flow rate described above. The relationship indicated by the line in Fig. 18 agrees fairly well with that of Eq. (28). This means that the irregular time series of liquid film thickness caused by the ripple waves still obeys a simple deterministic law, even when the disturbance waves appear.

On the other hand, the intervals of the two successive disturbance waves were obtained using the whole signal, including the disturbance wave portion. Figure 19 shows the Lorenz plot of the disturbance wave intervals. The points are scattered over a wide region. It is expected that the propagation of the disturbance waves cannot be explained by a deterministic law, but must be explained by a stochastic law. The scattered distribution of the points in the Lorenz plot of Fig. 17 may be caused by the disturbance waves and the

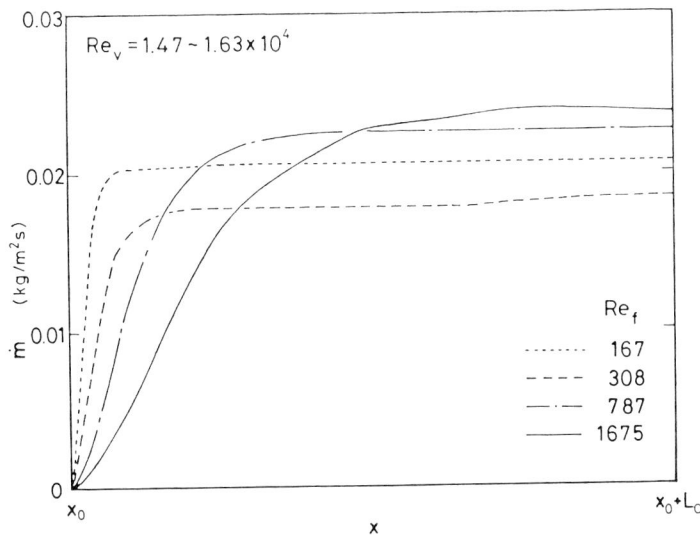

FIGURE 15. Distribution of local condensation rate

assumption was wrong, especially at larger liquid film Reynolds numbers.

INSTABILITY OF INTERFACE

To use the wave effect model mentioned above, the values of the two wave parameters (the wave height and the wave frequency) should be evaluated separately, for all the flow rate conditions where the calculation is made. However, no empirical or theoretical equation exists for the two wave parameters, which should be given as the functions of flow rate conditions, direction of each phase flow, and fluid properties. In this relation, some studies on the interface instability and on the wave characteristics have been initiated by one of the present authors. The results of such studies will be discussed in the following.

First, some results of the experimental approach will be discussed. Data were arranged in the form of a Lorenz plot to find if the random-like time series, both of liquid film thickness and of wall shear stress, have chaotic features (Hagiwara, 1988). Chaos is used here to describe a random-looking behaviour of a dynamic system obeying a deterministic law under a certain initial condition. Chaotic attractors must show some specific geometric forms in state space, and are used to characterize long term behaviours in the state space of the dynamic system (Crutchfield et al., 1986). The behaviour of the chaotic attractor is quite different from that of random or (quasi-)periodic motions. Some attempts have been made to reconstruct the chaotic attractor from experimentally obtained random-like signals (Dubounis et al., 1983). One of the effective methods for reconstructing trajectories of a chaotic attractor is the Lorenz plot.

WAVE EFFECT ON HEAT TRANSFER

The computed 'averaged' non-dimensional heat transfer coefficient over the short condensation test section is shown in Fig. 14, together with the experimental counterparts obtained by Blangetti. Re_f in the figure means the averaged film Reynolds number in the test section, defined by the following equation:

$$Re_f = (Re_l|_{x=x_0} + Re_l|_{x=x_0+L_c})/2, \qquad (26)$$

where x_0 is the starting point of the condensation test section, and L_c the test section length. The 'averaged' heat transfer coefficient in the test section is plotted in the non-dimensional form, defined as follows:

$$h^+ = \frac{1}{L_c}\int_{x_0}^{x_0+L_c} \frac{1}{\lambda_1} \frac{q_w}{T_i-T_w}\left(\frac{\nu_1^2}{g}\right)^{1/3} dx, \qquad (27)$$

where q_w is the wall heat flux, T_i the interfacial temperature, T_w the wall temperature and λ_1 the condensate thermal conductivity. According to Blangetti (1979), all the properties of the film are evaluated at the vapour flow temperature. The dash-dot line in the figure shows the Nusselt solution. Blangetti compared his data of the 'averaged' heat transfer coefficient with the local value of the Nusselt theory, because he regarded the test section as a simulation of a short segment of a long condenser tube. The bold line in the figure indicates the present result of h^+, taking into account both the wave effect and the shear stress. The broken line denotes the calculated result, disregarding the wave effect. Taking account of the interfacial shear stress leads to larger values of h^+, over the wide range of film Reynolds numbers $50 < Re_f < 2000$.

However, the result is still remarkably lower than the experimental results. The present results, accounting for both the effects of the interfacial wave and of shear stress, show fairly good agreement with the experimental results over the wide range of film Reynolds numbers. The same results were observed in the case of different vapour Reynolds numbers. Therefore, the present numerical analysis, employing the wave effect model, is found to be useful for predicting the forced-convection filmwise condensation heat transfer.

Another important feature of numerical computation is that it can supply information about the local heat transfer behaviour. Figure 15 shows how the local condensation rate distributes over the condensation test section, for several different liquid film Reynolds numbers. Blangetti assumed that this distribution should coincide with that obtained at the same point in a long condensation tube; therefore, that the condensation rate is almost uniformly distributed. However, the present computation indicates that his

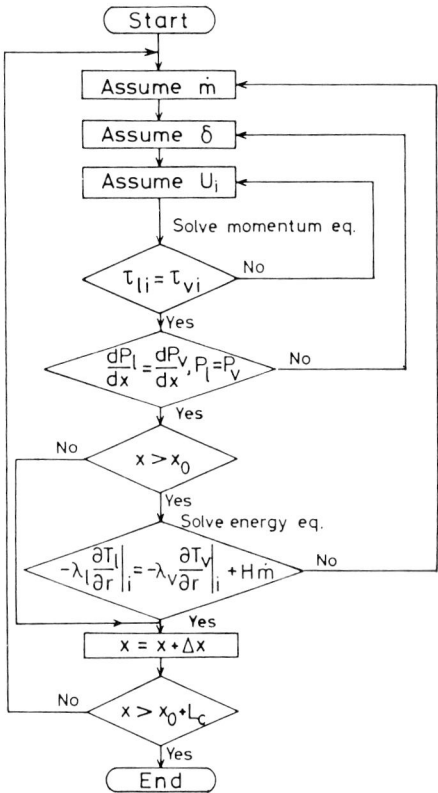

FIGURE 13. Computational procedure for prediction of forced-convection filmwise condensation

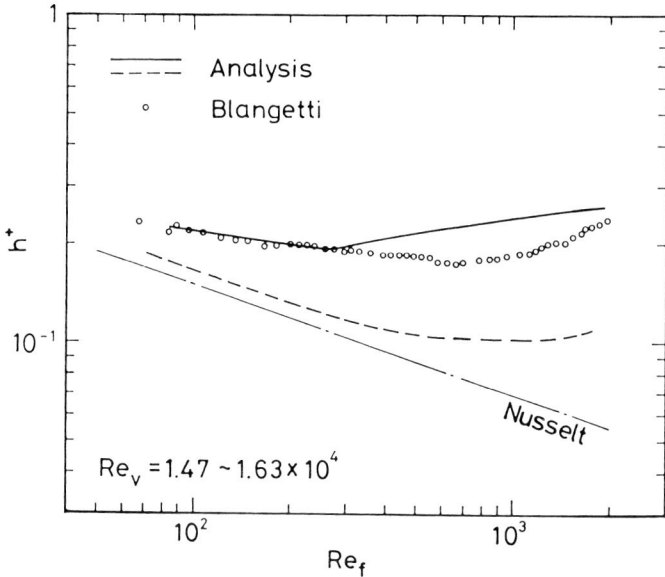

FIGURE 14. Non-dimensional heat transfer coefficient

WAVE EFFECT ON HEAT TRANSFER

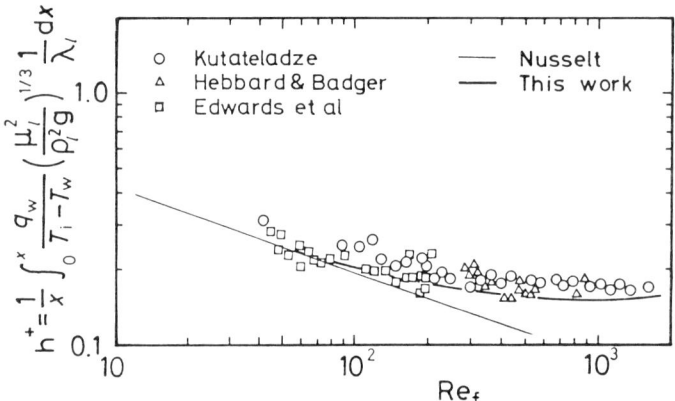

FIGURE 12. Local heat transfer coefficient (Body force convection condensation)

same form as Eqs (19) and (20) respectively. The energy conservation equation for the vapour flow is expressed as follows:

$$\rho_v Cp_v U_v \frac{\partial T_v}{\partial x} + \rho_v Cp_v V_v \frac{\partial T_v}{\partial r} = \frac{1}{r}\frac{\partial}{\partial r}[r(\frac{\mu_v Cp_v}{Pr_v} + \frac{\tilde{\mu}_v Cp_v}{P\tilde{r}_v} + \frac{\mu_t Cp_v}{Pr_t})\frac{\partial T_v}{\partial r}]. \quad (25)$$

The density of the vapour ρ_v was calculated from the total pressure P_v and the vapour temperature T_v, by means of the equation of state, assuming ideal gas.

Since no systematic experiment on the wave parameters for the condensate film exists, to the authors' knowledge, the experimental data obtained by Chu and Dukler (1975) for fully-developed adiabatic falling water film inside a vertical tube, was used for the values of wave parameters. Empirical relationships for the wave frequencies and for the wave amplitudes fitting their experimental data for different liquid film Reynolds numbers, were used in the calculation of wave apparent viscosity. The vapour Reynolds number for Blangetti's experiment was not very high, so such empirical equations may approximate the values of the wave parameters of the condensate film, under the effect of interfacial shear stress.

Figure 13 shows the flow chart of the computational process for filmwise condensation (Suzuki et al., 1988). The chart is almost the same as that seen in Fig. 8, except for the iterative process of evaluating the local condensation rate $\dot{m}$. The iterative process for solving the energy equation is continued until a guessed value of local condensate rate becomes equal to that satisfying the matching condition, within a predetermined accuracy.

heat and momentum transfer is not important in that case, only the interfacial wave is expected to affect the heat transfer mechanism of filmwise condensation.

The following governing equations for the condensate film were solved theoretically (Suzuki et al., 1983b).

$$(\mu_l + \tilde{\mu}_l)\frac{\partial^2 U_l}{\partial y^2} + g(\rho_l - \rho_v) = 0 \tag{23}$$

$$(\frac{\mu_l Cp_l}{Pr_l} + \frac{\tilde{\mu}_l Cp_l}{\tilde{Pr}_l})\frac{\partial^2 T_l}{\partial y^2} = 0. \tag{24}$$

The interfacial wave effect was taken into account with the model discussed above. The result of the calculation obtained for water vapour condensation shows fairly good agreement with the experimental data (Kutateladze, 1963; Hebbard and Badger, 1933; Edwards et al., 1948) (see Fig. 12). Thus, the interfacial wave effect on the filmwise condensation occurring in the body force convection is predictable with the present model.

Experimental data for the heat transfer coefficient of forced-convection filmwise condensation is well known to exceed the Nusselt solution (Nusselt, 1916) over a wide range of condensate film Reynolds numbers. Both the interfacial shear stress and the interfacial waves are the causes of larger heat transfer coefficient. However, suitable methods to directly predict the combined effects of these two causes on the condensation heat transfer are not available. Numerical analysis may be one of the methods useful for this purpose. Therefore, some numerical computations have been initiated to predict the forced-convection filmwise condensation heat transfer.

The computational results will be compared with the counterparts of the measurements by Blangetti (1979). He carried out his experiment in a vertical tube of 30mm I.D. He used a short condensation test section of 200mm length, which was connected to the downstream end of the adiabatic entry region of 3000mm length. The flow rates of both the water film and of the vapour were separately controlled at the upstream end of the entry region, independent from the phenomena occurring the test section. The main assumptions adopted in our computation are as follows: (1)The liquid film thickness is so thin that no turbulent motion exists inside the film in the whole condensate film Reynolds number range, but the waves appearing the interface can affect the momentum and heat transfer. (2)The droplet entrainment from the crests of the waves is neglected. (3)The fully developed state of the interfacial waves is assumed to be achieved well upstream of the test section.

The governing equations for the liquid film are the same as those of the two-component annular flow. The mass and momentum conservation equations for the vapour are of the

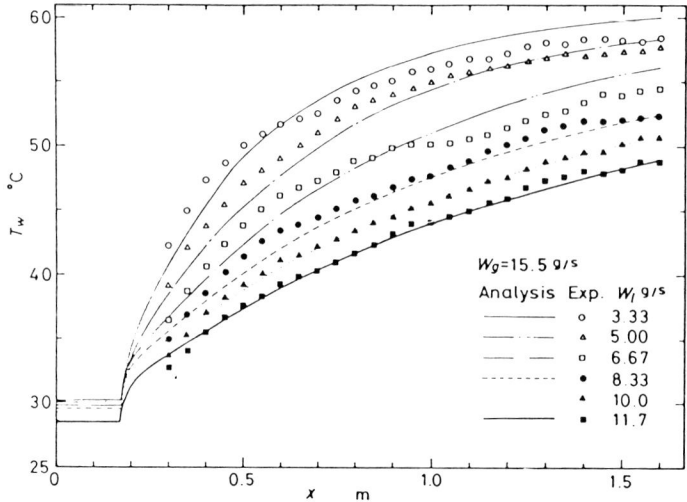

FIGURE 10. Axial distribution of tube wall temperature

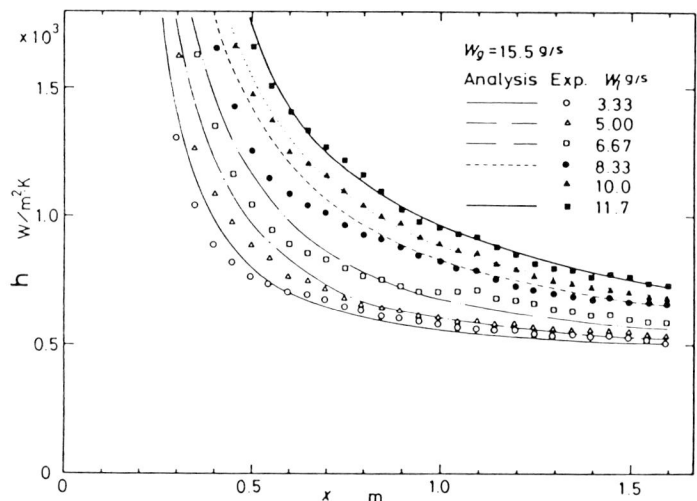

FIGURE 11. Local heat transfer coefficient

The results discussed here demonstrate that the model for the wave effects, and the method of accounting for the evaporation effect on heat transfer introduced by Suzuki et al. (1983a), are on the whole valid in predicting the heat and momentum transfer in the developing region of an annular two-phase flow. Also, they show that the present type of numerical analysis works well.

Case 2: Filmwise condensation

It is well known that the experimental heat transfer coefficient of filmwise condensation, occurring in the body force convection regime, exceeds the Nusselt solution (Nusselt, 1916), except at very low condensate film Reynolds numbers. Since the effect of the interfacial shear stress on

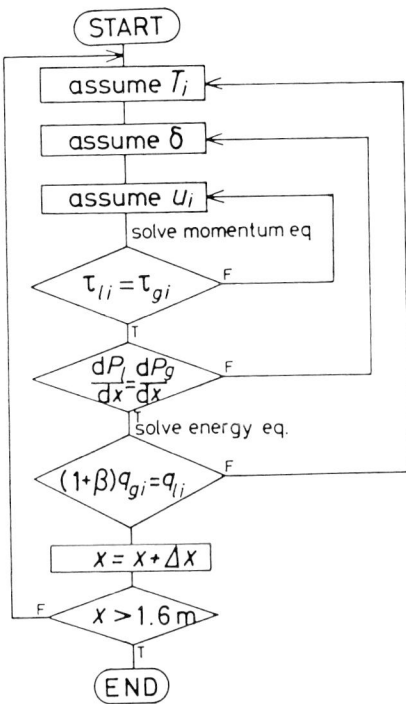

FIGURE 8. Flow chart of computational process

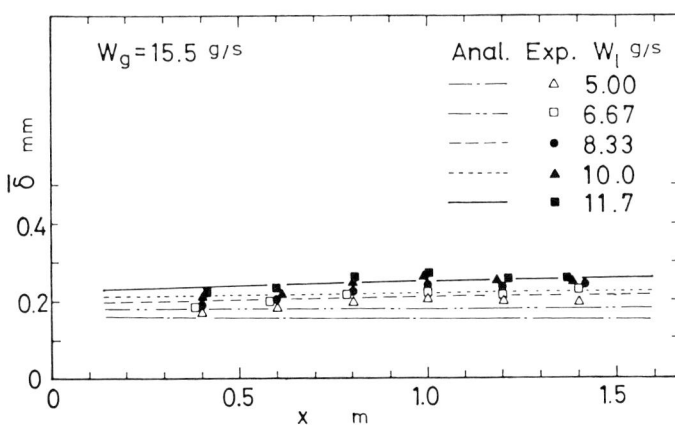

FIGURE 9. Streamwise variation of liquid film thickness

not always attained. Wave structure can be affected by heating. For better agreement, these factors should also be considered.

Local heat transfer coefficients based on the difference between the local wall temperature and the inlet flow temperature are shown in Fig. 11. The heat transfer coefficient is also predicted fairly well.

GENMIX, originally developed by Spalding (1975) for a single-phase flow, was used. The liquid film thickness is very thin, compared with the radius of the gas phase flow area. Therefore, grid nodes must be allocated very finely in the liquid film, but not in the gas phase flow domain. A too large non-uniformity in grid space is not preferable from the viewpoint of numerical instability. Reduction in grid space in the gas phase flow domain increases the total number of grid points. A too large number of grid points is also undesirable, because it is likely to cause numerical instability. Thus, the interface is regarded as one of the boundaries, and the governing equations for each phase are solved separately in each phase domain. With this method, iterative computation must be carried out at every streamwise location. This is because all the local interfacial values, and the local mean film thickness required to specify the boundary conditions at the interface, are not a priori known.

The iterative computation process used in this study is illustrated in Fig. 8. The values of T_i, δ and U_i are updated at each step of the iterative process, until the matching conditions at the interface are satisfied within a prescribed accuracy. After that, the iterative procedure is terminated, and the computation at the next streamwise location, $x+\Delta x$, is started. This iterative procedure was useful in securing good accuracy of the final results, but the computation time required was more than 10 times longer than that for a single-phase flow computation.

Figure 9 shows the streamwise variation of the computed mean liquid film thickness. The experimental results, obtained by processing the time trace of liquid film thickness mentioned above, are also shown in the figure. The mean liquid film thickness is found not to vary significantly with the change of x. Its variation is within 15% of its spatial mean. This agrees with the tendency of the experimental data.

In Fig. 10, the computed axial distribution of the tube wall temperature is compared with available experimental data obtained by Hagiwara et al. (1982). The present numerical analysis predicts the temperature distribution rather well, except for a few flow rate conditions of small W_l. There may be several reasons for this larger discrepancy found when W_l is small. When W_l is smaller, T_w is higher, so that the evaporation rate is larger. This follows from the fact that the value of β is then larger. A possible change in the gas phase flow dynamics, due to strong evaporation, was ignored in the present computation. Higher values of T_w means larger temperature inhomogeneity in the gas phase flow. Thus, the neglect of the radial variation of β becomes less satisfactory. Taking account of these evaporation effects may diminish the discrepancy found when W_l is small.

Moreover, the space difference between the liquid phase and gas phase flow areas becomes larger when the liquid film thickness is small. This may have slightly increased the numerical errors. The assumed thermodynamic equilibrium was

shown in Fig. 4. Empirical equations were obtained for the wave parameters as functions of axial distance. The local values of the effective viscosities for the developing two-component annular flow, $\tilde{\mu}_l$ and $\tilde{\mu}_g$, were calculated by making use of the developed empirical equations.

The numerical analysis of average liquid film thickness, wall temperature and heat transfer coefficient were carried out by solving the governing equations numerically, with a finite difference scheme. In the computation, the wave effect model and the mass transfer effect at the interface were taken into account. The conservation equations of mass, momentum and energy are expressed as follows: in the liquid film ($r_i < r < R$)

$$\frac{\partial U_l}{\partial x} + \frac{1}{r}\frac{\partial(rV_l)}{\partial r} = 0 \tag{16}$$

$$\rho_l U_l \frac{\partial U_l}{\partial x} + \rho_l V_l \frac{\partial U_l}{\partial r} = -\frac{dP_l}{dx} + \frac{1}{r}\frac{\partial}{\partial r}[r(\mu_l+\tilde{\mu}_l)\frac{\partial U_l}{\partial r}] + \rho_l g \tag{17}$$

$$\rho_l Cp_l U_l \frac{\partial T_l}{\partial x} + \rho_l Cp_l V_l \frac{\partial T_l}{\partial r} = \frac{1}{r}\frac{\partial}{\partial r}[r(\frac{\mu_l Cp_l}{Pr_l} + \frac{\tilde{\mu}_l Cp_l}{\tilde{Pr}_l})\frac{\partial T_l}{\partial r}], \tag{18}$$

and in the gas flow region ($0 < r < r_i$)

$$\frac{\partial U_g}{\partial x} + \frac{1}{r}\frac{\partial(rV_g)}{\partial r} = 0 \tag{19}$$

$$\rho_g U_g \frac{\partial U_g}{\partial x} + \rho_g V_g \frac{\partial U_g}{\partial r} = -\frac{dP_g}{dx} + \frac{1}{r}\frac{\partial}{\partial r}[r(\mu_g+\tilde{\mu}_g+\mu_t)\frac{\partial U_g}{\partial r}] \tag{20}$$

$$\rho_g Cp_g U_g (1+\beta)\frac{\partial T_g}{\partial x} + \rho_g Cp_g V_g (1+\beta)\frac{\partial T_g}{\partial r}$$
$$= \frac{1}{r}\frac{\partial}{\partial r}[r(1+\beta)(\frac{\mu_g Cp_g}{Pr_g} + \frac{\tilde{\mu}_g Cp_g}{\tilde{Pr}_g} + \frac{\mu_t Cp_g}{Pr_t})\frac{\partial T_g}{\partial r}]. \tag{21}$$

The last equation stands for the combination of the gas phase energy equation and the mass conservation equation of water vapour, as Eq. (12) does for the fully developed state. The boundary conditions are the same as those for the fully developed state. The matching conditions at the interface are

$$U_{li} = U_{gi} = U_i, \quad \tau_{li} = \tau_{gi} = \tau_i, \quad dP_l/dx = dP_g/dx,$$
$$T_{li} = T_{gi} = T_i, \quad q_{li} = (1+\beta)q_{gi}, \tag{22}$$

where all local interfacial values and the local mean film thickness are a priori unknown. They must be determined iteratively at each x step in the calculation.

For a numerical computation, a modified version of the code

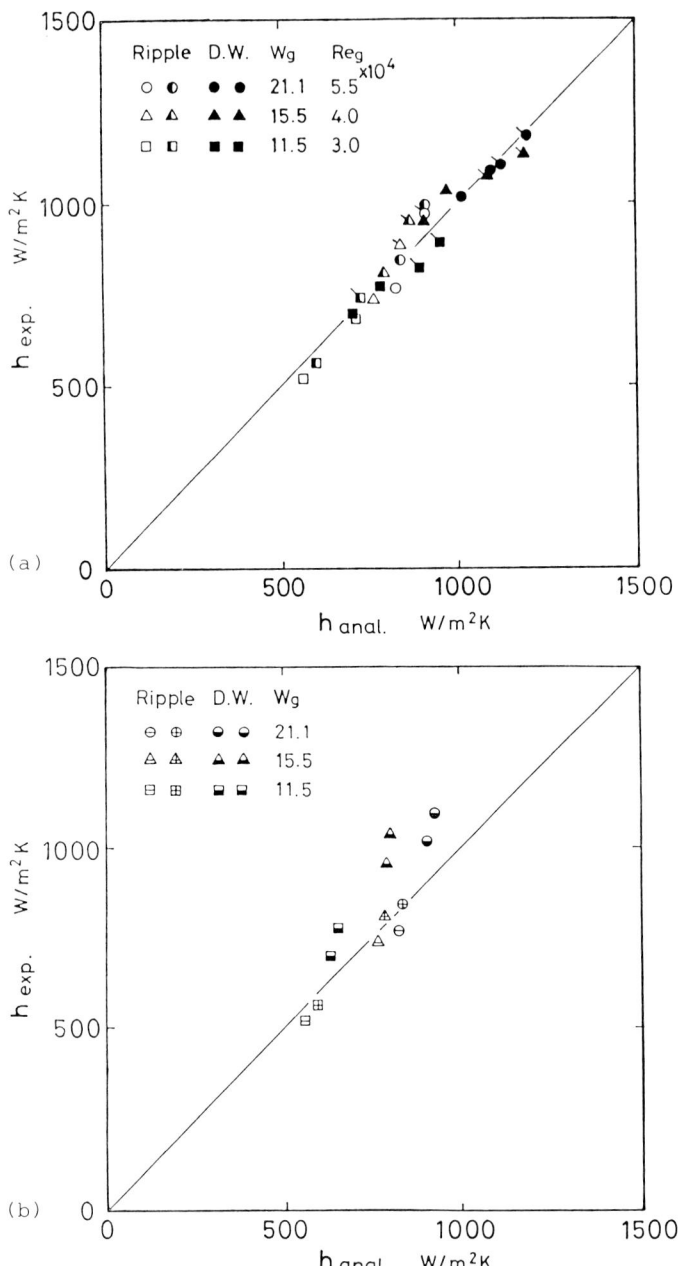

FIGURE 7. Augmentation of local heat transfer coefficient (a) taking account of wave effect; (b) disregarding wave effect

model plays an important role for disturbance waves.

In a developing region of the evaporative, two-component annular two-phase flow, both the wave structure and the thermal field develop simultaneously, along the tube axis. The wave parameters also vary with streamwise location, as

The term Hm_c, expressing the homogeneous evaporation effect, can be eliminated by making use of Eq. (11) (Suzuki et al., 1983). Introducing a non-dimensional parameter β, which expresses the degree of the mass transfer effect on heat transfer, a set of Eqs (10) and (11) reduces to a single combined equation for the gas temperature T_g. It is written as follows:

$$\rho_g Cp_g U_g (1+\beta) \frac{\partial T_g}{\partial r} = \frac{1}{r} \frac{\partial}{\partial r} [r(1+\beta)(\frac{\mu_g Cp_g}{Pr_g} + \frac{\tilde{\mu}_g Cp_g}{\tilde{Pr}_g} + \frac{\mu_t Cp_g}{Pr_t}) \frac{\partial T_g}{\partial r}]. \quad (12)$$

The parameter β is defined as

$$\beta = \frac{H^2 M_v}{M_a Cp_g PT_e v_{ve}}, \quad (13)$$

where M_v is the molecular weight of water vapour, M_a the molecular weight of air, T_e the equilibrium temperature introduced by Plotcher and McManus (1972), and v_{ve} the specific volume of vapour at the equilibrium temperature. In the above definition of β, its radial variation is ignored. Equation (12) is solved instead of solving both Eqs (10) and (11) with an iterative procedure. This combined form of the equation is the same as the energy equation for a single-phase flow. The difference between the problem under consideration and single-phase flow problems lies in the boundary condition for Eq. (12) at the interface. The gas side conductive heat flux at the interface, q_g, becomes $1/(1+\beta)$ times the total heat flux inside the liquid flow at the interface, q_{li}.

The boundary conditions on the axis were given by the axisymmetry of the gas phase velocity and temperature distributions. The boundary conditions at the wall are

$$U_l = 0, \quad -dT_l/dr = q_w/\lambda_l = \text{const}. \quad (14)$$

The matching conditions at the interface are

$$U_{li} = U_{gi} = U_i, \quad \tau_{li} = \tau_{gi} = \tau_i,$$
$$T_{li} = T_{gi} = T_i, \quad q_{li} = (1+\beta) q_{gi}, \quad (15)$$

where all interfacial values and the mean film thickness are a priori unknown. A set of Eqs (5)-(9) and (12) was solved numerically.

The heat transfer results of the analysis are compared with the experimental counterparts in Fig. 7. The wave effect was accounted for in (a), while it was neglected in (b). The figure demonstrates that the calculation disregarding the wave effect gives clearly lower values for the disturbance wave conditions. This may indicate that the wave effect

WAVE EFFECT ON HEAT TRANSFER

experimental data available. The effectiveness of both the wave model and of the analysis was examined, by comparing the predicted results and the experimental data.

Case 1: Evaporation in a downward air-water annular two-phase flow

A downward annular two-phase flow may be assumed to attain an almost fully developed state in the region far from the inlet. The flow and heat transfer characteristics in that region can be obtained theoretically, assuming thermally and hydrodynamically fully developed states (Suzuki et al., 1983). The governing equations to be solved are as follows: in the liquid film ($r_i < r < R$)

$$\frac{\partial U_l}{\partial x} = 0 \tag{5}$$

$$-\frac{dP_l}{dx} + \frac{1}{r}\frac{d}{dr}[r(\mu_l+\tilde{\mu}_l)\frac{dU_l}{dr}] + \rho_l g = 0 \tag{6}$$

$$\rho_l Cp_l U_l \frac{\partial T_l}{\partial x} = \frac{1}{r}\frac{d}{dr}[r(\frac{\mu_l Cp_l}{Pr_l} + \frac{\tilde{\mu}_l Cp_l}{\tilde{Pr}_l})\frac{\partial T_l}{\partial r}], \tag{7}$$

and in the gas flow region ($0 < r < r_i$)

$$\frac{\partial U_g}{\partial x} = 0 \tag{8}$$

$$-\frac{dP_g}{dx} + \frac{1}{r}\frac{d}{dr}[r(\mu_g+\tilde{\mu}_g+\mu_t)\frac{dU_g}{dr}] = 0 \tag{9}$$

$$\rho_g Cp_g U_g \frac{\partial T_g}{\partial x} = \frac{1}{r}\frac{d}{dr}[r(\frac{\mu_g Cp_g}{Pr_g} + \frac{\tilde{\mu}_g Cp_g}{\tilde{Pr}_g} + \frac{\mu_t Cp_g}{Pr_t})\frac{\partial T_g}{\partial r}] + Hm_c \tag{10}$$

$$\rho_g Cp_g U_g \frac{\partial \omega_v}{\partial x} = \frac{1}{r}\frac{d}{dr}[r(\frac{\mu_g}{Sc_g} + \frac{\tilde{\mu}_g}{\tilde{Sc}_g} + \frac{\mu_t}{Sc_t})\frac{\partial \omega_v}{\partial r}] - m_c$$

$$= \frac{1}{r}\frac{d}{dr}[r(\frac{\mu_g}{Pr_g} + \frac{\tilde{\mu}_g}{\tilde{Pr}_g} + \frac{\mu_t}{Pr_t})\frac{\partial \omega_v}{\partial r}] - m_c, \tag{11}$$

where ω_v is the vapour mass fraction, and m_c the homogenous vapour condensation rate. Pr_t is the turbulent Prandtl number, and $\tilde{Pr}_l$ and $\tilde{Pr}_g$ are its counterparts for the wave-induced motions in the gas and liquid phases, respectively. The relationships $Sc_g=Pr_g$ $\tilde{Sc}_g=\tilde{Pr}_g$ and $Sc_t=Pr_t$ have been assumed in the last equation.

correspond to the viscous sublayer in a single-phase turbulent flow,

$$0 < y_1^+ = (\rho_1|\tau_w|)^{1/2}(R-r)/\mu_1 < 5. \qquad (3)$$

Outside this layer, the apparent wave viscosity, $\tilde{\mu}_1$, is regarded to be constant in the liquid film. Therefore, the liquid film is assumed to consist of two layers of different constant apparent viscosities; in a thin layer close to the tube wall, the effective viscosity is same as the actual liquid viscosity μ_1, and in another, it equals $(\mu_1 + \tilde{\mu}_1)$.

The interfacial waves are also expected to affect the gas phase flow near the interface in various fashions, as was mentioned before. Only one of several possible effects is considered in the present wave effect model. The interfacial waves may intermittently break up the viscous sublayer in the gas phase flow, which should exist permanently near the interface, unless the waves appear. This effect can be taken into account by assuming a thin layer near the interface, where the following viscosity is effective:

$$\mu_g + \tilde{\mu}_g = \mu_g + C_d \rho_g L_d^2 F_d + C_r \rho_g L_r^2 F_r. \qquad (4)$$

In this expression, turbulence effect has been ignored. The thickness of this layer is determined by assuming that the effective viscosity given by Eq. (4) is equal to $(\mu_g + \mu_t)$ at the outer edge of this layer, where μ_t is the turbulent eddy viscosity. This is equivalent to assuming that the wave viscosity is ineffective outside this layer, where the turbulent viscosity prevails. The value of C in Eq. (4) is chosen so as to obtain good agreement with the experimental data.

μ_t is determined with the mixing length hypothesis in the present analysis. The value of mixing length was considered to vary radially according to the modified van Driest equation (van Driest, 1956), except for the region affected by waves, where μ_t is set equal to zero. Therefore, the gas core flow region is also assumed to consist of two parts: the thin layer affected by the interfacial waves where the apparent viscosity has a constant value of $(\mu_g + \tilde{\mu}_g)$, and the inner core region where the turbulent eddy viscosity expressed by the van Driest equation is predominant.

ANALYTICAL RESULTS AND DISCUSSIONS

Employing the wave effect model just discussed, numerical and theoretical analysis has been carried out for the following two different types of annular two-phase flow heat transfer. The values of the two wave parameters in the actual annular flow at any streamwise location were estimated with some

WAVE EFFECT ON HEAT TRANSFER

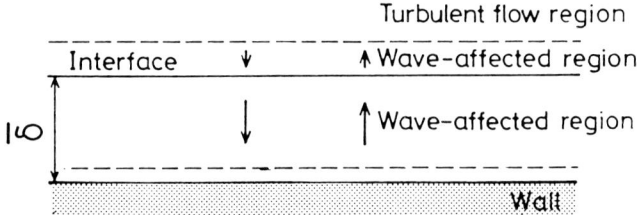

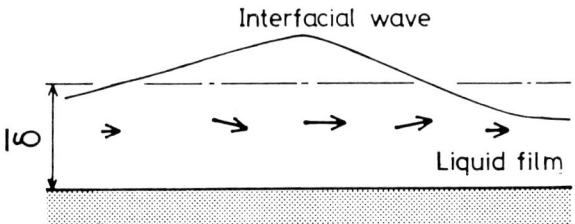

FIGURE 6. Motion of fluid particle inside liquid film

fluid thus cooled is then fed back toward the tube wall. This causes additional heat transport.

Although this wave-induced motion is not actually a turbulent one, the mechanism of heat and momentum transfer enhancement mentioned above is analogous to that caused by turbulent eddy motion. An effective turbulent eddy viscosity is expressed by the Prandtl-Kolmogoroff hypothesis (Launder and Spalding, 1972) as follows:

$$\mu^* = C\rho\Lambda u', \tag{1}$$

where C is an adjustable constant, and u' and Λ are the velocity and length scales of turbulent eddy motion, respectively. This expression is used as a basis in the development of an expression for the effective viscosity due to the wave-induced motion. Next, the length scale Λ and the velocity scale u' of turbulent eddy motion are replaced by the counterparts relevant to the wave-induced motion. The wave height L is used for Λ, and a velocity scale expressed by a combination of L and F is used for u'. A possible change in wave structure due to heating is not considered here. There are two types of waves on the interface: the ripple wave and the disturbance wave. Considering this, the final expression for the effective viscosity is as follows:

$$\mu_l + \tilde{\mu}_l = \mu_l + C_d \rho_l L_d^2 F_d + C_r \rho_l L_r^2 F_r, \tag{2}$$

where r denotes the ripple wave and d denotes the disturbance wave. The radial wave-induced motion must vanish at the wall, and should be suppressed close to the wall. Therefore, the apparent viscosity $\tilde{\mu}_l$ is set equal to zero, in a thin layer adjoining the wall. The thickness of the layer may

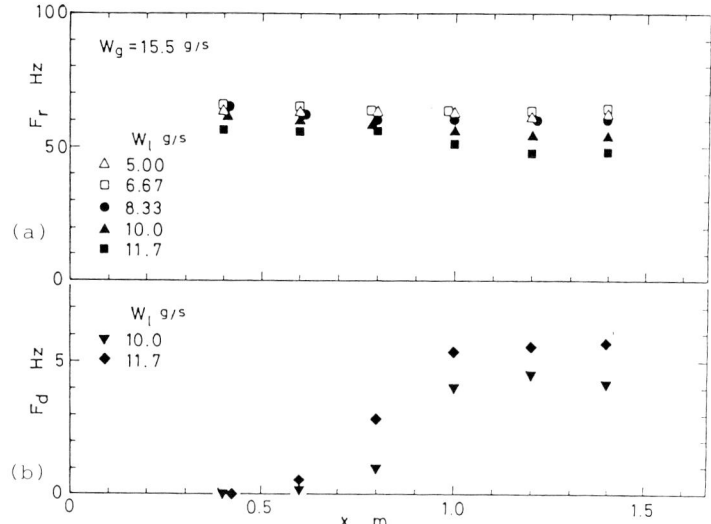

FIGURE 4. Streamwise change of wave frequency (a) ripple wave; (b) disturbance wave

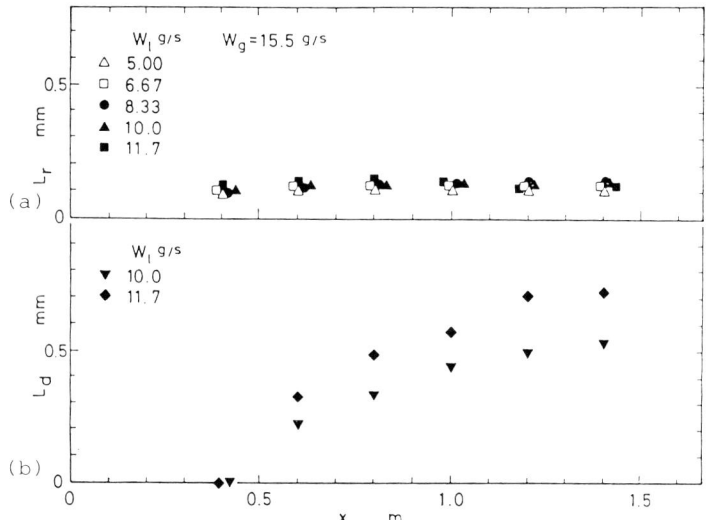

FIGURE 5. Streamwise change of wave height (a) ripple wave, (b) disturbance wave

of a wave (see Fig. 6). This means that a portion of liquid of relatively lower momentum may move inward and be accelerated there. Therefore, that portion of liquid thus accelerated, having higher momentum, is then fed back towards the tube wall. This radial motion is, therefore, expected to enhance the momentum transfer in the liquid film.

From the heat transfer viewpoint, this fluid motion causes additional heat transport. Considering a case when the flow is heated from the tube wall, hot fluid is brought toward the tube center line and is cooled there by exchanging heat. The

WAVE EFFECT ON HEAT TRANSFER

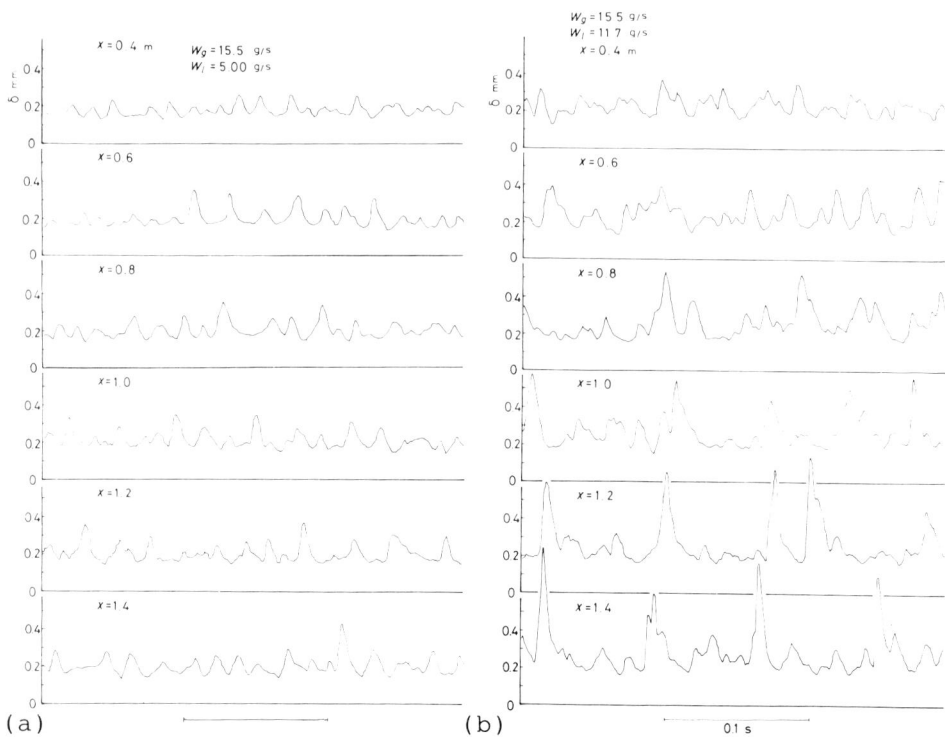

FIGURE 3. Time traces of liquid film thickness (a)W_g=15.5g/s (Re_g=4.0×10⁴), W_l=5.00g/s(Re_l=210);(b)W_g=15.5g/s (Re_g=4.0×10⁴), W_l=11.7g/s(Re_l=500)

disturbance waves also appeared. It is clearly observable, especially in (b), that the wave height increases downstream. The figure clearly indicates that the wave structure is developing within the examined region of x.

The time traces yield two important parameters of the wave structure: averaged wave frequency F and averaged wave height L. Some examples of the results for F and L are plotted in Figs 4 and 5. It is found that no disturbance wave appears in the first region of x down to x=0.6m (x/D=22.7), in any flow rate condition. While the value of F for ripple waves, F_r, does not seem to vary with x, its its value for disturbance waves, F_d, increases rapidly within a short distance after x=0.6m. The height of the ripple wave, L_r, does not seem to vary remarkably with x, irrespective of liquid flow rate. Its counterpart for disturbance waves, L_d, first increases almost linearly with x after x=0.6m, and attains an asymptotic value after x=1.2m (x/D=45.3).

WAVE EFFECT MODEL

A fluid particle in the liquid film is considered to experience an alternate radial inward and outward motion, during the time period between the arrival and the departure

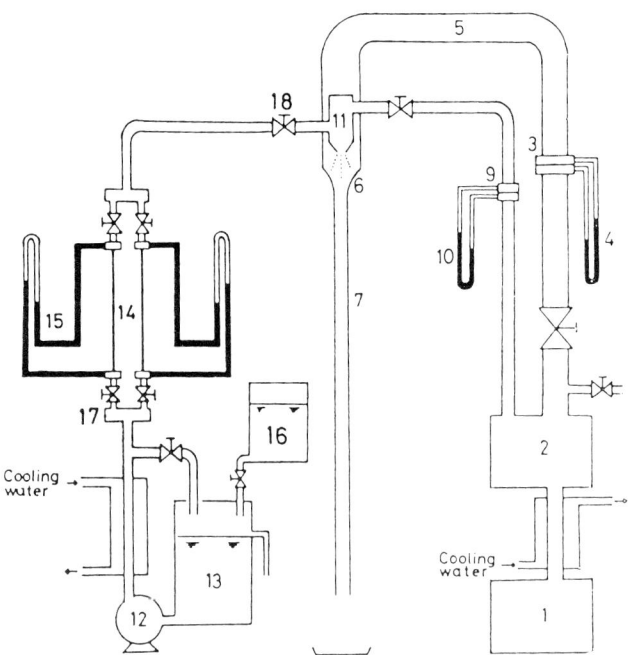

FIGURE 2. Experimental apparatus

phase fluid, by a pump 12. The water flow rate was measured by capillary flow rate meters. Almost all water droplets injected from the atomizer diffuse outward and deposit on the wall of the contraction nozzle 6. This leads to the formation of a water film on the inner surface of the tube from the tube inlet. The thickness of the film increases downstream due to the continuation of droplet deposition, but the rate of increase is small. Therefore, liquid flow rate in the film can be assumed to be approximately constant along the streamwise direction.

The liquid film Reynolds number is much lower than the critical Reynolds number. This means that the intrinsic turbulent motion can be ignored in the liquid film flow. Only the velocity fluctuation caused by the wave-induced motion should be taken into account in liquid phase flow, for the flow rate conditions considered. Incidentally, all the flow rate conditions chosen in the experiment stand above the inception criterion of droplet entrainment given by Ishii and Gromles (1975). However, it can be assumed that the amount of droplets to be entrained is small, and that the entrainment does not significantly affect the heat and momentum transfer, as in reference (Hagiwara et al., 1983). This is because the entrainment fraction is estimated to be at most 2% of the total liquid flow rate, according to reference (Ishii and Mishima, 1982).

Figure 3 demonstrates typical time traces of liquid film thickness obtained at six different axial locations respectively, in two typical conditions (a) and (b). Condition (a) represents the case when only ripple waves appeared, and condition (b) represents the case when

WAVE EFFECT ON HEAT TRANSFER

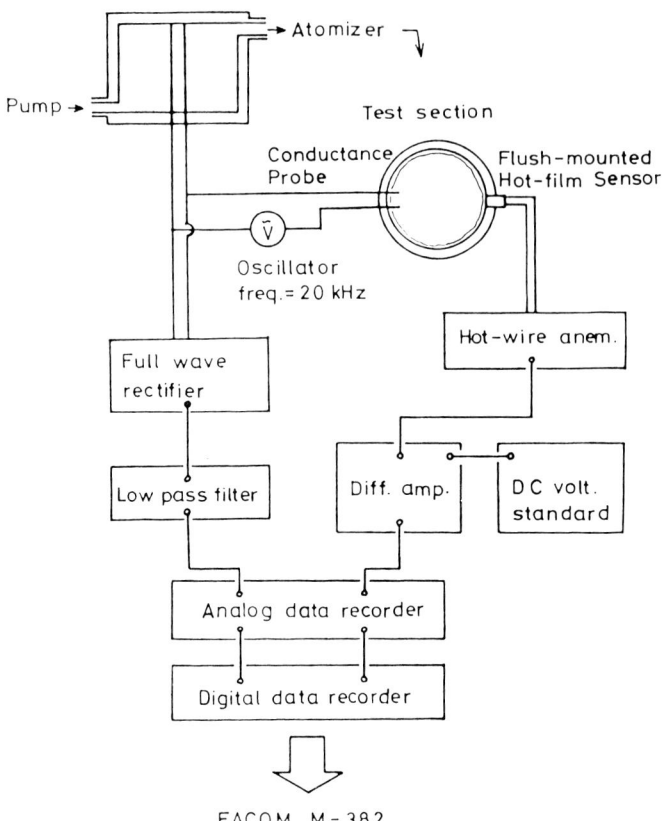

FIGURE 1. Signal recording circuit for time traces of liquid film thickness and wall shear stress

recording system used in the study. To avoid the polarizing action of water, a 20kHz alternating electric current was imposed between the two wire electrodes. This level of frequency is high enough to avoid any polarizing action of the water. Its rectified signal was low-pass-filtered and the obtained signal was used as a signal for the film thickness. To ensure the linearity between the final signal and the film thickness, a method similar to that used by Villeneuve and Ouellet (1978) was adopted —another two-wire conductance probe was inserted into the water feeding line. This was also useful to compensate for the water property's effect on the conductance probe. Static calibration of this probe system was made in a shallow open channel, comparing the output signal from the conductance probe with a film thickness measured by the needle contact method (Hagiwara et al., 1982).

Time-varying liquid film thickness was measured in downward air-water annular two-phase flow, in an acrylic tube of 26.4mm I.D. and 1850mm length (Hagiwara et al., 1985). Figure 2 shows the experimental apparatus used in the study. Air supplied by a compressor 1 serves as the gas phase fluid. It flows partly into an atomizer 11 and partly to a test section 7. Water is fed into the atomizer as the liquid

and leads to the incipience of interfacial waves. Instantaneous increase of wall shear stress was found to be closely correlated with the passing interfacial waves (Miya et al., 1971; Hagiwara et al., 1983; Hagiwara et al., 1989b), which means that the flow characteristics inside the liquid film are affected by the interfacial waves. The interfacial waves also affect the gas phase flow near the interface. The friction coefficient between the gas phase flow and the liquid film is larger than that for a single-phase flow flowing through a smooth pipe (Gill et al., 1964; Chien and Ibele, 1964; Anderson and Mantzouranis, 1960). Therefore, waves are sometimes considered to act as interfacial roughness elements (Ueda and Nose, 1973).

In theoretical approaches to the annular two-phase flow and related heat transfer, these wave effects are not usually distinguished from the intrinsic turbulence in each phase. However, in their paper on the filmwise condensation in a vertical tube, Blangetti and Schlünder (1978) tried to separate the wave effect from the turbulence effect. Brumfield et al. (1978) introduced a model for wave effects, and had fair success in correlating the gas absorption data in falling liquid films. Since the gas phase resistance is not important in this problem, their model does not provide good information about the wave effect on gas phase transport phenomena. Levy and Healzer (1981) reported a theory for annular flow assuming transition layer between the base and the top of waves, but the wave effect was not treated directly.

The complicated nature of the combined effects of interfacial shear stress, interfacial waves and turbulence has prevented us so far from developing good picture of the annular two-phase flow (Suzuki, 1986). Several types of annular two-phase flows were studied, in order to develop a new method distinguishing the interfacial wave effect from other complicated behaviours of each phase. A simple model for the interfacial wave effect is introduced, and its application will be discussed in the following.

TIME TRACES OF LIQUID FILM THICKNESS

The gas-liquid interface is covered with waves of different sizes (Hagiwara et al., 1984). A variety of experimental techniques is available for the determination of the characteristics of the interfacial wave structure. Liquid film thickness time trace may indicate characteristics of the interfacial wave structure. Four techniques have often been employed to obtain the time traces (Hewitt, 1978): conductance measurements, capacitance measurements, light absorption methods, and the fluorescence technique.

In the present study, we adopted the conductance probe techniques similar to those used by Brown et al. (1978) and Tsiklauri et al. (1979). The probe is composed of two electrodes, which are 0.5mm diameter stainless steel wires spaced 2.5mm circumferentially. Figure 1 shows the signal

INTERFACIAL INSTABILITY AND MODELING OF WAVE EFFECT ON HEAT TRANSFER IN ANNULAR TWO-PHASE FLOW

K. SUZUKI and Y. HAGIWARA
Department of Mechanical Engineering
Kyoto University
Kyoto, Japan

ABSTRACT

The streamwise development of interfacial wave structure of downward annular two-phase flow was studied experimentally, by measuring instantaneous liquid film thickness at six different locations. It was found that the wave height increased with axial distance, while the wave frequency did not change remarkably. Numerical and theoretical analyses were carried out to predict the characteristics of two typical types of annular two-phase flow heat transfer problems: evaporative heat transfer in a downward air-water annular two-phase flow, and forced-convection filmwise condensation. Governing equations for each phase were solved simultaneously with a finite difference scheme. A simple model, directly expressing the heat and momentum transfer augmentation by the interfacial waves, was employed in the analyses. The computed results of heat transfer coefficient and liquid film thickness showed fairly good agreement with available experimental counterparts. The effectiveness of both the model and the calculation process is suggested. Instability of interface and characteristics of interfacial waves are also discussed.

INTRODUCTION

The heat transfer process, accompanied with gas-liquid two-phase flows, is used in various industrial systems. To design such systems or facilities, numerous theories and empirical equations have been proposed for flow and heat transfer characteristics. Since almost all of these theories were obtained by modifying theories developed for single-phase flow, interaction between the two phases is not fully taken into account. Therefore, in order to use these theories for optimum design or optimum operation of heat transfer systems, it is still necessary to incorporate another approach, to evaluate separately the degree of interaction.

In annular two-phase flow, interfacial shear stress exerts an influence on liquid film thickness (Ueda and Tanaka, 1973),

10. Knoebel, D. H., Harris, S. D., and Crain, Jr. B., Forced-Convection Subcooled Critical Heat Flux. D_2O and H_2O Coolant with Aluminum and Stainless Steel Heaters, DP-1306, E. I. Dupont de Nemours and Company, U.S. AEC, February 1973.

11. Kroeger, P. G., and Zuber, N., An Analysis of the Effects of Various Parameters on the Average Void Fractions in Subcooled Boiling, International Journal of Heat Mass Transfer, vol. 11, pp. 211-233, 1968.

12. Levy, S., Forced Convection Subcooled Boiling - Prediction of Vapor Volumetric Fraction, International Journal of Heat Mass Transfer, vol. 10, pp. 951-965, 1967.

13. Lockhart, R. W., and Martinelli, R. C., Proposed Correlation of Data for Isothermal Two-Phase, Two-Component Flow in Pipes, Chemical Engineering Progress, vol. 45, no. 1, pp. 39-48, 1945.

14. Nariai, H., Inasaka, F., and Shimura, T., Critical Heat Flux of Subcooled Flow Boiling in Narrow Tube, Proc. ASME-JSME Thermal Engineering Joint Conf., Honolulu, Hawaii, vol. 5, pp. 455-462, 1987.

15. Ornatskii, A. P., and Vinyarskii, L. S., Heat Transfer Crisis in a Forced Flow Underheated Water in Small-Diameter Tubes, High Temperature, vol. 3, pp. 400-406, 1965.

16. Owens, W. L., and Schrock, V. E., Local Pressure Gradients for Subcooled Boiling of Water in Vertical Tubes, ASME Paper 60-WA-249, 1960.

17. Saha, P., and Zuber, N., Point of Net Vapor Generation and Vapor Void Fraction in Subcooled Boiling, Proc. 5th Int. Heat Transfer Conf., Tokyo, vol. 4, pp. 175-179, 1974.

18. Staub, F. W., The Void Fraction in Subcooled Boiling - Prediction of the Initial Point of Net Vapor Generation, Trans. ASME, Journal of Heat Transfer, vol. 90, no. 1, pp. 151-157, 1968.

19. Tarasova, N. Y., Leontiev, A. I., Hlopushin, V. I., and Orlov, V. M., Pressure Drop of Boiling Subcooled Water and Steam-Water Mixture Flowing in Heated Channels, Proc. 3rd Int. Heat Transfer Conf., Chicago, vol. 4, pp. 178-183, 1966.

20. Tong, L. S., Boundary-Layer Analysis of the Flow Boiling Crisis, International Journal of Heat Mass Transfer, vol. 11, no. 7, pp. 1208-1211, 1968.

21. Ueda, T., Kieki-Nisouryu, pp. 44, Yokendo, Tokyo, (in Japanese), 1981.

22. Zuber, N., and Findley, J., Average Volumetric Concentration in Two-Phase Flow Systems, Trans. ASME, Journal of Heat Transfer, vol. 87, pp. 453-468, 1965.

Subscript

ADB	Adiabatic single-phase flow (total tube length)
AV	Average
DB	Point of bubble detachment or net vapor generation
ex	Tube exit
exp	Experimental
F	Two-phase flow (total tube length)
f	Two-phase flow (bubble detachment region)
f0	Single-phase flow (bubble detachment region)
in	Tube inlet
l	Liquid
L	Condition given by Levy
OB	Point of incipient nucleate boiling
SAT	Saturation condition
S	Condition given by Saha-Zuber
v	Vapor

BIBLIOGRAPHY

1. Ahmad, S. Y., Axial Distribution of Bulk Temperature and Void Fraction in a Heated Channel with Inlet Subcooling, Trans. ASME, Journal of Heat Transfer, vol. 92, no. 4, pp. 595-609, 1970.

2. Bergles, A. E., and Rohsenow, W. M., Forced-Convection Surface-Boiling Heat Transfer and Burnout in Tubes of Small Diameter, EPL Rept. no. 8767-21, MIT, May 1962.

3. Bergles, A. E., and Rohsenow, W. M., The Determination of Forced-Convection Surface-Boiling Heat Transfer, Trans. ASME, Journal of Heat Transfer, vol. 86, no. 3, pp. 365-372, 1964.

4. Bergles, A. E., and Dormer, Jr. T., Subcooled Boiling Pressure Drop with Water at Low Pressure, International Journal of Heat Mass Transfer, vol. 12, pp. 459-470, 1969.

5. Dormer, Jr. T., and Bergles, A. E., Pressure Drop with Surface Boiling in Small-Diameter Tubes, EPL Rept. no. 8767-31, MIT, September 1964.

6. Gambill, W. R., Burnout in Boiling Heat Transfer (2) Subcooled Forced Convection Systems, Nuclear Safety, vol. 9, no. 6, pp. 467-480, 1968.

7. Griffel, J., Forced Convection Boiling Burnout for Water in Uniformly Heated Tubular Test Sections, NYO-187-7, Columbia University, May 1965.

8. Gunther, F. C., Photographic Studies of Surface Boiling Heat Transfer to Water with Forced Convection, Trans. ASME, vol. 73, no. 2, pp. 115-123, 1951.

9. Inasaka, F., and Nariai, H., Critical Heat Flux and Flow Characteristics of Subcooled Flow Boiling in Narrow Tubes, JSME International Journal, vol. 30, no. 266, pp. 1595-1600, 1987.

is almost unity for D = 3 mm and for D = 1 mm at G = 7000 kg/m²s. For D = 1 mm, at G = 13000 kg/m²s, α_{esti} is about 70 % of α_{Ahmad}, and at G = 20000 kg/m²s, α_{esti} is about 45 % of α_{Ahmad}.

Taking into account the decrease of void fraction in small inside diameters at high water mass velocities, a reevaluation of x_{ex} was conducted. The C values of the high heat flux region in Fig. 9 then agreed with those of the low heat flux region. This may be another verification that the void fraction for 1 mm inside diameter tubes is smaller than that for 3 mm tubes.(Inasaka, 1987)

CONCLUSIONS

The critical heat flux and flow characteristics of subcooled flow boiling with water in narrow tubes were investigated experimentally. The main results were as follows:
(1) With small inside diameter tubes at a high water mass velocity, the critical heat flux becomes very high. It is several times higher than the existing empirical correlations.
(2) With small inside diameter tubes at a high water mass velocity, the two-phase friction multiplier becomes very low.
(3) Characteristics of these results for narrow tubes are attributable to the decrease of void fraction due to the intense condensation effect by subcooled water at the core region.

NOMENCLATURE

English

C	Parameter in Tong CHF correlation	
D	Tube inside diameter	m
f	Single-phase friction factor	
G	Mass velocity	kg/m²s
H_{fg}	Latent heat of evaporation	J/kg
L	Tube length	m
ΔP	Pressure drop	Pa
q	Heat flux	W/m²
q_c	Critical heat flux	W/m²
T	Temperature	°C
ΔT_s	Wall superheat	K
ΔT_{sub}	Subcooling at tube exit	K
u	Water velocity	m/s
x	Quality	
x_{eq}	Thermal equilibrium quality	
X	Martinelli parameter	
Y	Value defined in equation (14)	
z	Distance along the heated channel	m

Greek

ζ	Form pressure drop factor	
μ	Viscosity	Pa.s
ρ	Density	kg/m³
ϕ_L	Two-phase friction multiplier	

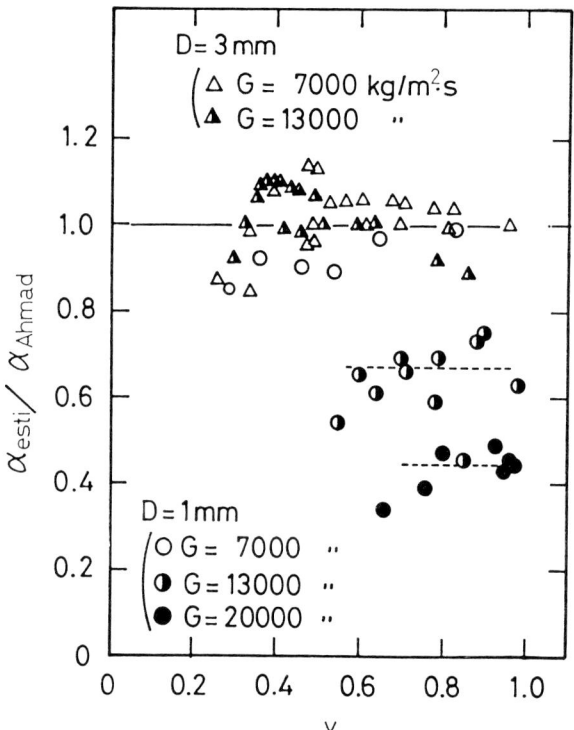

Figure 18. Comparison of Ahmad void fraction with estimated void fraction.

DISCUSSION

Void Fraction in Narrow Tubes

From the above mentioned experimental results, we reached the following conclusions. As tube inside diameter decreases and mass velocity increases, the diameter of the generated bubble or the thickness of the bubble boundary layer becomes smaller, due to the intense condensation effect by subcooled water at the core region. The void fraction also becomes smaller. Therefore, the CHF becomes larger and the two-phase multiplier becomes smaller. The Bergles-Rohsenow incipient nucleate boiling cerrelation agreed well with the experimental data. The effect of condensation by subcooled core water on the incipient nucleate boiling is very weak, since the bubble generation is governed only at the region near the wall in the thermal boundary layer, and depends on the heat flux to the heated wall.

Estimation of Void Fraction for Small-Diameter Tubes

Figure 18 shows the comparison of the void fraction α_{Ahmad} calculated by the Ahmad void fraction profile with the estimated fraction α_{esti}, calculated by assuming that the relation between α_{LAV} and X is represented by Eq. (17). $\alpha_{esti}/\alpha_{Ahmad}$

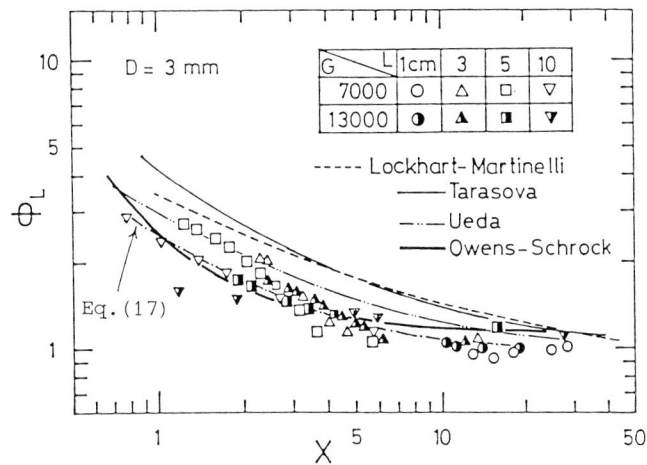

Figure 16. Average two-phase flow multiplier against Martinelli parameter.

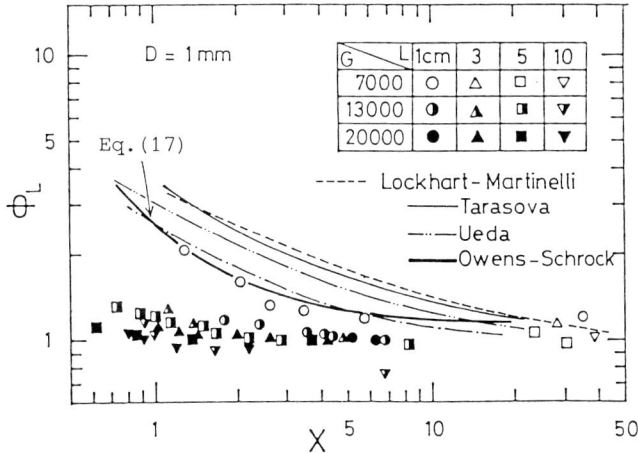

Figure 17. Average two-phase flow multiplier against Martinelli parameter.

assuming Saha-Zuber's bubble detachment correlation, as shown in Fig. 12.

Thus, the pressure drop correlation of subcooled flow boiling for 3 mm inside diameter tubes in the low heat flux region defined by the authors, is nearly equal to the existing pressure drop correlations for positive quality regions. However, for narrow tubes with 1 mm inside diameter and at high mass velocity, the pressure drop tends to differ widely from those correlations.

SUBCOOLED FLOW BOILING

length L_f from the bubble detachment point to the tube outlet, and ΔP_{f0} is the single-phase flow friction pressure drop for length L_f. ΔP_f and ΔP_{f0} were calculated respectively as follows:

$$\Delta P_f = \Delta P_{exp} - (fL_0/D + \zeta) G^2/2\rho_1 \tag{11}$$

$$\Delta P_{f0} = fL_f(1-x_{ex})^2 G^2/2D\rho_1 \tag{12}$$

where L_0 is the length from the tube inlet to the bubble detachment point, and ΔP_{exp} is the measured total pressure drop.

Owens-Schrock (Owens, 1960) derived the two phase friction multiplier for water above the ambient pressure with tubes of D = 3, 4, and 6 mm, and L = 38 and 41 cm. The average two-phase friction multiplier in the form of Eq. (10) is expressed as follows.

$$\phi_{LAV}^2 = 0.97 + 0.046 \,(\exp(6.13Y)-1)/Y \tag{13}$$

$$Y = 1 - \Delta T_{sub,exp}/\Delta T_{sub,DB} \tag{14}$$

Tarasova (1966) derived the multiplier for water at 1 to 20 MPa with D = 2.9 to 8.3 mm and G = 1400 to 3000 kg/m²s. The average multiplier is as follows:

$$\phi_{LAV}^2 = 1 + \left(\frac{q\rho_1}{H_{fg}G\rho_v}\right)^{0.7} \left(\frac{\rho_1}{\rho_v}\right)^{0.08} \left(\frac{26.3}{Y}\ln\frac{1.315}{1.315-Y} - 20\right) \tag{15}$$

Ueda (1981) made a correlation based on the Martinelli-Nelson correlation for steam water systems:

$$\phi_{LAV}^2 = 1 + 1.2\, x_{ex}^{0.75(1+0.01\sqrt{\rho_1/\rho_v})} \left(\left(\frac{\rho_1}{\rho v}\right)^{0.8} - 1\right) \tag{16}$$

Figure 16 shows a comparison of the empirical correlations with experimental data as the relation between the average two-phase friction multiplier ϕ_{LAV} and Martinelli's parameter X, for D = 3 mm. As shown in the figure, Owens-Schrock's correlation agrees well with the experimental data. The Lockhart-Martinelli and Tarasova correlations give higher values. Ueda's correlation comes between the two groups.

We made a simple correlation for our experimental data, which is shown as a chain line in the figure and is represented as follows:

$$\phi_{LAV} = 1 + \frac{2.42}{X} + \frac{2.80}{X^2} \tag{17}$$

Figure 17 shows the same comparison for a 1 mm inside diameter tube. At mass velocity G = 7000 kg/m²s, the relation between ϕ_{LAV} and X is similar to that for D = 3 mm. However, at G = 15000 abd 20000 kg/m²s, it is lower than the correlations and is approximately equal to unity. This suggests that the Ahmad quality profile correlation is larger for this case than the actual quality. Some experimental values for ϕ_{LAV} are smaller than unity. This is because the length of the two-phase flow region might be too large for D = 1 mm at high mass velocity,

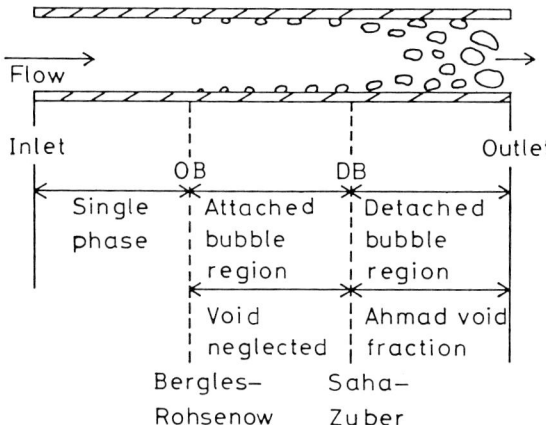

Figure 15. Modeling of subcooled flow boiling.

region from bubble detachment point DB to the outlet, could be calculated by assuming a suitable void fraction profile. As the actual quality profiles derived by assuming Zuber's (1965), Kroeger's (1968) and Ahmad's (1970) void fraction correlations were confirmed to be almost the same, the following actual quality correlation from Ahmad was used in this study:

$$x = \frac{x_{eq}(z) - x_{eqDB} \cdot \exp(x_{eq}(z)/x_{eqDB} - 1)}{1 - x_{eqDB} \cdot \exp(x_{eq}(z)/x_{eqDB} - 1)} \tag{7}$$

where x_{eqDB} is the equilibrium quality at the bubble detachment point, and $x_{eq}(z)$ is the equilibrium quality at point z downstream from point DB.

Lockhart-Martinelli (Lockhart, 1945) showed the friction pressure drop characteristics for adiabatic air-water systems in a positive quality region. They gave the two-phase friction multiplier ϕ_L as a function of Martinelli's parameter X:

$$X^2 = \left(\frac{1-x}{x}\right)^{2-n} (\rho_v/\rho_l)(\mu_l/\mu_v)^n \tag{8}$$

$$\phi_L = (\Delta P_f(x)/\Delta P_{f0}(x))^{0.5} \tag{9}$$

where n is 0.25 for turbulent flow.

The friction pressure drop obtained experimentally at the actual outlet quality x_{ex} corresponds to the value obtained by integrating the local pressure drop from x = 0 to x_{ex}. Therefore the average two-phase friction multiplier ϕ_{LAV} was defined as follows:

$$\phi_{LAV} = (\Delta P_f/\Delta P_{f0})^{0.5} \tag{10}$$

where ΔP_f is the two-phase flow friction pressure drop for

SUBCOOLED FLOW BOILING

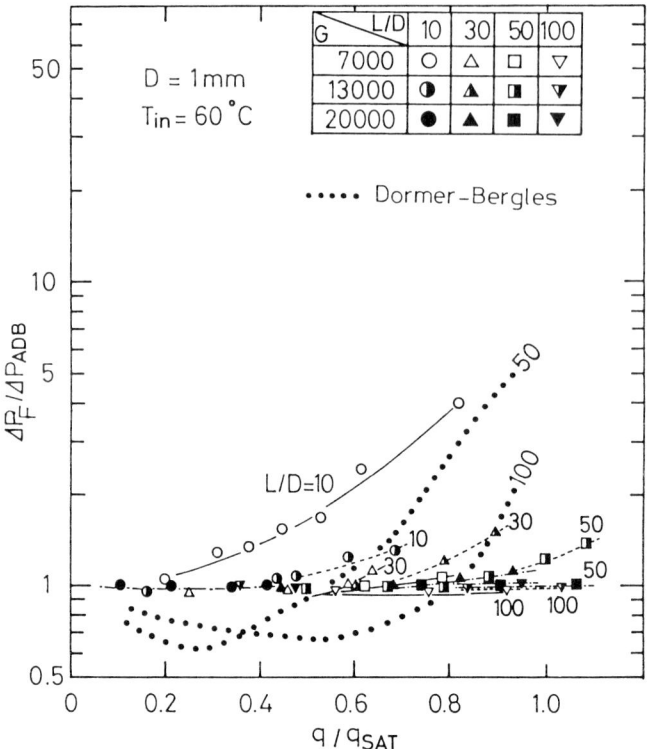

Figure 14. Friction pressure drop ratio against heat flux.

$\Delta P_F/\Delta P_{ADB}$ of 3 mm inside diameter correlated well in general to the length to diameter ratio, and agrees with that of Bergles.

Figure 14 shows the case of 1 mm inside diameter. The experimental pressure drop ratio for D = 1 mm is smaller than those of Dormer-Bergles, and the differences become larger with higher mass velocities.

As shown in the figure, the experimental data for the narrow tubes with 1 mm inside diameter at high mass velocity differ from the Saha-Zuber and Levy bubble detachment correlations, and from the pressure drop data given by Dormer-Bergles.

Two-phase Friction Multiplier

Two-phase friction multipliers were determined from the experimental pressure drop data of subcooled flow boiling, and compared with Lockhart's (1945), Owens's (1960), Tarasova's (1966) and Ueda's (1981) correlations. The flow model is shown in Fig. 15. Single-phase flow was assumed from the tube inlet to the point of incipient nucleate boiling (OB) determined by the Bergles-Rohsenow correlation. And at the attached bubble region existing from point OB to the point of bubble detachment (DB), the increases in void fraction and pressure drop were assumed to be negligible. The actual quality profile for the

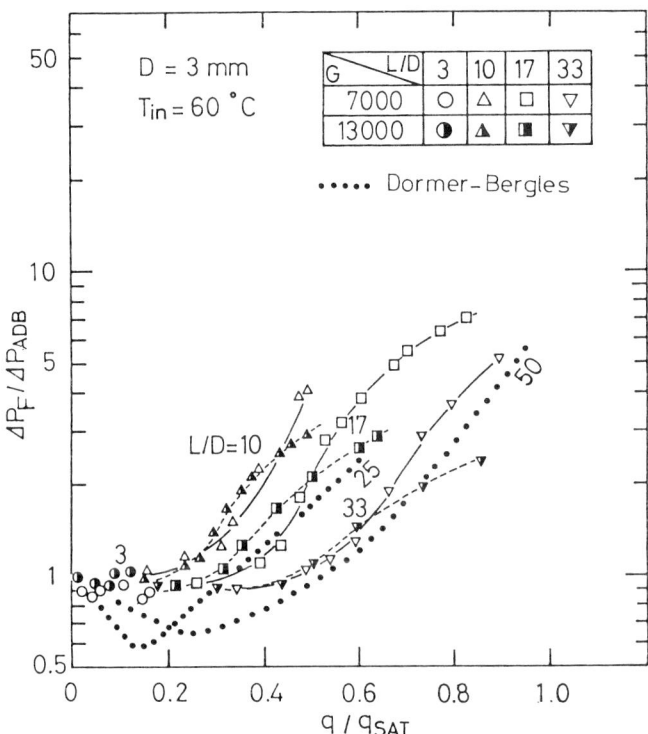

Figure 13. Friction pressure drop ratio against heat flux.

vapor generation correlation. Each experimental value is also indicated with a width. For D = 3 mm and D = 1 mm at G = 7000 kg/m²s, marked as open symbols, both heat fluxes agree well. For D = 1 mm at high mass velocity, the difference between the two becomes large, and the experimental heat flux is larger than that determined by the Saha-Zuber correlation. Comparison with the heat flux given by the Levy bubble detachment correlation gave almost the same result as that of the comparison with Saha-Zuber.

Dormer (1964) and Bergles (1969) conducted experiments on pressure drop for D = 1.57 to 5.0 mm. They selected two parameters to show the results. One is $\Delta P_F/\Delta P_{ADB}$, which is the ratio of the two-phase flow friction pressure drop in a heated tube to the single-phase flow friction pressure drop in a tube with similar geometry and inlet fluid conditions. The other is q/q_{SAT}, which is the ratio of the actual heat flux to the heat flux required for the exit condition to become saturated.

Figure 13 shows the comparison of experimental $\Delta P_F/\Delta P_{ADB}$ values of D = 3 mm with those given by Dormer-Bergles. (The inlet water temperature of our data is 60 °C, and that of Dormer-Bergles's data is 27 °C. Therefore, the rate of initial decrease of their ratio is larger than ours.) When $\Delta P_F/\Delta P_{ADB}$ for G = 13000 and 7000 kg/m²s are compared, the former becomes slightly smaller in the region of large q/q_{SAT}. However,

SUBCOOLED FLOW BOILING

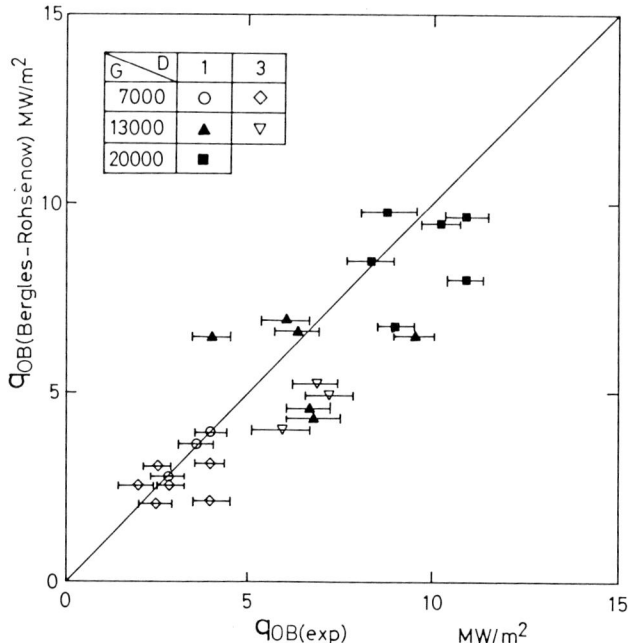

Figure 11. Comparison of Bergles-Rohsenow incipient nucleate boiling heat flux with experimental data.

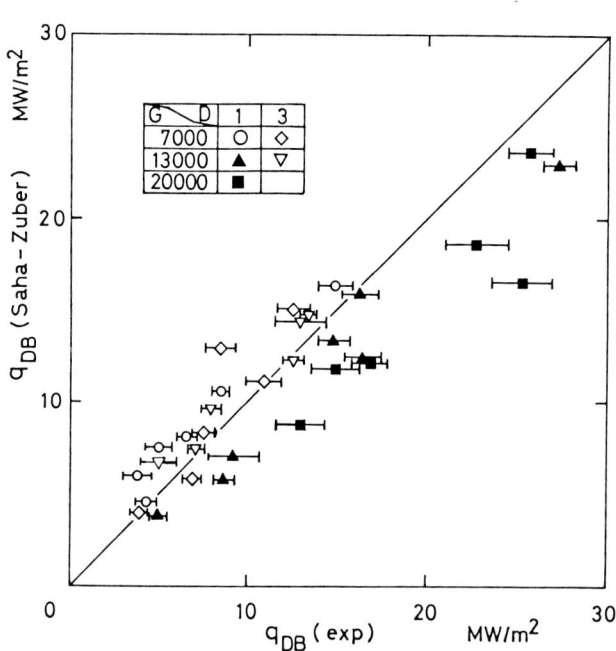

Figure 12. Comparison of Saha-Zuber net vapor generation heat flux with experimental data.

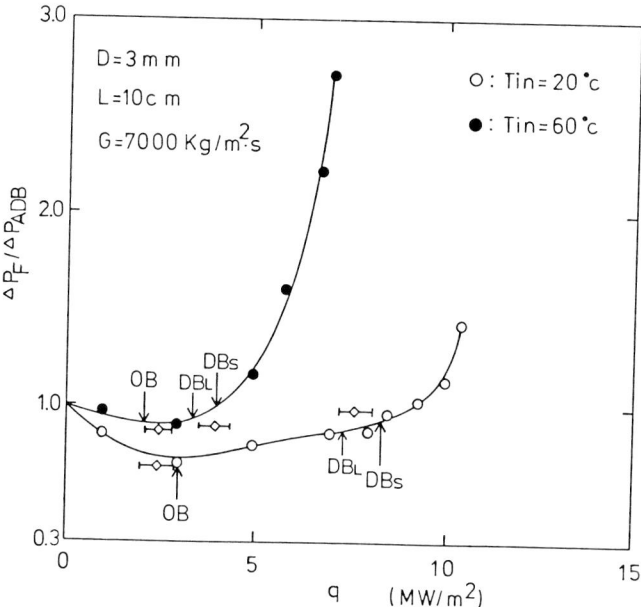

Figure 10. Friction pressure drop ratio against heat flux.

Bergles (1964) gave the heat flux q_{OB} at incipient nucleate boiling as a function of the wall superheat ΔT_s. Saha (1974) presented a correlation for the point of net vapor generation, based on the local thermal conditions. Levy (1967) presented a correlation for the point of bubble detachment, based on the hydraulic conditions. In Fig. 10, the symbol OB represents the heat flux of the incipient nucleate boiling at tube exit, determined by the Bergles-Rohsenow correlation. The symbol DB_S represents the heat flux of the net vapor generation at tube exit, determined by Saha-Zuber correlation. The symbol DB_L represents the heat flux determined by Levy's bubble detachment correlation. When the heat flux determined by Bergles-Rohsenow's, Saha-Zuber's and Levy's correlations is compared with the change of the experimental data, it becomes clear that the symbol OB is nearly the point where the friction pressure drop ratio becomes constant or begins to increase slightly. Also, the symbols DB_S and DB_L are nearly the point where this ratio begins to increase sharply.

Figure 11 shows the comparison of the heat flux, at the point where the experimental friction pressure drop changes from decrease to constant, with the heat flux determined by the Bergles-Rohsenow incipient nucleate boiling correlation. As it is very hard to determine the experimental heat flux as a value, the experimental values in Fig. 11 were indicated with a certain width (an example was shown near the symbol OB in Fig. 10). For a variety of tube diameters and mass velocities, both values agreed relatively well for a wide range.

Figure 12 shows the comparison of the heat flux, at the point where the frictional pressure drop ratio began to increase sharply, with the heat flux determined by the Saha-Zuber net

SUBCOOLED FLOW BOILING

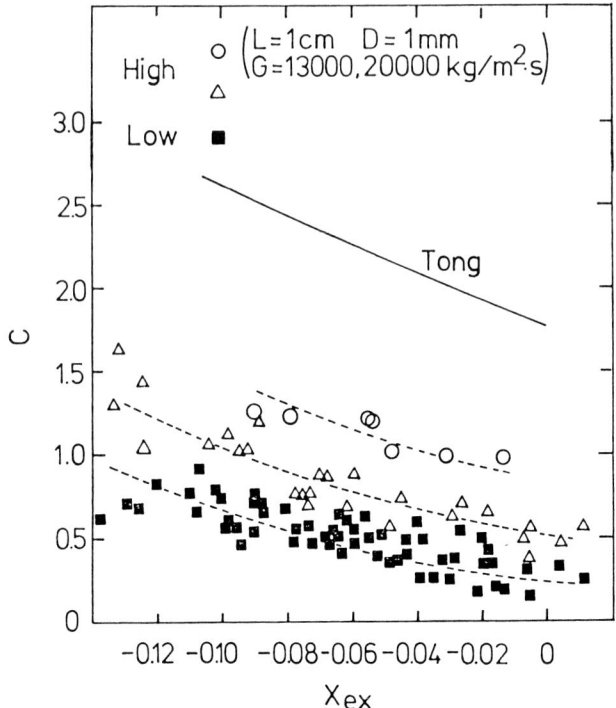

Figure 9. Coefficient C in Tong correlation for present data.

FLOW CHARACTERISTICS

Single-phase Pressure Drop

Friction pressure drop of single-phase flow was examined first. The friction pressure drop added the form pressure drop at the outlet and inlet flow areas, giving the measured pressure drop. The form pressure drop factor and the friction pressure drop factor were determined by the data for the different lengths and mass velocities. This form pressure drop factor was assumed to be the same for the subcooled flow boiling. Therefore, the two-phase flow friction pressure drop for subcooled flow boiling was determined by subtracting the form pressure drop from the measured total pressure drop.

Friction Pressure Drop and Flow Characteristics

Figure 10 shows an example of the ratio of subcooled flow boiling friction pressure drop to single-phase flow, with the increase of heat flux. The ratio decreases first because of the decrease of water viscosity, due to the temperature rise with the increase of heat flux. When the nucleate boiling begins, the ratio becomes constant or increases slightly, due to the stop of rising wall temperature and to generation of bubbles. When heat flux increases further, bubbles begin to detach and the pressure drop ratio begins to increase sharply, finally reaching the CHF condition.

It was compared with the present data, at G = 7000 kg/m²s. Considering that q_c decreases with the decrease of mass velocity, we can conclude that Bergles's data almost agrees with the present data. Further, we can find that the L/D at which q_c begins to increase sharply increases with the decrease of the tube inside diameter. This is explained by the fact that the values of L and D at the boundary of high and low heat flux regions are inversely proportional to each other, as shown in Fig. 4.

Comparison with Tong's Correlation

Tong (1968) assumed that the critical heat flux at the subcooled flow boiling occured at the separation of the bubble boundary layer along the tube wall, and derived the following equation:

$$\frac{q_c}{H_{fg}} = C \frac{G^{0.4} \mu_l^{0.6}}{D^{0.6}} \tag{2}$$

He determined the coefficient C, based on experimental data at water reactor conditions -- pressure: 7 - 14 MPa; mass velocity: 1400 - 4000 kg/m²s; and exit equilibrium quality: - 0.15 to 0.15. Tong then derived the following equation:

$$C = 1.76 - 7.433 \, x_{ex} + 12.22 \, x_{ex}^2 \tag{3}$$

Figure 9 shows the C values against x_{ex}. The solid line labeled Tong represents equation (3); the symbol ■ represents the experimental values at the low heat flux region; ∆ those at the high heat flux region; and ○ those at the particularly high heat flux region (L = 1 cm, D = 1 mm, and G = 13000 and 20000 kg/m²s). The C value by Tong is higher than that of present experimental data. This can be reduced to the difference of pressure, as the authors (Inasaka, 1987) showed in an other paper. However, the data are qualitatively well correlated, as shown in the figure. That means that the mechanism of the CHF of subcooled flow boiling might be based on the bubble boundary layer model. We tried to give the correlations for C values as follows.

For the low heat flux region,

$$C = 0.24 - 1.91 \, x_{ex} + 22.9 \, x_{ex}^2 \tag{4}$$

For the high heat flux region,

$$C = 0.52 - 2.61 \, x_{ex} + 26.2 \, x_{ex}^2 \tag{5}$$

For the particularly high heat flux region (L = 1 cm, D = 1 mm, G = 13000 and 20000 kg/m²s),

$$C = 0.84 - 3.42 \, x_{ex} + 30.4 \, x_{ex}^2 \tag{6}$$

These equations are presented in Fig. 9 as the broken lines.

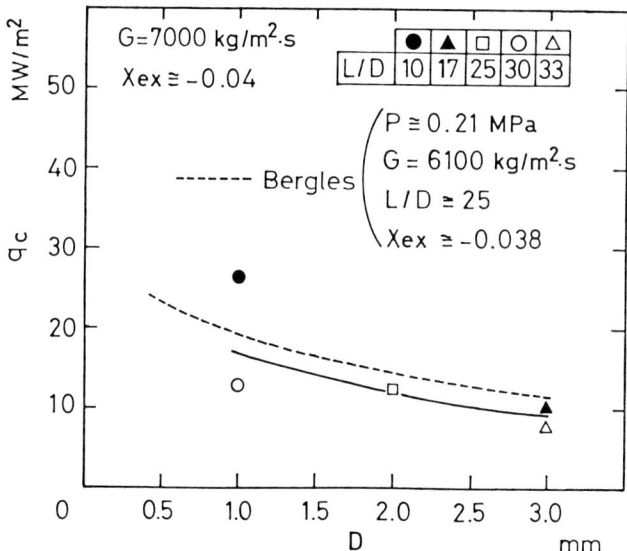

Figure 7. Effect of tube inside diameter on critical heat flux.

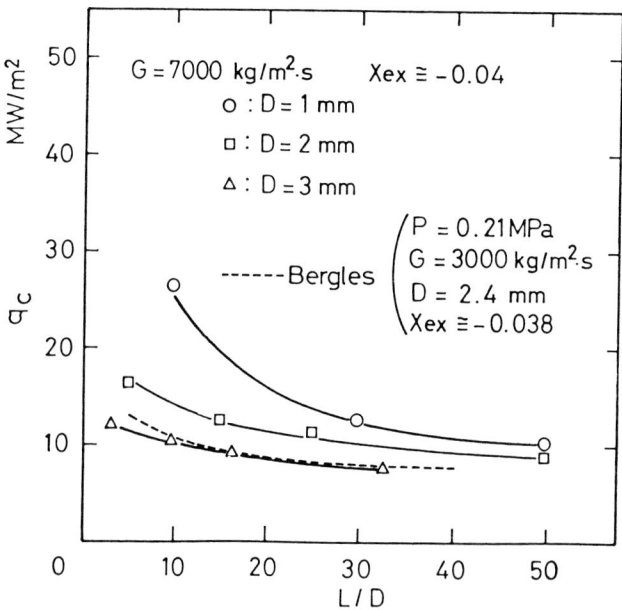

Figure 8. Effect of the ratio L/D on critical heat flux.

MPa, $G = 6100$ kg/m²s, L/D = 25, and $x_{ex} = -0.038$. Present data similar to these conditions were chosen. Both data agree well, as shown in the figure. Figure 8 shows the effect of L/D on q_c. Bergles showed the relation under the conditions $P = 0.21$ MPa, $G = 3000$ kg/m²s, $D = 2.4$ mm, and $x_{ex} = -0.038$.

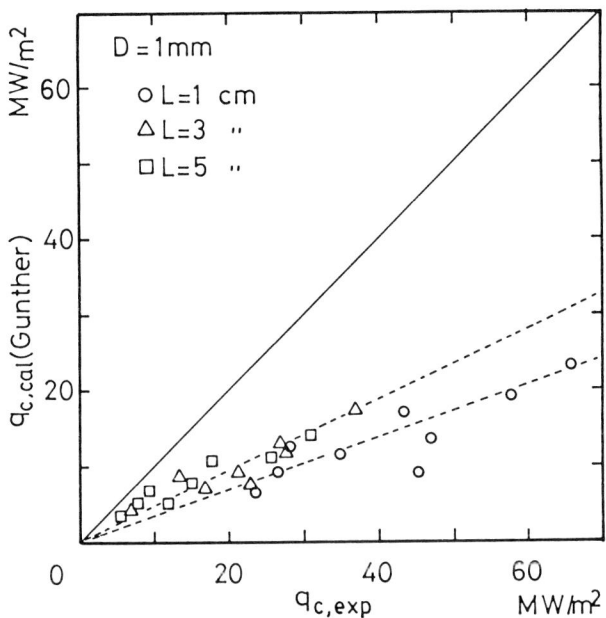

Figure 5. Comparison of experimental data with Gunther correlation.

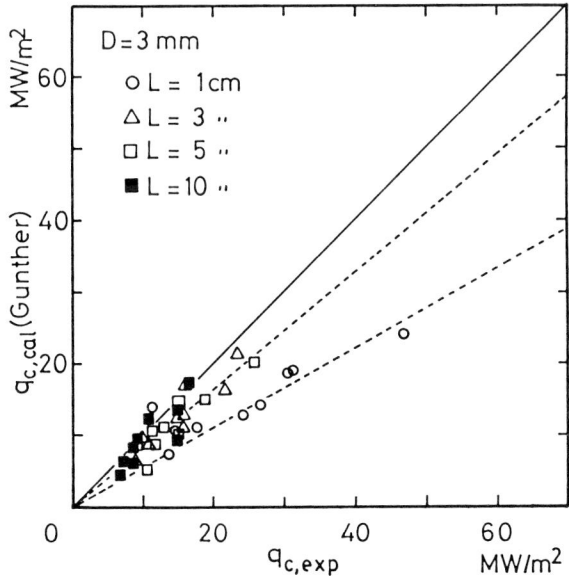

Figure 6. Comparison of experimental data with Gunther correlation.

SUBCOOLED FLOW BOILING

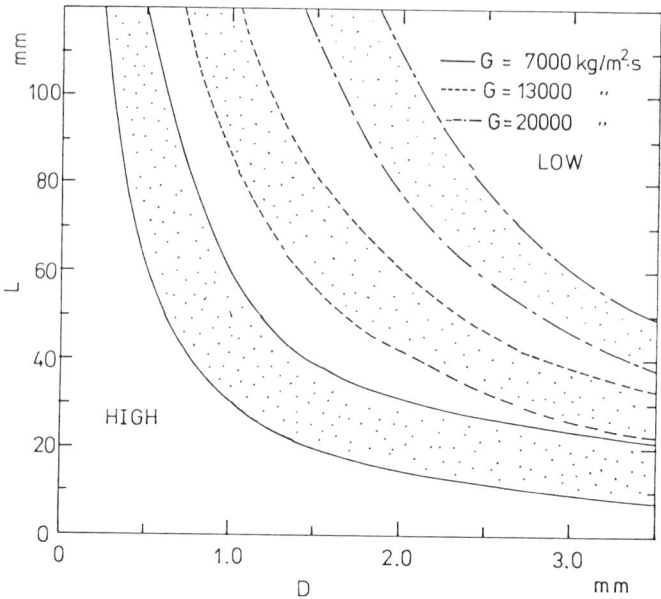

Figure 4. Boundary between high heat flux and low heat flux regions.

water velocity of 1.5 to 12 m/s, and water subcooling of 12 to 157 K.

$$q_c = 7.2 \times 10^4 \, u^{0.5} \Delta T_{sub} \tag{1}$$

Figure 5 shows a comparison of experimental data (for 1 mm inside diameter) with Eq. (1). The ordinate shows the prediction from Eq. (1), and the abscissa shows the experimentally derived q_c. With shorter tubes, the experimental q_c becomes larger than the predicted q_c. Figure 6 shows the case for 3 mm inside diameter. Both q_c agree fairly well for long tube.

The same kind of comparisons for Knoebel's (1973) and Griffel's (1965) correlations have already been reported by the authors (1987). Gambill's (1968) correlation is almost the same as Gunther's. As for the results, we can conclude that at the low heat flux region, or under the conditions of large inside diameter, long length and low water velocity, the experimental q_c agreed well with the empirical correlations. However, under the conditions at the high heat flux region, the experimental q_c was several times larger than the existing empirical correlations.

Comparison with Bergles's Data

Bergles (1962) studied the critical heat flux of subcooled flow boiling in narrow tubes. He investigated the effect of the tube inside diameter D and the ratio L/D on q_c, and showed that q_c increased with the decrease of D and L/D. Present data were compared with his data. Figure 7 shows the effect of D on q_c. Bergles's data are under the conditions that pressure P = 0.21

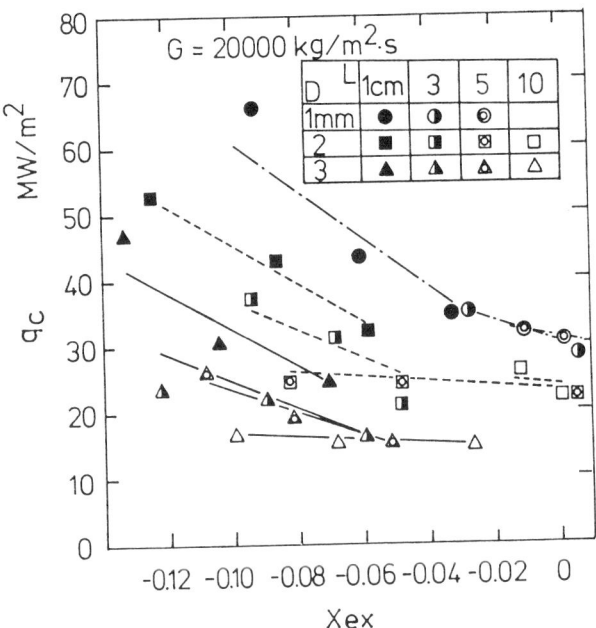

Figure 2. Critical heat flux for high water mass velocity.

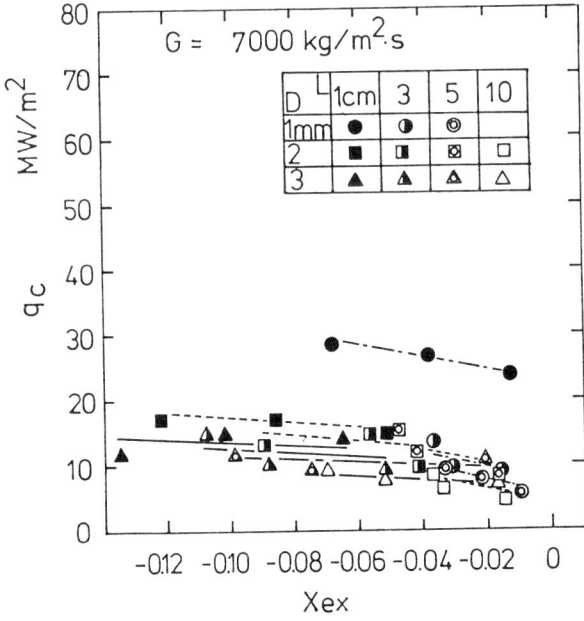

Figure 3. Critical heat flux for low water mass velocity.

SUBCOOLED FLOW BOILING

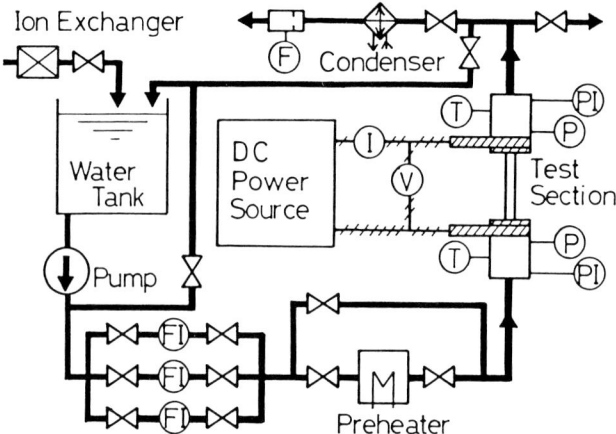

FIGURE 1. Schematic diagram of experimental apparatus.

inside diameter D = 1, 2, 3 mm; tube length L = 1, 3, 5, 10 cm; mass velocity of water G = 7000, 13000, 20000 kg/m²s; and inlet water temperature T_{in} = 20, 60 °C. The tube exit was kept almost at ambient pressure.

CRITICAL HEAT FLUX

High Heat Flux and Low Heat Flux Regions

Figures 2 and 3 show two examples of experimental results, representing the critical heat flux q_c against exit equilibrium quality x_{ex}, for the mass velocity G = 20000 and 7000 kg/m²s, respectively. As shown in Fig. 2, q_c increases with the decrease of both the tube inside diameter and the tube length, at high mass velocity. On the other hand, at low mass velocity, the effect of D and L on q_c disappears, if D and L are sufficiently large (see Fig. 3).

At the region where the tube inside diameter and the tube length affect q_c intensively, the critical heat flux is, in general, very high. We call this region the high heat flux region, and the region where they do not affect q_c so much we call the low heat flux region. The boundary between the two regions is shown in Fig. 4. In the figure, HIGH is the high heat flux region and LOW the low heat flux region. The boundary is the function of the mass velocity, and is shown in an appropriate width since the change from one region to another is gradual. The high heat flux region becomes wider at higher mass velocity.

Comparison with Existing Empirical Correlations

Gunther (1951) proposed an empirical correlation for the critical heat flux of subcooled flow boiling (Eq. (1)). Experiments were conducted with a rectangular channel of 7.6 cm length and 12.7 mm hydraulic diameter, with pressure of 0.1 to 1.1 MPa,

Kroeger (1968), and Ahmad (1970). For friction pressure drop, Lockhart's (1945), Owens's (1960), and Tarasova's (1966) correlations have been proposed. Ueda (1981) proposed a correlation based on Martinelli-Nelson correlation.

Almost all of above correlations were, however, derived for equivalent hydraulic diameters of larger than 3 mm. Present research is focusing on narrow tubes of less than 3 mm inside diameter. The CHF increases with a decrease of the tube inside diameter. Ornatskii (1965) derived CHF data of nearly 300 MW/m^2 with a tube of 0.4 mm inside diameter. Besides their data, a CHF of more than 100 MW/m^2 has not been reported. Bergles (1962) showed with small bore tube experiments that CHF increased with the decrease of tube inside diameter, and with the ratio of tube length to tube inside diameter. However, systematic study on the effect of tube dimension on CHF has been scarce.

On the friction pressure drop in narrow tubes, Dormer (1964) and Bergles (1969) conducted systematic studies and concluded that the pressure drop increased when bubbles in subcooled flow boiling grow and detach from the heated surface. However, there are few studies on the quantitative evaluation of the pressure drop of subcooled flow boiling in narrow tubes.

Present research aims to investigate the CHF and friction pressure drop of subcooled flow boiling, with a tube of 1 to 3 mm inside diameter. The CHF increases and the two-phase friction multiplier decreases with a decrease of tube inside diameter. That is attributed to the decrease of void fraction with the decrease of tube inside diameter, at the same average negative quality.

EXPERIMENT

Figure 1 shows a schematic diagram of the experimental apparatus. Water circulates through a pump whose maximum flow rate is 26 liter/min at a total head of 180 m, one of three float-type flow meters, a preheater and a test section, and returns to a water tank. The water is heated in the test section. As very low flow rates could not be measured by the flow meter, they were measured by a measuring cylinder downstream of the test section. The test section was made of a stainless steel tube, and two electrodes were silver brazed on the outside surface of the tube at the inlet and outlet portions. Tefron brocks, which served as an electric insulation, were connected with both electrodes. At the blocks, pressures were measured both by Bourdon tube-type pressure gauges and by strain gauge-type pressure transducers. Temperatures were measured by thermocouples. The test tube was directly heated by joule heating through electrodes, from a direct current power source.

After the test tube with a chosen length was installed, the flow rate and the inlet water temperature were set. Then the electric power was gradually increased step by step, and the electric power, pressures and temperatures were recorded at every suitable condition. The experimental conditions were selected by a combination of the following parameters: tube

CRITICAL HEAT FLUX AND FLOW CHARACTERISTICS OF SUBCOOLED FLOW BOILING WITH WATER IN NARROW TUBES

H. NARIAI
Institute of Engineering Mechanics
University of Tsukuba
Tsukuba, Ibaraki, Japan
F. INASAKA
Nuclear Technology Division
Ship Research Institute
Shinkawa, Mitaka, Tokyo, Japan

ABSTRACT

Critical heat flux (CHF) and flow characteristics of subcooled flow boiling with water in narrow tubes were investigated experimentally under uniform heat flux conditions. There were two experimental conditions: high heat flux regions, and low heat flux regions, based on the CHF characteristics. The CHF at high heat flux regions was higher than the predictions made by existing empirical correlations. They occured at small inside diameter, and short length tubes, as well as at high water mass velocity. The effect of tube inside diameter on the incipient boiling and the net vapor generation point, as well as on the two phase friction multiplier, was made clear from pressure drop characteristics. The decrease of friction pressure drop, and the increase of critical heat flux, with the decrease of tube inside diameter, are attributed to the decrease of void fraction in narrow tubes.

INTRODUCTION

Critical heat flux (CHF) and flow characteristics of subcooled flow boiling with water have been studied for more than 20 years, in the development of light water reactors. Many empirical correlations were proposed, both for subcooled flow boiling CHF and flow characteristics.

As for CHF, the correlations of Gunther (1951), Griffel (1965), Gambill (1968), and Knoebel (1973) are well known, for low pressure conditions. Tong's (1968) is one of the correlations used at high pressure (7 to 14 MPa), as the light water reactor condition.

Many correlations were also presented for flow characteristics and pressure drops of subcooled flow boiling. Bergles (1964) presented a correlation for incipient nucleate boiling. Concerning bubble detachment from heated surfaces or net vapor generation, Levy's (1967) and Staub's (1968) correlations based on force balance on a bubble, and Ahmad's (1970) and Saha's (1974) correlations based on local thermal conditions, have been proposed. Correlations for subcooled void fraction profiles along the heated surface were proposed by Zuber (1965),

Schrock, V.E. and Grossman, L.M., 1962, Forced Convection Boiling in Tubes, Nuc. Sci. Eng., Vol.12, pp.474-481.

Sekoguchi, K., Tanaka, O., Esaki, S. and Imasaka, T., 1980, Prediction of Void Fraction in Subcooled and Low Quality Regions, Bulletin of JSME, Vol.23, pp.1475-1482.

Sekoguchi, K., Tanaka, O., Ueno, O., Yamashita, M. and Esaki, S., 1982, Heat Transfer Characteristics of Boiling Flow in Subcooled and Low Quality Region, Proc. 7th Int. Heat Transfer Conf., Vol.4, pp.243-248.

Smith, S.L., 1969, Void Fraction in Two-Phase Flow : A Correlation Based upon an Equal Velocity Head Model, Proc. Instn. Mech Engrs., Vol.184, Pt.1, No.36, pp.647-664.

Thom, J.R.S., Walker, W.M., Fallon, T.A. and Reising, G.F.S., 1965, Boiling in Sub-cooled Water during Flow up Heated Tube or Annuli, Proc. Instn. Mech. Engrs., Vol.180, Pt.3C, pp.226-246.

Wright, R.M., 1961, Down Flow Forced Convection Boiling in Uniformly Heated Tubes, USAEC Rep., UCRL-9744.

REFERENCES

Beattie, D.R.H. and Whalley, P.B., 1982, A Simple Two-Phase Frictional Pressure Drop Calculation Method, Int. J. Multi-phase Flow, Vol.8 No.1, pp.83-87.

Bennet, J.A.R, Collier, J.G., Pratt, H.R.C. and Thornton, J.D., 1961, Heat Transfer to Two-Phase Gas-Liquid Systems, Pt.1 : Steam-Water Mixtures in the Liquid-Dispersed Region in Annulus, Trans. Instn. Chem. Engrs., Vol.39, pp.113-126.

Bergles, A.E. and Rohsenow, W.H., 1964, The Determination of Forced-Convection Surface-Boiling Heat Transfer, Trans. ASME, Ser.C, Vol.86, pp.365-372.

Chen, J.C., 1963, A Correlation for Boiling Heat Transfer to Saturated Fluids in Convective Flow, ASME Paper 63-HT-34.

Chisholm, D. and Laird, A.D.K., 1958, Two-Phase Flow in Rough Tubes, Trans. ASME, Vol.80, pp.276-286.

Collier, J.G., Lacey, P.M.C. and Pulling, D.J., 1964, Heat Transfer to Two-Phase Gas-Liquid Systems,Pt.2 : Further Data on Steam-Water Mixtures in the Liquid-Dispersed Region in Annulus, Trans. Instn. Chem. Engrs., Vol.42, pp.127-139.

Collier, J.G., 1981, Convective Boiling and Condensation, 2nd ed., p.212, McGraw-Hill Co, Ltd..

Davis, E.J. and Anderson, G.H., 1966, The incipience of Nucleate Boiling in Forced Convection Flow, A.I.Ch.E. Journal, Vol.12, No.4, pp.774-780.

Dengler, C.E. and Addmos, J.N., 1956, Heat Transfer Mechanism for Vaporization of Water in a Vertical Tube, Chemical Engng. Progress Symp. Ser., Vol.52, No.18, pp.95-103.

Jens, W.H. and Lottes, P.A., 1951, Analysis of Heat Transfer Burnout, Pressure Drop and Density Data for High Pressure Water, USAEC Rep., ANL-4627.

Martinelli, R.C. and Nelson, D.B., 1948, Prediction of Pressure Drop During Forced-Circulation Boiling of Water, Trans. ASME, Vol.70, pp.695-702.

Pujol, L. and Stenning, A.H., 1969, Effect of Flow Direction on the Boiling Heat Transfer Coefficient in Vertical Tubes, Symp. Ser. Canadian Soc. Chem. Engng., No.1, pp.401-453, Plenum Press N.Y..

Rohsenow, W.M., 1970, Nucleation with Boiling Heat Transfer, ASME paper 70-HT-18.

ACKNOWLEDGMENT

The authors wish to thank Messrs. N. Yoshimo, S. Hagihara, K. Takada, K. Irakai and K. Yamasaki for their assistance in the experiment. This work was supported by the Fund for Scientific Research granted by the Ministry of Education.

NOMENCLATURE

Bo	Boiling Number
C_1, C_2	constants in Eq. (7)
c_p	specific heat
D	tube diameter
G	mass velocity
g	acceleration due to gravity
h	heat transfer coefficient
K	parameter in Eq. (20)
k	thermal conductivity
m	exponent in Eq. (7)
n	exponent in Eq. (9)
P	pressure
Pr	Prandtl number
q_w	heat flux
T	temperature
u	velocity
x	quality
X_{tt}	Martinelli parameter
y	distance from wall
z	axial co-ordinate
dP/dz	pressure gradient

Greek

α	void fraction
δ	tube wall thickness
μ	viscosity
ρ	density
τ	shear stress
ϕ_ℓ^2	two-phase multiplier
ψ	constant, in Smith's correlation

Subscripts

a	actual value
b	bulk fluid
c	convection
g	gas
inc	incipience of boiling
ℓ	liquid
Lf	iquid-phase alone
o	main stream
sat	saturated
sup	suppression of boiling
TP	two-phase
w	wall

correspond to the decrease of film flow velocity, due to the increase of entrained liquid droplets. The declination of K becomes significant as the system pressure increases. The relation between K and $1/X_{tt}$ is fairly well expressed by a power low at each system pressure.

This supports the validity of the analogy between heat transfer and pressure drop, for two-phase forced convective heat transfer mechanism. Further detailed information on liquid and vapor phase distributions for various flow conditions is required to confirm the analogy.

CONCLUSIONS

An experiment on forced convective boiling heat transfer to water flowing vertically upward in a tube was carried out over the wide range of heat transfer regimes. This included single phase forced convection, subcooled boiling, saturated nucleate boiling and two-phase forced convection regions. Results obtained were as follows:

(1) The data clearly indicated transition boundaries between various heat transfer regimes. The heat transfer coefficient in the two-phase region was not correctly predicted by the empirical correlations of previous investigators.

(2) At low heat flux, particularly at low mass flow rate, the heat transfer coefficient evaluated by the usual definition was less than that of liquid single-phase flow in some regions at x>0. It was corrected by a modified empirical correlation and accounting for liquid superheat.

(3) For the two-phase convective heat transfer regime, including the low heat flux region, the following empirical correlation was obtained.

$$\frac{h_{CTP}}{h_{Lf}} = \left\{ 1 + 15.63 \left(\frac{1}{X_{tt}} \right)^2 \right\}^{1/3}$$

(4) Initiation conditions of nucleate boiling were compared with the previous works, but satisfactory results were not obtained. Boiling suppression conditions correlated well with the experimental conditions.

(5) Analogy between heat transfer and pressure drop was discussed, in particular for two-phase heat transfer mechanism in which the nucleate boiling was suppressed. Fairly good agreement was obtained between the correlating method by Beattie and Whalley for friction pressure drop and the present heat transfer data. This means that the present analogy is valid for two-phase flow. Introducing a parameter, K, which depends on quality and pressure, prediction of heat transfer coefficient by this analogy is improved.

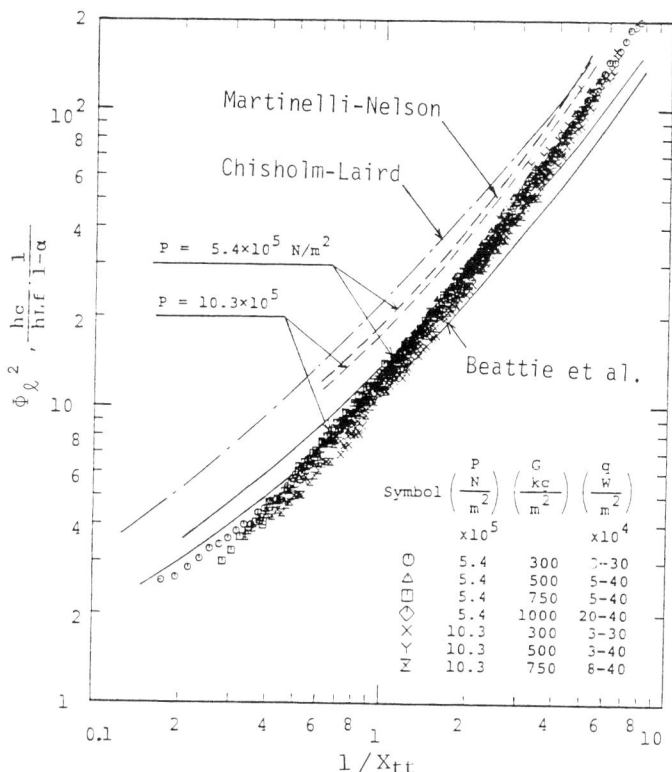

FIGURE 12 Comparison of the present analysis with frictional pressure drop correlations

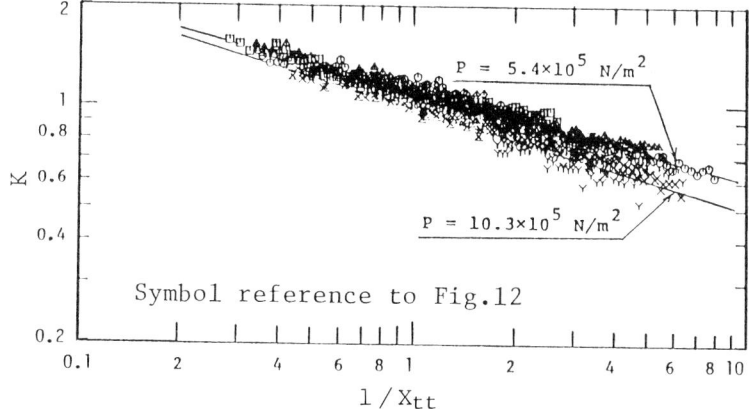

FIGURE 13 Value of K in comparison with correlation of Beattie et al. (1982)

In the single-phase flow the main stream velocity is approximately equal to the mean velocity. If the quality is evaluated as an actual one, x_a, u_{oLf} is given by

$$u_{oLf} = \frac{G(1-x_a)}{\rho_\ell} \quad (19)$$

For the two-phase film flow the main stream velocity might be different from its mean velocity. Therefore, multiplying a parameter K, u_o is described as follows:

$$u_o = K \frac{(1-x_a)G}{(1-\alpha)\rho_\ell} \quad (20)$$

From Eqs. (17) to (19) the relationship between heat transfer and friction pressure drop is derived.

$$K \frac{h_{CTP}}{h_{Lf}} \frac{1}{1-\alpha} = \frac{(dP/dz)_{TP}}{(dP/dz)_{Lf}} = \Phi_\ell^2 \quad (21)$$

where Φ_ℓ^2 is a two-phase multiplier and the parameter K may vary, depending mainly on flow pattern. At the present knowledge about K was not obtained: it will be discussed later.

Figure 12 shows comparisons between the present experimental results for K = 1 and frictional pressure drop correlations from various investigators. The coordinate represents $h_{CTP}/h_{Lf}(1-\alpha)$ and the abscissa represents Martinelli parameter $1/X_{tt}$.

Martinelli and Nelson (1948) presented a diagram to calculate friction pressure drop in a uniformly heated tube, up to the critical pressure for water. Chisholm and Laird (1958) presented a simplified correlation of two-phase multiplier as a function of Martinelli parameter, as expressed by Eq. (4). Recently Beattie and Whalley (1982) proposed another calculation method, which agreed well with experimental pressure drop data by many investigators, over a wide range of experimental conditions for adiabatic round tubes.

The present data are considerably lower than the correlations of Martinelli-Nelson and Chisholm-Laird. The correlation of Beattie and Whalley is rather close to our data. But the present results become slightly larger than the correlation of Beattie and Whalley for relatively large $1/X_{tt}$, and vice versa for smaller $1/X_{tt}$.

From Eq. (21) K is described as

$$K = \Phi_\ell^2 \frac{(1-\alpha)h_{Lf}}{h_{CTP}} \quad (22)$$

In Fig. 13 the parameter K calculated from the present data is plotted against $1/X_{tt}$, in which the two-phase multiplier by Beattie and Whalley is used. When $1/X_{tt}$ is small, i.e. in the low quality region, K is larger than unity and decreases with an increase in quality. The decrease of K might

ANALOGY BETWEEN HEAT TRANSFER AND FLUID FRICTION

In the single-phase forced convection heat transfer, an analogy between momentum and heat transfer has been well established both theoretically and experimentally. If the analogy is valid for the two-phase flow heat transfer problem, it might be a very useful means to predict the heat transfer coefficient. Collier (1981) developed a very simple model and derived the following relation:

$$\frac{h_{CTP}}{h_{Lf}} = \frac{1}{1-\alpha} \tag{12}$$

But no successful results were obtained. In the following section, a fundamental consideration is made and compared with the present experimental results.

For a single-phase fluid with Prandtl number of approximately unity, the velocity and temperature profiles in a cross section are similar. This relation is applied to the film flow in a gas-liquid two-phase system such as an annular flow.

$$\frac{1}{T_0-T_w}\left(\frac{\partial T}{\partial y}\right)_w = \frac{1}{u_0}\left(\frac{\partial u}{\partial y}\right)_w \tag{13}$$

where T_0 and u_0 are the temperature and the velocity of the so-called main stream, respectively. Substituting $Pr_\ell = c_{p\ell}\mu_\ell/k_\ell$ into Eq. (13) and rearranging, Eq. (13) yields

$$\frac{Pr_\ell}{c_{p\ell}}\frac{k_\ell}{T_0-T_w}\left(\frac{\partial T}{\partial y}\right)_w = \frac{\mu_\ell}{u_0}\left(\frac{\partial u}{\partial y}\right)_w \tag{14}$$

Heat flux and wall shear stress are respectively expressed as

$$q_w = -k_\ell\left(\frac{\partial T}{\partial y}\right)_w, \quad \tau_w = \mu_\ell\left(\frac{\partial u}{\partial y}\right)_w \tag{15}$$

Thus, Eq. (14) becomes

$$\frac{Pr_\ell}{c_{p\ell}}\frac{-q_w}{T_0-T_w} = \frac{Pr_\ell}{c_{p\ell}}\cdot h_{CTP} = \frac{\tau_{wTP}}{u_0} \tag{16}$$

where the heat transfer coefficient h_{CTP} is evaluated assuming $T_0 = T_{sat}$.

This relation is also valid when the flow in a tube is a liquid component alone.

$$\frac{Pr_\ell}{c_{p\ell}}\cdot h_{Lf} = \frac{\tau_{wLf}}{u_{OLf}} \tag{17}$$

From Eqs. (16) to (17)

$$\frac{h_{CTP}}{h_{Lf}} = \frac{u_{OLf}}{u_0}\frac{\tau_{wTP}}{\tau_{wLf}} = \frac{u_{OLf}}{u_0}\frac{(dP/dz)_{TP}}{(dP/dz)_{Lf}} \tag{18}$$

CONVECTIVE BOILING FLOW

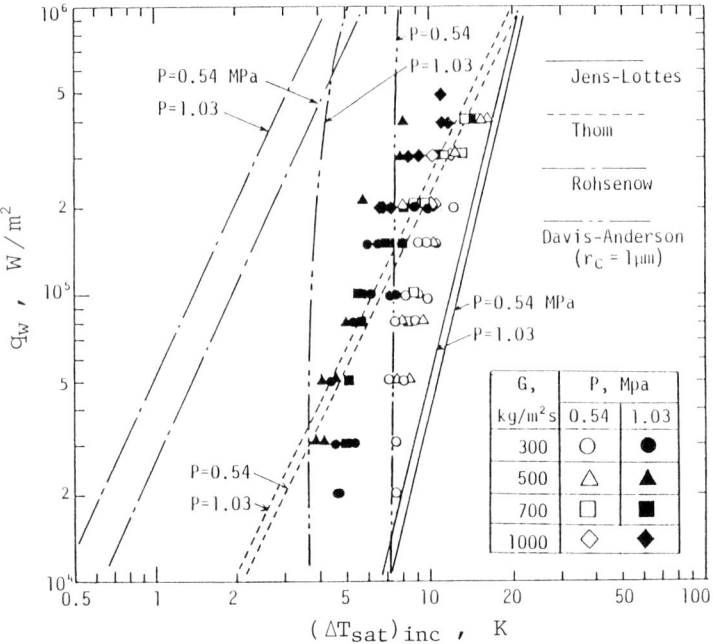

FIGURE 10 Comparison of boiling incipience condition with various correlations

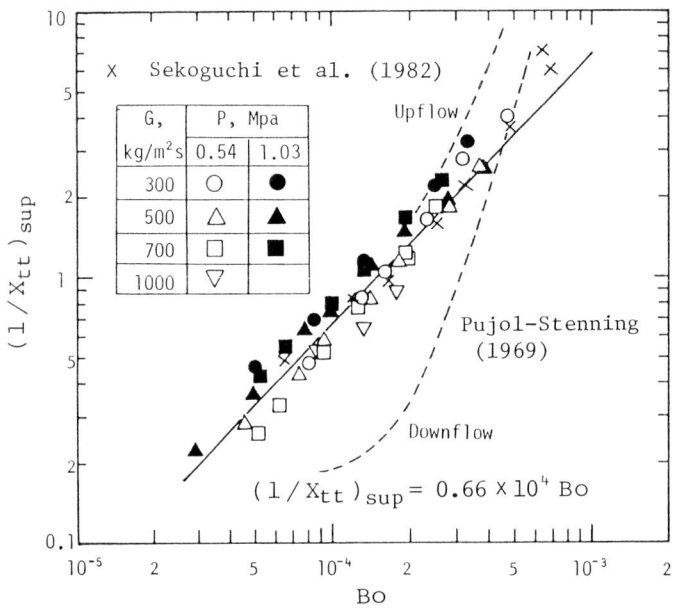

FIGURE 11 Boiling suppression condition in terms of Martinneli parameter and Boiling number

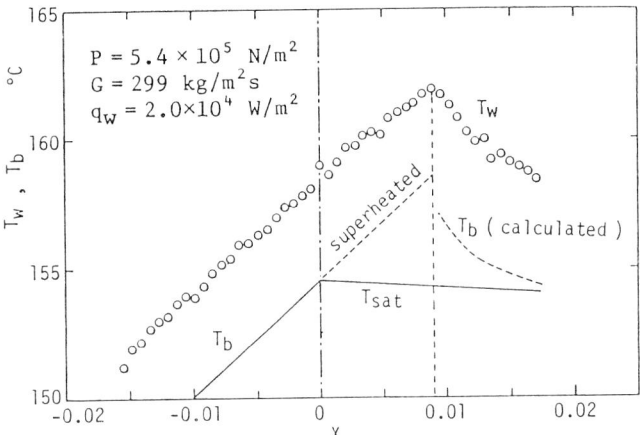

FIGURE 9 Relationship between wall and bulk temperatures at low heat flux

Incipience and Suppression of Nucleate Boiling

Conditions for the incipience of nucleate boiling were determined from the criterion in which the heat transfer coefficient began to be higher than that of liquid single-phase flow. The above criterion was not satisfied at low heat flux, but the maximum wall temperature which corresponded to the point immediately upstream of boiling initiation was detected with ease.

Figure 10 shows the relation between wall superheat ΔT_{sat} and heat flux q_w for incipience of boiling. The present data are compared with empirical correlations by Jens and Lottes (1951) and Thom et al. (1965), but agreements with both correlations are not satisfactory. Analytical predictions by Davis and Anderson (1966) and Rohsenow (1970) are also shown in this figure. Only the former calculations agree with the data for low heat flux conditions when available cavity size is assumed to be about 1 μm. At high heat flux, since the nucleate boiling initiates in the subcooled region, higher wall superheat will be required for vapor bubbles to grow and agitate the thermal boundary layer.

For a given mass flow rate, the quality at which nucleate boiling was suppressed became higher with an increase in heat flux. Figure 11 shows this relationship in terms of dimensionless groups, i.e. Martinelli parameter $(1/X_{tt})_{sup}$ and Boiling Number Bo. Experimental results by Pujol and Stenning (1969) for Freon-113, as well as data by Sekoguchi et al. (1982), are also shown in this figure. The present experimental data does not agree with them. A linear relation can be seen from our data. Although the data shifts to slightly higher quality as the pressure increases, almost all data are correlated with the following expression.

$$\left(\frac{1}{X_{tt}}\right)_{sup} = 0.66 \times 10^4 \, Bo \tag{11}$$

correlations proposed by various investigators. The present data fall in a straight line with little scattering. However, the correlations proposed by other investigators differ from each other and agreements with the present data are poor. From the present data for various experimental conditions, the following expression was obtained as the most correlative equation.

$$\frac{h_c}{h_{Lf}} = 2.5 \left(\frac{1}{X_{tt}}\right)^{2/3} \tag{8}$$

In Fig. 8 the whole of our data are compared with the above empirical correlation. It is found that almost all data falls in the above expression, within $\pm 10\%$ deviations.

Heat Transfer at Low Heat Fluxes

As previously described, the heat transfer coefficient evaluated by $h = q_w/(T_w-T_{sat})$ at $x > 0$ was somewhat less than that of liquid single-phase flow in the low heat flux region. The declination was significant as the mass flow rate decreased. Figure 9 shows an example of inside wall temperature T_w plotted against thermodynamic equilibrium quality x. Usually defined bulk fluid temperatures T_b and T_{sat} are shown by solid lines. After x exceeds zero the wall temperature continues to rise to some extent. This means that the nucleate boiling is absent and the liquid is superheated. Thus, the bulk temperature should be extended from the subcooled region. At the point where the wall temperature reaches the maximum value and begins to decrease, boiling might be initiated. However, the heat transfer mechanism is not considered to be the same as that for high heat flux. Taking this into account, the bulk temperature was corrected as shown by a dotted line, as follows:

Nucleate boiling is considered to be absent at low heat flux, unlike at high heat flux. For the transition region from single-phase to two-phase forced convection regions, correlation of heat transfer coefficient can be expressed as a combination of those for both regions, as follows:

$$\frac{h_{CTP}}{h_{Lf}} = \left\{1 + \left(\frac{h_c}{h_{Lf}}\right)^n\right\}^{1/n} \tag{9}$$

where h_{CTP}/h_{Lf} is given by Eq. (8) and n is an exponent. n was determined from the present data to be 3, and satisfactory results were obtained. As a result, Eq. (9) yields

$$\frac{h_{CTP}}{h_{Lf}} = \left\{1 + 15.63 \left(\frac{1}{X_{tt}}\right)^2\right\}^{1/3} \tag{10}$$

In Fig. 9 the bulk temperature corrected by Eq. (10) is shown by a dotted line. For comparison with other experimental conditions, the calculated result from Eq. (10) is also shown in Fig. 6. Except for the region immediately downstream of the maximum wall temperature point, reasonable results were obtained.

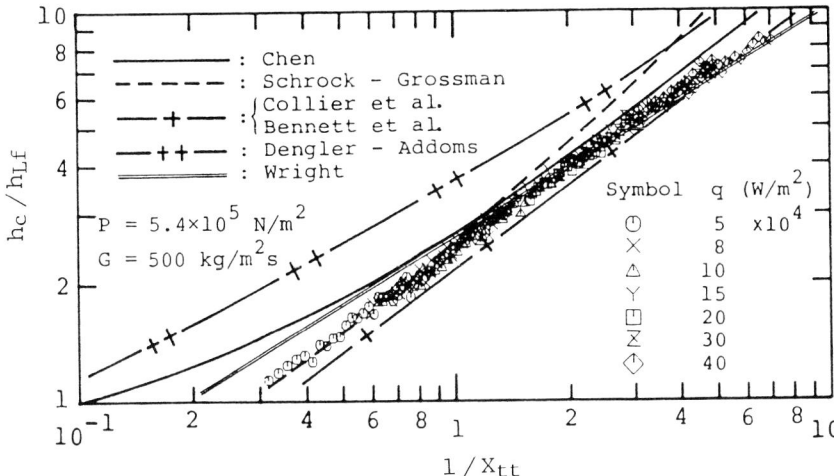

FIGURE 7 Comparison between the present results and empirical correlations for two-phase forced convective heat transfer region

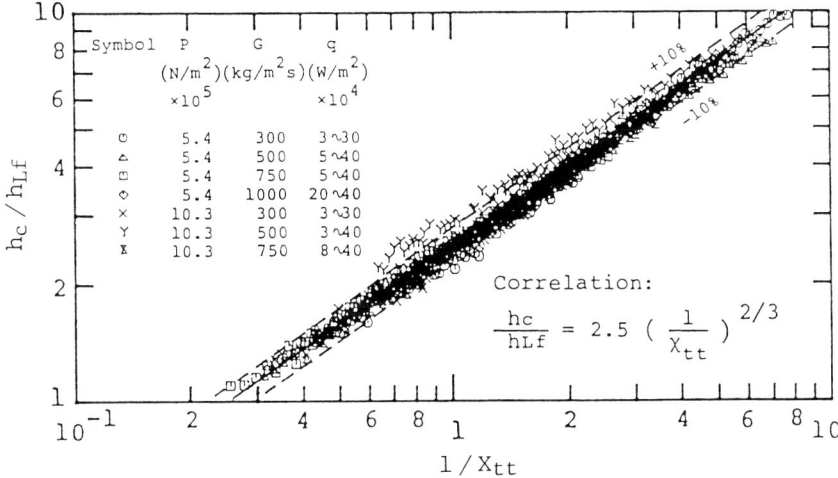

FIGURE 8 Comparison of Eq.(8) with experimental data for two-phase forced convective heat transfer region

In the two-phase forced convection region, heat transfer coefficient was found not to be influenced by heat flux for the present data. Therefore, the heat transfer coefficient in this region can be appropriately correlated only with a flow parameter, i.e. Martinelli parameter X_{tt}. Since the actual quality x_a was substantially equal to the thermodynamic equilibrium quality x in this region, $1/X_{tt}$ was used instead of $1/X_{tta}$.

Figure 7 shows a comparison between the present results and

CONVECTIVE BOILING FLOW

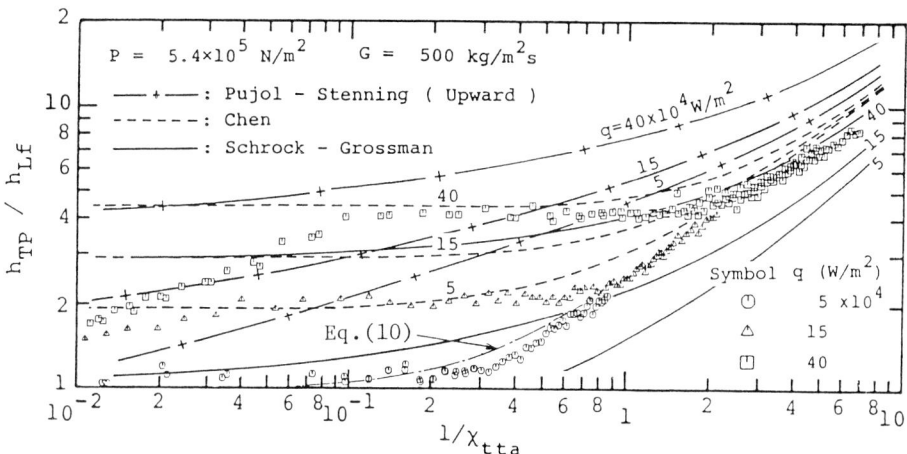

FIGURE 6 Comparison of the present experimental data with empirical correlations of other investigators

Comparison of Data with Empirical Correlations

Empirical correlations of two-phase heat transfer coefficient in the saturated boiling region, including nucleate boiling and two-phase forced convective heat transfer regimes, have been proposed by many investigators. According to Schrock and Grossman (1962) and Pujol and Stenning (1969), the correlation is generally expressed as

$$\frac{h_{TP}}{h_L} = C_1 \{Bo + C_2 (\frac{1}{X_{tta}})^m\} \tag{7}$$

where C_1 and C_2 are constants and m is an exponent. h_L is a liquid single-phase heat transfer coefficient, evaluated from mass flow rate either of liquid-phase alone (h_{Lf}) or of total flow (h_{Lo}).

In the present paper h_{Lf} was used and evaluated by Dittus-Boelter's equation. Chen (1963) proposed another correlating method, in which the two-phase heat transfer coefficient was composed of the forced convection term modified by the two-phase Reynolds number and the nucleate boiling term modified by a suppression factor.

Figure 6 shows a comparison of the present experimental data with the correlations described above. X_{tta} is the Martinelli parameter evaluated by the actual steam quality x_a. These correlations significantly differ from each other and do not agree with the present data, although Chen's correlation is rather close to our data for high heat flux. Transition from nucleate boiling to two-phase forced convection regions can be clearly detected from the present data, but other correlations give gradual transitions.

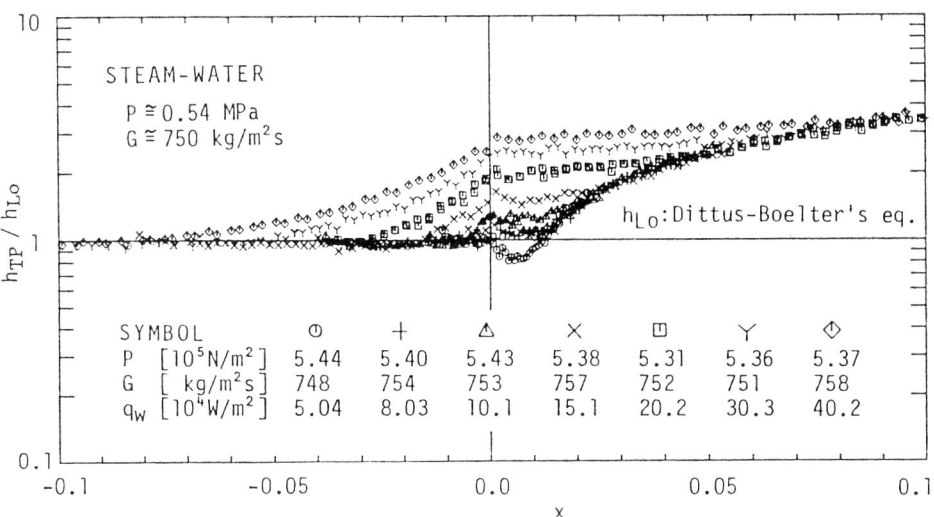

FIGURE 5 A typical example of relationship between heat transfer ratio h_{TP}/h_{LO} and quality x at various heat fluxes

x had a certain positive value. The reason why the ratio h_{TP}/h_{LO} becomes less than unity is from the definition of heat transfer coefficient at the quality region. Its correction will be discussed later. In the following, except for the case of low heat flux, heat transfer characteristics of various flow regimes are described.

Heat Transfer Characteristics Except for Low Heat Flux

When the liquid was highly subcooled, the heat transfer coefficient agreed with that of single phase liquid flow, and boiling was not considered to initiate. When the subcooling of the liquid decreased and the wall temperature rose above the saturation temperature to some extent, the heat transfer coefficient increased, which implied incipience of nucleate boiling. As the subcooling of the liquid decreased further, the heat transfer ratio h_{TP}/h_{LO} gradually increased. It became maximum at about x = 0. In the region where the quality was positive but relatively low, the heat transfer coefficient remained approximately constant.

As the quality increased further, nucleate boiling seemed to be gradually suppressed and two-phase convective heat transfer mechanism (evaporation) became dominant. In this region, the heat transfer coefficient increased with an increase in quality. Data plots of heat transfer coefficients at various heat fluxes in this region fall closely on a single curve for a given mass velocity, as shown in Fig. 5.

where

$$\frac{1}{X_{tta}} = \left(\frac{X_a}{1-X_a}\right)^{0.9} \left(\frac{\rho_\ell}{\rho_g}\right)^{0.5} \left(\frac{\mu_g}{\mu_\ell}\right)^{0.1} \tag{4}$$

As for the first approximation, the pressure drop was calculated successively along the tube axis from Eqs. (1) to (3). If the calculated total pressure drop over the whole length of the test section does not equal to the measured pressure drop, Eq. (3) was modified by multiplying a constant value proportional to its difference. This procedure was repeated until both values coincided. Very little difference in the result was seen when the correlation of Martinelli and Nelson (1948) was applied in place of Eq. (3). Thus the local saturation temperature of water was evaluated from the local static pressure.

Inside wall temperature T_w was evaluated from the measured outside wall temperature, by solving Fourier's equation for heat conduction. Thermal conductivity of the tube material is required in this calculation. It was measured for a sample piece cut off from the same lot as the test tube. The following relation was obtained:

$$k = 16.537 + 2.1253 \times 10^{-2} \, T \quad [W/mK] \tag{5}$$

where temperature T is given in °C.

Heat transfer coefficient is generally defined by the following expression.

$$h = q_w / (T_w - T_b) \tag{6}$$

For the two-phase region ($x > 0$), bulk temperature T_b is evaluated as equal to saturation temperature T_{sat}.

EXPERIMENTAL RESULTS

Figure 5 shows an example of experimental results obtained for various inlet fluid temperatures and heat fluxes, at a given mass flux and a given system pressure. The ordinate h_{TP}/h_{Lo} represents the ratio of two-phase heat transfer coefficient to liquid single-phase coefficient, calculated from Dittus-Boelter's equation, assuming that the total mass flow rate flows as a liquid phase alone. The abscissa represents the thermodynamic equilibrium steam quality x. At a given heat flux, the experimental data for various inlet subcooling conditions smoothly join with each other and fall in a curve with little scattering. For other experimental conditions of pressure and mass flow rate, similar results were obtained.

Except when the heat flux was low, the data covered liquid single-phase, subcooled partial boiling, fully developed nucleate boiling and two-phase forced convective heat transfer regions. When the heat flux was low, the heat transfer coefficient ratio was less than unity in the region where

Experimental conditions were as follows:
System pressure (P) : 5.4, 10.3 x 10⁵ N/m²
Mass velocity   (G) : 300 ~ 1000 kg/m²s
Heat flux   (q) : 2 ~ 50 x 10⁴ W/m²
Steam quality   (x) : -0.18 ~ 0.3

DATA REDUCTION

To obtain local heat transfer coefficient of boiling two-phase flow, the local value of pressure which corresponds to the saturation temperature is required. Since the local static pressure could hardly be measured in the present experiment, it was estimated by the following method:

Pressure drop along tube length is calculated from the conservation of momentum equation.

$$-\left(\frac{dP}{dz}\right)_{TP} = \{\alpha\rho_g + (1-\alpha)\rho_\ell\}g - \left(\frac{dP}{dz}\right)_{fTP}$$
$$+ G^2 \frac{d}{dz}\left\{\frac{x_a^2}{\alpha\rho_g} + \frac{(1-x_a)^2}{(1-\alpha)\rho_\ell}\right\} \qquad (1)$$

where actual quality x_a is calculated from net vapor generation, taking into account the temperature profile of the liquid in a cross section. Even in the subcooled region, the actual quality x_a becomes positive under the condition where nucleate boiling occurs, whereas the thermodynamic quality x has a negative value. x_a is usually higher than x until the bulk fluid slightly exceeds the enthalpy of the saturated liquid. Under the experimental conditions, x_a substantially equaled x for x > 0.002. The detail of the calculation procedure is presented in Sekoguchi et al. (1980). Average void fraction over the cross section was calculated from the equal velocity head model by Smith (1969).

$$\alpha = \left[1 + \frac{\rho_g}{\rho_\ell}\Psi\left(\frac{1}{x_a}-1\right) + \frac{\rho_g}{\rho_\ell}(1-\Psi)\left(\frac{1}{x_a}-1\right)\right.$$
$$\left. \times \left\{\frac{\rho_\ell/\rho_g + \Psi(1/x_a-1)}{1+\Psi(1/x_a-1)}\right\}^{1/2}\right]^{-1} \qquad (2)$$

where Ψ is the ratio of liquid entrained in vapor core to the total liquid flow rate. As Smith recommended, we assumed $\Psi = 0.4$.

Pressure gradient due to friction $(dp/dz)_{fTP}$ was calculated by the correlation of Chisholm and Laird (1958).

$$\Phi_\ell^2 = \frac{(dP/dz)_{fTP}}{(dP/dz)_{fLf}} = 1 + \frac{21}{X_{tta}} + \left(\frac{1}{X_{tta}}\right)^2 \qquad (3)$$

CONVECTIVE BOILING FLOW

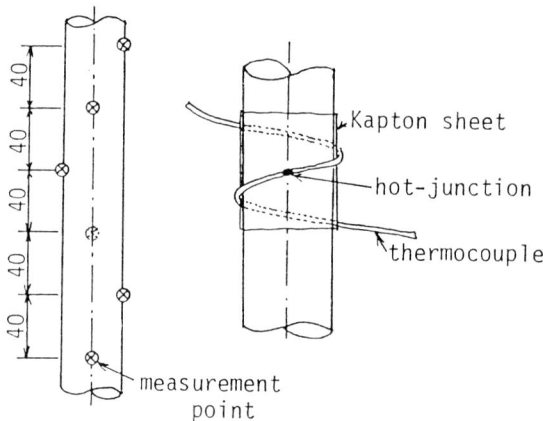

FIGURE 4 Details of thermocouple attachment method

Figure 3(a) shows the uniformity of outside diameters of the test section along the axis. The average value was 10.997 mm. Local electric conductivities, which corresponded to the wall thickness variation, were measured along the tube axis by a pair of electrodes spaced at 19.4 mm. In Fig. 3(b) three measurements are shown separately, and the maximum deviation is estimated as 0.023 mm for the wall thickness. This is smaller than that of the circumferential variation as shown in Fig. 2.

Outside wall temperatures were measured by 0.2 mm O.D. chromel-constantan thermocouples, carefully calibrated with an accuracy of 0.2°C, at 49 cross sections. The hot junctions of the thermocouples were arranged in locations every 90 degrees, rotated around the axis in subsequent cross sections at intervals of 40 mm along the axis, as shown in Fig. 4. This is to determine the average wall temperatures in cross sections as accurately as possible, since the tube wall thickness varied slightly around the circumference. The thermocouple was electrically insulated from the tube wall with a thin Kapton sheet. The test section was wrapped with ribbon heaters over thermal insulators to compensate for heat loss.

Flow rate was measured by a orifice flowmeter (O) and controlled by a throttling valve (F). After the preheaters (H_1, H_2), the water temperature was measured by three sheath thermocouples inserted in the inlet mixing chamber. Static pressures at the inlet and outlet of the test section were measured by a precision piston gauge specially manufactured for the present experiment, accurate to 20 N/m^2. During measurements steam generators ($G_1 \sim G_3$) were used to maintain system pressure, constantly feeding steam to a separator (S_1).

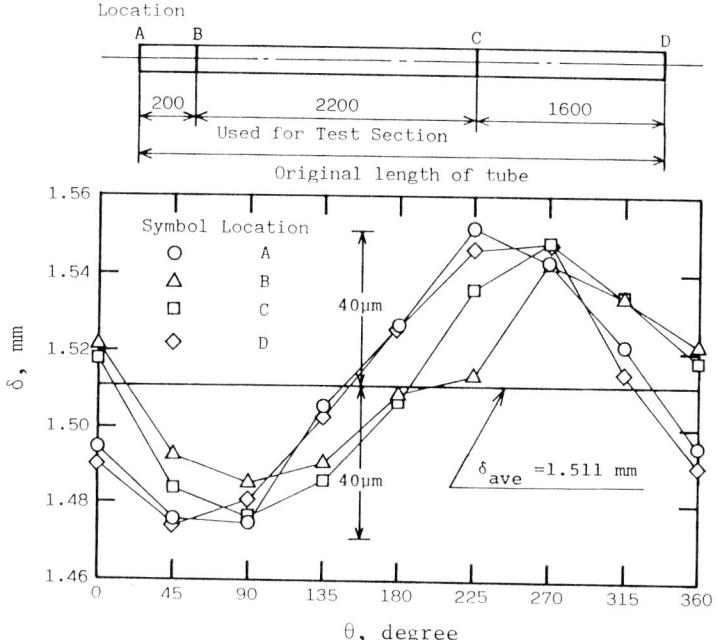

FIGURE 2 Circumferential wall thickness distributions at various cross sections of test tube

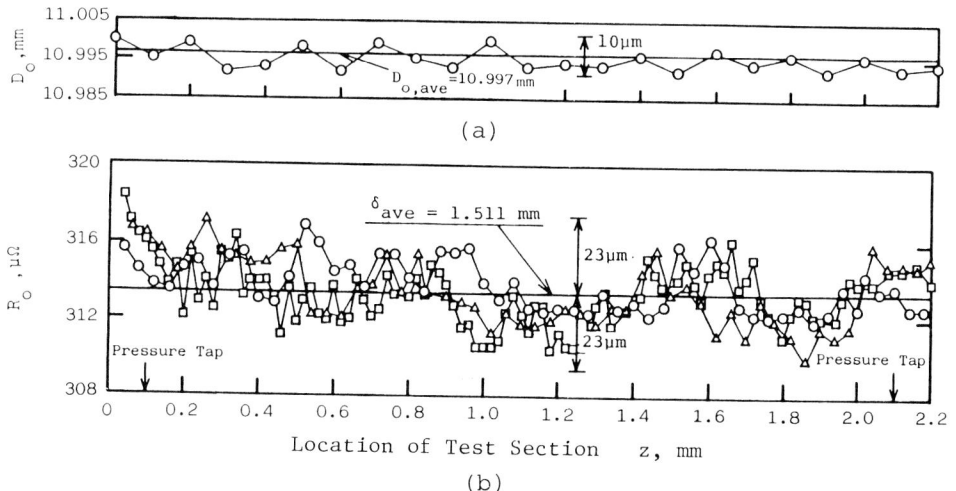

FIGURE 3 Variations of outside diameter and electric resistance along tube axis

CONVECTIVE BOILING FLOW

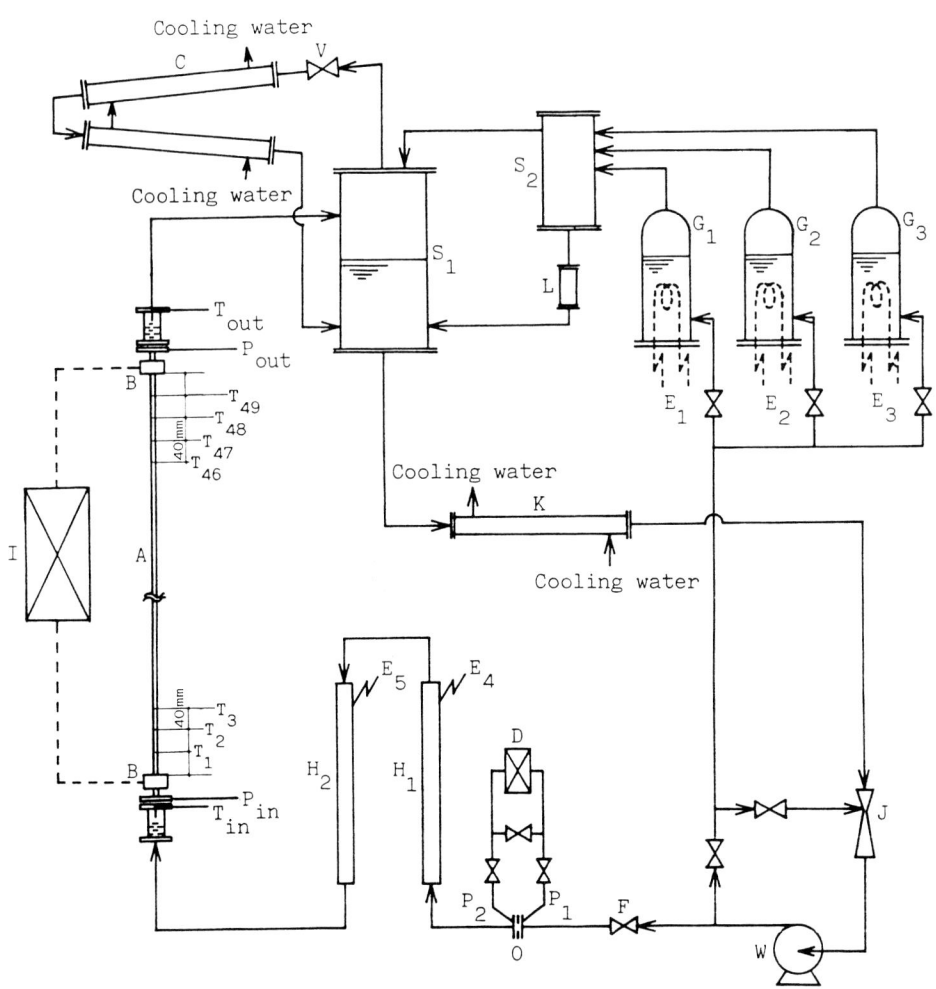

A : Test Section
B : Busbar
C : Condenser
D : Differential Pressure Transducer
E_1-E_5 : Electric Heater
F : Flow Rate Control Valve
G : Auxiliary Steam Generator
H_1,H_2 : Preheater
I : D.C. Electric Power Supplier
J : Ejector
K : Water Cooler
L : Steam Trap
O : Orifice Flow Meter
P : Pressure Tap
S_1,S_2 : Separator
T^2 : Thermocouple
V : System Pressure Control Valve
W : Water Circulation Pump

FIGURE 1 Schematic diagram of experimental apparatus

Figure 2 shows circumferential wall thickness distributions measured by a micrometer at different cross sections of the selected tube. The average wall thickness was 1.511 mm and the maximum deviation was about 0.04 mm. Both end parts were cut off and the middle part was used as the test section.

boiling and the two-phase forced convection (evaporation) regions.

In these investigations, attention is focused mainly on the boiling heat transfer characteristics at high heat fluxes, for the purpose of nuclear reactor safety design. On the other hand, precise prediction of heat transfer characteristics at lower heat fluxes and lower mass flow rates is indispensable for the effective use of thermal energy. It is uncertain whether existing empirical correlations are valid for these conditions. Moreover, the experimental data published so far do not wholly cover the regimes of flow boiling processes under various conditions, and considerable scattering of data is found among them. This results from the difficulties in the precise measurement of boiling heat transfer coefficient, in particular that of the extremely small temperature difference between the heated surface and the fluid.

The purpose of this paper is to investigate the boiling heat transfer coefficient, including lower heat and mass flux conditions for water flowing upward in a tube, measured as accurately as possible by experiment. For this purpose a selected test tube was used, and thermal conductivity of the tube material was measured. Carefully calibrated thermocouples were skillfully attached to the tube wall, and system pressure was measured by a precision piston gauge.

Systematic data from highly subcooled regions to relatively high quality regions, except for the occurrence of dryout, were obtained. When the heat flux was very low, commonly used definitions of boiling heat transfer coefficient gave unreasonable results, and a modifying method was proposed. An empirical correlation of heat transfer coefficient was derived in the high quality region where the boiling was suppressed, and the analogy between heat transfer and pressure drop was discussed.

EXPERIMENTAL APPARATUS AND PROCEDURE

Figure 1 shows a schematic diagram of the experimental apparatus. Purified water was circulated by a canned pump throughout the closed circulation loop. The test section was made of a stainless steel (AISI 304) tube of 7.974 mm I.D. and 1.511 mm thickness, and was heated by direct current. Its heated length was 2 m, and flow direction of the water in the test section was vertical upward.

Since the temperature difference between the heated wall and the fluid becomes extremely small in the boiling heat transfer as mentioned above, it is important to accurately estimate the inside wall temperature and the heat flux. To examine the accuracy of the evaluated heat transfer coefficient, which depends highly upon the uniformity of the tube wall, the wall thickness was inspected preliminarily.

AN ANALOGY BETWEEN HEAT TRANSFER AND PRESSURE DROP IN FORCED CONVECTIVE BOILING FLOW

K. SEKOGUCHI*, H. ZHEN-XING**, M. KAJI*, T. IMASAKA***, and Y. SUMIYOSHI****
*Dept. of Mechanical Eng., Faculty of Eng.
Osaka University, Suita, Osaka, Japan
**Beijing Institute of Aeronautics and Astronautics
Chen Hua Yuan, Beijing, China
***Toto Ltd., Kitakyushu, Fukuoka, Japan
****   Nagasaki Shipyard and Engine Works, Mitsubishi Heavy Industries Ltd., Akunoura, Nagasaki, Japan

ABSTRACT

An experiment on flow boiling heat transfer to water in a round tube was carried out, with emphasis on low heat flux and low mass flow rate conditions. Experimental data were compared with various existing empirical correlations, but satisfactory results were not obtained. A new empirical correlation of two-phase forced convective heat transfer in which nucleate boiling was absent, including the case of low heat flux, was presented. For a two-phase forced convection regime, an analogy between heat transfer and friction pressure drop was discussed. A fairly good agreement between our data and the prediction method by Beattie and Whalley (1982) was obtained.

INTRODUCTION

An increased requirement for the reliable prediction of boiling heat transfer coefficient of channel flow is needed in order to improve thermal performance of industrial equipment such as steam generators in nuclear reactors, conventional boilers, chemical plants and the cooling systems of electronic devices.

A number of investigations into flow boiling heat transfer have been done, and various empirical correlations to predict the heat transfer coefficient were presented. Bergles and Rohsenow (1964) proposed an interpolation method to evaluate heat flux in subcooled partial boiling regions, between single phase forced convection and fully developed nucleate boiling regions. Empirical correlations of heat transfer coefficient for the fully developed nucleate boiling water flow were presented by Jens and Lottes (1951) and Thom et al. (1965). For higher quality, including the region where the nucleate boiling is suppressed, Dengler and Addmos (1956), Wright (1961), Bennett et al. (1961), Schrock and Grossman (1962), Collier et al. (1964) and Pujol and Stenning (1969) presented similar correlations from their experimental data. Based on the data, Chen (1963) proposed a correlating method which covered both the saturated nucleate

FORCED CONVECTIVE AND POST-DRYOUT HEAT TRANSFER

Karasev, E.K., Vasinger, V.V., Mingaleyeva, G.S., Trubkin, E.I., 1977. "Investigation of Water Adiabatic Expansion from the Saturation Line." **Nuclear Energy, 42,** 6, pp. 478-481.

LASL, 1979. "TRAC-P1A. An Advanced Best-Estimate Computer Program for PWR LOCA Analysis." NUREG/CR-0665, LA-7777-MS.

Labuntzov, D.A., Kolchugin, B.A., Golovin, V.S., Zakharova, E.A., and Vladimirova, L.N., 1964. "High Speed Camera Investigation of Bubble Growth for Saturated Water Boiling in a Wide Range of Pressure Variations." **Thermophysics of High Temperature, 2,** 2, pp. 446-453.

McFadden, J.H., et al., 1981. "RETRAN-02. A Program for Transient Thermal-Hydraulic Analysis of Complex Fluid Flow Systems." EPRI Report NP-1850 CCM, V.1.

Richter, H.J., and Minas, S.E., 1981. "Separated Two-Phase Flow Model: Application to Critical Two-Phase Flow." EPRI report NP-1800.

Rivard, W.C., and Travis, J.R., 1980. "A Nonequilibrium Vapor Production Model for Critical Flow." **Nucl. Sci. and eng., 74,** pp. 40-48.

Schlichting, H., 1979. **Boundary Layer Theory,** McGraw Hill, New York.

Scriven, L.E., 1959. "On the Dynamics of Phase Growth." **Chem. Eng. Sci., 1,** pp. 1-13.

Shin, T.S., and Jones, O.C., 1990. "Nucleation and Flashing of Initially Subcooled Liquids in Nozzles: 1. A Distributed Nucleation Model. This volume.

Solbrig, C.W., McFadden, J.H., Lyczkowski, R.W., and Hughes, E.D., 1978. "Heat Transfer and Friction Correlations Required to Describe Steam-Water Behavior in Nuclear Safety Studies." **AIChE Sym. Ser. 74,** No. 174, pp. 100-128.

Soplenkov, K.I., and Blinkov, V.N., 1983. "Heterogeneous Nucleation in the Flow of Superheated Liquid." In **Multi-Phase Systems Transient flows with Physical and Chemical Transformation,** R.I. Nigmatulin and A.I. Ivandayev, Eds., Moscow State University, pp. 105-109.

Sozzi, G.L., and Sutherland, W.A., 1975. "Critical Flow of Saturated and Subcooled Water at High Pressure." General Electric Report NEDO-13418

Wu, B.J.C., Abuaf, N., and Saha, P., 1981. "A Study of Nonequilibrium Flashing of Water in a Converging-Diverging Nozzle. Vol. 2 - Modeling." NUREG/CR-1864, BNL-NUREG-51317.

REFERENCES

Abuaf, N., Jones, O.C., Jr., and Wu, B.J.C., 1980. "Critical Flashing flow in Nozzles with Subcooled Inlet Conditions." In **Polyphase Flow and Transport Technology**, R.A. Bajura, Ed., ASME, August.

Abuaf, N., Jones, O.C., Jr., and Wu, B.J.C., 1983. "Critical Flashing Flow in Nozzles with Subcooled Inlet Conditions." **Trans. ASME, J. Heat Trans., 105**, pp. 379-385.

Abuaf, N., Zimmer, G.A., and Wu, B.J.C., 1981. "A Study of Nonequilibrium Flashing of Water in a Converging-Diverging Nozzle. Vol. 1 - Experimental." NUREG/CR-1864, BNL-NUREG-51317.

Ardron, H.K., 1978. "A Two-Fluid Model for Critical Vapor-Liquid Flow." **Int. J. Multiphase flow, 4**, 3, pp. 323-337.

Beattie, D.R.H., 1973. "A Note on the Calculation of Two-Phase Pressure Losses." **Nucl. Eng. Des., 25**, pp. 395-402.

Chow, H., and Ransom, V.H., 1984. "A Simple Interphase Drag Model for Numerical Two-Fluid Modeling of Two-Phase Flow Systems." **Second Proc. Nucl. Thermal Hydraulics.** Summer Annual ANS Meeting, pp. 137-145

Dobran, F., 1985. "A Nonequilibrium Model for the Analysis of Two-Phase Critical Flows in Tubes." Presented at the 23rd National Heat Transfer Conference, Denver, Colorado, August 4-7.

Elias, E., and Chambre, P.L., 1984. "A Mechanistic Non-Equilibrium Model for Two-Phase Critical Flow." **Int. J. Multiphase Flow, 10**, 1, pp. 21-40

EPRI, 1982. "The Marviken Full Scale Critical Flow Tests. Vol. 1. Summary Report." EPRI report NP-2370.

EPRI, 1983. "Review and Application of the TRAC-PD2 Computer Code." Epri Report NP-2826.

Ishii, M., and Mishima, K., 1983. "Flow Regime Transition Criteria Consistent with Two-Fluid Model for Vertical Two-Phase Flow." NUREG/CR-3338, ANL-83-42.

Jones, O.C., Jr., 1984. "Thermal Design Concepts for the Rotating Fluidized Bed Reactor." **Nucl. Sci. Eng., 87**, pp. 13-27.

Jones, O.C., Jr., and Zuber, N., 1978. "Bubble Growth in Variable Pressure Fields." **Trans. ASME, J. Heat Trans., 100**, 3, pp. 453-459.

NOMENCLATURE

English

A_i	Interfacial area density
C	Specific heat
D_{hy}	Hydraulic diameter
f_f	Friction factor
Gi	Gibbs number
h	Heat transfer coefficient
Δi_{fg}	Latent heat of vaporization
$\dot{I}$	Nucleation rate
Ja	Jacob number
k	Thermal conductivity
L	Length
$\dot{m}$	Mass flow rate
N	Bubble number density
p	Pressure
P	Perimeter
$\dot{q}$	Heat flux
r_B	Bubble radius
S	Cross section area
t	Time
T	Temperature
ΔT	Temperature difference
u	Specific internal energy
w	Velocity
We	Weber number
x	Axial coordinate

Greek

α	Void fraction
Γ	Volumetric vapor generation rate
δ	Bubble diameter
μ	Dynamic viscosity
ν	Kinematic viscosity
ρ	Density
σ	Surface tension
χ	Quality

Subscripts

B	Bubble or bulk
d	Droplet
f	Saturated liquid
g	Saturated vapor
ℓ	Liquid (not at saturation)
m	Mixture
o	Stagnation
sat	Saturation
S	Re Taylor bubble
v	Vapor (not at saturation)
w	Wall
∞	Terminal

NUCLEATION AND FLASHING

In general, both qualitative and quantitative agreement was found between data and computations. For small geometry, bulk nucleation has negligible effect all data examined. For large geometry, bulk nucleation predominates.

Pressure profiles were best predicted by bubbly flow methods for all void fractions and the void profile were predicted best for a bubbly-churn-dispersed model. Some work obviously still needs to be accomplished to being the calculations into allignment with the magnitude and trends for all variables. The deficiency is thought to be in the computation of interfacial area density.

Void fractions at the throat in nozzles were found to be negligible in all cases, confirming previous hypotheses of Abuaf, Jones, and Wu [1980, 1983]. Both qualitative and quantitative agreement was found for downstream void development in comparison with existing data. It was shown that careful attention must be paid to modeling of interfacial area density.

Specific conclusions include:

1. Bubble nucleation density and resultant number density at the throat in nozzles with subcooled inlet may vary by many orders of magnitude;

2. Throat void fractions are negligible in all nozzle cases considered;

3. Void development downstream of the throat is dependent on the size and number density of nuclei at flashing inception, these values being provided accurately by the new wall nucleation model for small geometries. Assumptions of constant values for initial bubble size and/or number density are incorrect;

4. Bulk nucleation has negligible effect for small geometries;

5. Bulk nucleation becomes important for large geometries as the volume-to-surface ratio increases but insufficient data exist to determine the transition geometry.

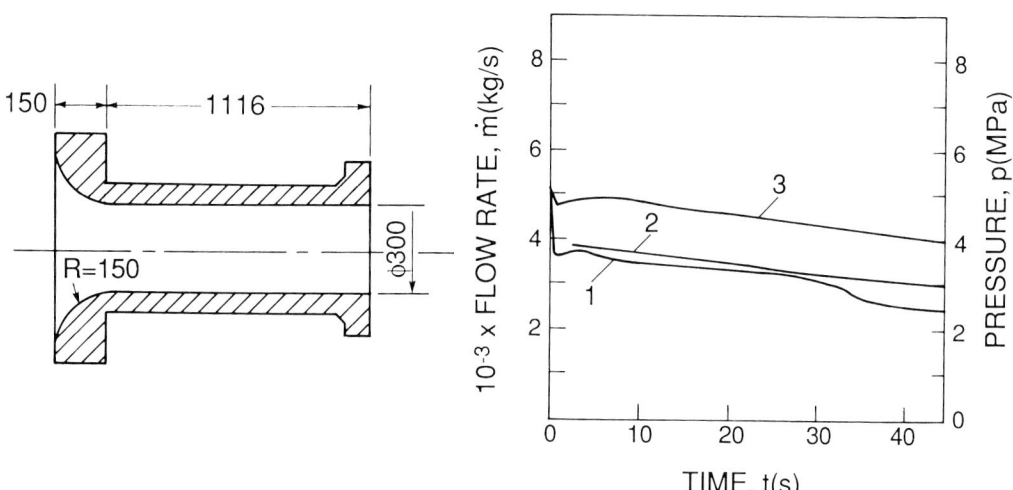

Figure 12. Blowdown of the Marviken experiment [EPRI, 1982] for a nozzle with L=1.266 m, D=0.3 m.
1 - $\dot{m}$ from experiment
2 - $\dot{m}$ from calculation
3 - experimental pressure history.

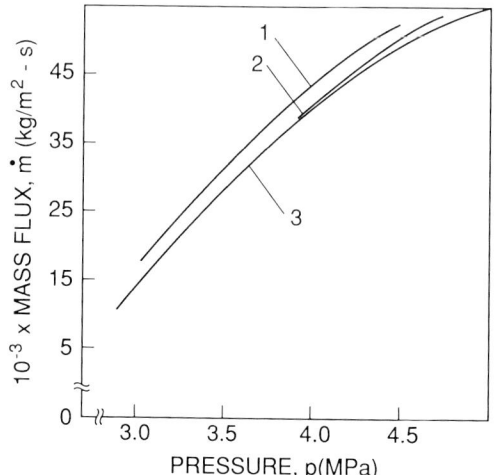

Figure 13. Critical flow behavior with stagnation pressure for different size Marviken nozzles [EPRI, 1982]

1 - Model presented, D=0.5 m, L=0.955 m
2 - Model presented, D=0.3 m, L=1.266 m
3 - Homogeneous equilibrium

NUCLEATION AND FLASHING

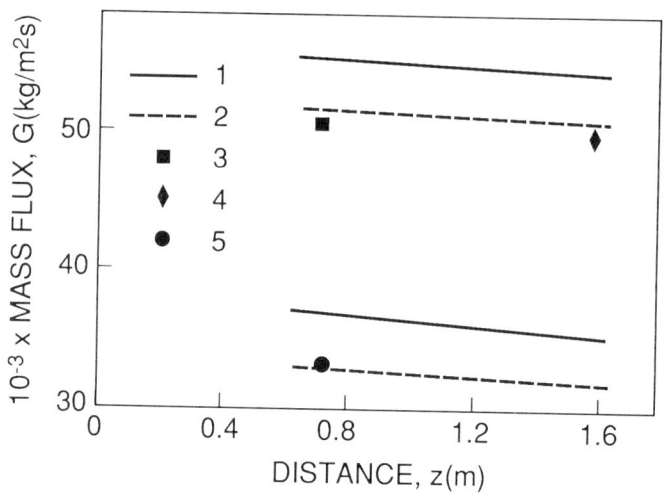

Figure 10. Comparison between calculated and experimental critical mass fluxes for the Marviken Nozzle [EPRI, 1982] large tube (D=0.509 m) having a round entrance.
1 - calculation with wall nucleation only;
2 - calculation with both wall and bulk nucleation;
3 - P_o=4.52 MPa, T_o=507.15K
4 - P_o=4.52 MPa, T_o=507.15K
5 - P_o=3.56 MPa, T_o=507.15K

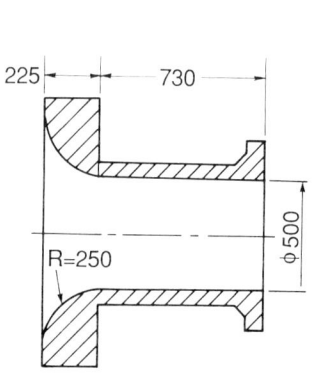 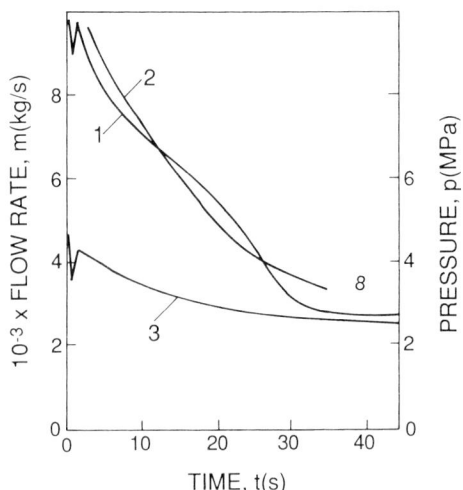

Figure 11. Blowdown of the Marviken experiment [EPRI, 1982] for a nozzle with L=0.955 m, D=0.5 m.
1 - $\dot{m}$ from experiment
2 - $\dot{m}$ from calculation
3 - experimental pressure history.

Figure 10 shows a comparison between calculated and experimental critical mass flow rates of the Marviken experiments when the water temperature is nearly constant. It is obvious from this case that the inclusion of bulk nucleation for this larger geometry substantially improves the computation of void development with resultant prediction of the critical flows. In this case, accurate calculation is crucial since the flow rate is void dominated.

The transient flow calculations for Marviken agree well with the experimental values (Figs. 11 and 12) up to about 30 ms after which the liquid temperature begins to become affected by the nonuniformity in the vessel temperature. Figure 13 shows that as the L/D ratio of the nozzle increases, the flow and the model approach that calculated by homogeneous equilibrium.

CONCLUSIONS

A computational framework for calculating the behavior of flowing, initially subcooled liquids in pipes and nozzles has been described for use on a microcomputer. The model uses the new distributed nucleation model of Shin and Jones [1990] coupled with a previous model for bulk nucleation developed by Sopolenkov and Blinkov [1983] to determine appropriate initial conditions for flashing and void development downstream of the throat. The model, a five-equation, mechanical equilibrium, thermal nonequilibrium model incorporates recent advances in the theory of distributed wall nucleation in small ducts, as well as a previously developed model for bulk nucleation on suspended particles in larger geometries.

The model consisted of mixture and vapor mass conservation equations, a mixture momentum equation, the liquid energy equation, and a bubble transport equation. Spherical bubble growth was calculated by traditional thermally-limited growth methods using local superheat.

For closure, a relatively simple wall friction model was utilized represented by a friction multiplier having different formulations in four different regions of void fraction.

Semi-implicit methods were used for differencing with all properties computed at cell centers and a donor cell method used to calculate convective flux effects with velocities computed at cell boundaries. The system was solved by Newton iteration. Typical computational times on a Hewlett-Packard 9816 microcomputer baed on the MC-68000 processor chip at 8 MHz were 40-45 seconds per time step and a total of 8-10 hours for a converged solution with all variables within 1 part in 10^3.

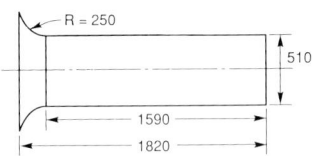

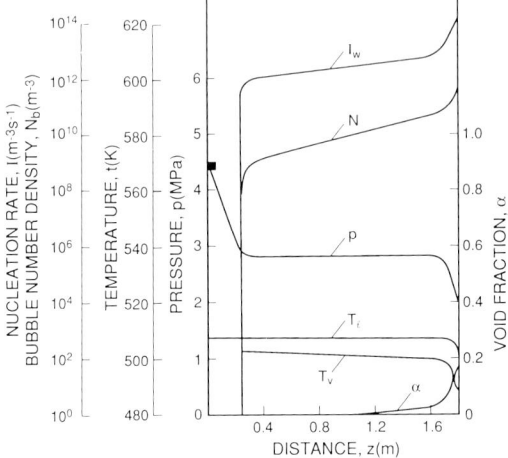

(a) Wall nucleation only

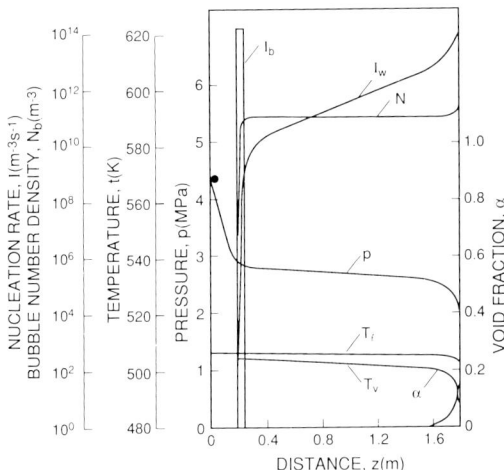

(b) Combination of wall and bulk nucleation

Figure 9. Calculated distributions of flashing flow parameters for the Marviken nozzle [EPRI, 1982]. P_o=4.52 MPa; T_o=507.15K; ΔT_o=23.85K; G_{exp}=51,000 kg/m²·s; G_{calc}=55,280 kg/m²·s.

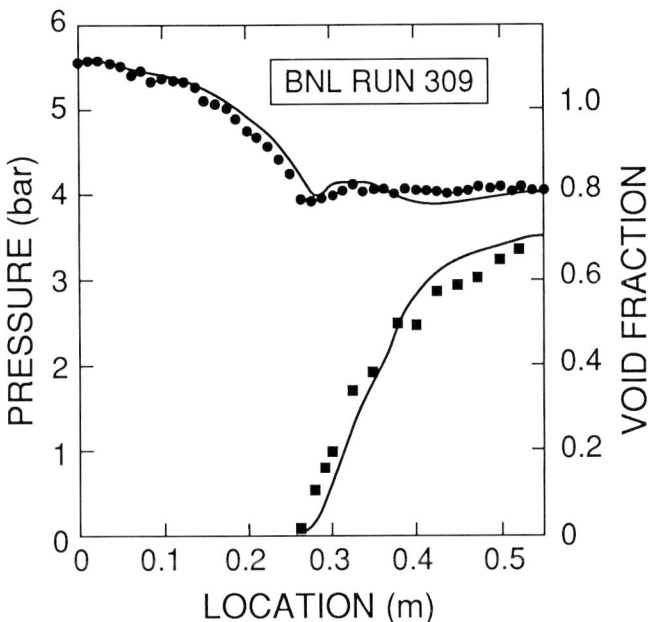

Figure 7. Comparison between experimental and calculated distributions of pressure and void fraction for the BNL nozzle run 309 [Abuaf et al., 1981]. P_{in}=5.559 bar; T_{in}=422.25K; ΔT_0=6.75K; $\dot{m}_{exp}$=8.8 kg/s; $\dot{m}_{calc}^{in}$=8.23 kg/s; p_∞=4.05 bar.

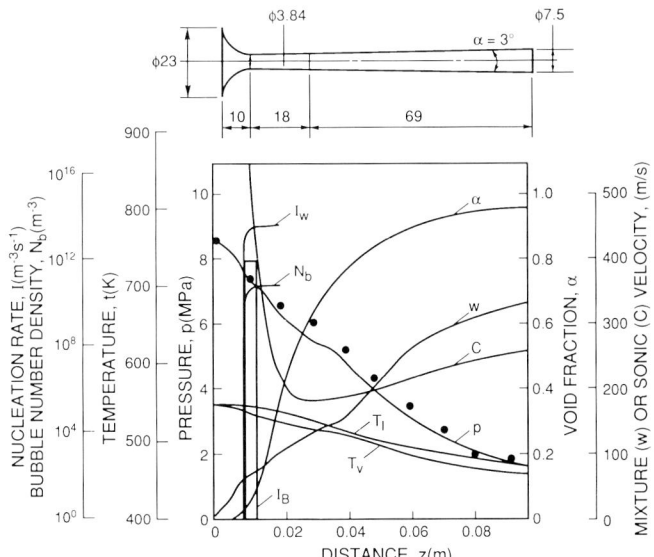

Figure 8. Calculated distributions of flashing flow parameters along the small scale Laval nozzle of Karasev et al. [1977. P_0=8.5 MPa; T_0=572K; ΔT_0=0K; $\dot{m}_{exp}$=0.535 kg/s; $\dot{m}_{calc}$=0.575 kg/s

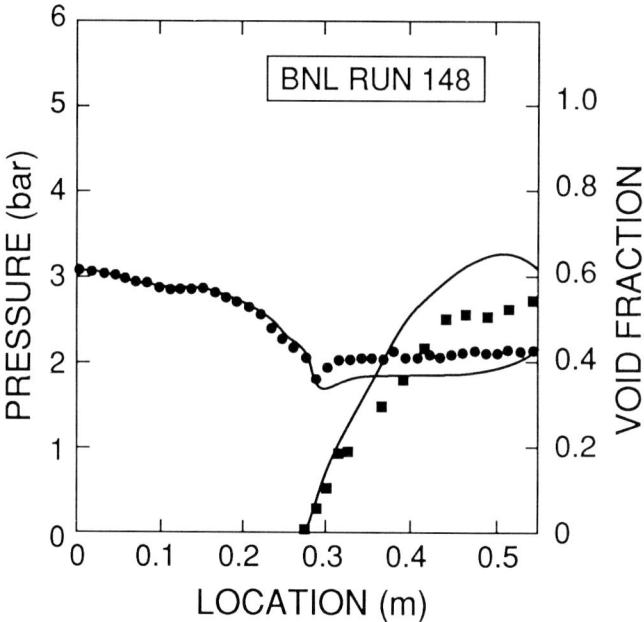

Figure 5. Comparison between experimental and calculated distributions of pressure and void fraction for the BNL nozzle run 148 [Abuaf et al., 1981]. P_{in}=3.05 bar; T_{in}=394.35K; ΔT_o=12.80K; $\dot{m}_{exp}$=7.5 kg/s; $\dot{m}_{calc}^{in}$=7.8 kg/s; p_∞=2.06 kPa.

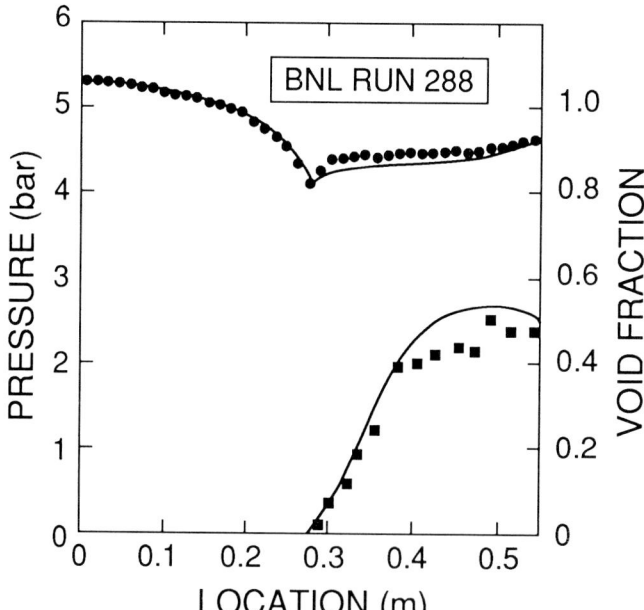

Figure 6. Comparison between experimental and calculated distributions of pressure and void fraction for the BNL nozzle run 288 [Abuaf et al., 1981]. P_{in}=5.3 bar; T_{in}=422.35K; ΔT_o=4.8K; $\dot{m}_{exp}$=7.25 kg/s; $\dot{m}_{calc}^{in}$=7.27 kg/s; p_∞=4.591 bar.

While the void fraction in the latter case is closer to the data, the pressure profiles are further away. The reasons for these results are unknown but show both the need for careful modeling of surface area density and for more definitive experiments to delineate the type of behavior to be expected.

Figures 5-7 show similar results for different BNL runs. In these cases, only the bubbly flow regime results are shown. Table 1 compares the exit void fraction for the bubbly flow calculation compared with that for the flow regime 1 results in comparison with the data. The void fractions calculated are: α_1, bubbly flow only; α_2, flow regime 1. It is seen that in all cases, the inclusion of a more realistic flow regime calculation which considers the reduction in interfacial area density due to agglomeration produces results closer to the data.

TABLE 1

RUN	EXIT α_1	EXIT α_2	Exit α_{data}
148	0.66	0.59	0.55
273	0.69	0.605	0.57
288	0.52	0.49	0.49
309	--	0.705	0.70

Citical flow of saturated water through a Laval nozzle having a throat diameter of 3.84 mm is shown in Fig. 8 for the data of Karasev et al. [1977]. In this case, wall nucleation dominates by two orders of magnitude the bulk nucleation, again affirming that this is the predominate mode of void formation in the nucleation zone in small geometies. It is seen that the flow becomes overexpanded and goes supersonic downstream of the throat. Substantial expansive cooling of both liquid and vapor are seen.

Void development downstream - large nozzles. Figure 9 shows calculations for a round-entrance short pipe used in Marvekin experiments [EPRI, 1982]. The single point at the inlet represents the measured inlet pressure used to drive the flow calculations. In Fig. 9a, results are shown with $I_B = 0$, no bulk nucleation. Figure 9b shows calculations where the bulk nucleation model is included. The number density is substantially larger in the early stages of decompression near the inlet when bulk nucleation is included and the voids begin to grow earlier in the nozzle but little effect on the temperatures or pressures is seen. However, near the exit where void development becomes significant, the number densities approach each other and void growth becomes similar. This shows that in the larger geometries, as the volume to surface ratio increases, the role of bulk nucleation can be expected to become increasingly important.

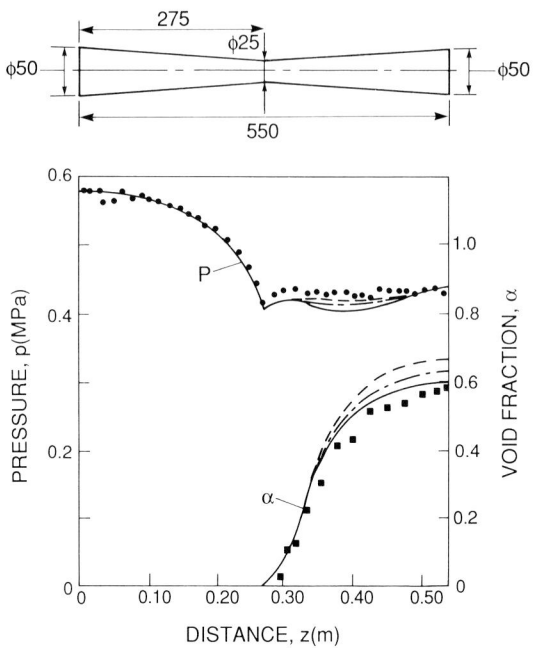

(a) Pressure and Void Distributions

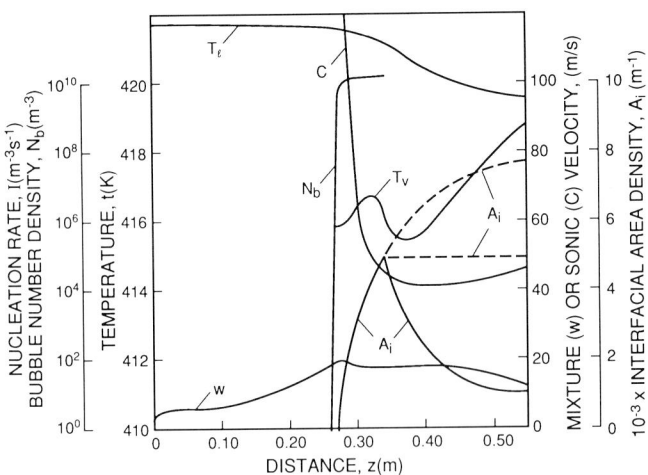

(b) Void Development Parameters

Figure 4. Comparison between experimental and calculated distributions of pressure and void fraction and other void development parameters for the BNL nozzle run 273 [Abuaf et al., 1981]. $P_0=0.573$ MPa; $T_0=421.85$K; $\Delta T_0=8.4$K; $\dot{m}_{exp}=8.71$ kg/s; $\dot{m}_{calc}=8.8$ kg/s; $p_\infty=0.442$ MPa. Solid lines, flow regime map 1; dot-dash lines, flow regime map 2; dash lines, bubbly flow over entire range.

computer bulk equivalent wall nucleation rate at the throat of slightly less than 5×10^{22} $m^{-3} \cdot s^{-1}$, 14 orders of magnitude larger than that of Ardron and Ackerman. In this case, the throat void fraction increased to 0.009. This value is also negligibly small and provides confirmation of the original hypothesis of Abuaf, Jones, and Wu [1980, 1983]. This further explains why the critical flow rates of all these runs can be accurately calculated (within ~3%) by correctly predicting the throat pressure through the superheat and then assuming single phase flow.

<u>Void development downstream - small nozzles</u>. Calculations for the vertical nozzle in the experiments of Abuaf et al. [1981] taken at Brookhaven National Laboratory (BNL) are shown in Fig. 4. In Fig. 4a, comparisons are shown between calculated void fraction and pressure profiles and those measured in the experiment. In Fig. 4b, curves are included showing the liquid temperature, T_ℓ, vapor temperature, T_v, interfacial area density, A_i, mixture velocity, w, bubble number density in the region up to 30% voids, log(N), and local frozen sonic velocity.

In all these calculations, it was confirmed that the calculated effect of bulk nucleation was negligible. This lends additional support to the experimental evidence that for nozzles of this small diameter wall nucleation predominates. Note also that the correlations which are used to calculate the wall nucleation rate were based on this assumption.

Three methods of calculation were used to make these calculations shown in Fig. 4:

(a)   Flow regime 1 consisting of bubbly, bubbly-slug, transitional, and dispersed flows;

(b)   Flow regime 2 consisting of bubbly, transitional churn turbulent, and dispersed flows;

(c)   Bubbly flow for any void fraction.

One can see that the three calculated values of void fraction in Fig. 4a differ slightly due to the differing interfacial area density for phase change. Bubbly flow gives the highest void calculated since the interfacial area density is the largest. Model flow regime 1 gives a smaller void fraction since the area density is lower and decreases after $\alpha = 0.3$ (Fig 4b). For the third model, flow regime 2, the interfacial area density is constant after $\alpha = 0.3$, but the void fraction is still lower. This is because as seen in the pressure profiles, this third model results in increased superheat offsetting the effect of reduced interfacial area density.

NUCLEATION AND FLASHING 653

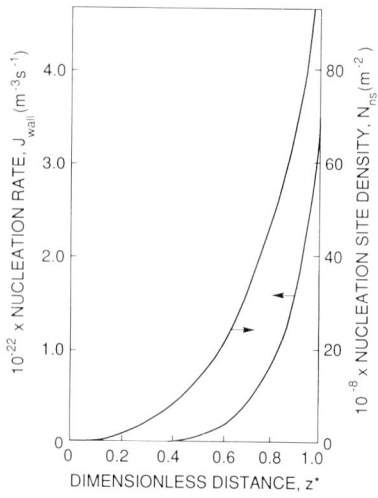

(a) Wall Nucleation Rate and Nucleation Site Density

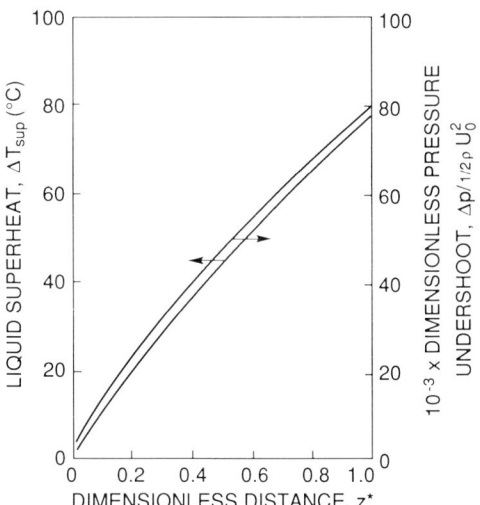

(b) Liquid Superheat and Pressure Undershoot.

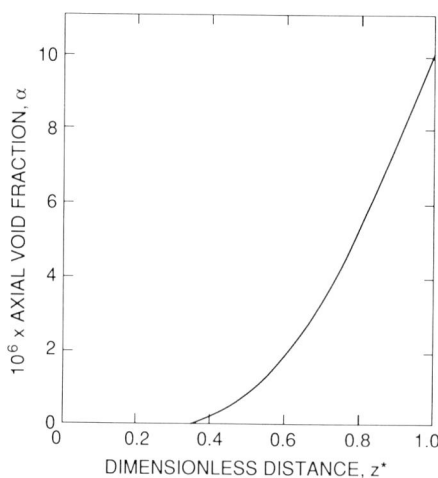

(c) Void Fraction in the Nucleation Zone.

Figure 3. Calculations for the nucleation zone for the conditions of Brown [1961], Run 39. T_{in}=280C p_{in}=1.59 bar; G=303 kg/m^2; ΔT_{sup}=81.6K

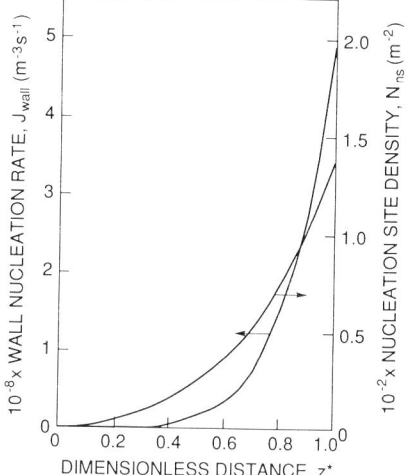

(a) Wall Nucleation Rate and Nucleation Site Density

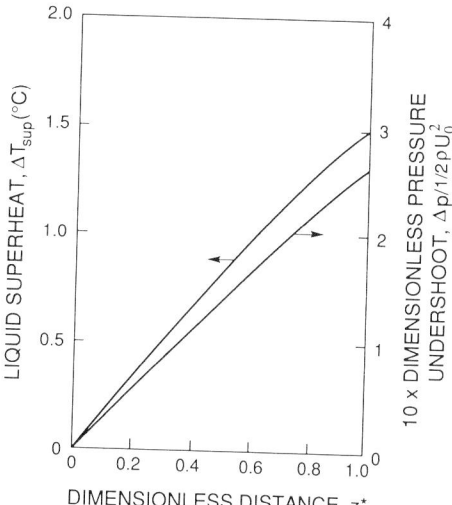

(b) Liquid Superheat and Pressure Undershoot.

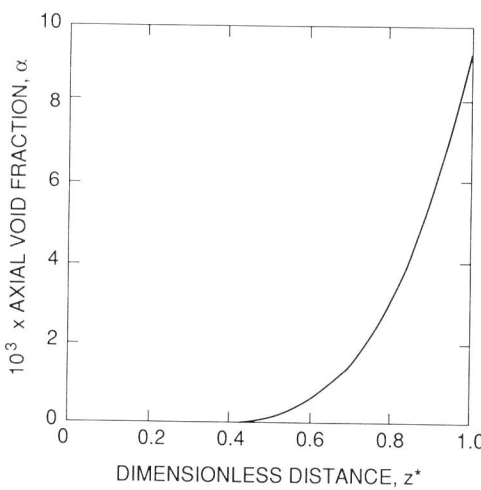

(c) Void Fraction in the Nucleation Zone.

Figure 2. Calculations for the nucleation zone for the conditions of Ardron and Ackerman [1978], Run C35. T_{in} = 111.5C, p_{in}=1.59 bar; G=7740 kg/m^2; ΔT_{sup}=1.66K

NUCLEATION AND FLASHING

RESULTS AND DISCUSSION

Numerical Results

The experience with the computations were reasonably good from the viewpoint of using a microcomputer for the calculations. The entire program required approximately 80 KB of main memory. Cell lengths for typical calculations were 5 mm for the BNL geometry [Abuaf et al., 1981] representing 110 cells, and ~2.5 mm for the Marvekin geometry [EPRI, 1982] (40-70 cells) and for the geometry of Sozzi and Sutherland [1975] (64 cells, results not shown).

Computational times required to reach a steady state solution usually took 200-300 time steps with each time step requiring ~4 internal iterations in the early stages of solution and 1-2 iterations near steady state. A single iteration required approximately 40 seconds so that calculational times ran about 8-10 hours. A typical case was set up during the day, and run overnight with the results ready the next morning. Since the computer was completely dedicated to the problem, there was no difficulty in undertaking computations in this manner. Moreover, it would be expected that newer and faster microcomputers would require less time to complete the same calculations.

Comparison with Experimental Results

Void development at the throat. In virtually all previous flashing models, the nucleation zone is treated as a single point of flashing inception. This has been justified since the zone of supersaturation in many cases is quite narrow. However, in this zone, the voids which develop from the nuclei form the basis for interfacial mass transfer and subsequent growth downstream. It is, therefore, important that both the size and number be determined so that accurate calculations of void development may be undertaken.

Calculations were made for all runs reported by Bailey [1951], Brown [1961], Sozzi and Sutherland [1975], Ardron and Ackerman [1978], Abuaf et al. [1981], and Celata [1982]. The two cases with the smallest and largest calculated throat void fractions are shown in Fig. 2. The smallest void fraction was computed for the Ardron and Ackerman run C25 having a superheat at the throat of 1.66K with 1.6 bar inlet pressure. As seen in Fig. 2a, while the nucleation site density increased to approximately 140 m^{-2} with overall bulk equivalent wall nucleation rates to about 5×10^8 $m^{-3} \cdot s^{-1}$, the void fraction increased only to $\sim 10^{-5}$ (Fig 2c).

The case with the largest computed throat void fraction was Run 39 from the data of Brown [1961] (Fig 3), having a throat superheat of 81.6K, an inlet pressure of 68.4 bar, and a

The first of the four equations involves the corrections Δp_{i-1}^{ν}, Δp_i^{ν}, Δp_{i+1}^{ν} only where

(33) $\quad K_i \Delta p_{i-1}^{\nu} + \Delta p_i^{\nu} + L_i \Delta p_{i+1}^{\nu} = M_i$

The system of equations in (33) written in each mesh cell from 1 to N is tridiagonal

(34) $\begin{pmatrix} 1 & L_1 & & & & & \\ K_2 & 1 & L_2 & & & & \\ & K_3 & 1 & L_3 & & & \\ & & \cdot & \cdot & \cdot & & \\ & & & K_{N-2} & 1 & L_{N-2} & \\ & & & & K_{N-1} & 1 & L_{N-1} \\ & & & & & K_N & (1+L_N) \end{pmatrix} \cdot \begin{pmatrix} \Delta p_1^{\nu} \\ \Delta p_2^{\nu} \\ \Delta p_3^{\nu} \\ \cdot \\ \Delta p_{N-2}^{\nu} \\ \Delta p_{N-1}^{\nu} \\ \Delta p_N^{\nu} \end{pmatrix} = \begin{pmatrix} M_1 \\ M_2 \\ M_3 \\ \cdot \\ M_{N-2} \\ M_{N-1} \\ M_N \end{pmatrix}$

The solution of Eq. (34) gives the corrections Δp_i^{ν}. Knowing these corrections, we get $\Delta \alpha_i^{\nu}$, $\Delta T_{\ell,i}^{\nu}$, ΔN_i^{ν} from Eq. (32). These corrections will then be added to p_i^{ν}, $T_{\ell i}^{\nu}$, α_i^{ν}, and N_i^{ν}, and the iterative process repeated until the condition $\text{Max}(\Delta p_i^{\nu}/p_i) < \varepsilon$ is satisfied, where ε is a sufficiently small number. The present case used $\varepsilon = 10^{-3}$ requiring 3-4 iterations per time step.

<u>Boundary Conditions</u>

Boundary conditions are achieved by adding additional cells with the subscripts "o" and "N+1." For subsonic flow through the inlet boundary the relations $K_1 \Delta p_o^{\nu} + \Delta p_1^{\nu} + L_1 \Delta p_2^{\nu} = M_1$, p_o = Const, and $\Delta p_o^{\nu} = 0$. The velocity w_o was determined by linear extrapolation upstream. For subsonic flow through the outlet boundary,

(35) $\quad \begin{cases} K_N \Delta p_{N-1}^{\nu} + \Delta p_N^{\nu} + L_N \Delta p_{N+1}^{\nu} = M_N \\ P_{n+1} = p_{\infty} = \text{Const} \quad \text{and} \quad \Delta p_{N+1}^{\nu} = 0 \end{cases}$

The values of $T_{\ell N+1}$, α_{N+1}, and N_{N+1} were determined by linear extrapolation downstream.

If the flow through the outlet boundary was supersonic, it was assumed that $\Delta p_{N+1}^{\nu} = \Delta p_N$, and linealy extrapolated for all other variables downstream.

NUCLEATION AND FLASHING

for all unknown variables at the new time step. The system is solved using Newton iteration shown in the following equation. The ν-superscript indicates successive iterative approximations to variables at the new time step with superscript "n+1." All equations are linearized by Taylor's expansion about the latest iteration values of the unknowns. The linearized form of Eq. (22) has the form:

$$
(30) \quad \begin{cases} \Delta p_i^\nu \left\{ \left(\alpha \frac{d\rho_v}{dp} \right)_i^\nu + \frac{1}{S} \frac{\Delta t}{\Delta x} \left(\alpha_{i+1/2}^n \rho_{v_{i+1/2}}^n S_{i+1/2} \beta_{i+1/2}^n + \cdots \right. \right. \\ \left. \cdots + \alpha_{i-1/2}^n \rho_{v_{i-1/2}}^n S_{i-1/2} \beta_{i-1/2}^n \right) - \left(\frac{\partial \Gamma_{\ell-v}}{\partial p} \right)_i^\nu \Delta t \right\} + \cdots \\ \cdots + \Delta \alpha^\nu \left[\rho_{v_i} - \left(\frac{\partial \Gamma_{\ell-v}}{\partial \alpha} \right)_i^\nu \Delta t \right] - \left[(1-\alpha)_i^\nu \Delta T_{\ell_i}^\nu \right] \cdots \\ \cdots \left(\frac{\partial \Gamma_{\ell-v}}{\partial T_\ell} \right)_i^\nu \frac{\Delta t}{(1-\alpha)_i^\nu} - (\Delta p_{i+1}^\nu \frac{1}{S_i} \frac{\Delta t}{\Delta x} \alpha_{i+1/2}^n \cdots \\ \cdots \rho_{v_{i+1/2}}^n S_{i+1/2} \beta_{i+1/2}^n) - (\Delta p_{i-1}^\nu \frac{1}{S_i} \frac{\Delta t}{\Delta x} \cdots \\ \cdots \alpha_{i-1/2}^n \rho_{v_{i-1/2}}^n S_{i-1/2} \beta_{i-1/2}^n) = \left\{ (\alpha^\nu \rho_v^\nu - \cdots \right. \\ \cdots - \alpha^n \rho_v^n) + \frac{1}{S_i} \frac{\Delta t}{\Delta x} \left[(\alpha^n \rho_v^n w^\nu S)_{i+1/2} - \cdots \right. \\ \left. \cdots - (\alpha^n \rho_v^n w^\nu S)_{i-1/2} \right] - \Gamma_{\ell-v,i}^\nu \Delta t \right\} \end{cases}
$$

In matrix form, the equation set may be written as

$$
(31) \quad \begin{pmatrix} B_1 \\ B_2 \\ B_3 \\ B_4 \end{pmatrix} \Delta p_{i-1}^\nu + A \begin{pmatrix} \Delta p_i^\nu \\ (1-\alpha)_i^\nu \Delta T_{\ell_i}^\nu \\ \Delta \alpha_i^\nu \\ N_i^\nu \end{pmatrix} - \begin{pmatrix} C_1 \\ C_2 \\ C_3 \\ C_4 \end{pmatrix} \Delta p_{i+1}^\nu = \begin{pmatrix} D_1 \\ D_2 \\ D_3 \\ D_4 \end{pmatrix}
$$

Where A is a 4×4 matrix. If Eq. (31) is multiplied by the inverse of A the result is

$$
(32) \quad A^{-1} \begin{pmatrix} B_1 \\ B_2 \\ B_3 \\ B_4 \end{pmatrix} \Delta p_{i-1}^\nu + \begin{pmatrix} \Delta p_i^\nu \\ (1-\alpha)_i^\nu \Delta T_{\ell_i}^\nu \\ \Delta \alpha_i^\nu \\ \Delta N_i^\nu \end{pmatrix} - A^{-1} \begin{pmatrix} C_1 \\ C_2 \\ C_3 \\ C_4 \end{pmatrix} \Delta p_{i+1}^\nu = A^{-1} \begin{pmatrix} D_1 \\ D_2 \\ D_3 \\ D_4 \end{pmatrix}
$$

The channel is divided into a number of cells with uniform spatial mesh spacing, Δx. All thermodynamic variables are cell-centered whereas velocities are computed at cell faces.

As an illustration, the finite-difference form developed for the vapor mass conservation equation is

$$(24) \quad (\alpha^{n+1}\rho_v^{n+1} - \alpha^n\rho_v^n)_i + \frac{1}{S_i}\frac{\Delta t}{\Delta x}\left[(\alpha^n\rho_v^n w^{n+1} S)_{i+1/2} \cdots\right.$$
$$\left.\cdots - (\alpha^n\rho_v^n w^{n+1} S)_{i-1/2}\right] = \Gamma_{\ell-v}^{n+1}\Delta t$$

Relationships between cell edge and cell center variables used are based on full donor cell differencing which is known to be extremely stable resulting in

$$(25) \quad \begin{cases} M_{i+1/2} = \begin{cases} M_i & \text{if } w_{i+1/2} \geq 0 \\ M_{i+1} & \text{if } W_{i+1/2} < 0 \end{cases} \\ M = \rho_v, \rho_\ell, u_\ell, \alpha, N \end{cases}$$

Other equations are treated identically. The equation of motion has the form

$$(26) \quad w_{i+1/2}^{n+1} = -\beta_{i+1/2}^n(p_{i+1}^{n+1} - p_i^{n+1}) + j_{i+1/2}^n$$

where $\beta_{i+1/2}$ and $j_{i+1/2}$ are calculated at the old time step and have the form

$$(27) \quad \beta_{i+1/2} = \frac{\Delta t}{\Delta x}\left[\rho_m^n + K^n \Delta t |w^n|\right]_{i+1/2}^{-1}$$

and

$$(28) \quad j_{i+1/2} = w_{i+1/2}^n \frac{\left[1 - \frac{\Delta t}{\Delta x}(w_{i+3/2}^n - w_{i-1/2}^n)\right]}{\left[1 + \frac{K^n \Delta t}{\rho^n}|w^n|\right]}$$

where

$$(29) \quad K^n = f_f(w^n|w^n|)^{-1}$$

Using (26) the number of variables is reduced by one where equations with the velocity at n+1/2 and i+1/2 appears.

The overall system of equations (1) written in a partially implicit form represents a nonlinear algebraic equation set

et al. [1987] The assumption was that the liquid always carried suspended particles whose size distribution is n(d). At nucleation sites, only supercritical particles (d > d*) can be active where d* is the diameter of the critical spherical vapor nucleus and depends on the physical properties of the liquid and degree of metastability. Thus the total number of nucleation sites where evaporation and bubble formation can occur is

$$(20) \quad N_B(G_i) = \int_{\infty}^{d^*} n(d)dd$$

where Gi is the Gibbs Number [Skripov, 1982]. An empirical correlation for $N_B = N_B(Gi)$ was obtained using experimental data on blowdown of initially subcooled water through short tubes ($4 \leq L/D \leq 10$, $L \leq 0.3m$) with sharp entrances where Gi ≳ 1500. The result was

$$(21) \quad \log(N_B) = 12.5 - 0.15 \log(Gi)$$

The nucleation source term for the numerical model is thus obtained as

$$(22) \quad \dot{I}_B = \frac{dN_B}{dt} = \frac{N_B(Gi)w}{\Delta x}$$

where Δx is the mesh spacing. The model applies the bulk nucleation zone to the mesh cell spanning the throat, consistent with maximum superheat at this location. Bulk nucleation thus vanishes everywhere except for the throat cell and in this cell is additive to the continuous wall nucleation determined from the model of Shin and Jones [1990]. For the computational model the characteristic time for bulk nucleation is thus taken as $\Delta t_n = \Delta x/w$.

Numerical Methods

For solving the equations given by Eq. (1), a semi-implicit method [EPRI, 1983] is used having the stability criterion

$$(23) \quad \Delta t \leq \frac{\Delta x}{w_{max}}$$

which significantly increases the allowable time step compared with explicit methods and makes the use of the microcomputer feasible.

for the nuclei allowed nucleation frequencies to be obtained from data. These were then correlated empirically with superheat as

(15) $$f_{max} = 10^4 \Delta T_{sup}^3$$

where $\Delta T_{sup} = (T_\ell - T_v)$. The nucleation site density is then determined by assuming there is a maximum energy available for nucleation in the layer of thickness $\delta = (\pi/6)R_c(\rho_v/\rho_\ell)$. Maximum nucleation site densities were thus determined for each data set and correlated empirically as

(16) $$\begin{cases} N_{nucl} = 0.25 \times 10^{-7} \dfrac{r_{dep}^2}{r_{cs}^4} \\ r_{cs} = \dfrac{2\sigma T_v}{\rho_v \Delta i_{fg}(T_\ell - T_v)} \end{cases}$$

The bubble departure radius was determined as

(17) $$\begin{cases} r_{dep} = 0.58\left[\left(\dfrac{\sigma r_c}{\rho_\ell}\right)^{1/2}\left(\dfrac{\mu_\ell}{\tau_o}\right)^{0.7}\left(\dfrac{\rho_\ell}{\mu_\ell}\right)^{0.3}\right]^{5/7} \\ \tau_o = 0.079\, Re^{-0.25}\, \dfrac{\rho_\ell w^2}{2} \end{cases}$$

Equations (16) and (17) were combined to obtain

(18) $$\begin{cases} N_{nucl} = 0.25 \times 10^{-7}\left[\left(\dfrac{T_v}{T_\ell - T_v}\right)\left(\dfrac{2\sigma}{\rho_v \Delta i_{fg}}\right)\right]^{-23/7} \cdots \\ \cdots \left\{0.58\left[\left(\dfrac{\sigma}{\rho_\ell}\right)^{1/2}\left(\dfrac{\mu_\ell}{\tau_o}\right)^{0.7}\left(\dfrac{\rho_\ell}{\mu_\ell}\right)^{0.3}\right]^{5/7}\right\}^2 \end{cases}$$

Combination of the departure frequency per site with the site density yields the wall nucleation rate $\dot{I}_w$ as

(19) $$\dot{I}_w = \frac{N_{nucl} f_{max} P(x)}{S(x)}$$

It should be noted that experiments in small geometries having diameters of the order of 10^{-2} m were used to develop the equations for wall nucleation rate identified above. Under these conditions, experiments have shown that wall nucleation dominates.

Bulk heterogeneous nucleation. A model for heterogeneous nucleation in the bulk fluid was developed by Soplenkov and Blinkov [1983] and further described in detail by Nigmatulin

area density, A_i. On the other hand, the coalescence and formation of large bubbles tends to reduce A_i. The variation of A_i with void fraction here may be small so it can be assumed that A_i remains constant having the value obtained at $\alpha=0.3$. The heat flux $\dot{q}_{\ell-v}$ for bubbly flow is obtained from Eq. (8).

For the case where $\alpha \geq 0.7$, the transition to dispersed droplet flow occurs and

$$(14) \quad A_i = \frac{3(1-\alpha)}{r_d} = (36\pi)^{1/3} N_d^{1/3} (1-\alpha)^{2/3}$$

Thus the variation with α is assumed to be symmetrical about $\alpha = 0.5$ ($A_i \sim \alpha^{2/3}$ if $\alpha < 0.3$, and $A_i \sim (1-\alpha)^{2/3}$ if $\alpha > 0.7$). The Nusselt number for the heat flux is determined by Eq. (12).

Nucleation Kinetics

Wall nucleation. The wall nucleation process is described in Shin and Jones [1990]. The formation, growth, and departure of a bubble from an active nucleation site are considered as a cyclical process consisting of two periods. During the bubble growth period the saturation temperature inside the bubble represents the boundary condition for a thermal wave to penetrate the wall.

During the waiting period after departure before the appearance of another nuclei at the site, superheated liquid contacts the wall resulting in a wall temperature between saturation and the superheat temperature. This wall temperature then relaxes causing both the solid and liquid to become increasingly closer to the temperature represented by the liquid superheat. The next bubble is nucleated when the wall-liquid contact line temperature reaches the temperature corresponding to the saturation temperature inside a bubble of critical radius equal to the cavity size.

Critical size increases as the waiting period, τ_w, decreases, corresponding to lower contact-line wall superheat. If τ_w vanishes, the maximum value of active cavity size is achieved, ($R^*_{c,max}$), which is independent of flow conditions. This cavity size is a function only of the thermodynamic state of the particular fluid-solid system, and results in minimum surface energy for the nuclei formed.

For the case where the dwell time, τ_w, vanishes, the nucleation frequency at the cavity is also maximized. The departure size is, therefore, essential to determining the growth time, and hence the nucleation frequency. This departure size is determined from a balance of drag and surface tension forces. Analysis of the activation criterion and the departure size where $\alpha_{Bmax} = 0.3$ and $\alpha_{Smax} = 0.8$. The heat transfer coefficient appropriate for Taylor bubbles is approximated by that given in the TRAC-P1A code [LASL, 1979] for slug flows as

$$(11) \quad \begin{cases} \dfrac{Nu_\ell}{Re_\ell Pr_\ell} = 0.0073 \\ \dot{q}_S = 0.0073 \rho_\ell w C_\ell (T_\ell - T_v) \end{cases}$$

(c) For the case of transitional and dispersed droplet flows where $\alpha > 0.8$, the difference is in the friction multiplier as expressed in Eq. (5). The heat and mass transfer occur on liquid droplets formed as a result of bubble coagulation and droplet entrainment from the lateral surface of the Taylor bubbles.

$$(12) \quad \begin{cases} A_i = \dfrac{3(1-\alpha)}{r_d} \quad \text{and} \quad \dot{q}_{\ell-v} = \dfrac{k_\ell Nu}{2r_d}(T_\ell - T_v) \\ Nu \cong 16 \quad \text{[Solbrig et al., 1978]} \\ r_d = \dfrac{We\sigma}{2\rho_v(w_v - w_\ell)^2} \quad ; \quad We = 5 \quad \text{[LASL, 1979]} \end{cases}$$

Of course, the value chosen for the Nusselt number is a time dependent approximation of a time independent value.

The slip, $\Delta w = w_v - w_\ell$, is needed for determination of the droplet diameter $2r_d$, and can be directly determined without iteration (Jones, 1984). In this case, the terminal velocity of single droplets is determined from the Reynolds number as

$$(13) \quad Re_\infty = \begin{cases} \text{for } Ar < 3.227 \times 10^5 \\ \dfrac{Ar}{18}\left[1 + 0.0487\left(\dfrac{4}{3}Ar\right)^{0.452}\right]^{-1} \quad \text{and} \\ 1.74\sqrt{Ar} \quad \text{for} \quad \begin{cases} Ar \geq 3.227 \times 10^5 \\ Re \leq 2 \times 10^5 \end{cases} \end{cases}$$

where Ar is the Archimedes number given by $Ar = g\rho_\ell \Delta\rho \delta^3/\mu_\ell^2$, and where the Reynolds number is based on the droplet size and terminal velocity, v_∞, taken identically as the slip velocity.

<u>Features of flow regime 2</u>. In bubbly flows there are no differences from flow regime map 1. The surface area density is proportional to $\alpha^{2/3}$ assuming $N^{1/3}$ is constant. If the void fraction exceeds 0.3, the surface area density is influenced by two opposing effects. On the one hand, the continuing bubble growth and distortion of their shape tend to increase the

(7) $$\begin{cases} A_i = 4\pi r_B^2 N \quad \text{and} \quad \alpha = (4/3)\pi r_B^3 N = (1/3)A_i r_B \\ A_i = (36\pi)^{1/3} N^{1/3} \alpha^{2/3} = 3\alpha/r_B \end{cases}$$

The growth of the bubbles is assumed to be controlled by transient conduction. An analytical solution of thermally-controlled bubble growth for constant values of liquid superheat due to Scriven [1959], expressed by an approximation given by Labuntsov et al. [1964] gives the heat input from the liquid to the bubble-liquid interface as:

(8) $$\begin{cases} \dot{q}_{\ell-v} = h\Delta T \quad \text{where} \quad h = \dfrac{k_\ell Nu}{2r_B} \\ Nu = \dfrac{12}{\pi} Ja \left[1 + \dfrac{1}{2}\left(\dfrac{\pi}{6Ja}\right)^{2/3} + \dfrac{\pi}{6Ja}\right] \\ Ja = \dfrac{c_\ell \rho_\ell (T_\ell - T_{sat})}{\Delta i_{fg} \rho_v} \end{cases}$$

While this expression in general is valid only for uniform superheat and not for variable pressure fields [Jones and Zuber, 1978], it has been shown approximately correct when the local superheat is used [Wu et al., 1981].

(b) For bubbly-slug flow in the range of $0.3 \leq \alpha < 0.8$, it is assumed that some of the bubbles coalesce to form larger (Taylor) bubbles while others continue to grow according to Eq. (8). Thus, two classes of bubbles are assumed to exist and grow at different rates.

Since vapor generation may take place on the surface of both kinds of bubbles, the total interfacial flux is given by

(9) $\quad \dot{q}_{\ell-v} = \dot{q}_S A_S + \dot{q}_B A_B$

The Taylor bubbles are assumed to be cylinders which, at $\alpha=0.8$ absorb smaller bubbles and merge with one another to form annular flow [Wu et al., 1981]. The interfacial area density is thus given by

(10) $$\begin{cases} \alpha = \alpha_S + \alpha_B \quad \text{where} \quad A_B = \dfrac{3\alpha_B}{r_B} \quad \text{and} \quad A_S = \dfrac{4\alpha^{2/3}}{\alpha_{Smax}^{1/6} D_{hy}} \\ \alpha_S = \dfrac{1}{1-\alpha_{Bmax}} \left\{\alpha - \alpha_{Bmax}\left[1 - \dfrac{(\alpha-\alpha_{Bmax})(1-\alpha_{Smax})}{\alpha_{Smax} - \alpha_{Bmax}}\right]\right\} \end{cases}$$

For two-phase flow, a friction multiplier, ϕ^2, is utilized so that

(4) $$\begin{cases} f_f = 2C_f \phi^2 \dfrac{\rho_\ell^2 w^2}{\rho_\ell D_{hy}} & \text{where } C_f = C_f(\text{Re}) \\ \text{Re} = \dfrac{\rho_m w D_{hy}}{\mu_\ell} \end{cases}$$

The multipliers used were taken from Beattie [1973] as included in RETRAN [McFadden, 1981], where a flow regime map similar to that utilized herein was used. The equations adopted for the two-phase friction multiplier are:

(5) $$\begin{cases} \text{for } \alpha < 0.3 \\ \phi^2 = \left[1+\chi\left(\dfrac{\rho_\ell}{\rho_v}-1\right)\right]^{0.8} \left[1+\chi\left(\dfrac{(3.5\mu_v+2\mu_\ell)\rho_\ell}{(\mu_v+\mu_\ell)\rho_v}-1\right)\right]^{0.2} \\ \text{for } 0.3 \leq \alpha < 0.8 \\ \phi^2 = \left[1+\chi\left(\dfrac{\rho_\ell}{\rho_v}-1\right)\right]^{0.8} \left[1+\chi\left(3.5\dfrac{\rho_\ell}{\rho_v}-1\right)\right]^{0.2} \\ \text{for } 0.8 \leq \alpha < 0.95 \\ \phi^2 = \left[1+\chi\left(\dfrac{\rho_\ell}{\rho_v}-1\right)\right]^{0.8} \left[1+\chi\left(\dfrac{\mu_v \rho_\ell}{\mu_\ell \rho_v}-1\right)\right]^{0.2} \\ \text{and for } 0.95 \leq \alpha \\ \phi^2 = \left(\dfrac{\mu_v}{\mu_\ell}\right)^{0.2} \left(\dfrac{\rho_v}{\rho_\ell}\right)^{0.8} \left[1+\chi\left(\dfrac{\rho_\ell}{\rho_v}-1\right)\right]^{1.8} \end{cases}$$

Interfacial heat and mass transfer. The rate of vapor generation is limited by the heat transfer rate and interfacial area according to the relation

(6) $$\Gamma_{\ell-v} = \dfrac{A_i \dot{q}_{\ell-v}}{\Delta i_{fg}}$$

which expresses the balance across the interface of the rate of energy loss due to vaporization and the rate of energy replenishment by heat transfer from the bulk.

(a) For bubbly flow, $\alpha<0.3$, the model assumes that the zone of intensive nucleation is vary narrow, $L_{nuc} \ll L$ and is located very close to the minimum area portion of the nozzle, which is confirmed by calculations and is consistent with the distributed nucleation model [Shin and Jones, 1990]. In this case, uniform sized bubbles may be assumed to exist at any cross section. From sphericity,

velocities and pressures. Vapor temperatures are assumed to be at local saturation conditions since the primary source of heat transfer is through the liquid continuum which may be subcooled, saturated, or superheated locally (according to the local pressure).

The model thus chosen is a quasi-one-dimensional, transient model which uses two continuity equations (mixture and vapor phase), one energy equation for the liquid, one momentum equation for the mixture, and one bubble transport equation. These are expressed below as:

$$(1) \begin{cases} \dfrac{\partial \rho_m}{\partial t} + \dfrac{1}{S}\dfrac{\partial}{\partial x}(\rho_m w S) = 0 \quad \text{and} \quad \rho_m = \alpha \rho_v + (1-\alpha)\rho_\ell \\[4pt] \dfrac{\partial \alpha \rho_v}{\partial t} + \dfrac{1}{S}\dfrac{\partial}{\partial x}(\alpha \rho_v w S) = \Gamma_{\ell-v} \\[4pt] \dfrac{\partial w}{\partial t} + w\dfrac{\partial w}{\partial x} = \dfrac{-1}{\rho_m}\dfrac{\partial p}{\partial x} - \dfrac{1}{\rho_m} f_f \\[4pt] \dfrac{\partial}{\partial t}\left[(1-\alpha)\rho_\ell u_\ell\right] + \dfrac{1}{S}\dfrac{\partial}{\partial x}\left[(1-\alpha)\rho_\ell u_\ell w S\right] + \\[4pt] \quad \ldots + \dfrac{p}{S}\dfrac{\partial}{\partial x}\left[(1-\alpha)wS\right] = p\dfrac{\partial \alpha}{\partial t} - \Gamma_{\ell-v} i_f - \dot{q}_{\ell-v} \\[4pt] \dfrac{\partial N}{\partial t} + \dfrac{1}{S}\dfrac{\partial}{\partial x}(NwS) = \dot{I}_W + \dot{I}_B \end{cases}$$

The unknowns in this equation set include void fraction, pressure, p, axial velocity, w, liquid temperature, T_ℓ, and bubble number density, N.

<u>Constitutive equations - flow regime map 1</u>. Constitutive equations for wall friction and interfacial heat and mass transfer must be provided for closure as well as those for nucleation.

Wall friction. For single-phase flow ($\alpha=0$) the shear stress is taken to be

$$(2) \begin{cases} \tau = 0.5 C_f \rho_\ell w^2 \quad \text{where} \quad C_f = C_f(Re) \\[4pt] \qquad\qquad\qquad \text{and} \quad Re = \dfrac{\rho_\ell w D_{hy}}{u_\ell} \end{cases}$$

The Blassius friction coefficient C_f is used for turbulent flow. Wall friction force per unit volume of the mixutre is

$$(3) \quad f_f = \tau \dfrac{P dx}{S dx} = 2C_f \dfrac{\rho_\ell w^2}{D_{hy}}$$

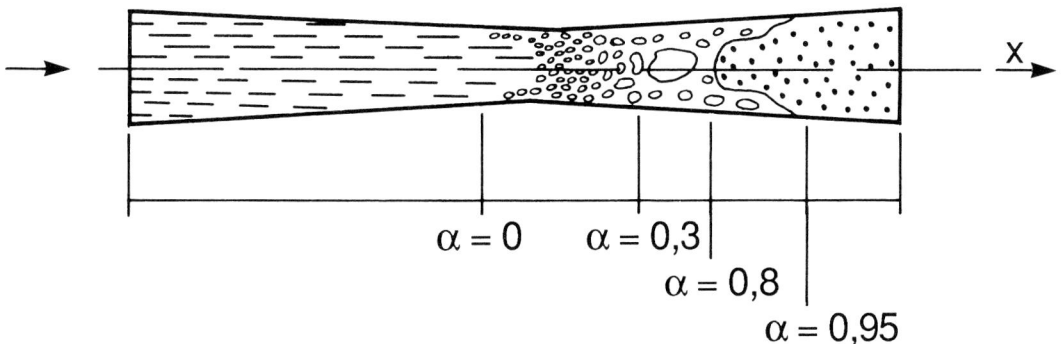

(a) Corresponding to flow regime 1

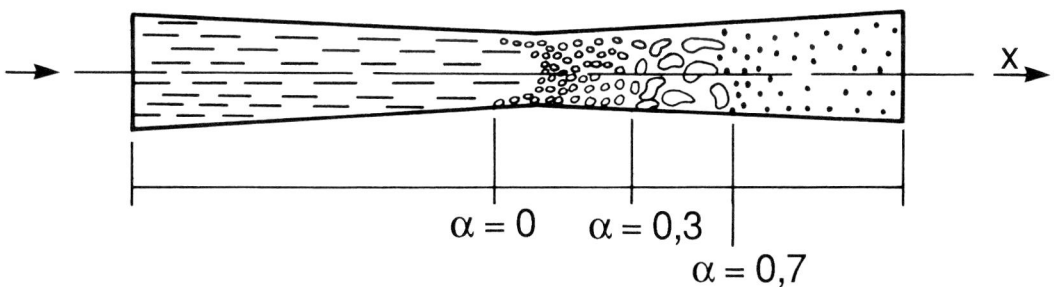

(b) Corresponding to flow regime 2

Figure 1. Assumed flow diagrams of flashing liquid in nozzles

void fraction has achieved recommended general usage [Wu et al., 1981, Ishii and Mashima, 1983] due to its simplicity. In what follows, two variants of this flow structure scheme are considered. A comparison of the resultant features of these two maps will be discussed following identification of the equations given particular to the two situations.

Flow regime 1 (Fig 1a). Flashing inception is assumed to result in a bubbly mixture. The limits of this flow regime vary depending on the rate of void development. For slowly developing systems, the transition is generally taken to be at approximately $\alpha=0.2$. For rapidly expanding systems, the transition to slug or churn flows may be inhibited until quite large void fractions up to over 0.7. For the purpose of this discussion, the bubbly mixture is assumed to exist up to $\alpha=0.3$, consistent with previous assumptions [Wu et al., 1981, and Dobran, 1985]. Note that Ishii and Mashima [1983] showed that for $\alpha>0.3$ spherical bubbles must touch. For bubbly flows to exist at larger void fractions, obviously the bubbles must distort. Such is the case observed in liquid metal MHD generators. In moderately accelerating systems, agglomeration begins to occur after bubbles begin to touch, thus leading to transition of the bubbly structure to the adjacent regime.

As void development continues past $\alpha>0.3$, it is assumed that this coalescence results in the formation of larger bubbles with the region between filled with small bubbles. This bubbly-slug regime is taken to exist throughout the region $0.3 \leq \alpha < 0.8$ whereupon the slugs have grown so long that they, in turn, coalesce to form annular flow or mist-annular. Dispersed droplet flow is assumed to exist for void fractions over 0.9. The region between $\alpha=0.8$ and $\alpha=0.95$ is considered herein to be a transitional zone coupling slug and dispersed liquid flows.

Flow regime 2 (Fig. 1b). In this case, the structure of slug flow is not developed and the zone between $\alpha=0.3$ and $\alpha=0.7$ is considered as a transitional zone. This zone may be characterized by intensive interaction or coalescense of bubbles and deviations from sphericity characteristic of the churn-turbulent regime [Solbrig et al., 1978]. In this case dispersed flow is assumed to occur as a breakdown of the continuous liquid filaments resulting in continuous vapor dispersed flows.

Two-Phase Flow Model

For the current case the interest is in the lower void fraction regions below dispersed flows. In such cases, the vapor is tightly coupled to the liquid from a mechanical viewpoint, relative velocities are small and any variations are due more to distribution than local slip. It is thus assumed that mechanical equilibrium exists and the phases have identical

INTRODUCTION

The first paper in this sequence described the technology relative to the initiation or inception of the flashing phenomena and showed development of a model for nucleation from the walls in pipes and nozzles. It was shown that negligible voids developed upstream of the throat, even in the most energetic of cases where throat superheats approached 100K. Predictions developed from the model of wall superheat were within 2K, and resultant predictions of critical discharge within 3%. The model provides the framework for the subsequent calculation of void development based on the superheat and area density for phase transformation thus determined. It is the purpose of this paper to provide a rational framework for the computation of void development downstream of the inception location in pipes, nozzles and restrictions.

The general framework used in the numerical development is not particularly new except for some specific details which will be delineated in the development. Rather, what is, perhaps, new is the use of such a detailed numerical model on a personal computer, in this case, a Hewlett Packard Model 9816. This machine is based on the 32-bit MC-68000 microprocessor so can be considered the forerunner of more modern 32 bit machines. As such, the use of such methods on the newer and faster 80386 microprocessor running at 16-24 MHz should be considerably faster than the experience reported herein. The practicality of such computational frameworks on "desktop" computers thus appears to be within reach.

DEVELOPMENT

Flow Regimes

The common approach to determination of flow structure is the use of flow regime maps. They are typically constructed with superficial velocities, flow rates and quality, and/or void fraction as coordinates. Among the variety of internal two-phase flow structures, the bubble, slug, churn, annular, dispersed-annular, and dispersed regimes may be identified. These classifications and transition criteria both have a qualitative nature so models constructed with the use of these flow maps may utilize existing transition criteria for comparison with data.

Advanced, best-estimate computer program for transient analysis of two-phase systems such as TRAC [LASL, 1979] and RELAP [Chow and Ransom, 1984] apply flow regime maps with void fraction, α, as the main transition criterion. Indeed, while other methods exist which are more detailed, and which are, perhaps, more accurate for specific transitions, the use of

NUCLEATION AND FLASHING OF INITIALLY SUBCOOLED LIQUIDS FLOWING IN NOZZLES: 2. COMPARISON WITH EXPERIMENTS USING A 5-EQUATION MODEL FOR VAPOR VOID DEVELOPMENT

V. N. BLINKOV
Kharkov Aviation Institute
Kharkov, USSR

O.C. JONES, JR.
Center for Multiphase Research
Department of Nuclear Eng. & Eng. Physics
Rensselaer Polytechnic Institute
Troy, New York 12180-3590

ABSTRACT

A quasi-one-dimensional, five-equation model has been developed on a microcomputer to calculate the behavior of flowing, initially subcooled, flashing liquids. Equations for mixture and vapor mass conservation, mixture momentum conservation, liquid energy conservation, and bubble transport were discretized and linearized semi-implicitly, and solved using a successive iteration Newton method. Closure was obtained through simple constitutive equations for friction and spherical bubble growth, and a new nucleation model for wall nucleation in small nozzles combined with an existing model for bulk nucleation in large geometries. Good qualitative and quantitative agreement with experiment confirms the adequacy of the nucleation models in determining both initial size and number density of nuclei. It is shown that bulk nucleation becomes important as the volume-to-surface ratio of the geometry is increased.

Powell, A.W., 1961. "Flow of Subcooled Water Through Nozzles," Westinghouse Electric Corporation, WAPD-PT-(V)-90, April.

Reocreux, M., 1976. "Experimental Study of Steam-Water Choked-Flow," Paper presented at the OECD/NEA Specialists' Meeting on Transient Two-Phase Flow, Toronto, Canada, Aug.

Richter, H.J., 1981. "Seperated Two-Phase Flow Model: Appilcation to Critical Two-Phase Flow," EPRI NP-1800, April.

Riebold, W.L., Reocreux, M., and Jones, O.C., 1981. "Blowdown Phase", in **Nuclear Reactor Safety Heat Transfer,** O.C. Jones, Jr., Ed., Hemisphere/McGraw-Hill, New York.

Rivard, W.C., and Travis, J.R., 1980. "A Non-Equilibrium Vapor Production Model for Critical Flow," **Nucl. Sci. Eng., 74,** 40.

Rohatgi, U.S., and Reshotko, E., 1975. "Non-Equilibrium One-Dimensional Two-Phase Flow in Variable Area Channels," in **Non-Equilibrium Two-Phase Flow,** R.T. Lahey and G.B. Wallis, Eds., 47.

Saha, P., 1978. "Review of Two-Phase Steam-Water Critical Flow Models with Emphasis on Thermal Nonequilibrium," NUREG/CR 0417, BNL-NUREG-50907.

Saha, P., Abuaf, N., and Wu, B.J.C., 1981. "A Non-Equilibrium Vapor Generation Model for Flashing Flows," ASME Paper No. 81-HT-84.

Schlichting, H., Boundary Layer Theory, McGraw-Hill, New York, 1979. Schrock, V.E., and Amos, C.N., 1984. "Two-Phase Critical Flow," **Proc. Japan-US Seminar in Two-Phase Flow Dynamics,** K. Akagawa and O.C. Jones, Eds., July.

Schrock, V.E., Starkman, E.S. and Brown, R.A., 1977. "Flashing Flow of Initially Subcooled Water Through Nozzles," **Trans. ASME, J. Heat Transfer, 99,** 263.

Scriven, L.E., 1959. "On the Dynamics of Phase Growth," **Chem. Eng. Sci., 10,** 113.

Shin, T.S., and Jones, O.C., Jr., 1986. "A Distributed Nucleation Model for Flashing of Initially Subcooled Liquids in Nozzles," submitted for publication in the Int. J. Heat and Mass Trans.

Shoukri, M.S.M., and Judd, R.L., "A Theoretical Model for Bubble in Frequency in Nucleate Pool Boiling Including Surface Effects," 6th International Heat Transfer Conference, Toronto, Canada, 1978.

Simoneau, R.J., 1975. "Pressure Distribution in a Converging-Diverging Nozzle during Two-Phase Choked Flow of Subcooled Nitrogen," in **Non-Equilibrium Two-Phase Flows,** R.T. Lahey and G.B. Wallis, Eds., 37.

Simpson, H.C., and Silver, R.S., 1962. "Theory of One-Dimensional Two-Phase Homogeneous Non-equilibrium Flow," **Proc. Inst. Mech. Eng. Symposium on Two-Phase Flow,** 45.

Sozzi, G.L., and Sutherland, W.A., 1975. "Critical Flow of Saturated and Subcooled Water at High Pressure," GE Rep. NEDO-13418.

Wegener, P.P., 1969. NonDEquilibrium Flows, Marcel Decker, New York. Wegener, P.P., 1975. "Non-Equilibrium Flow with Condensation," **Acta-Mechanica, 21,** 65.

Winters, W.S., and Merte, H., 1979. "Experiments and Non-Equilibrium Analysis of Pipe Blowdown," **Nucl. Sci. Eng., 69,** 411.

Wolfert, K., 1976. "The Simulation of Blowdown Processes with Condensation of Thermodynamic Non-Equilibrium Phenomena, Paper presented at the OECD/NEA Specialists Meeting on Transient Two-Phase Flow, Toronto, Canada, Aug.

Wu, B.J.C., Abuaf, N., and Saha, P., 1981. "A Study of Nonequilibrium Flashing of Water in a Converging-Diverging Nozzle," NUREG/CR-1864, BNL-NUREG-51317, Vol. 2.

Yang, J., Jones, O.C., Jr., and Shin, T.S., 1986. "Limitations in the Isentropic Homogeneous Equilibrium Model," **Nucl. Eng. Des., 95,** pg. 195.

Zimmer, G.A., Wu. B.J.C., Leonhard, W.L., Abuaf, N. and Jones, O.C., Jr., 1979. "Pressure and Void Distributions in a Converging-Diverging Nozzle with Non-Equilibrium Water Vapor Generation," BNL-NUREG-26003.

Zuber, N., Staub, F.W., and Bijwaard, G., "Vapor Void Fraction in Subcooled Boiling and in Saturated Boiling Systems," **Proc. 3rd Int. Heat Trans. Conf.,** Vol. V, pg. 25-38, 1966

Henry, R.E., Fauske, H.K., and McComas, S.T., 1970. "Two-Phase Critical Flow at Low Qualities, Part I: Experimental," **Nuclear Sci. and Eng., 41,** 79.

Hsu, Y.Y., and Graham, R.W., 1961. "An Analytical and Experimental Study of the Thermal Boundary Layer and the Ebullition Cycle in Nucleate Boiling," NASA TN-D-594.

Hsu, Y.Y., 1962. "On the Size Range of Active Nucleation Cavities on a Heating Surface," **Trans. ASME, J. Heat Trans., 94,** pg. 207.

Hsu, Y. Y., 1972. "Review of Critical Flow, Propagation of Pressure Pulse, and Sonic Velocity," NASA TND-6814.

Jones, O.C., Jr., 1980. "Flashing Inception in Flowing Liquids," **Trans. ASME, J. Heat Transfer, 102,** 439.

Jones, O.C., Jr., 1982. "Toward a Unified Approach for Thermal Nonequilibrium in Gas-Liquid Systems," **Nucl. Eng. Des., 69.,** 57.

Jones. O.C., Jr., and Saha, P., 1977. "Non-Equilibrium Aspects of Water Reactor Safety," in **Thermal Hydraulic Aspects of Nuclear Reactor Safety: Vol. 1. Light Water Reactors,** O.C. Jones, Jr. and S.G. Bankoff, Eds., ASME, New York.

Jones, O.C., Jr., and Shin, T.S., 1986. "An Active Cavity Model for Flashing," **Nuc. Eng. Des., 95,** pg. 185.

Jones, O.C. Jr., and Zuber, N., 1978. "Bubble Growth in Variable Pressure Fields," **Trans. ASME, J. Heat Transfer, 100,** 453.

Levy, S., and Abdollahian, D., 1982. "Homogeneous Non-Equilibrium Critical Flows."

Kochamustafaogullari, G., and Ishii, M., 1982. "Interfacial Area and Nucleation Site Density in Boiling Systems," **Int. J. Heat Mass Trans., 26,** 1377.

Malnes, D., 1975. "Critical Two-Phase Flow Based on Non-Equilibrium Model," in **Non-Equilibrium Two-Phase Flow,** R.T. Lahey and G.B. Wallis, Eds., 11.

Marvekin, 1979. "Marviken Full Scale Critical Flow Tests," Dec. Moody, F.J., 1966. "Maximum Two-Phase Vessel Blowdown from Pipes," **Trans. ASME, J. Heat Transfer, 88,** 285.

Plesset, M.S., and Zwick, S.A., 1954. "The Growth of Vapor Bubbles in Superheated Liquids," **J. Appl. Physics, 25,** 4.

Celata, C.P., Cumo, M., Farello, G.E., and Incalcaterra, P.C., 1982. "Critical Flow of Subcooled Liquid and Jet Forces," ENEA-RT/INC(8218).

Clark, H.B., Strenge, P.S., and Westwater, J.W., 1959. "Active Sites for Nucleate Boiling," **Chem. Eng. Prog. Symposium Ser., 55,** p. 103.

Collins, R.L., 1980. "Choked Expansion of Subcooled Water and I.H.E. Flow Model," **Trans. ASME, J. Heat Transfer, 100,** 275.

Edwards, A.R., 1968. "Conduction Controlled Flashing of a Fluid and the Prediction of Critical Flow Rates in One-Dimensional System," AHSB (S) R-147, UKAEA, AERE.

Fauske, H.K., 1964. "The Discharge of Saturated Water Through Tubes," Seventh National heat Transfer Conference, AICHE-ASME, Cleveland, Ohio, Aug.

Fincke, J.R., 1981. "The Correlation of Nonequilibrium Effects with Choked Flow with Subcooled Upstream Conditions,' Conference Paper, ANS Small Break Specialist Meeting, California, 4-1 - 4-30.

Fincke, J.R., Collins, D.R., and Wilson, M.L., 1981. "The Effects of Grid Turbulence on Non-Equilibrium Choked Nozzle Flow," NUREG/CR-1997, EGG-2088, Apr.

Forster, H.K., and Zuber, N., 1954. "Growth of a Vapor Bubble in a Superheated Liquid," **J. Applied Physics, 25,** 474.

Fritz, G., Riebold, W., and Sculze, W., 1976. "Studies on Thermodynamic Non-Equilibrium in Flashing Flow," Paper presented at the OECD/NEA Specialists Meeting on Transient Two-Phase Flow, Toronto, Canada, Aug.

Griffith, P., and Wallis, J., "The Role of Surface Conditions in Nucleat Boiling," Preprint 106, ASME-AIChE Heat Transfer Conf., Storrs, Conn., 1959.

Han, D., and Griffith, P., 1965. "The Mechanisms of Heat Transfer in Nucleat Pool Boiling," **Int. J. Heat Mass Transfer, 8,** p. 887.

Hendricks, R.C., Simoneau, R.J., and Barrows, R.F., 1976. "Two-Phase Choked Flow of Subcooled Oxygen and Nitrogen," NASA-TN-D-8169.

Henry, R.E. and Fauske, H.K., 1971. "The Two-Phase Critical Flow of One Component Mixtures in Nozzles, Orifices, and Short Tubes," **Trans. ASME J. of Heat Transfer, 93,** 179.

BIBLIOGRAPHY

Abdollahian, D., Healzer, J., Jansssen, E. and Amos, C., 1982. "Critical Flow Data Review and Analysis," EPRI NP-2192.

Abuaf, N., Zimmer, G.A., and Wu, B.J.C., 1981. "A Study of Nonequilibrium Flashing of Water in a Converging-Diverging Nozzle," NUREG/CR-1864, BNL-NUREG-51317, Vol. 1.

Abuaf, N., Jones, O.C., Jr., and Wu, B.J.C., 1980. "Critical Flashing Flow in Nozzles with Subcooled Inlet Conditions," in **Polyphase Flow and Transport Technology,** R.A. Bajura, Ed., Century-2 Engineering technology Conference, San Francisco, California, August 13-15.

Abuaf, N., Jones, O.C., Jr., and Wu, B.J.C., 1983. "Critical Flashing Flow in Nozzles with Subcooled Inlet Conditions," **Trans. ASME, J. Heat Transfer, 105,** 379.

Aguilar, F., and Thompson, S., 1981. "Non-Equilibrium Flashing Model for Rapid Pressure Transient," ASME Paper No. 81-HT-35.

Alamgir, M.D., and Lienhard, J.H., 1981. "Correlation of Pressure Undershoot During Hot-Water Depressurization," **Trans. ASME J. Heat Transfer, 103,** 52.

Ardron, K.H., 1978. "A Two-Fluid Model for Critical Vapor-Liquid Flow," **Int. J. Multiphase Flow, 4,** 323.

Ardron, K.H., and Ackerman, M.C., 1978. "Studies of the Critical Flow of Subcooled Water in a Pipe," **Proc. of 2nd CSNI Specialist Meeting,** June, Paris.

Bailey, J.F., 1951. "Metastable Flow of Saturated Water," Trans. ASME, 1109. Bankoff, S.G., 1958. "Entrapment of Gas in the Spreading of a Liquid Over a Rough Surface," **AIChE Journal, 4,** p. 24.

Bauer, E.G., Houdayer, G.R., and Sureau, H.M., 1976. "A Non-Equilibrium Axial Flow Model and Application to LOCA Analysis: the CLYSTERE System Code," Paper presented at the OECD/nea Specialists' Meeting on Transient Two-Phase Flow," Toronto, Canada, Aug.

Bergles, A.E., and Roshenow, W.M., 1962. "Forced Convective Surface Boiling Heat Transfer and Burnout in tubes of Small Diameter," MIT Report No. 8767-21, May.

Brown, R.A., 1961. "Flashing Expansion of Water Through a Converging-Diverging Nozzle," MS Thesis, Univ. of California, Berkeley, UKAEC Rep. UCRL-6665-T.

NUCLEATION AND FLASHING

We   Weber number
x   Axial coordinate

Greek

α Void fraction
Γ Volumetric vapor generation rate
δ Bubble diameter
μ Dynamic viscosity
ν Kinematic viscosity
ρ Density
σ Surface tension
θ Dimensionless temperature
χ Quality

Subscripts

B   Bubble or bulk
d   Droplet
f   Saturated liquid
g   Saturated vapor
ℓ Liquid (not at saturation)
m   Mixture
o   Stagnation
sat   Saturation
S   Re Taylor bubble
v   Vapor (not at saturation)
w   Wall
∞ Terminal

5. Determination of the maximum, energy-limited rate of nucleation.

Utilization of the nucleation model then allows the throat superheat to be calculated within 2% for existing data over the approximate range from less than 1K to nearly 100K. The range of data include pressures to almost 70 bar, and expansion rates from 0.2 bar/sec to over 1 Mbar/sec extending existing methods by four orders of magnitude.

Bubble sizes at departure upstream of the throat were calculated to be in the range of 1-100 μm and were not constant as has been here-to-fore assumed. Nucleation rates at the throat were also variable and were calculated to span the range of 10^8 to 10^{23} $m^{-3} s^{-1}$. Resultant calculations of throat number densities in all cases ranged between ~10^8 and 10^{11} m^{-3}.

Bubble transport calculations show that even for the cases with largest superheat near 100K, negligible (<1%) voids exist at the throat, confirming the previous hypotheses of Abuaf et al. [1980, 1983]. A result of this confirmation is the calculation of critical flow rates by single phase methods within 3% accuracy once the correct throat superheat (and thus pressure) is obtained.

NOMENCLATURE

English

A_i	Interfacial area density
C	Specific heat
D_{hy}	Hydraulic diameter
f_f	Friction factor
Gi	Gibbs number
h	Heat transfer coefficient
Δi_{fg}	Latent heat of vaporization
$\dot{I}$	Nucleation rate
Ja	Jacob number
k	Thermal conductivity
L	Length
$\dot{m}$	Mass flow rate
N	Bubble number density
p	Pressure
P	Perimeter
$\dot{q}$	Heat flux
r_B	Bubble radius
S	Cross section area
t	Time
T	Temperature
ΔT	Temperature difference
u	Specific internal energy
w	Velocity

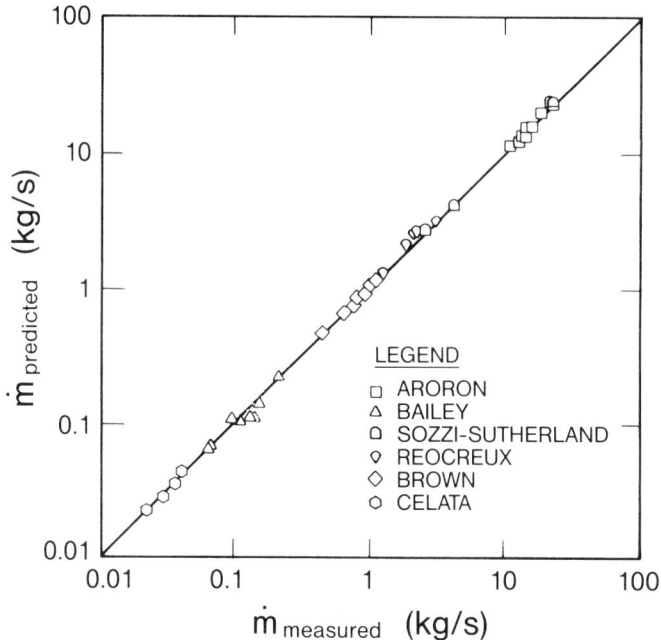

FIGURE 11. Comparison of Calculated Critical Flow Rates with Measured Values.

The result of these calculations is shown in Fig. 11 for the data cited previously. The standard deviation between predicted and measured critical flow rates is approximately 3%, a 40% improvement over earlier work [Abuaf et al, 1983] where 5% accuracy was obtained using the pressure undershoot correlation of Jones [1980].

CONCLUSIONS AND RECOMMENDATIONS

A distributed model for nucleation in the superheated zone upstream of the throat in nozzles during flashing has been described which has the following features:

1. Development of a stability criterion for active cavities.

2. Selection of a figure-of-merit for a nucleating surface which ties the stability criteria to an obtainable nucleation site density and cavity nucleation frequency in flashing flows.

3. Calculation of the departure size of nuclei in the nucleation zone.

4. Correlation of nucleation frequencies at a given site and surface density of nucleation sites as determined from existing data.

The volumetric vapor source for variable area geometry may be expressed as [Zuber et al., 1966]

$$(33) \quad \Gamma_v(z) = \frac{d}{dz}\left[\frac{1}{A_c}\int_0^z m(z,z')J_w(z')P_H(z')dz'\right]$$

where $m(z,z')$ is the mass of a bubble at z which was nucleated at z'. The mass of the bubble at any location may be determined through the departure size at the nucleation site (Eq. 19) and an analysis for bubble growth in a variable pressure field [Jones and Zuber, 1978].

RESULTS AND DISCUSSION

Void Development to the Throat

Calculations were made for all runs in the previously referenced data sets [Abuaf et al., 1981; Reocreux, 1974; Brown, 1961; Celata, 1982; Wu et al., 1981; Sozzi and Sutherland, 1975; Bailey, 1951; Ardron and Ackerman, 1978; Reocreux, 1976] and all had similar behavior.

The smallest throat void fraction was computed for Ardron and Ackerman's Run C25 having a throat superheat of 1.66K with 1.6 bar inlet pressure. While the nucleation site density increased to approximately 140 m^{-2} with overall wall nucleation rates of about 5×10^8 $m^{-3}s^{-1}$ (bulk equivalent), the throat void fraction was only 10^{-5}.

The largest throat void fraction of 0.9% was calculated for Brown's run 39 having a throat superheat of 81.6K, an inlet pressure of 68.4 bar, and a wall nucleation rate at the throat of 3×10^{23} $m^{-3}s^{-1}$, fifteen orders of magnitude larger than that found for Ardron and Ackerman.

Where measurements of void fraction exist, the agreement was within experimental accuracy of the experiment. This however, is no real test, since the data are only accurate to within 1-2% voids at best and the computed maximum throat void fraction when data existed was on the order of 6×10^{-5}. Nevertheless, the calculations and existing data support the original hypothesis of Abuaf et al. [1983].

Critical Mass Flows

The work described above shows that negligible voids exist at the throat for flashing of initially subcooled liquids and supports the original hypothesis of Abuaf et al. [1983]. Single-phase theory may thus be used to calculate critical flow rates under such conditions, where the correct throat pressure must be obtained from the calculation of the throat superheat (Eq. 30).

NUCLEATION AND FLASHING

where

(31) $$B = 8.41 \times 10^{-5} \left(\frac{\rho_g h_{fg}}{2\sigma T_{sat}}\right)^{23/7} \left\{\left(\frac{\sigma}{\rho}\right)^{1/2} \left(\frac{\mu}{\tau_o}\right)^{7/10} \left(\frac{1}{\nu}\right)^{3/10}\right\}^{10/7}$$

This dimensional coefficient has units of $(1/sK^3)$ because of the nature of the nucleation frequency correlation.

A comparison between the calculated and observed values is shown in Fig. 10. The standard deviation is $1.9°C$, indicating a reasonable accuracy in the calculation.

Void Development to the Throat

The void fraction at the throat may be found by integrating the vapor continuity equation to obtain

(32) $$\alpha_t = \frac{1}{\rho_g u_B} \int_0^{z^*} \Gamma_v(z) dz$$

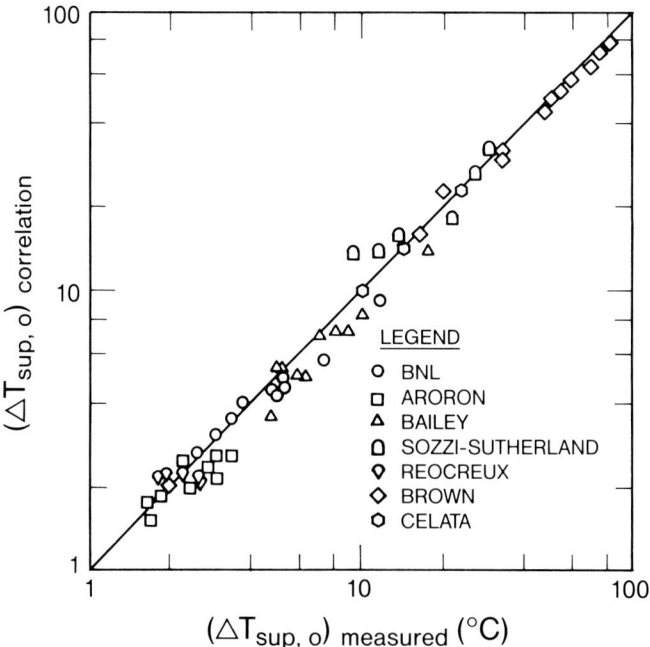

FIGURE 10. Comparison of Calculated Throat Superheats with Measured Values.

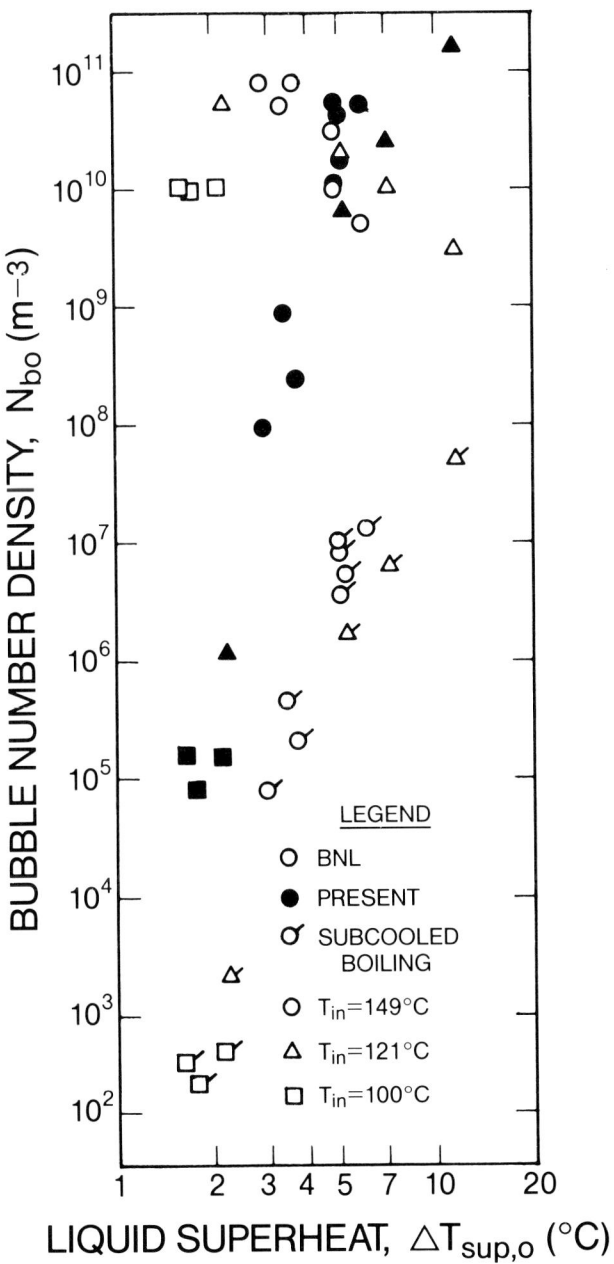

FIGURE 9. Comparison of Calculated Bubble Number Densities at the Nozzle Throat with Other Predictions.

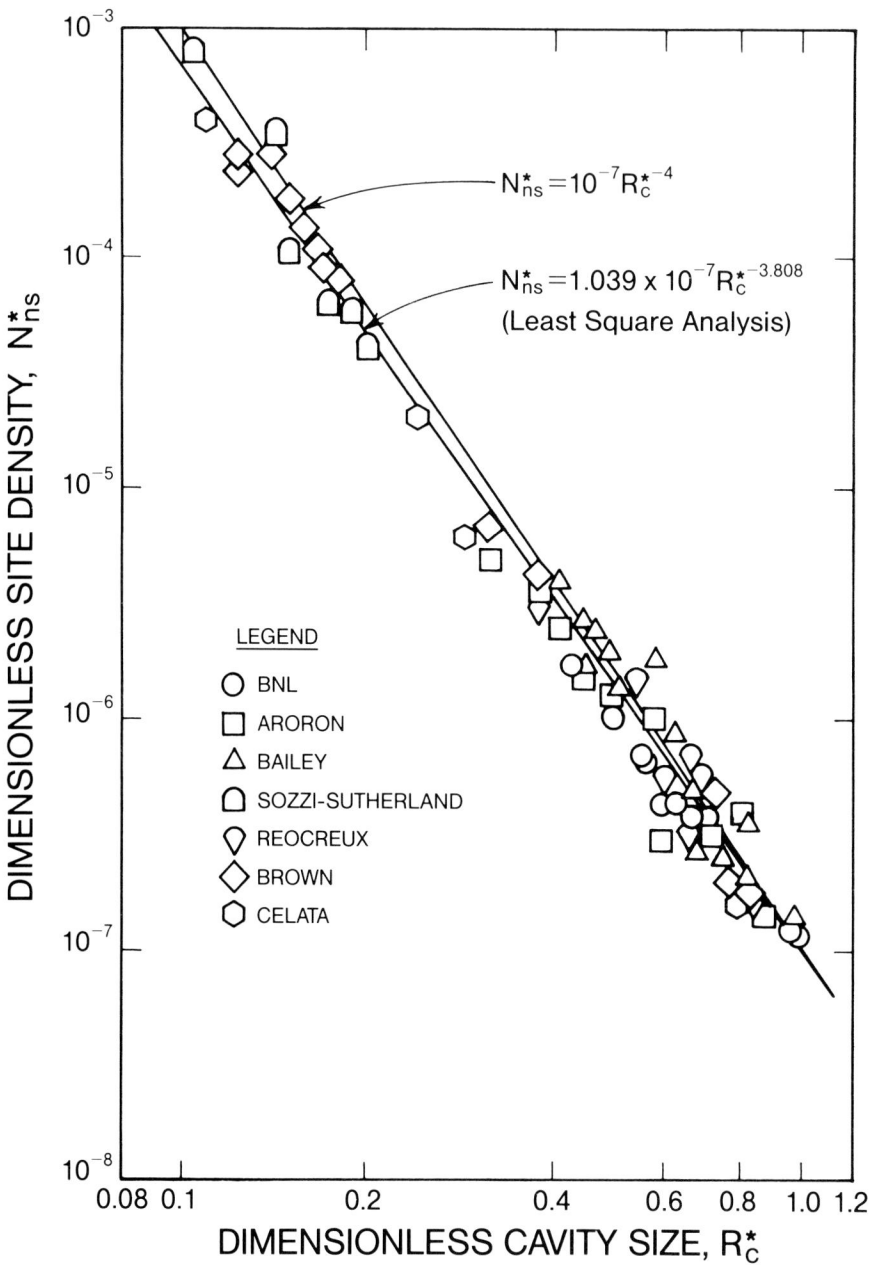

FIGURE 8. Correlation of Dimensionless Nucleation Site Density.

Equation (27) shows the density of active nucleation sites to be proportional to the square of the departure size and inversely dependent on the fourth power of the superheat-related cavity size. Thus, both thermodynamic and hydrodynamic states are important. Increasing the local velocity decreases the departure size and snuffs out active sites in accordance with the observations of Bergles and Rohsenow [1962] for subcooled boiling. An increase in frequency at remaining active sites would thus occur with decreasing pressure, increasing superheat, and increasing velocity.

Bubble Number Density

It is reasonable to expect that the nucleation site density at the throat is characteristic of the rest of the nucleation zone. The bubble number density at the throat may thus be determined by approximate integration of the bubble transport equation as

$$(28) \quad N_{b,o} \cong \frac{1}{u_B(L_n)A_c(L_n)} \int_0^{L_n} N_{ns}(z) f_{max}(z) P_h(z) dz$$

The results are shown in Fig. 9 in comparison with the estimates provided by Wu et al. [1981] on the basis of an assumed inception void fraction. Furthermore, the values predicted by the subcooled boiling model of Kocamustafaogullari and Ishii [1983], while not explicitly valid for flashing are also shown several orders lower than those predicted herein.

Superheat

The wall nucleation rate may be written in terms of the nucleation site density and site frequency as

$$(29) \quad J_w = \frac{N_{ns} f_{max} P_h}{A_c}$$

Recalling the expression for departure size, and using the cavity size in terms of the superheat through the Clausius Clapyron equation, the throat superheat is obtained as

$$(30) \quad \Delta T_{sup,o} = \left(\frac{J_{wm} A_c}{B P_h}\right)^{7/44}$$

$$(22) \quad N_{ns}^* = (2R_d)^2 N_{ns} = \frac{(2R_d)^2 J_{wm} A_c}{f_{max} P_h}$$

where J_{wm} is the maximum wall nucleation density which must be determined and A_c the local cross sectional area of the nozzle. Considering the nucleation wall layer, a convective energy balance yields

$$(23) \quad J_{wm}(z) = \frac{3\rho_\ell C_{p\ell} v_{fg}}{h_{fg} \pi R_N^2(z) R_C^3} \frac{d}{dz}\left[R_N(z) \Delta T_{sup}(z) \int_0^\delta u(y) dy\right]$$

δ is the thickness of a thin cylinder of liquid of radius equal to the cavity radius having the mass of the nuclei. Thus,

$$(24) \quad \delta = \frac{\pi}{6} R_c \left(\frac{\rho_g}{\rho_\ell}\right)$$

The velocity in (22) is obtained from the universal velocity profile over the thickness δ.

Equation (22) must, of course, be evaluated for each given geometry. For pipe flow, $R(z)$ is fixed and the evaluation of (23) is relatively straightforward. For the case where the radius varies linearly between R_1 at the inlet and R_2 at the throat, and where the acceleration pressure profile in the nucleation zone can be approximately linearized, it is found that

$$(25) \quad J_{wm} = 2.94 \times 10^{-3} \left(\frac{C_{p\ell} \dot{m}^{7/4}}{\sigma \mu^{3/4} T_{sat}}\right) \left(\frac{\rho_g \Delta T_{sup,o}}{\rho_\ell L_n}\right)^2 h(z)$$

where z is the distance from saturation ($0 < z < L_n$) and where

$$(26) \quad h(z) = \frac{z}{R_N^{19/4}(z)} \left[1 - 2.75 \left(\frac{\Delta R}{R}\right)\left(\frac{z}{L_n}\right)\right]$$

accounts for the nozzle geometry. Values for dimensionless nucleation site density can be obtained from the experimental data previously cited. It is found that the dimensionless nucleation site density correlates approximately with the superheat-based cavity size (Laplace equation) $R_c^* = R_{cs}/R_d$ (Fig. 8) as

$$(27) \quad N_{ns} = 10^{-7} R_c^{*-4}$$

The least squares coefficient and exponent were 1.039×10^{-7} and -3.808 respectively.

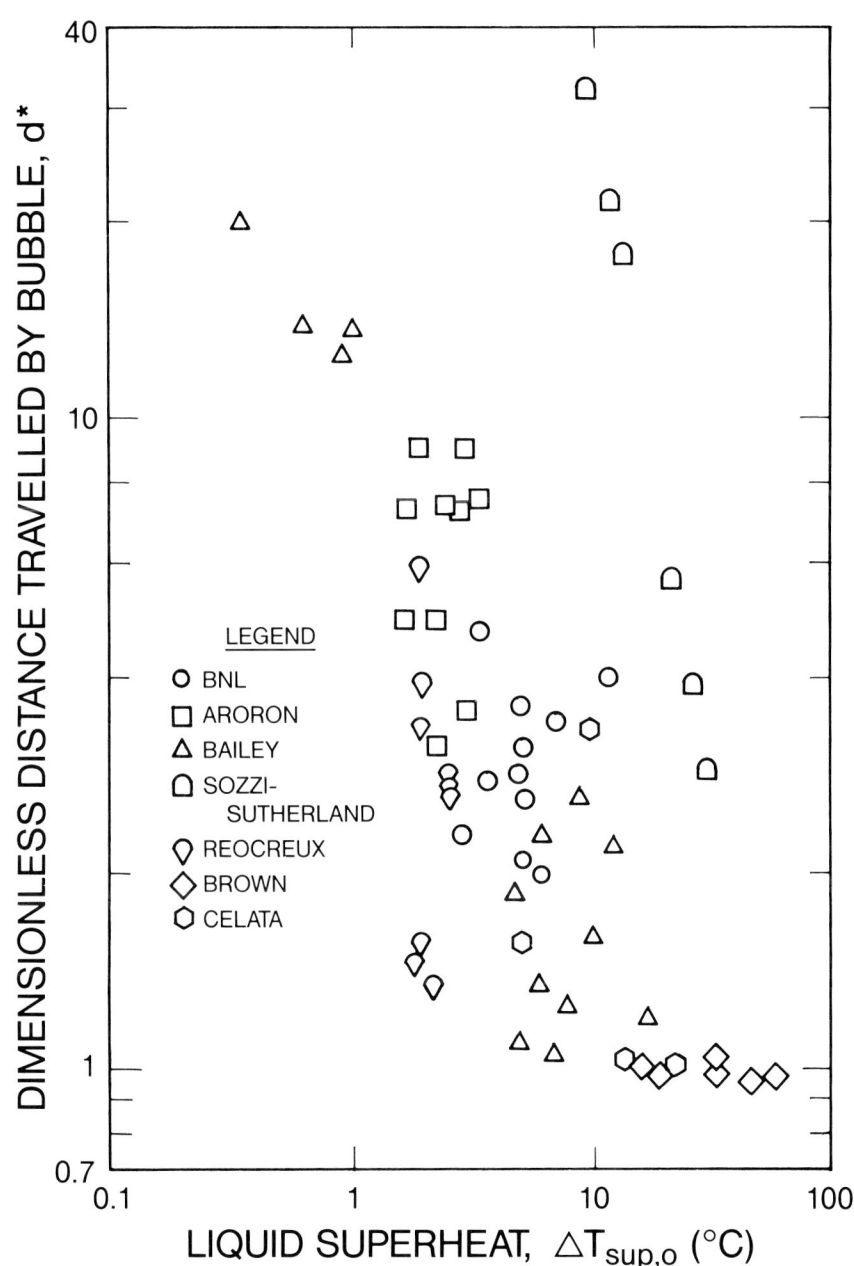

FIGURE 7. Distance Traveled by One Departing Bubble in One Nucleation Period Relative to Departure Diameter.

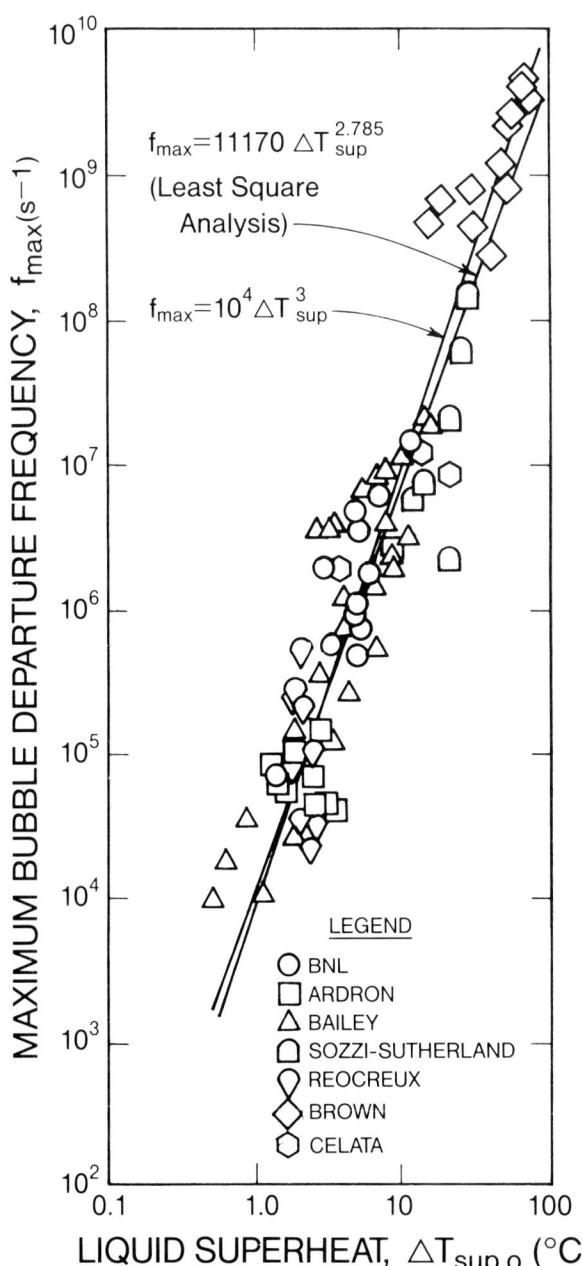

FIGURE 6. Throat Bubble Departure Frequencies for "Figure-of-Merit" Cavities.

Bubble Nucleation Frequency

For wall nucleation conditions, the nucleation frequency per site and the site density are the quantities which describes the system nucleation behavior. The cavity size distribution in the nozzle should be characterized by that in the throat where the superheat and nucleation frequency is at a maximum. Thus, the frequency in the throat will be determined and then used to obtain the site density. The latter will then be used for the balance of the nozzle.

Data which exist in the literature [Abuaf et al., 1981; Reocreux, 1974; Brown, 1961; Celata, 1982; Wu et al., 1981; Sozzi and Sutherland, 1975; Bailey, 1951; Ardron and Ackerman, 1978; Reocreux, 1976] were used to calculate the nucleation frequency for the figure-of-merit cavities (assumed size). Growth rates before departure were taken as the smaller of momentum controlled or thermal-diffusion-controlled [Plesset and Zwick, 1954] rates. Results are shown in Fig. 6. The liquid superheats span the range from less than $1^{\circ}C$ to close to $100^{\circ}C$. A reasonable fit close to the least squares condition of these data was found to yield

$$(21) \qquad f_{max} = 10^4 \Delta T_{sup}^3$$

Frequencies close to 10^{10} s^{-1} were calculated at the largest values of superheat near $100^{\circ}C$. These are unexpectedly large values characteristic of what one normally would consider at the lower bound for homogeneous nucleation. In spite of the facts that these are artificial values, it is seen in Fig. 7 that the ratio of distance traveled in one period after departure, u_ℓ/f, to the bubble size at departure, $2R_d$, is generally above unity. In only two cases is this ratio below unity having values of 0.9 and 0.95 respectively, well within the accuracy of the calculation of R_d.

The correlation for nucleation frequency is dimensional, a somewhat undesirable situation. However, at this time no satisfactory dimensionless method of correlating the data has been found.

Maximum Local Wall Nucleation Rate and Nucleation Site Density

It is typical of rapidly nucleating systems that the process of nucleation itself can turn off the nucleation process due to the energy absorbed or liberated by formation of the nuclei (Wegener, 1969, 1975). Thus, it is expected that with sufficiently rapid nucleation, the available superheat in the nucleation layer will be consumed. In order to calculate a maximum wall nucleation rate for each set of data, the nucleation site density can be calculated as

$$(16) \quad u_\ell = \frac{1}{2R_d} \int_0^{2R_d} \left(\frac{\tau_o y}{\mu}\right) dy = \frac{\tau_o R_d}{\mu}$$

The normal turbulent friction coefficient [Schlichting, 1979] is used to calculate the wall shear stress such that

$$(17) \quad C_f = 0.0791 \, Re_D^{-1/4}$$

This assumption will underestimate the wall shear in nozzles where the convergence suppresses the boundary layer thereby steepening the gradients. This will lead to late departure estimates and underestimated frequencies, tending to offset the vanishing dwell time estimate. The departure size thus becomes

$$(18) \quad R_d = \left[(K\mu/\tau_o)(4\sigma R_c/C_D \rho_\ell)^{1/2}\right]^{1/2}$$

The drag coefficient is taken to be [Schlichting, 1979]

$$(19) \quad C_D = \begin{cases} 24/Re_B & Re_B < 1 \\ 18.5 \, Re_B^{-0.6} & 1 \leqslant Re_B < 500 \\ 0.44 & 500 \leqslant Re_B < 2 \times 10^5 \end{cases}$$

so that the departure size becomes

$$(20) \quad R_d = 0.58 K^{5/7} \left\{ \left(\frac{\sigma R_c}{\rho_\ell}\right)^{1/2} \left(\frac{\mu}{\tau_o}\right)^{7/10} \left(\frac{1}{\nu}\right)^{3/10} \right\}^{5/7}$$

K accounts for the relative balance between drag and surface tension at departure is arbitrarily taken as unity in what follows.

Available data on flashing in pipes and nozzles [Abuaf et al., 1981; Reocreux, 1976; Brown, 1961; Celata, 1982; Wu et al., 1981; Sozzi and Sutherland, 1975; Bailey, 1951; Ardron and Ackerman, 1978] were examined and departure sizes calculated using Eq. (19) with K=1. For all these data, departure sizes in the range of 1 μm to 100 μm were obtained. The smallest values correspond to the highest flows and largest superheats near 100°C.

In all cases, the values of departure size calculated were approximately equal to or larger than the given value of $Max\{R_c, min\}$.

Figure of Merit

Actual determination of nucleation site density, even if possible, is patently impractical. An alternate approach would be to find a reasonable "figure-of-merit" which would be sufficiently definitive to allow analysis to proceed.

While there is no theoretical basis for a minimum energy principal for nucleation at cavities, it may be that preferential nucleation might take place at cavities in a way that the surface energy for production is minimized. If this were to be the case, nucleation would favor the maximum cavity size which will produce stable cavities.

A figure-of-merit for the system is thus obtained by considering the surface to be made up only of cavities producing the largest stable nuclei--those for which the dwell time vanishes and the wall temperature is at its minimum value.

Letting the waiting time vanish so that $\theta_w=-1$ yields the maximum value of minimum active cavity size from Eq. (13) as

$$(14) \quad \text{Max}(R_{C,min}^*) = \left\{1 - \left[1+\sqrt{(k\rho C)_\ell/(k\rho C)_s}\right]^{-1}\right\}^{-1}$$

which is shown in Fig. 5. Having assumed the surface to be made up of cavities of this (fictitious) size, one could proceed to calculate the needed quantities such as growth times, departure sizes, and nucleation frequencies, and see if the results proved useful. By considering this figure-of-merit, dwell time between bubbles has been neglected. Therefore, calculation of growth rate through standard methods, coupled with a departure size criteria, will then yield the nucleation frequency at a given site, this frequency being a maximum value.

Bubble Departure Size

By balancing drag and surface tension forces, the departure of a bubble is given by,

$$(15) \quad R_d = K \left(\frac{4\sigma R_c}{C_D \rho_\ell u_\ell^2}\right)^{1/2}$$

where K accounts for, among other things, the fraction of the surface tension forces acting in opposition to the drag, and u_ℓ is the average velocity acting over the bubble. It is assumed that all the bubbles grow entirely within the laminar sublayer. (This assumption has been confirmed by the result.) The average velocity is given by

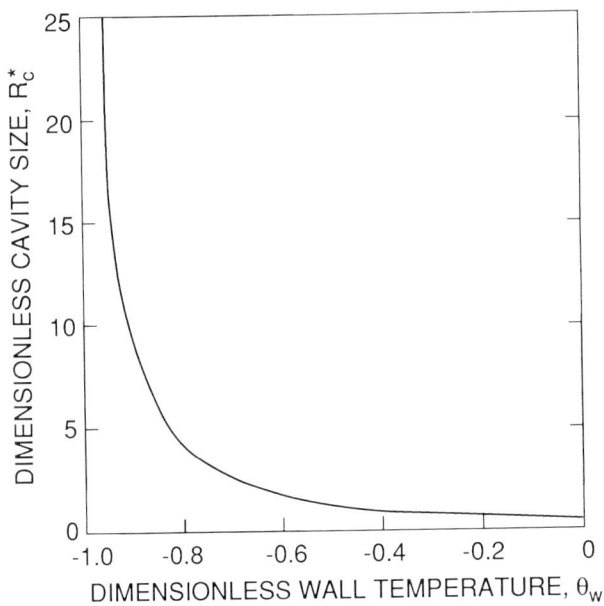

FIGURE 4. Minimum Active Cavity Size as a Function of Wall Temperature.

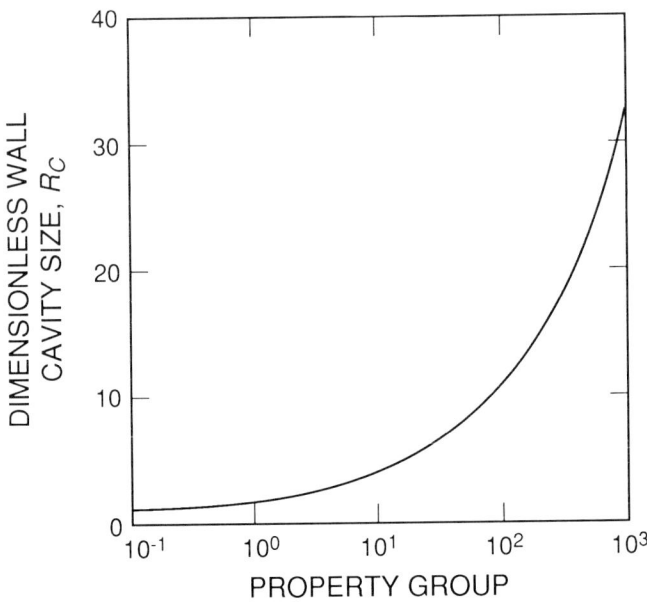

FIGURE 5. Figure-of-Merit for Nucleating Surfaces. Maximum Value of Minimum Active Cavity Size.

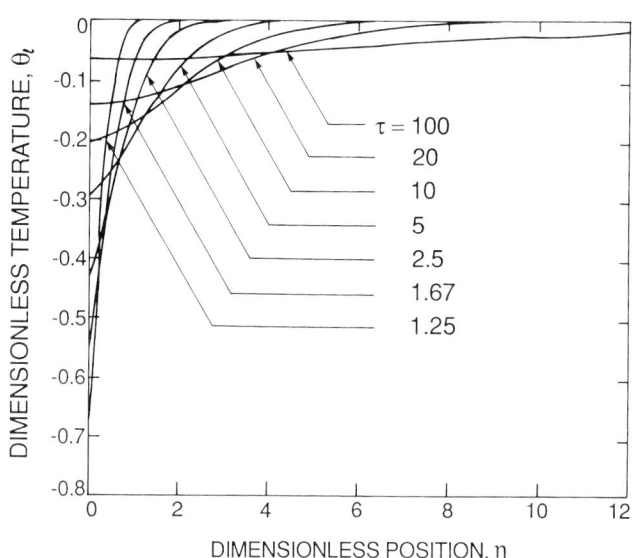

(a) Liquid Temperature

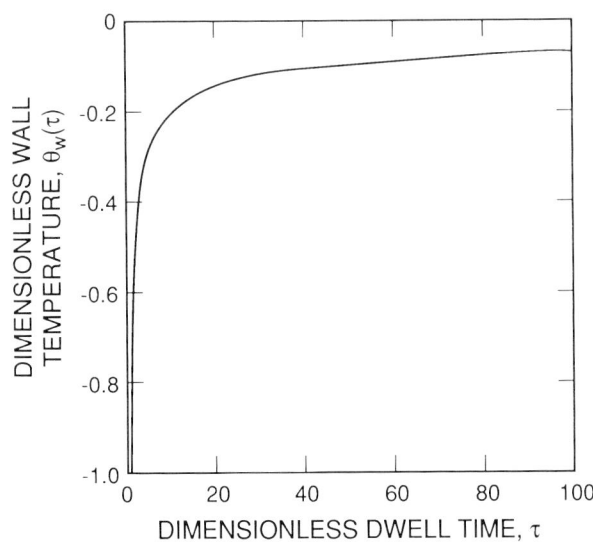

(b) Contact Line Temperature

FIGURE 3. Dimensionless Liquid Temperature Profiles and Contact Line Temperature During the Waiting Period.

NUCLEATION AND FLASHING

The transient temperature distribution is shown in Fig. 3a. Note that by the choice of parameters, the different combinations of material properties are automatically accounted for without change in dimensionless temperature profiles.

The contact line (wall) temperature time behavior is obtained from (9) by letting the distance from the wall, y, vanish to obtain

$$(11) \quad \begin{cases} \Theta_w(\tau) = \Theta_\ell(0,\tau) \\ \Theta_w(\tau) = -\frac{2}{\pi} \sum_{n=0}^{\infty} \frac{(2n)!}{(n!)^2 2^{2n}(2n+1)(\sqrt{(\tau)})^{2n+1}} \end{cases}$$

It is readily seen that as the dwell time progresses, the contact line temperature (Fig. 3b) increases back toward the superheat temperature. At some point, this temperature will be sufficiently large to sustain stable nucleation at a given cavity.

Following Hsu [1962], the expression for the minimum active cavity size comes directly from the Laplace equation for a metastable bubble combined with the Clausius-Clapyron equation to yield

$$(12) \quad R^*_{c,min} = \frac{R_{c,min} h_{fg} \Delta T_{sup}}{2\sigma T_{sat} v_{fg}}$$

so that

$$(13) \quad R^*_{c,min} = \left\{ 1 + \Theta_w(\tau)\left(1+\sqrt{(k\rho C)_\ell/(k\rho C)_s}\right)^{-1} \right\}^{-1}$$

This stability criterion is shown in Fig. 4. As the waiting time increases, Θ_w vanishes and the minimum active cavity size approaches that associated with the uniform superheat. As the dwell time vanishes, Θ_w approaches its limiting value of negative unity and only large cavities can nucleate since the wall superheat would be a minimum.

The difficulty in this analysis, of course, is identical to the difficulty in previous stability criteria developed for boiling. While this procedure is useful in qualitatively explaining observations, it requires quantitative information about the nucleating surface characteristics to be useful. In the next section, an approach shall be taken which will provide a useful quantitative method of using the results of the previous analysis.

(2) $$\frac{\partial^2 T_s}{\partial y^2} = \frac{1}{\alpha_s}\frac{\partial T_s}{\partial t} \quad \text{and} \quad \frac{\partial^2 T_\ell}{\partial y^2} = \frac{1}{\alpha_\ell}\frac{\partial T_\ell}{\partial t}$$

subject to the initial conditions

(3) At $t=0$, for $y<0$, $\left\{ T_s(y,0) = T_{sat} + \Delta T_{sup} \operatorname{erf}\left(\frac{-y}{2\sqrt{(\alpha_s t)}}\right) \right.$

and

(4) At $t=0$, for $y>0$, $\left\{ T_\ell(y,0) = T_{sup} \right.$

where t_g is the growth time of the precursor bubble. Boundary conditions are

(5) At $y=0$, $T_s(0,t) = T_\ell(0,t)$

and

(6) At $y=0$, $k_s\dfrac{\partial T_s}{\partial y} = k_\ell\dfrac{\partial T_\ell}{\partial y}$

and

(7) As $y \to \infty$, $T_\ell(y,t)$ is finite

and

(8) As $y \to -\infty$, $T_s(y,t)$ is finite

The solution for the liquid temperature distribution in dimensionless terms is

(9) $$\left\{ \begin{array}{l} \Theta_\ell(\eta,\tau) = \left\{ \operatorname{erf}\left(\dfrac{\eta}{\sqrt{(\tau-1)}}\right) - \operatorname{erf}\left(\dfrac{\eta}{\sqrt{(\tau)}}\right) - \cdots \right. \\[2ex] \cdots - \dfrac{2}{\pi} e^{-\eta^2/\tau} \displaystyle\sum_{n=0}^{\infty} \dfrac{(-1)^n H_{2n}(\eta/\sqrt{(\tau)})}{n! 2^{2n}(2n+1)(\sqrt{(\tau)})^{2n+1}} \end{array} \right.$$

where

(10) $$\left\{ \begin{array}{l} \eta \equiv \dfrac{y}{2\sqrt{(\alpha_\ell t_g)}} \;;\quad \tau \equiv \dfrac{t+t_g}{t_g} \\[2ex] \Theta_\ell(\eta,\tau) \equiv \left(\dfrac{T_\ell - T_{sup}}{T_{sup} - T_{sat}}\right)\left(1+\sqrt{(k\rho C)_\ell/(k\rho C)_s}\right) \end{array} \right.$$

NUCLEATION AND FLASHING

1. The problem is one dimensional and actual cavity geometry is ignored.

2. Properties of liquid and solid are constant.

3. There is no contact resistance between solid and liquid.

4. Liquid and solid are semi-infinite and stationary.

5. Liquid has uniform superheat far from the solid.

6. Convection of liquid is negligible.

The problem is broken up into two intervals: bubble growth period to departure; dwell time between departure and nucleation. A detailed solution to this problem may be found in Jones and Shin [1986].

Stage 1--growth period. During this time, the vapor temperature inside the bubble, and hence at the vapor-solid interface, is at saturation temperature according to the internal bubble pressure. It is assumed that this pressure decreases rapidly and has a value characteristic to that according to the local liquid pressure. The wall temperature, then, is assumed to start at uniform superheat and have an instantaneous decrease to local saturation temperature. The solution to this problem is well known to be

$$(1) \quad \frac{T_s(y,t)-T_{sat}}{T_{sup}-T_{sat}} = \frac{T_s-T_{sat}}{\Delta T_{sup}} = \mathrm{erf}\left(\frac{-y}{2\sqrt{(\alpha_s t)}}\right)$$

where the distance into the solid, y, is negative. Of course the assumption implies the wall equilibrates to the superheat temperature during the dwell times and is strictly applicable only to the case where the dwell times are long compared to the growth times. Nevertheless, this temperature profile will be used where such is not the case.

Stage 2--dwell period. During this period, the precursor bubble departs, uniformly superheated liquid impacts the solid, and an interfacial contact temperature is established providing in each a boundary layer. This layer then diffuses into both solid and liquid layers causing each to equilibrate towards the uniform superheat temperature.

The solution to stage-1 shall provide the initial conditions for the solution to this stage. In this case, both liquid and solid are considered with one dimensional conduction equations for each. The problem to be solved is thus,

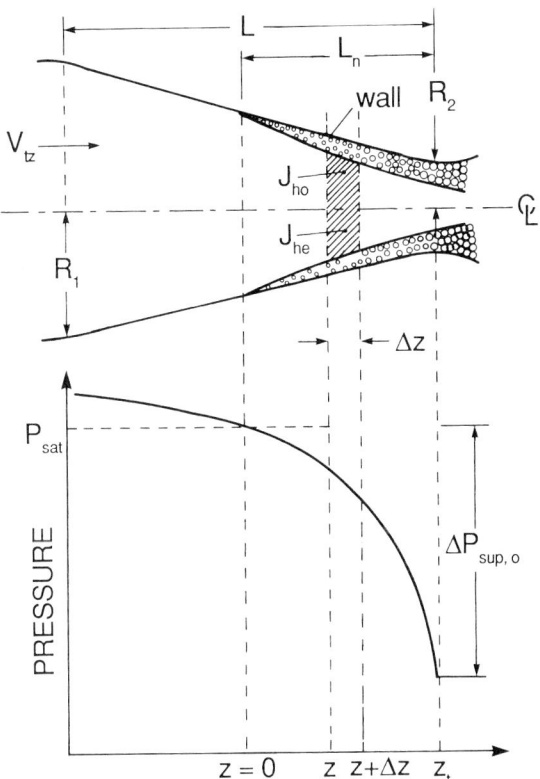

FIGURE 1. Schematic of nucleation zone in a converging Nozzle.

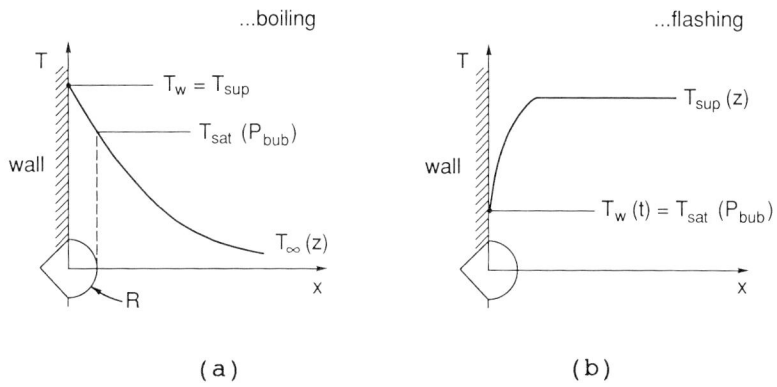

FIGURE 2. Stability Criteria for Cavity Nuclei. (a) Subcooled Boiling; (b) Flashing.

the formation of condensation nuclei in shock tubes [Wegener, 1969, 1975]. The net result is a limited nucleation zone wherein virtually all the bubbles which go into the subsequent void development process are formed.

Following the nucleation zone, bubbles will grow as they flow downstream, generally limited by conduction processes at liquid-vapor interfaces. It is expected that spherical bubble growth processes will dominate void development in the early stages of the process, and perhaps up to quite high void fractions owing to the rapid transport times and the limited times available for bubble agglomeration in high speed flows. This problem will be treated in two companion papers (Blinkov, et al., 1990a,b).

The analysis of this problem, therefore, will be broken up into three stages: nucleation, growth, and departure at a single cavity; overall nucleation zone and bubble transport to the throat; void development downstream of the throat. The first two will be discussed in turn. The third will be the subject of the second paper in this sequence (Blinkov et al., 1990a). Single cavity nucleation theory in uniform superheat has been previously reported [Jones and Shin, 1986] as well as the use of this theory for calculating continuous nucleation in the superheated upstream of the throat in nozzles [Shin and Jones, 1987]. These will only be summarized herein.

Single Cavity Nucleation

The first problem is to develop a criterion for determining when cavities are active. Consider the differences between flashing and boiling. One major difference was the way in which the thermal layers adjacent to the walls behaved. As shown in Fig. 2a, the wall layer in subcooled boiling is superheated and a bubble at a cavity would be stable and support growth if it were completely confined in the superheated layer (Hsu's criterion).

For flashing (Fig. 2b), the liquid is uniformly superheated away from the cavity. At the cavity during growth, the wall drops to saturation and then recovers as the wall and liquid start to equilibrate following departure of the vapor nucleus. The equivalent to Hsu's nucleation criterion for flashing is that a stable bubble will nucleate when the wall temperature increases to the saturation temperature inside a bubble of radius identical to that of the cavity.

An expression for the transient liquid temperature and hence the contact line temperature following departure can be determined by analyzing the combined fluid-solid conduction problem at a nucleation site but ignoring the geometry of the site itself. The following assumptions are made:

considerations to a nucleating surface to obtain bubble number densities.

Review Summary

Up until quite recently, then, the general state of knowledge relating to the flashing of initially subcooled liquids was as follows:

1. The general consensus is that thermal nonequilibrium plays a strong role in the behavior of flashing of initially subcooled liquids in nozzles.

2. Numerous models for nucleation inception have been suggested. None have been found to be adequate and all generally were found to rely on two or more empirical estimates such as bubble number density, initial void fraction, and the like. None have any general validity.

3. The initial flashing process has been largely treated as an inception phenomena. Little has been done to consider the distribution of nucleation over the entire superheated region even though it is well known that this is actually the case.

4. Flashing zone in nozzles is confined in practice to a region very close to the location of minimum pressure. No adequate explanation for this behavior exists.

5. The process of nonequilibrium vapor generation is an initial-value, path dependent process. Description of the initial value or initiation of this process has yet to be determined on other than an "ad hoc" basis.

ANALYSIS

The transition from liquid to a two-phase mixture through short pipes and nozzles by flashing usually takes place in several stages. Initially subcooled liquid encounters an acceleration-controlled pressure drop (Fig. 1) and is brought to a saturated state. A further decrease in pressure causes the liquid to become superheated. As liquid superheat is obtained, bubble nucleation starts. Nucleation is a strong function of the thermodynamic state of the superheated liquid.

Nucleation rates will first increase to a maximum value. Nucleation has been shown to occur on solid boundaries, at least in laboratory situations, and so heterogeneous wall nucleation shall be the mode considered herein. It is expected that the nucleation rate will subsequently decrease due to evaporative cooling of the liquid boundary layers adjacent to solid surfaces. In this respect it is similar to

the location or distribution of the heterogeneous nucleation sites. Simpson et al. [1962], Rohatgi and Reshotko [1975], Malnes [1975], and Ardron [1978] all assumed that heterogeneous nucleation sites occur in the bulk fluid. However, the BNL experimental work of Zimmer et al. [1979] and of Abuaf et al. [1981] showed conclusively that the important nucleation sites are at the perimeter of the channel, at least for small ducts. That this is generally the case in all geometries, however, has not yet been shown and indeed, is quite open to question.

Nucleation

Numerous studies have been undertaken with bubble nucleation on the surface in boiling. Among them, Bankoff [1958] studied the theoretical thermodynamic aspects of nucleation process. He found that the free energy difference required for nucleation on a solid phase can be smaller than, equal to, or larger than that for the homogeneous phase, depending on whether the solid geometry is a cavity, a perfect flat or a protruding point.

Later, Hsu [1962] analyzed the physics of bubble nucleation from cavities by using a model of Hsu and Graham [1961]. Hsu postulated that bubble growth from an incipient nucleation site would begin when the whole surface of the hemispherical bubble surmounting this site was at a temperature greater than the equilibrium temperature corresponding to a bubble of that radius. His simplified nucleation criterion for boiling was found to agree with the experimental results of Clark et al. [1959] and Griffith and Wallis [1959].

Han and Griffith [1965] proposed a similar nucleation criteria to Hsu's except that the constant was altered. Both of the proposed models for nucleation require knowledge of the thermal boundary layer thickness and were verified qualitatively by Bergles and Rosenhow [1962].

Shoukri and Judd [1978] developed a theoretical model for bubble departure frequency in boiling as a function of the cavity radius as well as the wall superheat and liquid subcooling. In their model, the time variations of the wall temperature through the bubble cycle are incorporated.

Jones and Shin [1986] used Hsu's concept but showed that nucleation at cavities in uniform superheat had somewhat different physics than in boiling. They developed a one sided nucleation activation criterion for such cases.

Shin and Jones [1987] reported on the use of this activation criterion together with a method which allowed characteristic cavity values for a given surface to be obtained. The result provided the first method of coupling active cavity void fraction data. The model also required a priori knowledge of the liquid superheat at flashing inception to obtain reasonable agreement with Zimmer et al.'s [1979] data for void fractions up to 0.2.

Abuaf et al. [1980, 1983] utilized a combination of the Alamgir-Lienhard and Jones approaches to determine the superheats at the throat in nozzles. They hypothesized that void development upstream of the throat was negligible. Critical flow rates were then based on single-phase flow equations using the throat pressure taken based on the predicted superheat. Results were within 5% of the experimental values for a wide range of experiments reported in the literature.

Levy and Abdollahian [1982] used a slight modification to the Alamgir Lienhard correlation based on the data of Reocreux [1976] and Zimmer et al. [1979]. Their final expression for critical flow rate was virtually identical to those proposed by both Abuaf et al. [1980, 1983] and Fincke [1981].

Schrock and Amos [1984] conducted experiments for flashing flows through slits. They found that the IHE model predicts the measured mass flux quite well in spite of considerable underprediction of the exit pressure. They also found that the prediction of Alamgir and Lienhard for flashing inception to be quite inaccurate in some cases, but no reasons were given.

Summary. In general, most models failed in prediction of critical flow rates outside the data base utilized for setting parameters in the correlation. Critical flow data for initially subcooled flows data up to about 1981 had one thing in common: flow rates are consistently lower than predictions based on equilibrium theory. The consensus was that significant thermal nonequilibrium must be present to account for the observations. Predictions of throat superheat improved calculations of critical flows but did little to advance the understanding of the process.

Analyses of these situations required two independent parameters at the "start" of the flashing process. These parameters generally were the equivalent of initial bubble number density and size. All bubble growth models used were applicable only to uniformly superheated liquids. Jones and Zuber [1978], hofwever, demonstrated that constant pressure theories of Plesset-Zwick [1954], Forster-Zuber [1954], or Scriven [1959] failed when applied to a transient pressure field.

All modeling efforts until recently were found to assume that the bubbles become uniformly distributed in the flow regardless of the nucleation site, but no consensus exists on

Several works on flashing phenomena with cryogenic liquids in a converging-diverging nozzle were reported by Simoneau [1975] and by Hendricks et al. [1976]. Their data included the axial pressure distribution along the nozzle and the critical flow rates for various subcooled inlet conditions. They found that the fluid appears to be metastable liquid upstream of the throat and that no model adequately describes the whole range of the experiment.

Zimmer et al. [1979] reported carefully controlled, degased, steady flow experiments with detailed pressure and two dimensional void profiles. Flashing inception superheats were found important in determining void development downstream of the inception point. For the first time, void data were obtained showing negligible voids at the throat.

Fincke [1981] and Fincke et al. [1981] conducted experiments to investigated the role played by grid-produced turbulence on the nonequilibrium choking process in short nozzles with subcooled inlet conditions. They reported pressure profiles, critical mass fluxes and turbulent intensities upstream of the nozzle throat.

Celata et al. [1982] investigated the critical flow of subcooled liquid through different channels. They reported data of critical mass flow rate at different degrees of inlet subcooling and static pressure profile along the channel.

Richter [1981] introduced a two-fluid model which required assumption of an initial bubble size and a fixed nucleation site density at onset of flashing. He chose an initial bubble number density of 10^{11} m^{-3}, and an initial bubble diameter of 2.5×10^{-5} m. Marviken test data were adequately predicted but prediction of other data such as that of Reocreux [1976] required adjustments.

The matter of initial superheats was first examined by Alamgir and Lienhard [1981] who developed a semi-empirical correlation motivated by classical nucleation theory to predict the pressure undershoot at the onset of flashing below the staturation pressure during a rapid depressurization in water covering the range from 0.004 to 1.8 Mbar/sec.

Jones [1980] incorporated the model of Alamgir and Lienhard into a Lagrangian model for flowing decompression in pipes and nozzles to include the effects of turbulence, but did not extend the dynamic decompression range.

Saha et al. [1981] developed a model for net vapor generation rate in each flow regime downstream of the throat. The bubble number density and initial void fraction at the flashing inception point was varied to obtain optimum fits with the

Malnes [1975] used a conduction-controlled bubble growth law similar to Simpson and Silver [1962]. Again, two dimensionless empirical constants were needed for data prediction. Those obtained based on data of Henry, et al. [1970], however, did not predict well the data of Fauske [1964].

Rohatgi and Reshotko [1975] also used a method similar to that of Simpson and Silver [1962], again needing two unknown parameters for closure which they determined from experimental data. Simoneau's [1975] nitrogen experiments were predicted by choosing a nucleation site density of about 10^6 m^{-3}.

Fritz, et al. [1976] pointed out that the bubble number density was the most uncertain factor in their model. They recommended bubble number densities between 10^8 and 2×10^{10} m^{-3} for water temperatures of 250°C to 325°C.

Wolfert [1976] again found that two independent parameters needed to be chosen to make comparisons with data. Different data required different initial void fractions and number densities of bubbles. Values of the order of 10^9 m^{-3} for number densities were chosen together with a minimum void faction of 10^{-6} corresponding to a minimum bubble diameter of 7.2 μm.

Aguilar and Thompson [1981] used number density and superheat as the two independent parameters and chose 10^9 m^{-3} as a bubble density. Constant superheat did not adequately predict the non-equilibrium flashing in rapid depressurization.

Ardron [1978], also introduced a nucleation model based on the kinetic theory and used a simple growth rate model for bubbles. The two empirical factors they chose were nucleation superheat (3°C) and nucleation site density (10^6 m^{-3}).

Winters and Merte [1979] formulated a model that treated the expanding two-phase fluid as a pseudo-homogeneous mixture of uniformly distributed, heat transfer dominated, spherical vapor bubbles surrounded by superheated liquid, but obtained poor agreement with his critical flow data.

Rivard and Travis [1980] based their model on a description of turbulence-enhanced, thermal diffusivity in the liquid and a Weber number criterion for bubble size. They chose an initial void fraction of 2×10^{-4} and initial bubble number density of 10^9 m^{-3}, which resulted in an initial bubble diameter of 7.2×10^{-5} m. Calculated critical flow rate deviated from the Semiscale blowdown tests up to 20% shortly after initiation of blowdown.

NUCLEATION AND FLASHING

Many experiments have been undertaken with critical flow in nozzles and orifices. Bailey [1951] conducted one of the earliest experiments to investigate the metastability of initially subcooled water finding only small superheats. Brown [1961] and Schrock et al. [1977] later observed throat superheats up to nearly 100 degrees centigrade reporting pressure profiles and critical flow rates. They noted that both number and size of bubbles are needed for prediction.

Powell's [1961] subcooled-inlet critical flow data having only inlet and exit pressures and inlet temperatures remained poorly predicted until Abuaf et al. [1980, 1983] obtained data indicating negligible throat void fraction and predicted the data based on the throat superheat model of Alamgir and Lienhard [1981].

Simpson and Silver [1962] used a kinetically-derived, nucleation rate model along with the Plesset and Zwick [1954] bubble growth model to calculate flashing critical flows. This model gave good agreement at small superheat but diverged from experimental data at large superheat.

Edwards [1968] used a conduction-controlled bubble growth law at constant superheat and constant pressure to formulate the mechanism of vapor generation during depressurization but needed to make an assumption regarding initial sizes and number densities.

A slip flow model was developed by Moody [1966] but is based on thermodynamic equilibrium. Its simplicity has resulted in its widespread usage in nuclear safety analysis and provides an upper bound prediction of critical flow rate for many two-phase flow conditions.

Henry and Fauske [1971] accounted for nonequilibrium effects at the choking plane as $x = Nx_e$ where N was a parameter. They matched their results well with the experiments for qualities above 1% but it was later found that N needed to be adjusted to predict other results such as the 1979 Marviken critical flow data adequately. A similar approach was used by Bauer et al. [1976].

Jones and Saha [1977], Riebold, Reocreux, and Jones [1981], and Jones [1982] have subsequently shown that linear relationship suggested by Henry and Fauske neglects the real relaxation phenomena. Fluids in thermal nonequilibrium tend toward the equilibrium state at rates governed by the interfacial processes. They showed that this relaxation was independent of the equilibrium path driving the phase change.

Sozzi and Sutherland [1975] found that geometry played an important role in critical flow phenomena, particularly for short nozzles. Critical mass flux was observed to decrease with increasing throat diameter.

INTRODUCTION

The discharge rate of subcooled liquids from pressurized containers is of interest in many situations which apply to the safety of chemical, process, and nuclear equipment. Critical flow phenomena limit the discharge rates, but the analysis of critical flow rate becomes difficult due to both mechanical and thermal nonequilibrium effects. An accurate knowledge of flashing phenomena is essential to the determination of critical flow rate and it has been previously shown to be the key to predicting such discharges when the inlet flows are subcooled [Abuaf, Jones, and Wu, 1980, 1983].

Abuaf, Zimmer, and Wu [1981] showed that the flashing inception of initially subcooled liquids was experimentally observed to be confined to regions very close to the throat so that single-phase calculations gave critical flow rates within a few percent. No theoretical explanation for these observations, however, was available and only the models of Alamgir and Leinhard [1981] and of Jones [1980] were available to predict "inception" superheat thus giving throat pressures.

It is well known that both initial size and number density of bubbles or equivalent are needed to provide closure for calculations of void development downstream of the throat in nozzles [c.f. Malnes, 1975; Rohatgi and reshotko, 1975; Wolfert, 1976; Ardron, 1978; Aguilar and Thompson, 1981; Ardron]. Typical past practice was to provide an initial estimate of initial void fraction (typical number much less than 1%) and number density (typically 10^8 to 10^{13} m-3). Void growth calculations have not been in good agreement with existing data in either trends or magnitude. It is the purpose of this paper to describe a composite model for continuous nucleation in the superheated zone upstream of the throat. It will be shown in two companion papers that these results lead to reasonable calculations of void development downstream.

BACKGROUND

Flashing Inception and Critical Flows

Early literature reviews of critical discharge of flashing flows have been done by Saha [1978] and by Hsu [1972]. Homogeneous equilibrium models were shown to underpredict critical discharge rates for short pipes and near saturation or subcooled upstream conditions due to the liquid superheat and resultant underprediction of void fraction. Abdollahian et al. [1982] reviewed two-phase critical flow models varying in complexity from simple correlation-type models to complex two-fluid models.

NUCLEATION AND FLASHING OF INITIALLY SUBCOOLED LIQUIDS IN NOZZLES: 1. A DISTRIBUTED NUCLEATION MODEL

T.S. SHIN and O.C. JONES, JR.
Rensselaer Polytechnic Institute
Troy, New York 12180-3590

ABSTRACT

It is well known that both number and size of bubbles must be accurately determined for initial calculation of flashing void development downstream of restrictions. This paper presents a new method of accurately determining both which results in accurate calculation of downstream void development.

A wall cavity model is described for use in the calculation of nucleation rates and bubble number densities at flashing inception in nozzles, and subsequently in the calculation of void development downstream of minimum area zones. The model is based on the physics of the nucleation phenomena in flashing and considers transient conduction to be the sole means of heat transfer from the superheated liquid to the vapor bubble.

The activation criterion developed for site nucleation is one sided due to the uniform superheat rather than two sided as in boiling. A figure of merit for the particular fluid solid combination is then determined which yields minimum nucleation surface energy per site. Characteristic site nucleation frequencies, and number densities of nucleation sites of given sizes are then obtained from data.

A bubble transport equation is used to predict the number density and size of bubbles at the throat. Throat superheats are calculated with a standard deviation of 1.9K for throat superheats up to ~100K and expansion rates between 0.2 bar/sec to over 1 Mbar/sec, extending previous correlations by more than three orders of magnitude. Throat void fractions for all data found in the literature are less than 1% confirming earlier assumptions and allowing nozzle critical flow rates to be calculated with an accuracy of ~3%.

[1]Currently with Korean Advanced Energy Research Institute, Daeduk Science Town, Korea

41. Schrock, V. E., Revankar, S. T., Mannheimer, R., Wang, C-H, and Jia, D.,1986c, Steam-Water Critical Flow Through Small Pipes from Stratified Upstream Regions, <u>HEAT TRANSFER 1986</u>, Ed. C. L. Tien, V. P. Carey and J. K. Farrell, Vol. 5, pp 2307-2311.

42. Schrock, V. E., Revankar, S. T., Lee, S. Y., and Wang, C-H, 1986d, A Computational Model for Critical Flow Through Intergranular Stress Corrosion Cracks, Lawrence Berkeley Laboratory, Engineering Div., Univ. of California, Report LBL-21967.

43. Simoneau, R. J., 1974, Two-Phase Choked Flow of Subcooled Nitrogen Through a Slit, NASA Lewis Research Center, NASA TM-71516.

44. Smoglie, C., 1984, Two-phase Flow Through Small Branches in a Horizontal Pipe with Stratified Flow, Kernforschungszentram Karlsruhe, KfK 3861.

45. Sozzi, G. L. and Sutherland, W. A., 1975, Critical Flow of Saturated and Subcooled Water at High Pressure, General Electric Co., NEDO-13418.

46. Uchida, H. and Nariai, H., 1966, Discharge of Saturated Water Through Pipes and Orifices, <u>Proc. of the Third International Heat Transfer Conference</u>, Vol. 5, Chicago, IL., pp 1-12.

47. Zuber, N., 1981, Problems in Modelling of Small Break LOCA, NUREG-0724.

27. Jeandey, Ch., Gros D'Aillon, L., Bourgine, L. and Barriere, G., 1981, Auto Vaporization D'Escoulements Eau/Vapeur, *Commisariat a L'Energie Atomique Report*, T. T. No. 163.

28. Jones, O. C. and Saha, P., 1977, Non-equilibrium Aspects of Water Reactor Safety, *Symposium on the Thermal and Hydraulic Aspects of Nuclear Reactor Safety*, Vol. 1, Light Water Reactors, Jones, O. C., Jr. and Bankoff, S. G., Editors, ASME.

29. Kroeger, P. G., 1978, Application of a Non-equilibrium Drift Flux Model to Two-phase Blowdown Experiments, CSNI Specialists' Meeting on Transient Two-phase Flow, Toronto.

30. Lee, S. Y. and Schrock, V. E., 1988, Homogeneous Non-Equilibrium Critical Flow Model for Liquid Stagnation States, *ASME Proc. 1988 Natl. Heat Transfer Conf.*, Vol. 2, pp 507-513.

31. Levy, S. and Abdollahian, D., 1982, Homogeneous Non-Equilibrium Critical Flow Model, *International Journal of Heat Mass Transfer*, Vol. 25, No. 6, pp 759-770.

32. Lubin, B. and Hurwitz, M., 1966, Vapor Pull-Through at a Tank Drain With and Without Dielecrophoretic Buffling, *Proc. Conf. Long Term Cryopropellant Storage in Space*, NASA Marshall Space Centre, Huntsville, AL, pp 173-180.

33. Mayfield, M. E. et al., 1980, Cold Leg Integrity Evaluation - Final Report, NUREG/CR-1319, Battelle Columbus Laboratories.

34. Reimann, J. and Smoglie, C., 1983, Flow Through a Small Break at the Top of a Large Pipe with Stratified Flow, Annual Meeting of the European Two-phase Flow Group: Zurich, Switzerland.

35. Reimann, J. and Khan, M., 1984, Flow Through a Small Break at the Bottom of a Large Pipe with Stratified Flow, *Nucl. Sci. Eng.* Vol. 88, pp 297-310.

36. Reocreux, M., 1978, Contribution to the Study of Critical Discharge of Steam-Water, Ph.D. Thesis, Medical Univ. of Grenoble, 1974, NUREG-tr-0002.

37. Rouse, H., 1956, Seven Exploratory Studies in Hydraulics, *J. Hydr. Div. Proc. ASCE*, Vol. 82, pp 1038-1-35.

38. Saha, P., 1978, A Review of Two-phase Steam-Water Critical Flow Models with Emphasis on Thermal Non-equilibrium, NUREG/CR-0417.

39. Schrock, V. E., Revankar, S. T. and Lee, S. Y., 1986a, Critical Flow Through Pipe Cracks, *Proc. 4th Miami International Symposium on Multiphase Transport and Particulate Phenomena*.

40. Schrock, V. E., Revankar, S. T., Mannheimer, R., and Wang, C-H, 1986b, Small Break Critical Discharge-The Roles of Vapor and Liquid Entrainment in a Stratified Two-phase Region Upstream of the Break, LBL-22024 and NUREG/CR-4671.

Analysis: The CLYSTERE System Code, presented at <u>OECD/NEA Specialists' Meeting on Transient Two-phase Flow</u>, Toronto.

13. Boivin, J. Y., 1979, Two-phase Critical Flow in Long Nozzles, <u>Nuclear Tech.</u>, Vol. 46, pp 540-545.

14. Celata, G. P., Cumo, M., Farello, G. E., and Incalcaterra, P. C., 1983, Critical Flow of Subcooled Liquid and Jet Forces, <u>Interfacial Transport Phenomena</u>, Ed. by J. C. Chen and S. G. Bankoff, p. 109, ASME, NY.

15. Chexal, B., Abdollahian, D., and Norris, D., 1984, Analytical Prediction of Single Phase and Two-Phase Flow Through Cracks in Pipes and Tubes, <u>AIChE Symp. Series - Heat Transfer, Niagara Falls</u>, 236, Vol. 80, p. 19.

16. Collier, R. P., Stulen, F. B. Mayfield, M. E., Pope, D. B. and Scott, P. M., 1980, Two-Phase Flow Through Intergranular Stress Corrosion Cracks and Resulting Acoustic Emission, EPRI Final Report NP-3540-LD, RP-T118-2.

17. Craya, A., 1949, Theoretical Research in the Flow of Non-Homogeneous Fluids, <u>La Houille Blanche</u>, pp 44-55.

18. Easton, C. R. and Catton, I., 1970, Nonlinear Free Surface Effects in Tank Draining at Low Gravity, <u>AIAA J.</u> Vol. 8, pp 2195-2199.

19. Edwards, A. R., 1968, Conduction Controlled Flashing of a Fluid, and the Prediction of Critical Flow Rates in a One-Dimensional System, UKAEA, ASHB(S) R-147.

20. Elias, E. and Chambre, P. L., 1984, A Mechanistic Nonequilibrium Model for Two-phase Critical Flow, <u>Int. J. of Multi-phase Flow</u>, Vol. 10, No. 1, pp 21-40.

21. Engineering Div. of Crane Co., 1976, Flow of Fluids Through Valves, Fittings and Pipes, Crane Co., Tech. Paper No. 410.

22. Fauske, H. K., 1965, The Discharge of Saturated Water Through Tubes, <u>Chemical Engineering Progress Symposium Series</u>, Vol. 61, No. 59, pp 210-216.

23. Gariel, P., 1949, Experimental Research on the Flow of Non-Homogeneous Fluids, <u>La Houille Blanche</u>, pp 56-64.

24. Hendricks, R. C., Simoneau, R. J. and Usu, Y. Y., 1975, A Visual Study of Radial Inward Flow of Subcooled Nitrogen, in: <u>Advances in Cryogenic Engineering</u>, Vol. 20, K. D. Timmerhaus, ed., Plenum Press, New York, pp 370.

25. Henry, R. E., 1970, The Two-Phase Critical Discharge of Initially Saturated or Subcooled Liquid, <u>Nuclear Science and Engineering</u>, Vol. 41, pp 336.

26. Ishimoto, S., Uematsu, M., Tanishita, I., 1972, New Equations for the Thermodynamic Properties of Saturated Water and Steam, <u>Bulletin of the JSME</u> Vol. 15, No. 88, p. 1278.

$$R = \frac{1}{v}\left[(v_g - v_\ell) - \frac{(h_g - h_\ell)}{c_{p_\ell}}\left(\frac{\partial v_\ell}{\partial T_\ell}\right)_p\right] \tag{B-30}$$

The parameter β accounts for non-equilibrium mass transfer, i.e.,

$$\beta = \beta_{HEM}\left\{1 + (x_E - x)/\tau^*\right\} \tag{B-31}$$

where β_{HEM} is the factor related to the equilibrium mass transfer due to local pressure change:

$$\beta_{HEM} = \left\{v - x\left(\frac{dh_g}{dp}\right)_{sat} - (1-x)\left(\frac{dh_f}{dp}\right)_{sat}\right\}/h_{fg} \tag{B-32}$$

REFERENCES

1. Abdollahian, D., Healzer, J., Janssen, E. and Amos, C., 1975, Critical Flow Data Review and Analysis, EPRI NP-13418.

2. Abdollahian D. and Chexal, D., 1983, Calculation of Leak Rates Through Cracks in Pipes and Tubes, EPRI Final Report NP-3395 RP-1757-19.

3. Abuaf, N., Jones, O. C. and Wu, B. J. C., 1983, Critical Flashing Flows in Nozzles with Subcooled Inlet Conditions, Transactions, ASME, Journal of Heat Transfer.

4. Agostinelli, A., Salemann, V. and Harrison, N. J., 1958, Predictions of Flashing Water Flow Through Find Annular Clearances, ASME Transactions, Vol. 80, pp 1138-1142.

5. Alamgir, M. D. and Lienhard, J. H., 1981, Correlation of Pressure Undershoot During Hot Water Depressurization, Journal of Heat Transfer, Vol. 103, No. 1, pp. 52-55.

6. Amos, C. N. and Schrock, V. E., 1983, Critical Discharge of Initially Subcooled Water Through Slits, NUREG/CR-3475, LBL-16363.

7. Amos, C. N. and Schrock, V. E., 1985, Two-Phase Critical Flow in Slits, Nuclear Science and Engineering, Vol. 88, No. 3, pp. 261-274.

8. Anderson, J. L. and Owca, W. A., 1985, Data Report for the TPFL Tee/Critical Flow Experiments, NUREG/CR-4164 Draft Report.

9. Anon., 1979, The Marviken Full Scale Critical Flow Tests, Summary Report, NUREG/CR-2671, MXC-301.

10. Ardron, K. H., 1978, A Two-Fluid Model for Critical Vapor-Liquid Flow, International Journal of Multiphase Flow, Vol. 4, pp. 323-337.

11. Arlotto, G. A., 1986, Leak-Before-Break Seminar Opening Remarks, pp. 1-5 in LEAK-BEFORE-BREAK: International Policies and Supporting Research, NUREG/CP-0077.

12. Bauer, E. G., Houdayer, G. R., and Sureau, H. M., 1976, A Non-equilibrium Axial Flow Model and Application to Loss-of-Coolant Accident

$$Ja = \frac{\rho_\ell cp_\ell \Delta T_{sub}}{\rho_g h_{fg}}$$

Three data sources were used to develop the following correlation. They are those of Amos and Schrock (1985), Schrock et al.(1986d) and Marviken (1979).

The data for saturated liquid stagnation states are represented by the correlation

$$f_1 = \begin{cases} 0.6385 Re^{0.1} - 1.689, & (Re \leq 5.5 \times 10^5) \\ 0.70, & (Re \geq 5.5 \times 10^5) \end{cases} \qquad (B-26)$$

and is shown in Fig. 9. The subcooling effect is represented by

$$f_2 = 1.40 - 0.40 \exp(-0.28 Ja) \qquad (B-27)$$

and is shown in Fig. 10. The predicted values of S is compared with the experimental values for the subcooled data in Fig. 11.

The S correlation $S = f_1(Re) f_2(Ja)$ was used in the model predictions applied to the various sources of experimental data.

Critical Flow Criterion

So far there are several criteria for critical flow condition advanced in the literature. Among these Kroeger's (1978) sound speed was for the choking criterion by Amos and Schrock and in the present model. Kroeger (1978) derived the sound speed for non-equilibrium two-phase flow. He used the method of characteristics applied to a four-equation drift flux model, which assumed the vapor phase to be saturated at the local pressure. For zero drift velocity (no slip) the sound speed was given by

$$a = a_{HF}(1 - R\beta a^2_{HF}/v)^{-1/2} \qquad (B-28)$$

In the above equation a_{HF} (homogeneous frozen sound speed) was given by

$$a_{HF} = v \left\{ \frac{1}{c_{p_\ell}} \left(\frac{\partial v_\ell}{\partial T_\ell} \right)_p \left[x \left(\frac{dh_g}{dp} \right) - v \right] - x \frac{dv_g}{dp} - (1-x) \left(\frac{\partial v_\ell}{\partial p} \right)_{T_\ell} \right\}^{-1/2} \qquad (B-29)$$

This sound speed corresponds to that of a particular situation in which no heat or no mass transfer takes place between vapor and liquid; thus the quality remains constant throughout the expansion. Frozen flow is a good approximation when the flow transit time inside a channel is less than the time constant of phase transition process, as may be the case in the high quality region, but not in flashing low quality flow. Therefore, the term in parentheses in Eq.(B-28) represents the effect of non-equilibrium due to flashing delay.

R is the non-dimensional two-phase mixture compressibility given by

CRITICAL FLASHING FLOW

b) When flashing occurs in the fully-developed flow region ($p_{F\ell} < p_{ent}$),

$$z_{F\ell} = L_e + (p_o - \frac{1}{2}(K_e + K_f)\rho_o u_m^2 - p_{sat}(T_o) + \Delta p_{F\ell})D_e / (\frac{1}{2} f \rho_o u_m^2) \quad (B-21)$$

where f = friction factor in fully-developed region.

The entrance length was given by Schlichting, i.e., for laminar flow

$$\frac{L_e}{D_e} = 2.875 \times 10^{-2} \, Re \quad (B-22)$$

for turbulent flow

$$\frac{L_e}{D_e} = 1.487 \, Re^{0.25} . \quad (B-23)$$

Criterion for Flashing Inception

Little work has been done examining the point of flashing inception. An experimental and theoretical study about flashing delay was performed by Alamgir and Lienhard (1981) and by Jones and Saha (1977). The Alamgir-Lienhard correlation was developed from the static depressurization data of Lienhard et al. in a semi-empirical way. Wall nucleation (heterogeneous nucleation) was assumed to be responsible for the flashing initiation in the metastable liquid.

Following Amos and Schrock, the Alamgir and Lienhard relation for pressure undershoot was used as a basis of flashing inception in the present model. However, as observed by Amos the pressure undershoot in channel flow was observed to be smaller than in Lienhard's static vessel experiments. This suggests that vapor/gas trapped in wall cavities and not molecular fluctuations is responsible for inception of flashing in channels. The pressure undershoot at flashing was taken as the Alamgir-Lienhard relation multiplied by the modifying factor, S, i.e.,

$$\Delta p_{F\ell} = S \frac{0.252 \, \sigma^{1.5} T_r^{13.73} (1 + 14 \Sigma^{0.8})^{0.5}}{(k_B T_c)^{0.5} (1 - \frac{v_\ell}{v_g})} \quad (B-24)$$

Pressure undershoot at flashing ($\Delta P_{F\ell}$) is defined as

$$\Delta p_{F\ell} = p_{sat}(T_\ell(z_{F\ell})) - p(z = z_{F\ell}) \quad (B-25)$$

Hence, the point at which flashing inception occurs directly affects the initial liquid superheat, and the amount of liquid superheat can be expected to play an important role in vapor generation under flashing critical flow conditions. The factor S (which ranges from 0 to 1) was correlated in terms of Reynolds number and Jakob number, which are related to fluid velocity, geometrical size, and initial subcooling, i.e.,
$S = f(Re, Ja) = f_1(Re) f_2(Ja)$.

where $Re = \dfrac{D_e \rho u}{\mu}$,

$$c_{p_\ell} = \frac{9.677258 \times 10^7 \, e^{T_r} - 7.752401 \times 10^6 \, T_r^4 - 1.081453 \times 10^7 \, T_r^3 - 5.193637 \times 10^7 \, T_r^2 - 9.562871 \times 10^7 \, T_r - 9.691202 \times 10^7 + 4.470480(1 - 1.2 \, p_r^2)}{\left[0.905 + 0.095 \left(\frac{T_{sat}(p)}{T_c}\right)^8 - T_r\right]^{\sqrt{3}}} \quad (B\text{-}18)$$

The solution of the differential equations described in this section depends on the flashing condition.

i.e., at $z = z_{F\ell}$,

$x = 0$, $\quad$ (B-19)

$p(z=z_{F\ell}) = p_{sat}(T_\ell(z_{F\ell})) - \Delta p_{F\ell}$, where $\Delta p_{F\ell}$ can be found from Eq.(B-24)

$t_r = 0$

$T_\ell = T_0$

$u = u_{F\ell}$.

In Eq.(B-4), the time constant (τ) related to the relaxation was chosen to give the best agreement between the pressure profile predictions and the experimental data. For the present model a value of 1.0 sec. for τ is recommended where Amos chose $\tau = 1.1 \times 10^{-3}$ seconds. The influence of τ upon the critical mass flux was found to be weak. When τ is changed from 10^{-5} sec. to 10^2 sec., only a 10% increase in mass flux is predicted. The lower values of τ used by Amos and Schrock resulted in poorer agreement between measured and predicted pressure profiles. It should be noted that for $\tau = 1$ the relaxation of the metastable condition is just balanced by the pressure gradient (which tends to create metastability).

In the above equations, the flashing inception position ($z_{F\ell}$) has to be evaluated. This position is evaluated for the given stagnation state (p_0, T_0) by the following procedure:

a) When flashing occurs in the entrance region ($p_{F\ell} > p_{ent}$),

$$z_{F\ell} = L_e(p_0 - \tfrac{1}{2} K_e \rho_0 u_m^2 - p_{sat}(T_0) + \Delta p_{F\ell})/(\tfrac{1}{2} K_f \rho_0 u_m^2) \quad (B\text{-}20)$$

where K_e = entrance loss coefficient associated with kinetic energy change.

$= 1 - (\text{area ratio})^2$

K_f = average distributed entrance loss coefficient in the entrance region

$= f_e \dfrac{L_e}{D_e}$

The average distributed friction factor (f_e) in the entrance region was evaluated using a relation of Schlichting, giving K_f of about 0.9.

CRITICAL FLASHING FLOW

$$\left.\frac{\partial v_\ell}{\partial p_\ell}\right|_{T_\ell} = -v_\ell \kappa \tag{B-10}$$

where κ = isothermal compressibility

$$= 6.570 \times 10^{-7} v_\ell$$

$$\left.\frac{\partial v_\ell}{\partial T_\ell}\right|_p = v_\ell \beta \tag{B-11}$$

where $\beta = 2.693455 \times 10^1 \, e^{T_r} - 9.360687 \times 10^1 \, e^{-T_r} - 1.672643 \times 10^1 \, T_r^3 + 3.125661 \times 10^1 \, T_r^2 - 1.198906 \times 10^2 \, T_r + 6.659029 \times 10^1$

For the evaluation of $\left.\frac{\partial h_\ell}{\partial p_\ell}\right|_{T_\ell}$, Gibb's equation can be used:

$$dh = T\,ds + v\,dp \tag{B-12}$$

$$\left.\frac{\partial h_\ell}{\partial p}\right|_{T_\ell} = T_\ell \left.\frac{\partial s}{\partial p}\right|_{T_\ell} + v_\ell \tag{B-13}$$

With the use of Maxwell relations, Eq.(B-13) becomes

$$\left.\frac{\partial h_\ell}{\partial p_\ell}\right|_{T_\ell} = v_\ell (1 - \beta T_\ell) \tag{B-14}$$

Finally, in order to provide closure to the differential equations, additional auxiliary equations must be specified.

In Eq.(B-2), the frictional pressure gradient, as proposed by Levy and Abdollahian (1982), is given by

$$\left(\frac{dP}{dz}\right)_f = -\frac{f}{(1-\alpha)^2} \frac{G^2(1-x)^2}{2D_e} v_\ell \tag{B-15}$$

f is the single-phase friction factor, i.e.,

$$\frac{1}{\sqrt{f}} = -2 \log_{10}\left(\frac{1}{3.7}\left(\frac{\varepsilon}{D_e}\right) + \frac{2.51}{Re\sqrt{f}}\right); \quad Re > 4.0 \times 10^3 \tag{B-16}$$

$$f = \frac{64}{Re}; \quad Re < 4.0 \times 10^3 \tag{B-17}$$

In the energy Eq.(B-3), a specific heat relation for metastable liquid is required. From curve fitting of Lienhard's graphical estimation of specific heat of superheated water we obtain

mass conservation:

$$\frac{1}{u}\left(\frac{du}{dz}\right) = \frac{(v_g - v_\ell)}{v_m}\left(\frac{dx}{dz}\right) + \frac{1}{v_m}\left\{x\left(\frac{dv_g}{dp}\right)_{sat} + (1-x)\left(\frac{\partial v_\ell}{\partial p}\right)_{T_\ell}\right\}\left(\frac{dp}{dz}\right) + \frac{(1-x)}{v_m}\left(\frac{\partial v_\ell}{\partial T_\ell}\right)_p\left(\frac{dT_\ell}{dz}\right) \qquad (B-1)$$

momentum conservation:

$$\frac{u}{v_m}\left(\frac{du}{dz}\right) = -\left(\frac{dp}{dz}\right)_f - \left(\frac{dp}{dz}\right) + \frac{g}{v_m} \qquad (B-2)$$

energy conservation:

$$\left\{x\left(\frac{dh_g}{dp}\right)_{sat} + (1-x)\left(\frac{\partial h_\ell}{\partial p}\right)_{T_\ell}\right\}\left(\frac{dp}{dz}\right) + (h_g - h_\ell)\left(\frac{dx}{dz}\right) + (1-x)c_{p_\ell}\left(\frac{dT_\ell}{dz}\right) + u\left(\frac{du}{dz}\right) = 0 \qquad (B-3)$$

The liquid superheat relation is given by the following equation:

$$\left(\frac{dp}{dT_\ell}\right)_{sat}\left(\frac{dT_\ell}{dz}\right) - \left(\frac{dp}{dz}\right) + \left(\frac{dp}{dz}\right)_{F\ell} \exp\left(-\frac{t_r}{\tau}\right) = 0 \qquad (B-4)$$

where t_r is the residence time of the two-phase mixture after flashing inception.

The residence time (t_r) is given by

$$t_r = \int_{z_{F\ell}}^{Z} \frac{dz'}{u(z')} \qquad (B-5)$$

In order to complete the homogeneous non-equilibrium model, the equations of state are required. They can be represented functionally as

$$v_g = v_g(p) \qquad (B-6)$$

$$h_g = h_g(p) \qquad (B-7)$$

$$v_\ell = v_\ell(p, T_\ell) \qquad (B-8)$$

$$h_\ell = h_\ell(p, T_\ell) \qquad (B-9)$$

The main dependent variables for this model are pressure (p), liquid temperature (T_ℓ), mixture fluid velocity (u), and flow quality (x).

In conservation equations for this model property gradients for the metastable liquid were obtained from the extrapolation of VDI data, i.e.,

CRITICAL FLASHING FLOW

Here n_2, n_3 are respectively, number per unit length of contractions and expansions, α is the angle of convergence, and β is the ratio of lower to higher diameter.

Close observation of photomicrographs of the IGSCC reveals that the bends, contractions and expansions each appear in the crack channel at intervals of approximately 10δ where δ is the average crack thickness (opening). For evaluating the loss coefficients K_i, the bends in the crack were approximated on the average as 45° bends. The ratio r/D was approximated a constant 0.60. Then the loss coefficient K_1 for bends can be calculated as

$$K_1 = 8f_p + \frac{7f_p}{n_1 L_1} \tag{A-15}$$

To evaluate the loss coefficient due to contraction and expansions, the angle of convergence was approximated to be 45 degrees and $\beta = 0.5$ was used.

The number of bends per unit length was evaluated as the

$$n_1 = \frac{1}{10\delta} \tag{A-16}$$

The number of contractions or expansions per unit length was evaluated as

$$n_2 + n_3 \frac{1}{10\delta} \text{ with } n_2 = n_3 . \tag{A-17}$$

The method of evaluating the equivalent friction factor involves determining the friction f_p for the crack and then the loss coefficients are determined using equations (A-13) through (A-15). Then using the equation (A-11) the equivalent friction factor f is determined.

To determine the friction factor f_p, Reynolds number has to be evaluated. As the inlet and outlet area are different the Reynolds number was taken as the average value of the Reynolds numbers at inlet and outlet plane of the crack. The surface roughness was taken as $\varepsilon = 1.78\mu$. This value for surface roughness was used in the BCL LEAK 00 code. Depending on the relative surface roughness ε/D and Re the friction factor was determined from the standard charts.

APPENDIX B

Improved HNEM Model for Pipes

General set of governing equations. Three mixture conservation equations are considered based on negligible interphase slip, i.e., homogeneous flow. The vapor phase is assumed to be saturated at the local pressure. The liquid phase is assumed to be superheated (metastable). Vapor properties were calculated from the steam tables, and properties of metastable liquid were evaluated by extrapolating the subcooled liquid properties into the metastable region.

Conservation equations were derived by summing the two-fluid conservation equations for each phase. The resulting mixture conservation equations do not contain any interphase transfer terms. Heat transfer at the two-phase interface is accounted for in the model by a liquid superheat relation since the vapor phase is assumed to be saturated. The conservation equations employed are as follows:

$$h = (1 - x)h_f(p) + x\, h_g(p) = h_f(p) + x\,[h_g(p) - h_f(p)] \tag{A-5}$$

(2) Single Phase Liquid

$$v_\ell \simeq \text{constant} = v_f(T_0) \tag{A-6}$$

$$h_\ell \simeq h_f(T_0) \tag{A-7}$$

Critical Flow Criterion

$$V_e = V_a = \left\{\left(-\frac{\partial p}{\partial \rho}\right)_s\right\}^{1/2} = v\left[-\left\{(1-x)\frac{\partial v_f}{\partial p} + x\frac{\partial v_g}{\partial p} + (v_g - v_f)\left(\frac{\partial x}{\partial p}\right)_s\right\}\right]^{1/2} \tag{A-8}$$

where

$$\left(\frac{\partial x}{\partial p}\right)_s = \frac{\left(\frac{\partial s_f}{\partial p}\right)(1-x) + x\left(\frac{\partial s_g}{\partial p}\right)}{(s_g - s_f)} \tag{A-9}$$

The wall shear term was represented by an equivalent friction factor in the usual form.

$$\left(\frac{dp}{dz}\right)_f = \frac{f}{D}\frac{\rho V^2}{2} \tag{A-10}$$

and

$$f = f_p + D\sum_i K_i \tag{A-11}$$

where f_p is the pipe friction coefficient, the K_i represent the loss factors for specific effects such as bends, contractions, expansion, and tortuosity, and the n_i represent the number per unit length in the flow direction of the type i. Details of the friction factor evaluation are presented below. A properties subroutine was developed using the equations for water given by Ishimoto et al.(1972). This is a very efficient and accurate package for saturation property evaluation.

Generalized prediction of equivalent friction factors. Choice of the optimum friction factor used to predict the results was based on the parametric study of BCL data. For general use of SOURCE a methodology to estimate the equivalent friction factor from global features of IGSCCs is required. Here a methodology, which has been tested against the results of the parametric study of the BCL data, is presented.

Assigning i = 1, for bends, i = 2 for contraction and i = 3 for expansions. The respective loss coefficients per unit length are given by [48](1976).

For bends $n_1 K_1 = (n_i - \frac{1}{L})(0.25\, f_p \cdot \pi \cdot (\frac{r}{D}) + 0.5\, K') + K'/L \tag{A-12}$

when n_1 is number of bends per unit length, r is the radius of the bend D is the hydraulic diameter, f_p is the tube friction factor, and $K' = 15\, f_p$ for 45° bends $K' = 30\, f_p$ for 90° bends.

For contractions $n_2 K_2 = (0.8 \sin\frac{\alpha}{2}(1-\beta^2))n_2 \tag{A-13}$

For expansions $n_3 K_3 = (2.6 \sin\frac{\alpha}{2}(1-\beta^2)^2) \cdot n_3 \; . \tag{A-14}$

CRITICAL FLASHING FLOW

ε	Roughness	-
K	Isothermal compressibility	Pa^{-1}
μ	Viscosity	kg/ms
τ	Relaxation time	s
ρ	Density	kg/m^3
Σ'	Decompression rate	Matm/s

Subscripts

c	Critical
E	Equilibrium
ent	Entrance
F_ℓ	Flashing
g	Vapor phase
i	ℓ or g
ℓ	Liquid phase

APPENDIX A

Crack Flow Model

Based upon the results of Amos's (1985), we do not believe that the flow entering the crack separates giving the type of phenomenon that Henry sought to represent by his model. Furthermore, the assumption of linear quality variation with length is not consistent with Fanno flashing flow in straight pipes and is probably not valid for crack flow. In view of the need to have a model that adds little to the running time cost of large codes into which it may be incorporated and also from the experience of Amos's calculations it appeared that the HEM is a good choice for the crack problem. Thus we have developed a computer code based upon the steady state form of the homogeneous equilibrium model. The geometry for the problem is a convergent channel with a "rough" surface. The governing equations are:

Continuity:

$$\frac{d}{dz}(\rho A V) = 0 \tag{A-1}$$

Momentum:

$$-\frac{dp}{dz} = \rho V \frac{dV}{dz} + \tau_w \frac{P_m}{A} \tag{A-2}$$

Energy:

$$\frac{d}{dz}\left(h + \frac{V^2}{2}\right) = 0 \tag{A-3}$$

State
(1) Two-Phase

$$v = (1 - x)v_f(p) + x\,v_g(p) = v_f(p) + x[v_g(p) - v_f(p)] \tag{A-4}$$

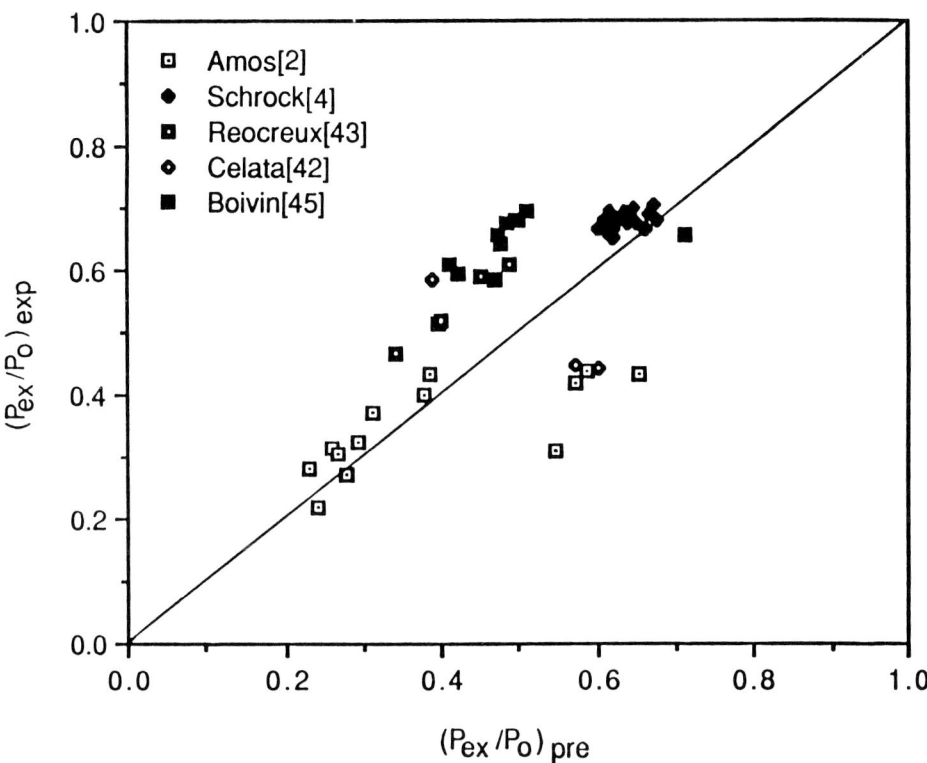

FIGURE 13. Comparison of predicted and experimental critical pressure ratio

Re	Reynolds number $\frac{\rho u D_e}{\mu}$	-
s	Specific entropy	J/kgK
T	Temperature	K
T_r	Reduced temperature, T/T_c	-
t	Transit time	s
t_r	residence time	s
u	Fluid velocity	m/s
v	Specific volume	m³/kg
V	Velocity	m/s
x	Quality	-
z	Spatial position	m
Greek		
α	Void fraction	-
β	Thermal expansion coefficient and Kroeger's non-equilibrium mass transfer parameter	
σ	Surface tension	N/m

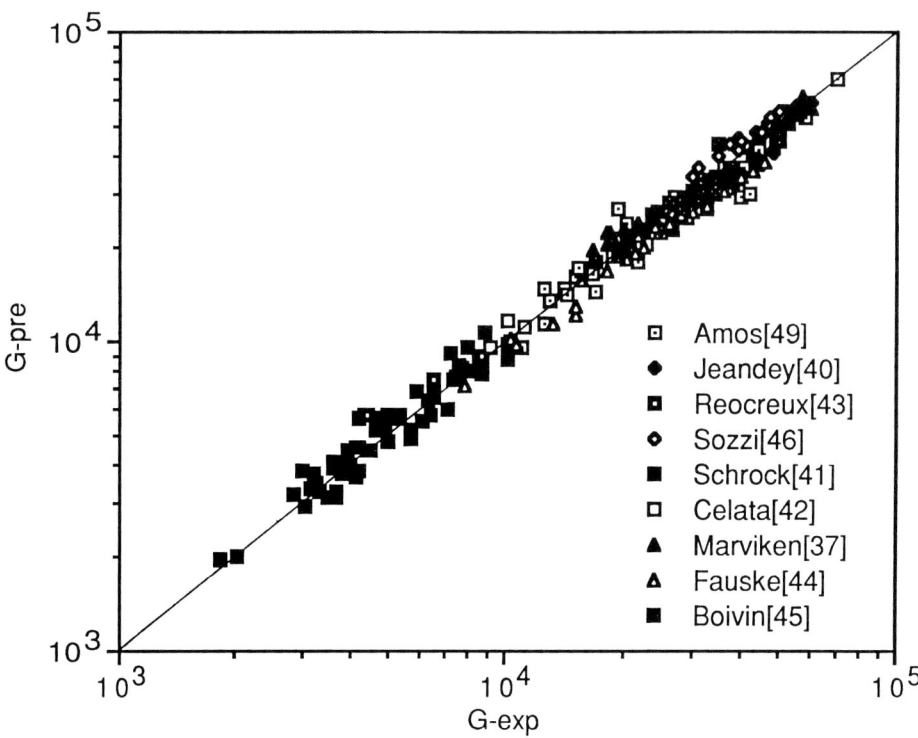

FIGURE 12. Comparison of predicted and experimental critical mass flux

C	Subcooling correction factor	-
c_p	Specific heat	J/kgK
D_e	Equivalent diameter	m
d	Diameter of branch line	-
Fr	Froude number, $V/(gd)^{0.5}$	-
f	Friction factor	-
G	Mass flux	kg/m²s
g	Gravitational acceleration	m/s²
h	Specific enthalpy	J/kg
Ja	Jakob number $\dfrac{\rho_\ell c_{p_\ell} \Delta T_{sub}}{\rho_g h_{fg}}$	-
L	Channel length	m
$\dot{m}$	Mass flowrate	kg/s
N	Viscosity number, $\mu_\ell \rho_\ell^{-0.5} \sigma^{-0.75} (g\Delta\rho)^{0.25}$	-
P	Pressure	Pa

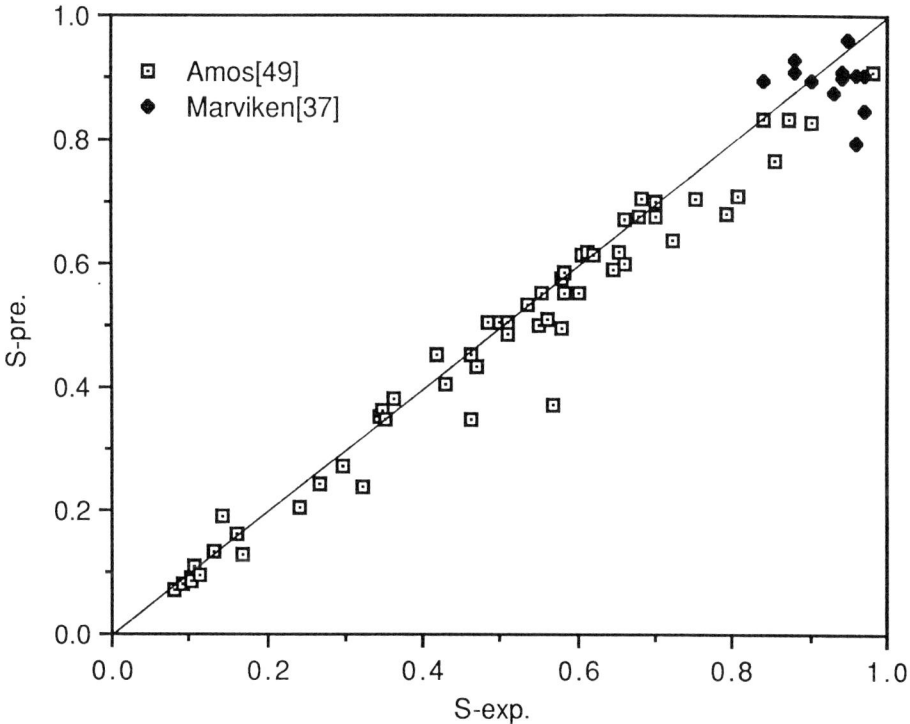

FIGURE 11. Pressure undershoot correction for subcooled data

It has been correlated in terms of Reynolds number and subcooling Jakob number, rather than just velocity, giving it a more physical basis.

In addition, the liquid superheat formulation was improved, and an improved interpretation of Kroeger's non-equilibrium sound speed was incorporated into the model.

The new model has been tested against nine data sets including experiments with thin slit geometry ($0.16 < D_e < 0.75$ mm), small diameter pipes ranging from 3.8 mm to 50 mm I.D. and large pipes ranging from 200 to 500 mm I.D. All the mass flux data are well represented by the new model within a standard deviation of 9.6%.

The modified Amos model gives excellent results over a very wide range of pipe size, of stagnation pressure and of subcooling.

NOMENCLATURE

Symbol	Meaning	Units
A	Cross sectional area	m^2
a	Sound speed	m/s
Bo	Bond number, $d\left[\dfrac{\sigma}{g\Delta\rho}\right]^{0.5}$	-

CRITICAL FLASHING FLOW

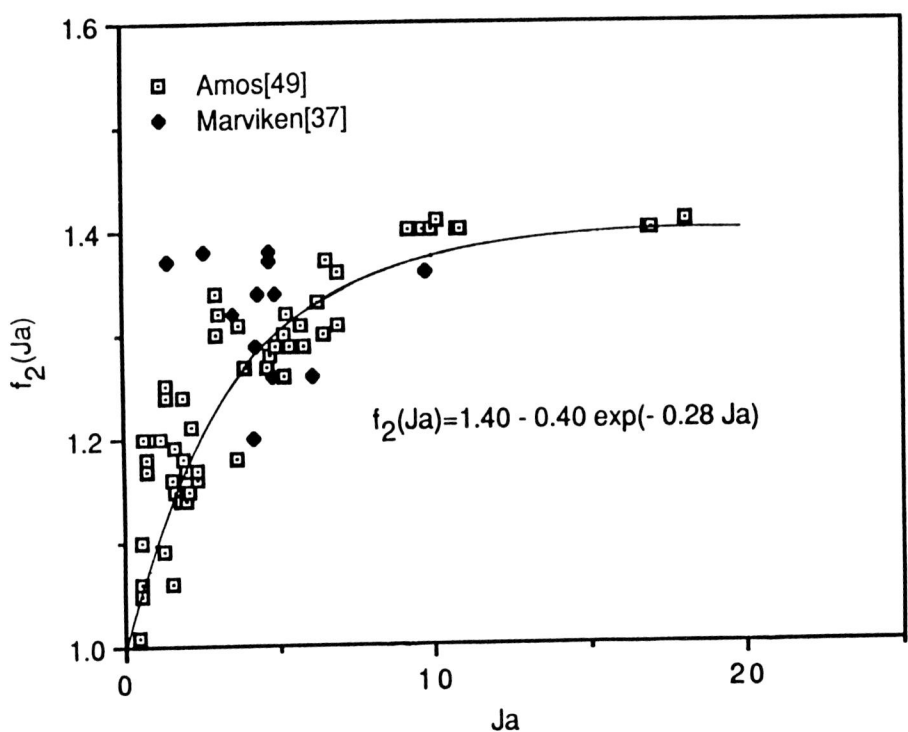

FIGURE 10. Subcooling factor for pressure undershoot

Entrainment From Stratified Upstream Regions

Critical flow through small breaks on horizontal pipes carrying stratified two-phase fluid depends strongly upon the quality of the fluid entering the break and therefore upon the phenomena of liquid entrainment and vapor pull-through. The present study of these phenomena compliments the results of air-water studies at KfK and steam-water tests done at higher pressure and larger scale at INEL. In the present study it was found that the steam-water level for incipient pull through was higher than that for air-water. A new correlation form for vapor pull-through has been developed which represents the data for both air-water and steam-water for all the data from the Berkeley, KfK, and INEL studies. The new correlation form uses the Bond and viscosity numbers as well as the Froude number to correlate the incipient height. Incipient liquid entrainment for steam-water was found to be independent of Bond and viscosity numbers and new correlations are presented in the form of Eq.(1). In addition quality correlations are presented for each break orientation for both liquid entrainment and vapor pull-through. The procedure for applying the results of the present study to the evaluation of critical flow in an LWR LOCA has been discussed.

Improved HNEM for Liquid Stagnation States

The homogeneous non-equilibrium model of Amos and Schrock has been improved by developing an improved method for predicting the S-factor which modifies the Alamgir-Lienhard pressure undershoot for incipient flashing.

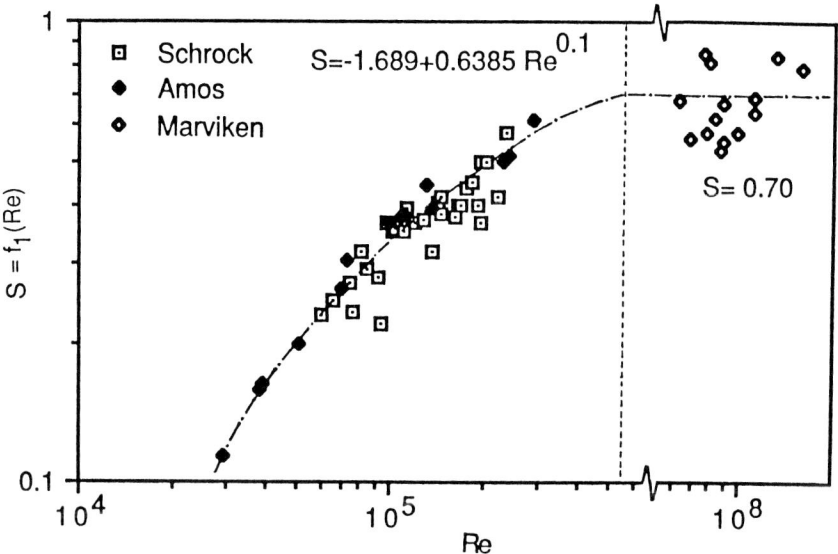

FIGURE 9. Pressure undershoot correction factor for saturated liquid (Ja = 0)

Critical pressure ratios are compared in Fig. 13. Although there is considerable scatter in the data, the improved model gives better results than did the original Amos-Schrock model.

CONCLUSIONS

Flow Through Cracks

A computer code (SOURCE) has been developed to estimate leak rates through IGSCC cracks. The basis of the code is the steady state Homogeneous Equilibrium Model with friction and area change. A general method was developed to estimate the equivalent friction factor, lumping the effects of wall friction, tortuosity of the flow path and expansions and contractions, from physical features of IGSCC. Experimental results of Collier et al. (BCL) were used to develop a subcooling dependent correction factor to apply to SOURCE critical flow predictions. Using optimum friction factors deduced from BCL data for each test section, corrected SOURCE predictions agree with the BCL measured flowrates for 61 qualified tests with a standard deviation of 1.35% and for all 82 tests with a standard deviation of 6.5%. Using the independently estimated friction factor (from the generalized method) the predictions agree with qualified data with a standard deviation of 15.9%. This is an improvement as compared with Chexal's method.

The code SOURCE is very fast running and should be adaptable to large systems codes without significant sacrifice in cost. The method developed is recommended for estimating leak rates through IGSCC.

Author (Data) [Ref.]	Stagnation Pressure (MPa)	Subcooling Jakob No.	Re	D_e (m x 10^3)	L/D_e	Standard Deviation of Exp. Data and Predictions (%)
Amos & Schrock (1983) [2]	2.73 to 15.70	0 to 18	2.0×10^4 to 5.0×10^5	0.159 to 0.748	85 to 400	10.4
Jeandey et al (1981) [40]	2.0 to 8.39	15 to 44	1.0×10^6 to 4.0×10^6	2.0	17	7.15
Celata et al. (1984) [42]	1.0 to 3.5	0 to 10	5.0×10^5 to 2.0×10^6	12.5	10 to 288	8.99
Reocreux (1974) [43]	0.2 to 0.34	9 to 25	2.0×10^5 to 6.0×10^5	20	134	13.2
Schrock et al (1986) [41]	0.2 to 1.07	0 to 0.1	2.0×10^4 to 2.0×10^5	3.76 to 6.32	19 to 33	10.2
Marviken (1979) [37]	2.6 to 5.2	0 to 10	6.0×10^7 to 3.0×10^8	200 to 509	0.3 to 3.6	10.4
Fauske (1965) [44]	0.79 to 12.07	0	2.0×10^5 to 2.0×10^6	6.35	8 to 40	7.09
Boivin (1979) [45]	1.96 to 10.1	1 to 7	5.0×10^6 to 2.0×10^7	12 to 50	37 to 58	6.08
Sozzi & Sutherland (1975) [46]	6.2 to 6.9	0 to 5	2.0×10^6 to 5.0×10^6	12.7	7 to 143	8.95

TABLE 1. Range of stagnation conditions and dimensions for comparison of experiments

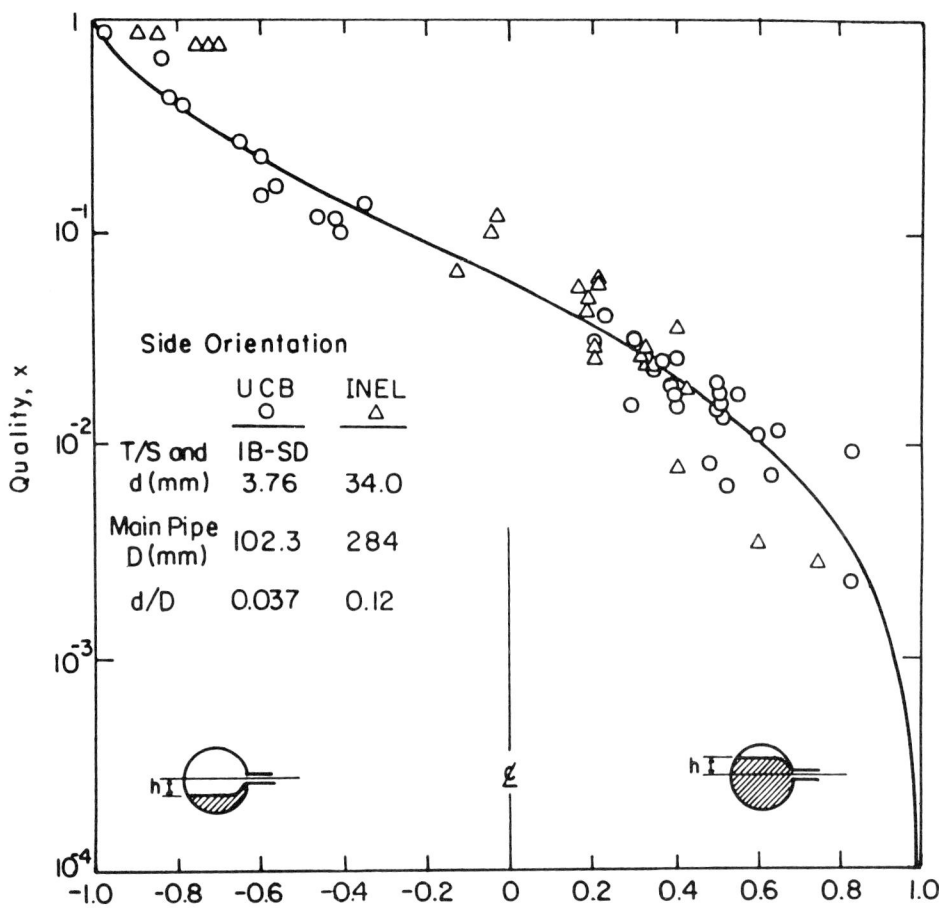

FIGURE 8. Side break quality correlation

$$f_1 = \begin{cases} 0.6385 \; Re^{0.1} - 1.689 \; ; & Re \leq 5.5 \times 10^5 \\ 0.70 \; ; & Re \geq 5.5 \times 10^5 \end{cases} \quad (11)$$

This correlation is compared with the data in Fig. 9. Subcooled stagnation state data from Amos and Schrock (1985) and Marviken (1979) are the basis for the correlation for f_2.

$$f_2 = 1.40 - 0.40 \; exp(-0.28 \; Ja) \quad (12)$$

This correlation is compared with the data in Fig. 10. Fig. 11 compares the predicted and experimental values of S.

The modified Amos model was then applied without modification to a total of nine data sets as listed in Table 1. As can be seen from the table the data span a very wide range of stagnation pressure, subcooling, Reynolds number, D_e and L/D_e. The standard deviation in critical mass flux for individual data sets ranges from 7 to 13%. For the whole group (256 data) the standard deviation is 9.6% as seen in Fig. 12.

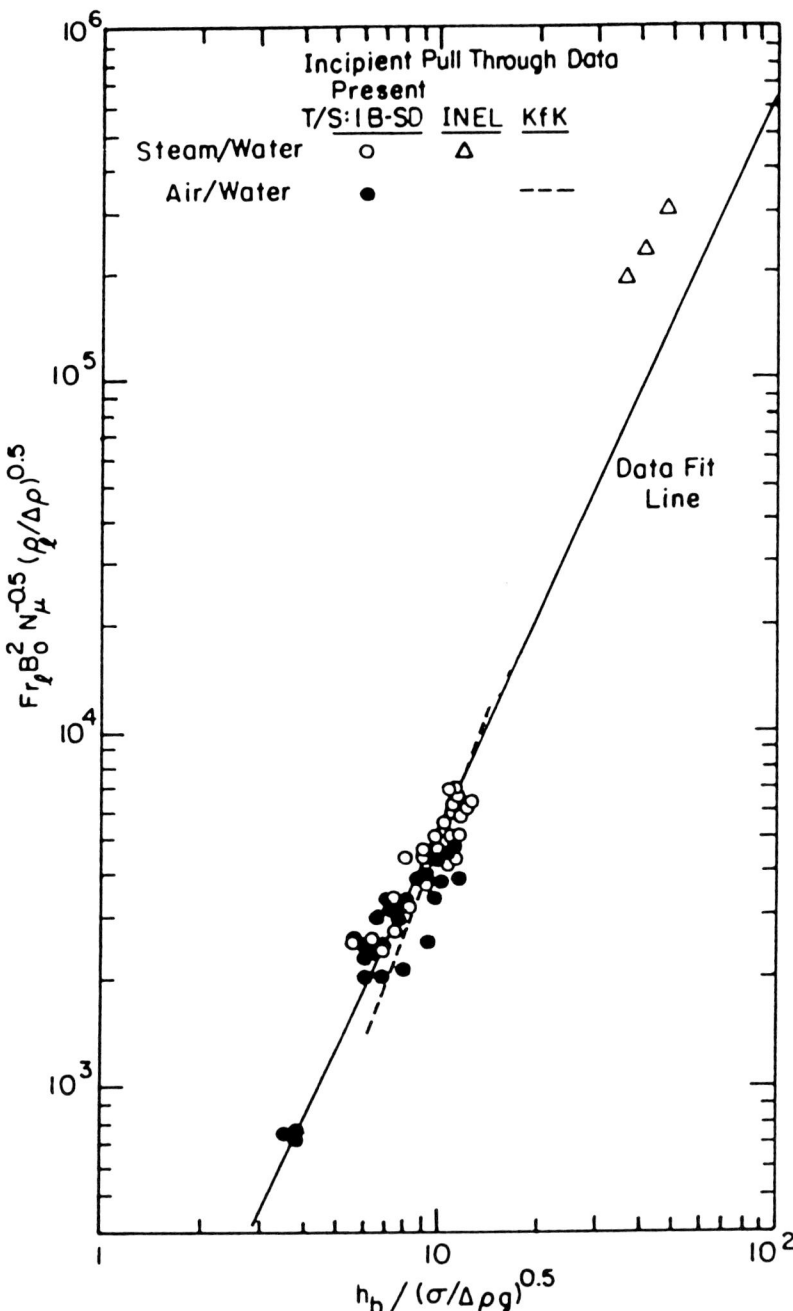

FIGURE 7. Side break incipient vapor pull-through

Side breaks. Similar to the bottom break case, the air-water and steam-water results for gas pull-through were different in the present study when presented in the form of Eq.(1). Thus similar procedures were followed leading to the incipient correlation

$$Fr_\ell \, Bo^2 \, N_\mu^{-0.5} \left(\frac{\rho_\ell}{\Delta\rho}\right)^{0.5} = 40.6 \left[\frac{h_b}{(\sigma/g\Delta\rho)^{1/2}}\right]^{2.1} \quad (6)$$

which is illustrated in Fig. 7 where it can be seen to represent well the data from Berkeley, KfK and INEL.

Liquid entrainment incipient heights for both side and top breaks were found to be independent of Bond and viscosity numbers. For the side break, liquid entrainment incipient heights are correlated by

$$Fr_g \left(\frac{\rho_g}{\Delta\rho}\right)^{0.5} = 3.25 \left(\frac{h_b}{d}\right)^{2.5} \quad (7)$$

The quality correlations for side breaks (see Fig. 8) are represented by a single expression

$$x = 0.06^{(1+h/h_b)^{0.7}} \left[1 - C_1 \frac{h}{h_b}\left(1 + \frac{h}{h_b}\right)\right] \quad (8)$$

where $C_1 = \begin{cases} 1/2 & \text{for } h/h_b \leq 0 \text{ (Liquid Entrainment)} \\ 0 & \text{for } h/h_b > 0 \text{ (Vapor Pull-Through)} \end{cases}$

where the incipient heights are from Eqs.(6) and (7) respectively for gas pull-through and liquid entrainment.

Top break. The incipient entrainment results from Berkeley data are correlated by

$$Fr_g \left(\frac{\rho_g}{\Delta\rho}\right)^{0.5} = 0.395 \left(\frac{h_b}{d}\right)^{2.5} \quad (9)$$

while the quality data are correlated by

$$x = \left(\frac{h}{h_b}\right)^{3.25(1-h/h_b)^2} \quad (10)$$

The incipient liquid entrainment data obtained in the present study did not agree with the data of KfK. No explanation has been found for this discrepancy.

Improved HNEM For Pipes

Experimental data for saturated liquid stagnation states were selected from Schrock et al.(1986a), Amos and Schrock (1985) and Marviken (1979) to obtain the basic dependence of the factor S upon Reynolds number.

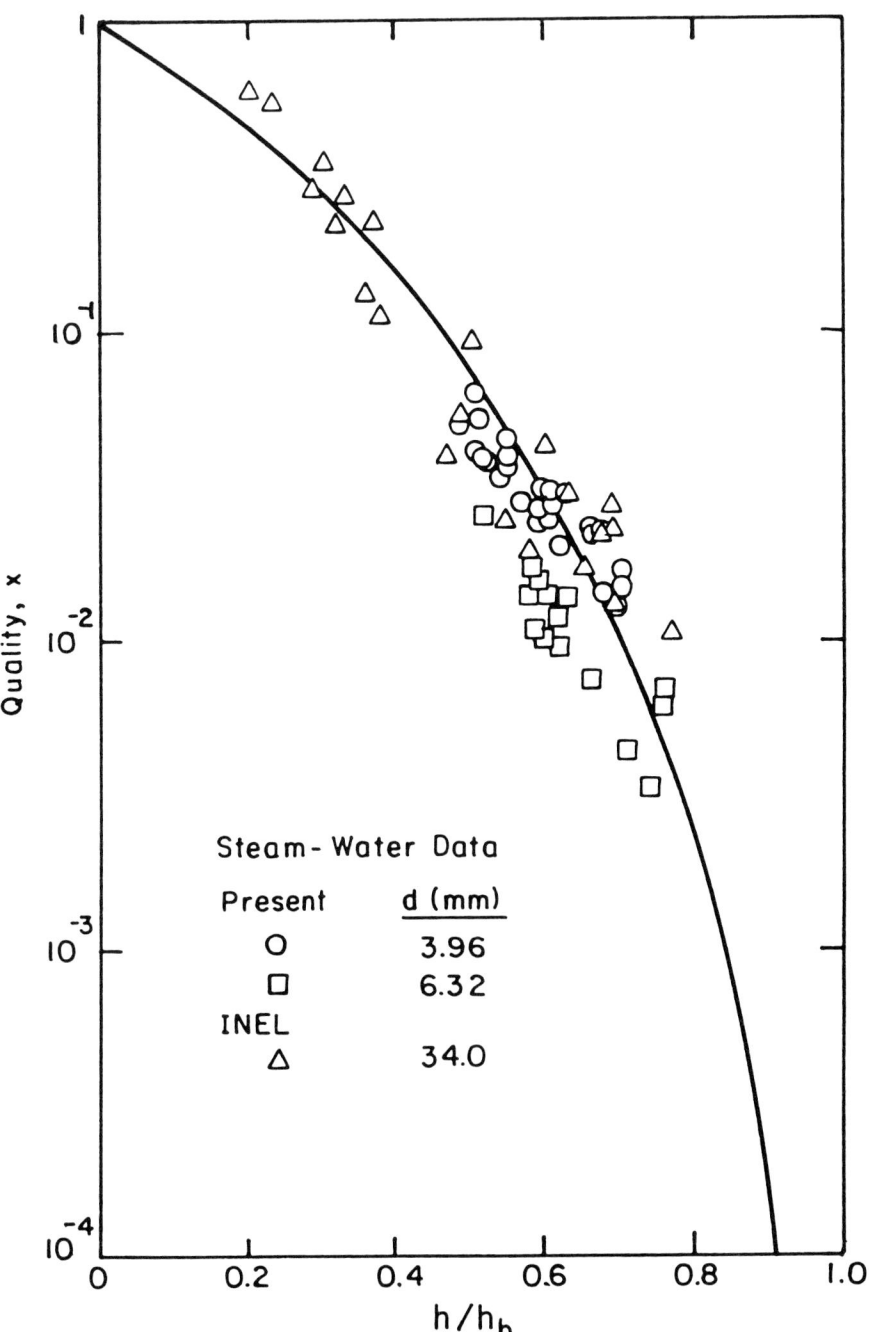

FIGURE 6. Bottom break quality correlation

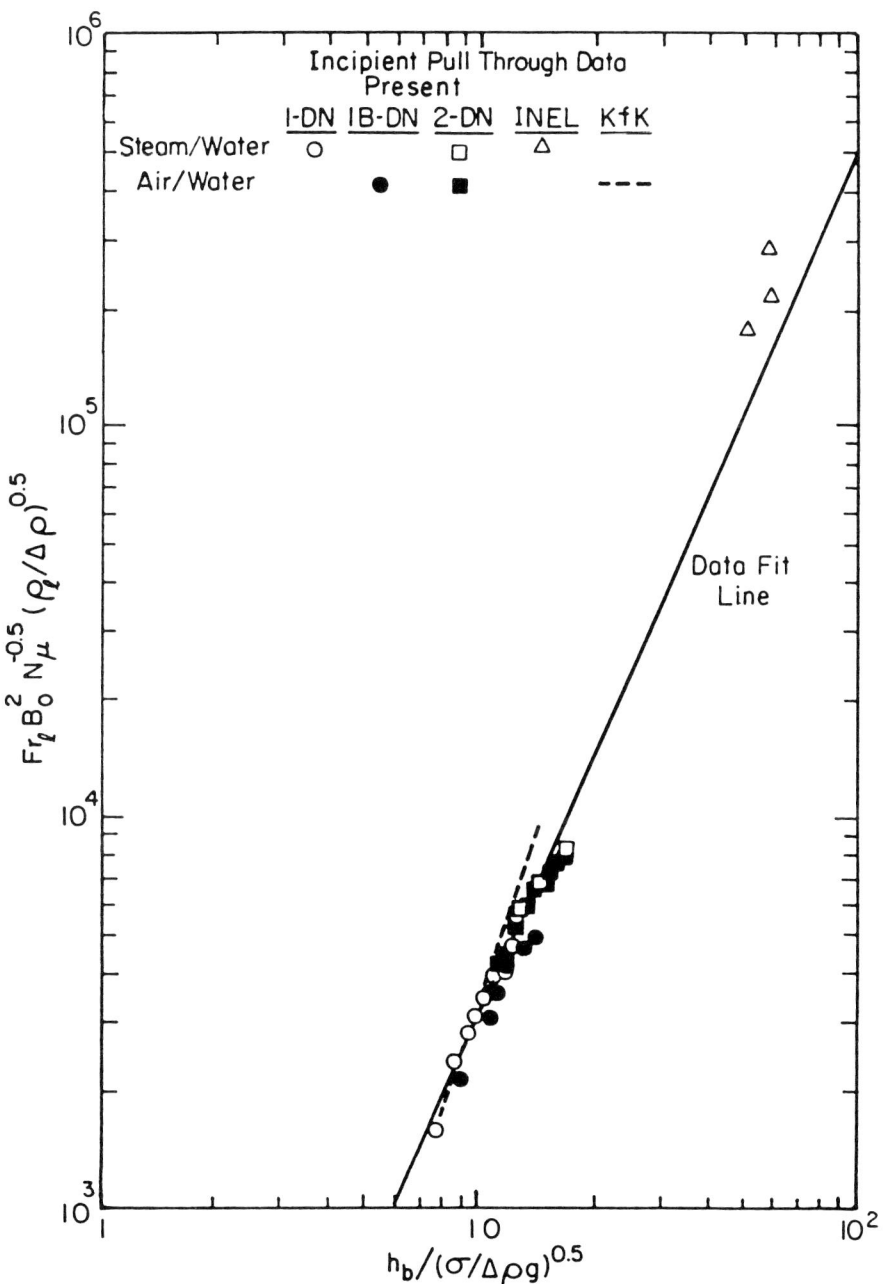

FIGURE 5. Bottom break incipient vapor pull-through

continued until the mass flowrate is found that corresponds to choking at the outlet. Details of the coding are presented elsewhere Schrock et al. (1986d). The code has been found to be very fast running and can complete a typical evaluation in about 3 seconds of CPU time on an IBM 3081.

Improved HNEM for Pipes

The improved version of Amos' model is presented in Appendix B. The major improvement is the development of a more general basis for obtaining the factor, S, which is a multiplier that modifies the Alamgir-Lienhard pressure undershoot. We have developed a correlation of the form

$$S = f_1(Re)f_2(Ja) \tag{2}$$

based upon the data Schrock et al.(1986a), Amos and Schrock (1985) and Marviken (1979). The pressure undershoot for high Reynolds number and high subcooling Jakob number approaches the Alamgir-Lienhard prediction. At low subcooling and low Reynolds number corresponding to low critical mass flux (long pipes and low stagnation pressure) the pressure undershoot becomes small.

Other improvements in the model involved the use of a larger relaxation time constant, improved use of Kroeger's sound speed and some more efficient numerical features.

RESULTS

Crack Flow Predictions

The code SOURCE was used to conduct a parametric study of the BCL data in order to determine the best fit equivalent friction factor for each of the five crack geometries. Of the 82 runs reported by BCL 21 were disqualified for this purpose because they showed major discrepancies with the general trends of data dependence upon stagnation pressure and subcooling. For each of the remaining 61 tests, SOURCE was used with a minimum of three trial values of the equivalent friction factor. Each computation produced a predicted mass flowrate which was compared to the measured value. In this way an optimum equivalent friction factor was found that best fit all the data for each test section. This involved a graphical interpolation procedure. Once the optimum equivalent friction factor was obtained it was used to again run SOURCE for each test condition. When these results were compared it was apparent that a systematic deviation existed between the prediction and the data that was dependent upon the stagnation subcooling. A correction factor was therefore developed as shown in Fig. 3. The predicted value then becomes the SOURCE result multiplied by the correction factor. The correction factor is given by

$$\begin{aligned} C &= 1.3015 - 5.3075 \times 10^{-3} \Delta T_{sub} \quad \text{for} \quad \Delta T_{sub} < 60C \\ &= 1.0 \quad \text{for} \quad \Delta T_{sub} > 60C \end{aligned} \tag{3}$$

The generalized friction factor methodology, described in Appendix A, was applied to each of the BCL test sections to evaluate their equivalent friction factors. SOURCE was again run using these equivalent friction factors to obtain new predictions, including subcooling correction, for

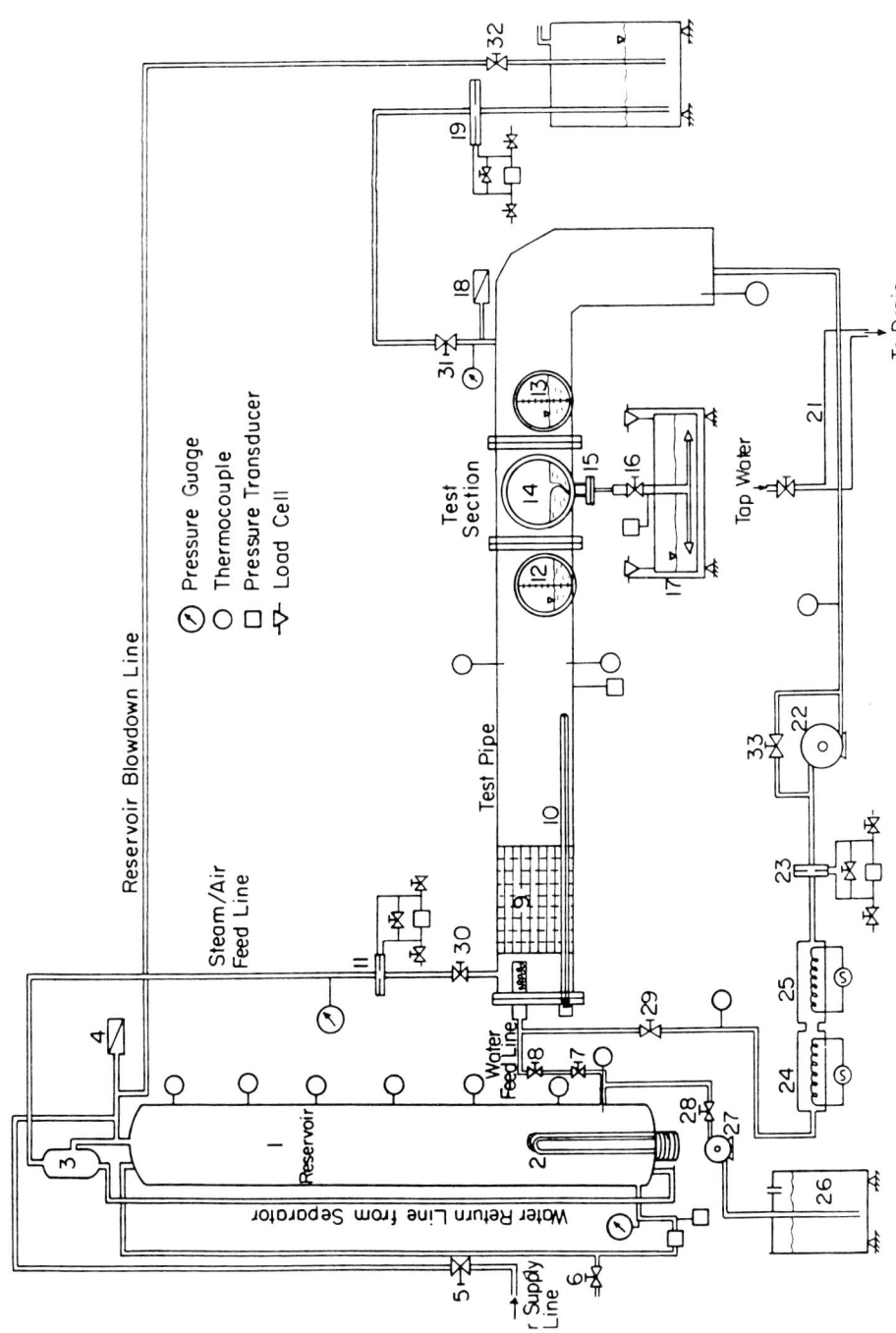

FIGURE 2. Test facility

CRITICAL FLASHING FLOW

for non-equilibrium sound speed in homogeneous flow. The conservation equations were derived from the two-fluid conservation equations with assumption of homogeneous flow. Model predictions showed good agreement with their experimental data for thin slits and with data of Jeandey et al.(1981).

EXPERIMENTS

Experiments done to study flow from a stratified upstream region were carried out in the test facility shown in Fig. 2. It consists of a pressure vessel constructed of 12 inch IPS Schedule 5 stainless steel pipe with welded end caps, 3m in height, which serves as a reservoir and a steam generator, a horizontal test pipe, a tee break section and a recirculation loop. The reservoir is heated using three adjustable 4kW immersion heaters during steam-water tests. By maintaining the reservoir pressure 15 to 35 kPa above the desired test section pressure, the steam, taken off the top and liquid flow taken off the bottom provided independently controlled flows to the test section. For air-water tests the reservoir was pressurized by the laboratory compressed air supply for pressures up to 650 kPa and using cylinders of compressed nitrogen for pressures up to 1065 kPa. The test pipe was equipped with a calming grid at the inlet. Viewing windows were placed at the break section and in the pipe just upstream and downstream of the break to allow observation visually and photographically, of the two-phase interface. The liquid passing the break was pumped back to the entrance in the recirculation loop. The steam passing the break was directed through a flow meter to a quench tank. The break flow went to a weigh tank. For air-water tests a separator and an air flow meter were also installed. In steam-water tests the steam in the discharged water was quenched in the weigh tank. Sufficient pressure and temperature measurements were provided to allow complete mass and energy balances to be performed. The data were recorded using an Auto Data Eight digital data acquisition system.

Data for incipient pull through and onset of entrainment were obtained visually for each break orientation and each fluid while simultaneously recording the system pressure and flowrates. Break flow was found from weigh tank measurements and also from level measurements in the reservoir and mass balance. Following incipient two-phase flow, the quality entering the break was obtained from mass flow measurements and the energy balance for steam-water. In the air-water tests water and air discharge rates were individually measured.

ANALYSIS

Critical Flow in Cracks

Based upon the experiments of Amos and Schrock (1985) the crack flow was modelled as homogeneous equilibrium flow with friction and area change. A subcooling dependent multiplier (1.0 to 1.3) corrects the model results. The model is presented in Appendix A. The equations were put into finite difference form and a Fortran code called SOURCE was developed. The calculation is initiated by assuming the mass flowrate and then the calculation marches downstream to obtain the fluid conditions in the channel. If choking is encountered before reaching the end of the channel the calculation is stopped and the mass flowrate is reduced for the next pass. If the calculation reaches the end of the channel with the flow still subcritical the flowrate is increased for the next pass. The procedure is

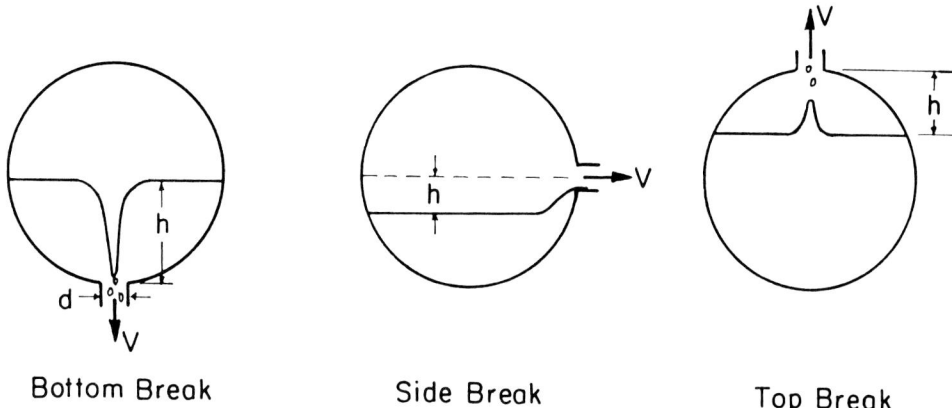

Bottom Break Side Break Top Break

FIGURE 1. Break geometries

In an earlier work on non-equilibrium, Henry (1970) introduced a simple correlation determined from the experimental data to include the lack of thermal equilibrium, and flashing was assumed to occur at L/D = 12 for all cases in an arbitrary way. He concluded that the slip ratio for low-quality single component critical flow should be near unity. It was shown that his model was in a good agreement with experimental data obtained by Fauske (1965) and Uchida and Nariai (1966).

Abuaf et al.(1983) presented a study of the non-equilibrium effect on the critical discharge of initially subcooled liquid through nozzles and orifices. They showed that flashing inception always occurs at or near the nozzle throat from experimental evidence. Pressure at the flashing inception was calculated from a modified Alamgir-Lienhard relation. Finally, they got a simple correlation for critical mass flux as a function of the pressure drop between the nozzle inlet and the flashing inception. Their empirical model agreed well with data for nozzles.

Levy and Abdollahian (1982) have proposed a model, which is similar to that of Henry in some respects but less arbitrary in considering the presence of a metastable liquid phase rather than specifying the fixed position of flashing inception as in Henry's model. In order to predict the critical flow rate, isentropic homogeneous flow was assumed. The isentropic assumption in their model makes the model easy to apply, but this cannot be true if a significant frictional loss is present within the channel. Their comparisons showed overall good agreement with the large scale Marviken (1979) data and some other small scale experimental data.

Recently, Amos and Schrock (1985) developed a new homogeneous non-equilibrium model (HNEM) with pressure undershoot at flashing inception. The pressure undershoot relation was obtained from the Alamgir-Lienhard correlation with a modification factor, which was correlated as a function of fluid velocity from the experimental data for their thin slit geometry. Relaxation of the metastable liquid phase produced by flashing delay was assumed to occur exponentially as was suggested by Bauer et al.(1976). The critical flow criterion was taken from Kroeger's (1978) expression

CRITICAL FLASHING FLOW

trance of a break channel connected to a horizontal upstream pipe carrying a stratified two-phase flow. The geometries of major concern are shown in Fig. 1. If no entrainment occurs, single phase steam enters the break if the break location is well above the free surface upstream and single phase liquid enters the break when it is well below the interface. When the break is in the proximity of the interface, entrainment may occur as illustrated in Fig. 1. Since the critical discharge depends significantly upon the stagnation state of the fluid entering the break, a method is needed for prediction of entrainment limits and the quality of the fluid entering the break after the onset of entrainment.

Early work on the draining of tanks (gas pull-through) by Lubin and Hurwitz (1966) and on liquid entrainment into gas streams by Rouse (1956), Craya (1949) and Gariel (1949) led to correlations of the form

$$Fr_i \left(\frac{\rho_i}{\Delta\rho}\right)^{0.5} = C \left(\frac{h_b}{d}\right)^n \tag{1}$$

where C and n are constants that depend upon the geometry. Such correlations are applicable to specific fluid properties. Lubin and Hurwitz did experiments with different fluids and noted a viscosity effect. Easton and Catton (1970) did analysis leading to a correlation form involving the Weber number (surface tension effect) but neglected the role of viscosity. Recent work for the LWR application by Reimann et al.(1984), Reimann and Smoglie (1983), and Smoglie (1984) at KfK (Kernferschungszentrum Karlsruhe) resulted in correlations of the above form for air-water data collected in a 206 mm diameter pipe with 6, 12, and 20 mm I.D. break tubes oriented at the top, side and bottom. The KfK work also recognized the need to obtain the quality of the mixture of the fluid entering the break as the level was modified following incipient two-phase flow. Steam-water data have been obtained by Anderson et al.(1985) at INEL (Idaho National Engineering Laboratory) using a 284 mm horizontal pipe and a 34 mm branch located either at the bottom or the side and at absolute pressures up to 6.2 MPa. Incipient two-phase flow was not observed visually but was inferred by the onset of noise in a pressure transducer output. The present study (Berkeley) used steam-water and air-water in a single apparatus and obtained data for a horizontal pipe diameter of 102 mm and break tubes of 4, 6, and 10 mm located at the bottom, side and top. The absolute pressure ranged up to 1.07 MPa.

Critical Flow Models for Liquid Stagnation States

Models for the case of liquid stagnation states must account for the nonequilibrium caused by delay in the onset of flashing and the rapid drop in pressure as the flow approaches the choking condition.

Edwards (1968) and Ardron (1978) are examples of authors who have contributed detailed mechanistic models. Both models involved some adjustable parameters such as initial bubble diameter, bubble number density in the liquid, and initial liquid superheat required to cause the bubble nucleation. Reviews of these non-equilibrium models have been presented by Jones and Saha (1977), Saha (1978), and Abdollahian et al.(1975).

Alamgir and Lienhard (1981) studied the dependence of flashing inception upon the depressurization rate and stagnation temperature. They proposed a semi-empirical correlation for pressure undershoot at the flashing inception point. This correlation was employed by several authors such as Abuaf et al.(1983) and Elias and Chambre´ (1984).

associated components of commercial boiling water reactors and steam generator tubes in pressured water reactors has attracted a considerable amount of attention over the past several years. Because of economic and safety considerations, it is highly desirable to determine if the failure of the piping system will occur in a leak-before-break mode. Leak-before break is demonstrated by establishing that postulated cracks in a pipe will be detected by leak detection methods before such cracks reach a critical size to cause unstable fracture. The prediction of leak rates has been identified as an important developmental area for implementation of a "leak-before-break" philosophy in the regulatory process (Arlotto, 1986).

Most reports on critical two-phase flow are related to flow in pipes, nozzles and orifices while there is little literature on two-phase flow in tight cracks. Agostinelli et al. (1958) studied flows of flashing water and steam through a smooth annular passages of constant area and with hydraulic diameters in the range of 0.15 to 0.43 mm. Test data were obtained with stagnation conditions of pressure from 3.50 to 20.51 MPa and sub-cooling from 9.3 to 67 C. Hendricks et al.(1975) made a qualitative study of radially inward flow of liquid nitrogen through a 0.076 mm gap between parallel glass plates. Flashing was seen to occur near the end of the 0.72 cm radial flow passage. Simoneau (1974) carried out an experimental study of two-phase nitrogen flow through a rectangular slit. The test section was 2.54 cm in length and width, with a gap of 0.292 mm. He concluded that a uniform two-phase flow pattern existed in most of the test runs and that flashing started at or near the exit plane. Amos and Schrock (1985) carried out experiments on rectangular slits 20 mm in width, gaps in 0.127 to 0.318 mm and L/D ratios from 83 to 400. Their data for subcooled water at pressures from 4.1 to 16.2 MPa were intended to simulate crack leakage at LWR conditions. The results showed that friction is a dominant factor in such channels and although the pressure profiles were not well predicted by the homogeneous equilibrium model, compensating effects result in the measured critical mass flux being less than 20% greater than the HEM prediction.

Recently, an experimental program was carried out by Collier et al. (1980). The experiments were done in two phases. In the first phase simulated cracks were used while in the second phase actual IGSCC cracks were used. Partially cracked pipes were machined on the outer surface to remove a portion of the uncracked wall material thus creating a through-crack which served as the test flow channel. five different crack channels were tested. Mayfield et al.(1980) developed an analytical model by extending Henry's (1970) non-equilibrium homogeneous model to account for flow area change and bends in the flow path. Further modifications were made to this model by Abdollahian and Chexal (1983) and Chexal et al. (1984) to improve its agreement with the data. Both versions of this model, coded into programs LEAK and LEAK 01 respectively, assumed that flashing always begins at an L/D of 12 and that the quality varies linearly with distance along the flow path. Quality was evaluated assuming an isentropic process in LEAK and an isenthalpic process in LEAK 01. The calculations were done by separately calculating channel pressure drop due to momentum and friction based upon length averaged conditions rather than solving the equations in a marching method to obtain the distribution of pressure and quality along the crack length.

Entrainment From Upstream Stratified Regions

Zuber (1981) called attention to the problem of entrainment at the en-

CRITICAL FLASHING FLOW IN PIPES AND CRACKS

V.E. SCHROCK
Department of Nuclear Engineering
University of California
Berkeley, CA

ABSTRACT

Critical flow occurs when, for a constant upstream stagnation state, the flow becomes independent of the pressure in the receiver. At critical flow, the downstream changes in pressure cannot propagate upstream beyond the location of the critical state unless the pressure there is exceeded. Critical flow has been studied extensively for the case of two-phase flashing fluids because of the importance of this phenomenon in the analysis of hypothetical accident sequences in nuclear reactor systems and many other industrial applications.

The present paper summarizes the results of an on-going study related to current issues in the nuclear applications. Leakage through pipe cracks is important as a potential means of detection and diagnosis of possible impending failure of the pressure envelope. In this case the modelling of the flow process becomes a key element in the "leak-before-break" philosophy. The work of Amos and Schrock (1985,1983) included experiments using thin rectangular slits to simulate cracks and the development of a homogeneous non-equilibrium model. A simpler homogeneous equilibrium model (with subcooling correction) was later presented by Schrock, et al. (1986a) for actual cracks. A study of entrainment phenomena at the entrance of a break channel connected to an upstream stratified region was carried out by Schrock, et al.(1986b,1986c). Correlations were obtained for incipient entrainment and for quality entering the break following onset of entrainment. Recently Lee and Schrock (1988) have completed some improvements in Amos' model to represent data from an exceptionally wide range of experimental conditions (size of break channel and stagnation states in the liquid region).

INTRODUCTION

Flow Through Cracks

The presence of intergranular stress corrosion cracks (IGSCC's) in weld heat affected zones of types 304 and 316 stainless steel piping and

3. Moody, F.J., 1966, Maximum Two-phase Vessel Blowdown from Pipes, *Trans. ASME J. Heat Transf.*, vol.88, pp.285-293.

4. Yagawa, G., 1985, Application of Leak-Before-Break Concept to Nuclear Plant, *J. Atomic Energy Soc. Japan*, (in Japanese), vol.27, pp.688-693.

5. Yano, T., Matsushima, E. and Okamoto, A., 1986, Leak Flow Rate from Through-Wall Crack of Pipe, *Proc. Nat. Heat. Transf. Conf. Japan*, (in Japanese), pp.268-270.

FLOW THROUGH A LABYRINTH

NOMENCLATURE

English

A	area
a	fraction of length occupied by a bubble ($\equiv \alpha^{1/3}$)
C_f	friction factor
c	specific heat
h	enthalpy
h_{fg}	latent heat of vaporization
L	length of the channel
L_{th}	total length of small gap region of the labyrinth channel
m	mass flux
n	number density
P	pressure
Pr	Prandtl number
Q	heat transferred to a bubble
S	slip ratio ($=w_v/w_\ell$)
T	temperature
ΔT_{sub}	inlet subcooling
t	time after a bubble is formed
t_h	gap height
w	velocity
x	quality
y	coordinate normal to the wall
z	coordinate along the flow direction

Greek

α	void fraction
κ	heat transfer coefficient
λ	thermal conductivity
μ	dynamic viscosity
ν	kinematic viscosity
ξ	coefficient of inlet pressure loss
ρ	density
τ_w	wall shear stress

Subscripts

c	critical or value at the center of the channel
cal	calculated result
in	inlet
ℓ	liquid
lab	labyrinth
m	mean value
v	vapor

BIBLIOGRAPHY

1. Amos,C.E. and Schrock,V.E., 1984, Two-Phase Critical Flow in Slits, *Nuclear Sci. Engng.*, vol.88, pp.261-274.

2. Hijikata,K., Nagasaki,T. and Nohata,K., 1987, Two-phase Critical Flow and Instability in a Capillary Tube, *Proc. Int. Seminar on Transient Phenomena in Multi-phase Flow.*

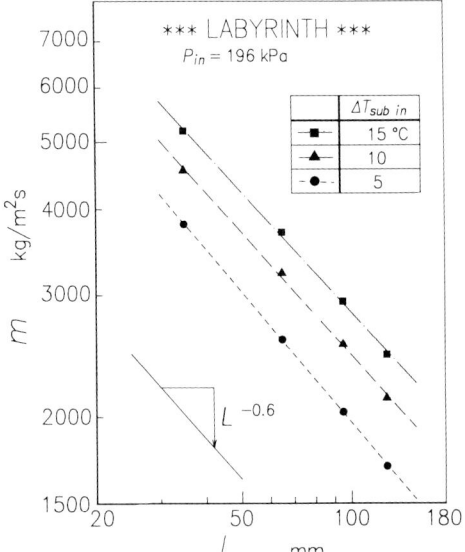

FIGURE 15. Calculated relation between critical mass flux and length of labyrinth channel

subcoolings. The weak dependence of the critical flow rate on the channel length is considered to be caused by a large portion of the pressure loss occurring at the last stage of labyrinth, as shown in Fig.13.

CONCLUSIONS

Characteristics of the single-component critical two-phase flow through a smooth rectangular channel, as well as through a labyrinth channel with small gap height, were investigated both experimentally and theoretically. The following conclusions were obtained:

(1) In the case of the smooth rectangular channel, thermal nonequilibrium between phases is small, and the critical flow rate is explained by the homogeneous equilibrium model or Moody's model, when the flow is in two-phase state at the inlet. However, when the inlet subcooling is positive, the critical flow rate increases rapidly with an increase of inlet subcooling, due to large thermal nonequilibrium. The present model, considering heat transfer from the liquid film between the bubble and the wall, predicts the critical flow rate well.

(2) The critical flow rate in a labyrinth channel is smaller than that in a smooth channel in a large flow rate region, due to the pressure loss caused by the acceleration. The critical flow rate is predicted by a model, where the thermal equilibrium in the cavity region and no phase change in the small gap region are assumed.

FLOW THROUGH A LABYRINTH

region, the following assumptions were made.
(1) In the cavity region, the two-phase flow is in thermal equilibrium due to long residence time, and pressure is uniform.
(2) In the small gap region, phase change is frozen due to rapid acceleration. Both acceleration and friction pressure loss are considered. Slip between phases is also considered.

Fundamental equations in the small gap region are given as follows.

$$\frac{d}{dz}(\rho_v \alpha\, w_v) = 0 \tag{16}$$

$$\frac{d}{dz}\{\rho_\ell(1-\alpha)w_\ell\} = 0 \tag{17}$$

$$\frac{d}{dz}\{\rho_v \alpha\, w_v^2 + \rho_\ell(1-\alpha)w_\ell^2\} + \frac{dp}{dz} = -\frac{2}{t_h}\tau_w \tag{18}$$

$$w_v = S\, w_\ell \tag{19}$$

In Eq.(19), the value of the slip ratio, S, is 2.3. This was determined by the matching of the calculated critical flow rate to the experimental one. In addition, isothermal change of vapor was assumed.

The above equations are written in a matrix form, and variations of flow parameters in each small gap region are calculated in the flow direction. The reduction of the pressure due to acceleration is considered at the entrance of each small gap region. The quality in each cavity is calculated from the energy conservation equation by assuming a thermal equilibrium. The critical flow rate is determined so that the flow is choked at the exit of the last stage of the labyrinth.

Calculated Results

The calculated critical flow rate is shown in Fig.12 by a solid line. Experimental data are also shown. The calculated rate agrees well with the experimental data under positive inlet subcooling, but it underestimates the flow rate when the inlet subcooling is negative. Therefore, the calculated pressure distribution agrees well with the experimental data for positive inlet subcooling, as shown in Fig.13. The disagreement between the theory and the experiment for negative inlet subcooling seems to be caused by the assumption of the constant slip ratio, because the actual slip ratio is considered to increase with an increase of the void fraction.

By using the present model, the critical flow rate was predicted for various lengths of labyrinth channel. These are shown in Fig.15. The geometry of the labyrinth is the same as in this experiment, but only the length of labyrinth was changed. Note that the critical flow rate is proportional to the -0.6 power of the length of the channel for various inlet

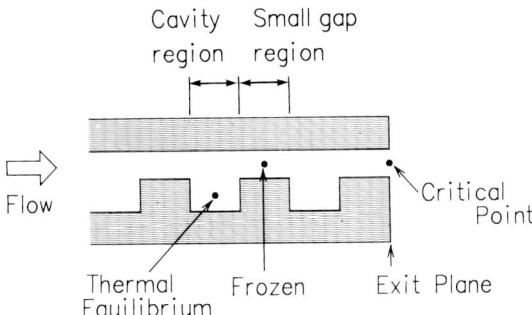

FIGURE 14. Modeling for labyrinth channel

direction with the calculated result, which is shown by the solid line. Stepwise reduction of pressure corresponds to the acceleration loss at the inlet of each stage of the labyrinth, and linear pressure decrease corresponds to the pressure loss caused by friction. The pressure is assumed uniform in each cavity region. It is shown that the pressure loss caused by the acceleration is nearly of the same order as that of the frictional loss.

Two-phase Critical Flow

The critical flow rate of the labyrinth channel is shown in Fig.12, along with that of the flat channel. The critical flow rate of the labyrinth channel is a little larger than that of the flat channel when the inlet subcooling is negative, because the flow rate in this region is small and frictional pressure loss is dominant. On the other hand, for positive inlet subcooling, the flow rate of the labyrinth channel is much smaller than that of the flat channel. The large flow rate of the flat channel is caused by large thermal nonequilibriums between phases. However, in the case of the labyrinth channel, thermal nonequilibrium is relaxed in each cavity. Therefore, the critical flow rate is not so large as that of the flat channel.

Pressure distribution in the labyrinth channel is shown in Fig.13 for positive and negative inlet subcooling. In Fig.13(a), the saturated pressure for the inlet liquid temperature is also shown. It is noted that flashing occurs just after the pressure becomes lower than the saturated value. This fact indicates that substantial thermal nonequilibrium doesn't occur in the labyrinth channel. Solid lines in these figures are the theoretical results presented in the next section.

ANALYSIS OF LABYRINTH CHANNEL

Modeling

The flow in the labyrinth channel is analyzed by separating the channel into two parts — small gap and cavity regions — as shown in Fig.14. According to the characteristics of each

FLOW THROUGH A LABYRINTH

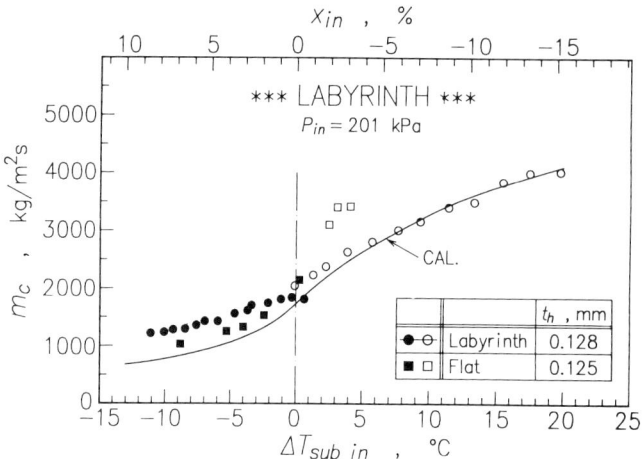

FIGURE 12. Relation between critical mass flux and inlet subcooling for labyrinth channel

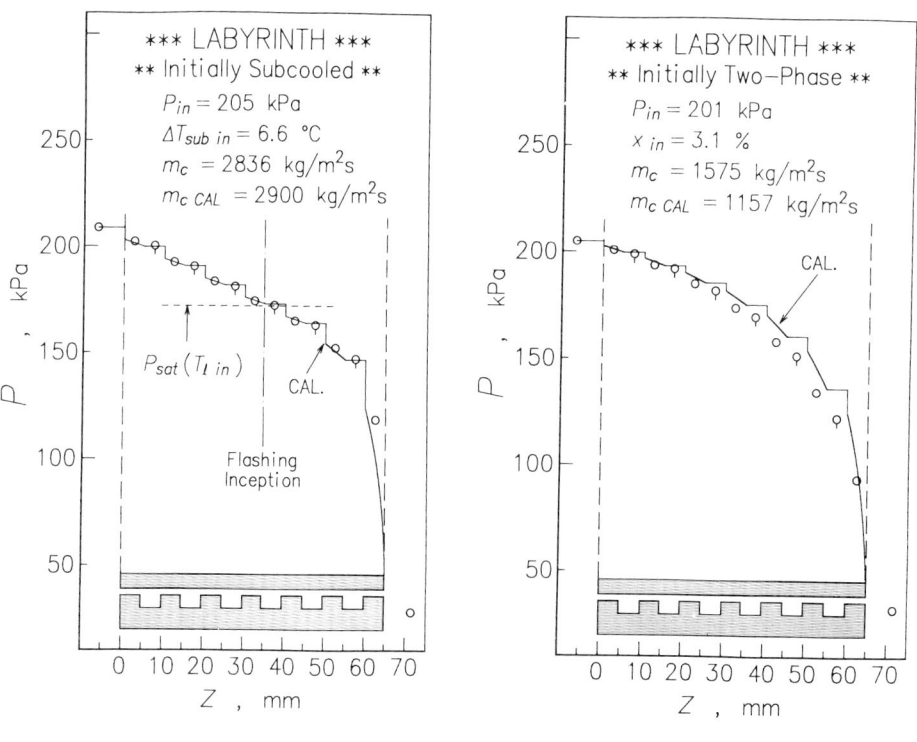

(a) Inlet subcooled liquid (b) Inlet two-phase

FIGURE 13. Pressure distribution in the labyrinth channel for critical two-phase flow

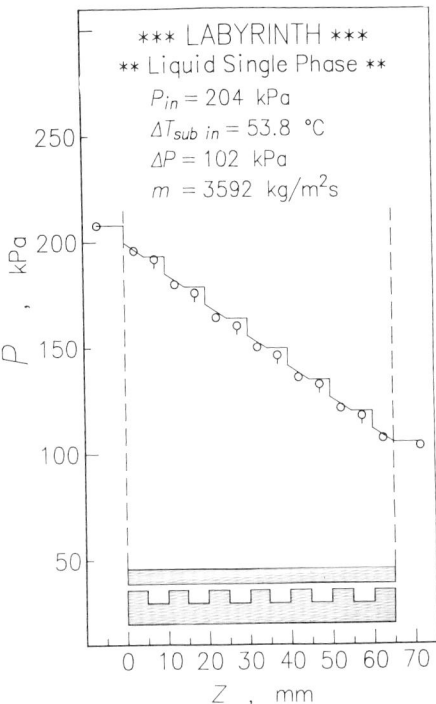

FIGURE 11. Pressure distribution in the labyrinth channel for liquid single-phase flow

channel and the flat channel, flow rates of liquid single-phase flow are compared in Fig.10, in which the abscissa, ΔP, is a pressure difference between the inlet and the exit of the channel. A calculated result obtained from the following equation is also shown by a solid line.

$$\Delta P = (2C_f \frac{L_{th}}{t_h} + N_{lab}\xi_{lab}) \frac{1}{2}\rho w_m^2 \tag{15}$$

where L_{th} is the total length of the small gap region in the channel, and N_{lab} is the number of the labyrinth. C_f is the friction factor for the fully developed flow, and ξ_{lab} is the coefficient of inlet pressure loss. The value of ξ_{lab} is 1.97, which was the best match with the experimental data. The broken line is the calculated result for the flat channel. In the small flow rate region, the flow rate of the labyrinth channel is larger than that of the smooth channel. This is because the pressure loss caused by the friction is larger than the inlet loss, and the total length of the small gap region in the labyrinth channel is smaller than the length of the flat channel. However, in the large flow rate region, the flow rate of the labyrinth channel is smaller than that of the flat channel, due to the increase of the acceleration pressure loss.

Figure 11 shows measured pressure distribution in the flow

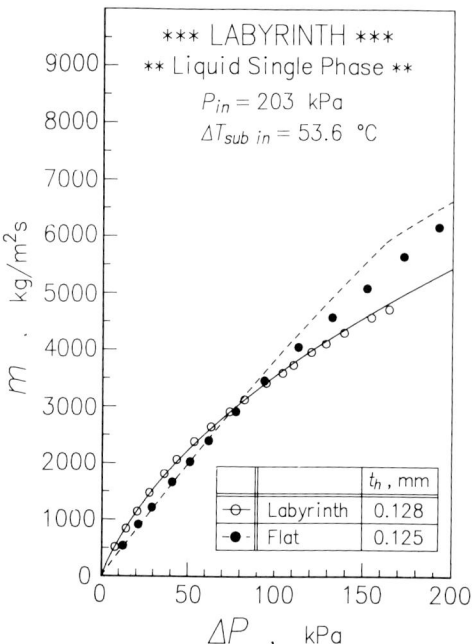

FIGURE 10. Relation between flow rate and pressure difference for liquid single-phase flow

experimental data. In Fig.9(b) the calculated mean velocity in the cross section is compared with the measured bubble velocity. The mean velocity is smaller than the bubble velocity, which suggests the large velocity difference between phases. In the present model, the velocity is assumed to be uniform in the lateral direction, but in the experiment, the velocity of the bubbly region is probably larger than that of the liquid region in the same lateral direction. Such a nonuniformity might be considered for more accurate analysis.

Calculated critical flow rates are shown by triangular symbols (N-E Model)in Figs.6 and 7. From these results, it was concluded that the present model can predict the critical flow rate for various inlet pressure and gap heights, when the inlet flow is subcooled. In this model, the location of the nucleation point is give by experiment. Therefore, further investigation is required for the criteria of the flashing inception, which depends upon the roughness of the wall, impurities in the working fluid, and other effects. However, the present model seems to explain well the structure of flow and heat transfer in the small gap channel.

EXPERIMENTAL RESULTS FOR LABYRINTH CHANNEL

Liquid Single Phase Flow

In order to clarify the difference between the labyrinth

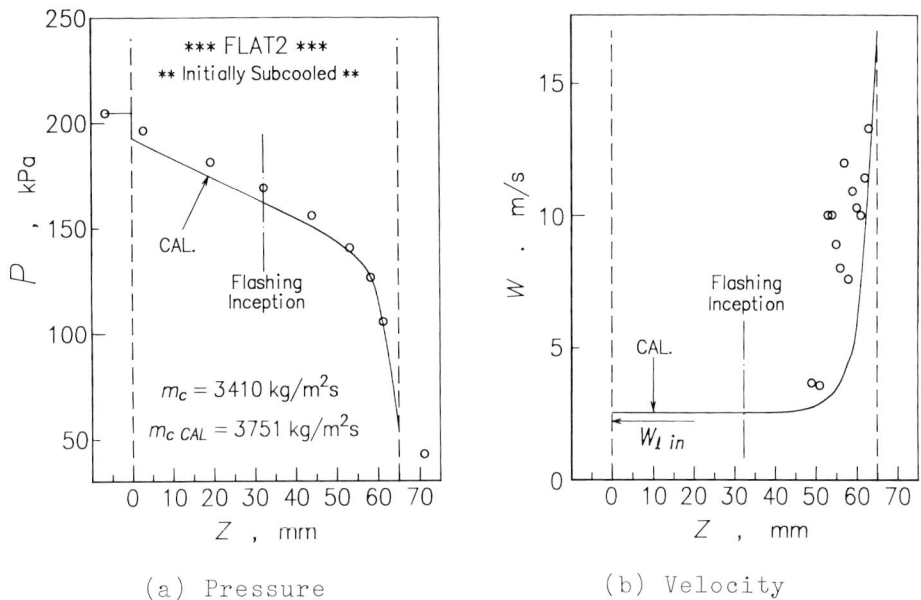

FIGURE 9. Changes of flow parameter in the flow direction
(P_{in}=201 kPa, $\Delta T_{sub\ in}$=3.1 °C)

$$Q_2 = A_2 \kappa_2 (T_\ell - T_v) \qquad (12)$$

$$\kappa_2 = \sqrt{\rho c_\ell \lambda_\ell / \pi t} \qquad (13)$$

where t is time after the bubble is formed.

Eqs (2),(3),(4) and (8) are written in a matrix form. Variations of flow parameters are calculated along the flow direction to the choking point, where the pressure gradient becomes a sufficiently large value (10^{20} Pa/m in this study). The location of the nucleation point in the flow direction is given by the experimental observation. Initial bubble radius and number density are also given by representative experimental values.

$$r_{bo} = 40\ [\mu m], \qquad n_{bo} = 1.7 \times 10^8\ [1/m^3] \qquad (14)$$

The above value of n_{bo} is measured from a photograph in which only one nucleation point exists in the cross section. The calculation is started by assuming the flow rate and is repeated until the choking point coincides with the exit of the channel. Hereafter, this flow model is called 'N-E model'.

Calculated Results

Calculated pressure distribution is shown and compared with experimental results in Fig.9(a). Since the calculated critical flow rate is about 10% larger than the measured value, the calculated pressure profile is slightly under the experimental data points, but it roughly agrees with the

$$\frac{d}{dz}\{a^2\int_0^{1-a}\rho_\ell(h_\ell+\frac{w^2}{2})wdy + a^2\int_{1-a}^1\rho_v(h_v+\frac{w^2}{2})wdy$$

$$+ (1-a^2)\int_0^1\rho_\ell(h_\ell+\frac{w^2}{2})wdy\} = 0 \tag{4}$$

In Eq.(3) mean wall friction, τ_w, is given by Blasius' friction law, using the following mean density and viscosity.

$$\rho_m = \alpha\rho_v + (1-\alpha)\rho_\ell \tag{5}$$

$$\mu_m = \alpha\mu_v + (1-\alpha)\mu_\ell \tag{6}$$

It is assumed that the nucleation occurs at one position in the flow direction, and the number of bubbles is conserved.

$$n_b \; w_m = \text{const.} \tag{7}$$

where n_b is the bubble number density in the cross section. The change of vapor mass flow rate is given by:

$$\frac{d}{dz}\{a^2\int_{1-a}^1\rho_v wdy\} = \frac{n_b Q}{h_{fg}} \tag{8}$$

where Q is transferred heat to one bubble from the surrounding superheated liquid. The heat is divided into two parts.

$$Q = 2Q_1 + Q_2 \tag{9}$$

Q_1 denotes heat from the upper and lower surfaces of the bubble as shown in Fig.8(b). Between each bubble surface and the channel wall, superheated liquid film exists. Since the mean velocity of the liquid film is much smaller than that of the bubble, it is assumed that a bubble is moving at velocity w_c on the stationary liquid film when Q_1 is estimated. As the thermal boundary layer develops in the liquid film from the edge of the bubble, Q_1 is expressed as follows.

$$Q_1 = A_1\kappa_1(T_\ell - T_v) \tag{10}$$

where A_1 is the upper surface area of a bubble. The vapor temperature, T_v, is the saturated temperature obtained from the pressure. The liquid temperature, T_ℓ, is obtained from the energy balance, Eq.(4). Heat transfer rate, κ_1, is given by that of the laminar boundary layer with characteristic length $\sqrt{A_1}$.

$$\frac{\kappa_1\sqrt{A_1}}{\lambda_\ell} = 0.66 \; Pr_\ell^{1/3}(\frac{w_c\sqrt{A_1}}{\nu_\ell})^{1/2} \tag{11}$$

On the other hand, the heat flow through the side surface of the bubble, Q_2, is estimated by unsteady heat conduction in the surrounding liquid. This is because the velocity difference in the lateral direction is not considered in this model.

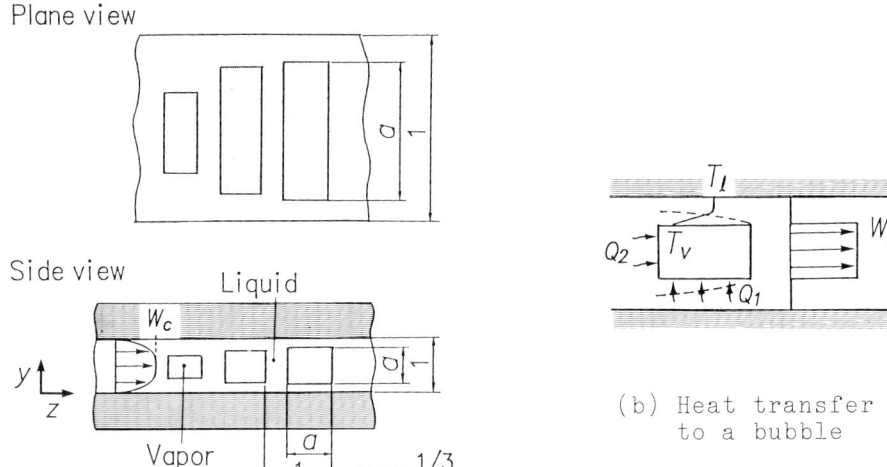

(a) Void distribution

(b) Heat transfer to a bubble

FIGURE 8. Modeling for flat channel

the distributions of void and velocity in the cross section were considered. Thermal nonequilibrium was also considered in our model. The distribution of void fraction in the channel is assumed, as shown in Fig.8(a). As the bubble grows three-dimensionally, the fractional length of vapor cross section in each direction, a, is assumed to be $\alpha^{1/3}$, where α is the averaged void fraction in a cross section. Further, the local velocities of vapor and liquid are equal, but a velocity profile is considered in the cross section by 1/7 power law.

$$w = w_c y^{1/7} \tag{1}$$

where w_c denotes velocity at the center of the cross section. y denotes nondimensional distance from the wall, normalized by half height of the channel.

Integrated conservation equations of mass, momentum and energy in the cross section are written as follows.

$$\frac{d}{dz}\{a^2\int_0^{1-a}\rho_\ell w\,dy + a^2\int_{1-a}^{1}\rho_v w\,dy + (1-a^2)\int_0^1 \rho_\ell w\,dy\} = 0 \tag{2}$$

$$\frac{d}{dz}\{a^2\int_0^{1-a}\rho_\ell w^2\,dy + a^2\int_{1-a}^{1}\rho_v w^2\,dy + (1-a^2)\int_0^1 \rho_\ell w^2\,dy\}$$

$$+ \frac{dP}{dz} = -\frac{2}{t_h}\tau_w \tag{3}$$

FLOW THROUGH A LABYRINTH

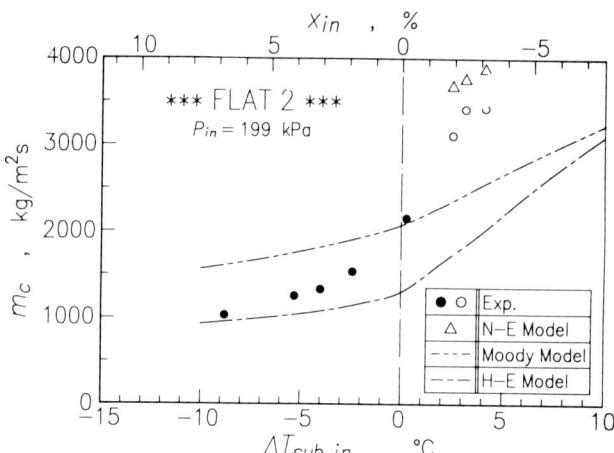

FIGURE 7. Relation between critical mass flux and inlet subcooling (flat channel with gap width 0.125mm)

pressure, which means that the effect of wall friction on the critical flow rate is important.

Calculated results by conventional models are also shown in Figs.6 and 7. The single-dotted line denotes results of a homogeneous equilibrium (H-E) model, in which equal velocity and thermal equilibrium between the phases are assumed. The double-dotted line denotes prediction by Moody's model (Moody,1966), which assumes thermal equilibrium and slip ratio being equal to $(\rho_\ell/\rho_v)^{1/3}$. Experimental data for the inlet flow in a two-phase state exist between H-E model and Moody's model, and approach the H-E model when increasing the inlet quality. This result is consistent with the observed flow pattern in Fig.5(d), which shows that the liquid and vapor phases are well mixed. On the other hand, when the flow is subcooled at the inlet, the critical flow rate deviates largely from either model when increasing the inlet subcooling. Such a large flow rate is caused by nonequilibrium between phases, due to the apparent separation of bubbly and liquid regions as shown in Fig.5(b).

ANALYSIS OF FLAT CHANNEL

Modeling

As shown in experimental results, in order to estimate the critical flow rate under inlet subcooled conditions it is necessary to take account of the thermal nonequilibrium between phases. Amos et al.(1984) have developed a homogeneous model, where pressure undershoot at a flashing point and an exponential relaxation to equilibrium were considered. Their model successfully predicted the critical flow rate by introducing some empirical constants in the model, but it failed to explain pressure distribution in the channel. The authors have proposed a theoretical model for critical two-phase flow in a capillary tube (Hijikata et al., 1987), where

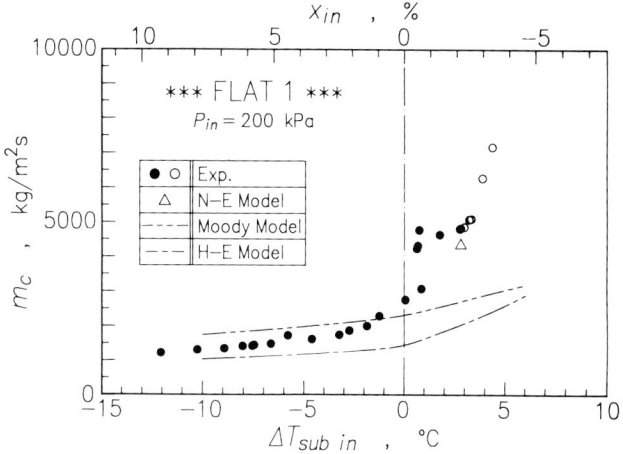

(a) P_{in} = 200 kPa

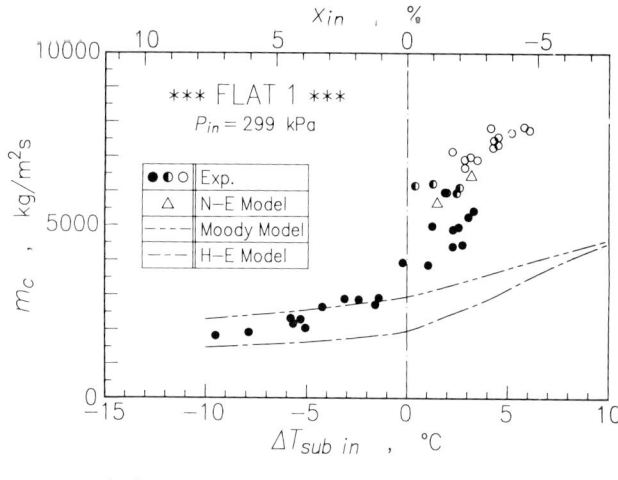

(b) P_{in} = 299 kPa

FIGURE 6. Relation between critical mass flux and inlet subcooling (flat channel with gap width 0.170mm)

On the contrary, when the inlet flow is a two-phase mixture, the dependence of critical flow rate on the inlet quality is small. The critical flow rate generally increases with an increase of inlet pressure. In these figures, the difference of symbol denotes the difference of nucleation pattern. ○ means that nucleation occurs in the middle of the channel, ◐ means that only one stream of bubbles exists at the inlet, and ● means that several streams of bubbles exist at the inlet. It is noted that the critical flow rate varies according to the number of nucleation points when the inlet subcooling is the same. Figure 7 shows results for the smaller gap height channel. The critical flow rate shown in Fig.7 is smaller than that in Fig.6(a) with the same inlet

(a) $\Delta T_{sub\ in}$ = 3.2 °C

(m_c = 6819 kg/m²s)

(b) $\Delta T_{sub\ in}$ = 1.5 °C

(m_c = 5512 kg/m²s)

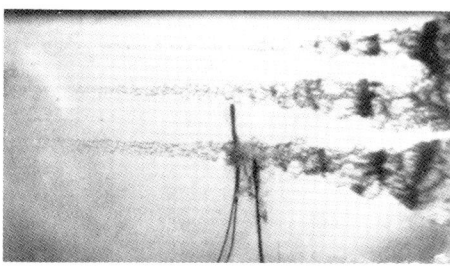

(c) $\Delta T_{sub\ in}$ = 1.3 °C

(m_c = 4868 kg/m²s)

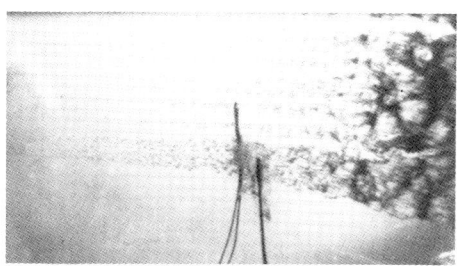

(d) x_{in} = 3.2 %

(m_c = 2585 kg/m²s)

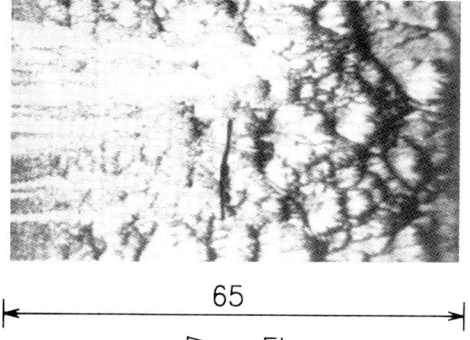

FIGURE 5. Flow pattern in flat channel
(gap width 0.170mm, P_{in}=293 kPa)

fraction, as shown in Fig.5(d).

Variations of critical mass flux with inlet subcooling are shown in Fig.6 for two different inlet pressures. Negative value of the inlet subcooling in the abscissa means a two-phase condition, and inlet qualities are indicated in the upper abscissa. As shown in these figures, the critical flow rate increases rapidly with the increase of inlet subcooling.

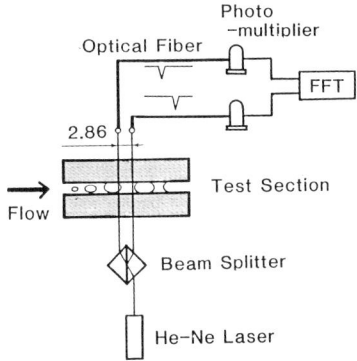

FIGURE 3. Method for bubble velocity measurement

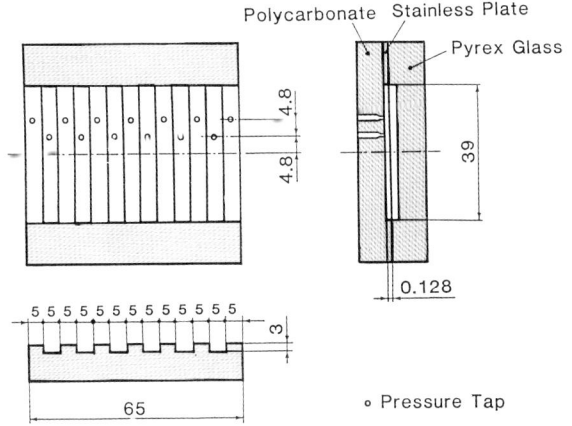

FIGURE 4. Test section (labyrinth channel)

height was nearly equal to that of FLAT2.

EXPERIMENTAL RESULTS FOR FLAT CHANNEL

Flow patterns in the flat channel are shown in Fig.5 for various inlet subcooling but with a constant inlet pressure. In the case of large inlet subcooling, as shown in Fig.5(a), a bubble nucleation occurs in the middle of the channel. On the other hand, in the case of smaller inlet subcooling, the nucleation occurs at the entrance of the channel due to acceleration pressure drop, as shown in Fig.5(b). Several streams of bubbles are formed whose widths grow in an exponential manner. On both sides of the bubble streams, large amounts of liquid phase exist in the same cross section. Therefore, it was considered that the thermal nonequilibrium between bubbles and liquid is large. The number of the bubble streams increases when decreasing the inlet subcooling, as shown in Fig.5(c). On the contrary, when the flow enters as a two-phase flow, the bubbles are distributed uniformly and the flow pattern becomes chaotic with the increase of void

FLOW THROUGH A LABYRINTH

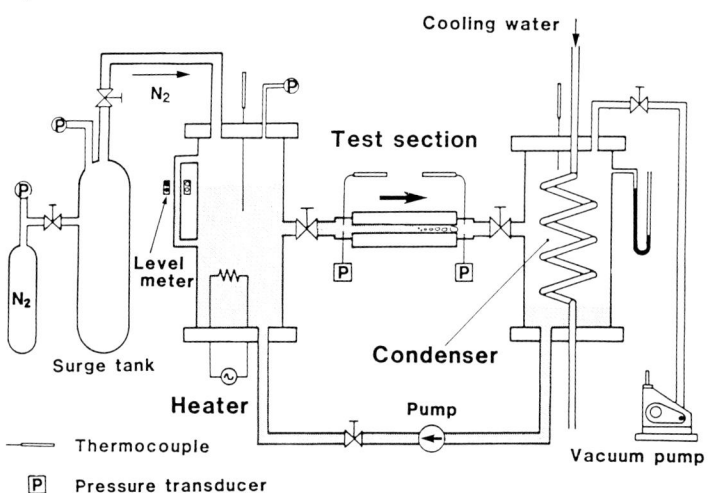

FIGURE 1. Experimental apparatus

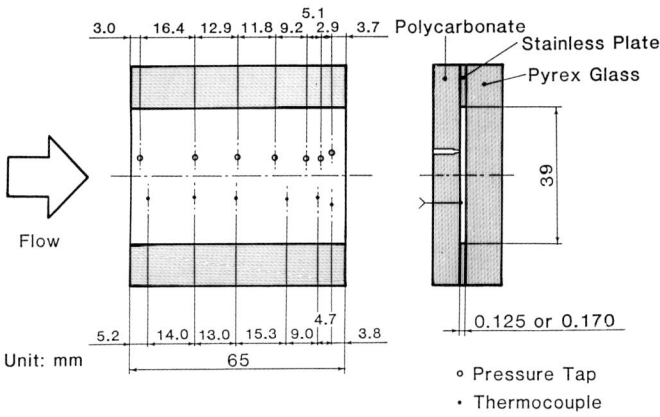

FIGURE 2. Test section (flat channel)

this channel was made of a polycarbonate plate, on which there were pressure taps and thermocouples as shown in Fig.2. This channel was named FLAT2 and was used mainly to measure the variation of pressure and temperature along the flow direction. In addition, bubble velocity was measured from a cross-correlation of intensity of two laser beams which passed in the cross section of the channel, as shown in Fig.3.

Detailed shape of the labyrinth channel is shown in Fig.4. One side of the channel was made of Pyrex glass, on which rectangular lateral grooves were made. The other side was made of a polycarbonate plate and had pressure taps. The length and lateral width of this channel were the same as those of the flat channel. The minimum and maximum gap heights were 0.128mm and 3.1mm, respectively. The minimum

the channel didn't agree with the experimental data (Amos et al., 1984), and detailed structure of the flow (such as a flow pattern in the channel) has not yet been clarified. Such a small leakage through a narrow channel is also important in the prediction of the flow leakage at the seal of a rotating machine, operating under two-phase condition. In this case, a labyrinth is more effective to reduce the leak flow rate.

From these points of view, the characteristics of one-component critical two-phase flow through a narrow rectangular channel and a labyrinth are investigated by the visual observation of the flow pattern and the measurement of the flow parameters for various inlet conditions. Based on the experimental results, a theoretical model was developed, and the results are compared with experimental data.

EXPERIMENTAL APPARATUS

A schematic diagram of the experimental apparatus is shown in Fig.1, which is essentially the same as that used in the previous study (Hijikata et al., 1987) except for the test section. The liquid R113 was stored in an inlet tank and maintained at a constant temperature. The inlet tank was connected to a surge tank containing highly pressurized N_2 gas. The tank was kept under a desired subcooled condition. An inlet valve was installed between the inlet tank and the test section, and the inlet two-phase flow condition was varied by changing the degree of flow flashing at this valve. When the flow was two-phase at the inlet, the inlet quality was calculated from the measured pressure and temperature before and after the flash valve, by assuming a thermal equilibrium. The exit pressure of the test section was kept at some constant value by adjusting a valve installed between the test section and exit tank. When a critical flow condition was desired, the exit pressure was held sufficiently low, so that the flow rate should not be influenced by the exit pressure under the critical flow condition. The two-phase flow was condensed in the exit tank, and the liquid R113 was returned to the inlet tank by a pump. The flow rate was determined from a measurement of the falling rate of the liquid level in the inlet tank, under a no-pumping condition. Pressures and temperatures at the inlet and the exit of the test section were measured.

Two types of test section were used; a flat channel and a labyrinth channel. Detail of the flat channel is shown in Fig.2. A small gap was formed between two transparent plates, using thin stainless strips placed between the plates to make the desired gap height. The length and width of the channel were 65mm and 39mm, respectively. Two flat channels with different gap heights were used. One of them had a 0.170mm gap height, and its walls were made of Pyrex glass. This channel had no pressure taps or thermocouples on the wall, in order to avoid the creation of a special artificial nucleation site. The gap height was estimated from the measured pressure loss of a liquid single-phase laminar flow through the channel. Hereafter, this channel is referred to as FLAT1. The other flat channel had a 0.125mm gap height. One side of

A STUDY ON CRITICAL TWO-PHASE FLOW THROUGH A LABYRINTH

K. HIJIKATA, T. NAGASAKI, and T. KANAYA
Department of Mechanical Engineering Science
Tokyo Institute of Technology
Ohokayama 2-12-1, Meguro-ku, Tokyo 152, Japan

ABSTRACT

Characteristics of critical two-phase flow through a very narrow rectangular channel and a labyrinth channel have been investigated both experimentally and theoretically. The channel was made of transparent glass plates, to observe the flow pattern and to measure the bubble velocity by an optical method. The critical flow rate was measured for various inlet conditions. For the rectangular channel, the measured flow rate under the subcooled condition of the inlet was larger than theoretical estimations by conventional models, because the thermal nonequilibrium between phases was significant in the experiment. By considering such a nonequilibrium effect, a new model was proposed, and the result was compared with experimental results. For the labyrinth channel, the measured flow rate was smaller than that for the rectangular channel, due to the pressure loss by repeated acceleration. The critical flow rate in the labyrinth channel was also predicted theoretically.

INTRODUCTION

One-component critical two-phase flow is important to the safety of nuclear reactors. Many experimental and theoretical works have been devoted to this problem. Considering the most severe accident, previous works have been concerned with critical flow through relatively wide bore pipes and nozzles. However, as it has been recently recognized that small leakage through pipe cracks can be used to predict the severe accident (Yagawa, 1985), the investigation of two-phase flow in small gaps is now prompted. Amos et al.(1984) studied critical two-phase flow of initially subcooled water through a narrow rectangular slit, with inlet pressure of 7 MPa, simulating the flow leakage through a pipe crack. Yano et al. (1986) also examined the critical flow rate and the effect of surface roughness, by using the channel where the minimum sectional area is at the outlet. In these studies, experimental data of the critical flow rate under typical working conditions of conventional reactors were obtained, and theoretical models for predicting the critical flow rate were presented. However, the calculated pressure distribution in

BIBLIOGRAPHY

1. Ball,L.J.,et.al., 1978, TREE-NUREG-1210

2. Crawford,T.J. and Weinberger,C.B., 1986, Two-Phase Flow Patterns and Void Fractions in Downward Flow. Part II: Void Fractions and Transient Flow Patterns, Int. J. Multiphase Flow, Vol.12 No.2, pp.219-236

3. Fukuda,K., Tanihira,M., Sakai,T., Hasegawa,S. and Kondo, T., 1987, Experimental Study on Two-Phase Flow Instability in System Including Downcomers, J. Nucl. Sci. Technol., Vol.24, No.4, pp.266-275

4. Fukuda,S., 1982, Pressure Variations due to Vapor Condensation in Liquid, (II) Phenomena at Large Vapor Mass Flow Flux, J. Atomic Energy Soc. Japan, Vol.24, No.6, pp.466-474

5. Ishii,M. and Mishima,K., 1984, Two-Fluid Model and Hydrodynamic Constitutive Relations, Nucl. Eng. Design, Vol.82, pp.107-126

6. Iwamura,T., Adachi,H. and Sobajima,M, Experimental Study of Two-Dimensional Thermal-Hydraulic Behavior in Core During Reflood Phase of PWR LOCA, J. Nucl. Sci. Technol., Vol.23, No.2, pp.123-135

7. Murase,M., Suzuki,H., Matsumoto,T. and Naitoh, M., 1986, Countercurrent Gas-Liquid Flow in Boiling Channels, J. Nucl. Sci. Technol., Vol.23, No.6, pp.487-502

8. Ueda,T., 1981, Gas-Liquid Two Phase Flow (in Japanese), Youkendou, p.57

9. Wallis,G.B., 1969, One-Dimensional Two Phase Flow, McGraw Hill

10. Wallis,G.B., Karlin,A.S., Clark,C.R. III, Bharathan and Hagi, Y., 1981, Countercurrent Gas-Liquid Flow in Parallel Vertical Tubes, Int. J. Multiphase Flow, pp.1-19

CONCLUSION

In downward two-phase flow experiments, two types of dryout, A(C) and B(D), and a static flow instability, were observed. The dryout A(C) occurs under low flow rate conditions, initiated by the accumulation of the uncondensed gas phase in the heated channel, while the dryout B(D) occurs when the flow rate is higher than 50 ml/s ($j_{l,in}$=7.3 cm/s), and is presumed to be caused by the flooding.

As the static ΔP-$W_{l,in}$ characteristic has a negative slope in this boiling downward flow system, a static instability --a flow excursion-- usually accompanies these dryouts. In multi-channel system this results in a large decrease in flow rate, or a reversed flow in some channels.

The onset of dryout A(C) was estimated with a two-fluid model, and its transition to dryout B(D) was predicted with Wallis' correlation. The results agreed well with the experiments.

NOMENCLATURE

English
a_i	interfacial surface ratio	(m^2/m^3)
D	equivalent diameter	(m)
f	friction factor	(-)
g	acceleration of gravity	(m/s^2)
h	enthalpy	(Joule/kg)
	heat transfer coefficient	(Watt/m^2)
h_{fg}	latent heat of evaporation	(Joule/kg)
j	superficial velocity	(m/s)
j^*	dimensionless superficial velocity	(-)
M	phase change mass rate	(kg/m^3s)
P	pressure	(Pa)
q	heat transfer rate	(Joule/m^3s)
T	temperature	(C)
u	velocity	(m/s)
v_{dj}	drift velocity	(m/s)
z	axial coordinate	(m)

Greek
α	void fraction	(-)
ρ	density	(kg/m^3)

Subscripts
BO boiling
CON condensing
g   gas phase
in inlet
l   liquid

ACKNOWLEDGEMENT

The authors would like to express their appreciation to Mr. K. Nakagawa for his help in manufacturing the test apparatus; and to Mr. Y. Miyake in assisting the experiments. This study was supported by research grant No.60460235, sponsored by the Ministry of Education, Japan.

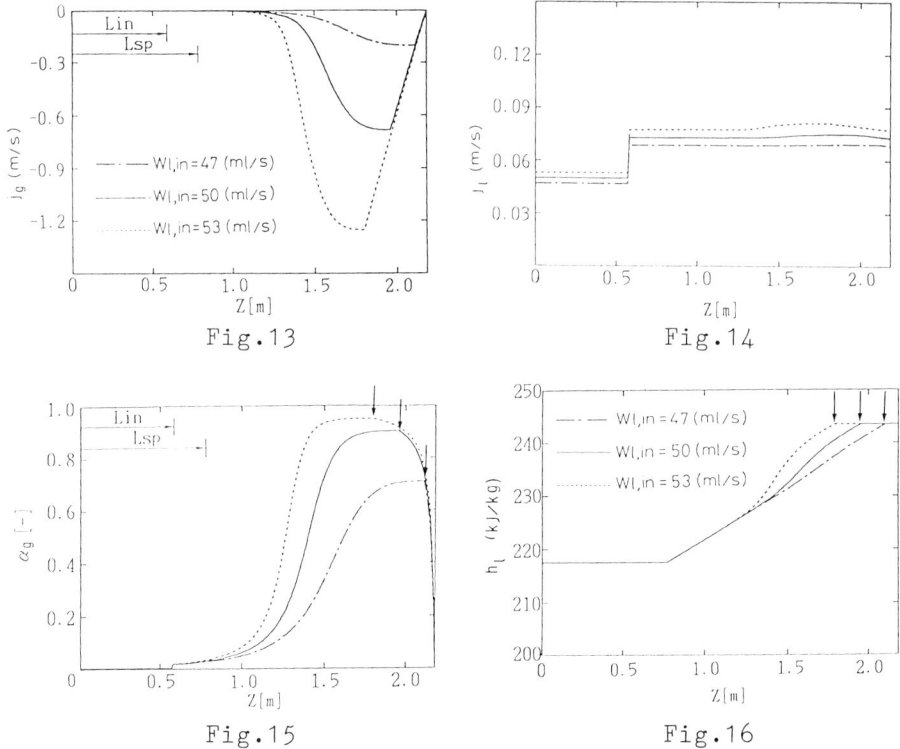

Fig.13 Analytical result for superficial gas-phase velocity
Fig.14 Analytical result for superficial liquid-phase velocity
Fig.15 Analytical result for void fraction
Fig.16 Analytical result for liquid phase enthalpy

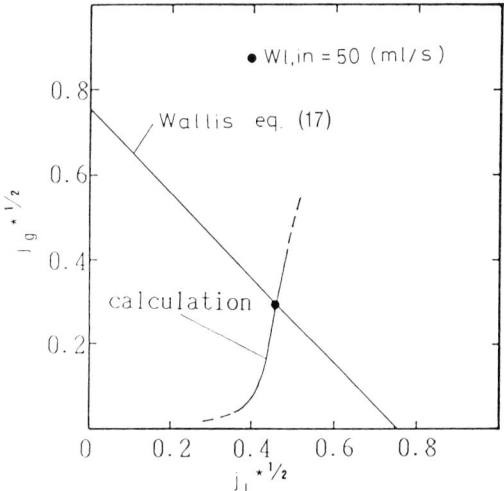

Fig.17 Plot of the analytical result in j_g^* vs. j_l^* plane. (Analytical j_k^*'s at their maxima are plotted, giving a curve which intersects with that from Eq.(17) at 50 ml/sec)

$$M_{CON} = q_{CON}/h_{fg} \tag{13}$$

$$q_{CON} = ha_i(T_{g,S} - T_l), \tag{14}$$

where Fukuda's correlation(1982) is taken for the heat transfer coefficient of condensation h, and Ishii's correlation (1984) for the interfacial surface rate a_i is assumed.

To evaluate the void fraction, the drift flux model is adopted, which is given by:

$$u_g = C_o(j_g + j_l) + v_{dj}, \tag{15}$$

where it is assumed that $C_o=1$, after the report by Crawford(1986) on experiments of downward flow with R-113. For v_{dj}, Nicklin's correlation(Ueda,1981) is taken.

Pressure drop is calculated with the equation for the homogeneous flow:

$$\frac{dp}{dz} = \{\alpha_g \rho_g + (1-\alpha_g)\rho_l\}g - \frac{2}{D}f\rho_l u_l^2,$$
$$f = 0.046(Du_l/\nu_l)^{-0.2}, \tag{16}$$

where the pressure drop due to the acceleration etc., is neglected.

Numerical analysis was carried out as follows. Assuming q_l, Eqs.(5),(6) and (8) are integrated with the Runge-Kutta method to obtain the boiling initiation point z_B, at which point the liquid enthalpy and j_g attains saturation. The heating power at the onset of dryout A should be searched such that j_g agrees with what is given by Eq.(12).

Analytical result

In Figs.13-16 are shown axial distributions of various parameters at the onset of dryout A and/or flow excursion, for three cases with different inlet flow rates. The superficial gas velocity j_g(Fig.13) increases linearly from the bottom, which decreases as the gas phase is condensed as it rises and vanishes to zero just at the top of the heated channel. On the other hand, the superficial velocity of the liquid phase does not change much (Fig.14). Since j_g and j_l take their maxima at the point of boiling initiation, it is there that the flooding is initiated, if it does occur.

If one increases the inlet flow rate as well as the heating power to meet the condition of the onset of dryout A, it is possible to expect the flooding to occur prior to dryout A.

The onset of flooding, followed by dryout B, is predicted with the Wallis' correlation(1969);

$$(j_g^*)^{1/2} + (j_l^*)^{1/2} = 0.75$$
$$j_k^* = j_k\{\rho_k/gD(\rho_l-\rho_g)\}^{1/2} \quad (k=l,g). \tag{17}$$

In Figs.17 and 6 $(j_l^*)^{1/2}$ vs. $(j_g^*)^{1/2}$ and $W_{l,in}$ vs. ΔP are plotted at the onset of dryout A, respectively. It is shown in Fig.17 that the flooding, and thus the consequent dryout B, is initiated at 50 ml/s. This is in good agreement with the experimental results.

DOWNWARD TWO-PHASE FLOW

$$\frac{d}{dz}(1-\alpha_g)\rho_l u_l = -M_{BO} + M_{CON}, \tag{2}$$

where M_{BO} and M_{CON} represent the evaporation and the condensation rate per unit volume, respectively. Conservation of energy for each phase is written by:

$$\frac{d}{dz}(\alpha_g \rho_g h_g u_g) = M_{BO} h_{BO} - M_{CON} h_{CON} + q_g, \tag{3}$$

$$\frac{d}{dz}(1-\alpha_g)\rho_l h_l u_l = -M_{BO} h_{BO} + M_{CON} h_{CON} + q_l. \tag{4}$$

If one introduces the superficial velocities $j_g = \alpha_g u_g$, $j_l = (1-\alpha_g) u_l$, the above equations are rewritten as:

$$\frac{dj_g}{dz} = \frac{1}{\rho_g}(M_{BO} - M_{CON}), \tag{5}$$

$$\frac{dj_l}{dz} = \frac{1}{\rho_l}(-M_{BO} + M_{CON}), \tag{6}$$

$$\frac{d}{dz}(j_g h_g) = \frac{1}{\rho_g}(M_{BO} h_{BO} - M_{CON} h_{CON} + q_g), \tag{7}$$

$$\frac{d}{dz}(j_l h_l) = \frac{1}{\rho_l}(-M_{BO} h_{BO} + M_{CON} h_{CON} + q_l). \tag{8}$$

With the assumptions (iv) and (v), q_g is defined such that Eq.(5) and (7) are equivalent, and the basic equations are Eq.(5),(6) and (8). These are solved with the initial conditions:

$$u_l = u_{l,in}, \quad h_l = h_{l,in}, \quad j_g = 0 \text{ at } z = 0 \tag{9}$$

$$j_g = 0 \text{ at } z = L, \tag{10}$$

where Eq.(9) represents a condition that, at the onset of dryout A, a small part of uncondensed gas phase begins to reach the top (z=0), while Eq.(10) means that the gas phase begins to be generated at the bottom (z=L) and flows upward.

In a subcooled zone, the gas phase which goes upward is condensed; the liquid phase in turn is heated. From the assumption (iii) $M_{BO}=0$ in the subcooled zone, whereas the assumptions (iv) yields $M_{CON}=0$ in the saturated zone.

It also results from the assumption (v) that all generation of heat from the heater is accounted as q_l. Thus, for the saturated zone, one obtains from Eqs.(6) and (8):

$$M_{BO} = q_l/h_{fg}, \tag{11}$$

and from integration of Eq.(5);

$$j_g = -q_l(L-z)/\rho_g h_{fg}, \tag{12}$$

where the negative sign corresponds to the upward flow.

The rate of condensation in the subcooled zone is given by:

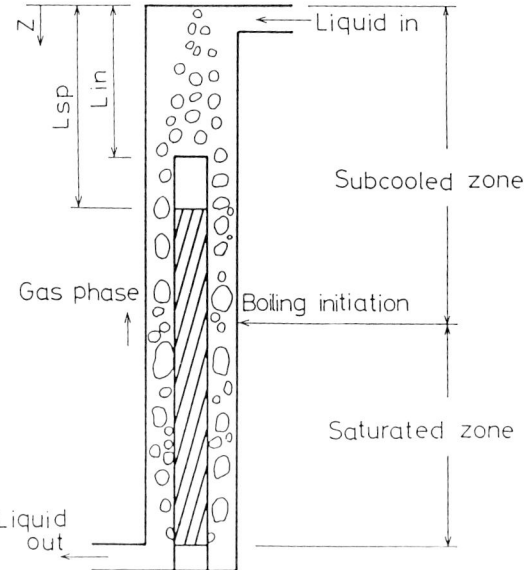

Fig.12 Model for the analysis of the onset of the dryout. (All the gas phase generated in the heated channel is assumed to rise, being condensed by the inflowing liquid phase, and to extinguish at the top of the heated channel.)

decreases; annular flow regime expands throughout the channel and the dryout D occurs.

Heating powers at initiation of dryout as well as those at onset of flow reversal and boiling are shown in Fig.11. It is found that the flow reversal in the first channel occurs easily as soon as the boiling is initiated. It should be noted that the flow rate at the transition from the Type C to the Type D dryout lies around 50 ml/s, which is consistent with those for experiments with a smaller number of channels.

ANALYSIS

Conditions for onset of the Type A dryout and/or flow excursion and for its transition to the Type B dryout were analysed.

Basic equations for the analysis of onset of dryout A

An analytical model is shown in Fig.12. Since the Type A dryout and/or flow excursion is assumed to be initiated when the uncondensed gas phase reaches the top of the heated section and begins to accumulate, void fraction there and at the bottom is zero. Analysis was carried out with the assumptions that (i) physical properties are constant, (ii) all the gas phase flows upward, (iii) subcooled boiling is not taken into account, (iv) in the saturated zone temperatures of gas and liquid phase are constant, and condensation of the gas phase will not happen, (v) superheat of the gas phase is not considered.

Conservation equations for liquid phase and for gas phase are given by:

$$\frac{d}{dz}(\alpha_g \rho_g u_g) = M_{BO} - M_{CON}, \qquad (1)$$

DOWNWARD TWO-PHASE FLOW

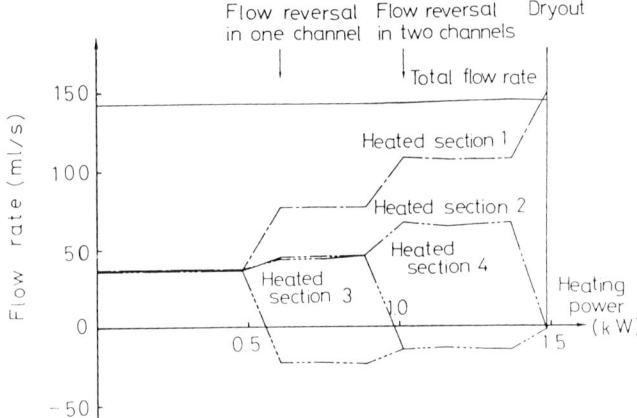

Fig.10 Schematic transients of flow rates for parallel four-channel experiments. (As the heating power is increased the flow reversal appears in one channel after another, successively.)

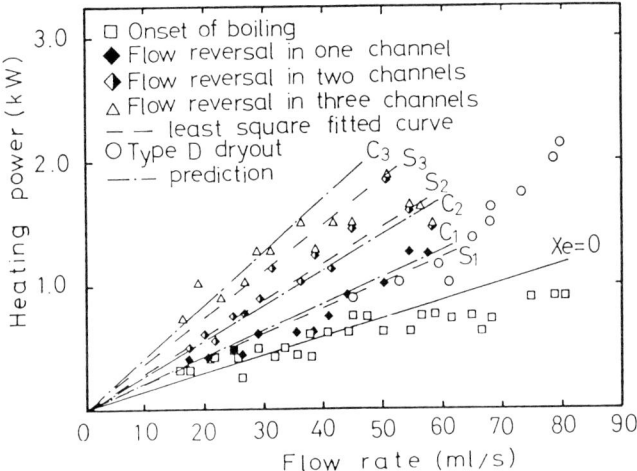

Fig.11 Heating power at initiation of dryout in parallel four-channel experiments. (In contrast with Fig.9 the change in heating limit at around 50 ml/sec is gradual. Each initiation of the flow reversal is predicted and is shown by C1-C3, based on the results of parallel two-channel experiments.)

Variation of distribution of flow rate among channels is shown schematically as a function of heating power in Fig.10. Because of the larger number of channels, the phenomenon leading to the dryout is more complex than the parallel two channel case. However, this is regarded as typical of the static instability resulting in the Type C dryout.
 In high flow rate conditions, it is observed that, after the initiation of boiling, the boiling in one channels becomes more violent and in the remaining three channel it becomes weaker. In the violent-boiling channel the flow rate

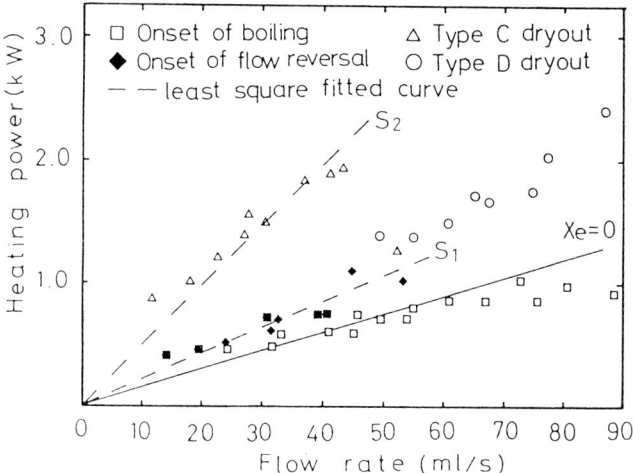

Fig.9 Heating power at initiation of dryout in parallel two-channel experiments. (A dramatic change in heating limit at 50 ml/sec is shown. The heating power and the flow rate are given per channel. Least square fitted curves for onset of the flow reversal S1 and for the Type C dryout S2 are shown.)

dryout D occurs at the bottom caused by stagnation, succeeding flooding, and the flow excursion. The only difference between them is that the critical heating power per heated section at the dryout D is lower than that for the dryout B(Fig.9). This is because the decrease in flow rate of the channel in question is larger for the parallel channel case than the single channel case, where it is prevented by the presence of the circulating pump. In Fig.9 it is shown that the flow reversal emerges at a flow rate less than 50 ml/s, and that a dramatic decrease in dryout heating power exists at this critical flow rate.

Parallel four channel experiment

The results of parallel four channel experiments are basically the same as the parallel two channel case: Type C dryout, with flow reversal at low flow rate conditions, and Type D dryout, without flow reversal under high flow rate conditions.

When the flow rate is low with the heating power being increased, the flow reversal (flow excursion) in a certain channel happens while the flow in the other three channels remains normal (downward flow). Since the total flow rate is kept constant throughout the process, flow rates in these three channels increase and boiling ceases. With the heating power increased further, flow in another channel becomes reversed -- i.e., normal flow in two channels, reversed flow in two channels. If the heating power is increased still further, flow in another channel is reversed, for normal flow in a single channel and reversed flow in three channels. However, in this case the dryout C at the top of a reversed flow channel follows.

DOWNWARD TWO-PHASE FLOW

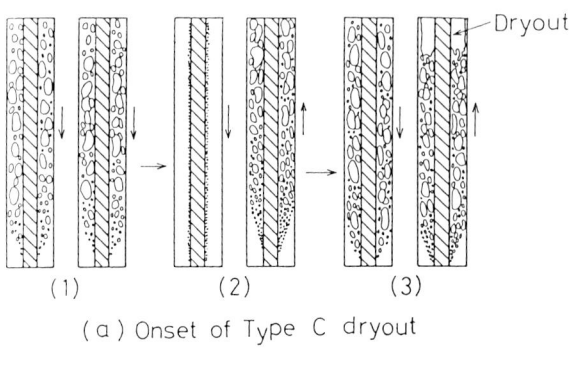

(a) Onset of Type C dryout

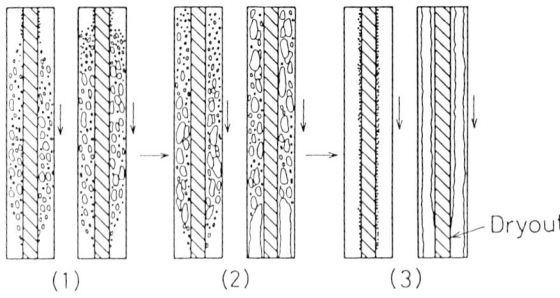

(b) Onset of Type D dryout

Fig.7 Observed flow regimes in parallel two channel experiments. (When flow rate is low the flow reversal occurs in one channel and the dryout C emerges at its top; when the flow rate is high the flooding causes a decrease in flow rate to nearly zero in one channel and the dryout D is initiated at its bottom.)

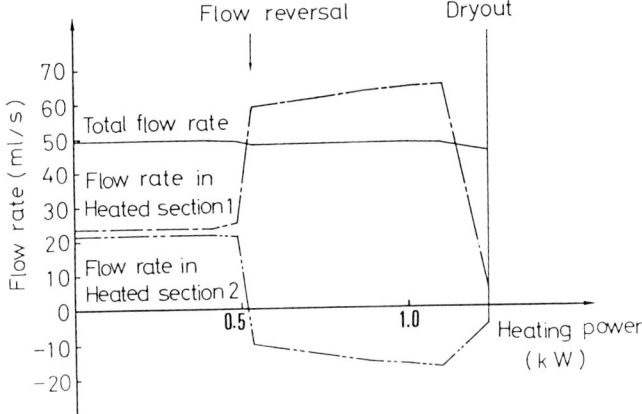

Fig.8 Schematic transients of flow rates showing an occurrence of the flow reversal. (As the total flow rate is kept constant, the flow rate in the normally-flowing-channel is more than the total flow rate.)

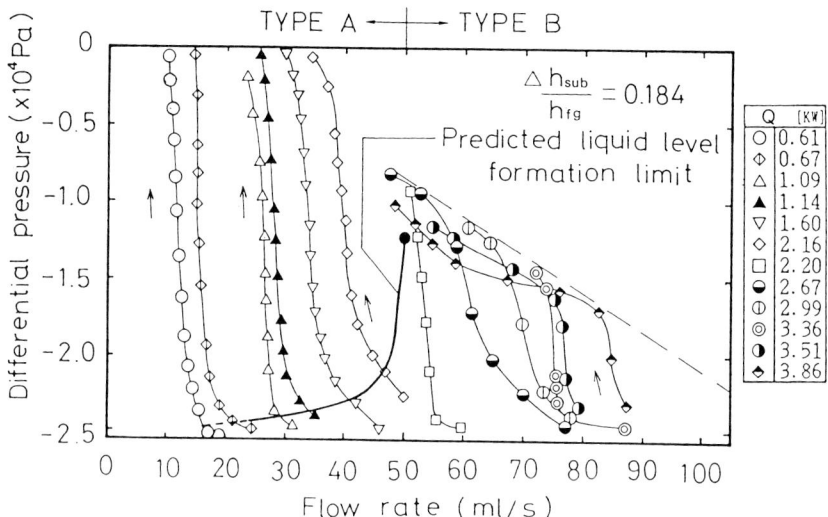

Fig.6 Differential pressure characteristics in single channel experiments. (The boundary between the Type A and the Type B dryout is about 50 ml/s. The pressure drop increases to zero in the Type A dryout case, or the annular flow regime expands throughout the heated channel, while its upper limit is given by the envelope curve along which the dryout B emerges.)

shift to Type B dryout) due to the onset of the flooding, is shown by a solid line and a solid circle. In the Type B dryout region, it seems that the sharp increase in pressure drop is caused by the flooding, and at the envelope shown by the dotted line the flow rate decreases, depending on the pump characteristics.

Parallel two channel experiment
    Similarly to the single channel experiment, two types of dryout (Type C and D) were observed(Fig.7). Under low flow rate conditions, once the uncondensed gas phase begins to accumulate at the top of the heated channel, the flow rates in two channels will not be balanced; they begin to deviate from each other. In a minute the flow in one channel is reversed (upward flow), the flow rate in another channel exceeds the total flow rate, and the boiling ceases. Fig.8 shows the transients of flow rates schematically, where one might find that the flow is reversed in the heated section 2, and that the dryout B occurs there. It is very different that the dryout emerges at the top of the reversed-flow channel, instead of at the bottom as in the case of Type A dryout for the single channel experiment. The mechanisms of the initiation of these dryouts, however, are considered to be the same: the arrival of the uncondensed gas phase and its accumulation at the top of one heated channel. The appearance of the reversed flow is also regarded as the manifestation of the static instability.
    Under high flow rate conditions, similar to dryout B,

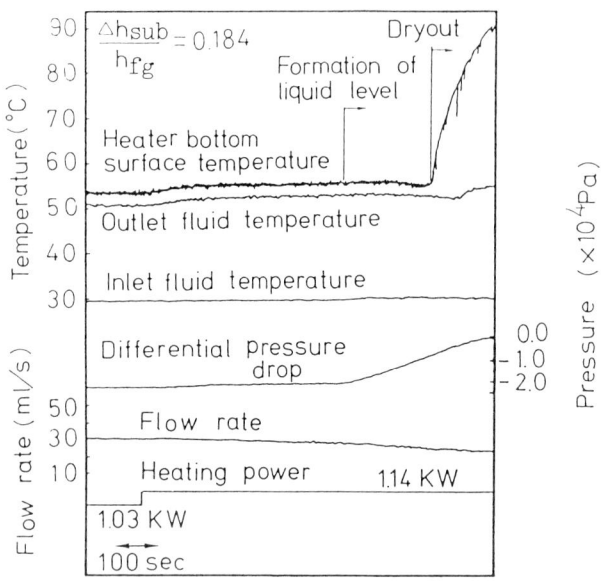

Fig.4 Transients of various parameters for the case of Type A dryout in single channel experiments. (As soon as the 'liquid level' is built up, differential pressure drop begins to increase and the flow rate begins to decrease. Decrease in flow rate is small because of the pump.)

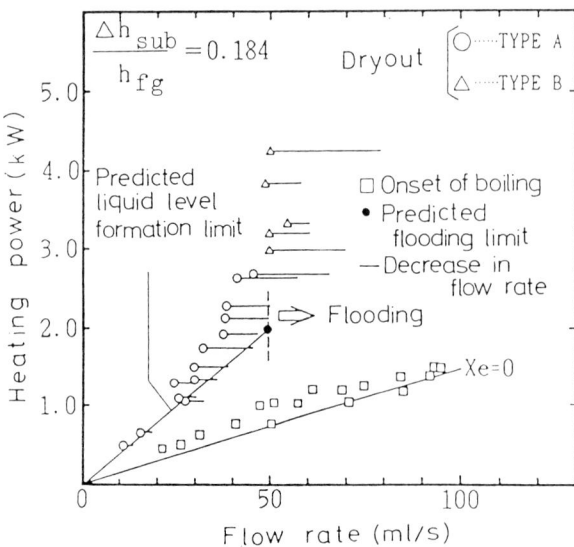

Fig.5 Heating power at onset of dryouts in single channel experiments. (After the formation of the 'liquid level' or the initiation of the flooding, the flow rate begins to decrease and dryouts occur.)

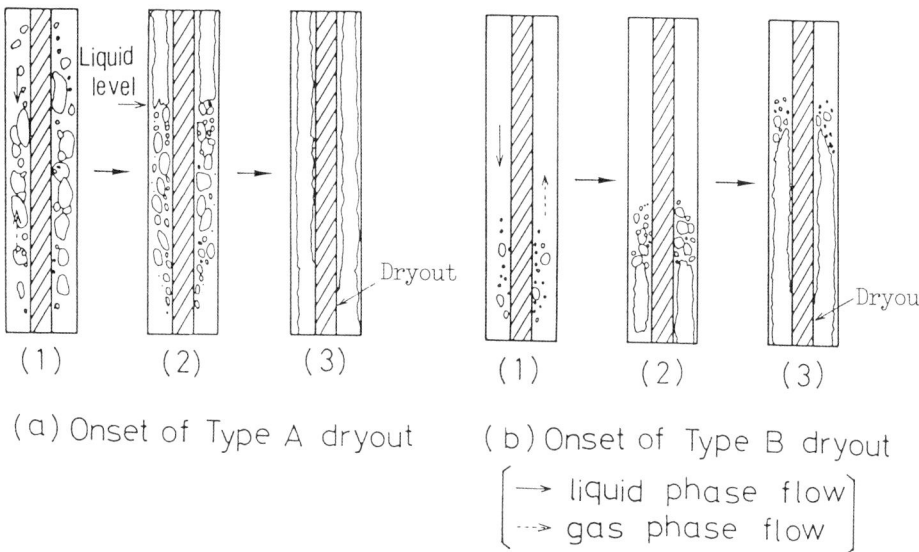

Fig.3 Observed flow regimes in single channel experiments (Dryout A occurs after the fall of the 'liquid level' to the bottom while the dryout B is caused by the flooding at the bottom)

heated channel. As a consequence, a liquid 'level' appears which gradually falls to the bottom and the dryout emerges there.

As the liquid 'level' falls, the pressure drop between the inlet (top) and the outlet (bottom) increases from a negative value to nearly zero; thus, the flow rate decreases slightly, depending on the pump characteristics(Fig.4). As the slope of the flow rate vs. the pressure drop characteristics is negative, this decrease in flow rate is regarded as the flow excursion -- a static instability.

The second dryout (Type B) was observed under high flow rate conditions, which are caused by the blockade of the gas phase due to its stagnation and succeeding flooding. When the downward liquid velocity exceeds the rising velocity of the gas phase, the gas phase begins to stagnate, and, with heating power increased, the bulk of the gas phase increases. This causes the flooding, as well as the consequent decrease in flow rate (flow excursion) and the dryout at the bottom.

The boundary between the Type A and the Type B dryout was at $W_{l,in}$ = 50 ml/s or $j_{l,in}$ = 7.3 cm/s. The decrease in flow rate during the transient from the initiation to the onset of dryout was larger for the Type B dryout case(Fig.5).

In a later section, the onset of the Type A dryout is predicted with the two-fluid model, and its results are shown in Fig.5. In Fig.6 are plotted the transients of the differential pressure against the inlet flow rate, where the phenomena proceed in the direction shown by arrows. The prediction of the onset of the Type A dryout and its limit (or

DOWNWARD TWO-PHASE FLOW

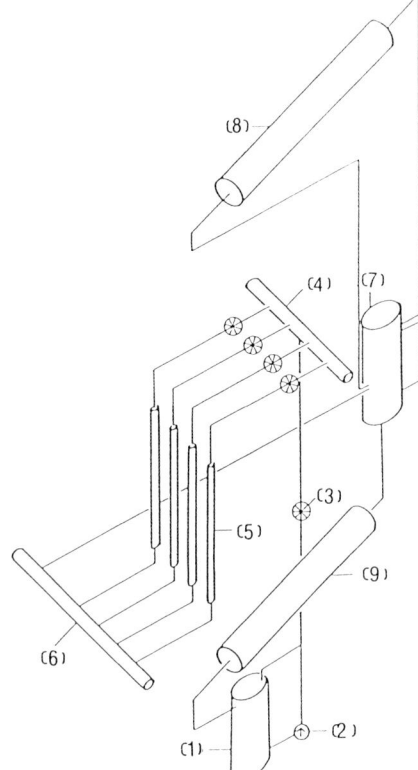

(1) Storage tank
(2) Pump
(3) Flow meter
(4) Inlet header
(5) Heated section
(6) Outlet header
(7) Steam drum
(8) Condenser
(9) Subcooler

Fig.1 Schematic flow diagram of the test apparatus

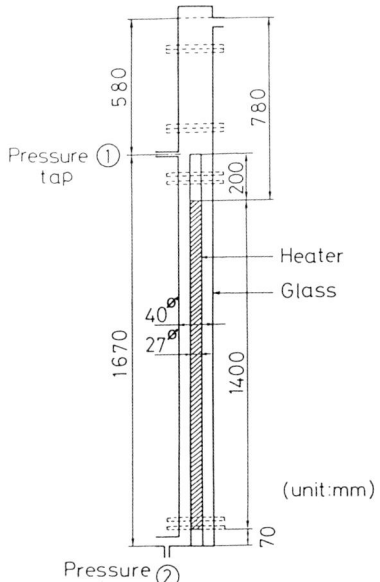

Fig.2 Details of the heated channel

by-pass flow during the injection of ECC-water at the upper plenum(Ball,1978) or the non-uniformity among the rewetting time(Iwamura,1986).

Therefore, it is important to dissolve the complicated phenomena into basic ones, and classify or investigate them with simple experimental apparatus. With an air-water system, Wallis(1981) and Murase(1986) pointed out the non-uniqueness inherent in downward multi-channel flow, as well as the discontinuous succession of various flow modes though a certain parameter, e.g. the upward gas phase flow rate was changed continuously.

The authors(Fukuda,1987) also noted some instabilities peculiar to the downward two-phase flow, and tried to classify them. Following this work, this paper presents experimental results showing another feature of the instabilities of downward two-phase flow, and their consequent heating limits. Analyses were carried out and compared with the experimental results.

EXPERIMENTAL APPARATUS AND PROCEDURE

A schematic flow diagram of the experimental apparatus is shown in Fig.1. The test fluid (R-113) is circulated by a pump (2) through a test loop which consists of turbine flow meters, an upper inlet header plenum, four parallel heated channels (5), a lower outlet header plenum (6), a steam drum (7) and a subcooler (9). In each heated channel, the test fluid flows downward in an annulus between a sheathed heater. (1400 mm heated length, 27 mm OD) and a Pyrex glass outer tube (2250 mm length, 40 mm ID) (Fig.2), through which the flow inside is visible, to make easier the understanding of the apparently complex phenomena. The pressure drop in the heated channel is measured by a differential pressure transducer, and the surface temperatures of each heater at six axial locations are measured with chromel-alumel thermocouples spot-welded on the surface.

Experiments were carried out with single or two parallel or four parallel heated channels. In single channel experiments, the flow rate and its inlet temperature were kept constant, then gradually the heating power was increased until the dryout emerged. In parallel multi-channel experiments, the flow rate in each channel was adjusted before the heating power was applied to give the same value by inlet valves; the heating power was then increased. Throughout the experiment the total flow rate, the inlet temperature and the inlet pressure were kept constant.

EXPERIMENTAL RESULTS

Single channel experiment

Two different types of dryout were observed visually. Their phenomena are shown schematically in Fig.3. The first one, termed Type A, emerged under low flow rate conditions. It was caused by the accumulation of unconfined steam in the heated channel. At low heat, the gas phase rises against the downward liquid phase and condenses to extinguish. However, with the heat increased, a part of the gas phase reaches the top without being condensed, and begins to accumulate in the

HEATING LIMITS AND INSTABILITY OF DOWNWARD TWO-PHASE FLOW

K. FUKUDA, T. KONDOH, H. SAKAI, and S. HASEGAWA
Department of Nuclear Engineering
Kyushu University
Fukuoka, Japan 812

ABSTRACT

Flow characteristics and heating limits of downward two-phase flow in single or parallel multi-channels were investigated experimentally and analytically. The heated channel used in the experiments was made of a glass tube with a heater rod inserted, and the flow regime inside was observed.

In single channel experiments under low flow rate conditions, it was found that the gas phase which flows upward against the downward liquid phase flow condenses and diminishes as it rises, being cooled by inflowing liquid. But as the heating power was increased, a portion of the gas phase reached the top and accumulated to form a liquid level, which eventually caused the flow excursion as well as the dryout. On the other hand, under high flow rate conditions, the flooding initiated at the bottom of the heated section was the cause of the dryout and the flow excursion.

In parallel multi-channel experiments, reversed (upward) flow leading to the dryout was observed in some of these channels with low flow rates. The situation was the same in the single-channel case under high flow rate conditions.

Analyses were carried out to predict the onset of the dryout using the drift flux model and the Wallis'(1969) flooding correlation. The analytical results agree well with the experimental results.

INTRODUCTION

To study flow characteristics and the heating limits of downward two-phase flow is of special importance, with respect to the LOCA (loss-of-coolant accident) analysis or the designing of the ECCS (emergency-core-cooling-system) of water-cooled nuclear reactors.

It was found that the downward flow, which might emerge in a postulated LOCA due to a break at an inlet leg, or a succeeding injection of ECC-water at the upper plenum, might be obstructed by upward steam flow. This could cause various complex phenomena such as non-uniform flow distribution, blockade of flow due to flooding, or flow instability.

Moreover, in the actual nuclear reactor core, regarded as a parallel multi-channel system with many channel boxes, far more complex 3-dimensional phenomena emerge --such as the

section. It was confirmed that the superheater did not cause it, since no effect on the oscillation was observed by releasing the two-phase mixture directly to the atmosphere without the superheater. Therefore, it was inferred that the pipeline upstream of the subcooler causes the oscillation. Fig. A-7 shows the periods of the oscillation of the single spiral channel as well as parallel spiral channels against the mass flux. Solid lines in Fig. A-7 are calculated residence times, assuming that boiling occured in the pipeline 2.1 m long between the liquid nitrogen bomb and the subcooler by heat penetration from the atmosphere, and density wave oscillation occured in it. It was seen that the oscillation was caused by the density wave oscillation in the pipeline upstream of the subcooler, since the experimental period was fairly equal to the residence time, assuming rather large penetration heat flux of 4×10^3 W/m^2. This result suggests that attention should be paid to heat penetration in non-heating area, to discuss the flow instabilities in a cryogenic fluid flow.

References

Takenaka, N., Fujii, T., Akagawa, K, and Nishida, K. 1989, Flow Pattern Transition and Heat Transfer of Inverted Annular Flow, Int. J. Multiphase Flow, to be published.
Wang, B. and Shi, D. 1985, A Semi-Empirical Theory for Forced-Flow Turbulent Film Boiling of a Subcooled Liquid along a Horizontal Plate, Int. J. Heat Mass Transfer Vol.28, 1499

FLOW INSTABILITIES IN PARALLEL CHANNELS

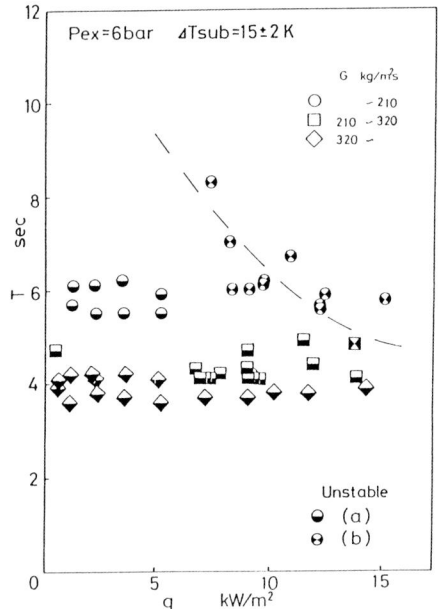

Fig.A-5 Periods of oscillation of single spiral channel

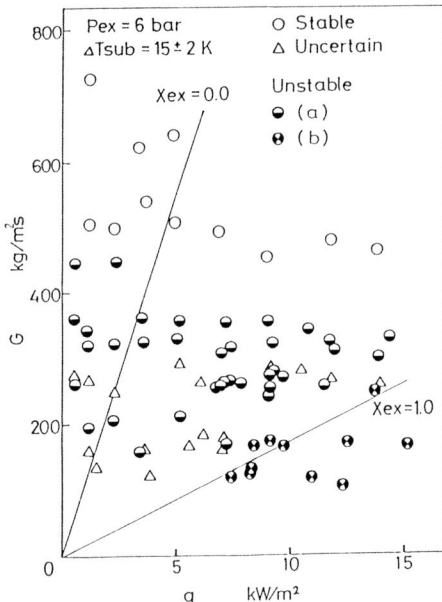

Fig.A-6 Stability map of single spiral channel

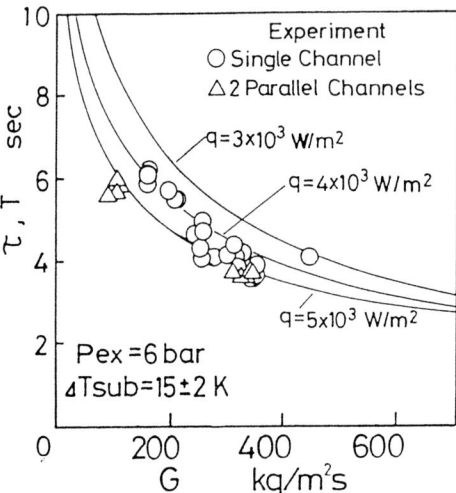

Fig.A-7 Residence time and period of oscillation in phase

affected by the mass flux. Fig. A-6 shows a stability map for a single spiral channel with nucleate boiling in the inlet section. The oscillations as shown in Fig. A-4 (a) were observed in a wide region, including the region where exit quality is smaller than 0, which indicates that this oscillation is caused by two-phase flow upstream of the test

Fig. A-3 shows the relations by the drift flux model between J and u_g by analyzing the experimental conditions. Though the analytical results were affected a little by the inlet mass flux and by the exit pressure, the distribution parameter and the drift velocity can be determined approximately as 2.4 and 0.309, respectively.

II. Oscillation in Phase

It was already mentioned that the oscillations in phase for parallel channels were considered to be caused by boiling in the upstream section of the heating test section. To confirm the cause of this oscillation, experiments were carried out, using a single spiral channel with nucleate boiling for the exit pressure of 6 bar.

Two typical oscillations were observed in a single channel, as shown in Figs. A-4 (a) and (b). Fig. A-4 (a) shows the oscillations similar to those observed in the parrallel channels as shown in Figs. 6-8 (b). Fig. A-4 (b) shows large oscillations of the hot wire signals, the inlet and outlet pressures, and the mass flux. Fig. A-5 shows the periods of the oscillations. The periods of the large oscillations decrease with the increase of the heat flux, similarly to the density wave oscillation for the parrallel channels. However, the periods of the oscillation as shown in Fig. A-4 (a) are not affected by the heat flux, but much

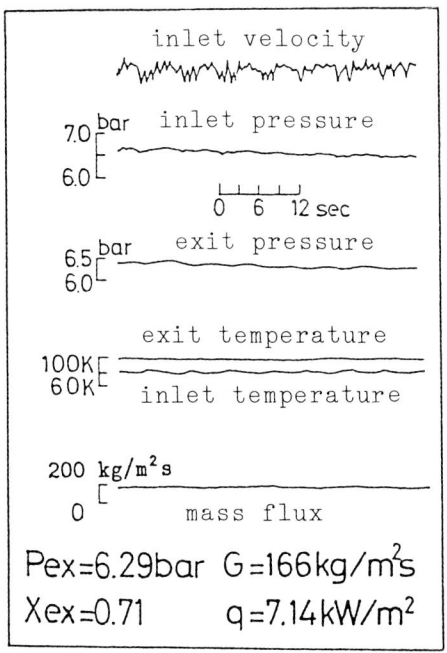
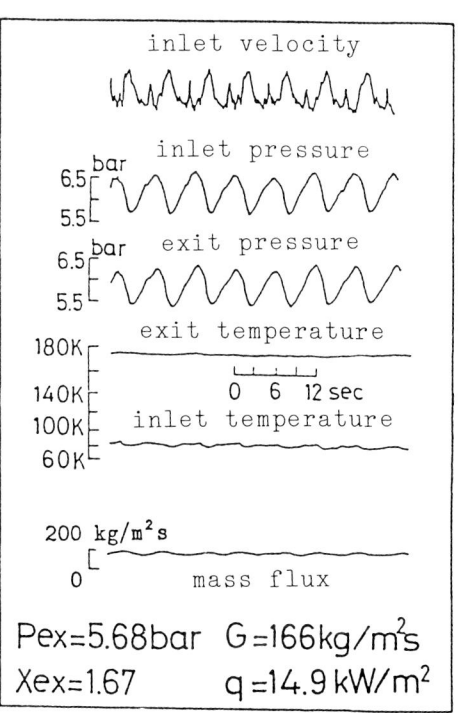

Fig. A-4 Recorder chart traces of single phase channel with nucleate boiling.

Fig. A-2 shows comparison of the experimental and analytical results for heat transfer plotted by Nusselt number Nu and film Reynolds number Re_f,

$$Nu = \frac{qD}{\lambda_g(T_w - T_s)} \qquad (A-12)$$

$$Re_f = \frac{Gx_a D}{\mu_g} \qquad (A-13)$$

Good agreements are obtained for heat transfer, therefore it is expected that these analyses predict well the void fraction which is not obtained experimentally.

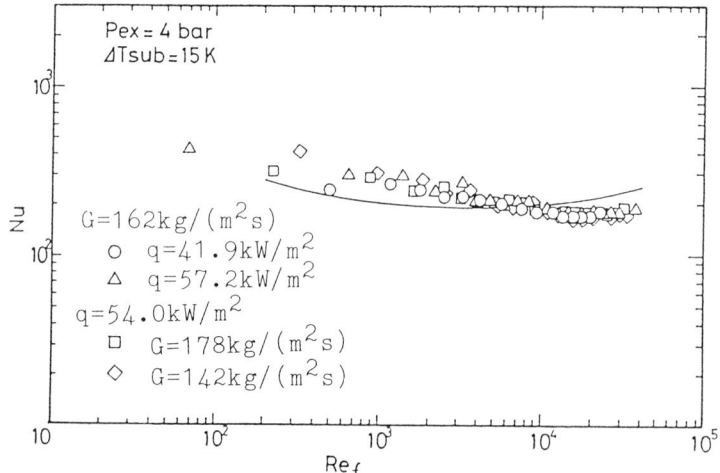

Fig.A-2 Heat transfer of inverted annular flow.

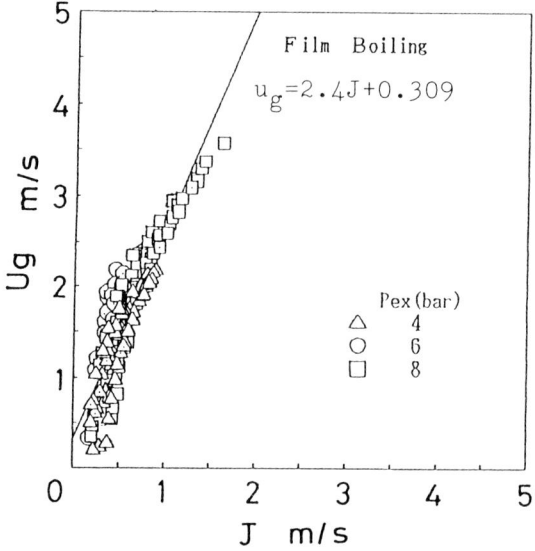

Fig.A-3 Drift flux model for inverted annular flow.

Substituting Eq. (A-5) into Eq. (A-1) and rearranging it with the thermodynamic equilibrium quality x_{eq} and that at inlet x_{eqin}, the actual quality can be obtained as

$$x_a = [x_{eq} - x_{eqin} \exp(-\frac{\pi D}{C_{p\ell} G} \int h_i dz)] / [\frac{H_{\ell gf}}{H_{\ell g}} - x_{eqin} \exp(-\frac{\pi D}{C_{p\ell} G} \int h_i dz)] \quad (A-6)$$

and h_i is given by the correlation of forced convective film boiling heat transfer from vapor to water reported by Wang and Shi (1985) as

$$h_i = 0.054 (\frac{k_\ell}{z})(\frac{Gz}{\mu_\ell})^{0.84} (\frac{C_{p\ell} \mu_\ell}{k_\ell}) \quad (A-7)$$

With the actual quality calculated with Eq. (A-5)-(A-7), heat and mass transfer is analyzed by the turbulent boundary layer theory.

Shear stress τ is determined with the shear stress at wall τ_w and at interface τ_i

in gas phase $\quad \tau = \tau_w - \frac{r_w - r}{r_w - r_i}(\tau_w - \tau_i) \quad (A-8)$

in liquid phase $\quad \tau = \frac{r}{r_i} \tau_i \quad (A-9)$

where r_w and r_i are the radius of the tube and the liquid column and

$$\tau = \rho(\nu + \varepsilon_M)\frac{du}{dy} \quad (A-10)$$

Eddy diffusivity ε_M is given by Reichardt's equations of a circular tube for both vapor and liquid phases. The shear stress in gas phase can be zero at a certain position y_d, therefore, ε_M is determined by Reichalt's equations symmetrically with respect to the position y_d.

Turbulent heat flux is determined as

$$q = C_p \rho (a + \varepsilon_H)\frac{dT}{dy} \quad (A-11)$$

and ε_H is assumed to be equal to ε_M by the Reynolds anology. Giving τ_w and $r_w - r_i$, and integrating Eqs. (A-10) and (A-11), the velocity and temperature profiles in both gas and liquid phases, as well as mass flux and actual quality, are calculated. Therefore, the heat transfer coefficient and the void fraction for a certain experimental condition are obtained by giving τ_w and $r_w - r_i$ by trial and error for calculated mass flux and the actual quality is to be equal to the experimental condition after calculating the actual quality by Eq. (A-5).

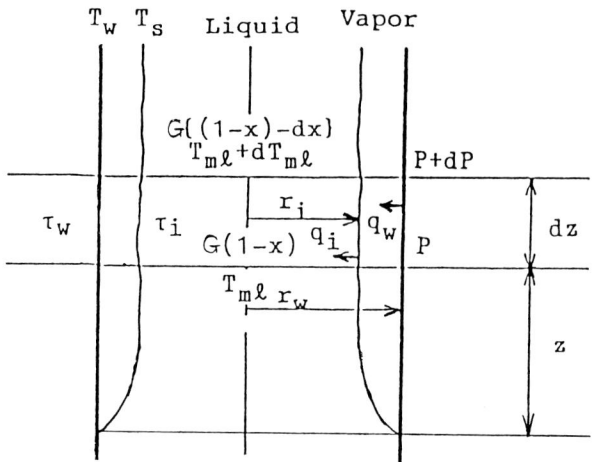

Fig.A-1 Model of inverted annular flow.

dimensional axial energy balance among wall heat inputs, latent heat and sensible heats of superheated vapor and subcooled liquid. Assuming that the axial gradients of the wall temperature T_w and averaged liquid temperature $T_{m\ell}$ are small, and that the actual quality x_a is small,

$$x_a = \frac{q \pi D z - C_{p\ell}(T_{m\ell}-T_{\ell in})G}{[H_{\ell gf}+C_{p\ell}(T_s-T_{m\ell})]G} \tag{A-1}$$

where $T_{\ell in}$ is the inlet temperature of subcooled liquid and $H_{\ell gf}$ is the effective latent heat by Kutateladze to take account of the sensible heat of the superheated vapor,

$$H_{\ell gf} = H_{\ell g} + \frac{1}{2} C_{pg}(T_w-T_s) \tag{A-2}$$

$T_{m\ell}$ is calculated with interfacial heat flux q_i from the vapor layer to the liquid column as

$$\frac{dT_{m\ell}}{dz} = \frac{q_i P_i}{C_{p\ell} G} \tag{A-3}$$

where the actual quality is small and P_i is the peripheral length of the liquid column, which is fairly equal to πD, and q_i is given as

$$q_i = h_i(T_s-T_{m\ell}) \tag{A-4}$$

Then $T_{m\ell}$ is integrated as

$$T_{m\ell} = T_s - (T_s-T_{\ell in})\exp\{-\frac{\pi D}{C_{p\ell}G} \int h_i dz\} \tag{A-5}$$

Subscript

    1   : liquid single flow,
    a   : actual,
    s   : saturation,
    eq : thermodynamic equilibrium,
    ex : exit,
    sub : subcool,
    g   : gas,
    i   : interface,
    w   : wall,
    in : inlet,
    ℓ   : liquid,
    m   : mean,
    ‾   : Laplace transformed.

REFERECES

Fukuda, K., Kato, A., Kondoh, T., and Hasegawa,S. 1986, Two-Phase Flow Instability in a Liquid Nitrogen Heat Exchanger, Nucl. Engng. and Design, Vol.95, pp.91-103 .
Hands, B.A. 1975, Pressure Drop Instabilities in Cryogenic Fluids, in Advances in Cryogenic Engineering Vol.20, pp. 355-369.
Hands, B.A. 1979, The Flow Instability of Liquid-Nitrogen Thermosiphon with 8 mm Bore Riser, AIChE Symp. Ser., Vol.75, No.189, pp.177-184.
Ishii, M. and Zuber, N. 1970, Thermally Induced Flow Instabilities in Two^{2}Phase Mixtures, Proc. 4th Int. Heat Transfer Conf., Paper No. B5.11.
Lahey, R.T. and Drew, D.A. 1980, An Assessment of the Literature Related to LWR Instability Modes, NUREG/CR-1414
Ozawa, M. 1976, Study on Flow Instabilities in Evaporator Tubes, Dr. thesis, Osaka Univ.. in Japanese
Saha, P. 1974, Thermally Induced Two-Phase Instabilities, Including the Effect of Thermal Non-Equilibrium between the Phases, Ph.D. thesis, Georgia Inst. Technol.
Spinks, N. 1971, Analysis of Flow Instability in Boiling Systems with the TOSCLE Code, AAEC/E217.
Zuber, N. and Findly, J.A. 1965, Average Volumetric Concentration in Two-Phase Flow System, Trans. ASME J. Heat Transfer, Vol.87, pp.453-468.

APPENDICES

I. Analysis of Inverted Annular Flow

   Turbulent boundary layer analysis was used to analyze inverted annular flow of liquid nitrogen, assuming a separated flow with a smooth gas-liquid interface as shown in Fig. A-1. The details of the analysis and the experimental and analytical results of Freon R-113 were reported by Takenaka et al. (1989).
   As the liquid column is subcooled and the vapor layer is superheated near the inlet, actual quality x_a is not equal to thermodynamic equilibrium quality. It is calculated by one-

FLOW INSTABILITIES IN PARALLEL CHANNELS

Acknowledgment

The authors wish to express their cordial acknowledgments to Dr. M. Ozawa in Kobe Univ. for his valuable suggestions and fruitful discussion and Messrs. K. Sugimoto and S. Sato for their help in preparation of this paper. Their acknowledgements are also extended to Nihon-Sanso Co. Ltd. for the support of the experimental equipment. This study was partly supported by Grant-in-Aid of Scientific Research of the Ministry of Education, Science and Culture of Japan.

NOMENCLATURE

<u>English</u>

- a : thermal diffusivity [m²/s],
- c_p : constant pressure specific heat [J/kg K],
- C : distribution parameter,
- D : diameter of tube [m],
- g : gravity acceleration [m/s²],
- G : mass flux [kg/m²s],
- h : heat transfer coefficient [W/m²],
- H : enthalpy [J/kg],
- $H_{\ell g}$: latent heat [J/kg],
- $H_{\ell gf}$: effective latent heat by Kutatelaze [J/kg],
- J : volumetric averaged velocity [m/s],
- k : heat conductivity [W/mK],
- K : restriction coefficient,
- P : pressure [Pa],
- ΔP : pressure drop [Pa],
- q : heat flux [W/m²],
- Q_v : heat input for unit volume [W/m³],
- r : radial distance [m],
- T : with suffix, temperature [°C],
- T : without suffix, period of oscillation [s],
- u : velocity [m/s],
- v : drift velocity [m/s],
- x : quality,
- y : distance normal to wall [m],
- z : axial distance [m],

<u>Greek</u>

- α : void fraction,
- θ : inclined angle of spiral tube,
- μ : viscosity [kg/ms],
- ρ : density [kg/m³],
- σ : surface tension [N/m],
- ϕ_ℓ : LM two-phase multiplier,
- χ_{tt} : LM parameter,
- ε_M : eddy diffusivity [m²/s],
- ε_H : thermal diffusivity [m²/s],
- ν : kinematic viscosity [m²/s],
- τ : residence time [s],

this feedback system is solved by Nyquist's diagram method on the open transfer function (G·H). The stability criteria predicted by using Ozawa's program mentionned above are shown by lines in Figs. 24-26, with the experimental results for cases [A], [B] and [C].

In Fig. 24 (c) for the spiral channels with nucleate boiling in the inlet section, the calculated stability criteria by Eq. (5) for slug flow model and Eq. (6) for churn flow model without and with taking into acount the wall capacity are shown. The latest model agrees fairly well with the experimental results. Therefore, the calculated results by the latest model are shown in Figs. 24 (a) and (b).

In Figs. 25 (a), (b) and (c) for the vertical channels with nucleate boiling in the inlet section, the calculated results by the homogenious model, the slug model and the churn model are shown. In this case, the homogenious model fits well the experimental results for every exit pressure condition.

In Figs. 26 (a), (b) and (c) for the vertical channels with film boiling, i.e. inverted annular flow in the inlet section, the calculated results by the inverted annular flow model are shown. This agrees well with the experimental results.

With the results it is concluded that the analytical method described above is effective to predict the stability criteria in cryogenic fluid flow. Also, as the stability analysis indicated that the flow is more stable for the model with larger slip, it is estimated that the flow with film boiling is more stable than that with nucleate boiling, because of the larger slip in the inverted annular flow.

CONCLUSIONS

Experimental and analytical studies were carried out on the flow instabilities of cryogenic fluid in parallel channels, consisting of spiral tubes and vertical tubes with nucleate boiling and with film boiling in the inlet section, using liquid nitrogen as a working fluid.

Concluding remarks are as follows:
1. Density wave oscillations, which cause alternative increases and decreases of flow rate out of phase in parallel channels, were observed in every evaporator in certain experimental conditions.
2. Stability maps were obtained and the stability criteria can be approximately expressed by constant exit qualities.
3. Flows with film boiling in the inlet section were stable for higher exit quality than those with nucleate boiling.
4. Linear stability analysis predicts well the stability criteria with nucleate boiling, by using previous drift flux models, and with film boiling, by using the drift flux model formulated by the turbulent boundary layer analysis for inverted annular flow.

FLOW INSTABILITIES IN PARALLEL CHANNELS

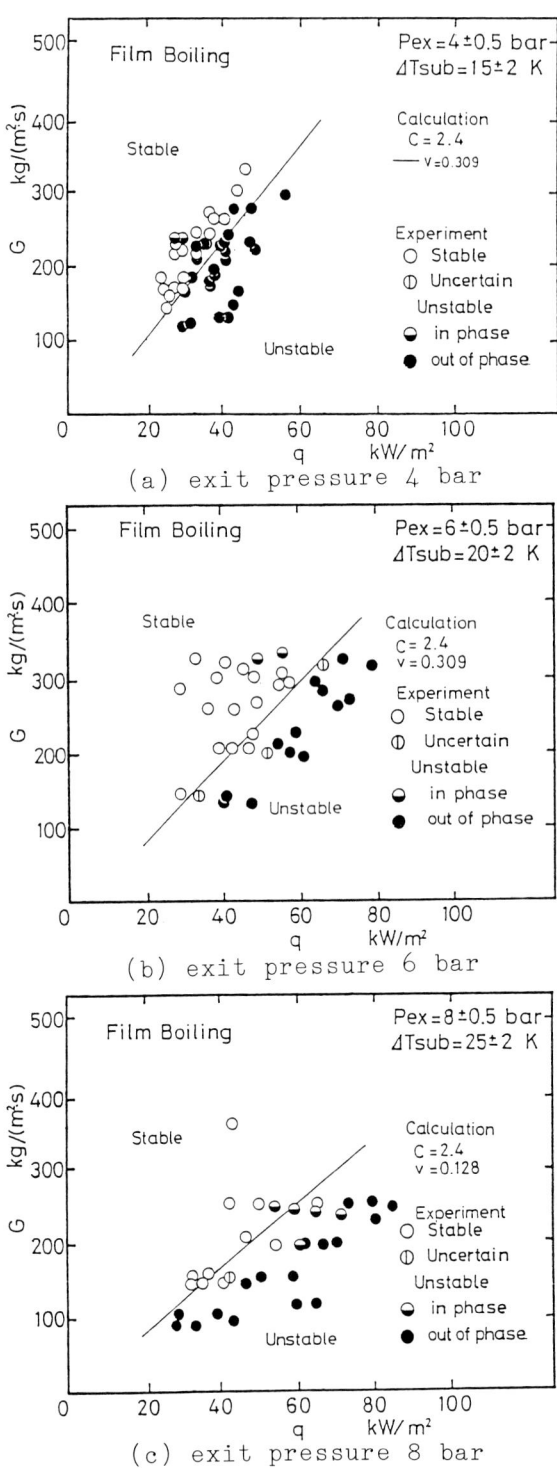

Fig.26 Stability maps with analytical prediction, vertical channels with film boiling case [C].

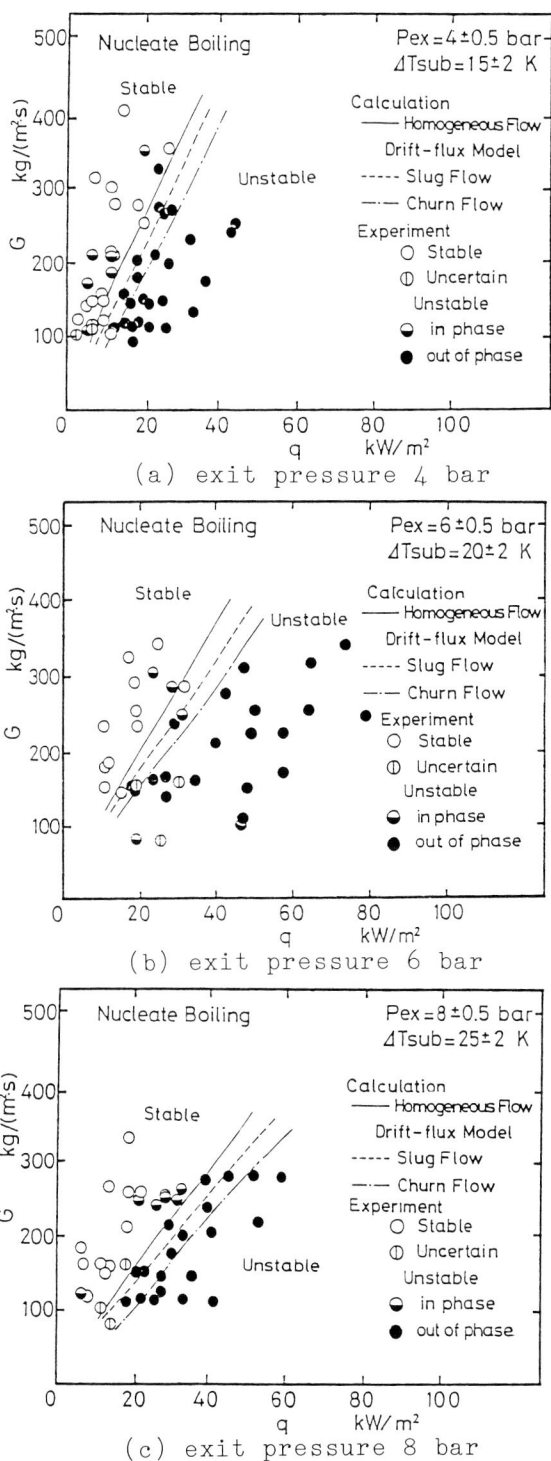

Fig.25 Stability maps with analytical prediction, vertical channels with nucleate boiling, case [B].

FLOW INSTABILITIES IN PARALLEL CHANNELS

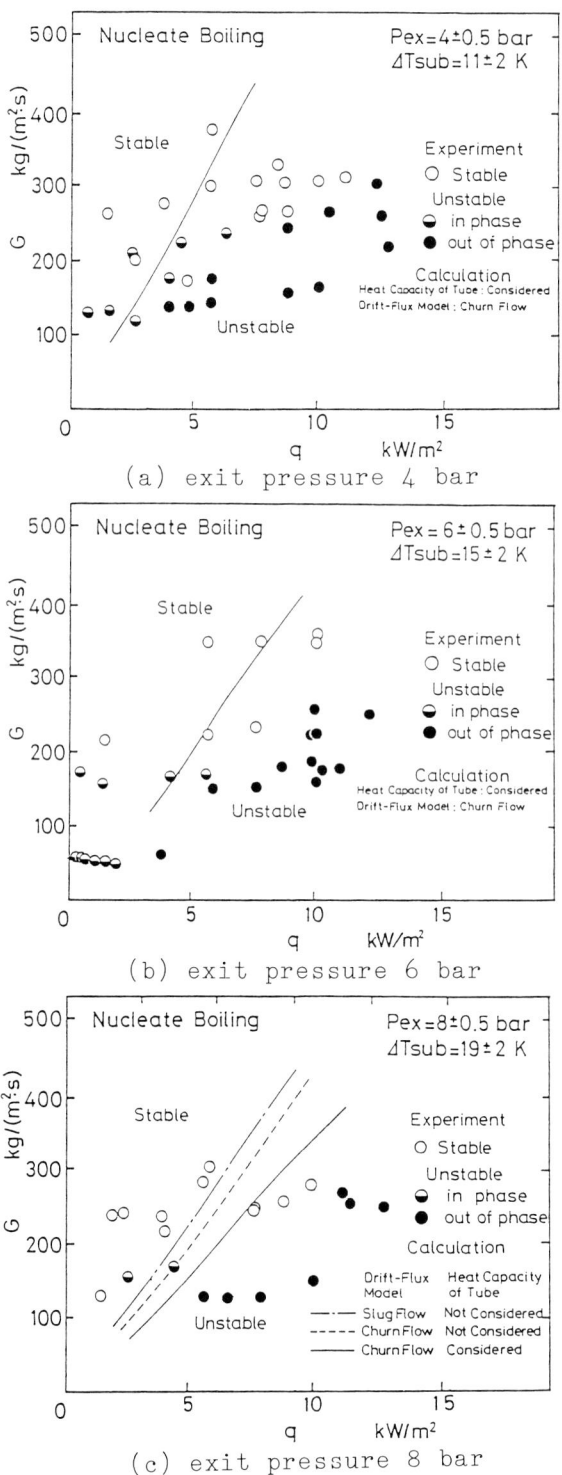

Fig.24 Stability maps with analytical prediction, spiral channels with nucleate boiling, case [A].

when the exit quality is larger than 1,

$$\Delta P_{ex} = \frac{K_{ex}}{2\rho_g} G^2 \tag{11}$$

where K_{in} and K_{ex} are the inlet and exit restriction coefficients, and are determined $K_{in}=8.75$ and $K_{ex}=9.74$ for the spiral channels, and $K_{in}=3.60$ and $K_{ex}=2.19$ for the vertical channels with the experiments using air flow.

The calculated values of the total pressure drop ΔP_{cal} between the inlet and exit headers fairly agree with the experimental values ΔP_{ex} as shown in Figs. 20, 21 and 22. For case [A] in Fig. 20, Eq. (6) for churn flow is used; for case [B] in Fig. 21, Eq. (4) for homogenious flow is used; and for case [C] in Fig. 22, Eq. (7) for inverted annular flow is used. The over-estimation of 30% in case [C] is because the flow pattern changes from inverted annular flow to dispersed flow in the downstream section of the test channel. The accuracy of the pressure drop estimation is sufficient for the flow stability analysis.

The stability analysis is carried out by a Nyquist diagram method program coded by Ozawa (1976). Linearization and Laplace transformation of the equations give a block diagram, as shown in Fig. 23. Here the transfer functions of G and H are expressed by Eqs. (12) and (13), respectively. The transfer function G is for the inlet restriction and the liquid single phase flow region, and H is for the two-phase flow region and the exit restriction.

$$G = [\frac{\overline{u}_1}{u_1}] / [\frac{-\overline{\Delta P}_{a1}}{\rho_\ell u_1^2} + \frac{-\overline{\Delta P}_{f1}}{\rho_\ell u_1^2} + \frac{-\overline{\Delta P}_{in}}{\rho_\ell u_1^2}] \tag{12}$$

$$H = [\frac{-\overline{\Delta P}_g}{\rho_\ell u_1^2} + \frac{-\overline{\Delta P}_a}{\rho_\ell u_1^2} + \frac{-\overline{\Delta P}_f}{\rho_\ell u_1^2} + \frac{-\overline{\Delta P}_{ex}}{\rho_\ell u_1^2}] / [\frac{\overline{u}_1}{u_1}] \tag{13}$$

where subscript 1 means the terms for the liquid single phase flow region and hat ¯ means the Laplace transformed perturbation terms. For the flow with film boiling, the term of the liquid single phase flow is not necessary. Therefore, G only consists of the inlet restriction. The stability criterion of

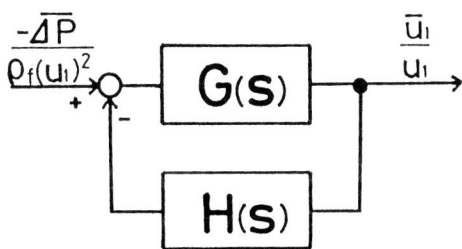

Fig.23 Block diagram.

FLOW INSTABILITIES IN PARALLEL CHANNELS 523

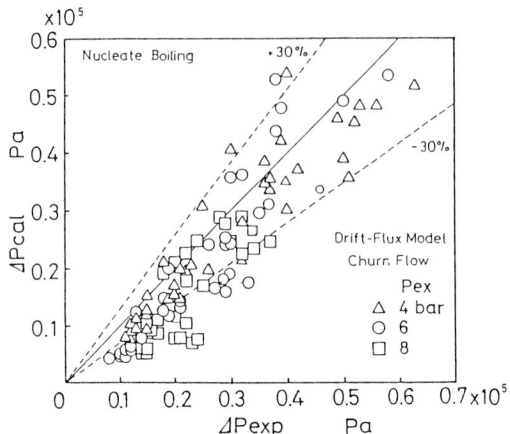

Fig.20 Spiral channels with nucleate boiling, case [A].

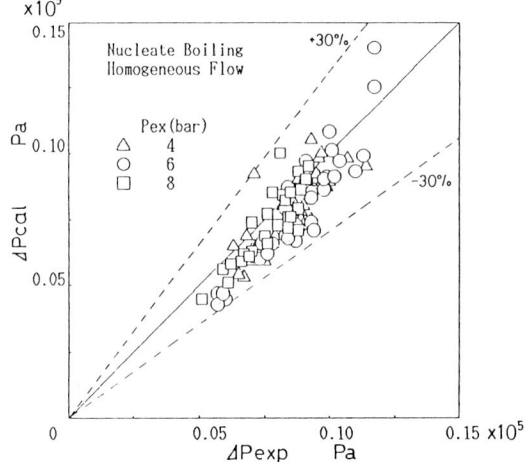

Fig.21 Vertical channels with nucleate boiling, case [B].

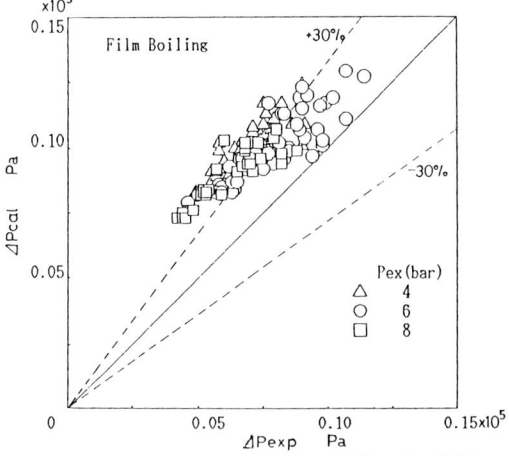

Fig.22 Vertical channels with film boiling, case [C].

Fig.20-22 Pressure drop between inlet and exit headers.

drop, which is given by a constitutive equation. L.S.H of Eq. (2) is an accerational pressure drop ΔP_a and the third term of R.S.H of Eq. (2) is a gravitational pressure drop ΔP_g.

Gas velocity u_g is determined by a drift flux model, $u_g = CJ + v$, where J is the volumetric averaged velocity $(1-\alpha)u_\ell + \alpha u_g$. C and v are a distribution parameter and a drift velocity. Giving ΔP_f, C and v as constitutive equations, Eqs. (1)-(3) are closed and the solutions can be obtained.

The values of C and v for the flow with nucleate boiling used in this calculation are those proposed by Zuber and Findly (1965) as follows:

for homogenious flow;
$$C=1 \quad , \quad v=0 \tag{4}$$

for slug flow;
$$C=1.2 \quad , \quad v=0.35[g(\rho_\ell-\rho_g)D/\rho_\ell]^{1/2} \tag{5}$$

and for churn flow;
$$C=1.35, \quad v=1.53[g(\rho_\ell-\rho_g)\sigma/\rho_\ell]^{1/4} \tag{6}$$

The values of C and v for the flow with film boiling are obtained from the analytical results by turbulent boundary layer model, described in Appendix I as follows:

for inverted annular flow;
$$C=2.4 \quad , \quad v=0.309 \tag{7}$$

The frictional pressure drop ΔP_f is estimated from the two-phase multiplier Φ_ℓ of the LM method of Chisholm's correlation Eq. (8) both for nucleate boiling and film boiling.

$$\Phi_\ell = 1 + \frac{21}{X_{tt}} + \frac{1}{X_{tt}^2} \tag{8}$$

Steady solutions of u_ℓ, u_g and α can be obtained by Eqs. (1) and (3), neglecting the time varying terms with Eqs. (4), (5), (6) or (7) and the steady total pressure drop is obtained with Eq. (8), inserting the solutions into Eq. (2).

The frictional pressure drop in single phase flow is calculated with the Blasius equation. No liquid single phase flow region is considered for film boiling.

The pressure drops in non-heating pipe lines beetween the inlet header and the heating channels, and between the heating channels and the exit header, are assumed as local inlet and exit pressure drops ΔP_{in} and ΔP_{ex}, and are expressed by:

$$\Delta P_{in} = \frac{K_{in}}{2\rho_\ell} G^2 \tag{9}$$

$$\Delta P_{ex} = \frac{K_{ex}}{2\rho_\ell} G^2 \Phi_\ell^2 (1-x_{ex})^2 \tag{10}$$

FLOW INSTABILITIES IN PARALLEL CHANNELS 521

When the heat flux is increased gradually under a constant mass flux, the flow oscillation out of phase; i.e., density wave oscillation, occur at a definite heat flux. The stability maps using heat flux and mass flux as coordinates obtained by the experiments are shown in Figs. 17, 18 and 19 for cases [A], [B] and [C], respectively. Figures (a),(b) and (c) are for the exit pressures of 4, 6 and 8 bar, respectively. Solid lines in the figures are constant lines of exit thermodynamic equilibrium quality. It can be seen from the figures that the boundary between stable and unstable regions of the density wave oscillation is approximately expressed by a constant exit thermodynamic equilibrium quality line. For instance, x_{ex} is approximately 0.6 in case [A], and is approximately 0.1 in case [B], and is from 0.3 to 0.6 in case [C]. It is shown that the flow with film boiling, i.e. inverted annular flow, in the inlet section is more stable than that with nucleate boiling.

ANALYTICAL RESULTS AND DISCUSSION

The stability criteria is predicted by using a linear stability analysis with a drift flux model, according to previous reports. Assumptions used in this study are as follows:

1. One-dimensional two-phase flow is assumed.
2. The spiral tube is treated as an inclined straight tube, neglecting the centrifugal force.
3. Physical properties are constant of the values at the saturation temperature.
4. Internal energy is assumed to be equal to enthalpy and kinematic energy is small enough to neglect, compared to thermal energy.
5. Thermodynamic equilibrium is assumed for the flow with nucleate boiling, and thermodynamic non-equilibrium in inverted annular flow is considered, by calculating the actual quality as described in Appendix I.

The continuous, momentum and energy equations for the two-phase flow region are expressed, as well known, with the assumptions 1-5,

$$\frac{\partial}{\partial t}[\rho_\ell(1-\alpha)+\rho_g\alpha] + \frac{\partial}{\partial z}[\rho_\ell(1-\alpha)u_\ell+\rho_g\alpha u_g]=0 \tag{1}$$

$$\frac{\partial}{\partial t}[\rho_\ell(1-\alpha)u_\ell+\rho_g\alpha u_g] + \frac{\partial}{\partial z}[\rho_\ell(1-\alpha)u_\ell^2+\rho_g\alpha u_g^2]$$
$$= -\frac{\partial \Delta P}{\partial z} - \frac{\partial \Delta P_f}{\partial z} + [\rho_\ell(1-\alpha)+\rho_g\alpha]g\sin\theta \tag{2}$$

$$\frac{\partial}{\partial t}[\rho_\ell(1-\alpha)H_\ell+\rho_g\alpha H_g] + \frac{\partial}{\partial z}[\rho_\ell(1-\alpha)u_\ell H_\ell+\rho_g\alpha u_g H_g]=Q_v \tag{3}$$

where θ is the inclined angle of the spiral tube, Q_v is the heat input for unit volume and ΔP_f is the frictional pressure

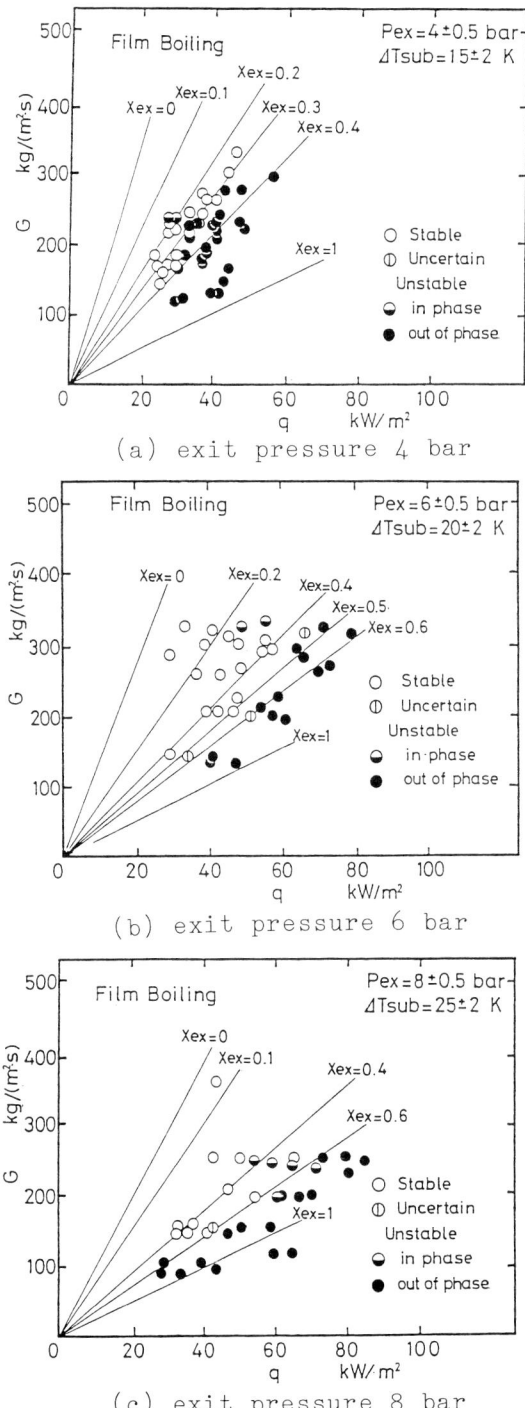

Fig.19 Stability maps, vertical channels with film boiling, case [C].

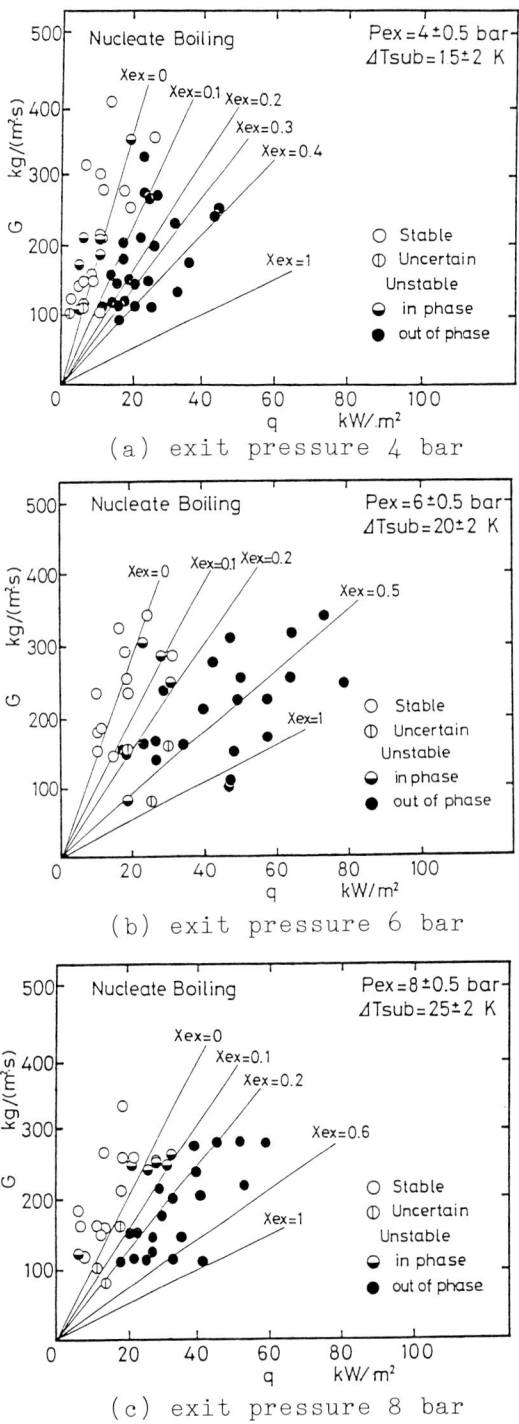

Fig.18 Stability maps, vertical channels with nucleate boiling, case [B].

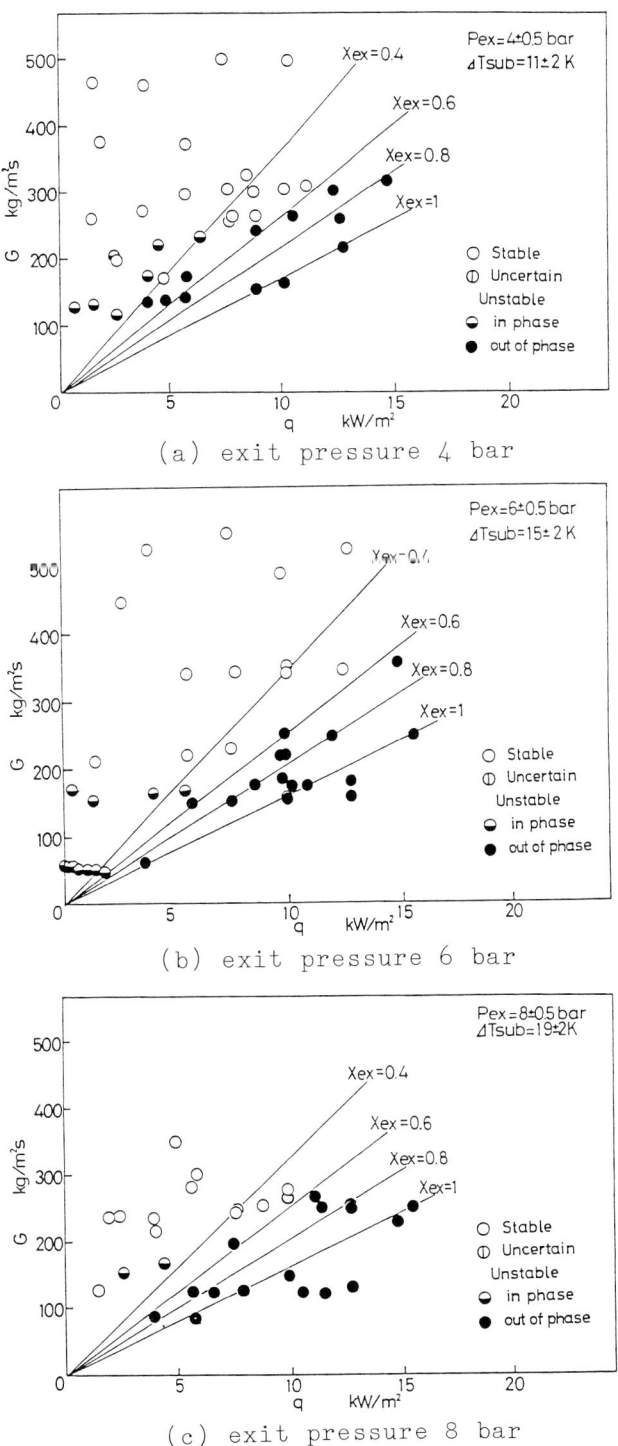

Fig.17 Stability maps, spiral channels with nucleate boiling, case [A].

FLOW INSTABILITIES IN PARALLEL CHANNELS

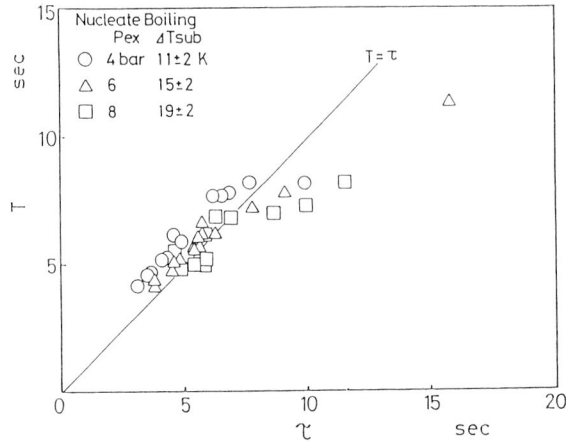

Fig.14 Spiral channels with nucleate boiling, case [A].

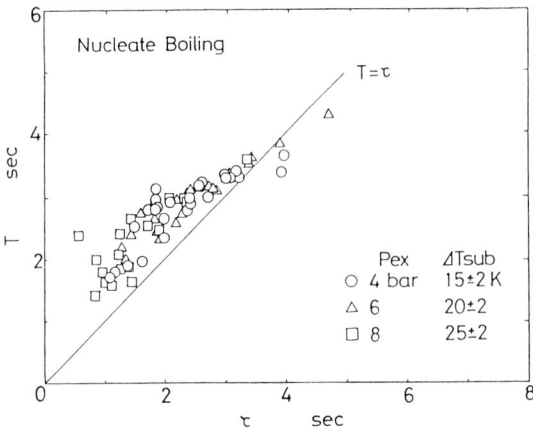

Fig.15 Vertical channels with nucleate boiling, case [B].

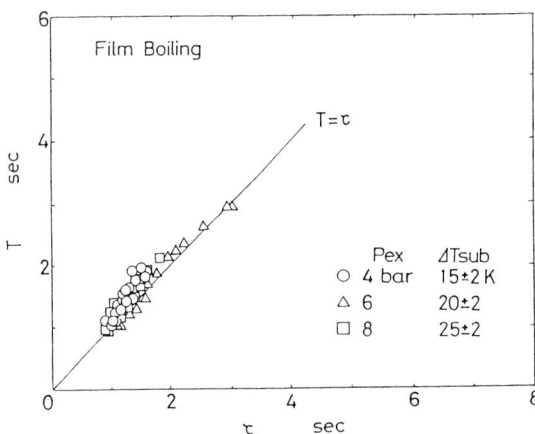

Fig.16 Vertical channels with film boiling, case [C].

Figs.14-16 Residence time and period of oscillation.

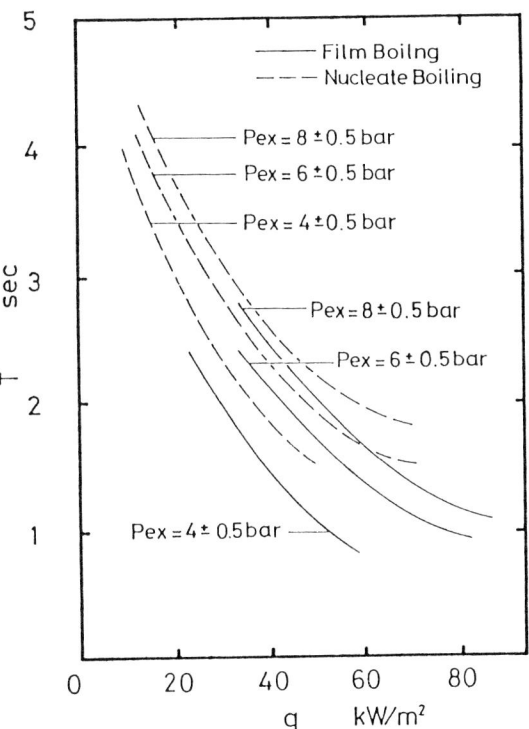

Fig.13 Period of oscillation out of phase in vertical channels, case [B] and case [C].

and that both periods increase with an increase in exit pressure.

In order to investigate the type of flow instability out of phase in this experiment, the periods are compared with the residence time τ in the heating test section. The residence time with nucleate boiling is calculated assuming that boiling occurs at $x_{eq}=0$ and that two-phase flow is homogenious and is perfectly vaporized at $x_{eq}=1$. With film boiling, inverted annular flow is in a thermodynamic non-equilibrium condition near the inlet, as the liquid column is subcooled while the vapor near the wall is superheated. Actual quality x_a is estimated by the energy balance, details of which are described in Appendix I. τ is calculated assuming the two-phase flow is homogeneous. Comparisons between the period and the residence time are shown in Figs. 14, 15 and 16 for cases [A], [B] and [C] respectively, and for exit pressures of 4, 6 and 8 bar. It is seen that the period is fairly equal to the residence time for every case, although the data for nucleate boiling are somewhat scattered. This result indicates that the oscillation out of phase is a density wave oscillation, and the difference of the periods between case [B] and case [C] is due to the difference of actual quality; i.e., the existence of thermodynamic non-equilibrium in inverted annular flow.

FLOW INSTABILITIES IN PARALLEL CHANNELS 515

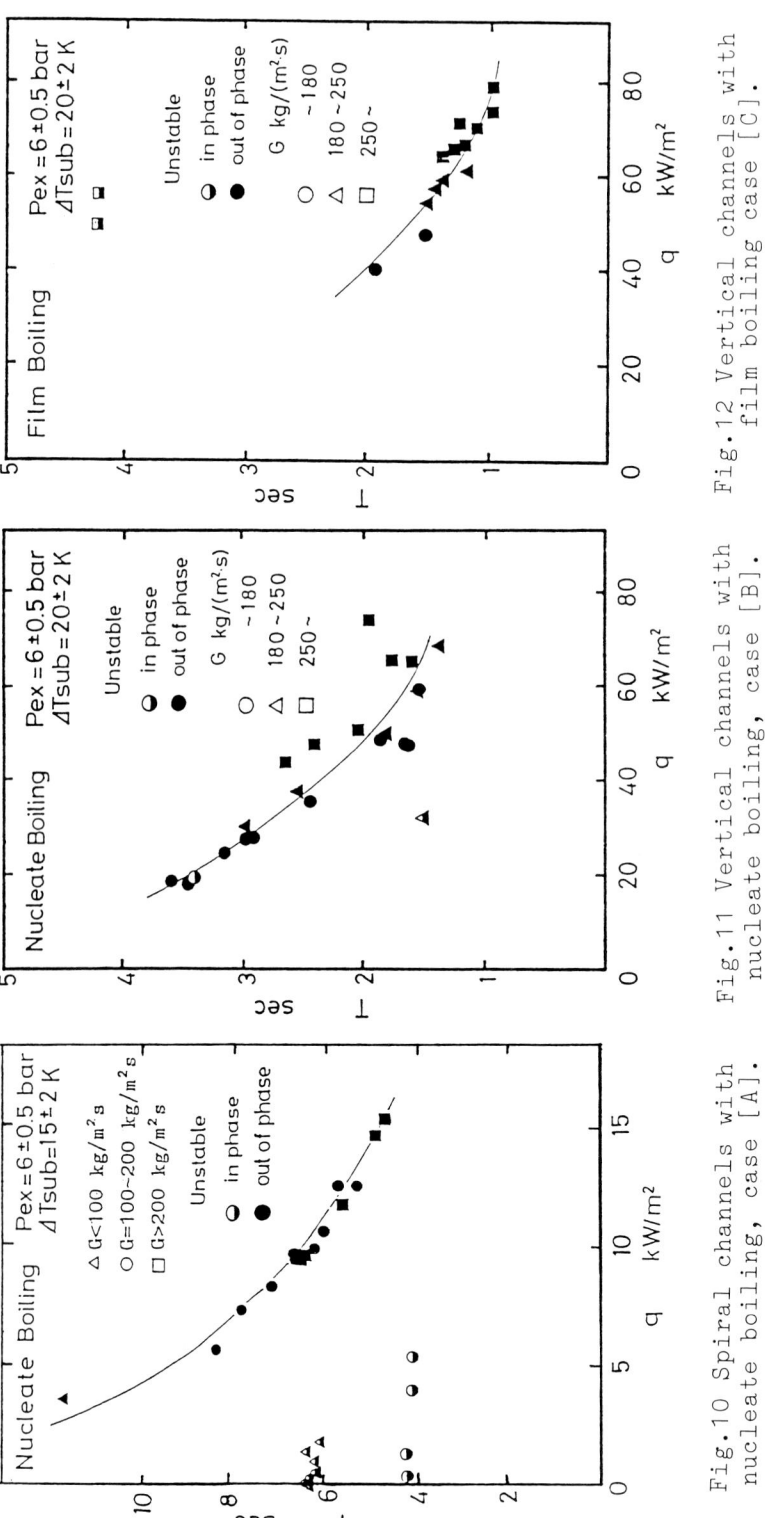

Fig.10 Spiral channels with nucleate boiling, case [A].

Fig.11 Vertical channels with nucleate boiling, case [B].

Fig.12 Vertical channels with film boiling case [C].

Fig.10-12 Periods of oscillations.

Examples of the wall temperature traces along the channel are shown with the hot wire signal in Figs. 9 (a) and (b), for cases [B] and [C] respectively. The wall temperature oscillations are out of phase for the flow rate oscillation; i.e., the wall temperature decreases when the flow rate increases. The wall temperature oscillation is attenuated with the distance from the inlet.

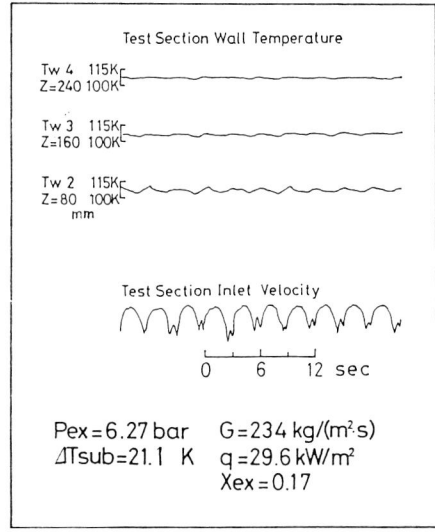
(a) with nucleate boiling, case [B].

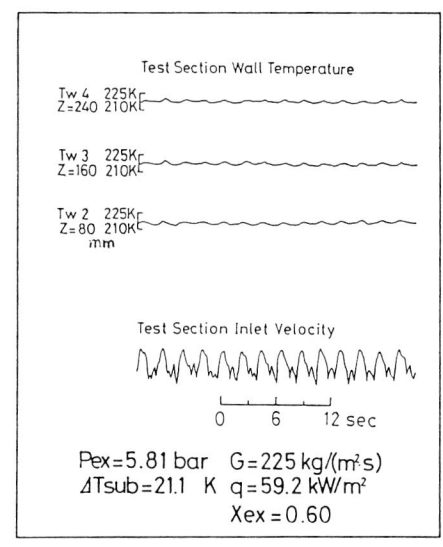
(b) with film boiling, case [C].

Fig.9 Wall temperature oscillation.

Figures 10, 11 and 12 show the oscillation periods of cases [A], [B] and [C] against heat flux, for the exit pressure of 6 bar. The periods of the oscillation in phase, as shown in Figs. 6-8 (c), are determined by the period of a pair of the high and low peaks. The periods of the oscillation out of phase decrease with increasing the heat flux, and are little affected by the inlet mass flux. On the other hand, those in phase are not affected by the heat flux, but are affected by the inlet mass flux. Therefore it is concluded that the oscillation out of phase is caused by flow instabilities in the test section itself, but that in phase is caused by flow instabilities in the whole system including the test section, the supply pipeline and the nitrogen bomb. Therefore, the instability in phase is not an original flow instability in evaporators. Consequently the oscillation in phase is not the subject of this report, but will be discussed in Appendix II.

The periods of the oscillation out of phase in vertical channels are shown in Fig. 13, in order to make clear the effects of the flow pattern and the exit pressure. The values for nucleate boiling are shown by broken lines and those for film boiling by solid lines. It is seen that the periods with film boiling are shorter than those with nucleate boiling,

FLOW INSTABILITIES IN PARALLEL CHANNELS

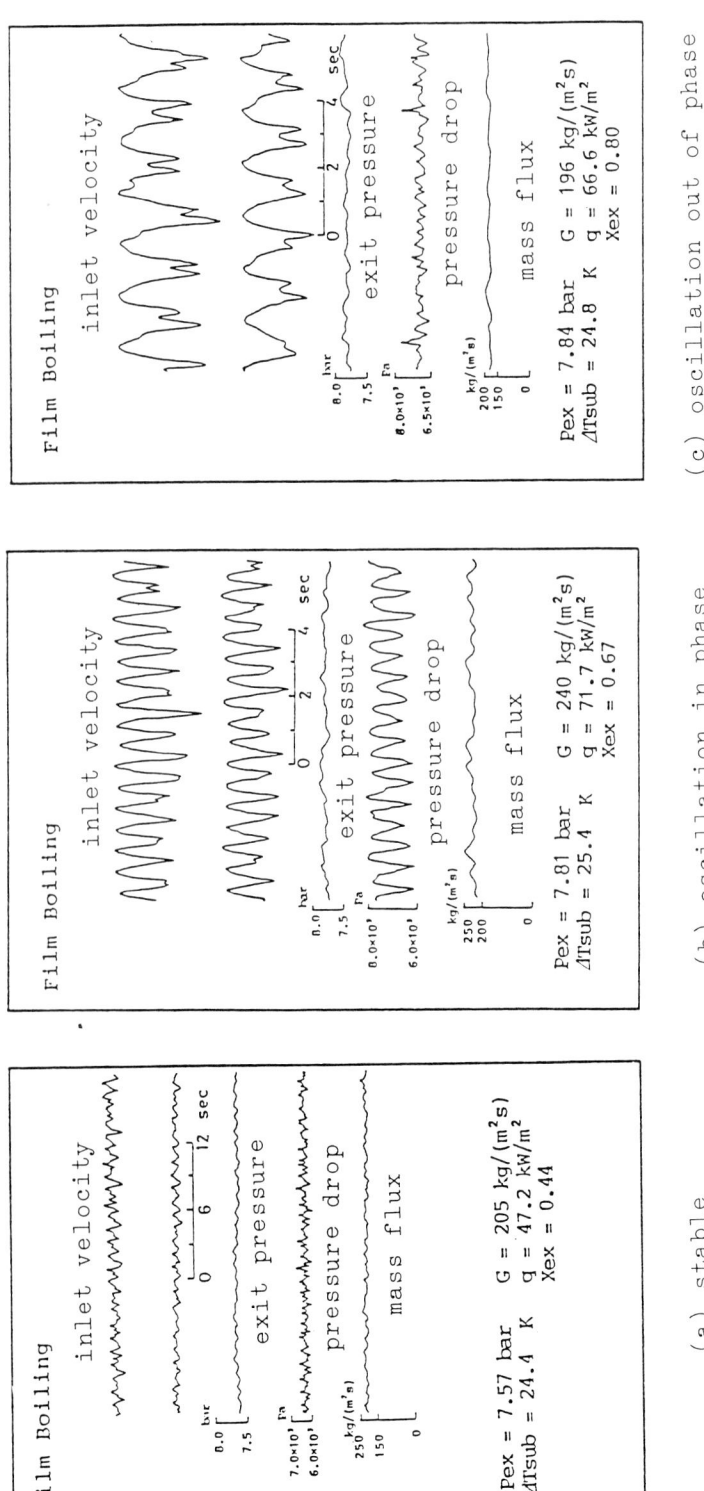

Fig.8 Recorder chart traces, vertical channels with film boiling, case [C].

(a) stable (b) oscillation in phase (c) oscillation out of phase

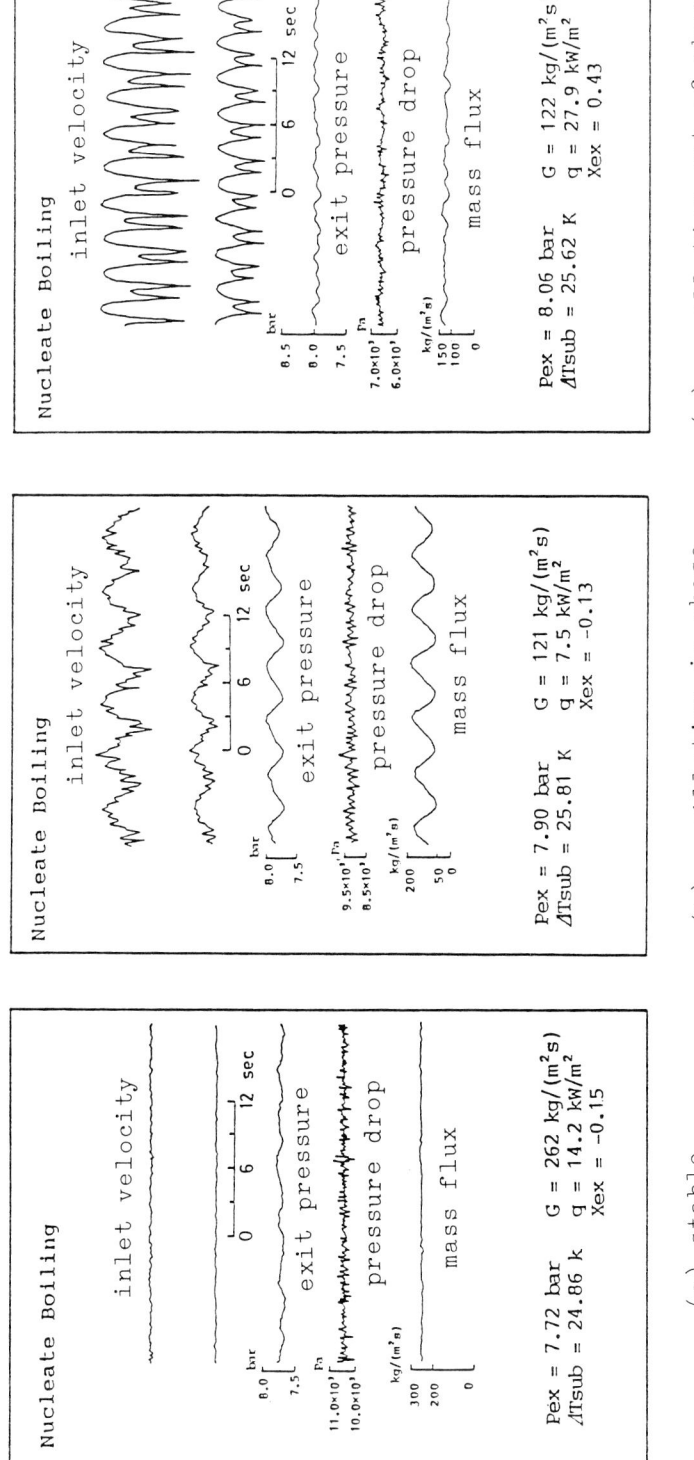

Fig.7 Recorder chart traces, vertical channels with nucleate boiling, case [B].

(a) stable   (b) oscillation in phase   (c) oscillation out of phase

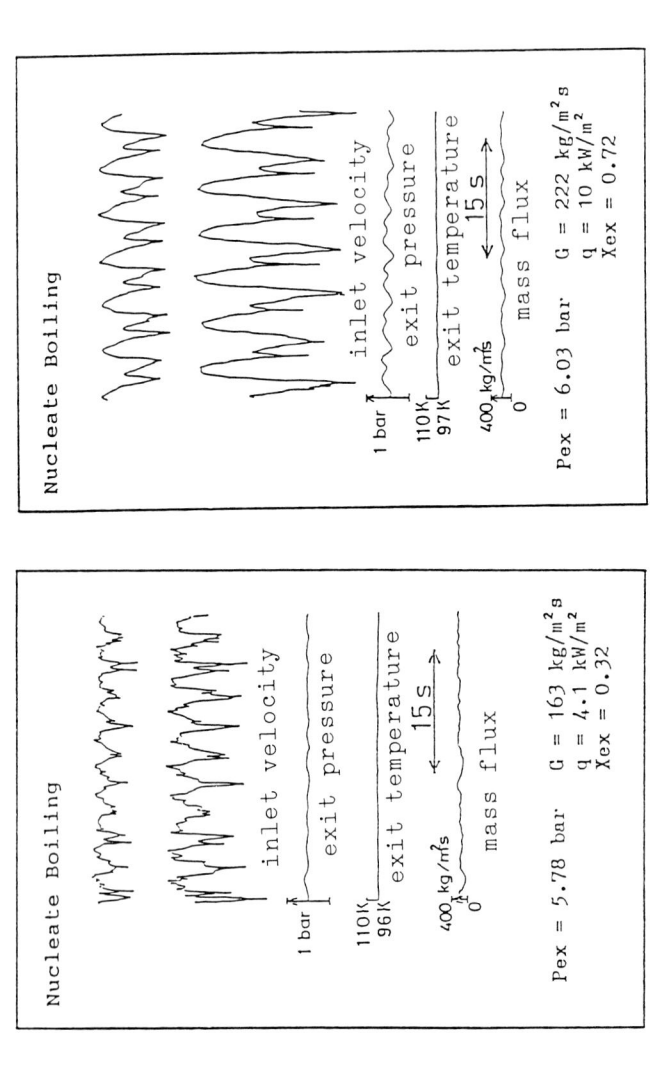

Fig.6 Recorder chart traces, spiral channels with nucleate boiling, case [A].

are repeated alternatively in each channel, as shown in Fig. 4 (b).

Examples of the axial distributions of wall temperature in cases [B] and [C] for the fairly same experimental conditions, are shown in Fig. 5. The wall superheat in case [C] is much higher than that of [B]. It can be seen that the whole wall is surely dried out in the case of [C].

Typical examples of recorder chart traces of inlet velocities of each channel, measured by the hot wire anemometry, exit header pressure, inlet mass flux measured by exit orifice etc. are shown in Figs. 6, 7 and 8 for the cases [A], [B] and [C], repectively. The experimental conditions of P_{ex}, G, q, ΔT_{sub} and x_{ex} are noted in the figures. The exit qualities x_{ex} in the figures are the values calculated as thermodynamic equilibrium qualities at the exit. In Figs. 6-8 (a) small and irregular fluctuations of the hot wire signals, the pressure and the mass flux can be seen. This state is denoted as stable flow. In Figs. 6-8 (b) the oscillations of flow rates in each channel are in phase, and the pressure and the mass flux also oscillate periodically. In Figs. 6-8 (c) the hot wire signals in each channel oscillate out of phase periodically with a pair of high and low peaks, while the oscillation amplitudes of the pressure and the mass flux are small.

The oscillation in phase indicates that the whole flow oscillates from the liquid nitrogen bomb to the exit of the superheater, while that out of phase indicates that the flow rate increases and decreases alternatively in two channels, and that the total flow rate is fairly constant. It can be seen from calibrated examples as already shown in Fig. 2 (a), for the oscillation out of phase, the flow velocity after a lower peak reaches toward zero but after a higher peak does not. Therefore, it is inferred that downward flow should not occur. As shown in Fig. 2 (b) the flow velocity reaches toward zero for the oscillation in phase.

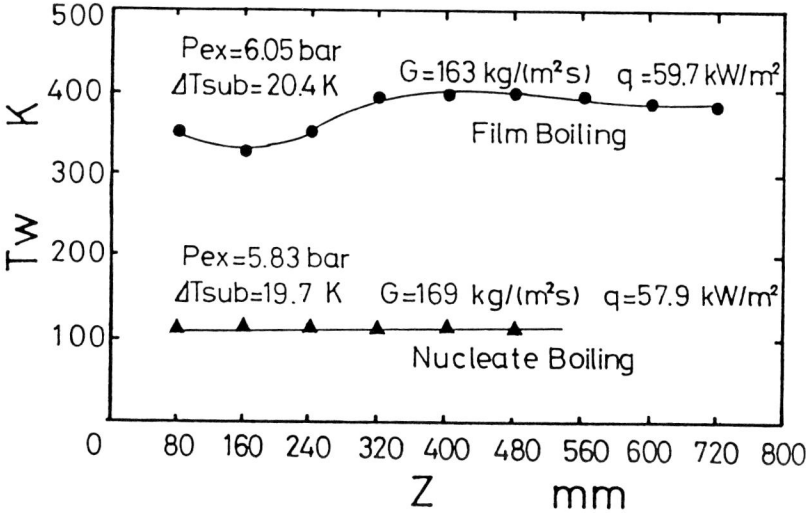

Fig.5 Axial wall temperature distributions.

FLOW INSTABILITIES IN PARALLEL CHANNELS

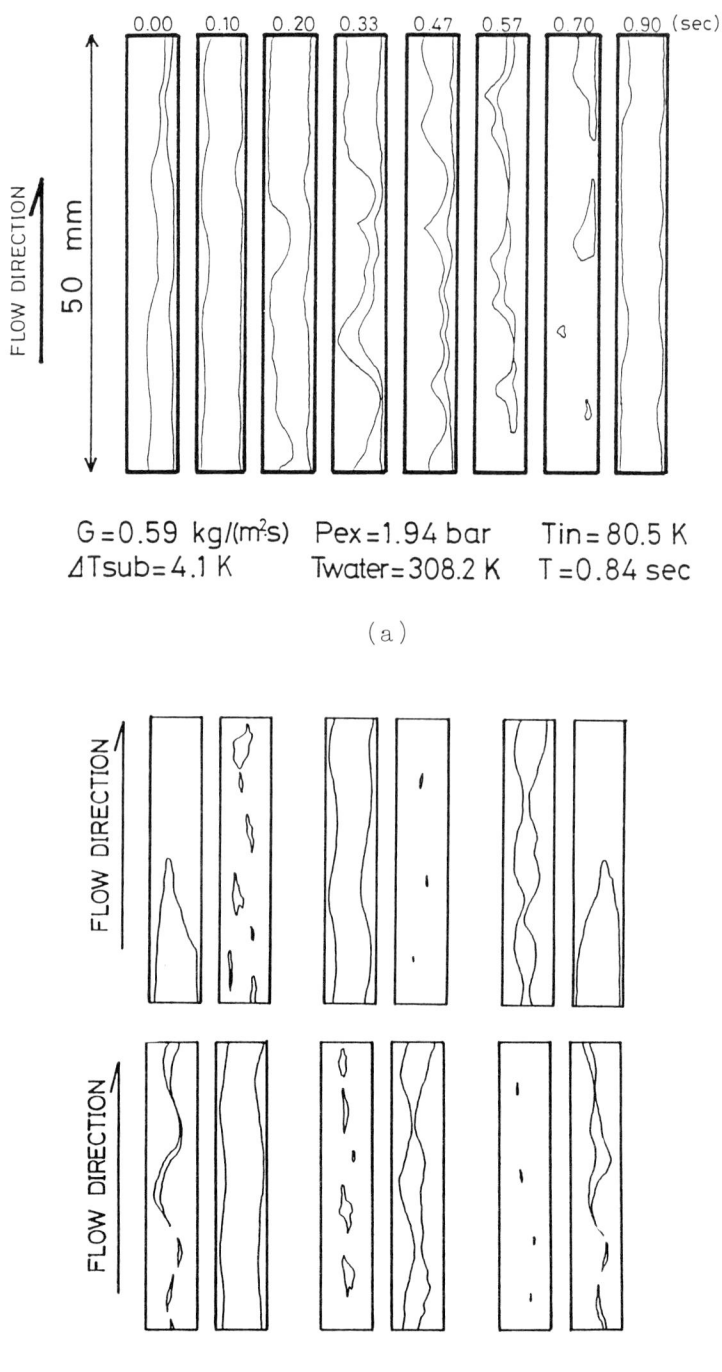

Fig.4 Flow pattern of inverted annular flow in oscillation out of phase.

low heating ratio. Figures. 3 (a) and (b) show examples of calibrated hot wire signals at typical oscillating flow in phase and out of phase in case [A]. The mass flux calculated by integrating the calibrated flow velocity agrees approximately with the mass flux measured by the exit orifice. In this report, however, the absolute values of flow rates in each channel are not discussed, while the variation characteristics and the oscillation period of the flow rate are discussed. Therefore, the calibration of the hot wire is conducted for some experimental runs.

Every signal is recorded by a multi-pen-recorder and a data logger.

Experimental procedures. For the experimental run with nucleate boiling in the inlet section, the pipeline and the channels are cooled by supplying liquid nitrogen without heat supply, before the start. For the experimental run with film boiling in the inlet section, the heating sections and the inlet electrode are heated at temperatures higher than the rewetting temperature, and the upstream pipeline is cooled by supplying liquid nitrogen before the start. Then the experiments are carried out, confirming by the hot wire signals that the fluid in the inlet of the channels is liquid state. Experimental conditions are summarized in Table 1.

Table 1 Experimental conditions

case	configuration of test channels	boiling state	exit pressure (subcooling)	max. mass flux $kg/m^2 s$	max. heat flux kW/m^2
[A]	spiral	nucleate	4bar(11K) 6bar(15K) 8bar(19K)	500	15
[B]	vertical	nucleate	4bar(15K) 6bar(20K) 8bar(25K)	400	80
[C]	vertical	film	4bar(15K) 6bar(20K) 8bar(25K)	400	80

EXPERIMENTAL RESULTS AND DISCUSSION

The flow instability criteria and the characteristics of flow instabilities are experimentally studied for three cases [A], [B] and [C].

Before the series of the experiments, flow patterns of inverted annular flow were observed. The flow patterns in parallel vertical channels are schematically shown in Figs. 4 (a) and (b) when flow oscillations occur. While the flow rate is decreasing, the liquid column surrounded by the vapor annulus becomes thin and is torn to liquid filaments. The thick liquid column rises when the flow rate begins to increase, as shown in Fig. 4 (a). These flow pattern changes

annular flow pattern in the inlet section, as shown in Fig. 1 (b). Visualizations of flow patterns in case [C] were carried out, using a test section consisting of acrylic resin tubes of 6 mm I.D., 8 mm O.D., and 800 mm long, heated by city water of room temperature. The flow patterns were observed by a high-speed video camera.

In case [A], the value of the frictional pressure drop is much higher than the gravitational and accelerational pressure drops; that is, the frictional pressure drop is dominant in comparison with other pressure losses. On the other hand, in cases [B] and [C], the gravitational pressure drop is dominant. The length of the pipeline from the liquid nitrogen bomb to the subcooler is 2.1 m for case [A] and 0.5 m for cases [B] and [C].

<u>Measuring system.</u> Pressures at the inlet and the exit headers are measured by using Buldon tubes and pressure transducers, and the pressure drop between them is measured by using a differential pressure transducer. Temperatures at the headers are measured by using sheathed CC thermocouples of 1mm diameter, and the wall temperature is measured at ten locations along the tube by using CC thermocouples of 0.1mm diameter welded at the outside of the tube. The inlet mass flux is determined by measuring the released superheated vapor flow rate by the orifice, as already described above. To measure the flow rate of each channel, constant temperature-type hot wire anemometry, composed of a tungsten wire of 5μm in diameter and 1mm in sensing length is inserted at the center of each tube cross section. Both void detection and velocity measurement can be accomplished by using hot-wire anemometry. The calibration is carried out by rotating the hot wire in a liquid nitrogen pool under atmospheric pressure; King's calibration equation is then obtained. The equation is then modified for the liquid temperature measured during each experimental run. However, the calibration at rest state, i.e. velocity zero, is not possible because of boiling of the liquid nitrogen on the hot wire, even with a

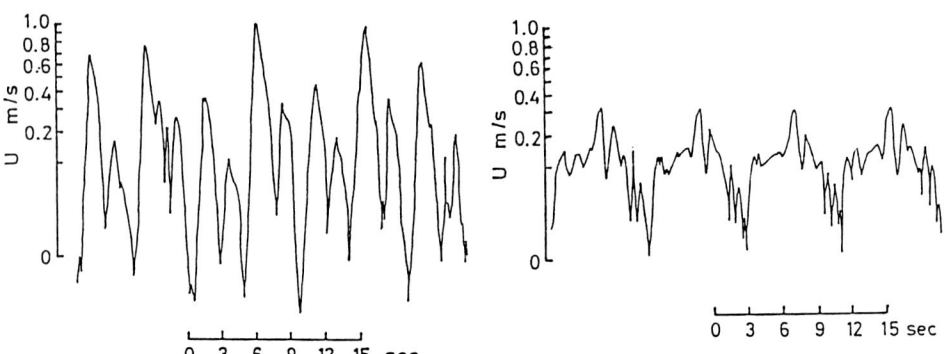

(a) oscillation out of phase, P_{ex}=6.15bar, G=145kg/m²s, x_{ex}=1.71, q=13.6kW/m².

(b) oscillation in phase, P_{ex}=5.82bar, G=148kg/m²s, x_{ex}=1.71, q= 2.2kW/m².

Fig.3 Calibrated signals of hot wire anemometry.

pipeline upsream of the inlet header is immersed in a subcooler tank 3, where liquid nitrogen is supplied from another nitrogen bomb under atmospheric pressure. The liquid nitrogen flowing in the conduit is cooled to near saturation temperature for atmospheric pressure. The flow rate is controlled by a valve V7. The subcooled liquid is supplied to the inlet header, and is divided into the two channels of the test section. The values of inlet subcooling ΔT_{sub} depend on the pressure of the test section, as the inlet liquid temperature is kept near the saturation temperature for atmospheric pressure. The two-phase mixture generated in each channel flows into the exit header. The exit pressure is ajusted by a valve V8. The two-phase mixture downstream of the valve V8 is superheated in a heat exchanger 5 which uses room temperature city water. The vapor then proceeds through an orifice where the flow rate measured, and then is released to atmosphere. Therefore, the inlet mass flux of the liquid nitrogen can be calculated by the superheated vapor flow rate.

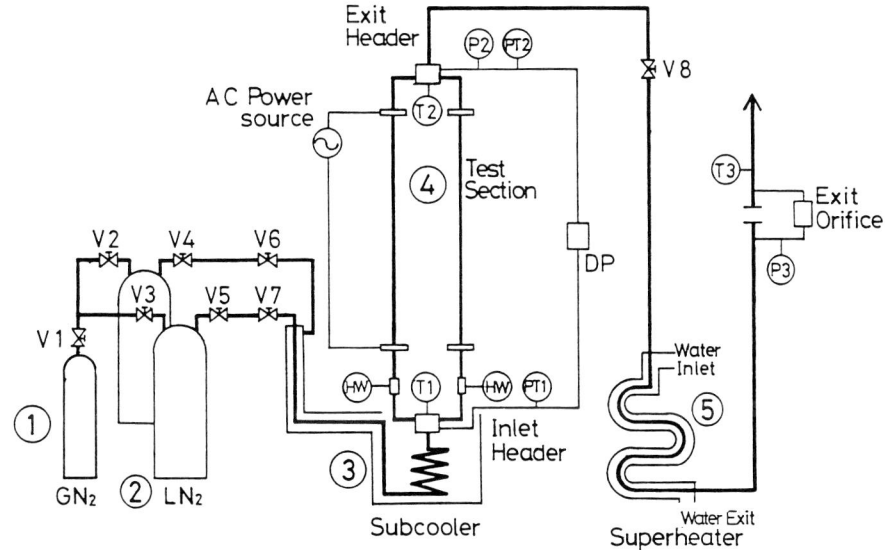

Fig.2 Schematic diagram of experimental apparatus.

Test sections. In order to study the effects of channel configuration on the flow instabilities, two types of test sections were used. One consisted of two spiral channels, used in case [A]; type 304 stainless steel tubes of 6 mm I.D., 8 mm O.D., and 5200 mm heating length, coiled into ten turns, 150 mm in coil diameter and 700 mm in height. The other channel consisted of two parallel vertical channels, used in the cases [B] and [C]; type 304 stainless steel tubes of 6 mm I.D., 8 mm O.D., and 800 mm heating length. The tubes are equipped with brozen electrodes connected to an AC power supply. The inlet electrodes are wrapped by sheathed heaters to hold high temperature, to keep the quench front at the inlet of the test section for the experiments of the inverted annular-mist flow along the tube. The flow pattern transition shown in Fig. 1 (b) is encounted when the tube temperature is higher than the rewetting temperature. Film boiling occurs in the inlet section, and the flow pattern transfers from inverted annular flow to dispersed flow along the tube. When the cryogenic fluid is evaporated in a heat exchanger using water or steam as a higher temperature fluid, the flow pattern transition is that as shown in Fig. 1 (b), since the temperature of water or steam is much higher than the rewetting temperature. When heat flux at the tube's inner surface is constant, both flow pattern transitions can occur depending on the initial wall temperature. This is similar to pool boiling.

3) Boiling may occur at non-heating section:

Cryogenenic fluid can be heated by external heat penetration from the atmosphere without intensional heating, and may become a saturated liquid or two-phase flow mixture as the heat capacity and the latent heat is low. Therefore, the possibility of flow instabilities should be considered in non-heating sections in cryogenic fluid equipments.

In this study three series of experiments were conducted on the flow instabilities of nitrogen evaporators, consisting of two parallel channels heated with constant heat flux: (1) long spiral channels, case [A]; and (2) short vertical channels, case [B] with nucleate boiling in the inlet section as shown in Fig. 1 (a) and short vertical channels, and case [C] with film boiling in the inlet section as shown in Fig. 1 (b). The effects of the channel configuration, flow pattern, pressure, heat flux and mass flow rate on the flow instabilities were investigated.

One of the purposes of this study is to investigate the estimation of occurence of the flow instabilities in the cryogenic fluid channels. There are many investigations on flow instabilities of ordinary fluids. The linear stability analysis using a drift flux model was first reported by Ishii and Zuber (1970) and was also reported by Spinks (1971), Saha (1975) and Ozawa (1976). The effect of subcooled boiling on flow instabilities was investigated by Saha (1975). The effect of heat capacity of the tube wall was by studied Ozawa (1976). According to these studies, the stability analysis can well predict the flow stability criteria, for experiments using water and Freon R-113 as the working fluids.

In this report, the linear stability analysis, using a drift flux model, is conducted for the cryogenic fluid flow. As the result, the method can be applied to predict the stability criteria, with reasonable assumptions of the distribution parameter and the drift velocity.

EXPERIMENTAL APPARATUS AND PROCEDURES

Experimental Apparatus

A schematic diagram of the experimental apparatus is shown in Fig. 2. Liquid nitrogen bombs 2 are pressurized at about 10 bar by nitrogen gas bomb 1. Liquid nitrogen is supplied from the liquid nitrogen bomb 2 to the test section 4 through a pipeline insulated by urethane foam. The coiled

Flow instabilities are also important for operational problems. Recently, many cryogenic fluids, such as LNG for city gas, liquid oxygen and hydrogen for rocket fuel, liquid helium for superconductor electro-magnets and so on, are being used, and in the future they will be used much more if high temperature superconductivity equipment is realized, which at first will be kept at the saturation temperature of liquid nitrogen. Therefore, it is necessary to understand the fundamental characteristics of the flow instabilities of cryogenic fluids and to test whether the analytical method developed to predict the instabilities of ordinary fluids such as water can be applied to those of cryogenic fluid flows.

The characteristics of a cryogenic fluid flow, except those of a superfluid, are summarized (compared to ordinary fluid flow) as follows:

1) Physical properties are different:

The liquid to vapor density ratio of cryogenic fluid is lower than that of water for the same pressure if the temperature is low. Latent heat and heat capacity are much lower, and viscosity and surface tension are also lower than that of water.

2) Film boiling may occur and the flow pattern may be different:

The flow pattern transitions in an evaporator tube when the inlet fluid is a subcooled liquid are divided into two cases, as shown schematically in Fig. 1 (a) and (b). In the case of an ordinary flow pattern transition shown in Fig.1 (a), nucleate boiling occurs in the inlet section, and the flow pattern transfers from bubbly flow to annular or

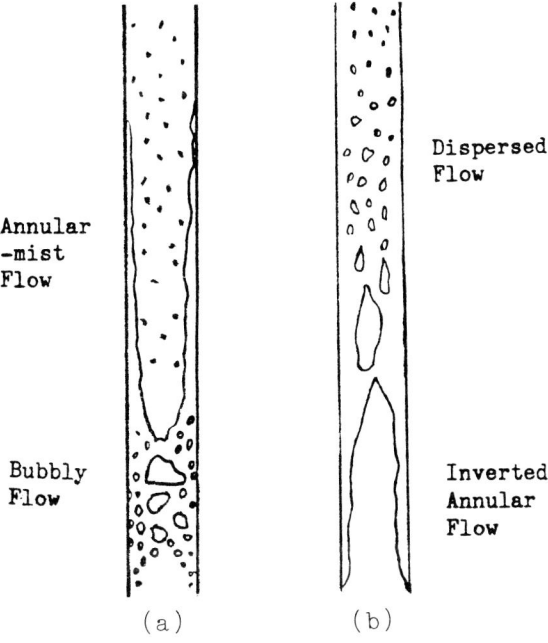

Fig.1 Flow pattern transition in evaporator tube.

FLOW INSTABILITIES IN PARALLEL CHANNELS WITH NUCLEATE BOILING AND FILM BOILING

K. AKAGAWA
Department of Mechanical and System Engineering
Ryukoku University
Yokotani, Oe-cho Seat, Ohtsu 520-21, Japan
N. TAKENAKA, T. FUJII, and Y. OKU
Department of Mechanical Engineering
Kobe University
Rokkodai, Nada, Kobe 657, Japan

ABSTRACT

Experimental and analytical results are presented on thermally induced flow instabilities of cryogenic fluid in parallel channels, using liquid nitrogen as a working fluid with nucleate boiling in the inlet section. The fluid flow pattern transfers from bubbly flow to annular flow with film boiling, and from inverted annular flow to dispersed flow, respectively. Two types of channels, consisting of two spiral tubes and two vertical tubes, are tested. Stability criteria are obtained in each experiment and the criteria are approximately given by constant exit thermodynamic equilibrium qualities. The flows with film boiling in the inlet section are stable for higher exit thermodynamic equilibrium qualities than those with nucleate boiling. Linear stability analysis for density wave oscillation is adopted to predict the stability criteria using a drift flux model. The stability criteria for nucleate boiling are well predicted by using the previous drift flux models, and those for film boiling are also well predicted by using the drift flux model formulated by analytical results obtained with a turbulent boundary layer theory for the inverted annular flow. These results indicated that the criteria of the flow instabilities should be estimated by the slip between liquid and vapor for both boiling states.

INTRODUCTION

Thermally-induced flow instabilities in channels flowing two-phase mixtures are important problems in the operation and safety of power generating systems, heat exchangers and chemical process systems. Experimental and analytical studies have been reported by many researchers the safety problems of light water reactors and steam generators in fast breeder reactors, Lahey and Drew (1980). However, studies of flow instabilities of cryogenic fluids are limited (Hands, 1975; Hands, 1979; Fukuda et al., 1986).

In cryogenic equipment, since atmospheric temperature is much higher than the boiling temperature, boiling can occur in almost every component in the equipment, and thermal and hydraulic problems of the two-phase flow are very important.

t	time
u	velocity
x	flow direction
X	vector of variables

Greek

α	void fraction
θ	$k\Delta x$
φ	$1-e^{-i\theta}$
λ	wavelength
μ	eigenvalue
ρ	density
σ	surface tension

Subscripts/Superscript

g	gas phase
i	interface
j	spatial mesh number
k	phase index
l	liquid phase
m	index used in Eq.(3)
n	time step

BIBLIOGRAPHY

Akimoto M. et al. (1986), "Development of Transient Two-Phase Flow Analyzer: MINCS", 2nd Int. Topical Meeting on Nucl. Power Plant Thermal-Hydraulics and Operations, Tokyo, April, Vol.1, p.72-79.

Hancox W.T. et al. (1980), "One-Dimensional Models for Transient Gas-Liquid Flows in Ducts", Int. J. Multiphase Flow, Vol.6, 25-40.

Hewitt G.F. et al. (1987), Multiphase Science and Technology, Vol.3, Hemisphere, 477-479.

Milne-Thomson L.M. (1960), Theoretical Hydrodynamics, 395.

Ransom V.H. and Trapp J.A. (1983), "Applied Mathematical Methods in Nuclear Thermal Hydraulics", 2nd Int. Topical Meeting on Nucl. Reactor Thermal-Hydraulics, Santa Barbara, January, Vol.1, p.99-110.

Roache P.J. (1976), Computational Fluid Dynamics, Hermosa.

Rousseau J.C. and Ferch R.L. (1979), "A Note on Two-Phase Separated Flow Models", Int. J. Multiphase Flow, Vol.5, 489-493.

stability limit determined by the eigenvalue defined by
Eqs.(11) or (13) strongly depends on each phasic velocity.
The wave growth phenomena shown in Fig.16 can be well
described by the eigenvalues in Fig.17. It is concluded that
the numerical wave growth phenomena are not described by the
classical theory or the well-posed limit, but described by the
eigenvalues of amplification matrix. This fact indicates that
problems such as the wave growth problem should be carefully
treated when we use some numerical procedures. Otherwise,
nothing can be concluded from the numerical solution.

We used the fully-implicit upwind scheme in this problem, and
the explicit upwind scheme in the accelerated flow problem.
The staggered mesh system was applied for both problems.
However, the selection of numerical schemes was found to be
not essential for the above discussions, since similar
expressions of eigenvalues can be obtained for another scheme.
While the sample problems calculated here were both cocurrent
flow problems, the same procedure can be applied for
countercurrent flow problems. Through several applications,
the stability analysis mentioned above has been found to be
effective for better understanding of numerical results.

CONCLUSION

The stability of numerical analysis on stratified two-phase
flow was studied for the accelerated flow problem and the wave
growth problem. Governing equations for the two-fluid model
with two phasic pressures were solved by the finite difference
method. The stability analyses based on Neumann's method were
carried out for the explicit and implicit finite difference
schemes. The stability criterion for the numerical procedure
was shown in terms of the eigenvalue of amplification matrix
for two sample problems. The numerical instability and wave
growth phenomena in the numerical solutions were demonstrated
to be well described not by the well-posed condition or the
classical wave growth theory for the differential governing
equations, but by the stability limit determined by the
eigenvalue of amplification matrix for the discretized
equations.

NOMENCLATURE

a	sound speed
A	coefficient matrix
B	coefficient matrix
C	$\Delta t/\Delta x$
Co	Courant number
C_g	Cu_g
C_l	Cu_l
C_a	$C\alpha$
C_{1-a}	$C(1-\alpha)$
g	gravitational acceleration
H	channel height
I	imaginary unit
K	wave number ($2\pi/\lambda$)
M	variable used in Eq.(12)
p	pressure

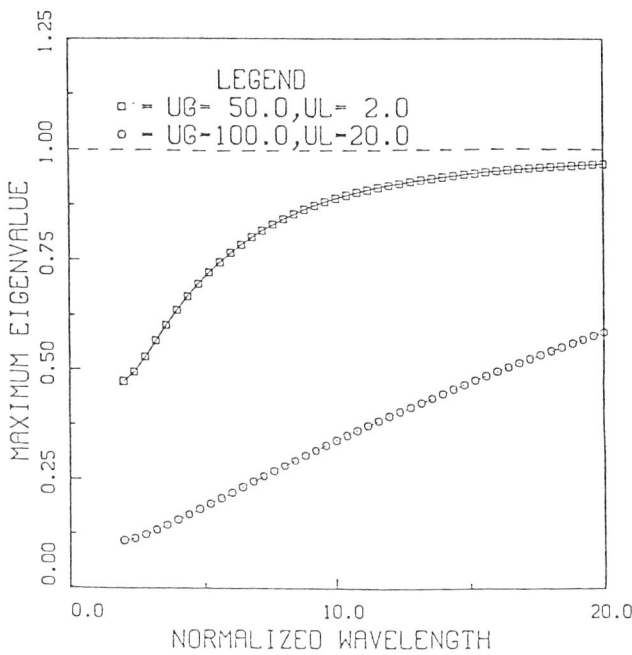

FIGURE17. Maximum eigenvalues (with large relative velocity)

The void fraction transients at the mid-length of channel are shown in Fig.16 for another two cases: one is with gas velocity $50.0 m/s$ and liquid velocity $2.0 m/s$, the other is gas $100.0 m/s$ and liquid $20.0 m/s$. The time step size is $0.01s$ for both cases. In the former case, waves should grow from the classical theory shown in Fig.9, since the relative velocity is larger than the stability limit. Moreover, in the latter case, waves should also grow from the classical theory. The relative velocity in the latter case is larger than the Kelvin-Helmholtz stability limit, that is, larger than the well-posed limit. However, the initial waves are damped in both cases, as shown in Fig.16. It is suggested that the numerical wave growth phenomena does not depend on the classical theory or ill-posed condition.

The eigenvalues for both cases are shown in Fig.17. It is found that the eigenvalues are smaller than unity in all regions for both cases. The eigenvalue for the latter case is seen to be smaller than that for the former case. The latter case is thus regarded to be more stable than the former case, in spite of the large relative velocity. In fact, the initial wave of the latter case is damped faster than the former case, as shown in Fig.16.

The well-posed limit defined as Eq.(4) or the wave growth limit defined as Eq.(12) depend not on each phasic velocity, but on the relative velocity. However, the numerical

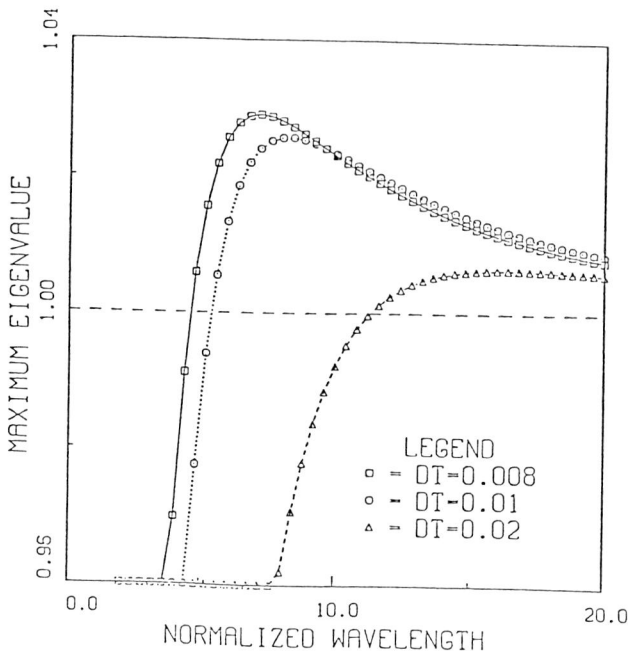

FIGURE15. Maximum eigenvalues (u_g=50.0m/s, u_l=0.001m/s)

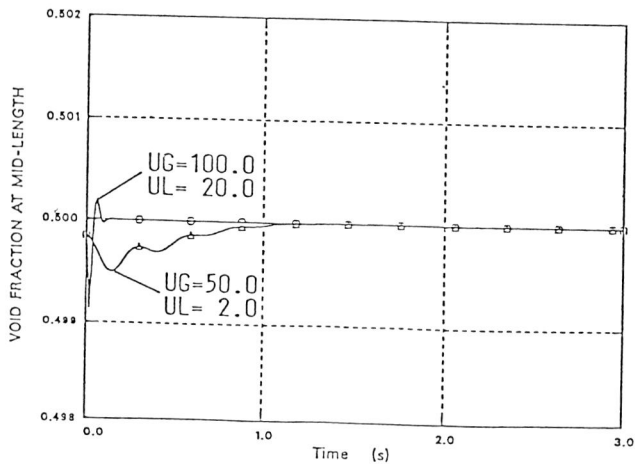

FIGURE16. Void fraction transients at mid-length (with large relative velocity)

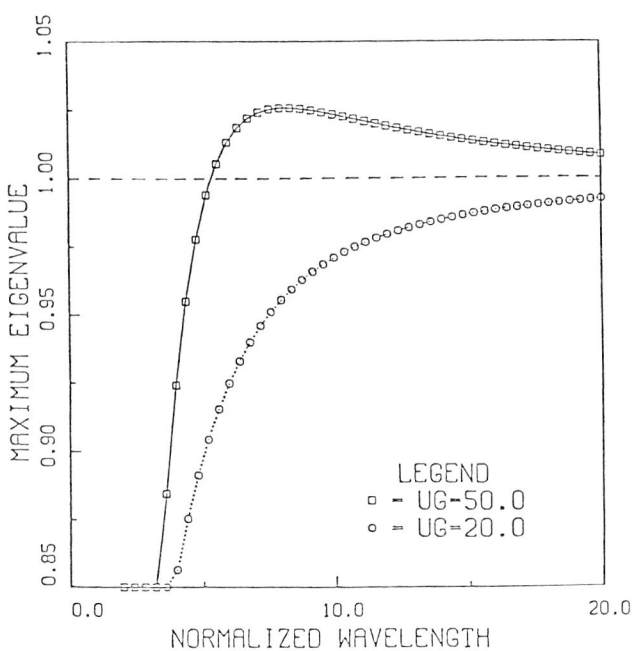

FIGURE13. Maximum eigenvalue (u_l=0.001m/s)

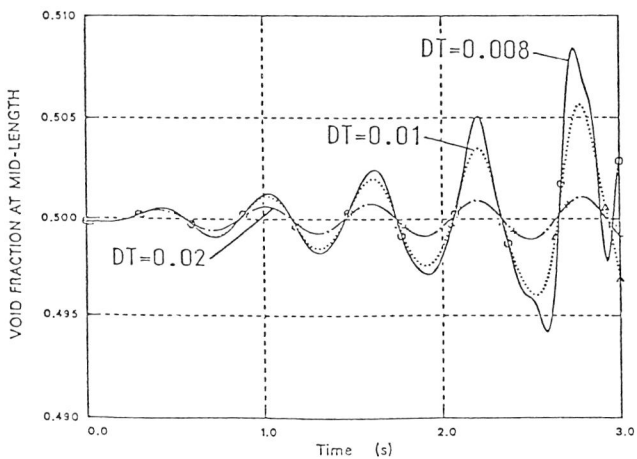

FIGURE14. Void fraction transients at mid-length (u_g=50.0m/s, u_l=0.001m/s)

ANALYSIS ON VOID WAVES

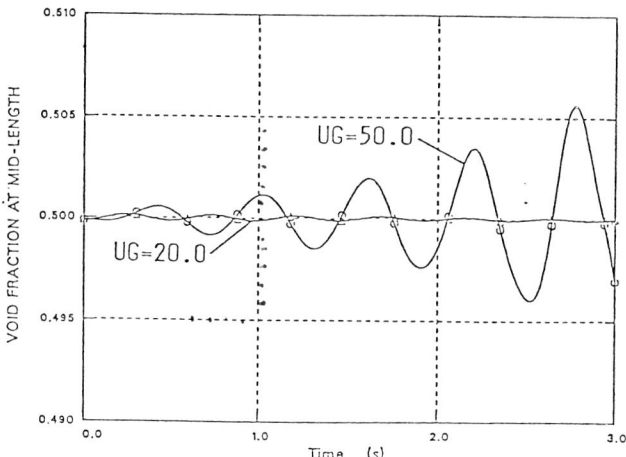

FIGURE12. Void fraction transients at mid-length
(u_l=0.001m/s)

performed based on Neumann's method. The discretized equations are similar to Eq.(7), however, variables in spatial difference terms are all in the time step $n+1$. We can obtain a matrix form after substituting the difference terms by components of Fourier series. Finally, the eigenvalue is given by

$$\{\alpha\rho_l(1+C_l\varphi)^2 + (1-\alpha)\rho_g(1+C_g\varphi)^2 - \alpha(1-\alpha)\Delta\rho g H e^{i\theta}C^2\varphi^2\}\mu^2$$
$$- 2\{\alpha\rho_l(1+C_l\varphi) + (1-\alpha)\rho_g(1+C_g\varphi)\}\mu + \{\alpha\rho_l+(1-\alpha)\rho_g\} = 0 \quad, \tag{13}$$

for the fully-implicit scheme.

The maximum eigenvalues in terms of the normalized wavelength are shown in Fig.13. The eigenvalue is smaller than unity in all regions of wavelength with the gas velocity $20.0 m/s$, while it is larger than unity in the region of large wavelengths with velocity $50.0 m/s$. This result corresponds to the wave growth phenomena shown in Fig.10 to 12.

The void fraction transients at the mid-length of channel obtained by using different time step sizes are shown in Fig.14. The initial velocities are $50.0 m/s$ for the gas and $0.001 m/s$ for the liquid, respectively. The time step sizes are chosen to be $0.008s$, $0.01s$ and $0.02s$. The initial waves are found to grow in all cases; however, the large amplitude is obtained by using the small time step size. The eigenvalues are shown in Fig.15 for three time step sizes. The eigenvalues exceed unity for all cases, and it is clearly shown that the system goes to more instability when the smaller time step size is used in this numerical scheme. If the time step size or Courant number becomes infinitely large, the eigenvalue is zero from Eq.(13). It is suggested that the stability of numerical solutions increases by using large time step sizes, so long as the implicit numerical procedure is used.

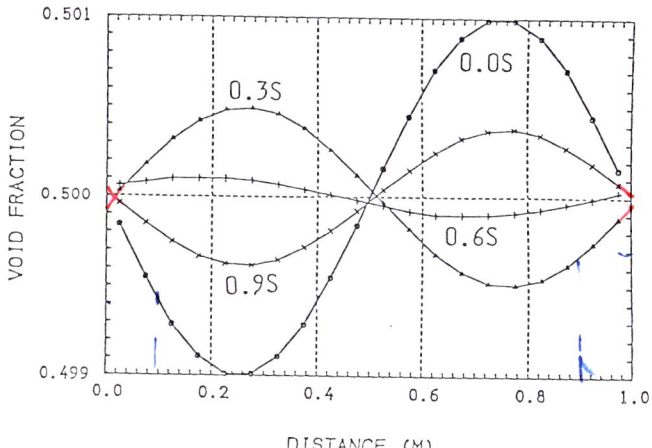

FIGURE10. Overall void fraction profiles
(u_g=20.0m/s, u_l=0.001m/s)

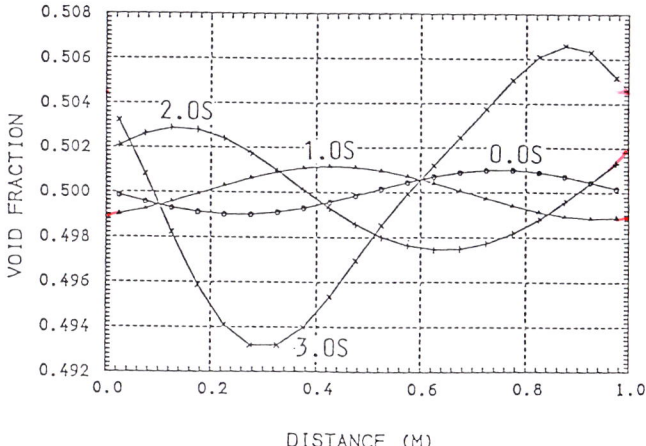

FIGURE11. Overall void fraction profiles
(u_g=50.0m/s, u_l=0.001m/s)

The initial gas velocities are $20.0 m/s$ and $50.0 m/s$, and the initial liquid velocity is $0.001 m/s$. The transients of overall void fraction profiles are shown in Figs.10 and 11. The initial sinusoidal wave is damped with time as shown in Fig.10, while it grows in Fig.11. The transients of void fraction at the mid-length of the channel are shown in Fig.12. We can easily find the wave growth phenomena in Fig.12. These calculated results seem to be in good agreement with the classical theory.

The stability analysis for the fully-implicit scheme was also

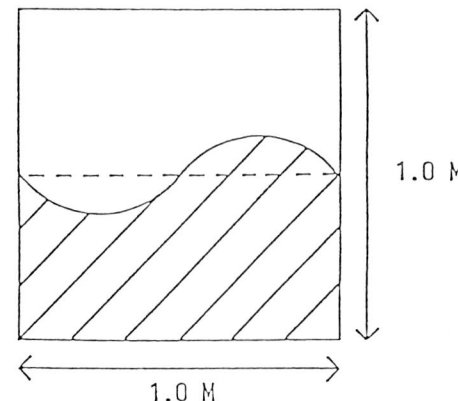

FIGURE 8. Flow channel of wave growth problem

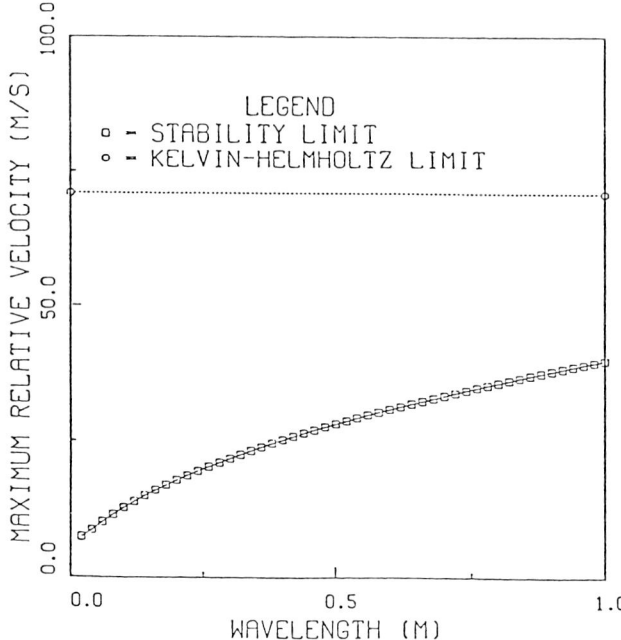

FIGURE 9. Maximum relative velocity as stability limit by potential flow theory

et al., 1986). The numerical method of MINCS is based on the fully-implicit donor cell scheme with the staggered mesh system, and the large time step size can be used. We basically used 20 spatial mesh cells and the time step size of 0.01s. The Courant number was thus about 68. Of course, such a large time step calculation can not be performed by the explicit method.

WAVE GROWTH

Problem description

The growth of small amplitude sinusoidal waves is discussed. In this problem, governing equations and fluid properties are the same as in the accelerated flow problem. The flow channel is depicted in Fig.8. A stratified flow with void fraction of 0.5 flows in a channel which is $1.0m$ in length and $1.0m$ in height. The inlet and outlet of the channel are assumed to be connected. A small amplitude sinusoidal wave is put on the stratified flow and initial velocities for both phases are given. It should be noted that this problem is based on the benchmark problem of International Two-Phase Fundamentals (Hewitt et al., 1987).

The growth of small amplitude sinusoidal waves has been discussed by Milne-Thomson (1960). According to the literature, the two-dimensional stratified flow can be theoretically studied by using the potential flow theory, and the relative velocity which indicates the wave growth is obtained in terms of the wavelength and flow depth:

$$(u_g - u_l)^2 \leq \frac{\rho_g M_g + \rho_l M_l}{\rho_g \rho_l M_g M_l} \left(\sigma K + \frac{g(\rho_l - \rho_g)}{K} \right) . \tag{12}$$

where M_g and M_l are defined as

$M_g = \coth(K\alpha H)$,
$M_l = \coth(K(1-\alpha)H)$.

In Eq.(12), σ, which is the surface tension, is considered to be negligibly small in comparison with the gravity term for waves with small amplitude and long wavelength. In this problem, the initial wavelength and wave amplitude are set equal to $1.0m$ and $0.002m$, respectively. The several hypotheses used in the potential theory are satisfied by such a small wavelength-amplitude ratio. The maximum relative velocity which gives the damping of waves in this case is shown in Fig.9 in terms of the wavelength. Equation (12) is identical to what is called the Kelvin-Helmholtz stability condition when the wavelength goes to infinity. The relative velocity strongly depends on the channel height, the void fraction and the wavelength. The difference between the Kelvin-Helmholtz limit and the limit given by Eq.(12) in this case is large, as shown in Fig.9. The stability limit for the wave with $1.0m$ wavelength is about $39.8m/s$, as shown in Fig.9. That is, the wave can grow if the initial relative velocity is larger than $39.8m/s$, or the wave is damped if the initial relative velocity is smaller than $39.8m/s$.

Calculated results and discussion

The implicit finite difference scheme was used for numerical analysis, since we have to perform several seconds of calculations to study the wave growth. We used the MINCS code developed at Japan Atomic Energy Research Institute (Akimoto

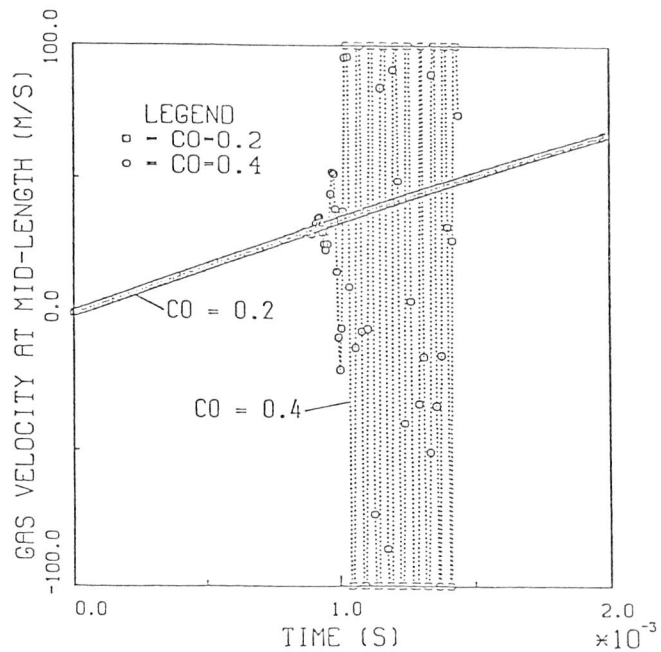

FIGURE 6. Gas velocity transients at mid-length (H=1.0m, 200meshes)

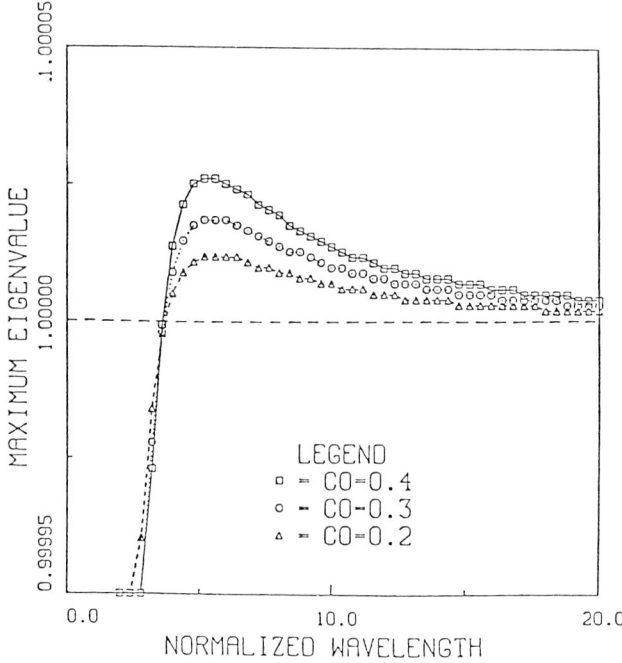

FIGURE 7. Maximum eigenvalues (H=1.0m, 200meshes)

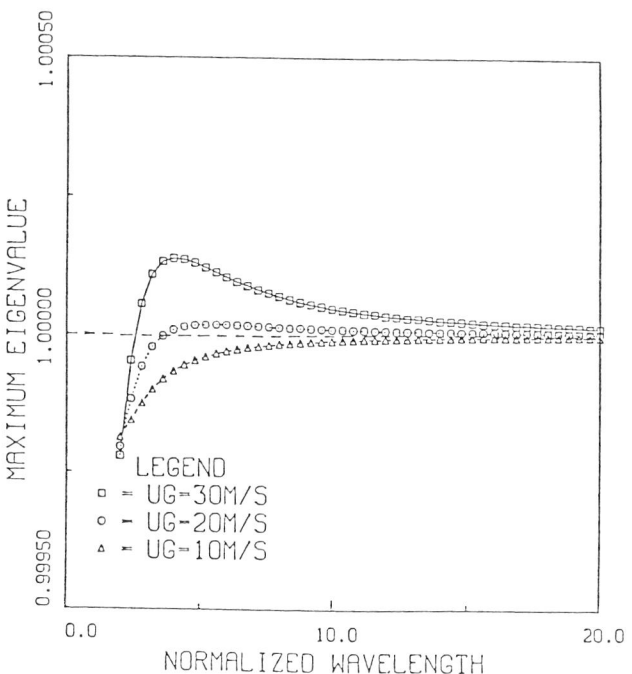

FIGURE 4. Maximum eigenvalues (H=1.0m, Co=0.3)

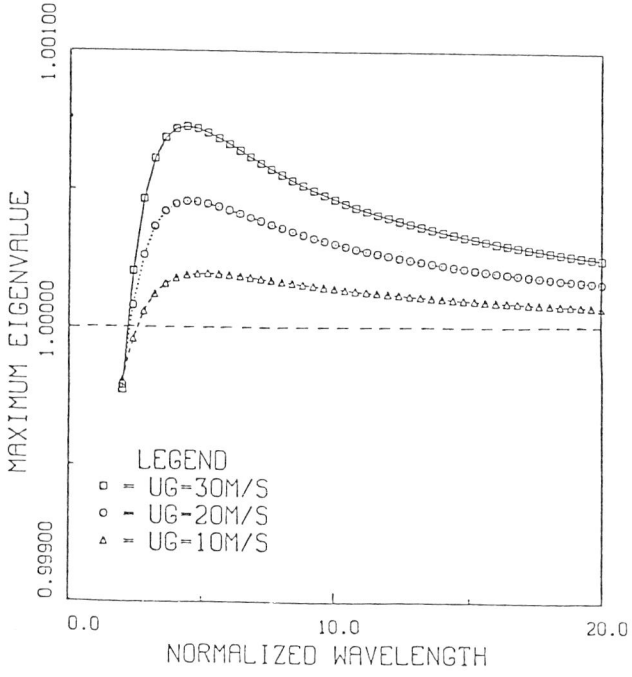

FIGURE 5. Maximum eigenvalues (H=0.0m, Co=0.3)

ANALYSIS ON VOID WAVES

calculating the maximum absolute value of eigenvalue of the amplification matrix $A^{-1}B$. The eigenvalue μ is defined as the root of a parabolic equation:

$$\{\alpha\rho_l + (1-\alpha)\rho_g\}\mu^2 - 2\{\alpha\rho_l(1-C_l\varphi) + (1-\alpha)\rho_g(1-C_g\varphi)\}\mu$$
$$+ \alpha\rho_l(1-C_l\varphi)^2 + (1-\alpha)\rho_g(1-C_g\varphi)^2 - \alpha(1-\alpha)\Delta\rho g H e^{i\theta}C^2\varphi^2 = 0 \ . \quad (11)$$

The finite difference scheme is considered to be stable against the small perturbation if the maximum absolute value of eigenvalue is smaller than unity (Roache 1976).

The maximum eigenvalues in terms of the normalized wavelength ($\lambda/\Delta x$) are shown in Figs.4 and 5. In this problem, as the gas velocity increases from $0.0 m/s$, the eigenvalues at three different gas velocities are shown. In Figs.4 and 5, u_l is assumed to be constant at $0.1 m/s$ since, in every case, the maximum u_l reached in the time scale of this problem is smaller than $0.1 m/s$. The spatial increment is set equal to $0.005 m$, that is, the number of mesh cells is 200. Figures 4 and 5 correspond to the case of $H=1.0 m$ and $H=0.0 m$, that is, to Figs.2 and 3, respectively. The maximum eigenvalues are larger than unity in some region of the normalized wavelength, even if the gas velocities are small as shown in Figs.4 and 5. It is suggested that the discretized system may be ill-posed, even if the differential system is well-posed. The maximum eigenvalues come closer to unity when H becomes larger. Such a qualitative tendency is about the same as the meaning of Eq.(4).

In this problem, as the gas is accelerated with time, the eigenvalue of the amplification matrix may exceed unity. The system is thus almost unstable even if the relative velocity is much smaller than the well-posed limit. The instability is, however, concealed in the coarse mesh cases, since the numerical damping has some stabilizing effects as shown in Figs.2 and 3.

The dependency of numerical solutions on the time step size is discussed here while the Courant number was set equal to 0.3 in Figs.2 and 3. The gas velocities obtained by using different time step sizes are shown in Fig.6. The channel height is $1.0 m$ and the number of mesh cells is 200 in Fig.6. The Courant numbers 0.2 and 0.4 are used, while 0.3 is used in Figs.2 and 3. We can find that as the Courant number becomes larger, the oscillation occurs earlier. The maximum eigenvalues with the gas velocity of $20.0 m/s$ for three cases of Courant number are shown in Fig.7. The discretized system is found to become more and more stable as the Courant number becomes smaller. If the Courant number becomes an infinitesimal value, the maximum eigenvalue is unity from Eq.(11). It is suggested that the stability of the numerical solution increases by using sufficiently small time step sizes, even if the eigenvalue of the amplification matrix is larger than unity. It is concluded that the coarse mesh cell and small time step size can give more stable solutions, so long as the explicit numerical procedure is used.

A stability analysis for the numerical procedure is performed in the following, based on Neumann's stability analysis method (Roache 1976). Three equations are considered: two mass conservation equations and the difference of two momentum equations. For simplicity, the incompressible and cocurrent flow is assumed. The discretized equations for the explicit upwind scheme are described as

$$\frac{a_j^{n+1} - a_j^n}{\Delta t} + a_j^n \frac{u_{g,j}^n - u_{g,j-1}^n}{\Delta x} + u_{g,j-1}^n \frac{a_j^n - a_{j-1}^n}{\Delta x} = 0 ,$$

$$\frac{a_j^{n+1} - a_j^n}{\Delta t} - (1-a_j^n) \frac{u_{l,j}^n - u_{l,j-1}^n}{\Delta x} + u_{l,j-1}^n \frac{a_j^n - a_{j-1}^n}{\Delta x} = 0 ,$$

$$\rho_g \frac{u_{g,j}^{n+1} - u_{g,j}^n}{\Delta t} - \rho_l \frac{u_{l,j}^{n+1} - u_{l,j}^n}{\Delta t} \qquad (7)$$

$$+ \rho_g u_{g,j}^n \frac{u_{g,j}^n - u_{g,j-1}^n}{\Delta x} - \rho_l u_{l,j}^n \frac{u_{l,j}^n - u_{l,j-1}^n}{\Delta x} + \Delta\rho g H \frac{a_{j+1}^n - a_j^n}{\Delta x} = 0 ,$$

where the subscript j and the superscript n denote the spatial mesh number and the time step, respectively, and $\Delta\rho$ is the density difference.

We suppose that small perturbations — δa_j, $\delta u_{g,j}$ and $\delta u_{l,j}$ — are superposed on the solution of Eqs.(7), and that Eqs.(7) are linearized around the solution. The variables in the difference terms are then substituted by components of the Fourier series, that is, $\delta u_{g,j}^n$ is substituted by $\delta u_{g,j}^n e^{IKj\Delta x}$, where I is an imaginary unit, K the wave number defined as $(2\pi/\lambda)$ and λ the wavelength. We obtain

$$\delta a_j^{n+1} = \delta a^n - C_a \varphi \delta u_{g,j}^n - C_g \varphi \delta a_j^n ,$$

$$\delta a_j^{n+1} = \delta a^n + C_{1-a} \varphi \delta u_{l,j}^n - C_l \varphi \delta a_j^n ,$$

$$\rho_g \delta u_{g,j}^{n+1} - \rho_l \delta u_{l,j}^{n+1} = \qquad (8)$$

$$\rho_g \delta u_{g,j}^n - \rho_l \delta u_{l,j}^n - \rho_g C_g \varphi \delta u_{g,j}^n + \rho_l C_l \varphi \delta u_{l,j}^n - \Delta\rho g H C e^{I\theta} \varphi \delta a_j^n ,$$

where

$$\begin{aligned}
C &= \Delta t / \Delta x , \\
C_a &= C a , \\
C_{1-a} &= C(1-a) , \\
C_g &= C u_g , \\
C_l &= C u_l , \\
\varphi &= 1 - e^{-I\theta} \quad \text{and} \\
\theta &= K \Delta x .
\end{aligned} \qquad (9)$$

In Eq.(9), a, u_g and u_l represent unperturbed values. Equation (8) can be rewritten in a matrix form:

$$A X^{n+1} = B X^n , \qquad (10)$$

where X is a vector composed of δa_j, $\delta u_{g,j}$ and $\delta u_{l,j}$. We can find whether the small perturbation can grow or not by

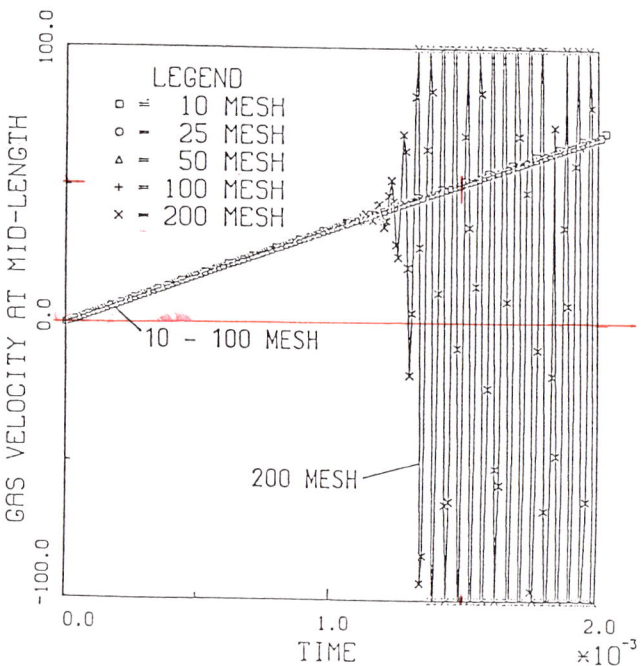

FIGURE 2. Gas velocity transients at mid-length (H=1.0m, Co=0.3)

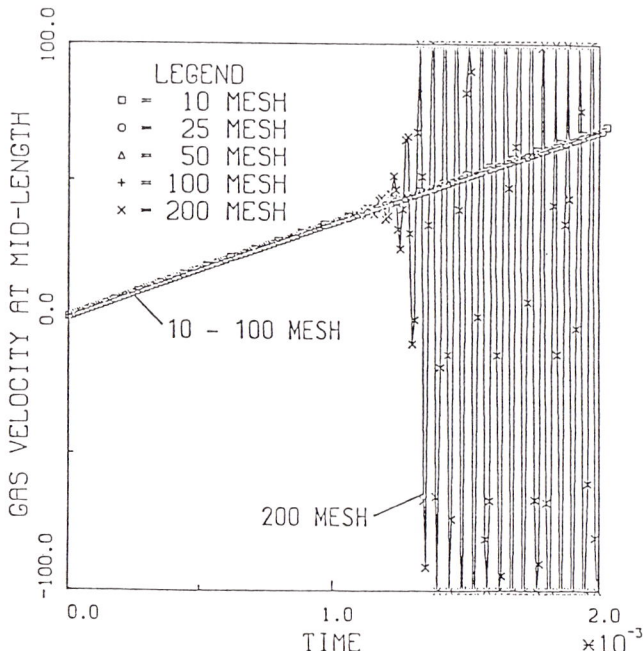

FIGURE 3. Gas velocity transients at mid-length (H=0.0m, Co=0.3)

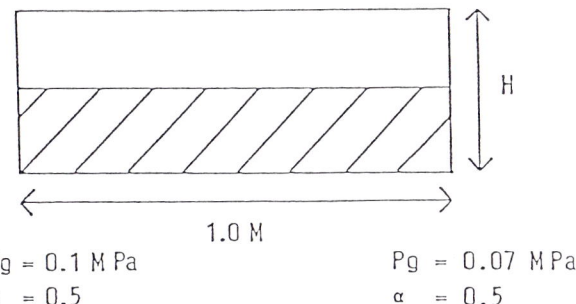

FIGURE 1. Flow channel of accelerated flow problem

Courant number. The Courant number is defined as:

$$Co = \frac{\Delta t}{\Delta x} \left\{ \frac{\alpha_g \rho_l + \alpha_l \rho_g}{(\alpha_l \rho_g / a_l^2 + \alpha_g \rho_l / a_g^2)} \right\}^{1/2} \quad (6)$$

where a_l and a_g are the sound speeds in liquid and gas phases, respectively. For simplicity, the sound speeds are set equal to constants: $a_l=1500 m/s$ and $a_g=340 m/s$. Although a detailed description was not given in the paper of Hancox et al., the problem definition and the numerical procedure mentioned above are about the same as they used (Hancox et al., 1980).

Calculated results and discussion

The channel heights are chosen as $1.0 m$ and $0.0 m$. Thus, the maximum relative velocities which maintain the system hyperbolicity are $70.7 m/s$ and $0.0 m/s$, respectively. The channel height of $0.0 m$ has no physical meaning; however, it can be considered as the limiting case and corresponds to the one-pressure model. The transients of gas velocity at the mid-length of the channel are shown in Figs.2 and 3. In Figs.2 and 3, the Courant number is set equal to 0.3.

The gas velocity linearly increases and, only in the 200 mesh cell case, the numerical instability occurs as shown in Fig.2. It is suggested that the numerical damping effect associated with the coarse mesh may stabilize the solution. In the 200 mesh cell case, the oscillation occurs at the gas velocity of about $30 m/s$. In the time scale of this problem, the liquid is not accelerated to more than $0.1 m/s$. The relative velocity at which the oscillation occurs is thus much smaller than the stability limit ($70.7 m/s$) determined by the well-posed condition. Moreover, even if the channel height is $0.0 m$ and the corresponding maximum relative velocity which gives the system hyperbolicity is $0.0 m/s$, about the same result is obtained as seen in Fig.3. It is concluded from Figs.2 and 3 that the numerical instability cannot be fully described by the stability condition defined by Eq.(4).

In governing equations, we neglect interfacial mass and momentum transfer terms in order to simplify the equation system. For this reason, we also omit energy equations. In other words, adiabatic two-phase flows are assumed in this study.

The system defined by Eqs.(1) to (3) is hyperbolic only when (Rousseau et al., 1979):

$$(u_g - u_l)^2 < (\frac{\alpha}{\rho_l} + \frac{(1-\alpha)}{\rho_g})(\rho_l - \rho_g)gH. \tag{4}$$

This hyperbolicity condition is referred to as the well-posed condition. That is, if the condition defined by Eq.(4) is violated, the system is non-hyperbolic and constitutes an ill-posed problem. Equation (4) is also known as the Kelvin-Helmholtz stability condition for waves with long wavelength (Rousseau et al., 1979).

Governing equations (1) to (3) are numerically solved under appropriate conditions by using the finite difference method. The relation between the instability associated with numerical solutions and the well-posed condition defined by Eq.(4) is discussed.

ACCELERATED FLOW

Problem description

The stratified flow is suddenly accelerated from the rest, due to the constant pressure difference between the inlet and outlet. In this study, the channel length is set equal to $1m$ and the channel height is varied to change the well-posed condition. The schema of flow channel for this problem is depicted in Fig.1.

Initial conditions are:

$$\alpha = 0.5 ,$$
$$u_g = u_l = 0.0 \ (m/s) ,$$
$$\rho_g = 1.0 \ (kg/m^3) ,$$
$$\rho_l = 1000.0 \ (kg/m^3) \text{ and} \tag{5}$$
$$dp_g/dx = p_{g,outlet} - p_{g,inlet} .$$

$p_{g,outlet}$ and $p_{g,inlet}$ are set equal to $0.07MPa$ and $0.1MPa$, respectively, and are assumed to be constant throughout the calculation as the boundary conditions. The void fractions at the inlet and outlet are also assumed to be constant at 0.5. The gravitational acceleration is $10.0 m/s^2$.

Equations (1) to (3) are solved under the above conditions by a simple explicit finite difference procedure, with the advective terms approximated by upwind differencing. We use a staggered mesh system for spatial discretization: velocities are defined at the cell edge, and other state variables defined at the cell center. The numbers of space mesh cells are varied from 10 to 200 and the mesh cells are assumed to be equally spaced. The time step size is determined by the

pointed out that the numerical damping effect stabilized the solution. One numerical example obtained by an explicit upwind scheme was shown; however, a detailed discussion about the numerical solution procedure was not carried out.

Some possible approaches for formulating a well-posed numerical problem for an ill-posed differential model have been investigated and discussed by Ransom and Trapp (1983). They applied Neumann's stability analysis to some finite difference schemes, and discussed the stability of discretized systems in terms of amplification factors. They used advection-diffusion equations as model equations. The implicit scheme with large Courant numbers was shown to be almost stable for ill-posed equations. They concluded that the ill-posed differential initial-value problem could be rendered well-posed for finite interval by a numerical process. However, they used simple equations and did not apply such a method to the usual two-phase flow equations. Numerical results or examples were not shown.

In this paper, the accelerated stratified flow problem, such as was used by Hancox et al. (1980), is numerically solved. The numerical instability of a one-dimensional finite difference scheme is studied, based on Neumann's stability analysis method. A stability criterion is derived in terms of an eigenvalue of amplification matrix. It is shown that the numerical instability does not depend on the ill-posed condition of differential equations, but rather depends on the stability condition determined by the eigenvalue. The growth of void waves such as the Kelvin-Helmholtz instability problem is also numerically calculated. The growth of waves with long wavelengths is shown to be determined not by the classical wave growth condition, but rather by the numerical stability condition.

GOVERNING EQUATIONS

The governing equations for one-dimensional stratified two-phase flows are described as (Hancox et al., 1980):

$$\frac{\partial (\alpha_k \rho_k)}{\partial t} + \frac{\partial (\alpha_k \rho_k u_k)}{\partial x} = 0 , \tag{1}$$

$$\frac{\partial (\alpha_k \rho_k u_k)}{\partial t} + \frac{\partial (\alpha_k \rho_k u_k^2)}{\partial x} + \alpha_k \frac{\partial p_k}{\partial x} + (p_k - p_i)\frac{\partial \alpha_k}{\partial x} = 0 , \tag{2}$$

where α, ρ, u and p are the volume fraction, the density, the velocity and the pressure, respectively. Subscripts k and i refer to each phase and interface, respectively. The pressure differences between each phase and interface are defined, assuming static force balance:

$$p_k - p_i = (-1)^m \alpha_k \rho_k g \frac{H}{2} , \tag{3}$$

where g and H are the gravitational acceleration and the height of flow channel, respectively. A superscript m indicates an index: $m=1$ for $k=g$ and $m=2$ for $k=l$.

STABILITY OF NUMERICAL ANALYSIS ON VOID WAVES IN STRATIFIED TWO-PHASE FLOW

T. WATANABE and M. HIRANO
Department of Reactor Safety Research
Japan Atomic Energy Research Institute
Tokai-mura, Ibaraki-ken, 319-11, Japan

ABSTRACT

The stability of numerical analysis on stratified two-phase flow has been studied. The stability analysis based on Neumann's method was carried out for one-dimensional finite difference schemes. A stability criterion was shown in terms of an eigenvalue of amplification matrix. The numerical instability of accelerated flow and the growth of small amplitude sinusoidal void waves were studied. It was found that the instability and wave growth did not depend on the stability limit derived by the classical potential theory or on the ill-posed condition, but depended only on the stability limit determined by the eigenvalue of amplification matrix.

INTRODUCTION

An accurate prediction of two-phase flow phenomena is quite important in safety analyses of nuclear reactors. Among the several two-phase flow models, the two-fluid model is considered to be appropriate for the most general and detailed description of transient two-phase flows. In the two-fluid model, each phase is separately described in two sets of conservation equations. In practical concerns, the conservation equations are averaged equations of local instantaneous conservation equations. Although the detailed information is lost by averaging, the averaged model is quite important from the engineering point of view.

It is considered that phasic pressures should be separately described in the averaged equations of stratified flow, since the basic equation system exhibits non-hyperbolic behavior when the equality of two pressures is assumed. The system hyperbolicity condition has been shown by Rousseau and Ferch (1979), assuming hydrostatic force balance. A numerical example has been shown by Hancox et al. using a finite difference method (Hancox et al., 1980). The numerical instability was shown to occur when the hyperbolicity condition was violated. Further, the instability was concealed when the coarse mesh was adapted, even if the hyperbolicity condition was violated. Thus, Hancox et al.

Plesset, M.S. and Hsieh, D.Y., "Theory of Gas Bubbly Dynamics in Oscillating Pressure Fields", *The Physics of Fluids*, Vol. 3, pp. 882-892, 1960.

Prosperetti, A., "Thermal Effects and Damping Mechanisms in the Forced Radial Oscillations of Gas Bubbles in Liquids", *J. Acoust. Soc. Am.*, Vol. 61, pp. 17-27, 1977.

Rschevkin, S.N., *The Theory of Sound*, Pergamon Press, 1963.

Ruggles, A.E., "An Experimental and Analytical Investigation of the Propagation of Pressure Perturbations in Bubbly Air/Water Flows", Ph.D. Thesis, Rensselaer Polytechnic Institute, Troy, NY, 1987.

Ruggles, A.E., Scarton, H.A. and Lahey, R.T., Jr., "An Investigation of Pressure Perturbations in Bubbly Air-Water Flows", *J. of Heat Transfer*, Vol-110, 1988.

Silberman, E., "Sound Velocity and Attenuation in Bubbly Mixtures Measured in Standing Wave Tubes", *J. Acoust. Soc. Am.*, Vol-29, 925, 1957.

Stuhmiller, J.H., "The Influence of Interfacial Pressure Forces on the Character of Two-Phase Flow Model Equations", *J. Multiphase Flow*, Vol-3, 551-560, 1977.

Tournaire, A., "Detection et Etude des Ondes de Taux de Vide en Ecoulement Diphasique a Bulles Jusqu'a la Transition Bulles-Bouchons", Docteur Engenieur Thesis, L'Université Scientific et Médicale et L'Institut National Polytechnique de Grenoble, 1987.

van Wijngaarden, L., "Linear and Non-Linear Dispersion of Pressure Pulses in Liquid-Bubble Mixtures", Proc. 6th Symp. on Naval Hydrodynamics (ed. R.D. Cooper & S.W. Doroff), U.S. Govt. Printing Office, pp. 129-135, 1966.

Wallis, G.B., *One-Dimensional Two-Phase Flow*, McGraw-Hill, 1969.

Zuber, N. and Hench, J., "Steady-State and Transient Void Fraction for Bubbling Systems and Their Operating Limit: Part 1. Steady-State Operation", General Electric Report No. 62 GL 100, 1962.

Bernier, R.N.J., "Unsteady Two-Phase Flow Instrumentation and Measurement", Report E200.4, Division of Engineering and Applied Science, California Institute of Technology, 1982.

Biesheuvel, A. and van Wijngaarden, L., "Two-Phase Flow Equations for a Dilute Dispersion of Gas Bubbles in Liquid", *J. Fluid Mech.*, 168, pp. 301-318, 1984.

Bouré, J.A., Fritte, A.A., Giot, M.M. and Reocreux, M.L., "A Contribution to the Theory of Critical Two-Phase Flow: On the Links Between Maximum Flow Rates, Choking, Sonic Velocities, Propagation and transfer Phenomena in Single and Two-Phase Flow", *Energie Primaire*, Vol. 11, pp. 1-27, 1975.

Carstensen, E.L. and Foldy, L.L., "Propagation of Sound Through a Liquid Containing Bubbles", *J. Acoust, Soc. Am.*, Vol. 19, No. 3, p. 481-501, 1947.

Cheng, L-Y., Drew, D.A., and Lahey, R.T., Jr., "An Analysis of Wave Dispersion, Sonic Velocity, and Critical Flow in Two-Phase Mixtures", NUREG/CR-3372, 1983.

Cheng, L-Y., Drew, D.A. and Lahey, R.T., Jr., "An Analysis of Wave Propagation in Bubbly Two-Component Two-Phase Flow", *Journal of Heat Transfer*, Vol. 107, pp. 402-408, May, 1985.

Delhaye, J.M., Giot, M., Riethmuller, M.L., *Thermohydraulics of Two-Phase Systems for Industrial Design and Nuclear Engineering*, McGraw-Hill, 1981.

Hall, P., "The Propagation of Pressure Waves and Critical Flow in Two-Phase Mixtures", Ph.D. Thesis, Heriot-Watt University, Edinburgh, U.K., 1971.

Harmathy, T.Z., "Velocity of Large Drops and Bubbles in Media of Infinite or Restricted Extent", *AIChE Journal*, Vol-6, No. 2, 1960.

Ishii, M., *Thermal-Fluid Dynamics of Two-Phase Flow*, Eyrolles, 1975.

Ishii, M. and Zuber, N., "Relative Motion and Interfacial Drag Coefficient in Dispersed Two-Phase Flow of Bubbles, Drops and Particles", *AIChE Journal*, Vol. 25, No. 5, 1979.

Landau, L.D. and Lifshitz, E.M., *Fluid Mechanics*, Pergamon Press, New York, 1959.

Mercadier, Y., "Contribution a L'Etudes des Propagations de Perturbations de Taux de Vide dans les Ecoulements Diphasiques Eau/Air a Bulles", Docteur Engenieur, Thesis, L'Université Scientifique et Médicale & L'Institut National Polytechnique de Grenoble, 1981.

Nigmatulin, R.I., "Spatial Averaging in the Mechanics of Heterogeneous and Dispersed Systems", *J. of Multiphase Flow*, 5, 353-385, 1979.

Nishihara, H. and Michiyoshi, I., "Acoustic Velocity and Attenuation in an Air/Water Two-Phase Medium", *Two-Phase Flow Dynamics*, Hemisphere Publishing Corp., 1981.

Pauchon, C. and Banerjee, S., "Interphase Momentum Interaction Effects in the Averaged Multifield Model, Part I: Void Propagation in Bubbly Flows", *Int. J. of Multiphase Flow*, 16, 4, pp. 559-573, 1986.

Pauchon, C. and Banerjee, S., "Interphase Momentum Interaction Effects in the Averaged Multifield Model, Part II: Kinematic Waves and Interfacial Drag in Bubbly Flows", *Int. J. Multiphase Flow*, Vol-14, No. 3, 253, 1988.

k	Wavenumber
M_k	Interfacial momentum transfer
n	Drift-flux parameter
$1/L_s$	Interfacial area density
p	Pressure
q_{ki}''	Interfacial heat flux
R	Radius
STP	One atmosphere, 23°C
T	Temperature
t	Time
u	Velocity
v	Eigenvalue
V	Parameter defined in Eq. (67a)
z	Axial location

Greek

α	Void fraction
$\delta(\ )$	Perturbed quantity
$\eta\alpha$	Attenuation coefficient
λ	Wave length
μ	Dynamic viscosity
ω	Angular frequency
ρ	Density
κ	Reynolds stress parameter
ψ	State vector
$\overline{\sigma}$	Surface tension
θ	Angle of inclination of flow from vertical
υ	Kinematic viscosity or parameter defined in Eq. (67c)
τ	Stress or parameter defined in Eq. (67b)
ξ	Interfacial pressure distribution parameter

Subscripts

b	Bubble
2ϕ	Two-phase
1ϕ	One-phase
g	Gas
H	Hydraulic
i	Interfacial
k	Phase indicator (g = gas, ℓ = liquid)
ℓ	Liquid
o	Equilibrium (steady-state) value
VM	Virtual mass
w	Wall
∞	Terminal rise (velocity)

Symbols

$\delta(\ )$	Perturbation
$(\ )^T$	Transpose of a vector matrix
∇	Gradient
$\dfrac{\partial(\ )}{\partial(\ )}$	Partial derivative
$\dfrac{d(\ )}{d(\ )}$	Total derivative

REFERENCES

Bendat, J.S., Pierson, A.G., *Random Data: Analysis and Measurements Procedures*, Wiley-Interscience, 1971.

WAVE PROPAGATION

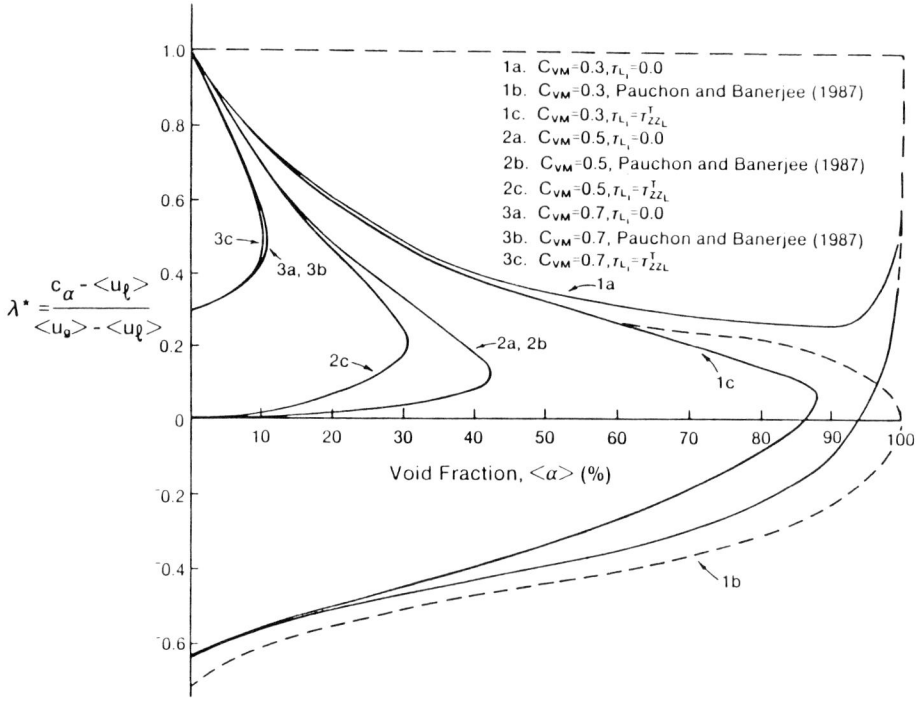

FIGURE 19. Sensitivity of Eigenvalue Model to Changes in C_{VM} (with Reynolds Stress.

Pressure wave propagation involves rapid local accelerations and thus the predicted celerities were very sensitive to virtual mass effects but not to interfacial drag laws nor Reynolds stress models. In contrast, the void wave celerities predicted by the complete dispersion relation showed that they were not as sensitive to the virtual mass model but were very sensitive to interfacial drag. It thus appears that wave propagation phenomena in two-phase flow represent an excellent means of isolating effects and thus assessing he interfacial closure laws of two-fluid models. It is hoped that this review paper will help stimulate further research of this type.

NOMENCLATURE
Latin

a	Acceleration
a_{VM}	Virtual mass acceleration
c	Speed of propagation (celerity)
C_{VM}	Virtual volume coefficient
D	Diameter
$\frac{D_k(\ )}{Dt}$	Material derivative, $\frac{\partial(\ )}{\partial t} + u_k \frac{\partial(\ )}{\partial z}$
f	Frequency; friction factor
F	Force
g	Gravitational acceleration
g_c	Gravitational constant
h	Enthalpy
i	Imaginary number, $\sqrt{-1}$
j_k	Superficial velocity of phase-k
$j_{g\ell}$	Drift-flux

implies that the other terms (eg, wall shear) in the matrix $\underline{\underline{C}}_o$ of Eq. (63) are relatively small compared to interfacial drag.

Finally, it is interesting to note in Fig. 18 that the kinematic drift-flux model for distorted bubbly flow (n = 7/4) coincides with the corresponding dispersion results up to fairly large void fractions. Interestingly, the drift-flux parameter n = 2 agrees reasonably well with the dispersion results for Newton's regime drag law. Thus we find that the void wave celerities given by a drift-flux model (which is based only on the quasi-static closure model) are closely linked (via the interfacial drag law) to the corresponding eigenvalues and linear dispersion results of the two-fluid model which is valid for low frequencies. This clearly shows the importance of interfacial drag on void wave propagation phenomena. Moreover, it implies that a lot of physics associated with interfacial momentum transfer (eg: virtual mass effects, turbulence, etc.) is implicit in the drift-flux parameter, n.

It appears that the linear dispersion relation of a typical two-fluid model is capable of predicting kinematic void wave propagation. In contrast, the void eigenvalues generally underpredict such data since these results imply zero interfacial drag, among other things (ie, they imply setting $\underline{\underline{C}}_o = 0$). Moreover, it has also been shown that kinematic drift-flux models of void wave propagation are closely linked to the dispersion relation in which the corresponding interfacial drag law is used.

It can be noted in Fig. 19 that the eigenvalues of the model may yield increasing values for λ^* as the void fraction approaches unity. This occurs because the primary contribution to the liquid pressure gradient is the hydrostatic head due to the mixture density. Thus, the relative velocity, $\langle u_g \rangle - \langle u_\ell \rangle$, approaches zero as the mixture density approaches that of the gas phase (ie, bubble buoyancy goes to zero). This result is somewhat artificial since the Newton's regime drag law used in the eigenvalues shown in Fig. 19 is not appropriate for very high void fractions, nor are some of the other closure laws which have been used. While the same comments apply, this behavior is not seen in the model of Pauchon and Banerjee [1988] since no explicit expression for relative velocity was included.

Finally, it should be noted that all the two-fluid models shown on Fig. 18 were run with the virtual volume coefficient, C_{VM}, equal to 0.5 in order to be consistent with the eigenvalue models of Pauchon and Banerjee [1986; 1988]. However, it can be noted in Fig. 19 that the eigenvalues are significantly affected by the value of C_{VM}. In particular, the domain of hyperbolicity of the two-fluid model is reduced when C_{VM} is greater than 0.5 and increased when it is less.

Including a finite interfacial shear stress, τ_{Li}, also reduces the domain of hyperbolicity. Indeed, a significant change can be noted when $C_{VM} \leq 0.5$. It is also interesting to note that for $C_{VM} < 0.5$, the lower branch of λ^* can be negative, implying, $c_\alpha^- < \langle u_\ell \rangle$.

SUMMARY AND CONCLUSIONS

It has been shown that a properly formulated one-dimensional two-fluid model can predict the propagation of pressure and void perturbations in two-phase systems. It has also been shown that a two-fluid model which is only valid for low frequencies can predict the so-called two-phase critical flow velocity. Indeed, it was verified that the local sonic velocity at the choking plane given by such models is the choking velocity for critical two-phase flows.

WAVE PROPAGATION

$$\lambda^\star \overset{\Delta}{=} \frac{c_\alpha^+ - <u_\ell>}{<u_g> - <u_\ell>} = <\alpha> + \frac{1}{\left(<u_g> - <u_\ell>\right)} \frac{\partial j_{g\ell}}{\partial <\alpha>} \tag{70}$$

Wallis [1969] has also proposed a drift-flux relation of the form,

$$j_{g\ell} = <\alpha>(1-<\alpha>)\left(<u_g> - <u_\ell>\right) = u_\infty <\alpha>(1-<\alpha>)^n \tag{71}$$

Thus Eqs. (69), (70) and (71) yield,

$$\lambda^\star = 1 - n<\alpha> \tag{72}$$

The value of the drift-flux parameter (n) depends on the flow conditions. For example, in steady, fully-developed, vertical bubbly two-phase flow Eqs. (9) yield:

$$\left(<u_g> - <u_\ell>\right) = \left[\frac{4}{3} \frac{g(\rho_\ell - \rho_g)(1-<\alpha>)D_b}{\rho_\ell C_D}\right]^{1/2} \tag{73}$$

For the distorted bubble regime, Eq. (13) can be combined with Eq. (73) to yield,

$$\left(<u_g> - <u_\ell>\right) = 1.414\left[\frac{gg_c(\rho_\ell - \rho_g)\sigma}{\rho_\ell^2}\right]^{1/4}(1-<\alpha>)^{3/4}$$

That is,

$$\left(<u_g> - <u_\ell>\right) = u_\infty(1-<\alpha>)^{3/4} \tag{74}$$

Hence Eqs. (71) and (74) imply n = 7/4 for distorted bubbly flow. Similarly, for other flow conditions the value of the drift-flux parameter recommended by Wallis [1969], n = 2, may be appropriate.

Finally, it should be noted that, using an entirely different analytical approach, Pauchon and Banerjee [1988] have deduced a result for the case of a constant interfacial drag coefficient, which yields, n = 3/2. This result is consistent with the use of a constant C_D in Eq. (73) and with the result proposed previously by Zuber and Hench [1962]. Moreover, they also derived the void dependence of neutrally stable void waves and found agreement with the Newton regime drag model [Ishii and Zuber, 1979] up to a void fraction of about 30%. It will be shown later that this model corresponds to an 'n' which is slightly less than 2.0.

The appropriate roots ($\xi_j^\star = c_\alpha^\pm$) of the linear dispersion model, Eq. (63), are compared with the simplified eigenvalue models due to Pauchon and Banerjee [1986; 1988], and the kinematic drift-flux model of Wallis [1969] in Fig. 18. It can be seen that the dispersion model gives results which are larger than the eigenvalue models, with the deviation increasing as the interfacial drag coefficient (C_D) increases. Also, it was found that for zero interfacial drag (ie, C_D = 0) that the dispersion model gave results which were close to the eigenvalue model. This

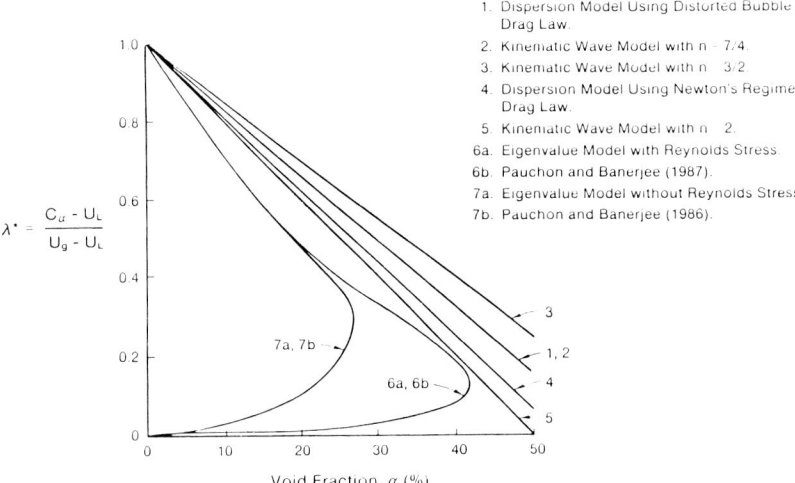

FIGURE 18. Comparison of Void Wave Models, $C_{VM} = 0.5$

necessarily indicative of flow regime transition. Indeed flow regime transition involves phenomena (eg, bubble coalescence) which are not in the model presented herein. Thus flow regime transition should not be inferred from the results presented in this paper.

The quasi-static two-fluid model, Eqs. (55), can also be treated as an eigenvalue problem,

$$\det\left\{\underline{\underline{A}}_s \xi - \underline{\underline{B}}_s\right\} = 0 \qquad (68)$$

where, ξ, is a system eigenvalue.

These eigenvalues include the effect of compressibility, vapor inertia, and bubble pulsation induced forces (F_R). However, it is seen in Fig. 18 that the two-fluid model gives virtually identical values for ($c_\alpha^\pm$) as the simplified eigenvalue models due to Pauchon and Banerjee [1986, 1987]. This is a useful check on consistency and verifies that, as expected, the effects of compressibility, bubble pulsation, surface tension, and gas phase inertia are negligible for void wave propagation.

A kinematic drift-flux model for void waves has been proposed by Wallis [1969]. Significantly, this model is based entirely on steady-state considerations. The celerity of void perturbations is given by,

$$c_\alpha^+ = \left(\frac{\partial j_g}{\partial <\alpha>}\right)_j = j + \frac{\partial j_{g\ell}}{\partial <\alpha>} \qquad (69)$$

since,

$$j = <\alpha><u_g> + (1-<\alpha>)<u_\ell>$$

we have,

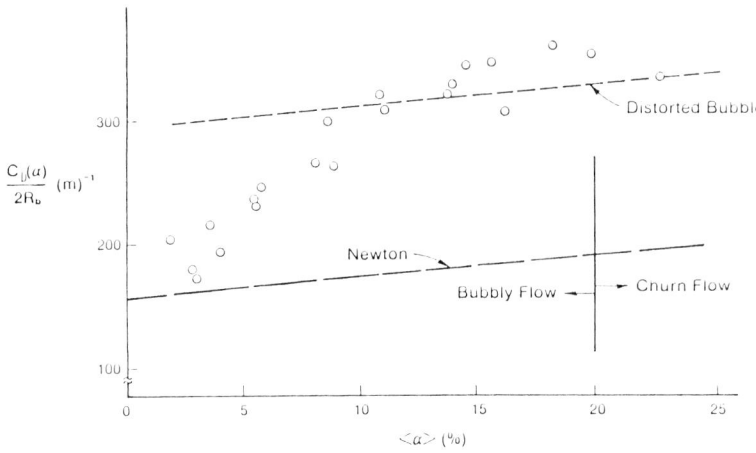

FIGURE 17. Drag Coefficient vs. Data (Mercadier, 1981)

where,

$$V = \frac{(1-<\alpha>)\left[C_{vm} - \xi - \kappa <\alpha>\right]}{<\alpha>(1-<\alpha>) + C_{vm}} \tag{67a}$$

$$\tau = <\alpha>(1-<\alpha>) + C_{vm} \tag{67b}$$

$$\upsilon = (1-<\alpha>)^2 \frac{\left[C_{vm} - \xi - \kappa <\alpha>\right]^2}{\left[<\alpha>(1-<\alpha>) + C_{vm}\right]} + <\alpha>(1-<\alpha>)\left[\xi + \kappa - C_{vm}\right]$$

$$+ 2(1-<\alpha>)^2 \left[\xi - C_{vm}/2\right] \tag{67c}$$

It should be noted that when $\xi = 1/4$ the expression given in Eq. (21) is implied. Similarly, when $\kappa = 1/5$, Eq. (25) is implied. Correspondingly, $\kappa = 0$ implies zero Reynold's stress. The larger value of λ^* in Eq. (55), $c_{\alpha'}^+$ is associated with kinematic void wave propagation [Pauchon and Banerjee, 1988].

It is interesting to note in Fig. 18 that the eigenvalues are quite sensitive to how the Reynolds stress is modeled. Moreover, we can note in Eq. (67c) that the two void wave eigenvalues become complex, and cause the mathematical system to become elliptic, for void fractions larger than that where these two eigenvalues are equal. Significantly, the point at which they coincide (ie, the "nose" of the eigenvalue curves in Fig. 18) shifts dramatically with Reynolds stress. This change in the nature of the system of differential equations implies that the assumed constitutive relationships may not be appropriate for void fractions beyond where the eigenvalues are real (ie, beyond where the system model is well-posed). This is not surprising since the models used for virtual mass effects and Reynolds stress are based on single bubble analyses. Appropriate models for closely packed bubbly flows will require improved constitutive models. Moreover, flow regime changes are expected as the void fraction is increased. Nevertheless, it should not be assumed that the mathematical transition from a hyperbolic to non-hyperbolic system is

shown in Fig. 15 for the dispersion model with no interfacial drag (ie, $C_D = 0$). It can be seen that the drag models for distorted bubbles, Eq. (13), and zero interfacial drag ($C_D = 0$) tend to bracket the data, while the Newton's regime drag model, Eqs. (14), falls in between. It is evident that the predictions are quite sensitive to the interfacial drag law used. It should be noted that the most probable reason that the scatter in the data shown in Fig. 15 is so great is that only naturally occurring void waves were measured. A better experimental technique appears to be one in which the void perturbations are harmonically forced [Tournaire, 1987].

The linear dispersion model, Eq. (63), is compared to the data of Mercadier [1981] in Fig. 16. Only the distorted bubble drag law is shown since it was considered to be the most appropriate for the test conditions examined. This choice of drag law is further supported by comparing the interfacial drag data from Mercadier [1981], given in Fig. 17, with the two interfacial drag models given in Eqs. (13) and (14). Note that while neither of these drag laws is appropriate for the entire range of void fractions investigated, the distorted bubble drag law does the best job for $(<\alpha>) \geq 10\%$.

Comparison of the Dispersion Model to Other Void Wave Models

Pauchon and Banerjee have proposed a model which is based on the continuity and momentum equations for the liquid and gas, however, phasic compressibility, surface tension, bubble pulsation induced forces (F_R) and gas phase inertia were neglected. The resulting expression for the eigenvalues of their model is [Pauchon and Banerjee, 1988]:

$$\lambda^* = \frac{c_\alpha^{\pm} - <u_\ell>}{<u_g> - <u_\ell>} = V \pm \sqrt{\upsilon/\tau} \tag{66}$$

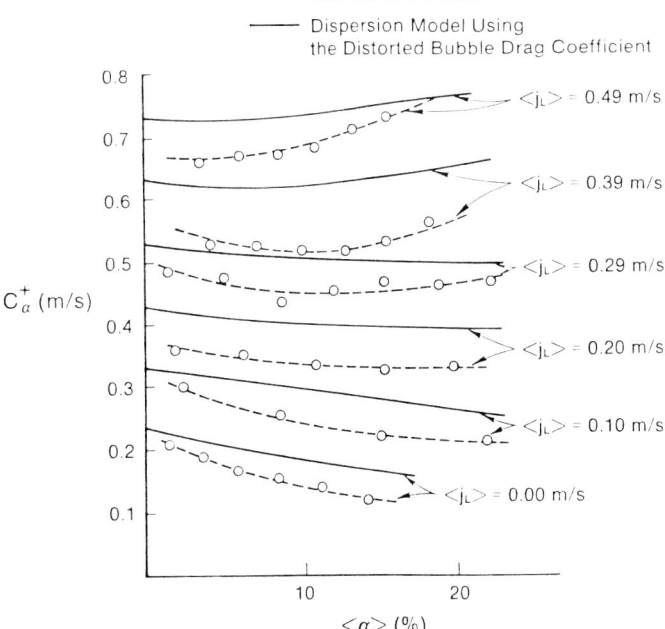

FIGURE 16. Void Wave Data Due to Mercadier (1981)

WAVE PROPAGATION

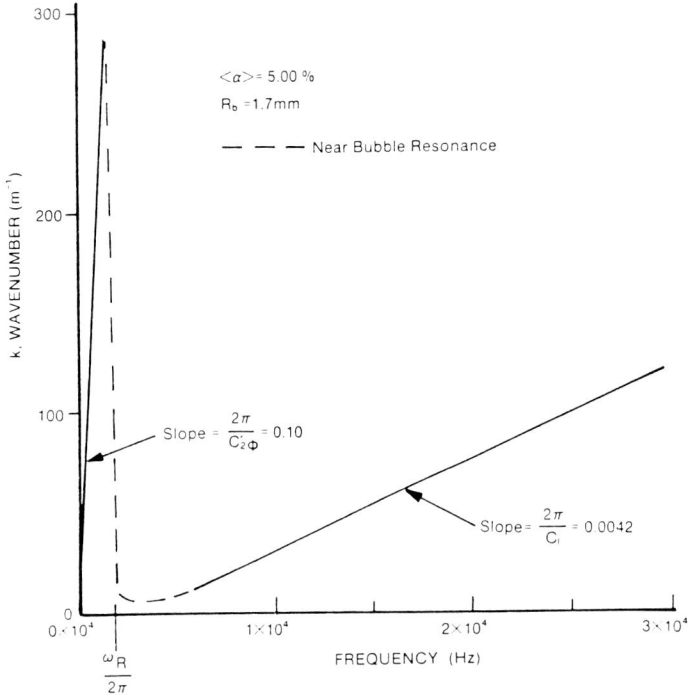

FIGURE 14. Dispersion Relation as Predicted by a Linear Model

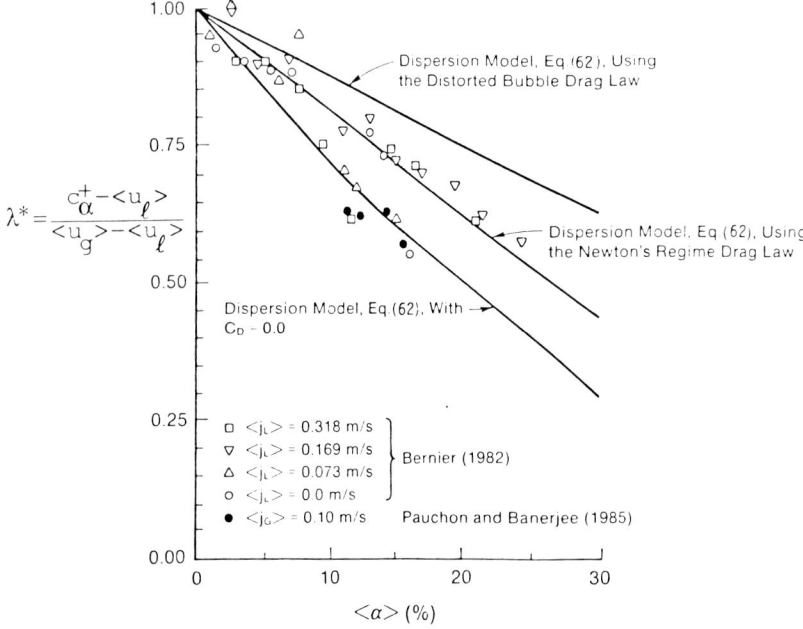

FIGURE 15. Void Wave Data, λ^* vs $\langle\alpha\rangle$

Thus we find that the wave propagation model associated with low frequency phenomena predicts the so-called two-phase choking velocity, $c'_{2\phi}$.

The reader is cautioned that while adiabatic models may be adequate for predictions of the two-phase sound speed they are not normally adequate for predictions of the spatial attenuation.

It is also interesting to note that the terms resident in matrix $\underline{\underline{C}}'_s$ are due to constitutive laws often used to model phenomena such as bubble drag, wall shear, and heat transfer. These phenomena are actually based on local gradients, but for convenience they are normally modeled algebraically. Hence their appearance on the right hand side of Eq. (55), and of Eq. (36), are a result of procedures used in averaging and modeling. Fortunately, they are not important if one is only interested in predictions of the sound speed.

The critical flow model just discussed was found to be consistent with the commonly used two-fluid model given by Eq. (55). This model uses the quasi-steady pressure relation of Eq. (50) that assumes that the gas phase pressure perturbations follow the liquid phase pressure perturbations. This model was shown to be appropriate for pressure perturbations having frequency content well below bubble resonance, and yields the low frequency two-phase sound speed, $c'_{2\phi}$.

However, it has been shown herein, and has been previously observed [van Wijngaarden, 1966], that the two-fluid media is also capable of transmitting high frequency ($\omega > \omega_r$) pressure disturbances at velocity c_ℓ. It is significant to note that if one chooses to use the dynamic pressure relationship between the gas and liquid phases appropriate for these high frequencies, Eq. (34), then the high frequency critical flow velocity would have corresponded to c_ℓ. It appears, however, that these modes can only be excited for situations involving rapid transients (eg, sudden pressure changes).

Finally, it is interesting to note that a general spatial pressure field can be constructed either from frequencies above resonance (fast waves), or from frequencies below resonance (slow waves). This non-unique representation is possible because, as shown in Fig. 14, a significant range of frequencies above bubble resonance, and below bubble resonance, produce an identical range of wave numbers. That is, the two-phase dispersion relation is multi-valued.

The discussion to date has had to do with the prediction of the propagation of pressure perturbations. Let us now turn our attention to the propagation of void perturbations.

THE ANALYSIS OF VOID WAVE PROPAGATION

The void wave propagation data of Pauchon and Banerjee [1986] and Bernier [1982] are compared in Fig. 15 with the roots (ξ^*_j) of Eq. (63) associated with void wave propagation ($c_\alpha^{\pm}$). For consistency with previous investigations [Pauchon and Banerjee, 1986; 1987] the virtual volume coefficient, C_{VM}, was taken to be 0.5 in these evaluations.

The two interfacial drag coefficients given in Eq. (13) and Eqs. (14) were used in conjunction with the two-fluid model. Predictions are also

WAVE PROPAGATION

$$\det\left[\underline{\underline{B}}_s\right] = 0 \tag{60}$$

The choking condition can be determined by solving Eq. (59) as an initial value problem, while simultaneously testing for the condition of Eq. (60). A detailed discussion of how this type of analysis is performed has been given previously [Bouré et al, 1975] and thus will not be repeated here.

The velocity associated with the choked state (ie, the critical flow velocity) is calculated from the characteristics (ξ) of Eq. (57) by [Delhaye et al, 1981]:

$$\det\left\{\underline{\underline{A}}_s \xi - \underline{\underline{B}}_s\right\}\bigg|_{choked} = 0 \tag{61}$$

Obviously Eq. (61) is just a special case of Eq. (57). One of the eigenvalues of Eq. (61), $\xi_c = 0$, determines the choking condition. This eigenvalue is the only one capable of transmitting information on pressure variations downstream of the throat to positions upstream of the throat. Once the discharge velocity at the throat reaches sonic conditions, further change of downstream pressure is not capable of affecting conditions upstream of the choking plane (ie, the throat). Moreover, since Eq. (61) is valid for all eigenvalues, and in particular for the one that vanishes, this equation implies Eq. (60) when choking occurs.

Equation (61) is somewhat different from the corresponding linear dispersion relation for the low frequency pressure pulses,

$$\det\left\{\left(\underline{\underline{A}}_s \xi^* - \frac{i}{k}\underline{\underline{C}}_s'\right) - \underline{\underline{B}}_s\right\} = 0 \tag{62}$$

where,

$$\xi^* = \frac{\omega}{k}.$$

However, Eq. (62) can be put in the form of Eq. (61) if one takes the limit as the wave number, k, grows large (ie, the high frequency limit):

$$\det\left\{\lim_{k \to k_m}\left\{\left(\underline{\underline{A}}_s \xi^* - \frac{i}{k}\underline{\underline{C}}_s'\right) - \underline{\underline{B}}_s\right\}\right\} \equiv \det\left\{\underline{\underline{A}}_s \xi^* - \underline{\underline{B}}_s\right\} \tag{63}$$

where k_m is the maximum wavenumber for which the low frequency pressure relationship, Eq. (50) holds. In practice the celerities predicted by Eqs. (61), (62) and (63) are almost identical for all frequencies where the bubble behavior is essentially adiabatic. That is, for [Prosperetti, 1977],

$$\frac{R_b}{\lambda_g} \gg 1 \tag{64}$$

where,

$$\lambda_g = \left(\frac{a_g}{2\omega}\right)^{1/2} \tag{65}$$

As noted previously [Ruggles et al, 1988] two of the eigenvalues of Eq. (55) yield the $c'_{2\phi}$ shown in Fig. 10. Interestingly, the linear dispersion relation, Eq. (43), which results from Eq. (55), also yields $c'_{2\phi}$. This is apparently because, for bubbly air/water flows, viscous effects are negligible for frequencies below the resonance frequency [Prosperetti, 1977].

The 'fast' portion of the original square wave pulse is composed of frequencies above bubble resonance. For perturbations of high frequency (ie, $\omega \gg \omega_r$), the full Rayleigh equation, Eq. (34), yields,

$$R'_b \cong 0 \tag{56}$$

Hence, as shown in Fig. 10, when Eq. (43) is used, the resulting prediction for the liquid sound speed is, c_ℓ. In contrast the dispersion relation corresponding to Eq. (55) always yields $c'_{2\phi}$ irrespective of frequency.

Let us now determine how $c'_{2\phi}$ is related to the eigenvalues of Eq. (55). The eigenvalues (ξ_j) of this nonlinear system of equations can be determined from:

$$\det\left[\underline{\underline{A}}_s \xi - \underline{\underline{B}}_s\right] = 0 \tag{57}$$

where,

$$\text{Re}\left[\xi_j\right] = \left(\frac{dz}{dt}\right)_j \tag{58}$$

The seven roots of Eq. (58) are not frequency dependent. Subtracting $\xi_p^+/2$ from $\xi_p^-/2$ we obtain the speed of the propagating pressure pulse $\left(c'_{2\phi}\right)$. The pressure pulse data was compared to the data of Ruggles et al (1988) for a range of C_{VM}. As can be seen in Fig. 9 the agreement was excellent when C_{VM} values corresponding to Eq. (16) were used.

The Relationship to Critical Flow

Let us now consider the relationship of $c'_{2\phi}$ to the critical flow velocity. The form of the one-dimensional two-fluid model commonly used for critical two-phase flow analysis is identical to Eq. (55). In particular, for steady choked flow, Eq. (55) reduces to,

$$\underline{\underline{B}}_s \frac{d\underline{\psi}}{dz} = \underline{\underline{C}}_s \underline{\psi} \tag{59}$$

A necessary condition for choking is [Bouré et al, 1975]:

WAVE PROPAGATION

$$\frac{p_g(R_b,t) - p_\ell}{\Delta p_{QS}} = \frac{\rho_\ell R_b R_b' \omega^2}{\Delta p_{QS}} \left(\ddot{R}_b^*\right) + \frac{\frac{3}{2}\rho_\ell {R_b'}^2 \omega^2}{\Delta p_{QS}} \left(\ddot{R}_b^*\right)^2 + \frac{4\mu_\ell R_b' \omega}{R_b \Delta p_{QS}} \dot{R}_b^* + 1 \qquad (49)$$

where,

$$\Delta p_{QS} \overset{\Delta}{=} p_g - p_\ell = \frac{2\sigma}{R_b} - \frac{\rho_\ell}{4}\left(<u_g> - <u_\ell>\right)^2 \qquad (50)$$

and Δp_{QS} is the quasi-steady relationship between the phasic pressures.

Equation (50) accurately describes the relationship between the liquid and gas phase pressures if the coefficients of the derivative terms in Eq. (49) are much less than unity. This is true when the following three inequalities are satisfied,

$$R_b' \omega^2 \ll \frac{\Delta p_{QS}}{\rho_\ell R_b} \qquad (51)$$

$$R_b'^2 \omega^2 \ll \frac{\Delta p_{QS}}{\frac{3}{2}\rho_\ell} \qquad (52)$$

and,

$$R_b' \omega^2 \ll \frac{\Delta p_{QS} R_b}{4\mu_\ell} \qquad (53)$$

These inequalities were evaluated for an air/water mixture, at standard temperature and pressure conditions (STP), assuming that the amplitude of the perturbed bubble radius, R_b', was less than 0.1% of the steady-state value, R_{b_0}. Equation (51) gives the most stringent limitation, requiring that:

$$\omega < 0.5 \omega_r \qquad (54)$$

This scaling exercise shows that the quasi-steady pressure relationship of Eq. (50) is sufficient for modeling the propagation of frequencies well below bubble resonance. This indicates that the gas phase pressure follows the liquid phase pressure for low frequencies, and thus low frequency pressure perturbations are strongly effected by gas phase compressibility.

It has been previously shown that the bubble's thermal response is essentially adiabatic, over a wide range of frequencies which satisfy Eq. (54) [Prosperetti, 1977]. Thus when the quasi-steady phasic pressure relationship, Eq. (50), is substituted for Eq. (34) in the two-fluid model we obtain:

$$\underline{\underline{A}}_s \frac{\partial \underline{\psi}}{\partial t} + \underline{\underline{B}}_s \frac{\partial \underline{\psi}}{\partial z} = \underline{\underline{C}}_s \underline{\psi} \qquad (55)$$

where the subscript, s, denotes the matrices appropriate for quasi-static (ie, low frequency) disturbances.

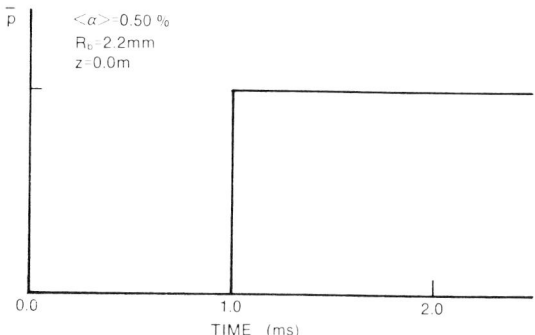

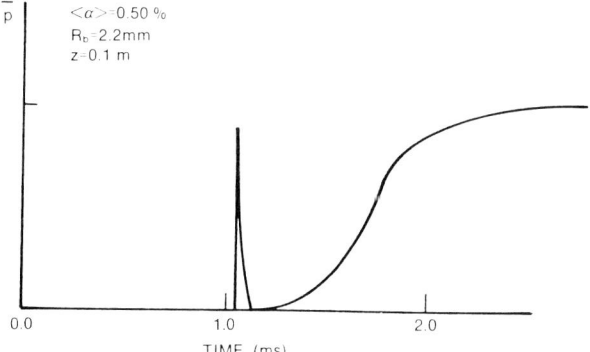

FIGURE 13. Predicted Dispersion of Step Input

The rapid separation of a square pulse into 'slow' and 'fast' portions indicates that the physical behavior that governs sound propagation for frequencies below bubble resonance is quite different from that which governs the propagation of sound for frequencies above bubble resonance.

The physics governing the frequencies below bubble resonance can be examined by returning to the modified Rayleigh equation, Eq. (35). This equation can be scaled by introducing the following non-dimensional variables,

$$\dot{R}_b^* = \frac{\dot{R}_b}{R_b' \omega} \tag{48a}$$

and,

$$\ddot{R}_b^* = \frac{\ddot{R}_b}{R_b' \omega^2} \tag{48b}$$

where R_b' is one of the perturbed amplitude components of the vector, $\underline{\Psi}'$. The generalized Rayleigh equation, Eq. (34), may now be written as,

$$g_p(\omega) = \frac{1}{2\pi} \int_{t=-\infty}^{\infty} \delta p_\ell(0,t) e^{i\omega t} dt \qquad (47b)$$

where the wave number of the pressure perturbation traveling in the negative z direction (ie, in the flow direction), $\bar{k}_p^-$, has been used.

Equations (47) have been employed to predict the propagation speed and dispersion of the pressure perturbation shown in Fig. 8, thus relating the standing wave and propagating pressure perturbations. The time trace of the pressure perturbation (measured at the lower transducer) was used to generate the function $g_p(\omega)$ using a FFT algorithm. The time trace of the pressure pulse at the upper transducer was then constructed through a numerical evaluation of Eq. (47a) with the values for $\bar{k}_p^-(\omega)$ taken from the model's dispersion relation, given by Eq. (43). The results are shown in Fig. 12. It can be seen that the model predictions are in close agreement with the data.

It is also interesting to use the model to predict the dispersion of a hypothetical square pressure pulse. The results are shown in Fig. 13. Notice that the pressure pulse rapidly separates into two distinct pulses, one that is traveling relatively slowly, at the low frequency two-phase speed of sound $\left(c'_{2\phi}\right)$, and one that is traveling much more rapidly, at the speed of sound in liquid (c_ℓ). This is easily understood by examining the predicted values of $c_{2\phi}$ in Fig. 10. All the frequency components of the pulse having values less than bubble resonance (ω_r) are traveling relatively slowly, at the low frequency two-phase speed of sound $\left(c'_{2\phi}\right)$. In contrast, all components for frequencies above bubble resonance are traveling much faster, at the speed of sound in liquid (c_ℓ). Thus, the high frequency components ($\omega > \omega_r$) which comprise the leading edge of a square pulse, rapidly separate from of the low frequency components ($\omega < \omega_r$). All of this is accompanied by strong attenuation of all components of frequency near bubble resonance (see Fig. 11).

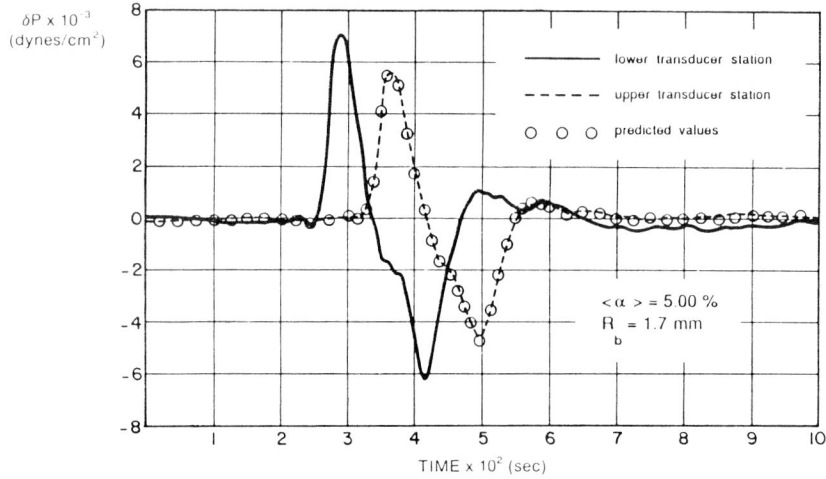

FIGURE 12. Measured and Predicted Pressure Traces

$$\underline{\delta\psi}(z,t) = \int_{\omega=-\infty}^{\infty} \sum_{j=1}^{7} g_j(\omega)\underline{\delta\psi}_j \, d\omega$$

$$= \int_{\omega=-\infty}^{\infty} \left[\sum_{j=1}^{7} g_j(\omega)\underline{\psi}'_j e^{i(k_j(\omega)z)} \right] e^{-i\omega t} d\omega$$

$$= \int_{\omega=-\infty}^{\infty} \left[\sum_{j=1}^{7} \underline{G}_j(\omega) e^{i(k_j(\omega)z)} \right] e^{-i\omega t} d\omega \tag{45a}$$

where,

$$\sum_{j=1}^{7} \underline{G}_j(\omega) = \sum_{j=1}^{7} \left[g_j(\omega)\underline{\psi}'_j \right] = \frac{1}{2\pi} \int_{t=-\infty}^{\infty} \underline{\delta\psi}(0,t) e^{i\omega t} dt \tag{45b}$$

and,

$$\underline{g}_j(\omega) = \frac{1}{2\pi} \int_{t=-\infty}^{\infty} \underline{\delta\psi}_j(0,t) e^{i\omega t} dt \tag{45c}$$

The $\underline{\delta\psi}_j$ in Eq. (45c) is the portion of $\underline{\delta\psi}$ due to the summation of Fourier components of eigenmode, $\underline{\delta\psi}_j$, such that

$$\delta\psi_j = \int_{t=-\infty}^{\infty} g_j(\omega)\underline{\delta\psi}_j \, d\omega \tag{46a}$$

and,

$$\underline{\delta\psi} = \sum_{j=1}^{7} \underline{\delta\psi}_j \tag{46b}$$

Normally, the perturbed pressure, δp_ℓ, is the dominant eigenmode. Equation (45a) can thus be simplified to give the approximate Fourier integral representation of the single state variable p_ℓ in the propagating eigenmode associated with the traveling pressure wave

$$\delta p_\ell(z,t) = 2 \int_{\omega=0}^{\infty} \left[g_p(\omega) e^{i\bar{k}_p z} \right] e^{-i\omega t} d\omega \tag{47a}$$

where,

WAVE PROPAGATION

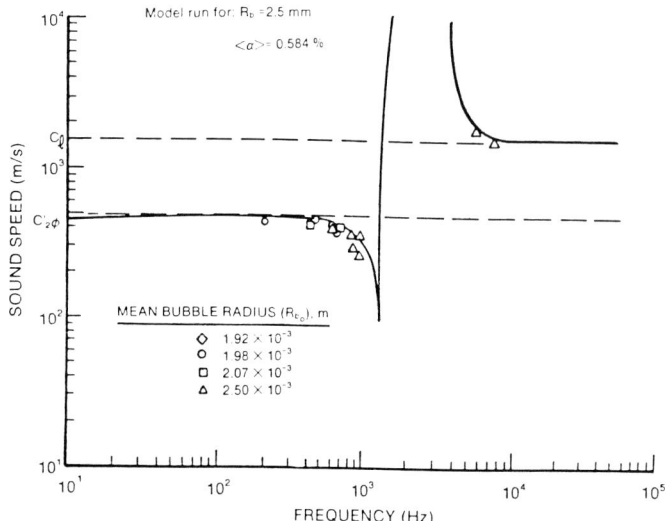

FIGURE 10. Sound Speed vs. Frequency (Data due to Silberman [1957])

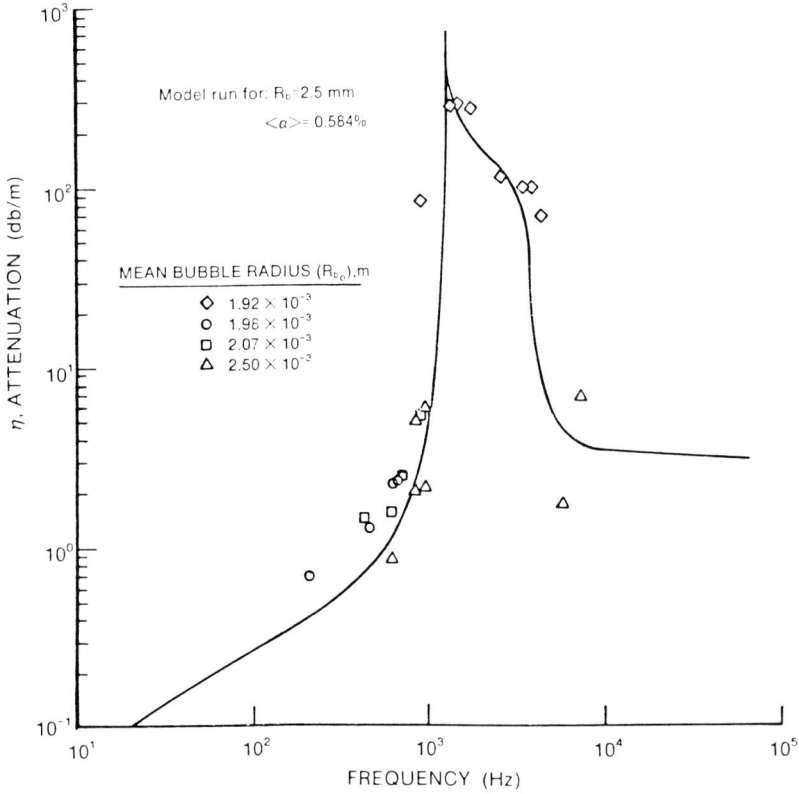

FIGURE 11. Attenuation vs. Frequency (Data due to Silberman [1957])

$$\left\{ \underset{=o}{A}(\underline{\psi})[-i\omega] + \underset{=o}{B}(\underline{\psi})[ik] - \underset{=o}{C'}(\underline{\psi}) \right\} \underline{\psi}' = 0 \quad . \tag{42}$$

Equation (42), in conjunction with the requirement that $\underline{\psi}'$ be finite, implies a *dispersion relationship* of the form,

$$\det\left\{ \left(\underset{=o}{A}(\omega/k) - (i/k)\underset{=o}{C'} \right) - \underset{=o}{B} \right\} = 0 \quad . \tag{43}$$

For standing waves, Eq. (43) gives a relationship between the angular frequency, ω, and wavenumber, k. The wavenumber is in general a complex number, with its real part corresponding to 2π divided by the wavelength, and its imaginary part corresponding to a spatial attenuation coefficient (η).

The dispersion relation normally gives seven roots, (ω/k), for each value of angular frequency, ω. Three of these roots yield celerities typical of the convective velocities of the liquid and vapor phase, $<u_\ell>$ and $<u_g>$. Two other roots are associated with the celerity of void perturbations ($c_\alpha^\pm$). The wavelengths associated with these roots are too short to allow the two-fluid media to be treated as a continuum when frequencies go above about ten to twenty Hertz. Thus these roots are only meaningful for low frequencies.

The remaining two roots travel at celerities typical of the so-called speed of sound in the two-phase medium. One of these acoustic roots is positive (ie, it travels with the flow). The other is negative, and travels against the flow. These roots have velocities that vary in absolute value by an amount associated with a two-phase convective velocity. The difference of these two roots has been used to predict the standing wave previously discussed. That is, the speed of sound was given by,

$$c_{2\phi} = \frac{\left| [\omega/\text{Re}(k)]_p^+ - [\omega/\text{Re}(k)]_p^- \right|}{2} \tag{44}$$

It can be seen in Figs. 1-6 (the RPI data) and 10-11 (Silberman's data [1975]) that this model predicts the standing wave data very well. It is also interesting to note in Fig. 10 that the two-phase speed of sound is essentially nondispersive at both low and high frequencies, but varies dramatically near the resonant frequency. Moreover, it should be noted that $c_{2\phi}$ becomes c_ℓ above the resonant frequency. This is because the inertia of the liquid does not allow the bubbles to respond to high frequency excitation.

Prediction of Propagating Pressure Pulses

Let us now consider the analytical relationship between standing waves and propagating disturbances.

A linear perturbation of the form $\delta\underline{\psi}$ can be represented in terms of the harmonic perturbations given in Eq. (41) through the superposition of Fourier integrals:

WAVE PROPAGATION

$$\dot{R}_b = \frac{\Delta \frac{D_g R_b}{Dt}}{Dt}. \tag{35}$$

A microscopic thermodynamic analysis of a bubble suspended in a liquid media can be used instead of the gas thermal energy equation, Eq. (10). This type of analysis has been done by previous investigators [Prosperetti, 1977; Plesset et al., 1960; Cheng et al., 1983] and results in being able to write the perturbed form of Eq. (34) as a damped harmonic oscillator.

The Dispersion Relation

Equations (8), (9), (10)*, and (35) can be written in matrix form as,

$$\underline{\underline{A}} \frac{\partial \underline{\psi}}{\partial t} + \underline{\underline{B}} \frac{\partial \underline{\psi}}{\partial z} = \underline{\underline{C}} \underline{\psi} \tag{36}$$

where,

$$\underline{\psi} = \left[<\alpha>, p_\ell, <u_g>, <u_\ell>, h_\ell, R_b, \frac{D_g R_b}{Dt} \right]^T \tag{37}$$

Equation (36) can be perturbed as follows,

$$\left[\underline{\underline{A}}_o + \delta\underline{\underline{A}}\right]\left[\frac{\partial \delta\underline{\psi}}{\partial t}\right] + \left[\underline{\underline{B}}_o + \delta\underline{\underline{B}}\right]\left[\frac{\partial \underline{\psi}_o}{\partial z} + \frac{\partial \delta\underline{\psi}}{\partial z}\right] = \left[\underline{\underline{C}}_o + \delta\underline{\underline{C}}\right]\left[\underline{\psi}_o + \delta\underline{\psi}\right] \tag{38}$$

The steady-state equation describing the unperturbed fully developed two-phase flow is given by,

$$\underline{\underline{B}}_o \frac{\partial \underline{\psi}_o}{\partial z} = \underline{\underline{C}}_o \underline{\psi}_o \tag{39}$$

Assuming that the partial derivatives of the steady-state solution are of order δ, the linearized equation set describing the response of the system to small perturbations can be expressed as,

$$\underline{\underline{A}}_o \frac{\partial \delta\underline{\psi}}{\partial t} + \underline{\underline{B}}_o \frac{\partial \delta\underline{\psi}}{\partial z} = \underline{\underline{C}}'_o \delta\underline{\psi} \tag{40}$$

where,

$$\underline{\underline{C}}'_o = \underline{\underline{C}}_o + \underline{\psi}_o \frac{\partial \underline{\underline{C}}}{\partial \underline{\psi}}\bigg|_o$$

The perturbation of the state variables is assumed to be of the form,

$$\delta\underline{\psi} = \underline{\psi}' e^{i(kz-\omega t)} \tag{41}$$

This transforms Eq. (40) into the algebraic equation,

---

*   Only used for liquid phase.

where the velocity potential of a stationary pulsating bubble is given by [Landau & Lifshitz, 1959],

$$\phi(r) = -\frac{R_b^2 \dot{R}_b e^{ik_b(r-R_b)}}{r(1 - ik_b R_b)} \tag{30}$$

where the wave number, k_b, is given by,

$$k_b = \frac{\omega}{c_{2\phi}(\omega)} \tag{31}$$

Other investigators have defined the wavenumber, k_b, as ω/c_ℓ. This assumed the surrounding media transmitted the radiated pressure disturbance at celerity c_ℓ (ie, the speed of sound in liquid). However, if the original sinusoidal excitation of frequency ω is propagating at celerity $c_{2\phi}(\omega)$, then it is more consistent to assume that a radiated pressure disturbance of the same frequency, ω, will also be transmitted through the media at velocity $c_{2\phi}$.

It is significant to note that the radiated acoustic energy is negligible for frequencies below bubble resonance, where $c_{2\phi}$ is much less than c_ℓ, thus the differences in k_b are of no consequence. Moreover, the propagation speed of pressure perturbation, $c_{2\phi}$, is approximately equal to c_ℓ for frequencies above bubble resonance, thus the old and new definitions for k_b are the same. Near bubble resonance, however, the definition of k_b can have a significant effect.

It is also interesting to note the expression for bubble resonance. If an inviscid isentropic analysis is performed one obtains [Reschevkin, 1963],

$$\omega_r = \omega_{r1\phi} \frac{\rho_\ell}{\rho_{2\phi}} \tag{32}$$

where $\omega_{r1\phi}$ is the value for the resonant frequency if only a single bubble were imbedded in a liquid.

If we combine Eqs. (28), (29) and (30) we obtain for the assumption $k_b R_b \ll 1$, the classical Rayleigh equation,

$$p_g(t) - p_\ell = \rho_\ell R_b \ddot{R}_b + \frac{3}{2} \rho_\ell \dot{R}_b^2 + \frac{4\mu_\ell \dot{R}_b}{R_b} + \frac{2\sigma}{R_b} \tag{33}$$

where we have recognized that the average pressure of the liquid phase, p_ℓ, is the appropriate far field liquid measure (p_{ℓ_∞}).

In general, when there is relative motion between the bubble(s) and the liquid, Eqs. (21) and (33) yield,

$$p_g(t) - p_\ell = \rho_\ell \left[R_b \ddot{R}_b + \frac{3}{2} \dot{R}_b^2 + \frac{4\upsilon_\ell \dot{R}_b}{R_b} \right] + \frac{2\sigma}{R_b} - \frac{\rho_\ell}{4} \left(<u_g> - <u_\ell> \right)^2 \tag{34}$$

where,

$$C_1 = \frac{3}{20} \qquad C_2 = \frac{1}{20}. \tag{23}$$

It should be noted that for one-dimensional pipe flow, Eqs. (22)-(23) reduce to:

$$\tau^T_{zz_\ell} = -\frac{1}{5} <\alpha> \rho_\ell \left(<u_g> - <u_\ell>\right)^2 \tag{24}$$

Hence, the Reynolds stress gradient term in Eq. (9) can be written as,

$$\frac{\partial}{\partial z}\left[(1-<\alpha>) \tau^T_{zz_\ell}\right] = -\frac{\partial}{\partial z}\left[\frac{1}{5}<\alpha>(1-<\alpha>)\rho_\ell\left(<u_g>-<u_\ell>\right)^2\right] \tag{25}$$

This is basically the same expression as developed by Biesheuvel and van Wijngaarden [1984] and used by Pauchon and Banerjee [1988].

The interfacial shear, τ_{k_i}, is often set to zero. However, one may also assume that it is equal to $\tau^T_{zz_k}$. When this is done the third and fourth terms on the right hand size of Eq. (9) combine to yield,

$$-\tau^T_{zz_k} \frac{\partial <\alpha_k>}{\partial z} + \frac{\partial}{\partial z}\left[<\alpha_k> \tau^T_{zz_k}\right] = <\alpha_k> \frac{\partial}{\partial z}\left[\tau^T_{zz_k}\right] \tag{26}$$

In this study, all results concerning sonic velocity were obtained using $\tau_{k_i} = 0.0$. This was done since, unlike for void waves, the sonic velocity is insensitive to the model used for τ_{k_i} and interfacial drag (C_D).

For quasi-static conditions we note from Eq. (21), and the well known Laplace equation, that:

$$p_g - p_\ell = 2\sigma/R_b - \frac{\rho_\ell}{4}\left(<u_g> - <u_\ell>\right)^2 \tag{27}$$

Let us next develop a dynamic relationship between the variables p_g and p_ℓ. This can be done in a manner similar to that of Prosperetti [1977] and Cheng et al [1985]. To accomplish this, a single bubble is considered to be surrounded by an infinite liquid medium, and excited by sinusoidal pressure oscillations. The bubble response is assumed to be spherically symmetric and without translatory oscillations. These assumptions are valid for pressure excitations having wavelengths, λ, which are much greater than the bubble radius, R_b.

Continuity of normal stress at the bubble surface gives,

$$p_g(R_b, t) - p_\ell(R_b, t) = \frac{2\sigma}{R_b} - 2\mu_\ell \frac{\partial u_\ell}{\partial r}\bigg|_{r=R_b} \tag{28}$$

The Bernoulli equation for the liquid phase can be written as,

$$\frac{p_\ell(R_b, t)}{\rho_\ell} + \frac{1}{2}\left[\nabla\phi(R_b)\right]^2 + \frac{\partial \phi(R_b)}{\partial t} = \frac{p_\ell(t)}{\rho_\ell} \tag{29}$$

$$a_{VM} = \left[\frac{\partial <u_g>}{\partial t} + <u>_g \frac{\partial u_g}{\partial z}\right] - \left[\frac{\partial <u_\ell>}{\partial z} + <u_\ell> \frac{\partial u_\ell}{\partial z}\right]$$

or, equivalently,

$$a_{VM} = \left[\frac{D_g <u_g>}{Dt} - \frac{D_\ell <u_\ell>}{Dt}\right] \tag{17}$$

The axial reaction force due to bubble pulsation, F_R, results from the interaction of a spherical bubble with the flow field around the bubble. This flow field is due to both bubble translation relative to the liquid phase and to radial bubble pulsations. This force is given by [Cheng et al, 1985],

$$F_R = \frac{3}{R_b} C_{VM} \rho_\ell \left(<u_g> - <u_\ell>\right) \frac{D_g R_b}{Dt} \tag{18}$$

Since the gas phase is assumed to be dispersed within the liquid phase, the wall shear stress on the gas phase is,

$$\tau_{g_w} = 0 \tag{19a}$$

and the liquid phase wall shear stress (when walls are present) is,

$$\tau_{\ell_w} = \frac{1}{2} \frac{f}{D_H} \rho_\ell <u_\ell> \left|<u_\ell>\right| \tag{19b}$$

For low pressure air/water flows, the interfacial pressure in the gas phase is related to the average pressure of the gas phase by,

$$\Delta p_{gi} = p_{gi} - p_g \cong 0 \tag{20}$$

This is normally a very good assumption and implies that one is dealing with situations in which the bubble has essentially a uniform internal pressure.

The difference between the interfacial pressure and the mean pressure in the liquid phase is, for a non-pulsating bubble, given by [Stuhmiller, 1977],

$$p_{\ell i} - p_\ell = \Delta p_{\ell i} = -\frac{\rho_\ell}{4}\left(<u_g> - <u_\ell>\right)^2 \tag{21}$$

The Reynolds stress ($\underline{\underline{\tau}}_k^T$) is negligible in the gas phase but not in the liquid phase. It is beyond the state-of-the-art to model this term in general, however, the Reynolds stress in the liquid phase due to "bubble-induced" turbulence has been given by Nigmatulin (1979) as,

$$\underline{\underline{\tau}}_\ell^T = -\alpha \rho_\ell \left[C_1 \left|<\underline{u}_g> - <\underline{u}_\ell>\right|^2 \underline{\underline{I}} + C_2 \left(<\underline{u}_g> - <\underline{u}_\ell>\right)\left(<\underline{u}_g> - <\underline{u}_\ell>\right)\right] \tag{22}$$

It can be shown [Biesheuvel & van Wijngaarden, 1984] that the coefficients C_1 and C_2 are given by,

WAVE PROPAGATION

where the subscript k denotes either the gas (k=g) or the liquid (k=ℓ) phase.

In this model it is assumed that both phases are compressible and that, $\rho_k = \rho_k(h_k, p_k)$.

Appropriate constitutive equations must be used to model the interaction between the flow components. The momentum transfer between the gas and liquid phases can be written as the sum of three forces,

$$M_\ell = -M_g = <\alpha>\left[F_D + F_{VM} + F_R\right] \tag{11}$$

where,

$$<\alpha> \stackrel{\Delta}{=} <\alpha_g>.$$

The interfacial drag force, F_D, for bubbly flow can be modeled as,

$$F_D = \frac{3}{8}\rho_\ell \frac{C_D}{R_b}\left(<u_g> - <u_\ell>\right)\left|<u_g> - <u_\ell>\right|$$

The drag coefficient, C_D, for distorted bubbly flows is given by [Harmathy, 1960] as:

$$C_D = \frac{4}{3}R_b\left[\frac{g(\rho_\ell - \rho_g)}{g_c \sigma(1-<\alpha>)}\right]^{1/2} \tag{13}$$

where R_b is expressed in meters. In contrast, for some of the other bubble sizes used in this study, the Newton's regime drag law [Ishii & Zuber, 1979] is given by,

$$C_D = 0.45\left[\frac{1+17.67[f(<\alpha>)]^{6/7}}{18.67f(<\alpha>)}\right]^2 \tag{14a}$$

where,

$$f(<\alpha>) = (1-<\alpha>)^{1/2}\left(1-<\alpha>/<\alpha_M>\right)^{2.5\alpha_M\left(\mu_g+0.4\mu_\ell\right)/\left(\mu_g+\mu_\ell\right)} \tag{14b}$$

and, the maximum packing fraction, $<\alpha_M>$, can be assumed to be unity. The virtual mass force, F_{VM}, is given by,

$$F_{VM} = \rho_\ell C_{VM} a_{VM} \tag{15}$$

The virtual volume coefficient, C_{VM}, can be expressed as an empirical function of the global void fraction [Ruggles et al, 1988] as,

$$C_{VM} = 0.5\left[1 + 12<\alpha>^2\right] \qquad [(<\alpha>) \leq 20\%] \tag{16}$$

and the virtual mass acceleration, a_{VM}, is given by,

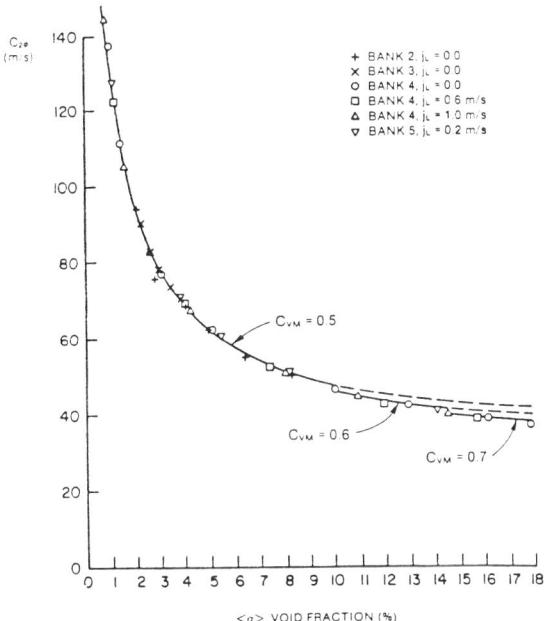

FIGURE 9. Pressure Pulse Propagation Speed vs. Global Void Fraction

DISCUSSION OF ANALYSIS

The space-time averaged one-dimensional two-fluid conservation equations for adiabatic air/water flow in a constant area duct are given by [Ishii, 1975]:

CONSERVATION OF MASS

$$\frac{\partial}{\partial t}\left(<\alpha_k>\rho_k\right) + \frac{\partial}{\partial z}\left(<\alpha_k>\rho_k<u_k>\right) = 0 \tag{8}$$

CONSERVATION OF MOMENTUM

$$\frac{\partial}{\partial t}\left(<\alpha_k>\rho_k<u_k>\right) + \frac{\partial}{\partial z}\left(<\alpha_k>\rho_k<u_k>^2\right) = -<\alpha_k>\frac{\partial p_k}{\partial z} + \Delta p_{ki}\frac{\partial<\alpha_k>}{\partial z}$$

$$-\tau_{k_i}\frac{\partial<\alpha_k>}{\partial z} + \frac{\partial}{\partial z}\left[\alpha_k \tau^T_{zz_k}\right] - <\alpha_k>\rho_k g_k \cos\theta + M_k - \tau_{k_w}/D_H \tag{9}$$

CONSERVATION OF ENERGY

$$\frac{\partial}{\partial t}\left(<\alpha_k>\rho_k h_k\right) + \frac{\partial}{\partial z}\left(<\alpha_k>\rho_k<u_k>h_k\right) = <\alpha_k>\left(\frac{\partial p_k}{\partial t} + <u_k>\frac{\partial p_k}{\partial z}\right)$$

$$-<u_k>\tau_{k_w} + \frac{q''_{ki}}{L_s} \tag{10}$$

WAVE PROPAGATION

$$\delta\underline{\psi}(t,z) = \delta\underline{\psi}\left(t - \frac{z}{c_{2\phi}}\right) \tag{4}$$

If we Fourier transform Eq. (4) to the frequency domain we obtain:

$$\delta\underline{\psi}(\omega,z) = e^{i\left[\frac{\omega z}{c_{2\phi}} + \theta\right]} |\delta\underline{\psi}(\omega)| \tag{5}$$

where θ is the phase angle of frequency component ω. The CPSD may be written as,

$$\text{CPSD}(\omega) = \delta\underline{\psi}(\omega,z_1)\delta\underline{\psi}^*(\omega,z_2) = e^{-\frac{i\omega}{c_{2\phi}}(z_2 - z_1)} |\delta\underline{\psi}(\omega)|^2, \quad z_1 \leq z_2 \tag{6}$$

This indicates the phase of the CPSD of a propagating perturbation is locally linear with frequency, thus the slope $\left(\frac{d\phi}{d\omega}\right)$ of the CPSD phase angle (ϕ) versus frequency plot yields,

$$\frac{d\phi}{d\omega} = \frac{(z_2 - z_1)}{c_{2\phi}}$$

In Ruggles data [1987], $z_2 = z_{lower}$ and $z_1 = z_{upper}$, thus we have:

$$c_{2\phi} = \left[(z_{lower} - z_{upper})360°\right]\left(\frac{d\theta}{df}\right)^{-1} \tag{7}$$

The CPSD's were calculated using a Hewlett-Packard 3562A dynamic signal analyzer. This technique was a useful independent verification that the time-of-flight measurements accurately indicated the pressure pulse propagation speed.

The pressure pulse propagation data taken in this study are presented in Fig. 9.

DISCUSSION OF EXPERIMENTAL RESULTS

The standing wave dispersion data given in Fig. 2 clearly shows the strong dependence of the propagation speed on the interfacial heat transfer between the two phases. The larger bubbles have less interfacial area available for heat transfer and thus exhibit a more-nearly adiabatic process, and thus a greater celerity. It can also be noted that the celerity also increases with frequency. More time exists for heat transfer at the lower frequencies thus promoting a more-nearly isothermal process and a lower celerity.

The pulse data given in Fig. 9 clearly shows the strong dependence of two-phase pulse propagation speed with global void fraction in the region of low void fraction. The pulse propagation speed data given here is consistent with, but more tightly controlled than, those of previous researchers [eg, Hall, 1971].

Let us now consider the analysis of wave propagation phenomena. The model used will be a one-dimensional two-fluid model of the two-phase flow.

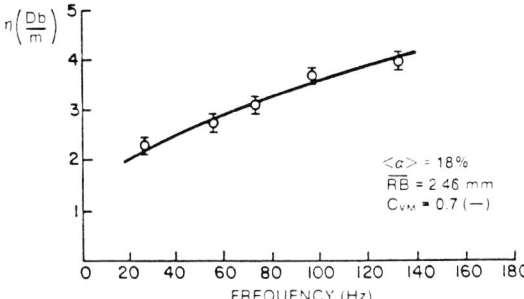

FIGURE 7. Attenuation vs. Frequency

show that the spatial attenuation coefficient, η, is a strong function of frequency.

The apparatus used to take the pressure pulse propagation speed measurements was identical to that in Fig. 1. The pressure pulses were introduced by driving the electromechanical shaker with a square-wave generator. For these experiments, the upper end of the waveguide was connected to a phase separation tank and the lower plenum was fitted with a metered water flow to facilitate variation of the superficial liquid velocity, $<j_\ell>$.

A typical pressure pulse is given in Fig. 8 as it appeared at the lower transducer station of Fig. 1. It was noted that the pressure pulses exhibited some attenuation as they passed the upper transducer stations. However, no significant distortion of the leading edge of the pulse was observed, thus time-of-flight measurement techniques are appropriate.

Small amplitude pressure pulse speed measurements were made using two independent techniques. The first of these was a time-of-flight method. This technique used a Tektronix 7854 digital oscilloscope with the peak positive pressure chosen as the discrete time feature of each pulse. The second pressure pulse speed measurement technique involved taking the slope of the phase angle vs. frequency plot of the cross power spectral density (CPSD) function [Bendat et al, 1971] between a lower and upper side-mounted transducer set. This slope can be related to the propagation speed of a pulse by noting that,

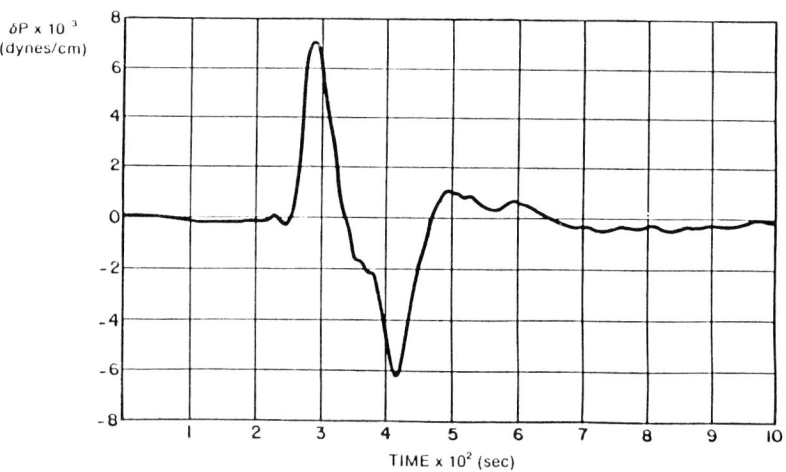

FIGURE 8. Pressure Pulse (Typical)

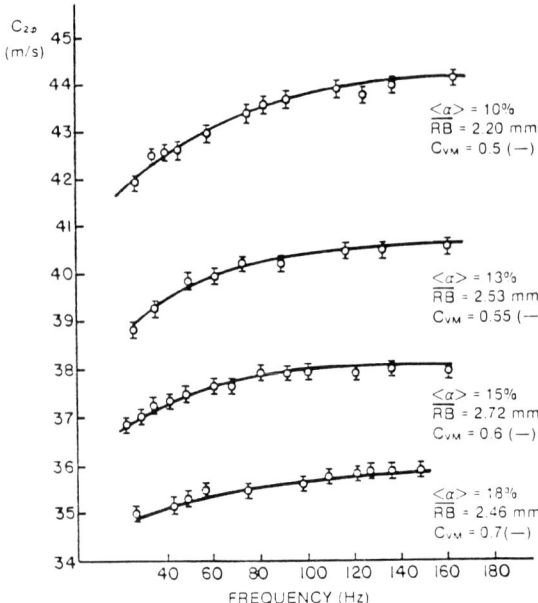

FIGURE 4. Propagation Speed vs. Frequency

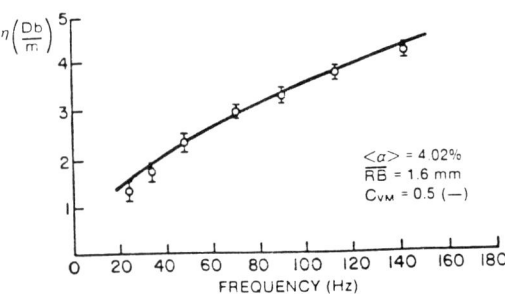

FIGURE 5. Attenuation vs. Frequency

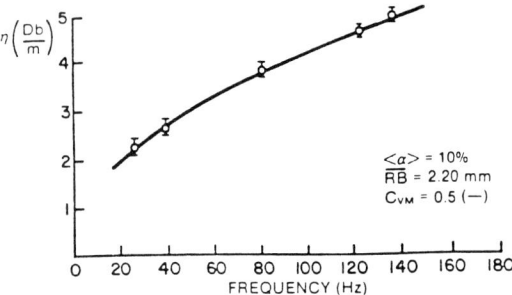

FIGURE 6. Attenuation vs. Frequency

number of half wavelengths was then measured using the traversing hydrophone and the propagation speed was calculated using Eq. (1). These redundant measurement techniques were used to assure the accuracy of the data and to verify that the presence of the traversing hydrophone did not affect the data.

An error analysis was performed [Ruggles, 1987] using "propagation of error" techniques. The measurement uncertainties are given along with representative standing wave data in Figs. 2-7. It can be seen in Fig. 2 that the sonic speed ($c_{2\phi}$) is a strong function of both frequency and bubble radius. In addition, Figs. 3 & 4 show that the sonic speed decreases markedly as void fraction increases. In addition, Figs. 5-7

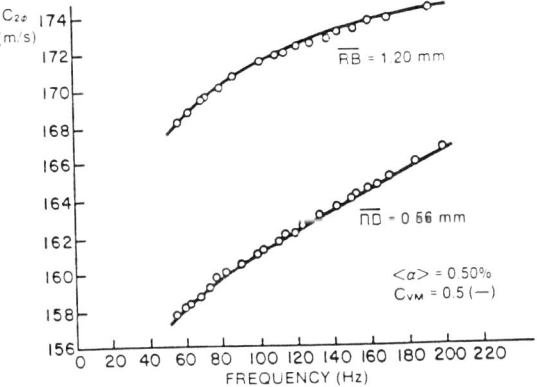

FIGURE 2. Propagation Speed vs. Frequency

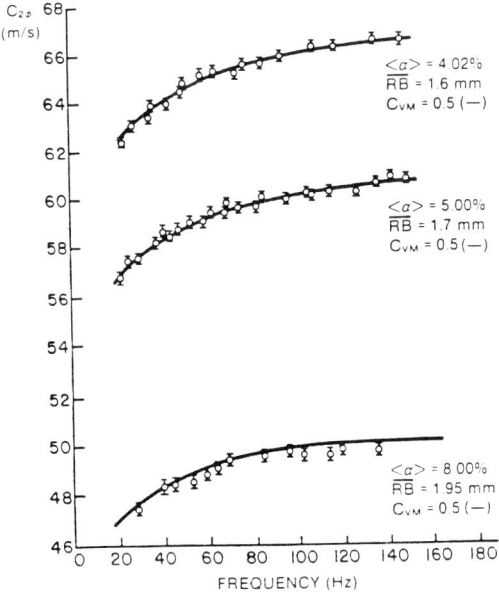

FIGURE 3. Propagation Speed vs. Frequency

WAVE PROPAGATION

High speed photographs of the flow in the acrylic section were also taken to allow the bubble size and distribution to be determined. These observations accounted for any bubble coalescence that occurred at higher void fractions.

It is well known that, at low global void fractions, sonic velocity is a strong function of void fraction. This is especially true for global void fraction less than 3%. Thus accurate measuring techniques were needed for void fraction. For the low, or no, liquid flow runs, one of three manometers was used to measure the variation in hydrostatic pressure due to void fraction. The first of these manometers measured global void fractions, $<\alpha>$, from 0% to 1%, the second measured $<\alpha>$ from 1% to 8%, and the third measured $<\alpha>$ from 8% to 20%. This manometer bank arrangement provided sufficient range and accuracy in the measurement of global void fraction.

In the standing wave experiments the global void fraction was varied from 0.5% to 18% and the bubble radii, R_b, varied from 0.5 mm to 2.5 mm. The forcing frequency was varied from 20 Hz to 200 Hz to allow measurement of the dispersion and attenuation curves for each flow situation. The propagation speed and attenuation of the pressure perturbations associated with standing waves was measuring using three independent techniques. In the first of these techniques a hydrophone was traversed through the waveguide and the locations and amplitudes of the pressure nodes and anti-nodes were recorded. The propagation speed ($c_{2\phi}$) and attenuation (η) could then be calculated, since the distance between the nodes (L_N) is half of the wavelengths, λ. That is [Cheng et al, 1983]:

$$c_{2\phi} = \lambda f = 2 L_N f \qquad (1)$$

and,

$$\eta = \left(\frac{4\pi}{\lambda}\right) \sinh^{-1} \frac{\left|\delta p_{min}(n+1)\right| - \left|\delta p_{min}(n)\right|}{2\left|\delta p_{max}\left(n+\frac{1}{2}\right)\right|} \qquad (2)$$

where, n is the node number, counting from the top of the waveguide.

The frequency, f, was imposed, and was independently verified using a Tektronix 7854 digital oscilloscope with a waveform calculator.

The second measurement technique used the root mean square pressure amplitude readings, δp_{RMS}, from three side mounted transducers to infer the wavelength and attenuation of the standing wave. That is [Cheng et al, 1983]:

$$\frac{\delta p_{RMS}}{\left|A^+\right|} = 2 e^{-\eta \ell} \left[\cosh^2(\eta \ell) - \cosh^2\left(\frac{2\pi}{\lambda}\ell\right)\right]^{1/2}, \qquad (3)$$

where in Fig. 1 the distance from the free surface is

$\ell = L - z$

and A^+ is the amplitude of the upward traveling wave at $z = L$.

The third measurement technique involved varying the angular frequency (ω) until a pressure node was situated over one of the side-mounted transducers. This indicated an integer number of half wavelengths existed between that transducer and the bubbly air/water interface. The

DISCUSSION OF EXPERIMENTAL PROGRAM

Let us begin by discussing the experimental findings of Ruggles [1987]. The apparatus he used for standing wave and pressure pulse measurements is shown in Fig. 1. The waveguide used to generate the standing wave pattern was constructed from a 63.5 mm ID, 76.2 mm OD, stainless steel tube two meters in length. This tube was fitted with three side mounted pressure transducers and a hydrophone mounted on a traversing mechanism. Sinusoidal pressure oscillations were introduced through a wye, using an electromechanical shaker and piston arrangement. An isolation system prevented sound energy from the shaker and piston from entering the waveguide walls and disturbing the side mounted transducers. In addition, an air cushion isolated the entire wave guide and lower plenum from laboratory floor vibrations.

All standing wave data were taken for conditions of no liquid flow, while some of the data for the pulse propagation experiments involved finite liquid flows. Air bubbles were introduced into the lower plenum using one of four banks of hypodermic needles. The bubble radius produced by each needle was inferred from the measurement of the volumetric flow rate through the individual needles and measurement of the corresponding bubble departure frequency. Bubble radii were also measured directly using high speed strobe photographic techniques. The global void fraction ($<\alpha>$) was normally determined from the pressure drop. However, some experiments were performed in which the pressure drop was not just hydrostatic. For these runs a special optically clear acrylic section was used to replace the waveguide in the neighborhood of side-mounted transducer-B. Some of the pulse propagation runs, to be described shortly, had a finite $<j_\ell>$ and friction pressure drop, thus the global void fraction was measured by simultaneously closing the two quick-closing valves in the acrylic section. This avoided errors in trying to infer the global void fraction from measurements of the hydrostatic head.

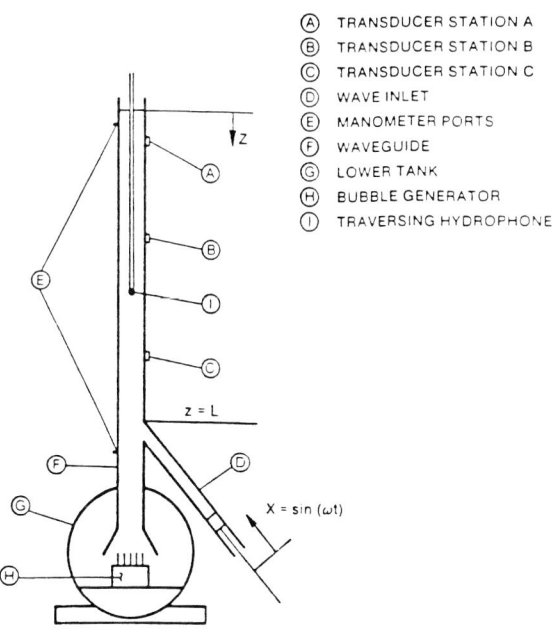

FIGURE 1. Measurement system for sound progagation in a bubbly air-water mixture.

AN ANALYSIS OF WAVE PROPAGATION PHENOMENA IN TWO-PHASE FLOW

R.T. LAHEY, JR.
Rensselaer Polytechnic Institute
Troy, New York 12180-3590

ABSTRACT

This paper reviews the state-of-the-art in our understanding of wave propagation phenomena in two-phase flows. In particular, both pressure and void perturbation waves are discussed. It is shown that the analysis of wave propagation phenomena is an excellent means of two-fluid model assessment.

INTRODUCTION

A good physical understanding of transient phenomena in two-phase flow is important if one is to properly model sonic and void wave propagation using two-fluid models. Moreover, since propagating perturbations cause relatively large local acceleration of the fluid, wave propagation data can be used to assess some postulated interfacial momentum transfer laws. Indeed, it will be shown in this review paper that the combination of pressure and void wave data are an excellent means of assessing two-fluid models.

Let us begin by considering propagating pressure perturbations. Previous investigators have taken data on the speed of sound in a two-phase mixture ($c_{2\phi}$) using both standing wave and pulse propagation (ie, time-of-flight) techniques [eg, Carstensen & Foldy, 1947; Silberman, 1957; Hall, 1971; Nishihara & Michiyoshi, 1981]. In these studies both side mounted and traversing pressure transducers (ie, hydrophones) were used for standing wave experiments. Unfortunately no one has taken data using both of these techniques, thus it was not possible to determine what bias may have occurred. For example, what effect the acoustic impedance mismatch associated with an intrusive probe (ie, a traversing hydrophone) might have on the data. Moreover, no one has taken both standing wave and pulse propagation data in the same facility and the analytical linkage between the two-phase speed of sound measured in both ways was not previously demonstrated.

The purpose of this paper is to address these shortcomings. Moreover, new data which was taken in a special air/water waveguide will be presented. This test facility was constructed such that both standing wave and pressure pulse propagation data could be taken using several different intrusive and nonintrusive measuring techniques. In addition, two-fluid analytical models were developed and assessed against these data and other data in the literature. In particular, both pressure perturbation and void wave celerities were predicted with this model. These data and analyses have given new insight into wave propagation phenomena in two-phase flow and into the proper formulation of two-fluid models.

Subscripts
G: gas
L: liquid
0: steady flow

Superscripts
$^-$: phasic average, Laplace transform
$\sim$: Fourier transform
', *: perturbation

BIBLIOGRAPHY

Bouré, J.A., 1987, Properties and Modeling of Kinematic and Pressure Wave in Two Phase Flow, Int. Seminar on Transient Phenomena in Multiphase Flow, Dubrvnik.

Drew, D.A. and Lahey, R.T., 1979, Application of General Constitutive Principles to the Derivation of Multidimensional Two-Phase Flow Equations, Int. J. Multiphase Flow, Vol. 5, pp. 243-264.

Drew, D.A., 1983, Mathematical Modeling of Two-Phase Flow, Ann. Rev. Fluid Mech., Vol. 15, pp. 261-291.

Liu, J.T.C., 1982, Note on a Wave-Hierarchy Interpretation of Fluidized Bed Instabilities, Proc. R. Soc. London, A380, pp. 229-239.

Monji, H., Onisawa, M., Watanabe, T. and Morioka, S., 1986, Visual Simulation Experiment of Two-Phase Flow in Riser for Heavy Liquid Metal MHD Generation Loop, Proc. 9th Int. Conf. MHD EPG, Tsukuba, pp. 733-742.

Monji, H. and Onisawa, M., 1987, Observation of Wave-Hierarchy Instability in Bubbly Liquid Flow in Decompressed Riser, J. Phys. Soc. Jpn., Vol. 56, pp. 1909-1910.

Morioka, S. and Nakajima, T., 1987, Modeling of Gas and Solid Particles Two-Phase Flow and Application to Fluidized Bed, J. Méc. Théo. Appl., Vol. 6, pp. 77-88.

Murray, J.D., 1965, On the Mathematics of Fluidization, J. Fluid Mech., Vol. 21, pp. 465-493.

Toma, T. and Morioka, S., 1986, Acoustic Waves Forced in Flowing Bubbly Liquid, J. Phys. Soc. Jpn., Vol. 55, pp. 512-520.

van Wijngaarden, L. and Biesheuvel, A., 1987, Voidage Waves in Bubbly Flows, Int. Seminar on Transient Phenomena in Multiphase Flow, Dubrvnik.

Whitham, G.B., 1974, Linear and Nonlinear Waves, Wily-Interscience, pp. 339-359.

Thus, the lower-order wave plays the role almost over the riser. The smallness of the diffusion coefficient means that the transition from bubble flow to slug flow is not so fast. In fact, the diffusion distance for the time t may be estimated as $(4Dt)^{1/2}$, on the basis of the characteristic solution of the diffusion equation. The diffusion distance 0.01 m for 9 seconds, for which the bubbles move through 2 m length of riser with their velocity 0.22 m/s, seems to be sufficient for bubbles to gather and coalesce.

CONCLUSIONS

It has been shown that the transition from bubble flow to slug flow observed in a decompressed vertical riser can be interpreted by the instability of voidage wave on the basis of a dilute bubbly liquid model. The instability has been characterized by the negative diffusion of the voidage, that is by the concentration of the denser and larger bubbles. The experimental result has shown that the diffusion distance estimated from the diffusion coefficient and the time up to the coalescence reasonably agrees with the initial inter-bubble distance, together with the concentrating process of bubbles.

NOMENCLATURE

c_1, c_2: propagation velocities of higher-order waves
c_0: propagation velocity of lower-order wave
c_r: real part of c_1 and c_2
c_i: imaginary part of c_1
g: gravitational acceleration
k: wave number
p: pressure
s: variable in Laplace transform
s_1, s_2: poles
t: time
u: flow velocity of water
v: velocity of bubbles
x: length along riser
x_i: spatial coordinates
C_D: drag coefficient
D: diffusion coefficient
f(): initial profile of voidage perturbation
L: length of riser
Res(): residue at pole
V: volume of a bubble

α: void fraction
β: reciprocal of square Froude number
η: nondimensional time
μ: reciprocal of velocity slip ($=u_0/v_0$)
ξ: nondimensional length along riser
ρ: density
τ: characteristic time
ω: angular frequency
Ω: nondimensional drag coefficient

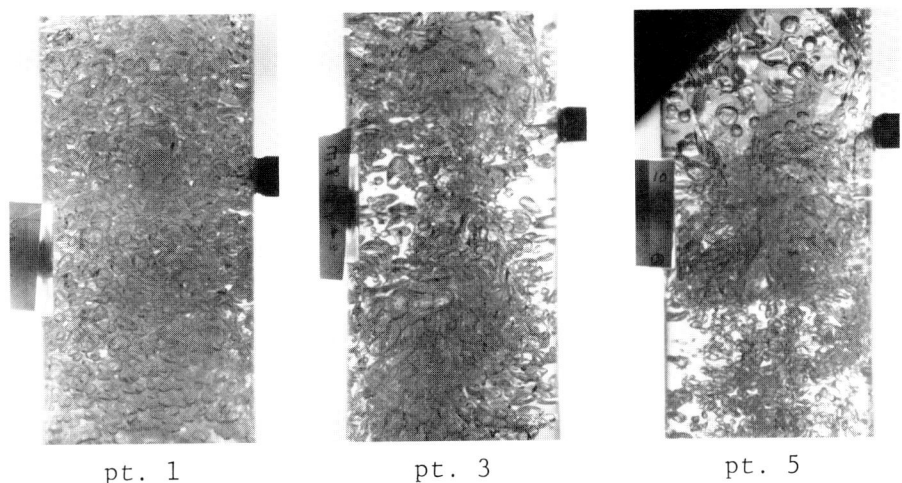

pt. 1 pt. 3 pt. 5

FIGURE 4. Photographs of flow pattern: Inlet void fraction 0.27.

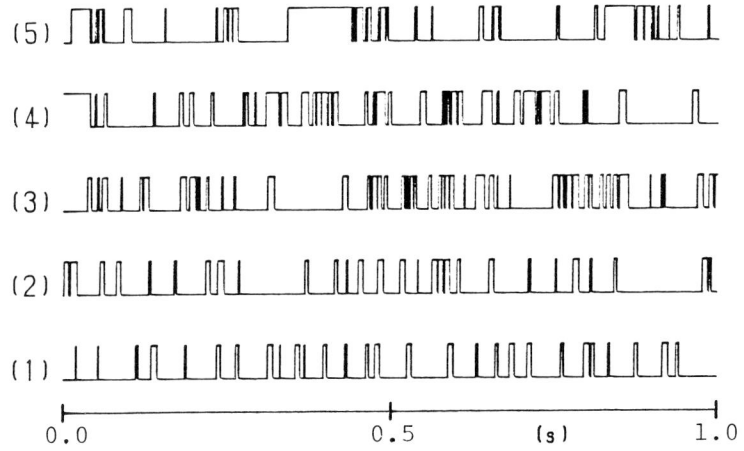

FIGURE 5. Process signal delivered by probes.

$$c_0 = 0.19 \text{ m/s},$$
$$c_r = 0.20 \text{ m/s}, \tag{38}$$
$$c_i = 0.022 \text{ m/s},$$

according to Eq. (14), Eq. (16) and Eq. (17), respectively. On the other hand, the negative diffusion coefficient $-D$ and the transition distance $c_r \tau$ are

$$-D = -3.46 \times 10^{-6} \text{ m}^2/\text{s},$$
$$c_r \tau = 0.0012 \text{ m}, \tag{39}$$

according to Eq. (21) and Eq. (22), respectively.

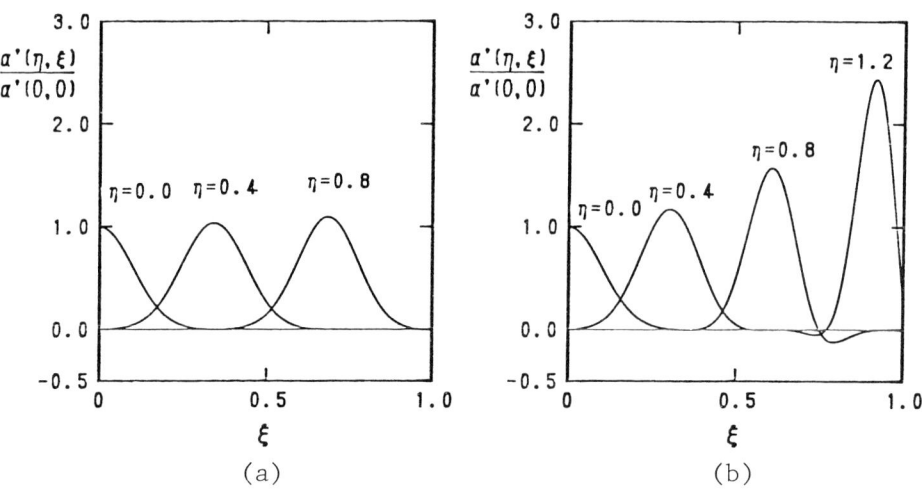

FIGURE 2. Propagation of voidage perturbation: $u_0 = 0.16$ m/s, $\alpha_0 = 0.27$, (a) $v_0 = 0.22$ m/s and (b) $v_0 = 0.32$ m/s.

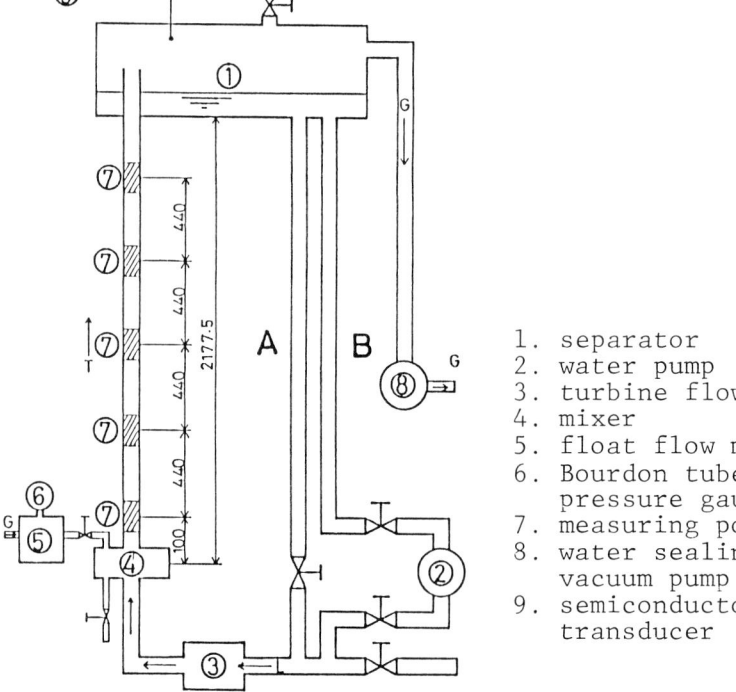

1. separator
2. water pump
3. turbine flow meter
4. mixer
5. float flow meter
6. Bourdon tube pressure gauge
7. measuring points
8. water sealing vacuum pump
9. semiconductor pressure transducer

FIGURE 3. Experimental facility

that is

$$(s+ikc_1)(s+ikc_2) + \sigma(s+ikc_0) = 0 . \tag{36}$$

If the form Eq. (29) for $f(\xi)$ is assumed, then one obtains

$$\tilde{f}(k) = (\pi/p)^{\frac{1}{2}}\exp(-k^2/4p) . \tag{37}$$

Thus, Eq. (31) can be reduced to a real integral with respect to k ($0 \leq k < \infty$) and it can be evaluated for given values of η and ξ by using a personal computer. Figure 2 shows the results of the numerical integration. The step and the range of the integration have been taken as $\Delta k = 0.01$ and $0 \leq k \leq 50$, respectively. The perturbation profile given by Eq. (29) at $\eta = 0$ moves downstream and is deformed steeper and steeper with increasing η. Apparently this result demonstrates the fact that the bubbles are gathered by the instability of voidage wave. By comparing Figs. 2(a) and (b), one notes that the gathering of bubbles may be promoted with increasing velocity slip in the main stream.

COMPARISON WITH EXPERIMENTAL RESULT

The detail of the experimental facility, based on a natural circulation loop of water with injected air bubbles, has been described in references by Monji et al. (1986) and Monji & Onisawa (1987) (Fig. 3). The riser was of about 2 m length and 70 mm diameter. The pressure was reduced below 10 kPa at the exit of riser, with the intention of simulating a loop of heavy liquid-metal at normal pressure.

The photographs (Fig. 4) show the flow patterns at three measuring points downstream of the mixer. At point 1, the bubbles were of almost same size and distributed uniformly. At point 3, the dense and dilute portions appeared in the distribution of bubbles and the coalescence of bubbles began to occur. The coalescence of bubbles increased more downstream from point 3, and the flow pattern changed into slug flow at point 5.

Measurement was made first on the pressure at each measuring point by semiconductor pressure transducers, and the linear decrease of the pressure in the upward direction was checked. Second, the void fraction, the velocity of bubbles and the chord length of bubbles was measured by double resistivity probes. Figure 5 shows the signal profiles delivered by the probe at each measuring points. The distribution of bubble chord length at each measuring point was almost homogeneous at point 1, but dispersed distinctly at the subsequent point.

The experimental conditions at the inlet of riser were the flow velocity of water 0.16 m/s, the velocity of bubbles 0.22 m/s and the void fraction 0.27. Thus, the propagation velocity c_0 of the lower-order wave and the real part c_r and the imaginaly part c_i in the propagation velocities of the higher-order waves were estimated as

where $u_0/v_0 = \mu$, $gL/v_0^2 = \beta$ and $C_D L/\rho_L V v_0 = \Omega$.

The initial and boundary conditions are assumed as follows,

$$\alpha' = f(\xi), \quad u^* = v^* = 0 \text{ at } \eta = 0, \tag{27}$$

$$\alpha' = u^* = v^* = 0 \text{ as } \xi \to \pm\infty, \tag{28}$$

where the function $f(\xi)$ denotes the initial profile of the voidage perturbation, for instance, it is assumed as

$$f(\xi) = \exp(-p\xi^2). \tag{29}$$

The Laplace transform with η and the Fourier transform with ξ are taken, for instance,

$$\bar{\alpha}' = \int_0^\infty \alpha' \exp(-s\eta) d\eta, \tag{30}$$

$$\tilde{\bar{\alpha}}' = \int_{-\infty}^\infty \bar{\alpha}' \exp(-ik\xi) d\xi, \tag{31}$$

taking account of the conditions Eq. (27) and Eq. (28). As a result three algebraic equations for $\tilde{\bar{\alpha}}'$, $\tilde{\bar{u}}^*$ and $\tilde{\bar{v}}^*$ are obtained. Eliminating $\tilde{\bar{u}}^*$ and $\tilde{\bar{v}}^*$ and solving for $\tilde{\bar{\alpha}}'$ results in

$$\tilde{\bar{\alpha}}' = \frac{(s+ikF+\sigma)\tilde{f}(k)}{(s+ikc_1)(s+ikc_2) + \sigma(s+ikc_0)} \tag{32}$$

where

$$F = \frac{1+2\alpha_0(1-\alpha_0)\mu}{1+2\alpha_0(1-\alpha_0)}, \quad \sigma = \frac{1}{\tau}, \tag{33}$$

and $\tilde{f}(k)$ is the Fourier transform of $f(\xi)$, and c_0, c_1, c_2 and τ have been presented by Eq. (14) to Eq. (18).

The inversion of Eq. (32) can be expressed as

$$\alpha' = \frac{1}{2\pi}\int_{-\infty}^\infty \exp(ik\xi) [\frac{1}{2\pi i}\int_{\delta-i\infty}^{\delta+i\infty} \tilde{\bar{\alpha}}' \exp(s\eta) ds] dk. \tag{34}$$

As seen in Eq. (32), the integrand $\tilde{\bar{\alpha}}'\exp(s\eta)$ has two poles in s-plane. They are denoted by s_1 and s_2. Then Eq. (34) can be written as

$$\alpha' = \frac{1}{2\pi}\int_{-\infty}^\infty \exp(ik\xi) [\text{Res}\{\tilde{\bar{\alpha}}'\exp(s\eta)\}_{s=s_1} + \text{Res}\{\tilde{\bar{\alpha}}'\exp(s\eta)\}_{s=s_2}] dk \tag{35}$$

where "Res" denotes the residue at their poles. The s_1 and s_2 are given by the roots of the denominator in Eq. (32),

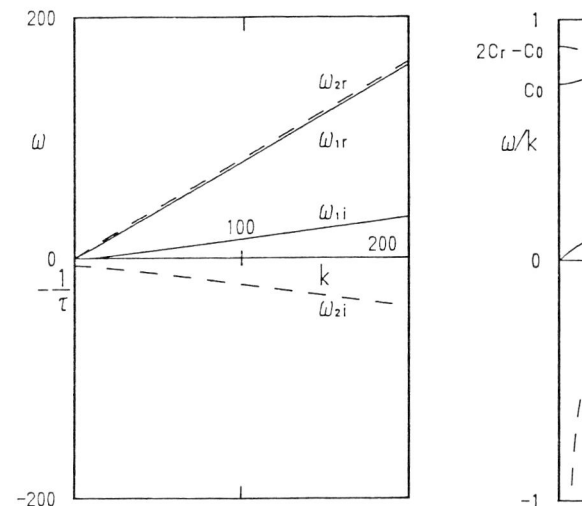 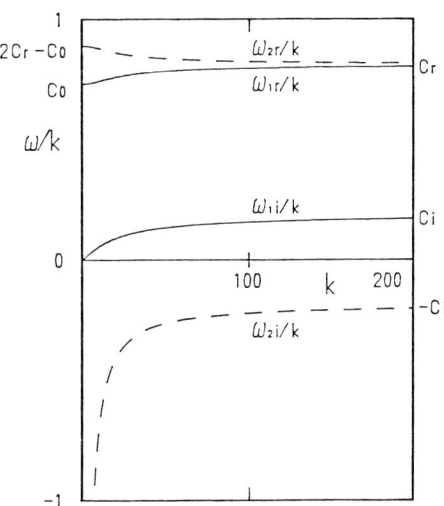

FIGURE 1. Solution of dispersion relation: $\alpha_0 = 0.27$, $v_0 = 0.32$ m/s, $u_0 = 0.16$ m/s.

third power of the velocity slip in the main stream. Thus, the effect of negative diffusion becomes so small that it is not appreciable over the riser if the bubbles are of small size and hence of small velocity slip.

The region where the lower-order wave plays the role may appear at the distance $c_r \tau$ downstream of the mixer, where $c_r \tau$ is given by

$$c_r \tau = \frac{(2-\alpha_0)v_0 + \alpha_0(5-4\alpha_0)u_0}{4(1-\alpha_0)} \frac{(v_0-u_0)}{g} . \tag{22}$$

SOLUTION OF PERTURBATION EQUATIONS

Now we consider an initial-boundary value problem for perturbation equations given by Eq. (6), Eq. (7) and Eq. (10). For this purpose, the variables are made nondimensional:

$$x = L\xi, \quad t = L\eta/v_0, \quad u' = v_0 u^*, \quad v' = v_0 v^*, \tag{23}$$

where L denotes a characteristic length of the flow field, for instance, the length of riser.

Substituting Eq. (23) into Eq. (6), Eq. (7) and Eq. (10), one obtains

$$\frac{\partial \alpha'}{\partial \eta} + \mu \frac{\partial \alpha'}{\partial \xi} - (1-\alpha_0)\frac{\partial u^*}{\partial \xi} = 0 , \tag{24}$$

$$\frac{\partial \alpha'}{\partial \eta} + \frac{\partial \alpha'}{\partial \xi} + \alpha_0 \frac{\partial v^*}{\partial \xi} = 0 , \tag{25}$$

$$(3-2\alpha_0)\frac{\partial u^*}{\partial \eta} + [1+2\mu(1-\alpha_0)]\frac{\partial u^*}{\partial \xi} - \frac{\partial v^*}{\partial \eta} - \frac{\partial v^*}{\partial \xi} = \beta\alpha' + \Omega(v^*-u^*), \tag{26}$$

where

$$c_0 = (1-2\alpha_0)v_0 + 2\alpha_0 u_0 , \tag{14}$$

$$\left.\begin{matrix}c_1\\c_2\end{matrix}\right\} = c_r \pm ic_i , \tag{15}$$

$$c_r = \frac{(2-\alpha_0)v_0 + \alpha_0(5-4\alpha_0)u_0}{2[1+2\alpha_0(1-\alpha_0)]} , \tag{16}$$

$$c_i = \frac{[\alpha_0(8-17\alpha_0+8\alpha_0^2)]^{\frac{1}{2}}}{2[1+2\alpha_0(1-\alpha_0)]}(v_0-u_0) , \tag{17}$$

$$\tau = \frac{1+2\alpha_0(1-\alpha_0)}{2(1-\alpha_0)} \frac{(v_0-u_0)}{g} . \tag{18}$$

Figure 1 shows the solution of the dispersion relation

$$(\omega-c_1 k)(\omega-c_2 k) + \frac{i}{\tau}(\omega-c_0 k) = 0 \tag{19}$$

provided by substituting $\alpha = \hat{\alpha}\exp[i(kx-\omega t)]$ into Eq. (13). In this figure, the wave number k and the angular frequency ω has been nondimensionalized by L and L/u_0, respectively, where L is the length of riser and u_0 the flow velocity of water, and c_0, c_1, c_2 and L/τ have been made nondimensional by u_0. The solution with $\omega_i > 0$ denotes an unstable mode.

Equation (13) represents a wave equation, which includes the higher-order wave operators with complex conjugate propagation velocities c_1 and c_2 and the lower-order wave operator with real propagation velocity c_0. The fact that the propagation velocities of the higher-order waves are complex conjugate means that one of them increases exponentially with time. The growth rate is represented as $c_i k$ and it is comparatively small proportional to the velocity slip and the void fraction in the main stream.

The real part of the propagation velocities of the higher-order waves is different from the propagation velocity of the lower-order wave. The former is larger than the latter for the present case, in contrast to the case of fluidized bed of fluid and solid particles. However, the negative diffusion of the voidage may occur in the region where the lower-order wave plays the role. In fact, if we approximate $\partial/\partial t$ in the higher-order terms of Eq. (13) by $-c_0(\partial/\partial x)$ for $t \gg \tau$, we have

$$\frac{\partial \alpha'}{\partial t} + c_0 \frac{\partial \alpha'}{\partial x} = -D\frac{\partial^2 \alpha'}{\partial x^2} \tag{20}$$

where the coefficient of negative diffusion can be expressed as

$$D = \tau[(c_r-c_0)^2 + c_i^2]$$
$$= \frac{\alpha_0(1-4\alpha_0+8\alpha_0^2-4\alpha_0^3)}{(1-\alpha_0)} \frac{(v_0-u_0)^3}{g} . \tag{21}$$

It is noticed that the coefficient is proportional to the

wave in a fluidized bed of gas and solid particles (Murray, 1965; Morioka & Nakajima, 1987) and in bubbly liquid flow (Bouré, 1987; van Wijngaarden & Biesheuvel, 1987). This assumption may be resonable for low flow velocity and low frequency disturbance. The coordinate x in the upward direction along the riser is taken, and it is assumed that the motion depends only on the coordinate x and the time t. Then, the basic equations presented in the previous section are reduced as follows,

$$\frac{\partial}{\partial t}(1-\alpha) + \frac{\partial}{\partial x}(1-\alpha)u = 0 \quad , \tag{6}$$

$$\frac{\partial \alpha}{\partial t} + \frac{\partial}{\partial x}\alpha v = 0 \quad , \tag{7}$$

$$(1-\alpha)(\frac{\partial u}{\partial t} + u\frac{\partial u}{\partial x}) = -\frac{1}{\rho_L}\frac{\partial p}{\partial x} - (1-\alpha)g \quad , \tag{8}$$

$$\frac{1}{2}(\frac{\partial}{\partial t} + v\frac{\partial}{\partial x})(v-u) = -\frac{1}{\rho_L}\frac{\partial p}{\partial x} - \frac{C_D}{\rho_L V}(v-u) \quad . \tag{9}$$

Eliminating $\partial p/\partial x$ between Eq. (8) and Eq. (9), one obtains

$$(3-2\alpha)\frac{\partial u}{\partial t} + [2(1-\alpha)u + v]\frac{\partial u}{\partial x} - \frac{\partial v}{\partial t} - v\frac{\partial v}{\partial x} = -2(1-\alpha)g + \frac{2C_D}{\rho_L V}(v-u) \quad . \tag{10}$$

Equations (6) to (9) have a steady flow solution in simple form, which represents a steady bubble flow in vertical riser, that is

$$\left.\begin{array}{l}\alpha = \alpha_0, \quad u = u_0, \quad v = v_0, \\ -\frac{\partial p_0}{\partial x} = \frac{C_D}{V}(v_0-u_0) = \rho_L(1-\alpha)g \quad , \end{array}\right\} \tag{11}$$

where α_0, u_0 and v_0 are constant. The pressure p_0 linearly decreases in the upward direction.

EQUATION FOR VOIDAGE PERTURBATION

The perturbation around the above steady flow solution is considered and taken as

$$\alpha = \alpha_0 + \alpha', \quad u = u_0 + u', \quad v = v_0 + v', \tag{12}$$

where the primed quantities denote their perturbations. Substituting Eq. (12) into Eqs. (6), (7) and (10) and retaining only the first order terms in the perturbations α', u' and v', the linearized equations are obtained. Furthermore, eliminating the variables u' and v', the equation for the voidage perturbation α' is obtained:

$$(\frac{\partial}{\partial t} + c_1\frac{\partial}{\partial x})(\frac{\partial}{\partial t} + c_2\frac{\partial}{\partial x})\alpha' + \frac{1}{\tau}(\frac{\partial}{\partial t} + c_0\frac{\partial}{\partial x})\alpha' = 0 \tag{13}$$

$$\frac{\partial}{\partial t}\alpha\bar{\rho}_G + \frac{\partial}{\partial x_i}\alpha\overline{\rho_G v_i} = 0 ,\qquad(2)$$

where α is the void fraction, u_i the velocity of liquid, v_i the velocity of bubbles, ρ_G the gas density, and the bar denotes the phasic average. Here the evaporation and condensation have been neglected.

On the other hand, taking the phasic average from the local instant equations of motion for liquid and gas phase, and adding the resultant equations to cancel out their interaction terms, the following equation is obtained:

$$\frac{\partial}{\partial t}(1-\alpha)\bar{u}_i + \frac{\partial}{\partial x_j}(1-\alpha)\overline{u_i u_j} = -\frac{1}{\rho_L}\frac{\partial \bar{p}}{\partial x_i} + (1-\alpha)g_i \qquad(3)$$

where ρ_L is the liquid density, p the pressure, and g_i the gravitational acceleration. Here the viscous term and the inertia term of the gas phase have been neglected. Furthermore, it is assumed that $\overline{\rho_G v_i} = \bar{\rho}_G \bar{v}_i$ and $\overline{u_i u_j} = \bar{u}_i \bar{u}_j$ in Eq. (2) and Eq. (3), respectively.

As a constitutive equation which presents the interaction between liquid and gas phase, the equation of slip motion for a single spherical bubble placed in the averaged flow is taken:

$$(\frac{\partial}{\partial t} + v_j\frac{\partial}{\partial x_j})[\frac{1}{2}\rho_L V(v_i-u_i)] = -V\frac{\partial p}{\partial x_i} - C_D(v_i-u_i) \qquad(4)$$

where V is the volume of a bubble and $-C_D(v_i-u_i)$ the drag force on a bubble. Here it is assumed that $u_i = \bar{u}_i$, $v_i = \bar{v}_i$, $p = \bar{p}$ and $V = \bar{V}$. Equation (4) obeys a constitutive principle that in the limit as $\alpha \to 0$ the equation of motion for gas phase approaches the appropriate single bubble equation, with the equation for liquid phase approaching the correct equation for liquid (Drew & Lahey, 1979).

Here it is noticed that the relation of the same form as Eq. (4) can be derived from the kinetic equation of bubble with the virtual mass $0.5 \rho_L V$ and the external force $-V\partial p/\partial x_i - C_D(v_i-u_i)$ (Morioka & Nakajima, 1987).

Furthermore, the state equation of an ideal gas is supplemented, assuming isothermal variation,

$$\bar{p}/\bar{\rho}_G = \text{constant} . \qquad(5)$$

Equations (1) to (5) present a dilute bubbly liquid model.

STEADY SOLUTION FOR BUBBLE FLOW IN RISER

In order to obtain a steady solution in simple form, which represents a bubble flow in the riser, it is assumed that the gas is imcompressible, as used in the argument of the voidage

is examined in this paper.

The modeling of bubbly liquid flow and the corresponding possible wave modes, particularly the voidage wave, have been discussed in detail in recent papers (Toma & Morioka, 1986; Bouré, 1987; van Wijngaarden & Biesheuvel, 1987). Here, a classical dilute bubbly liquid model is assumed, which takes account of a pressure gradient appreciable over the interbubble distance. It is assumed that the gas is incompressible for low velocity and low frequency disturbance, as in the voidage wave (Bouré, 1987; Murray, 1965; van Wijngaarden & Biesheuvel, 1987). Then, a solution in simple form presenting a steady bubble flow in the riser is found. The perturbation around this steady flow solution yields a wave equation governing the voidage disturbance.

The wave equation includes the higher-order wave operator with complex conjugate propagation velocities and the lower-order wave operator with real propagation velocity. The imaginary part of the complex conjugate propagation velocities is comparatively small, proportional to velocity slip (the bubble velocity relative to the liquid flow velocity) and the void fraction in the main stream, meaning the gradually growing voidage disturbance. On the other hand, the real part of the complex conjugate propagation velocities of the higher-order waves is larger than the propagation velocity of the lower-order wave for the present case, but it also provides the effective negative diffusion in the lower-order wave. The explicit expressions are obtained for the propagation velocities, the growth rate, the characteristic distance in which the lower-order wave becomes important, and the coefficient of negative diffusion.

This linear wave equation can be solved for given initial and boundary conditions, using Laplace and Fourier transform techniques. Then the solution in a real definite integral form can be obtained from the inversion formula, and can be evaluated numerically by using a personal computer. As expected, the result shows that a given perturbation profile of voidage moves downstream in course of time, with increasing height and decreasing width.

Lastly the experimental result is discussed in the light of the present theoretical result. It is shown that the diffusion distance predicted from the diffusion coefficient and the coalescence time is comparable with the interbubble distance at the inlet of riser.

BASIC EQUATION

When the method of averaging (Drew, 1983) is applied to the local instant continuity equations of liquid and gas phase, respectively, the following relations are obtained:

$$\frac{\partial}{\partial t}(1-\alpha) + \frac{\partial}{\partial x_i}(1-\alpha)\bar{u}_i = 0 \ , \tag{1}$$

FLOW PATTERN TRANSITION DUE TO INSTABILITY OF VOIDAGE WAVE

S. MORIOKA
Department of Aeronautical Engineering
Kyoto University, Kyoto, Japan
F. JOUSSELLIN and H. MONJI
Institute of Engineering Mechanics
University of Tsukuba, Tsukuba, Ibaraki, Japan

ABSTRACT

In order to show that the transition from bubble flow to slug flow observed in a vertical riser is due to the instability of voidage wave, an anlysis and a calculation are performed on the basis of a dilute bubbly liquid model. The result demonstrates such a possibility, together with expressions for propagation velocities, growth rate and coefficient of negative diffusion. A comparison is made with the experimental result.

INTRODUCTION

Recently an experiment was performed by the authors using water and air to simulate the two-phase flow in the riser of a heavy liquid-metal MHD generation loop (Monji et al, 1986; Monji & Onisawa, 1987). In order to have common similarity paramenters with the liquid-metal case, the pressure in the riser has been reduced. The flow pattern changed from bubble flow to slug flow under the condition with such values of parameters. Careful observation of the transition process showed that the bubbles were distributed uniformly just after the air was injected at the mixer, but they changed to a non-uniform distribution in the middle of the riser, and the dense part coalesced to develop into the slug flow. This phenomenon strongly suggests an instability of voidage wave.

This kind of instability has recently attracted the attention of researchers as a mechanism of bubble generation in fluidized bed of fluid and solid particles (Liu, 1982). In that case, the wave equation for the voidage disturbance consists of two higher-order waves and one lower-order wave, and the propagation velocities of both of the higher-order waves are smaller than that of the lower-order wave. Thus, the voidage disturbance generated at the inlet first grows along with the higher-order waves and then is gathered by the negative diffusion effect along with the lower-order wave. Such an instability has been known as the violation of the stability condition in the wave hierarchies (Whitham, 1974). Whether or not such an instability is possible in bubbly liquid flow

WAVE AND SHOCK PHENOMENA AND CRITICAL FLOW

For the X-ray CT scanner, if sufficient quantities of projection data from all directions around the object could be obtained at the same time, very detailed two-phase flow structure (such as the behaviour of liquid film and droplets) will clearly be visualized. Therefore useful information on two-phase flow will be obtained.

In this sense, these two methods have a great promise for the future.

REFERENCES

(1) Martin, R., "Measurements of the Local Void Fraction at High Pressure in a Heating Channel," BT269-38, 1972, Centre d'Etudes Nucleaires de Granoble, France.

(2) Smith, A. V., "Fast Response Multi-Beam X-Ray Adsorption Technique for Indentifying Phase Distributions during Steam-Water Blowdowns," Journal of British Nuclear Energy Society, Vol. 14, No. 3, 1975, pp. 227-234.

(3) Lassahn, G. D., "Two-Phase Fluid Density Measurement with a Two-Beam Radiation Attenuation Densitometer," IEEE, 1975 Nuclear Science Symposium.

(4) Lecroart, H., and Lewi, J., "Localized Measurements and their Statistical Interpretation in Two-Phase Flow at High Velocity and Large Void Fraction," French Hydrotechnic Society, Douziemes Journees de l'Hydraulique, Question IV, Report 7, 1972.

(5) Miller, N., and Mitchie, R. E., "Measurement of Local Voidage in Liquid/Gas Two-Phase Flow Systems," Journal of British Nuclear Energy Society, Vol. 9, No. 2, 1970, pp. 94-100.

(6) T. Narabayashi, et al., "Measurement of Transient Flow Pattern by High Speed Scanning X-Ray Void Fraction Meter," Measuring Techniques in Gas-Liquid Flows, 259-280, Springer (1984).

(7) Iizuka, M., Morooka, S., Kimura, M., and Kagawa, T., "Measurement of Distribution of Local Void Fraction in Two-phase Flow by X-ray Computed Tomography", Journal of Atomic Energy Society Japan, Vol. 26, No. 5, 1984, pp. 63-71.

(8) Zuber, N., and Findlay, J. A., "Average Volumetric Concentration in Two-Phase Flow System", Trans. ASME, J. Heat Transf., 87-(1965), pp. 453-468.

PHASE DISTRIBUTION PHENOMENA 437

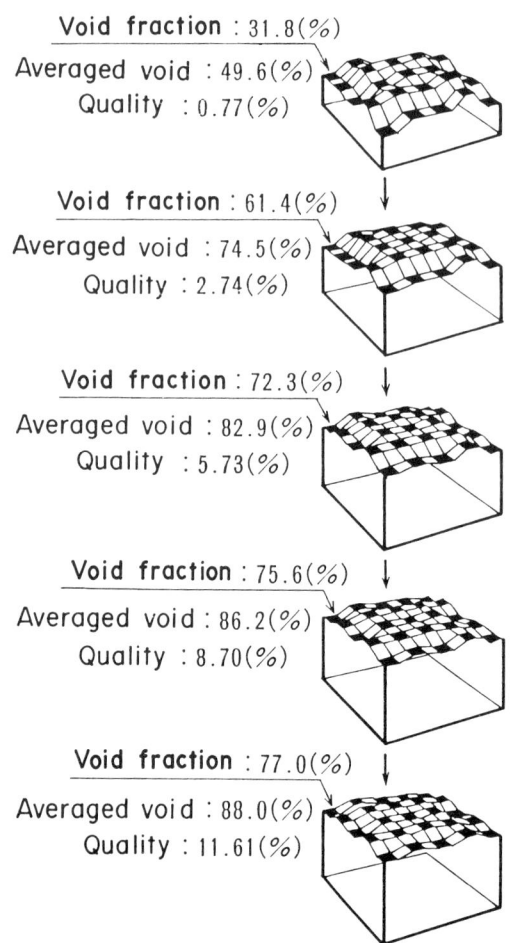

Fig. 19 Subchannel Void Fraction Distribution in
        4 x 4 Rod Bundle

in a tube. This is presumably attributable to the thick liquid film
covering the channel wall, and the tendency of the vapor (void) to move
into the flow area at a higher flow speed (void drift)[8]. In addition,
the void fraction increases in all subchannels, gradually developing
homogeneity with increasing quality.

4. CONCLUDING REMARKS

In this report, two new two-phase flow measurement methods by X-ray were
presented.

The X-ray beam scanning method can be improved for measuring more rapid
two-phase flow phenomena, using higher intensity X-ray sources. Also,
this method was shown to be very promising for accurate measurements of
averaged two-phase flow velocity.

$$\frac{J_g}{\alpha} = C_0 J + V_{gj} \tag{18}$$

In Fig. 18, the data represented by the solid line was obtained through the least squares method, and can be expressed as the following equation.

$$\frac{J_g}{\alpha} = 1.08J + 0.45 \tag{19}$$

From Fig. 18, the following results were obtained:

a. The data follows the linear relation given by Eq. (18). Therefore, it was considered that the cross-sectional averaged void fraction data for a rod bundle can be closely correlated as simple flow channel data (tube or annuli), using the Drift Flux model.

b. As the distribution parameter Co (the solid line slope in Fig. 18) is larger than unity, the void distribution across the rod bundle is not uniform. Also, the void fraction channel wall is lower than the central void fraction. This result agrees with the measured void distribution (see Fig. 19).

(4) Subchannel void fraction distribution

The rod bundle flow was divided into multiple subchannels, enclosed by rods, and the average void fraction for the subchannels (the subchannel void fraction) was examined. Figure 19 shows a typical subchannle void fraction distribution. In this figure, a solid square represents a subchannel void fraction. The void fraction for the center subchannel is larger than those for external (corner and side) subchannels. Their distribution resembles the void fraction distribution in an annular flow

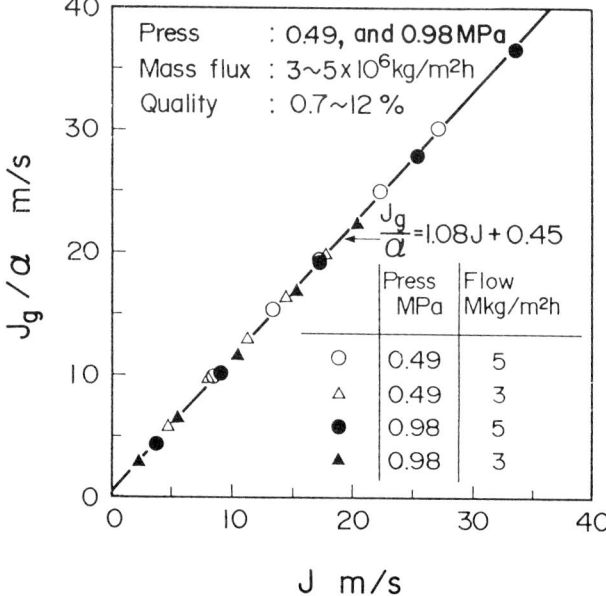

Fig. 18 Velocity-Flux Plane for Rod Bundle Data

PHASE DISTRIBUTION PHENOMENA

Table 2 Specifications for X-Ray CT Scanner (TCT-30)

Scanning method	Translate-rotate method
Scanning time (s)	105
Counting time per one beam (ms)	4
Sampling time (ms)	4
X-ray tube voltage (kV)	120
Current (mA)	25
Detector	BGO scintillator
Detector response time (μs)	5.3
Reconstruction element dimensions (mm)	1.0 × 1.0

(2) Two-phase flow CT picture

Figure 17 shows a typical CT scan image for two-phase flow inside a rod bundle. The void fraction distribution is shown in color.

(3) Drift flux model evaluation

Figure 18 summarizes the rod bundle void fraction data in one of the general forms for the Drift Flux model [8], as explained below.

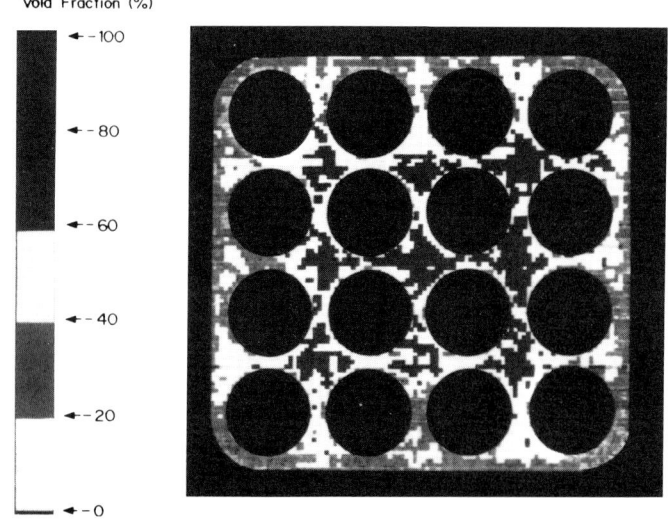

PRESS=0.98(MPa), Tin=447(K), QUALITY=0.800(%), MASS FLUX=3.00×10^6(kg/m^2·h)

Fig. 17 Typical CT Picture

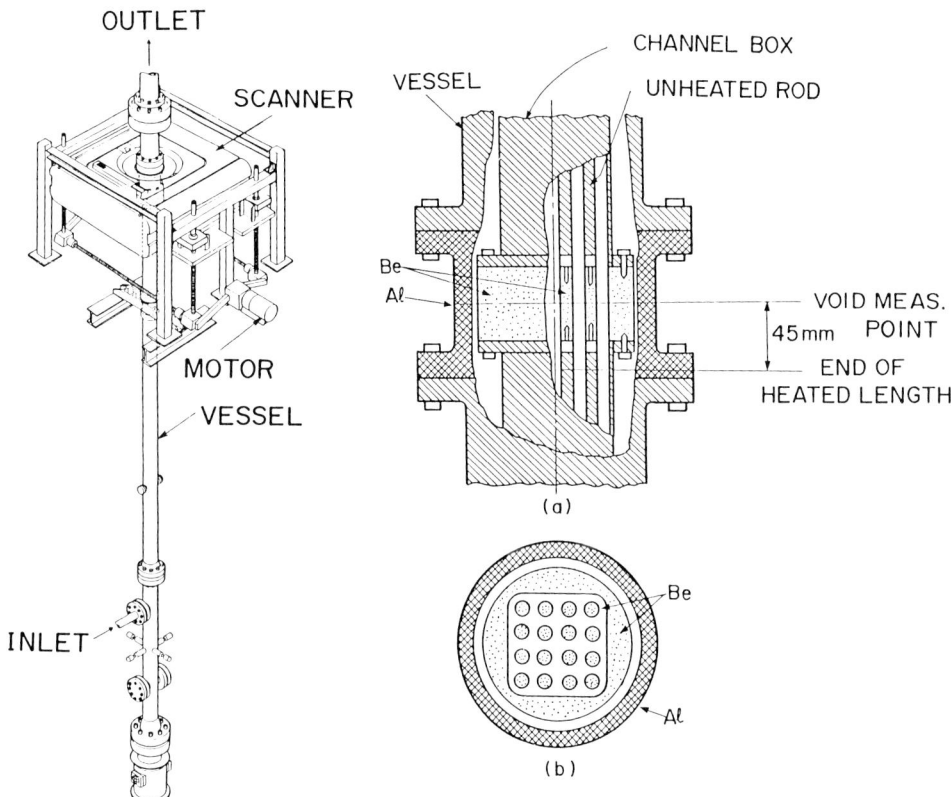

Fig. 15 Test Section

Fig. 16 Void Measuring Section

This test section is installed in a high-pressure heat transfer loop with the X-ray CT scanner.

Figure 15 shows the test section. The rod bundle is inserted in the test vessel. As shown in the figure, an X-ray CT scanner is placed on a stand that can be moved upward and downward.

Figure 16 shows the void measuring section. The void fraction is measured at a place about 4 cm above the end of the heated length. X-ray paths for the rods and the channel walls are made of beryllium (Be). The pressure vessel is made of aluminum to minimize the X-ray attenuation in the structural material. Basic specifications for the X-ray scanner used for this experiment are shown in Table 2.

3.3 Test Results and Discussion

(1) Experimental variables range

The experiment has been carried out in the following range:

Pressures: 0.49 and 0.98 MPa
Mass Flux: 3 to 5 x 10^6 kg/m^{2}h
Quality : 0 to 12 %

PHASE DISTRIBUTION PHENOMENA

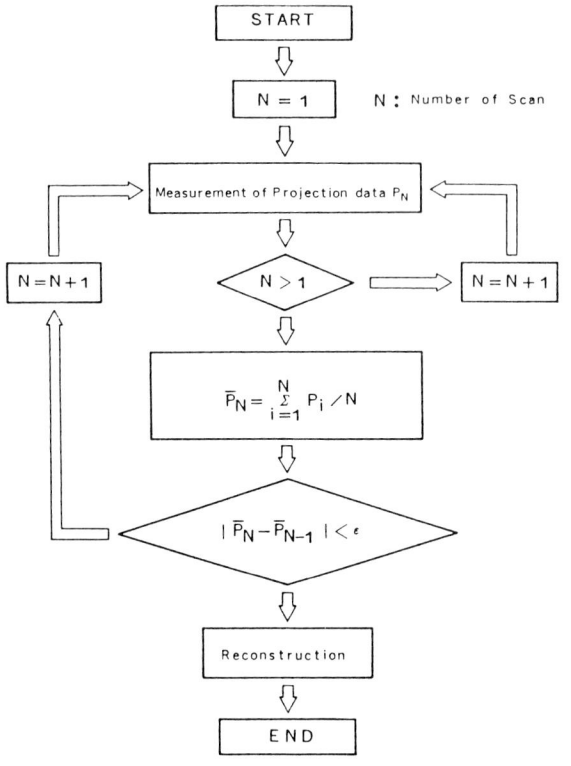

Fig. 13 Time-Averaging Technique Flow Chart

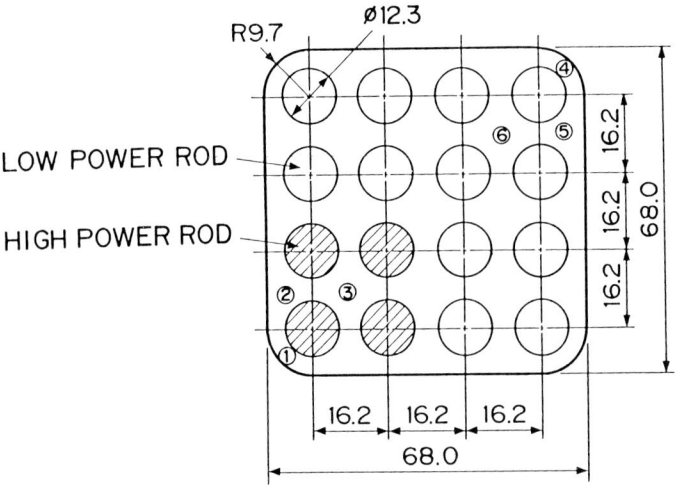

Fig. 14 Test Bundle Cross-Sectional View

The reconstruction method applies to the object. The object is divided into several reconstructed elements (i,j). Thus, the f(x,y) projection is given by

$$P(x,\theta) = \int f(x,y)\,dy = \sum_{ij} q_{ijk}\mu'_{ij} \tag{16}$$

where μ'_{ij} is a linear attenuation coefficient for an element (i,j) and q_{ijk} is a weighted coefficient, proportional to the area through which X-ray beam $k(\theta)$ passes for each element. Subscript k shows that an element (i,j) is on the line for the X-ray beam $k(\theta)$. Element dimensions are 1 mm x 1 mm.

Equation (16) can be solved for the unknown function. The solution has already been developed; analytic reconstruction methods, involving filtered back projection, have been widely used. In a two-phase flow system, Eq. (16) is rewritten by

$$\bar{\alpha}(x,\theta) = \int \alpha(x,y)\,dy / \int dY = \sum_{ij} h_{ijk}\alpha_{ij} \tag{17}$$

where $\bar{\alpha}(X,\theta)$ is the so-called 'chordal void fraction', α_{ij} is a local void fraction for an element (i,j), and h_{ijk} is a quantity proportional to q_{ijk}.

With reference to Eq. (17), it is evident that the accuracy of reconstructed local void fraction α_{ij} is dependent upon the sensitivity achieved in measuring chordal void fraction $\bar{\alpha}(X,\theta)$. In general, the void fraction for a steady state flow is a time-averaged quantity. However, the time per X-ray beam (for the X-ray CT scanner for medical application) is about 4 milliseconds. This is not adequate for void fraction measurement. A new time-averaging technique involving the X-ray CT scanner was developed, called ACTS (Advanced CT System). In this technique, the measurement (of a projection data set) is repeated N times and an average is obtained of the projection data $\overline{P_N}^*$. If the difference between $\overline{P_N}$ and $\overline{P_{N-1}}$ is within limit ε, the $\overline{P_N}$ value is taken as a chordal void fraction.

A flow chart for the time-averaging technique is shown in Fig. 13. The local void fraction for an element (i,j) is reconstructed from the above projection data $\overline{P_N}$.

3.2 Test Section and Test Apparatus

Figure 14 shows a test bundle cross-sectional view. The test section consists of electrically heated rods in a 4 x 4 array configuration. These are indirect heater rods and are supported by spacers to keep the rod gap constant. The heated length and the outer diameter of the heater rod, as well as the geometry and the axial locations of the spacers, are the same as those for a BWR fuel bundle. The axial profile for the heat flux is uniform, while the radial power distribution is biased to 4 rods near the corner (shaded in Fig. 14).

PHASE DISTRIBUTION PHENOMENA

Figure 11 shows average void fraction $\bar{\alpha}_1$ and $\bar{\alpha}_2$. The time difference between $\bar{\alpha}_1$ and $\bar{\alpha}_2$ is 22.5 ms (9 scans). Therefore, vapor phase slug velocity u_g is 2.7 m/s, and quality x from Eq. (12) is 0.039, using $\bar{\alpha}^t$ = 26 %.

As mentioned above, using the dual beam scanning X-ray void fraction meter, void fraction and vapor mean velocity were measured at the same time under slug flow conditions. It is believed that the method can be applicable for other flow patterns, except for stratified flow.

3. X-RAY CT SCANNER METHOD

3.1 Principle

This X-ray CT scanner consists of an X-ray tube and 8 detectors, which carry out a linear scan over a two-phase flow test section with unknown two-dimensional property. It is a linear attenuation coefficient. An outline of the CT scanner principle is described below, with reference to Fig. 12. In the linear scan, a finely collimated X-ray beam is attenuated by the object and sensed by the detector. The number of photons sensed by the detector is given by

$$I(x,\theta) = I_0 \exp[-\int f(x,y)\,dY] \tag{15}$$

where f(x,y) is the linear attenuation coefficient for the object.

The line integral for the function f(x,y), along an X-ray beam oriented by X, θ, is called the projection data.

In the linear scan, one projection data set for the object, for a given angle θ, is obtained. A second projection data set is obtained by rotating the source and detector about the center of the object in 6° increments. After each rotation, the next linear scan is made. All projections were obtained through a 180° arc around the object. About 100 seconds were required to obtain all projections.

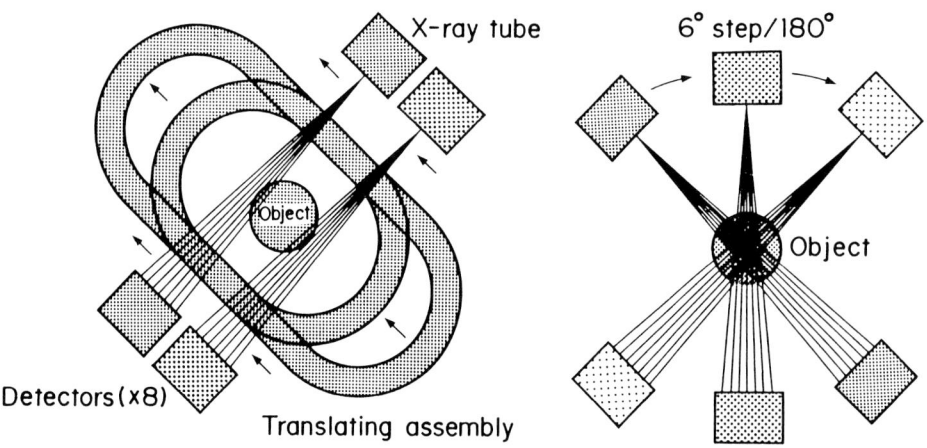

Fig. 12 Scanning Method of Scanning Translate/Rotate

2.3 Test Results

Figure 9 shows one example of the measured flow pattern on the color CRT. Its pattern is annular flow. Vapor phase mean velocity u_g could not be obtained with the stratified flow pattern.

Figure 10 shows one example of measured flow patterns for slug flow at measuring cross sections 1 and 2, respectively. Pressure was 7 MPa and mass flow rate was 0.13 kg/s. Vapor slugs are located near the upper wall of the measuring pipe, because of their buoyancy.

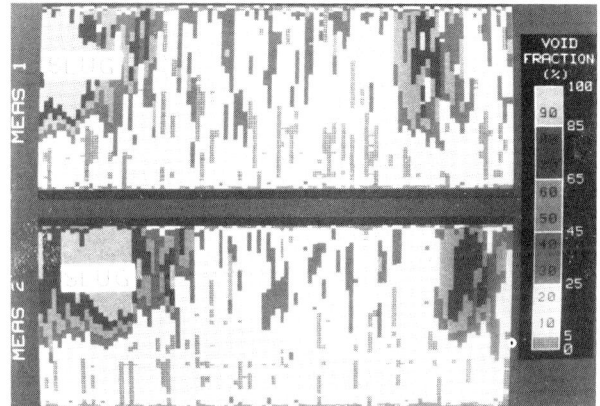

Fig. 10 Measured Flow Patterns (Slug Flow Showing Time Difference between Cross-sections 1 and 2)

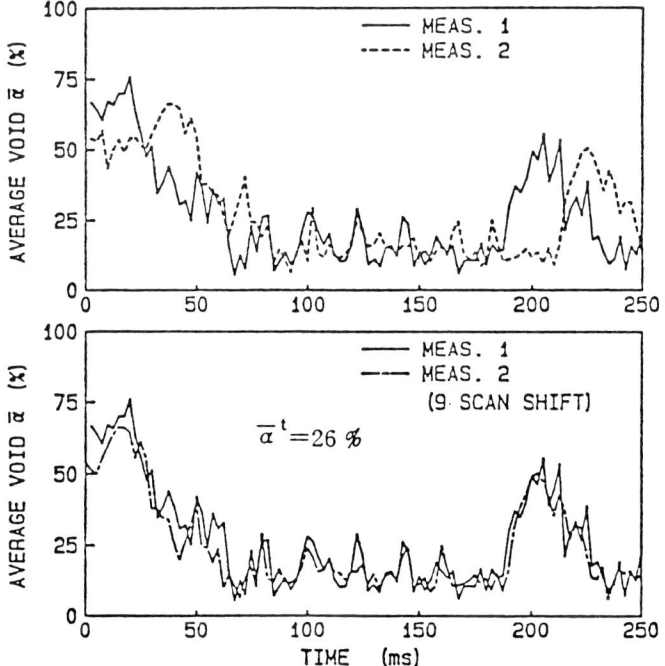

Fig. 11 Cross-Sectional Average Void Fraction for Calculating Vapor-Phase Mean Velocity

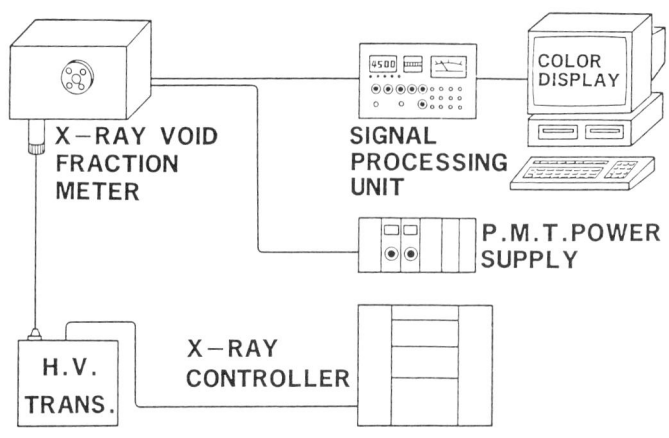

Fig. 5 Void Fraction Meter System

Fig. 6 Dual Beam X-Ray Scanner*

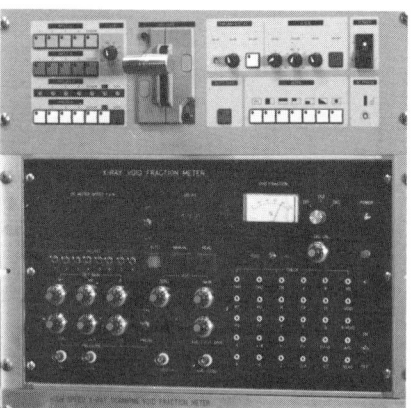

Fig. 7 High-Speed Analog Processing Circuit

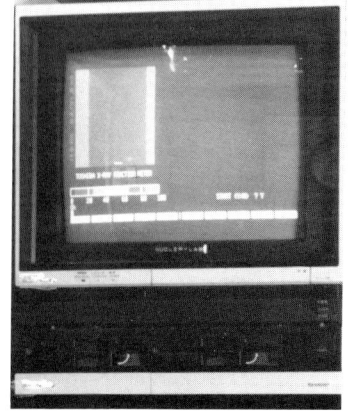

Fig. 8 Color Display

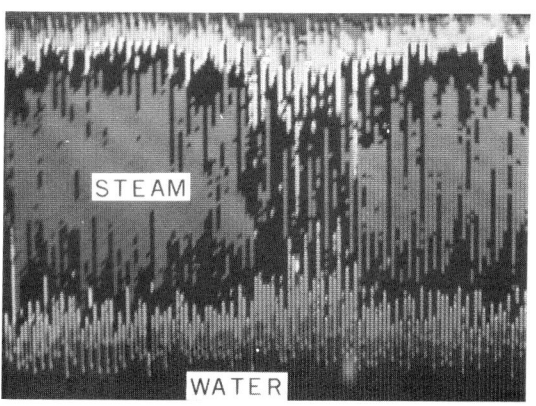

Fig. 9 A Measured Flow Patterns (Annular Flow)

* Figures 6 through 10 and 17 appear in color following page 312.

When the average void fraction at cross section 1 and 2 are $\bar{\alpha}_1$ and $\bar{\alpha}_2$, respectively, mean velocity for vapor phase u_g is calculated using time difference τ. This is obtained from the cross-correlation between $\bar{\alpha}_1$ and $\bar{\alpha}_2$.

$$u_g = 60/\tau \tag{10}$$

Mass flux G is given by:

$$G = \dot{m}/A \tag{11}$$

Quality x is given by:

$$x = u_g \rho_g \bar{\alpha}^t / G \tag{12}$$

where $\bar{\alpha}^t$ is the cross-sectional and short time averaged void fraction.

Figure 4 shows a block diagram of the high-speed analog/digital processing circuit for calculating Eqs. (5) and (9). Eq. (5) is changed into the following equation for the circuit.

$$\alpha(x) = F(p) \, \frac{\log I_X(x) - W(x)}{V(x)} - G(p) \tag{13}$$

where

$$W(x) \equiv \log I_w(x) \, ,$$

$$V(x) \equiv \log[I_A(x)/I_w(x)] \tag{14}$$

Figure 5 shows a block diagram for the total X-ray void fraction meter system. Photographs of the dual beam X-ray scanner, the high speed analog processing circuit, and the color display are shown in Fig. 6, Fig. 7 and Fig. 8, respectively.

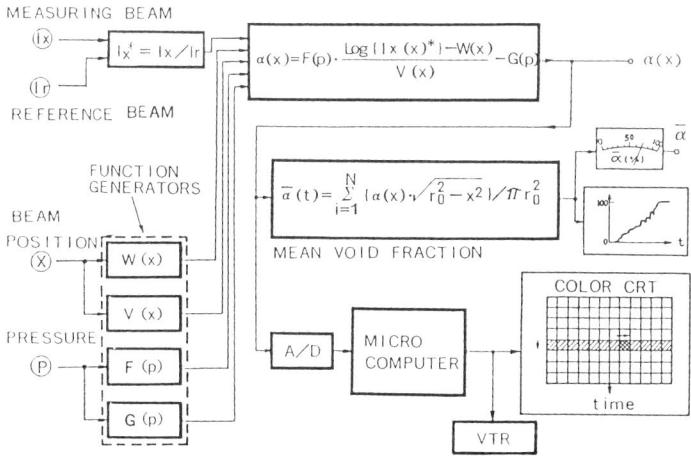

Fig. 4 Block Diagram of High-Speed Analog Signal Processing Circuit

Table 1 Disc Speed and Scanning Period

DISC SPEED N (rpm)	SCANNING FREQUENCY (Hz)	SCANNING PERIOD (msec)	DRIVING AND SAMPLING PULSE (kHz)
300	40	25.0	5.0
600	80	12.5	10.0
1500	200	5.0	25.0
3000	400	2.5	50.0
4500	600	1.66	75.0
6000[1)]	800	1.25	100.0

where R is the radius for the concentric circle on 8 hole-slits and θ is the rotation angle for the disc, which ranges from $-22.5°$ to $+22.5°$.

Figure 3 shows the dual beam X-ray void fraction meter and the test apparatus used in this study. Dual scanning beams from the anode of the X-ray tube pass through cross sections 1 and 2 and are measured by detectors 1 and 2. The test section pipe inner radius is 16.1 mm, and the distance between cross sections 1 and 2 being measured is 60 mm. As shown in Table 1, when the disc speed is 3000 rpm and flow mean velocity is 1 m/s, the flow patterns at cross sections 1 and 2 are displayed on the CRT, with a difference of 24 scans. As the sampling frequency is 50 kHz at the disc speed of 3000 rpm, sampling time is 30 μs for beam position x. The fluid moves only 20 μm during this time.

The test apparatus, shown in Fig. 3, can measure mass flow rate within a residual mass by ΔP cell or a venturi flow meter.

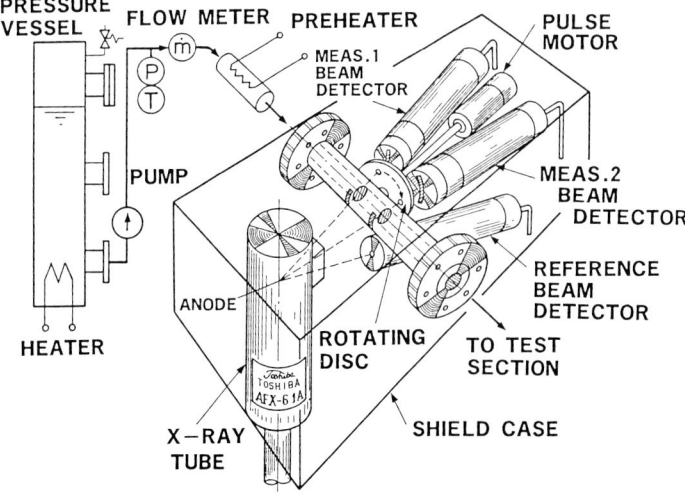

Fig. 3 Dual Beam X-Ray Scanner and Test Apparatus

$$F(p) = \frac{\rho_w - (\frac{\mu_A}{\mu_w})\rho_A}{\rho_w' - \rho_v'} \qquad (6)$$

$$G(p) = \frac{\rho_w - \rho_w'}{\rho_w' - \rho_v'} \qquad (7)$$

where, in Eq. (6), $(\mu_A/\mu_w)\rho_A$ is negligible, since it is very small compared with ρ_w. Both the functions present approximately straight lines for a 2 MPa or greater pressure, thus providing practical convenience. In Eq. (5), $I_A(x)$ and $I_w(x)$ are determined by the shape of the beryllium pipe, so only one measurement need be made. The void fraction $\alpha(x)$ distribution form can thus be obtained immediately by measuring only $I_x(x)$. The average sectional void fraction $\bar{\alpha}$ can be calculated from $\alpha(x)$, using the following equation. $\bar{\alpha}$ is an important factor for comparison with calculation codes analysis results.

$$\bar{\alpha} = \int_{-r_0}^{+r_0} 2\sqrt{r_0^2 - x^2}\, \alpha(x)\, dx / \pi r_0^2 \qquad (8)$$

where r_0 is the beryllium pipe inner radius.

2.2 Test Section and Test Apparatus

Figure 2 shows the high speed scanning X-ray void fraction meter principle. The scanning X-ray beam is collimated by 8 hole-slits on a rotating disc. The x-ray beam scans across the measuring pipe from high to low, according to hole-slit movement. The disc is driven by a pulse motor and its maximum speed is 3000 rpm. Therefore, the maximum scanning frequency is 400 Hz, and a void distribution curve can be obtained every 2.5 milliseconds. The relations between disc speed and scanning frequency are shown in Table 1. Due to the geometric arrangement for the X-ray source, the test section, the disc and the detector, as shown in Fig. 2, beam position x is calculated as follows:

$$X = \frac{L}{L+D} H = \frac{L}{L+D} R \sin\theta \qquad (9)$$

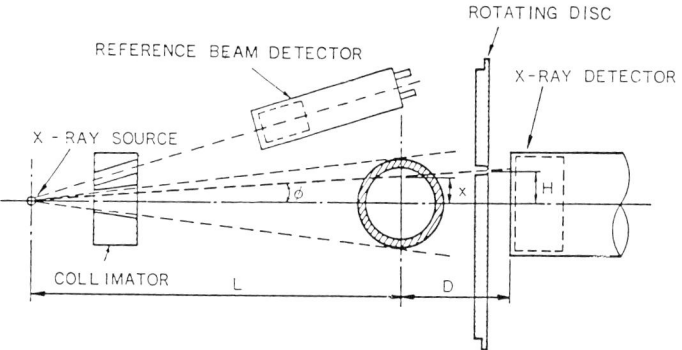

Fig. 2 Geometric arrangement for X-ray source, test section, disc and detector

PHASE DISTRIBUTION PHENOMENA

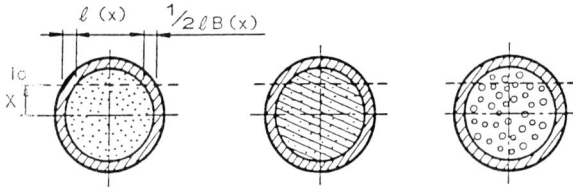

(a) FILLED WITH AIR, (b) FILLED WITH WATER, (c) FILLED WITH TWO PHASE MIXTURE

Fig. 1 Measuring X-ray void fraction meter principle

is expressed by the following equation.

$$I = I_0 \exp(-\rho \mu \ell) \tag{1}$$

where ρ is the density of fluid through which the X-ray beam passes, ℓ is the mass X-ray absorption factor, and μ is the path length.

When a pipe is filled with air, as shown in Fig. 1(a), the X-ray at a height or x is attenuated by the pipe wall and air between the X-ray source and the X-ray detector, excluding the inside of the test section pipe. In this study, a metallic beryllium pipe, having a small X-ray attenuation factor, was used for the test section. Accordingly, the normalized intensity $I_A(x)$ for the X-ray at a height of x is expressed, with subscripts B and A corresponding to beryllium and air respectively, as follows:

$$I_A(x) = \frac{I}{I_0} = \exp\{-\rho_B \mu_B \ell_B(x) - \rho_A \mu_A \ell_T(x) - \rho_A \mu_A \ell_A(x)\} \tag{2}$$

When the pipe is filled with water at normal temperature, as shown in Fig. 1(b), the intensity is expressed, with the subscript w corresponding to water, as follows:

$$I_w(x) = \frac{I}{I_0} = \exp\{-\rho_B \mu_B \ell_B(X) - \rho_w \mu_w \ell_T(x) - \rho_A \mu_A \ell_A(x)\} \tag{3}$$

When the pipe is filled with a two-phase water/vapor flow of high temperature and high pressure, the intensity is expressed, with a void fraction of an X-ray path as $\alpha(x)$, as follows:

$$I_X(x) = \frac{I_X}{I_0} = \exp\{-\rho_B \mu_B \ell_B(x) - \rho_A \mu_A \ell_A(x) - \mu_w \ell_T(x) [\rho_w'\{1-\alpha(x)\} - \rho_v'\alpha(x)]\} \tag{4}$$

where ρ_w' and ρ_v' are saturation densities for water and vapor under high temperature and high pressure, respectively.

Putting Eqs. (2) - (4) in order, with respect to the X-ray void fraction $\alpha(x)$ at a beam height of x, the following equation can be obtained:

$$\alpha(x) = F(p) \frac{\log\{I_X(x)/I_w(x)\}}{\log\{I_A(x)/I_w(x)\}} - G(p) \tag{5}$$

where $F(p)$ and $G(p)$ are functions of pressure. They can be calculated easily by using the steam tables, as indicated by the following equations:

$P(X, \theta)$ = projection data along an X-ray beam oriented by X, θ

V_{gj} = drift velocity

u = velocity

X = distance or quality

Greek

α : void fraction

μ : mass attenuation coefficient

μ' : linear attenuation coefficient

ρ : density

Subscript

A : Air

B : Beryllium

g : Gas (vapor)

W : Water

1. INTRODUCTION

Many studies have been reported concerning void fraction measurement in two-phase flow. The methods used to measure void fraction are probe, X-ray, and γ-ray attenuation (1) ∿ (5).

The γ-ray attenuation method is difficult to use when measuring a transient void fraction due to a limited radioactive dose. The probe has disadvantages: it disturbs the flow and has less reliability under high temperature conditions.

The X-ray, which was used in the authors' methods (X-ray beam scan and X-ray CT scanner) has these advantages: ① no flow disturbance ② high resolution ③ good time response ④ applicable for high pressure conditions (by appropriately selecting structural material for pressure vessel) ⑤ easy handling and shielding.

This report presents the new X-ray measurement methods.

2. X-RAY BEAM SCANNING METHOD (6)

2.1 Principle

As is generally known, the intensity of X-ray I for initial intensity I_0

MEASUREMENT OF PHASE DISTRIBUTION PHENOMENA IN HIGH PRESSURE STEAM-WATER TWO PHASE FLOW

T. NARABAYASHI, S. MOROOKA, T. KAGAWA, T. ISHIZUKA, A. HOSHIDE, and T. ISHIYAMA
Nuclear Engineering Laboratory
Toshiba Corporation, Japan

ABSTRACT

To improve the accuracy of BWR fuel design and safety evaluation, it is necessary to obtain a better understanding of high pressure two-phase flow phenomena. For this purpose, the authors developed a new two-phase measurement apparatus using X-ray beam scanning and X-ray CT (Computed Tomography) scanner methods, and measured the high pressure steam-water two-phase phenomena in a horizontal tube and a simulated BWR fuel rod bundle.

X-ray beam scanning and X-ray CT scanning showed the effectiveness of measuring the transient two-phase phenomena, as well as the local void distribution in the complicated flow channel.

These new two-phase flow measurement methods can also be used to visualize two-phase flow phenomena under high pressure conditions.

NOMENCLATURE

Co = concentration parameter

$f(x,y)$ = linear attenuation coefficient

G = mass flux

g_{ijk} = weighted coefficient for an element (i,j) at X-ray beam line $k(\theta)$

h_{ijk} = weighted coefficient for an element (i,j) at X-ray beam line $k(\theta)$

I_0 = initial X-ray intensity

I = X-ray intensity

j = cross-sectional average volumetric flux

j_g = cross-sectional average volumetric flux for vapor

La Grange Park, ILL.

Lahey, R.T.Jr., 1986, Subchannel Measurements of the Equilibrium Quality and Mass Flux Distribution in a Rod Bundle, *Proc. 8th Int. Heat Transfer Conf.*, *San Fransisco*, NR-01, pp.2399-2404.

Macbeth, R.V., 1974, Odds and Evens — A Formula for Enhancing the Dry-Out Power in Boiling Water Reactor Fuel Channels, AEEW-R-954, 21 pages.

Sadatomi, M., Sato, Y. and Saruwatari, S., 1982, Two-Phase Flow in Vertical Noncircular Channels, *Int. J. Multiphase Flow*, Vol.8, No.6, pp.641-655.

Sato, Y., Sadatomi, M. and Mine, T., 1984, Flow Distribution and Pressure Drop for Parallel Flow through Rod Bundles, *Bulletin of JSME*, Vol.27, No.224, pp.180-187.

Sato, Y. and Sadatomi, M., 1985, Data on Two-Phase Gas-Liquid Flow Distribution in Multiple Channels, *Proc. 2nd Int. Conf. on Multi-Phase Flow*, *London*, BHRA, pp.27-37.

Sato, Y., Sadatomi, M. and Tsukashima, H., 1987, Two-Phase Flow Characteristics in Interconnected Subchannels with Different Cross-Sectional Areas, *Proc. 1987 ASME·JSME Therm. Eng. Joint Conf.*, *Honolulu*, Vol.5, pp.389-395.

Shoukri, M., Tawfik, H. and Midvidy, W.I., 1982, An Experimental Investigation of Two-Phase Flow Interactions in Horizontal Rod Bundle Geometries, *Proc. 7th Int. Heat Transfer Conf.*, *Munich*, Vol.5, pp.355-360.

Shoukri, M., Tawfik, H. and Chan, A.M.C., 1984, Two-Phase Redistribution in Horizontal Subchannel Flow — Turbulent Mixing and Gravity Separation, *Int. J. Multiphase Flow*, Vol.10, No.3, pp.357-369.

Smith, S.L., 1969-70, Void Fraction in Two-Phase Flow — A Correlation Based upon an Equal Velocity Head Model, *Proc. Instn. Mech. Engs.*, Vol.184, Pt.1, No.36, pp.647-664.

Taitel, Y., Barnea, D. and Dukler, A.E., 1980, Modelling Flow Pattern Transitions for Steady Upward Gas-Liquid Flow in Vertical Tubes, *AIChE J.*, Vol.26, No.3, pp.345-354.

Tapucu, A., Ahmad, S.Y. and Gencay, S., 1982, Behaviour of Two-Phase Flow in Two Laterally Interconnected Subchannels, *Proc. 7th Int. Heat Transfer Conf.*, *Munich*, Vol.5, pp.361-366.

van der Ros, T., 1970, On Two-Phase Flow Exchange between Interacting Hydraulic Channels, Report WW015-R160, Eindhoven University of Technology, The Netherlands.

S = gap clearance
u = mean velocity in single-phase flow
x = quality
α = void fraction
ρ = density
ϕ_L^2 = two-phase frictional multiplier defined by Eq.(6)
χ = Lockhart-Martinelli parameter defined by Eq.(8)

Subscripts

A,B,E,F,I,J,a,b,c,i,j = subchannel identifiers
ij = identifier considering two adjacent subchannels i and j as a pair
G = gas
L = liquid

In addition, all parameters without a subchannel idenifier refer to a multiple channel as a whole.

REFERENCES

Bergles, A.E., 1969, Two-Phase Flow Structure Observations for High Pressure Water in a Rod Bundle, in *Two-Phase Flow and Heat Transfer in Rod Bundles*, pp.47-55, ASME Booklet, New York.

Chisholm, D. and Laird, A.D.K., 1958, Two-Phase Flow in Rough Tubes, *Trans. ASME*, Vol.80, No.2, pp.276-286.

Collier, J.G., 1979, Two-Phase Gas-Liquid Flows within Rod Bundles, in *Turbulent Forced Convection in Channels and Bundles —— Theory and Applications to Heat Exchangers and Nuclear Reactors*, ed. Kakaç, S. and Spalding, D.B., Vol.2, pp.1041-1055, Hemisphere, London.

Gonzalez-Santalo, J.M. and Griffith, P., 1972, Two-Phase Flow Mixing in Rod Bundle Subchannels, *ASME Paper* 72-WA/NE-19, 13 pages.

Lahey, R.T.Jr. and Schraub, F.A., 1969, Mixing, Flow Regimes and Void Fraction for Two-Phase Flow in Rod Bundles, in *Two-Phase Flow and Heat Transfer in Rod Bundles*, pp.1-14, ASME Booklet, New York.

Lahey, R.T.Jr. et al., 1972, Out-of-Pile Subchannel Measurements in a Nine-Rod Bundle for Water at 1000 PSIA, in *Progress in Heat and Mass Transfer*, Vol.6, Proc. Symposium on Two Phase Systems, ed. Hetzroni, G. et al., pp.345-363, Pergamon, Oxford.

Lahey, R.T.Jr. and Moody, F.J., 1979, *The Thermal-Hydraulics of a Boiling Water Nuclear Reactor*, 2nd ed., pp.122-138, ANS,

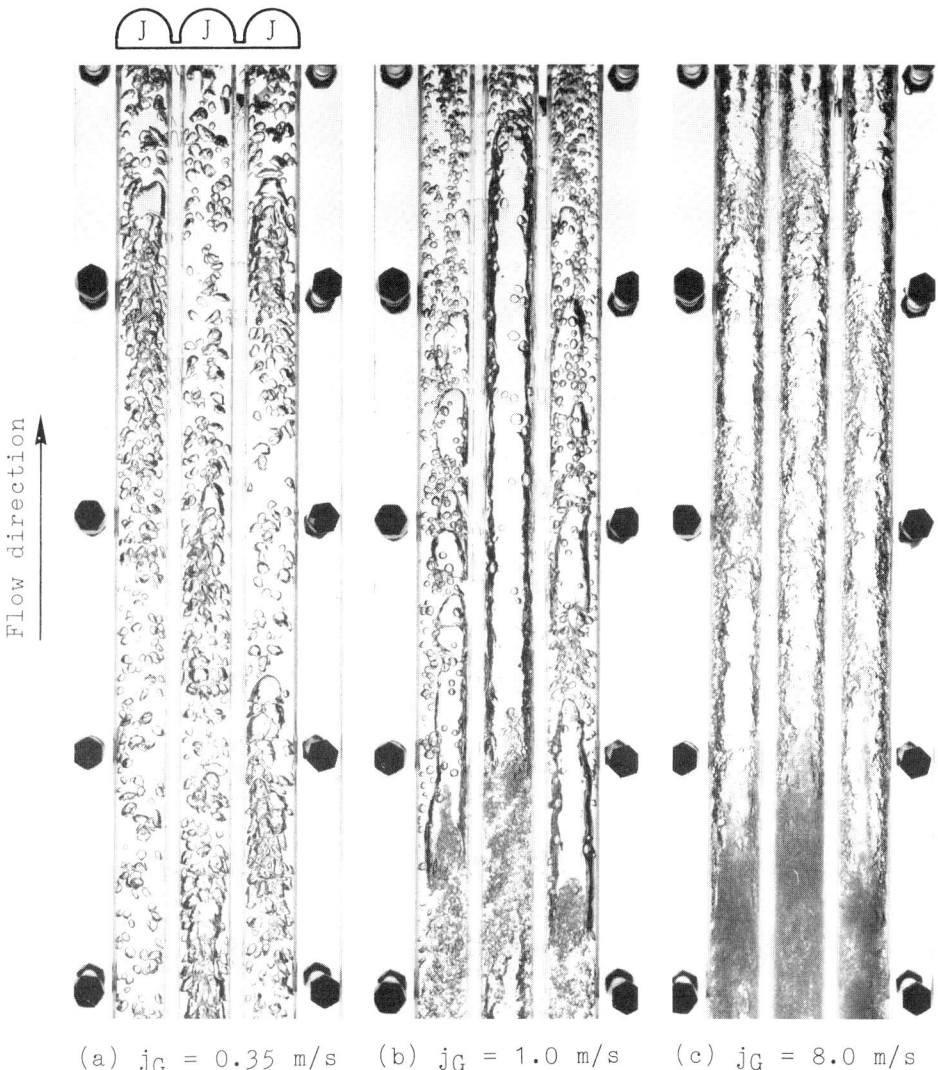

(a) j_G = 0.35 m/s   (b) j_G = 1.0 m/s   (c) j_G = 8.0 m/s

FIGURE A3. A series of flows in CH.J-J-J at j_L = 1.0 m/s

NOMENCLATURE

A = cross-sectional area
C = geometry factor of noncircular subchannel
D = hydraulically equivalent diameter
j = volumetric flux (= Q/A)
dp/dz = pressure gradient
Q = volume flow rate

FLOW DISTRIBUTIONS

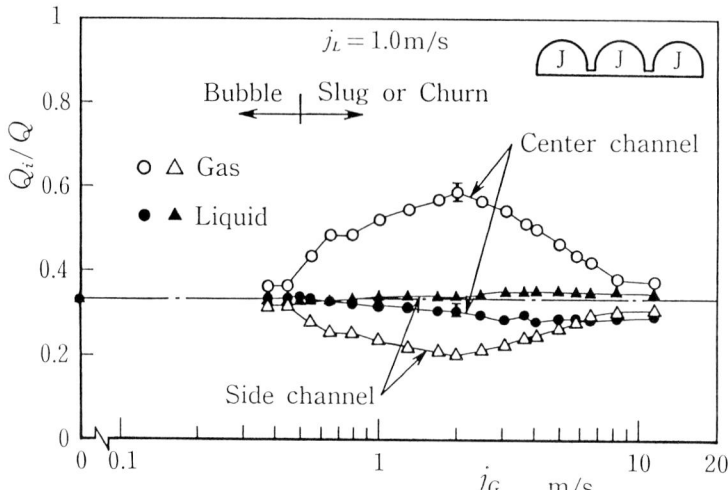

FIGURE A2. Flow distributions in Ch.J-J-J at j_L = 1.0 m/s

flow rate equally, as can be expected. When j_G < 0.5 m/s and the flow pattern is bubble flow, the flow distributions are similar to those in a single phase flow. But, when j_G > 0.5 m/s and the flow pattern becomes slug or churn flow, non-uniform flow distributions take place, resulting in a higher air flow rate in the center subchannel. At $j_G \simeq 2$ m/s where the non-uniformity of flow is the most pronounced, the proportion of air flow in the central subchannel amounts to 60%. This value corresponds to three times that in each side subchannel. The above-mentioned phenomena can be seen in the pictures shown in Fig. A3 (on the next page).

If the side subchannels were interconnected with each other, such non-uniform flow distributions would never have happened. That is, the distributions are not due to any difference in cross-sectional area of subchannels, as discussed in the text. Therefore, it can be said that the non-uniform flow observed in this test section is attributable to the subchannel arrangement.

Perhaps Ch.J-J-J should be considered to be uncommon because multiple channels of this kind are not frequently used in industries. However, in designing a multiple channel, attention should be paid to the fact that subchannel arrangement can be a cause of two-phase non-uniform flow distributions. Related to this, there is a similar statement in the report of Macbeth (1974).

Paying attention to the gas-phase flow distribution between two adjacent subchannels, an empirical correlation — Eq. (1) — has been proposed, in which non-uniform flow characteristics are taken into account. Equation (1) is applicable where the cross-sectional area ratio of the larger subchannel to the smaller ranges from 1.4 to 3.

A method for the numerical calculations of the flow distributions has been presented in Eq. (1). The agreement between the calculated values and the available data is reasonable.

It is felt that more systematic and reliable experiments for multiple channels having more than two subchannels are required for the progress of the research in this area.

APPENDIX: EFFECT OF SUBCHANNEL ARRANGEMENT

Different cross-sectional areas between adjacent subchannels causes non-uniform flow distributions, as mentioned in detail in the text. In addition, another supplementary experiment showed that non-uniform flow can occur even if the subchannels' geometries are identical, due to the arrangement of the subchannels. This experimental finding is described briefly below.

The cross section of the test channel used in this experiment is shown in Fig. A1. It had three semicircular subchannels, geometrically identical, arranged in a row as seen in the figure. The simplicity of the channel geometry made its fabrication and the visual observation easy. This test section is labeled as Ch.J-J-J since its channel geometry is identical to that of the subchannel-J in Ch.I-J shown in Fig. 1(c).

Figure A2 represents the flow distributions obtained from this Ch.J-J-J at j_L = 1.0 m/s. The data acquisition and presentation were similar to those in the case of Ch.a-b-c shown in Fig. 5. The following results can be seen from Fig. A2: In a single phase flow, the three subchannels divide the

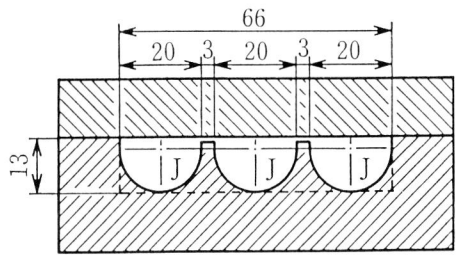

$D = D_J = 14.8$ mm, $S_{JJ} = 1.3$ mm

FIGURE A1. Cross section of Ch.J-J-J

FLOW DISTRIBUTIONS

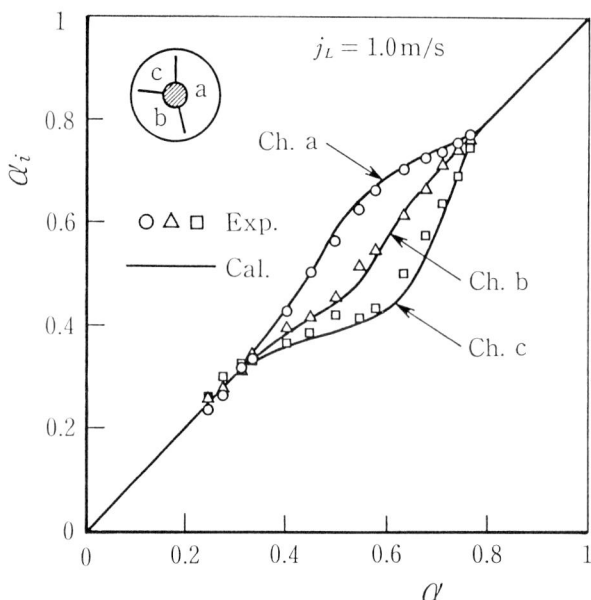

FIGURE 11. Comparison of the predicted void fraction distribution with the data obtaind from Ch.a-b-c

against the mean value. Each data point was calculated from the measured flow rates of both phases using Smith's correlation, Eq. (9).

As for the available experiments, it can be said, on the whole, that the prediction gives fairly good results for the subchannel flow distributions of both phases. However, there are a few cases in which a noticeable disparity has been observed between the predicted and measured values. The explanation for this discrepancy probably lies not only in the flow modeling but also in the measurement. The major errors found in the experiments may be attributable to the method of isokinetic flow splitting.

CONCLUSIONS

Experiments with air-water mixtures through a multiple channel having two or three subchannels have provided data on two-phase flow distributions under fully developed flow conditions. These data are available for the examination of a subchannel analysis.

The data showed that the flow distributions between two adjacent subchannels depend substantially on the volumetric flux of each phase. This considers the two subchannels as a pair and considers the ratio of the respective cross-sectional areas, independent of the subchannel geometry and the presence of other neighboring subchannels.

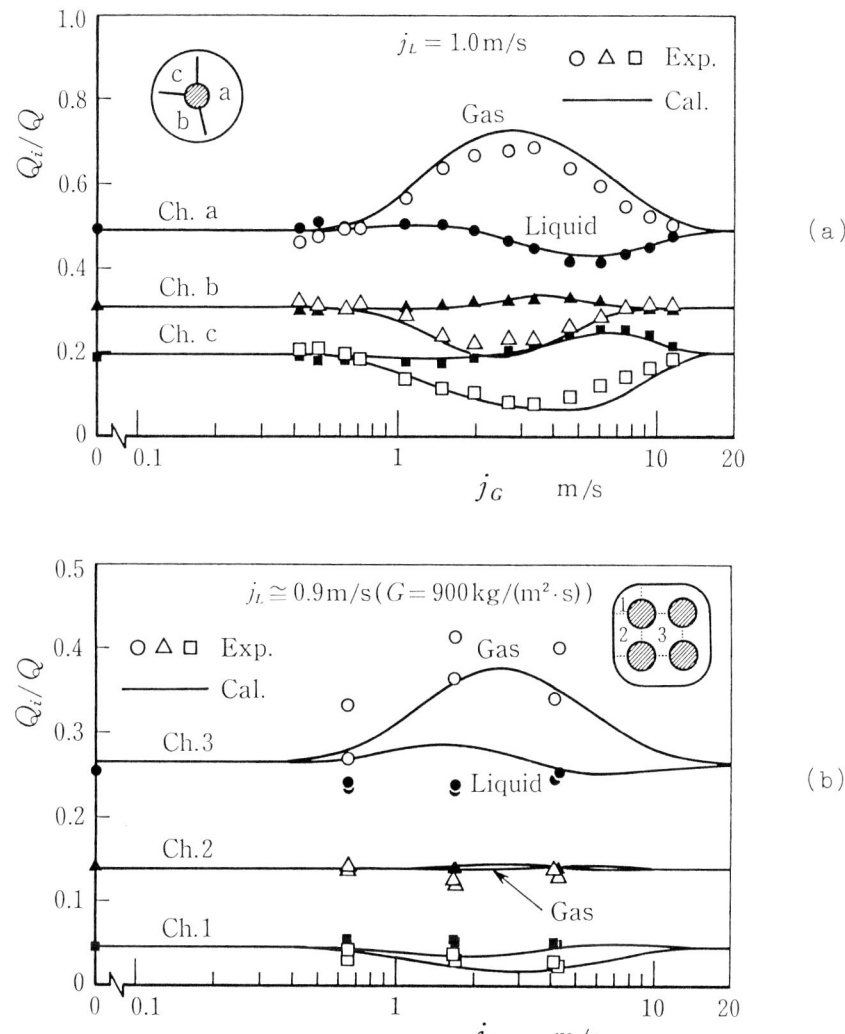

FIGURE 10. Comparisons of the predicted flow distributions with the data of (a) the present experiment and (b) Lahey (1986)

Figures 10(a) and (b) relate to a multiple channel having more than two subchannels: Figure (a) is concerned with Ch. a-b-c of the present experiment, whereas Fig.(b) is concerned with Lahey's 2×2 rod array channel, whose dimensions are twice those of a typical square pitched BWR fuel rod bundle (Lahey, 1986). Figure 11 shows a comparison of the subchannel void fraction distribution predicted by the proposed method (solid lines) with the experimental data obtained in Ch.a-b-c. The source experiment is the same as that shown in Fig. 10(a). The void fraction of each subchannel is plotted

FLOW DISTRIBUTIONS

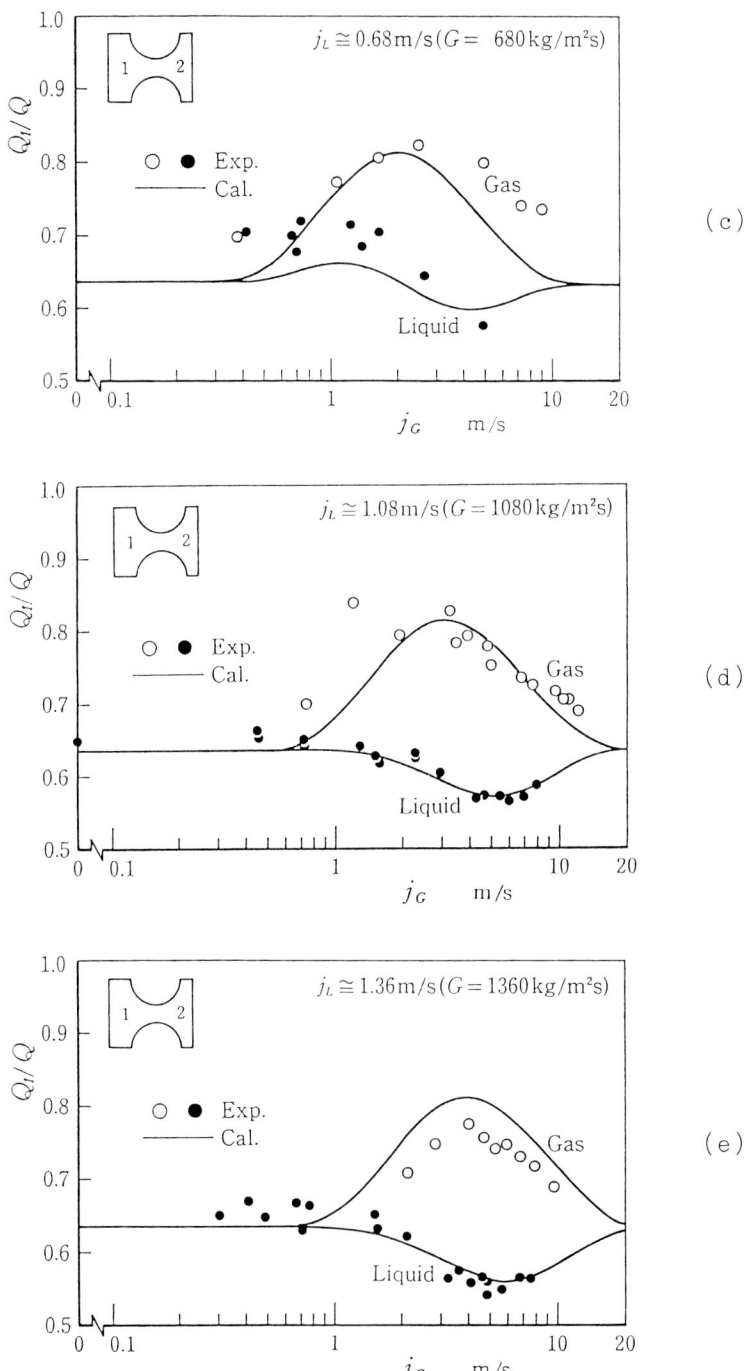

FIGURE 9. Comparisons of the predicted flow distributions with the data relating to 2-subchannel test sections. (a) and (b): the present experiment, (c)-(d): Gonzalez and Griffith (1972)

Comparison between predictions and experiments. Figures 9 (a) ∿ (e) show comparisons of the flow distributions predicted by the proposed method (solid lines) with those obtained experimentally in 2-subchannel test sections. Figure (a) is the case of the highest mass flux of the available data, obtained from Ch.I-J at j_L = 2.0 m/s. Figure (b) is an extreme case regarding the cross-sectional area ratio of the subchannels; i.e., the presented data relate to A_A/A_B = 3 in Ch.A-B. Figures (c) ∿ (e) are concerned with the experiment of Gonzalez and Griffith (1972). The geometry of their test section is shown by the figure in the upper left corner, with D_1 = 12.8 mm and D_2 = 9.8 mm in hydraulically equivalent diameter, and $A_1/A_2 \simeq 1.5$.

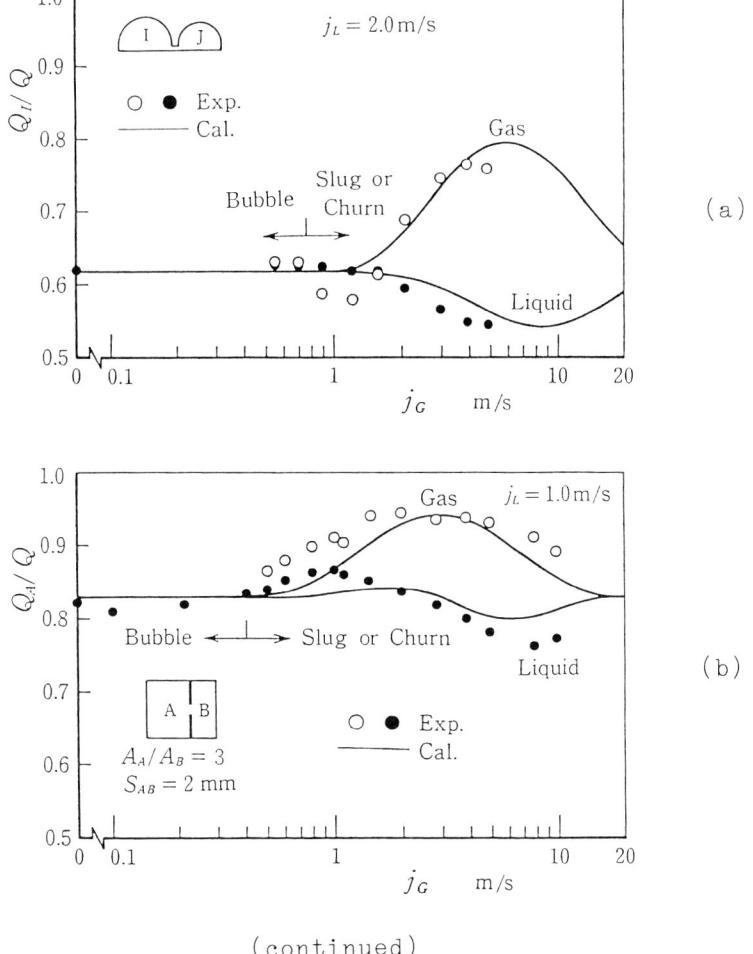

(continued)

FIGURE 9. Comparisons of the predicted flow distributions with the data relating to 2-subchannel test sections. (a) and (b): the present experiment, (c)-(d): Gonzalez and Griffith (1972)

FLOW DISTRIBUTIONS

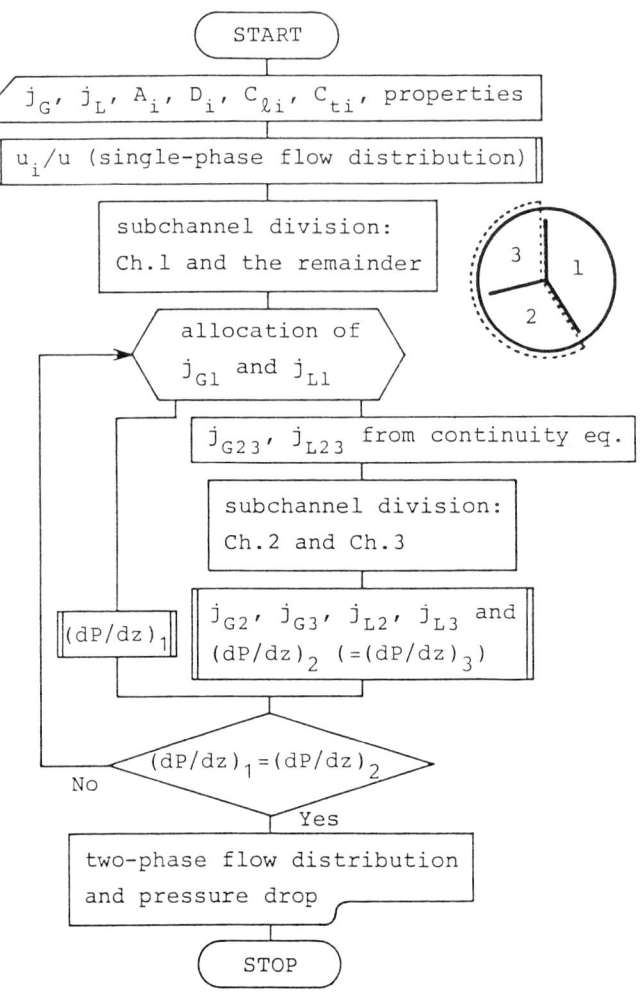

FIGURE 8. Explanatory diagram for the calcuration of flow distributions

5. Assume the volumetric flux of liquid phase in subchannel-1, j_{L1}, and calculate the corresponding pressure gradient $(dp/dz)_1$.
6. Determine the flow rates of the remainder from the continuity.
7. Divide this part into subchannels-2 and 3.
8. Determine j_{G2}, j_{G3}, j_{L2} and j_{L3} to satisfy Eqs. (1) $\sim$ (4), and obtain $(dp/dz)_2 = (dp/dz)_3$.
9. Compare $(dp/dz)_2$ with $(dp/dz)_1$ calculated at step 5.
10. Repeat the process in steps 4-9, with newly revised volumetric fluxes j_{G1} and j_{L1}, until consistent flow distributions result. The corresponding values of void fraction and quality in each subchannel can also be determined.

where $(dp_f/dz)_{Li}$ is the frictional pressure gradient for the liquid phase flowing alone within the same (i-th) subchannel. As to the multiplier ϕ_{Li}^2, the Chisholm-Laird correlation (Chisholm and Laird, 1958) Eq. (7), is used, since its additional applicability to a noncircular channel has been verified by Sadatomi et al (1982):

$$\phi_{Li}^2 = 1 + \frac{21}{\chi_i} + \frac{1}{\chi_i^2} \tag{7}$$

in which χ_i is the Lockhart-Martinelli parameter defined by

$$\chi_i = \sqrt{\frac{(dP_f/dz)_{Li}}{(dP_f/dz)_{Gi}}} \tag{8}$$

At this stage, where $(dp_f/dz)_{Li}$ and $(dp_f/dz)_{Gi}$ are calculated, noncircular subchannel geometry is taken into account by means of the friction factor, as well as the hydraulically equivalent diameter (Sadatomi et al., 1982).

In the calculation of the gravitational and accelerational terms, the void fraction α_i is estimated by use of Smith's correlation (Smith, 1969-70),

$$\alpha_i = \left[1 + 0.4 \frac{\rho_{Gi}}{\rho_{Li}} \left(\frac{1}{x_i} - 1 \right) + 0.6 \frac{\rho_{Gi}}{\rho_{Li}} \left(\frac{1}{x_i} - 1 \right) \left\{ \frac{\frac{\rho_{Li}}{\rho_{Gi}} + 0.4 \left(\frac{1}{x_i} - 1 \right)}{1 + 0.4 \left(\frac{1}{x_i} - 1 \right)} \right\}^{\frac{1}{2}} \right]^{-1} \tag{9}$$

in which x_i is the quality. This correlation has been found to be satisfactory for the determination of void fraction in a noncircular subchannel (Sato et al., 1987).

Numerical calculation. An explanatory diagram for the calculation of flow distributions is shown in Fig. 8, in which a multiple channel having three subchannels is considered as an example. The iteration method is used to determine the flow rates of both phases in each subchannel, for Eqs. (1) $\sim$ (4) to be satisfied. The step-by-step procedure is as follows:

1. Input prescribed values such as total flow rates of both phases, geometrical parameters of the channel, and fluid properties.
2. Calculate the flow distribution in a single-phase flow case, u_i/u.
3. Divide the channel into two parts; e.g. the subchannel-1, and the remainder enclosed by a dashed line, as seen in the attached figure. (There is no restriction to this subchannel division. That is, dividing into subchannel-3 and the remainder does not make any difference.)
4. Assume the volumetric flux of gas phase in subchannel-1, j_{G1}.

FLOW DISTRIBUTIONS

Prediction of Flow Distributions

Equations. Using the knowledge of the present experiment, an attempt was made to predict the flow distributions of a fully developed two-phase flow in a multiple channel. The essential point is that, if attention is paid to two adjacent subchannels from among the constituent subchannels, the flow distributions between the two become similar to those where these two subchannels alone are isolated from the others. To be more precise, Eq. (1) is used as one of the main equations for the flow in these two subchannels.

When the fluids are assumed to be incompressible and isothermal, the continuity equations for both phases are written as

$$j_G A = \sum_{i=1}^{k} j_{Gi} A_i \qquad (2)$$

$$j_L A = \sum_{i=1}^{k} j_{Li} A_i \qquad (3)$$

where k is the number of subchannels under consideration. Equations (2) and (3) mean that the sum of the flow rate of every subchannel is equal to that of the multiple channel as a whole.

In a fully developed flow region the equi-pressure gradient condition must be satisfied; i.e., the pressure drop in any subchannel must be equal to that in the multiple channel:

$$\frac{dP}{dz} = \left(\frac{dP}{dz}\right)_i, \quad (i = 1, 2 \cdots\cdots k) \qquad (4)$$

From a force balance the right hand of Eq. (4) can be written in the form,

$$\left(\frac{dP}{dz}\right)_i = \left(\frac{dP_f}{dz}\right)_i + \left(\frac{dP_h}{dz}\right)_i + \left(\frac{dP_a}{dz}\right)_i \qquad (5)$$

The three terms on the right side of this equation are the frictional, gravitational and accelerational components respectively. Axial momentum change due to turbulent mixing is assumed to be negligible in Eq. (5), because in a fully developed flow, gap clearance has no considerable effect on the flow distributions of both gas and liquid phases (Sato et al., 1987).

The frictional term is determined by use of the two-phase frictional multiplier ϕ_{Li}^2,

$$\left(\frac{dP_f}{dz}\right)_i = \phi_{Li}^2 \left(\frac{dP_f}{dz}\right)_{Li} \qquad (6)$$

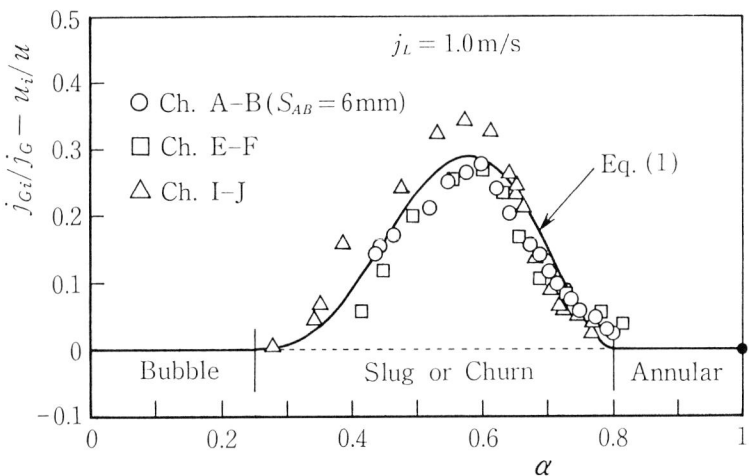

FIGURE 7. Comparison of Eq. (1) with experimental data

a result, the data on the gas phase flow distribution between two adjacent subchannels, say subchannels i and j ($A_i > A_j$), can be correlated well by

$$\frac{j_{Gi}}{j_G} = \begin{cases} \dfrac{u_i}{u} & : \alpha < 0.25 \\ \dfrac{u_i}{u} + 250 \dfrac{A_j}{A_i} (\alpha - 0.25)^3 (0.8 - \alpha)^2 & : 0.25 \leq \alpha \leq 0.8 \\ \dfrac{u_i}{u} & : 0.8 < \alpha \end{cases} \quad (1)$$

where j_{Gi}/j_G is the volumetric flux of the larger subchannel referred to the mean value, u_i/u the ratio of the larger subchannel to the mean velocity in a case of homogeneous fluid flow, and α the mean void fraction. Figure 7 is presented in connection with Eq. (1) for Ch.A-B, Ch.E-F, and Ch.I-J at $j_L = 1.0$ m/s. The data points relating to the deviation from the homogeneous case, $j_{Gi}/j_G - u_i/u$, are plotted in terms of the mean void fraction of both subchannels, $\alpha = (\alpha_i A_i + \alpha_j A_j)/(A_i + A_j)$. u_i/u was calculated from the knowledge of single-phase flow in a noncircular channel and in a multiple channel (Sadatomi et al., 1982; Sato et al., 1984). α_i and α_j were determined from the measured flow rates of both phases by means of Smith's correlation (Smith, 1969-70), which has been found to be satisfactory in the determination of void fraction in noncircular subchannels (Sato et al., 1987). The applicability of Eq. (1) will be discussed below in more detail.

FLOW DISTRIBUTIONS

flux in this part to 1 m/s, i.e., $j_{Lab} = (Q_{La} + Q_{Lb})/(A_a + A_b) = 1.0$ m/s, the air flow rate was gradually increased in small steps. Then, both air and water flow rates pertaining to the subchannel-a and the subchannel-b were measured, for each value of air flow. (In those runs, of course, a two-phase mixture flowed through the remaining subchannel-c, but no attention was paid to it.) Similar runs were also conducted, giving attention to subchannels 'b' and 'c' alone.

In Fig. 6, the data are plotted in terms of the volumetric flux in the larger subchannel referred to the mean value. As can be seen in the figure, similar flow distributions have been obtained from the three considered channels, 'a-b', 'b-c', and Ch.A-B, although there are slight differences between the gas phase data. Thus, it can be concluded that fully developed two-phase flow distributions between two adjacent subchannels depend substantially on the volumetric fluxes of both phases and the ratio of cross-sectional area, independent of the subchannel geometry and the presence of other neighboring subchannels.

As to fully developed two-phase flow distributions among interconnected subchannels, the above-mentioned experimental finding suggests the following idea: Consider two adjacent subchannels in a certain multiple channel, as well as two adjacent subchannels in a separate multiple channel, under the condition that the ratio of cross-sectional area of the respective subchannels of the former is equal to that of the latter. The flow distributions between the two adjacent subchannels will then become similar to each other, providing that the respective volumetric fluxes of gas and liquid phases are identical, considering the two subchannels as a pair. A multiple channel having only two subchannels can also be included in this idea. (The description of this paragraph is not necessarily applicable to any kind of multiple channel regardless of subchannel arrangement. An exceptional case will be represented in the APPENDIX.)

DISCUSSION

Empirical Correlation of Gas-Phase Flow Distribution

Theoretical prediction of fully developed two-phase flow distributions in any multiple channel has been desirable, but the complexity of the phenomena makes such approaches difficult. An alternative, more practical method is an empirical one. Along this line, Lahey (1986) has suggested correlations of subchannel void and mass flux distributions, based on his data of a 2×2 square rod array channel. However, the applicability of his correlations is limited to specific channel geometries. Another empirical correlation is proposed in this section.

The characteristics of two-phase distribution to each subchannel were examined thoroughly in the present experiment and in the previous experiment (Sato and Sadatomi, 1985). As

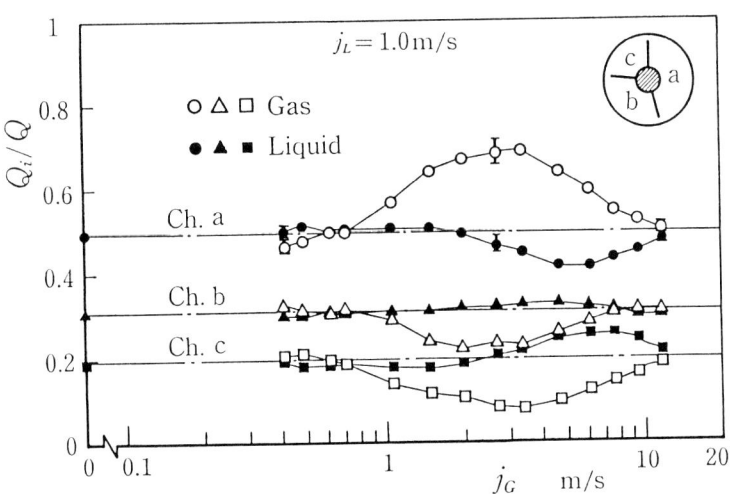

FIGURE 5.   Flow distributions in Ch.a-b-c at j_L = 1.0 m/s

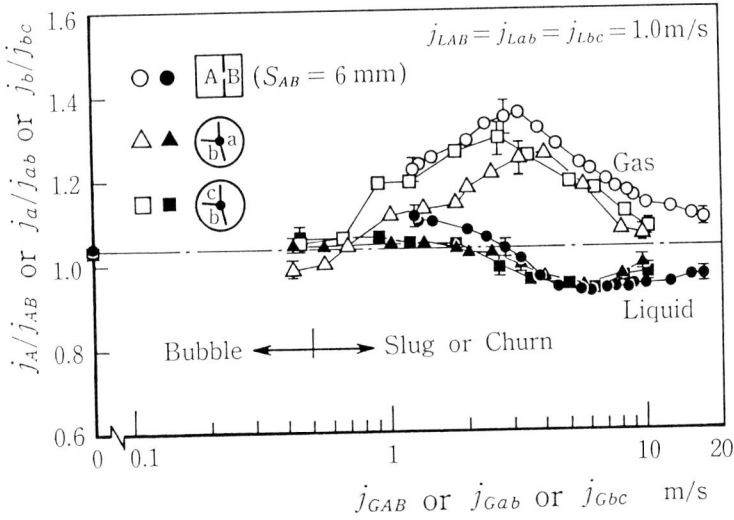

FIGURE 6. Comparison of flow distributions in 'a and b', 'b and c', and Ch.A-B

In addition, special test runs were conducted to study the interrelation between two adjacent subchannels in more detail, giving attention to the respective combinations of 'a and b' and 'b and c.' The results are presented in Fig. 6, together with those for Ch.A-B to facilitate comparison. Regarding cross-sectional area, the ratio of the adjacent subchannels was nearly identical in these test channels; i.e., $A_a/A_b = A_b/A_c = A_A/A_B = 1.5$. When the subchannels a and b alone were considered as a pair, the data acquisition was done in the following way: Adjusting the water volumetric

FLOW DISTRIBUTIONS

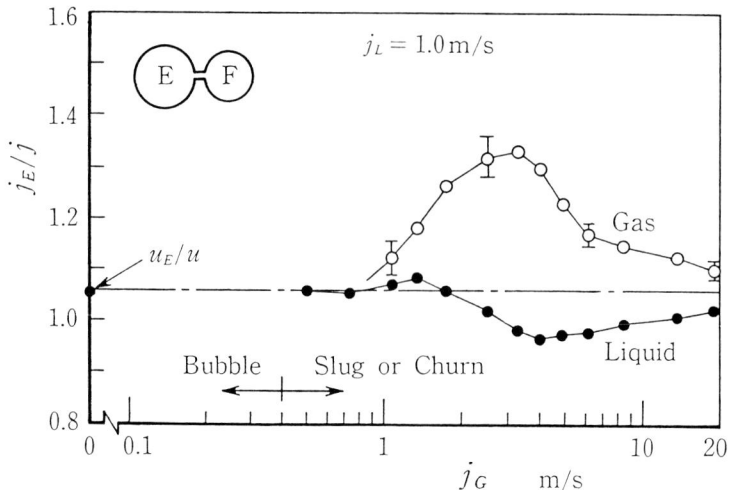

FIGURE 4. Flow distributions in Ch.E-F at j_L = 1.0 m/s

gas phase to drift toward the more open, higher-velocity subchannel as mentioned in the INTRODUCTION. A non-uniform flow depends greatly on the flow pattern of two-phase flow. The following trends can be observed when an air flow rate is increased gradually at a fixed water flow rate. Until the air flow rate corresponding to the bubble flow-slug flow transition is reached, non-uniform flows are indiscernible and the flow distribution of both phases is similar to that of a homogeneous fluid flow. But, once the air flow rate exceeds the above value and the flow pattern becomes a slug or churn flow, a void drift takes place with increasing air flow rate. This results in a higher air flow rate and a lower water flow rate in the larger subchannel-E, and vice versa in the smaller subchannel-F. After reaching the ultimately non-uniform flow distribution, the magnitude of non-uniformity decreases gradually with further increase of air flow rate. And finally, when the mixture velocity becomes high enough, the flow distributions are near the value of a homogeneous flow.

Flow Distributions in Ch.a-b-c

The results for Ch.a-b-c at j_L = 1.0 m/s are presented in Fig. 5. The data presentation is similar to that in Fig. 4, but in this figure the ratio of the volume flow rate of each subchannel to that of the entire channel (Q_i/Q) is plotted against j_G. The three dot-dash lines represent the flow rate ratio of the subchannels a, b and c respectively, if the mixture were homogeneous. Non-uniform flow distributions can also be seen in this 3-subchannel test section. At $j_G \simeq 3.5$ m/s, where the non-uniformity of flow is the most pronounced, the subchannel-a (A_a/A = 0.46) takes an excessive 70% share of air flow rate. Subchannels b and c (A_{bc}/A = 0.54) share the residual 30%.

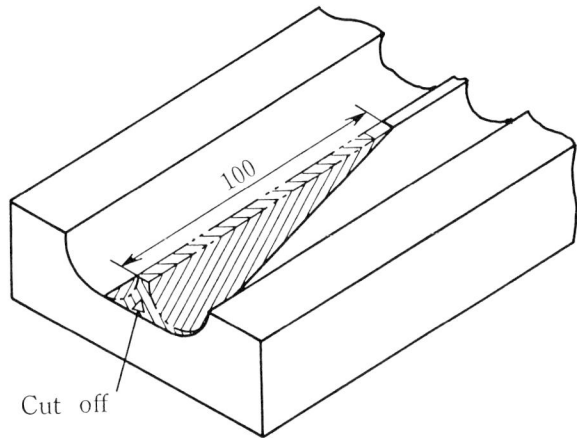

FIGURE 3. Inlet of Ch.E-F and Ch.I-J

surge tank, while the drainage led to a metering tank via a weir which controlled a water level in the separator.

The flow rate of each phase was measured at two locations, before entering the entry section and after leaving the subchannels. These two values were not always similar, particularly the values for air. Regarding the air flow rate, differences of less than 5% were found between the sum of the exit flow rate of the respective subchannels and the corresponding inlet flow rate. Since before entering a channel the air flow was steady and the measurement of it was considered to be correct to within ±1%, the major error involved in the determination of the air flow rate was attributed to flow fluctuations at the exit of the surge tank. Repeated measurements of flow rates of both phases were taken for each run. It was expected that the results would be generally reproducible to ±5% for air and ±2% for water, and accurate to 7% for air and 3% for water.

RESULTS

Flow Distributions in 2-Subchannel Test Sections

Figure 4 shows a typical example of the flow distributions obtained from Ch.E-F at $j_L = 1.0$ m/s. The ordinate is the volumetric flux ratio of the larger subchannel-E to the entire channel, and the abscissa is the volumetric flux of gas for the entire channel. The open and darkened symbols indicate gas and liquid phases respectively. The dot-dash line represents the flow distribution in a single-phase flow case. If the two-phase mixture were homogeneous, all data points would have been on the dot-dash line.

In this figure, non-uniform flows can be seen as deviations from the dot-dash line. Such flows result from the so-called 'void drift,' which is characterized by the tendency for the

FLOW DISTRIBUTIONS

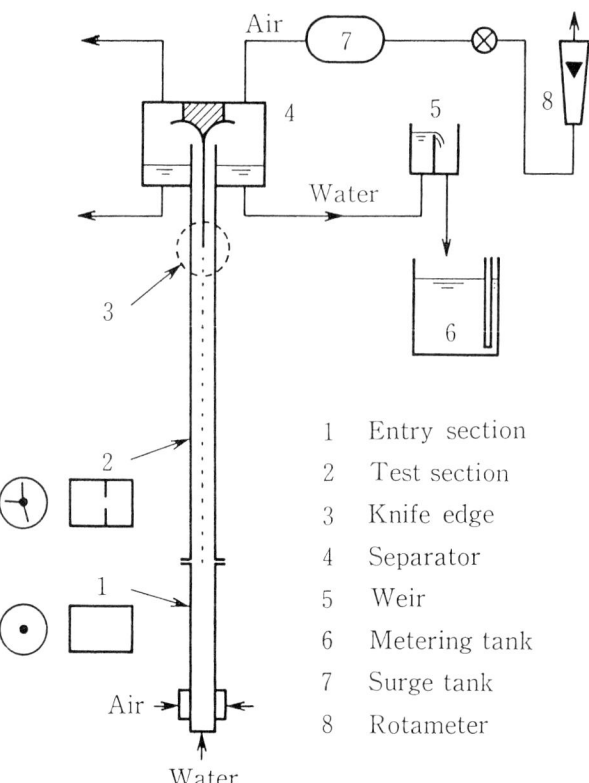

FIGURE 2. Test rig

and Ch.I-J. For the latter test sections, the ridge dividing the channel was cut off and tapered over 100 mm at the inlet of the test section to diminish the contraction effect. This is shown in Fig. 3.

After leaving a calibrated orifice, water flowed into the entry section at the bottom end. Following a calibrated rotameter, air was injected into the water stream through a number of holes drilled on the channel wall. The diameter of these air holes was either 0.3 mm or 2 mm, depending on the magnitude of the air flow rate; i.e., either $j_G < 2$ m/s or $j_G > 2$ m/s. This two-phase mixture flowed upward in the entry section, $(30 \sim 40)D$ in length, and entered into the test section, $(260 \sim 320)D$ in length. At the top end of the test section, the two-phase mixture was split into each subchannel by a knife edge under isokinetic conditions. The mixture then passed through each leading duct, $(10 \sim 20)D$ in length, which was the same subchannel as the test section but not interconnected with each other. The mixture then finally issued into each separator. As to the flow development, it had been confirmed by the previous experiment that an equilibrium state can be fully attained at the knife edge location (Sato and Sadatomi, 1985). In order to make the measurement of flow rate, air was then led to a rotameter via a

experiments (Sato and Sadatomi, 1985; Sato et al., 1987).

Ch.A-B was basically a 20.2×23.2 mm rectangular duct. A straight brass fin, 1 mm in thickness, was attached axially to each side wall to split the passage into subchannels A and B. The gap clearance or the area of both subchannels was variable, by changing the height or position of these fins.

Ch.E-F consisted of two circular subchannels, 20 and 16 mm in diameter respectively, connected by a gap clearance. The cross-sectional area ratio of the larger subchannel to the smaller, A_E/A_F, was about 1.5. This test section was made in the following way: The two semicircular troughs were first grooved 21 mm apart on a acrylic resin slab, and then the two identical slabs were put together.

Ch.I-J had two semicircular subchannels, made in a similar way. The two troughs, having respective diameters of 25 and 20 mm, were grooved 25.5 mm apart, and a flat slab was then affixed to cover them, resulting in the subchannels I and J. The cross-sectional area of each subchannel was nearly equal to that of Ch.E-F; $A_I \simeq A_E$ and $A_J \simeq A_F$.

Ch.a-b-c had three annular sector subchannels. This test section was basically a concentric annular channel made up of an acrylic resin pipe (33 mm ID) and a brass core rod (11 mm OD). Three straight brass fins, 2 mm in thickness, were rooted axially on the core rod to split the annular space into the three subchannels, so that one subchannel was connected to the other two by a gap clearance. These three subchannels were labeled 'a', 'b' and 'c' respectively, from the largest to the smallest. The angles between the fins were determined for the cross-sectional areas so the areas would differ from each other by a factor of 1.5. Also, the resulting areas of 'a' and 'b' became nearly identical to those in Ch.E-F and Ch.I-J; $A_a \simeq A_E \simeq A_I$, $A_b \simeq A_F \simeq A_J$. In order to hold the core rod concentrically and to keep the gap clearance constant, spacers were arranged axially every 0.5 m interval. These spacers were brass pieces (1 mm in height and 4 mm in length) soldered as a protuberance on top of each fin.

Test Rig

Measurement of two-phase flow distributions was made for the above-mentioned test sections, aligned vertically, using air and water as the working fluids.

A sketch of the essential part of the test rig is shown in Fig. 2. At the most upstream part of each test channel there was an entry section of (30 ∼ 40)D in length whose cross section was either nonsplit rectangular or concentric annular; the dimensions were 20.2×23.2 mm for Ch.A-B, 20×39 mm for Ch.E-F, 15.5×48 mm for Ch.I-J, and 11 mmOD and 33 mmID for Ch.a-b-c respectively. The geometries of both test section and entry section were identical in Ch.A-B and Ch.a-b-c (except for the fins) but were different in Ch.E-F

FLOW DISTRIBUTIONS

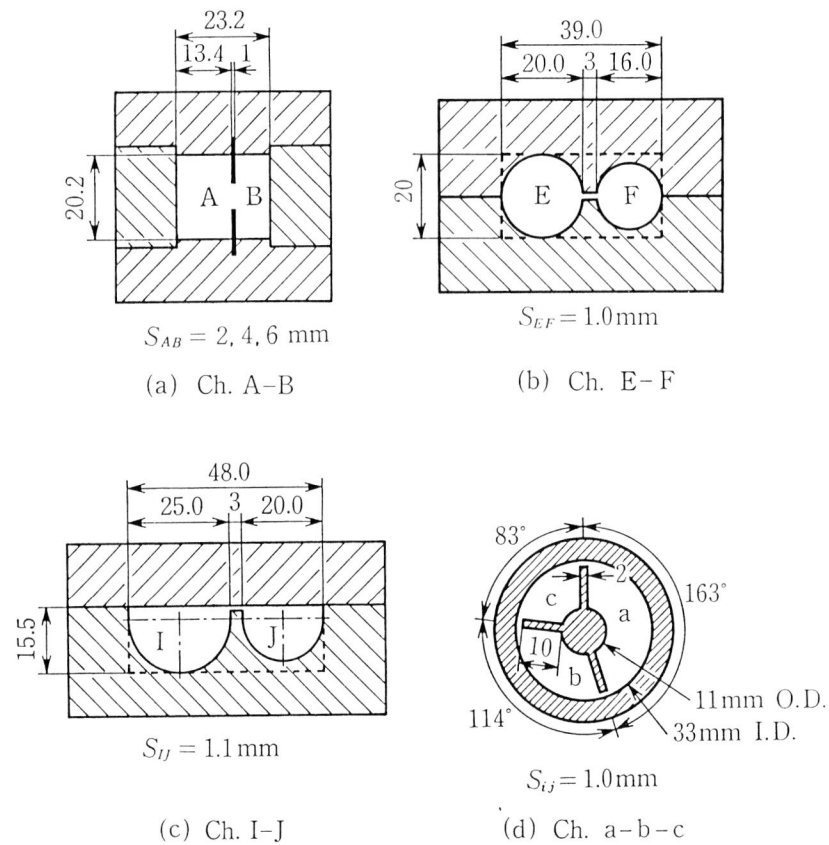

FIGURE 1. Cross section of the test sections

TABLE 1. Cross-sectional areas, hydraulically equivalent diameters, and gap clearances of the test sections

test section	Ch.	A mm^2	D mm	S_{ij} mm
A-B	A	276	17.7*)	2,4,6
	B	181	13.6*)	
	whole	457	15.2*)	
E-F	E	317	19.5	1.0
	F	204	15.6	
	whole	521	17.7	
I-J	I	317	17.7	1.1
	J	221	14.8	
	whole	538	16.4	
a-b-c	a	324	15.7	1.0
	b	221	13.8	
	c	155	11.9	
	whole	700	14.1	

*) In the case of S_{AB} = 6 mm

developing flow. Accordingly, a subchannel analysis for the two-phase flow must be able to predict accurately the non-uniformity of the flow. As yet, few satisfactory models exist which would describe such flow behavior. About ten years ago, Lahey and Moody (1979) and Collier (1979) pointed out that the understanding and quantification of void drift was one of the most important and difficult problems in two-phase flow. This problem remains unsolved today.

In order to gain accurate and useful information on the subchannel flow, it is more effective to perform experiments by multiple channels, using much simpler geometries than rod bundles. A considerable number of experiments has been done from this point of view (for example, van der Ros, 1970; Gonzalez-Santalo and Griffith, 1972; Tapucu et al., 1982; Shoukri et al., 1982; Shoukri et al., 1984; Lahey, 1986). However, there is a scarcity of reliable and systematic data on the two-phase flow, particularly when the constituent subchannels are geometrically dissimilar (Gonzalez-Santalo and Griffith, 1972; Lahey, 1986).

The studies of this series are based on the above-mentioned view and are concerned with flow characteristics of fully developed two-phase flows in multiple channels consisting of two or three subchannels in different cross-sectional areas. It was found in the previous experiments (Sato and Sadatomi, 1985; Sato et al., 1987) that void drift due to difference of subchannel area occurs even in a horizontal flow; that both gap clearance and geometry of the subchannel cross section have no considerable effect on the flow distributions; that as to the determination of void fraction in a noncircular subchannel, the Smith's correlation (Smith, 1969-70) is satisfactory, and that the flow pattern — even for a non-circular subchannel — can be predicted by the method of Taitel et al. (1980). The present study aims to increase systematic data on the flow distributions, in order to find main system parameters having significant effects on non-uniform flow phenomena, as well as to propose a practical method for the calculation of the flow distributions in a multiple channel.

EXPERIMENT

Test Sections

The purpose of the experiment was to observe fully developed two-phase flow in interconnected subchannels, when the cross-sectional area of the respective subchannels is different. Figures 1(a) ∿ (d) show the cross section of the four test sections used in the experiment. Three consisted of two subchannels while the fourth had three subchannels. From here on these test sections will be called 'Ch.A-B', 'Ch.E-F', 'Ch.I-J' and 'Ch.a-b-c' respectively, as shown in the figure. The cross-sectional areas, hydraulically equivalent diameters, and gap clearances are summarized in Table 1. Ch.A-B and Ch.I-J were the same channels used in the previous

TWO-PHASE GAS-LIQUID FLOW DISTRIBUTIONS IN MULTIPLE CHANNELS

Y. SATO and M. SADATOMI
Department of Mechanical Engineering
Kumamoto University
Kumamoto 860, Japan

ABSTRACT

To provide useful data for the improvement of subchannel analysis, experiments were performed for two-phase flow of air-water mixtures through a multiple channels. Four test channels were used. Three channels consisted of two interconnected subchannels, while the fourth consisted of three interconnected subchannels. In each channel, the cross-sectioal area of the respective subchannels was different. Data on the fully developed flow distributions to each subchannel are presented. Based on the data, a practical method to calculate the flow distributions in a multiple channel is proposed. Comparisons are then made between the predictions and the available data.

INTRODUCTION

In describing a flow in a complex channel geometry, the flow passage is usually divided into pertinent subchannels. This kind of conduit, made up of subchannels, is termed a 'multiple channel.' The most sophisticated examples in industry are water-cooled nuclear reactors and shell-and-tube heat exchangers.

For a quarter of a century, two-phase flow in rod bundles has drawn the attention of nuclear reactor designers, to establish a rational basis for predicting the local heat flux and void fraction. This explains the chief difficulty of the problem — the complexity of the flow. One of the complex flow phenomena making the problem difficult is phase separation. Early experiments on two-phase flow in rod bundles have shown that there is a strong tendency for the gas phase to flow in the more open, higher velocity central region of the bundle (Lahey and Schraub, 1969; Bergles, 1969; Lahey et al., 1972). Lahey et al. (1972) called this phenomenon 'void drift' and gave an explanation of it.

Void drift results in non-uniform distributions of the flow rates of both phases and the void fraction over the subchannels in a fully developed equilibrium flow, as well as in a

Transient Phenomena in Multiphase Flow, May 24-30, 1987, Dubrovnik, Yugoslavia.

Zun, I. (1986), Pressure Drop Prediction for Bubbly Flow Based on Local Turbulent Characteristics, Proc. of 4th Miami Int. Symposium on Multiphase Transport & Particulate Phenomena, Miami Beach, Florida.

Zun, I., Kljenak, I. and Serizawa, A. (1988), Bubble Coalescence and Transition from Wall Void Peaking to Core Void Peaking in Turbulent Bubbly Flow, Proc. of the Japan-U.S. Seminar on Two-Phase Flow Dynamics, July 15-20, 1988, Ohtsu, Japan.

u	:	ℓ_s/D_b
$\vec{v}_b$	:	bubble velocity
v_{sz}	:	passing velocity of a bubble
x	:	coordinate
y	:	coordinate
z	:	coordinate

Greek

α : angle defined in Fig. 5
α_o : maximum α defined by Eq.(12)
α_v : local void fraction
β : angle defined by Fig. 5
μ : angle defined by Fig. 5
ν : angle defined by Fig. 5
σ_{sz} : standard deviation of ensemble-average bubble velocity v_{sz}

Subscripts

g : gas
j : j-th bubble
ℓ : liquid

Symbols

$-$: ensemble average
$\rightarrow$: vector

REFERENCES

Anderson, T. T. (1964), Comments on Two-Phase Measurements Using a Resistive Probe, AIChE J., Sept. pp. 776-791.

Bankoff, S. G. (1964), Bubble Radius Distribution Functions from Resistivity Probe Measurements, AIChE J., Sept., pp. 776 - 792.

Gakuhari, K. (1984), Bubble size distribution in Two-Phase Flow, B.S. Thesis, Kyoto University (in Japanese).

Kataoka, I., Ishii, M. and Serizawa, A. (1986), Local Formulation of Interfacial Area Concentration in Two-Phase Flow, Int. J. Multiphase Flow, 12-4, pp. 505-529.

Herringe, R. A. and Davis, M. R. (1976), Structural Development of Gas-Liquid Mixture Flows, J. Fluid Mech., Vol.73, part 1, pp. 97-123.

Sekoguchi, K., et.al. (1984), Velocity Measurements with Electrical Double-Sensing Devices in Two-Phase Flow, in Measuring Techniques in Gas-Liquid Two-Phase Flow, edited J.M. Delhaye, G. Cognet, Springer-Verlag.

Serizawa, A. (1984), Improvement of Resistivity Probe Method (unpublished).

Serizawa, A. and Kataoka, I. (1987), Phase Distribution in Two-Phase Flow, Proc. of the 1987 ICHMT Int. Seminar on

CONCLUSIONS

The effect of bubble size on phase distribution in a vertically upward two-phase flow in a round tube was studied experimentally, using a specially designed nozzle-type bubble generator. Results showed very clearly a strong dependence of the phase distribution on the bubble size distribution. Even a small change in bubble size which is difficult to recognize by visual observation may result in a considerable change of phase distribution pattern. This fact might explain the well-known discrepancies existing among phase distribution data obtained by different investigators.

Within the range of bubble size covered in this experiment (1 ∿ 8 mm in bubble diameter), there was a trend for larger bubbles to rise in the flow core, as expected. However, this general trend appears to vary dependently on the liquid velocity j_ℓ. Under certain conditions, bubble coalescence process is likely to be important in determining phase distribution. A formula was derived to calculate the bubble-size distribution from the probability density function for the chord length measured by a double-sensor probe method.

This paper is the first of a series dealing with the bubble-size effect, particularly on bubbly flow structure. For this reason, not much has been included in this paper. However, the authors hope some useful results have been obtained for further steps.

Acknowledgement
The authors acknowledge Mr. Z. Kawara for his help in computations.

NOMENCLATURE

D_b : bubble diameter

j : volumetric flux

ℓ_j : chord length for j-th bubble

ℓ_s : apparent chord length defined by Eq.(2)

$\vec{n}$: normal unit vector

$P_k(k)$: probability density function for quantity k

$P_u^N(u)$: normalized probability density function of u, defined by Eq.(14)

r : radial coordinate

$s_{\alpha\mu}$: projected area of the surface onto α-μ plane in (α, μ, β - ν) space

Δs : probe distance

t_c : contact time

Δt : transit time

BUBBLE SIZE EFFECT

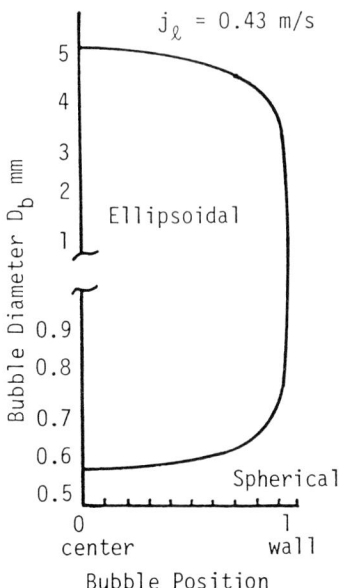

Fig. 11 Observation of bubble lateral motion (Zun, 1986)

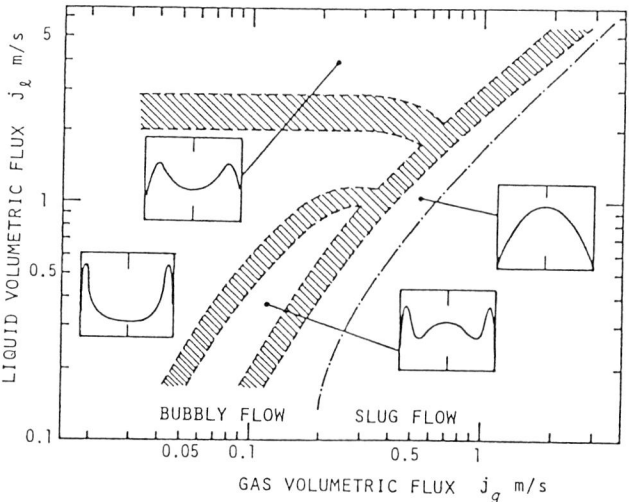

Fig. 12 Phase distribution pattern
(Serizawa and Kataoka, 1987)

diameter distribution. However, solving numerically this integral equation based on measured chord-length histogram is not an easy task, and, perhaps, needs some approximations. Therefore, in this paper, the authors demonstrated just the result of chord-length distribution instead of bubble-size distribution.

very sensitive to the method of bubble injection, and not uniquely determined by j_ℓ, j_g and physical properties of the fluids alone.

By refering to the bubbly flow phase distribution data obtained in this measurement, it is understood that the effect of increasing the nozzle velocity j_n is to increase the local void fraction in the wall region and to decrease core void fraction, that is, a change from core void-peaking flow to wall void-peaking. In view of smaller bubbles being generated at higher j_n, this result can be interpreted in such a way that smaller bubbles tend to be collected near the wall, so far as the bubble size covered in this experiment is concerned. This observation is qualitatively consistent with Zun's observation of single bubble lateral migration, indicating that bubbles with intermediate size tend to rise near the wall (Fig. 11; Zun, 1986).

Bubble coalescence was clearly observed in most cases of both slug flow and the transition from slug to bubbly flow. Typical examples of this can be seen in Photo. 3 (j_ℓ = 2.0 m/s, j_g = 0.1 m/s : j_n = 0 m/s, slug flow : j_n = 0.4 m/s, transition flow : j_n = 0.6 m/s, bubbly flow). This bubble coalescence is important particularly in the transition from wall void-peaking to core void-peaking and vice versa. This process should be taken into consideration in modelling the phase distribution phenomena (Zun et al., 1988).

Using the technique to change bubble size like as adopted in this work, the phase distribution can be thus, in some sense, easily changed for fixed j_ℓ and j_g. However, there appears still to exist a general rule regarding the j_ℓ-dependence of the phase distribution patterns. Figure 12 represents a bubbly flow regime map obtained by summarizing the existing phase distribution data reported by many investigators with different bubble injection methods (Serizawa and Kataoka, 1987). According to this map, at low j_g, the bubbly flow shows wall void-peaking distribution at low liquid velocity (but not very low). On the other hand, at higher liquid velocity, it does not have a sharp void fraction-peak near the wall with the maximum at a point nearer to the tube center. The present result shows a similar trend when comparing Fig. 6(b) (j_ℓ = 1.0 m/s) and Fig. 6(d) (j_ℓ = 4.0 m/s).

The effect of bubble shape is another important factor determining phase distribution phenomena as described previously in Fig. 1. However, this effect is very difficult to evaluate, since we have no means at present to define the bubble deformation quantity in an appropriate way.

Finally, a knowledge of bubble-size distribution is also quite important to understand in more detail the phase distribution and turbulence mechanisms. For this reason, the authors have derived an integral equation relating the probability density function for the chord-length to bubble

BUBBLE SIZE EFFECT

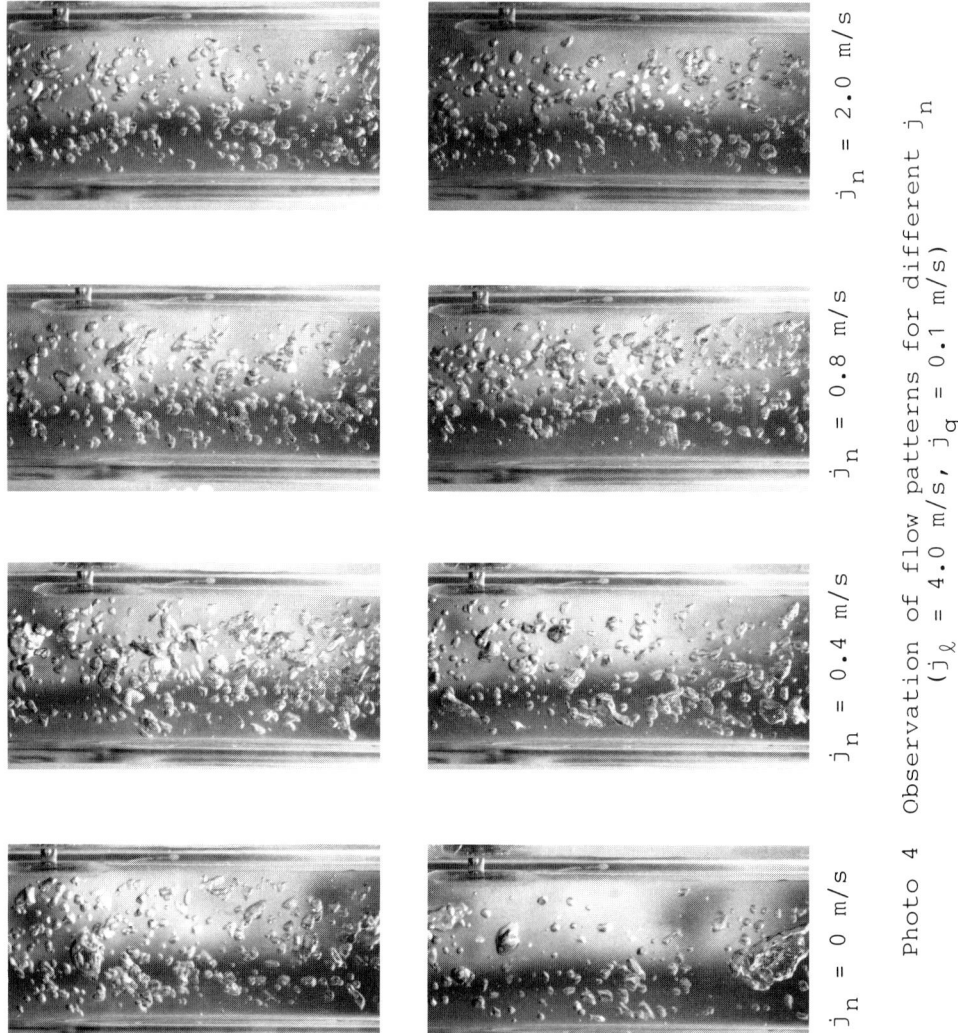

Photo 4 Observation of flow patterns for different j_n
(j_ℓ = 4.0 m/s, j_g = 0.1 m/s)

394 DISTRIBUTION PHENOMENA

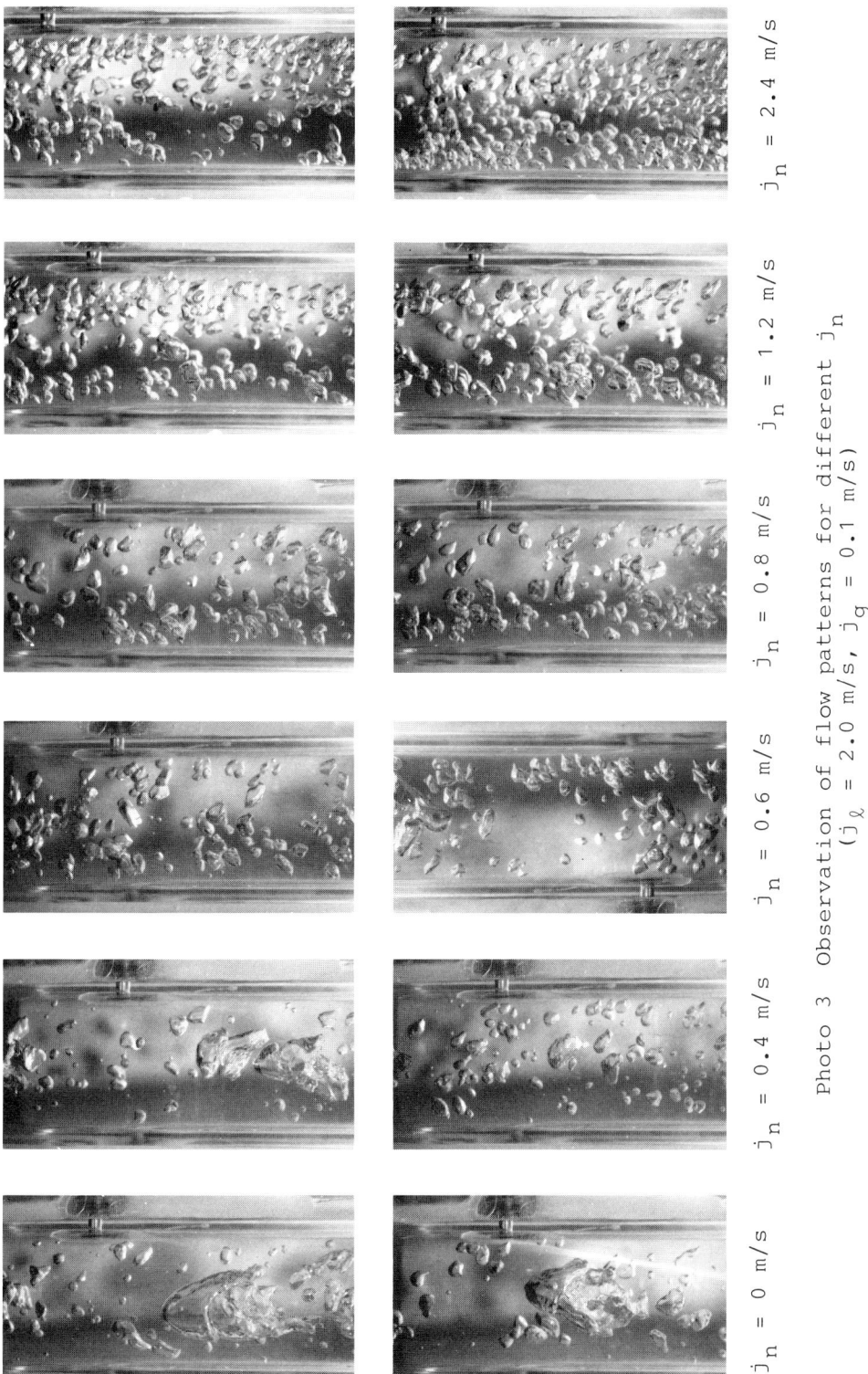

Photo 3 Observation of flow patterns for different j_n
($j_\ell = 2.0$ m/s, $j_g = 0.1$ m/s)

BUBBLE SIZE EFFECT

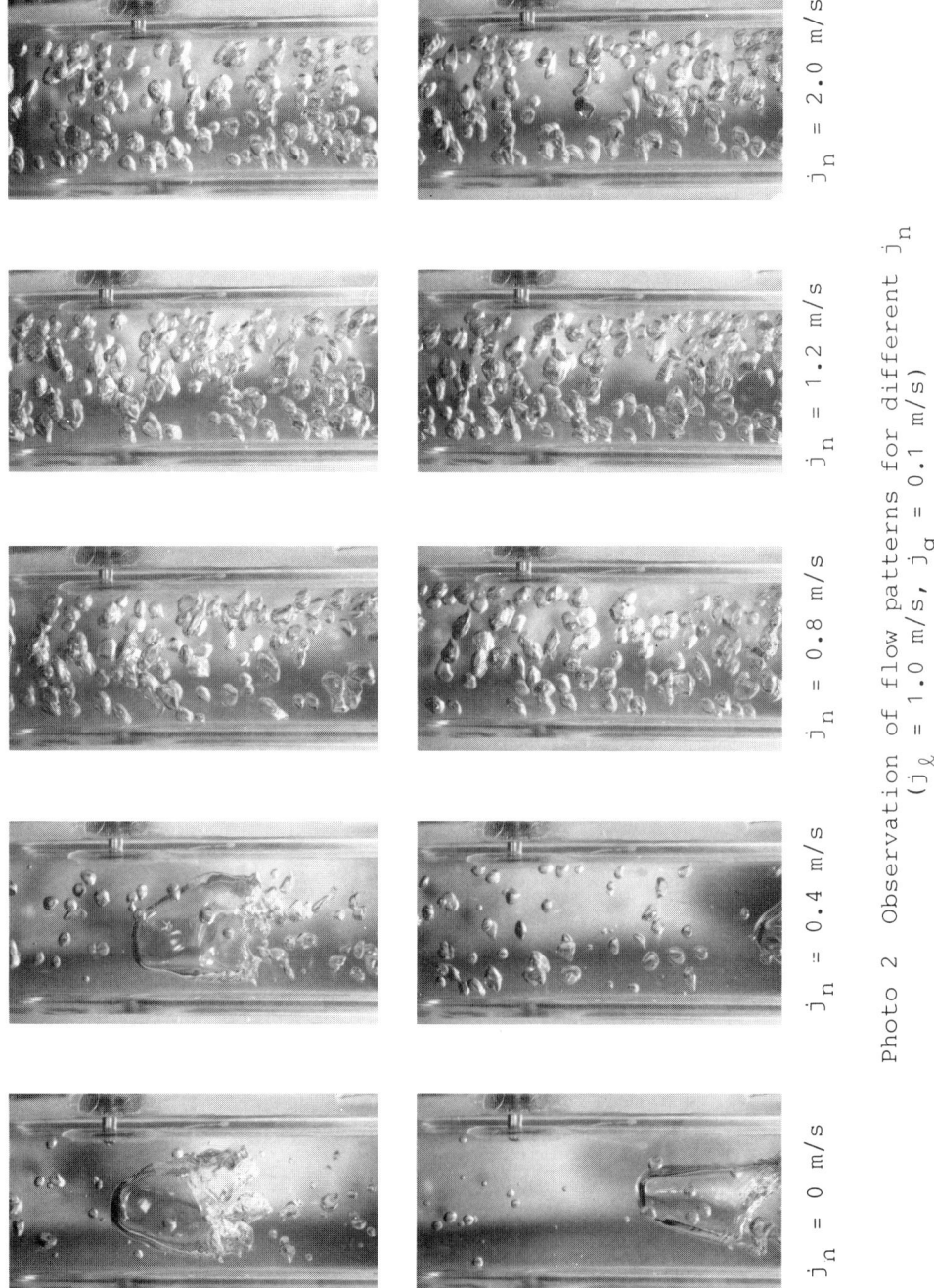

Photo 2 Observation of flow patterns for different j_n
($j_\ell = 1.0$ m/s, $j_g = 0.1$ m/s)

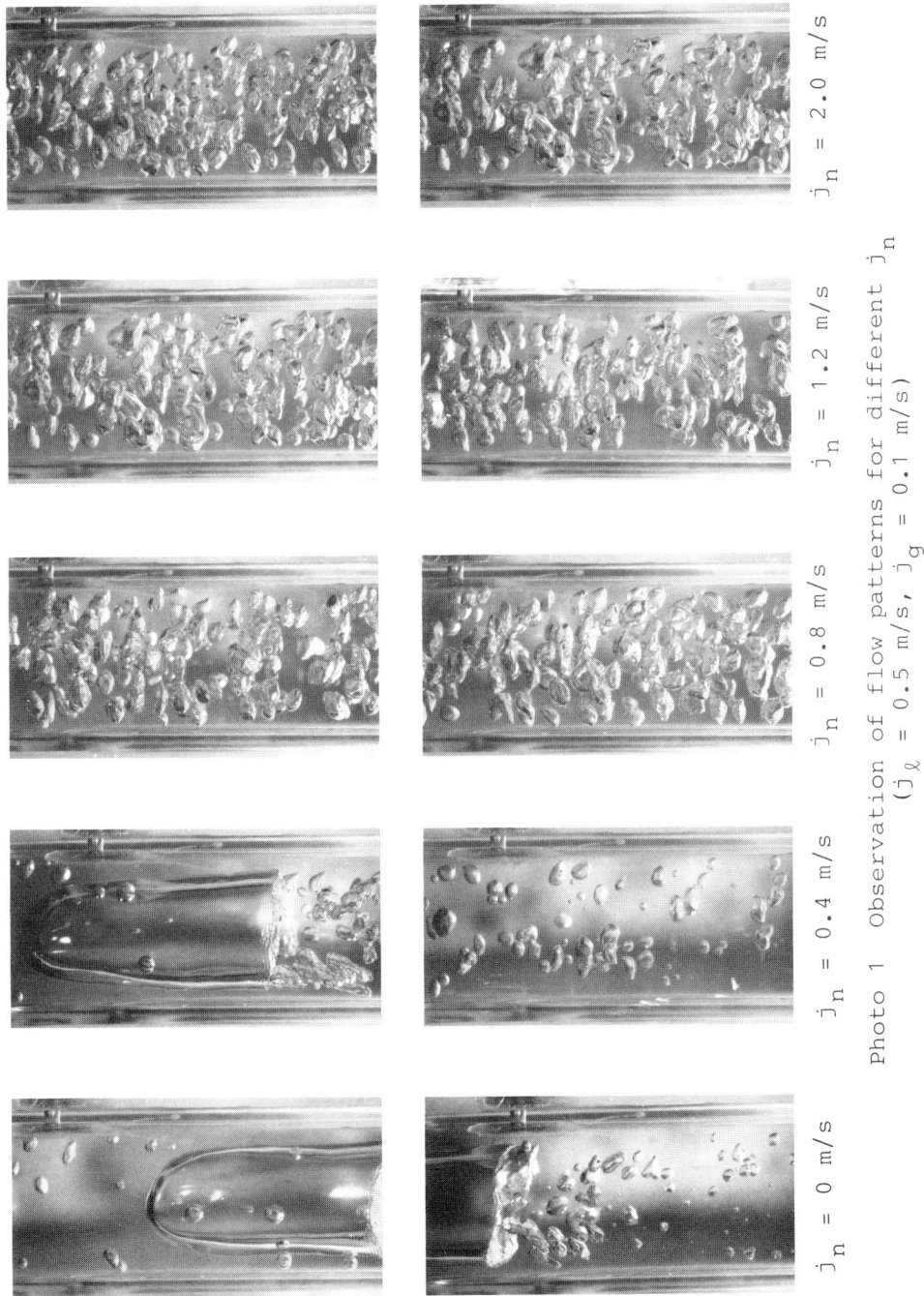

Photo 1 Observation of flow patterns for different j_n
(j_ℓ = 0.5 m/s, j_g = 0.1 m/s)

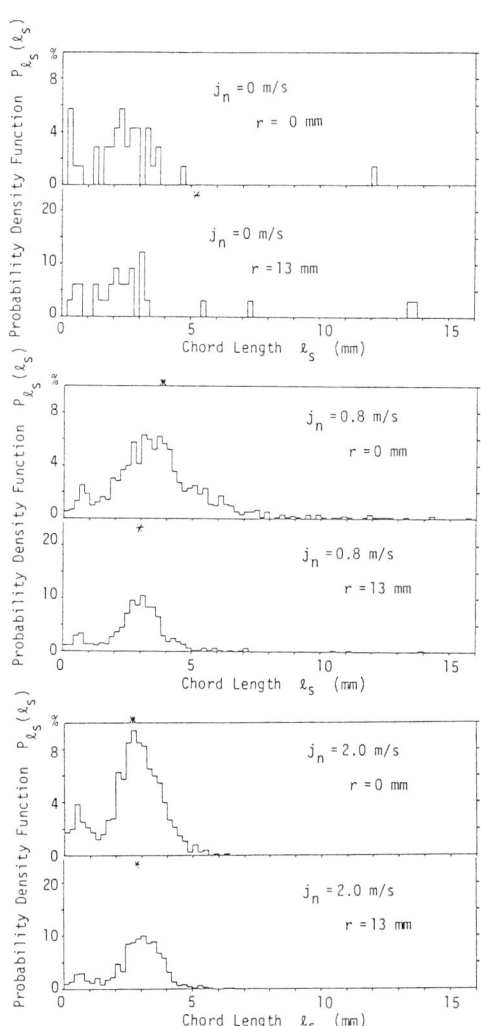

Fig. 10 Chord length distribution (j_ℓ = 4.0 m/s, j_g = 0.1 m/s)

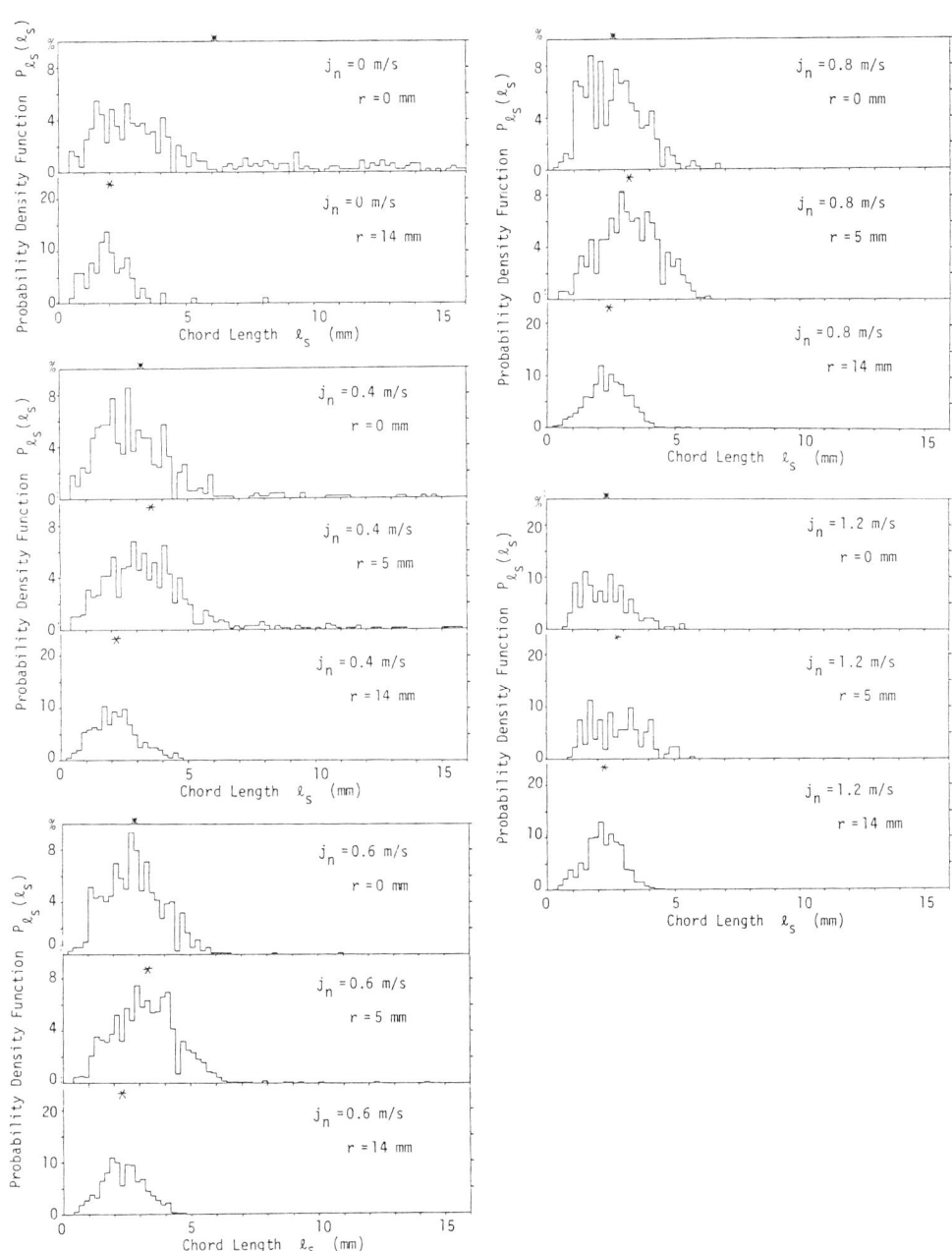

Fig. 9 Chord length distribution (j_ℓ = 2.0 m/s, j_g = 0.1 m/s)

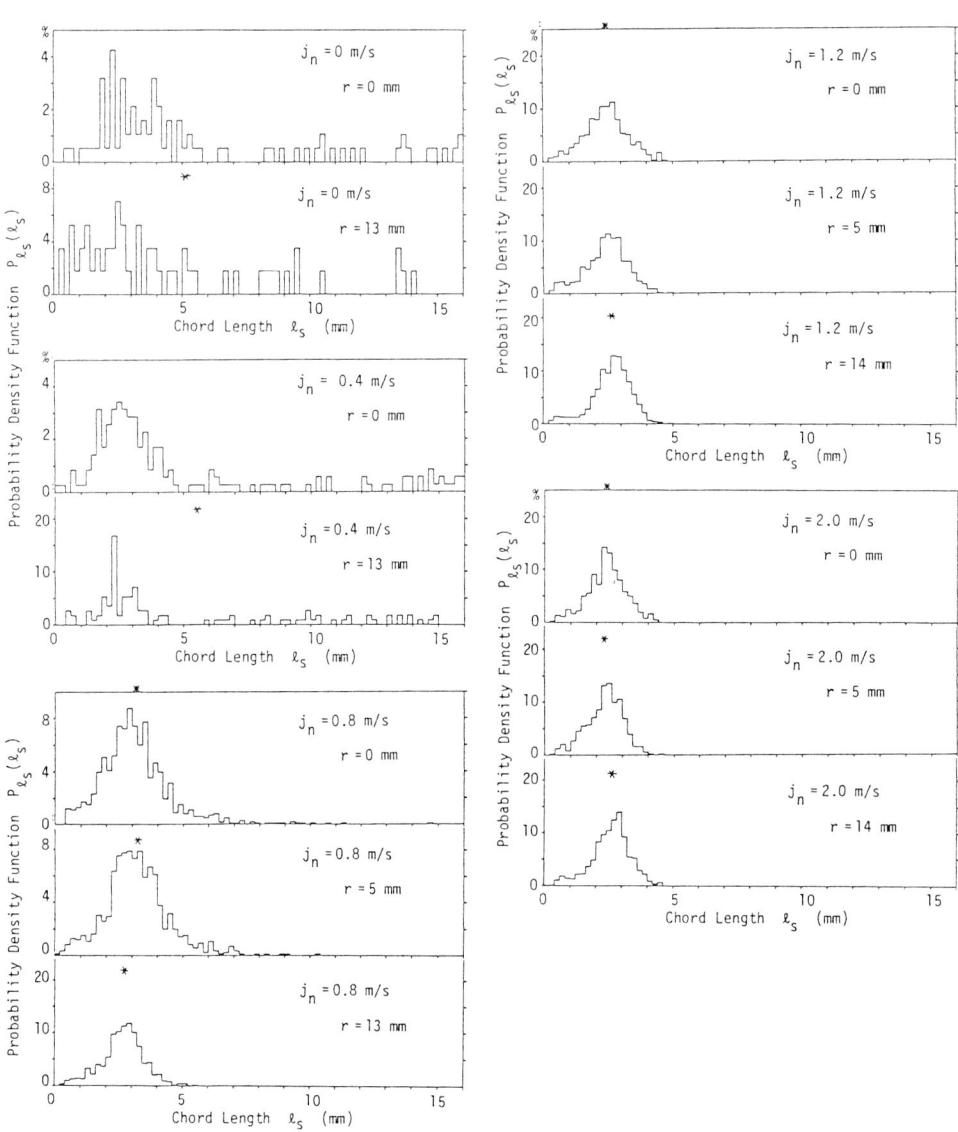

Fig. 8 Chord length distribution (j_ℓ = 1.0 m/s, j_g = 0.1 m/s)

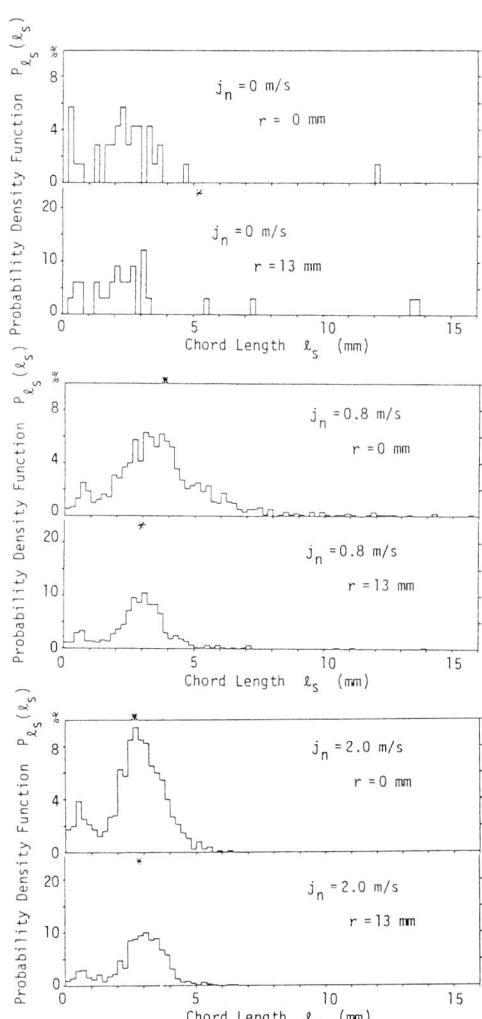

Fig. 7 Chord length distribution (j_ℓ = 0.5 m/s, j_g = 0.1 m/s)

BUBBLE SIZE EFFECT

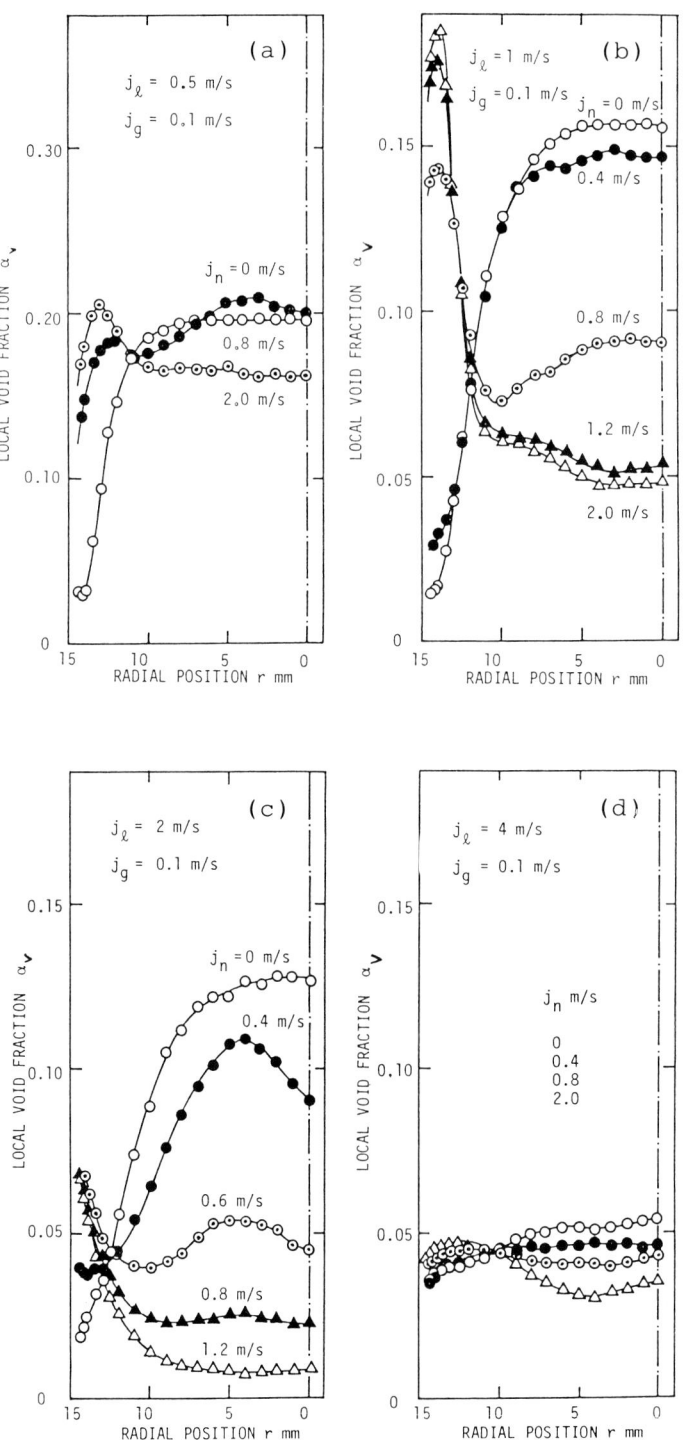

Fig. 6 Variation of phase distribution due to a change of j_n

$$P_u^N(u) = P_u(u) / \int_0^\infty P_u(u)\, du \tag{14}$$

As is clear from the discussions above, the function $P_u^N(u)$ represents the probability density function of ℓ_s/D_b defined for constant bubble diameter D_b. Therefore, the probability density function for ℓ_s for constant D_b is given by

$$P_{\ell_s}^{D_b}(\ell_s) = P_u^N\left(\frac{\ell_s}{D_b}\right) \frac{1}{D_b} \tag{15}$$

Then the probability density function for bubble chord length P_{ℓ_s} for all D_b is given as follows, using Eq.(14) and the probability density function $P_{Db}(D_b)$ for the bubble diameter D_b.

$$P_{\ell_s}(\ell_s) = \int_0^\infty P_{\ell_s}^{D_b}(\ell_s)\, P_{Db}(D_b)\, dD_b$$

$$= \int_0^\infty P_u^N\left(\frac{\ell_s}{D_b}\right) \frac{1}{D_b} \cdot P_{Db}(D_b)\, dD_b \tag{16}$$

From Eq.(16) the bubble-diameter distribution $P_{Db}(D_b)$ is thus calculated by solving the integral equation given by Eq.(16) for given (measured) chord-length distribution $P_{\ell_s}(\ell_s)$, since $P_u^N(\ell_s/D_b)$ is now known from Eqs.(11), (12), (13) and (14).

RESULTS AND DISCUSSIONS

Phase distribution, and measured chord-length distribution were given through Fig. 6 to Fig. 10 together with visual observations (Photos. 1 ~ 4) for $(j_\ell, j_g) = (0.5, 0.1), (1.0, 0.1), (2.0, 0.1)$ and $(4.0, 0.1)$ m/s. Measurement was performed for different bubble-size distributions obtained by changing nozzle velocity j_n, where j_n was defined as the superficial liquid velocity within the nozzle tube (6-mm i.d.), not in the test section (30-mm i.d.). Since the bubble size changed significantly in low nozzle velocity region, the experiment was restricted to rather low nozzle velocities.

Two photographs of each experimental condition are shown to give a realistic view of the flow situation. It is understood from these photographs that, in general, smaller bubbles were brought about by higher nozzle flow j_n as defined above.

It is surprising that, as shown in these photographs, only a small addition to the nozzle flow can change two-phase flow pattern from slug flow to bubbly flow for fixed values of j_ℓ and j_g. Although the experiment was performed under limited flow conditions, it should be concluded from the results reported here that the two-phase flow pattern boundary is

$$P_u(u) = \iint_{S_{\alpha\mu}} \frac{P(\alpha, \mu, \beta-\nu)}{\left|\frac{\partial}{\partial(\beta-\nu)} P(\alpha, \mu, \beta-\nu)\right|} \, d\alpha d\mu \tag{8}$$

where $s_{\alpha\mu}$ is projected area of the surface defined by Eq.(9) onto the $\alpha-\mu$ plane in ($\alpha, \mu, \beta-\nu$) space.

$$\frac{\{\cos\alpha\cos\mu + \sin\alpha\sin\mu\cos(\beta - \nu)\}^2}{\cos\mu} = \text{const.} \ (\equiv u) \tag{9}$$

Assuming that bubble motion is random in manner, and that the probe strikes any element of the projected bubble frontal area with equal probability, $P(\alpha, \mu, \beta-\nu)$ is given by

$$P(\alpha, \mu, \beta-\nu) = \frac{1}{\pi} \cos\mu \sin\mu \, P_\alpha(\alpha) \tag{10}$$

$$0 \leq \alpha, \mu \leq \pi/2$$

$$0 \leq \beta - \nu \leq 2\pi$$

Substituting Eq.(10) into Eq.(8), we have

$$P_u(u) = \frac{1}{2\pi} \iint_{S_{\alpha\mu}} \frac{P_\alpha(\alpha) \cos^2\mu \, d\alpha d\mu}{\{\cos\alpha\cos\mu + \sin\alpha\sin\mu\cos(\beta-\nu)\} \sin(\beta-\nu)\sin\alpha} \tag{11}$$

In the above equation, the probability density function for the angle α should be known based on the observation of bubble velocity in three-dimensional measurement. Since the authors' knowledge is quite limited at present, the following approximation will be made (Kataoka et.al., 1986).

$$P_\alpha(\alpha) = \frac{1}{\alpha_0} \quad 0 \leq \alpha \leq \alpha_0$$

$$= 0 \quad \alpha_0 \leq \alpha \leq \pi/2 \tag{12}$$

where α_0 is given by

$$\frac{\sin 2\alpha_0}{2\alpha_0} = \frac{1 - \{\sigma_{sz}^2 / (|\overline{v_{sz}}|)^2\}}{1 + 3\{\sigma_{sz}^2 / (|\overline{v_{sz}}|)^2\}} \tag{13}$$

Bubble Diameter Distribution

The following normalized probability function of $P_u(u)$ is defined.

probe (passing velocity of bubbles) is given by

$$|v_{szj}| = \Delta s/\Delta t_j \tag{1}$$

Therefore, the apparent chord length ℓ_{sj} becomes

$$\ell_{sj} = t_{cj} \cdot |v_{szj}| = s(t_{cj}/\Delta t_j) \tag{2}$$

The problem was how to deduce the bubble-diameter probability density function from that for the measured ℓ_{sj}.

By referring to Figs. 4 and 5, the true chord length which should be pierced by the upstream sensor is

$$\ell_j = D_{bj} \cos\phi_j \tag{3}$$

where, $\vec{v}_{bj}$ and $\vec{n}_j$ are the velocity of the j-the bubble and the normal unit vector at the point of contact made by the upstream sensor and the interface. And the angle ϕ_j which is made by $\vec{v}_{bj}$ and $\vec{n}_j$ is given by

$$\cos\phi_j = \cos\alpha_j \cos\mu_j + \sin\alpha_j \sin\mu_j \cos(\beta_j - \nu_j) \tag{4}$$

When the probe distance Δs is small, the following approximation can be made (Kataoka et.al., 1986).

$$|v_{szj}|\cos\mu_j = |\vec{v}_{bj}|\cos\phi_j \tag{5}$$

Since t_{cj} is defined by

$$t_{cj} = \ell_j / |\vec{v}_{bj}| \tag{6}$$

we have

$$\ell_{sj}(\equiv t_{cj}|v_{szj}|) = \ell_j \cos\phi_j / \cos\mu_j$$
$$= D_{bj} \frac{\{\cos\alpha_j \cos\mu_j + \sin\alpha_j \sin\mu_j \cos(\beta_j - \nu_j)\}^2}{\cos\mu_j} \tag{7}$$

Probability Density Function for Chord Length

The authors next derived a formula relating the probability density function of the apparent chord length ℓ_s to the bubble-diameter distribution function on the assumption of constant bubble size, that is, D_b = const..

Letting the probability density functions of the event for the random variables, α, μ and $\beta - \nu$, given by $P(\alpha, \mu, \beta - \nu)$, the authors obtain the probability density function for variable ℓ_s/D_b ($\equiv u$) as follows, consulting a statistical theory.

BUBBLE SIZE EFFECT

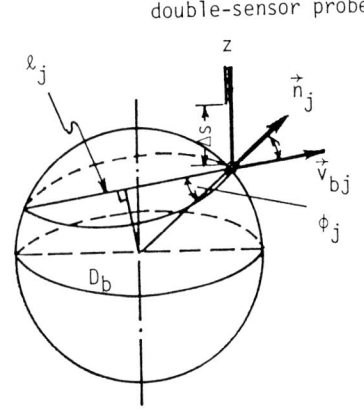 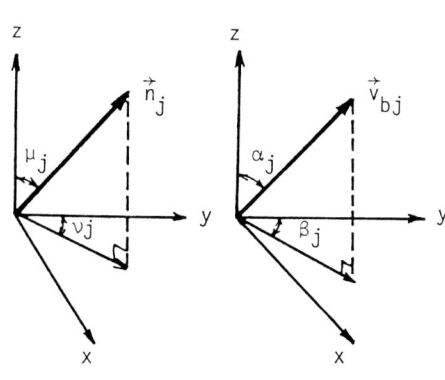

Fig. 4 Double-sensor probe and j-th interface

Fig. 5 Coordinate systems

CALCULATION PROCEDURES OF BUBBLE-SIZE DISTRIBUTION

In general, bubble-size distribution is related to the probability density function for bubble chord length or contact time from resistivity probe measurements (for example, Bankoff, 1964; Anderson, 1964; Herringe and Davis, 1976). Bankoff (1964) and Anderson (1964) obtained an integral equation which related bubble-size distribution to the bubble-contact time distribution by assuming uni-directional bubble movement.

Herringe & Davis (1976) proposed a similar theory and applied their results to an experiment. However, they did not solve the integral equation. Instead, they calculated an accumulative function of bubble size (integration of bubble-size distribution function) with several assumptions and simplifications.

The method to be described in this paper is, in principle, similar to that proposed by Bankoff (1964). However, it considers a variation of the direction of bubble movements, assuming that the bubble is not deflected or slowed down by the probe.

Bubble Chord Length

The following additional assumptions were made:
(1) bubbles are spherical,
(2) bubble velocity is defined as the velocity of the gas-liquid interface, and
(3) two probe sensors are located at a certain distance apart in z-direction.

The measurable quantities associated with j-th bubble are the contact time t_{cj} with the upstream sensor and the time Δt_j required for the bubble to travel over a distance Δs between the two sensors. Then, the bubble velocity measured by the

were at temperatures close to 20 °C.

Bubble Injector

The bubble injector used in this work was designed to realize bubbles as uniform as possible and also to change bubble size independently of the flow conditions of j_ℓ and j_g.

As schematically shown in Fig. 3, the bubble generator was equipped at the bottom of the test section. The air was injected through a 6-mm i.d. x 20 mm long porous cylinder wall of nominal porosity 5 μm into the water and was blown off by the high speed nozzle flow up to 20 m/s. The bubble size was thus easily controlled by setting the nozzle flow and the pressure difference between gas and liquid flows at appropriate values.

The nozzle flow containing thus generated small bubbles entered the test section coaxially and mixed with the main liquid flow. The ratio of the nozzle flow to the total liquid flow was varied from 0 to 1.0 depending on the experimental conditions. And, the mean bubble size was changed from 1 mm to 15 mm, also depending on the experimental conditions required. The bubble-size range mentioned above covered wall-peaking void and core-peaking void distributions under a fixed flow condition of j_ℓ and j_g.

The air flow was regulated and metered by a thermal gas flow controller/meter with ± 0.2-% error band, and the volume of air housing was about 230 cc with additional accumulator, which guarantees a constant air supply to the nozzle.

Instrumentation

The local void fraction and bubble chord length were measured with a double-sensor resistivity probe, using 0.1-mm o.d. electrically insulated platinum wires with 10-μm naked tips. The distance between the two sensors was about 5 mm in axial direction. In order to minimize the surface tension effects, the authors adopted the method recognizing the change of phase based on derivatives of the probe signals (Serizawa, 1984).

In order to calculate the bubble chord length, the bubble residence time and velocity must be known, since the bubble chord length is defined as a product of both quantities. To identify the probe signals originating from the same bubble, Sekoguchi's method was adopted (Sekoguchi et.al., 1984; Gakuhari, 1985). The bubble-size distribution was then computed from bubble chord-length histogram in a way described in the next section[1].

---

[1] Since the time left before this seminar was unfortunately limited, the authors were not able to demonstrate the results in this paper.

BUBBLE SIZE EFFECT

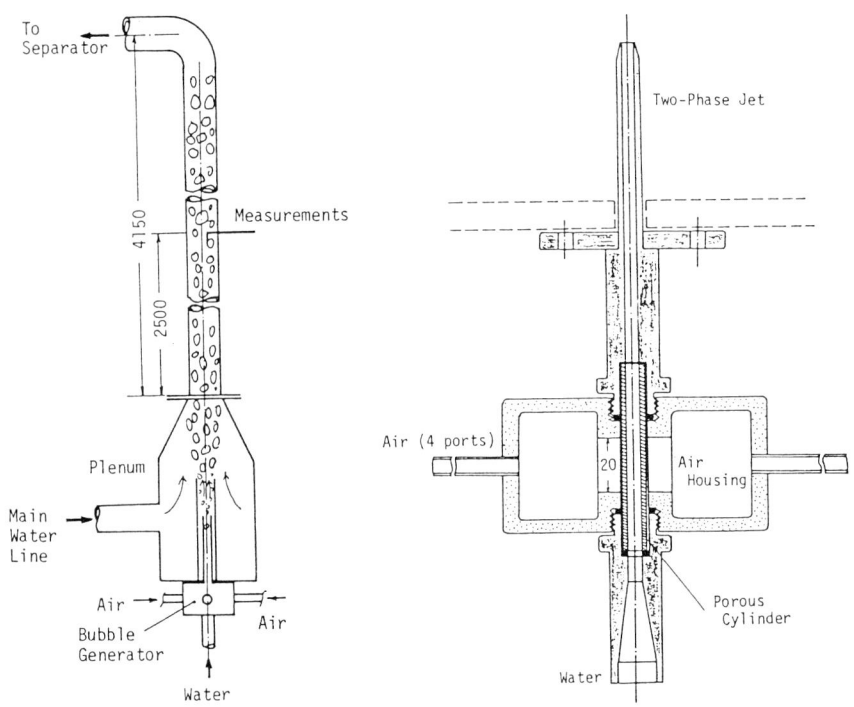

Fig. 2 Test section and lower plenum

Fig. 3 A schematic of bubble generator

length or contact time measured by a double-sensor probe is presented.

This paper, the first in a series deals with the effects of bubble size on bubbly flow structures. It presents some experimental results of bubble size effect on phase distribution. This study is a part of the joint work with Ljubljana University, Yugoslavia.

EXPERIMENT

Experimental Apparatus

The phase distribution measurement was carried out with a low pressure air-water flow test facility (Fig. 2). The air was injected from the bottom of the test section into deionized water through a specially designed nozzle (Fig. 3) to form vertically upward bubbly two-phase flow in a 30-mm i.d. round acrylic tube. The test section length was 4,150 mm, and the measurements were carried out at z = 2,500 mm downstream of the test section inlet which was enough to attain a developed flow condition. The water velocity j_ℓ ranged up to 6 m/s and the gas velocity j_g up to 0.3 m/s. The pressure at the measurement station was around 110 ∼ 130 kPa, and the fluids

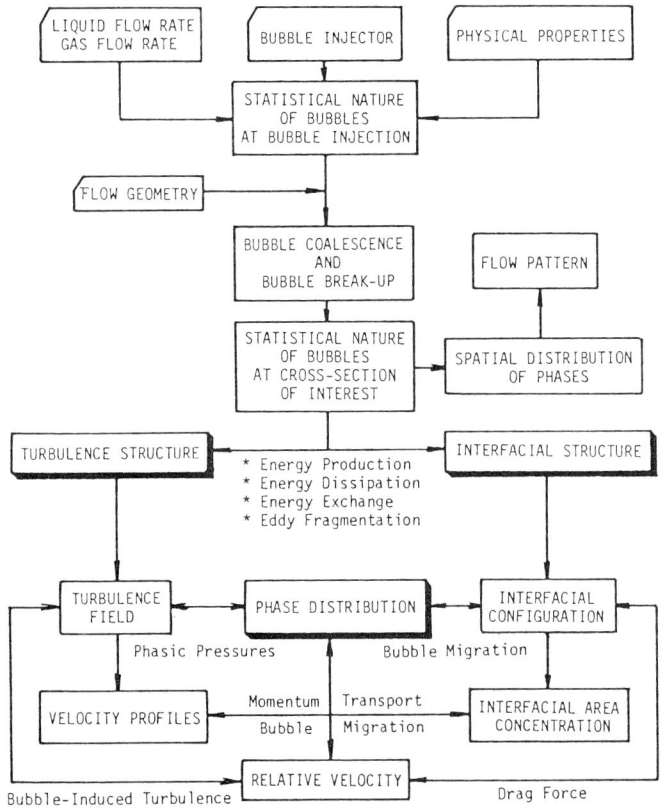

Fig. 1 Physical picture of gas/liquid flow

one of the most difficult points in experimental techniques is determining how to vary the size of bubbles in the flow with one bubble generator under a fixed flow rate condition of gas and liquid phases. Use of a surfactant such as ethanol or salt may be the easiest way to obtain small bubbles of the order 1 ∿ 2 mm in diameter. However, this method is obviously not appropriate to study the effect of bubble size on the complex nature of the close triangular relation among three structures mentioned above. This is shown in Fig.1. Here the turbulence and phase distribution are closely inter-related to the energy required to maintain the interfacial structure. This is very difficult to evaluate for a (liquid + surfactant) - gas system. The authors therefore adopted a mechanical method for obtaining the different size of bubbles using a high velocity water jet in a small cylinder nozzle.

Another problem is how to know the bubble size and size distribution. A conventional method is to take photographs of bubbles. However, this method gives rise to a problem concerning the focalized plane, and does not give the size distribution of bubbles passing at a particular point in the flow. In this paper a method to deduce bubble-size distribution from the probability density function for bubble chord

BUBBLE SIZE EFFECT ON PHASE DISTRIBUTION

A. SERIZAWA and I. MICHIYOSHI
Department of Nuclear Engineering, Kyoto University
Yoshida, Sakyo, Kyoto 606, Japan
I. KATAOKA
Institute of Atomic Energy, Kyoto University
Gokasho, Uji, Kyoto 611, Japan
I. ZUN
Faculty of Mechanical Engineering, University of Ljubljana
Murnikova 2, 61000 Ljubljana, Yugoslavia

ABSTRACT

The effect of bubble size on phase distribution in vertically upward two-phase flow in a round tube is discussed based on measurement and visual observations. It turned out that, for fixed values of j_ℓ and j_g, the phase distribution was drastically changed even with a small change of bubble-size distribution. The two-phase flow pattern transition was also affected very much by the bubble-size distribution. A formula was proposed to derive bubble-size distribution from the probability density function for bubble chord length, which takes into account the effect of multi-directional bubble movement.

INTRODUCTION

It is currently understood that in bubbly two-phase flow the local flow field is defined by a triangular linkage among phase distribution mechanism, eddy, and interfacial structures. Their interactions are complicated, as illustrated in Fig. 1. The close linkage among these three structures is very sensitive to bubble shape and size distribution. The bubble-size effect becomes, therefore, one of the most important problems to be solved in order to clarify the bubbly flow structure. And, in fact, both at the International Workshop on Two-Phase Flow Fundamentals held at the Rensselaer Polytechnic Institute in 1987 and at the 1987 ICHMT International Seminar on Transient Phenomena in Multiphase Flow in Dubrovnik, Yugoslavia, the importance of the effect of bubble size was stressed as a key parameter in explaining discrepancies that exist among phase distribution data obtained by different investigators who used similar experimental rig and conditions. However, the details of the effect of bubble size were left to be explained in future studies. Unfortunately, only a limited amount of experimental information was available at that time.

In studying the effect of bubble size on phase distribution,

9. Sato, Y., Sadatomi, M., Horita, K., and Sekoguchi, K., 1980, Momentum and Heat Transfer in Two-Phase Bubble Flow (1st Report), Trans. JSME(B), vol. 46, no. 409, pp. 1780-1789.

10. Sekoguchi, K., Nakasatomi, M., Sato, H. and Tanaka, O., 1980, Heat Transfer in Vertical Air-Water Bubble Flow, Trans. JSME(B), vol. 46, no. 402, pp. 291-298.

11. Sekoguchi, K., Fukui, H., and Sato, H., 1981, Flow Characteristics and Heat Transfer in Vertical Bubble Flow, Two-Phase Flow, ed. A. E. Bergles and S. Ishigai, pp. 59-74.

12. Serizawa, A., Kataoka, I. and Michiyoshi, I., 1975, Turbulence Structure in Air-Water Bubbly Flow— (II) Local Properties., Int. J. Multiphase Flow, vol. 2, no. 3, pp. 235-246.

13. Theofanous, T. G. and Sullivan, J., 1982, Turbulence in Two-Phase Dispersed Flows, J. Fluid Mech. vol. 116, pp. 343-362.

14. Zun, I., 1980, The Transverse Migration of Bubbles Influenced by Walls in Vertical Bubbly Flow, Int. J. Multiphase Flow, vol. 6, no. 6, pp. 583-588.

Subscripts, Superscripts, and Other

```
a    = Air
anal = Analytically obtained
b    = Bulk mean
c    = Condensation
d    = Disturbance wave
e    = Equilibrium
exp  = Experimentally obtained
f    = Condensate film
g    = Gas phase
i    = Interface
l    = Liquid phase
max  = Maximum or predominant
ND   = Non-disturbance wave
n    = n-th
r    = Ripple wave
t    = Turbulent flow
v    = Vapour
w    = Wall
0    = Upstream boundary or initial state
‾    = Mean
+    = Non-dimensional
∿    = Affected by wave
```

BIBLIOGRAPHY

Anderson, G. H., and Mantzouranis, B. G., 1960, Two-phase (gas-liquid) Flow Phenomena, I. Pressure Drop and Hold-up for Two-phase Flow in Vertical Tubes, Chem. Eng. Sci., Vol. 12, pp. 109-126.

Blangetti, F., and Schlünder, E. U., 1978, Local Heat Transfer Coefficients on Condensation in a Vertical Tube, Proc. 6th Int. Heat Transfer Conf., Montreal, Vol. 2, pp.437-442.

Blangetti, F., 1979, Lokaler Wärmübergang beider Kondensation mit Überlagerter Konvektion im Vertikalen Rohr, PhD Thesis, Universität Karlsruhe.

Brown, R. C., Andreussi, P., and Zanelli, S., 1978, The Use of Wire Probes for the Measurement of Liquid Film Thickness in Annular Gas-liquid Flows, Canadian J. Chem. Eng., Vol. 56, pp. 754-757.

Brumfield, L. K., House, R. N., and Theofanous, T. G., 1975, Turbulent Mass Transfer at Free Gas-liquid Interfaces with Applications to Film Flow, Int. J. Heat Mass Transfer, Vol. 18, pp. 1077-1081.

Chien, S. F., and Ibele, W., 1964, Pressure Drop and Liquid Film Thickness of Two-phase Annular and Annular-mist Flow, J. Heat Transfer, Vol. 86, pp. 89-96.

Chu, K. J., and Dukler A. E., 1975, Statistical Characteristics of Thin, Wavy Films III.Structure of the Large Waves and Their Resistance to Gas Flow, AIChE J., Vol. 21, pp. 583-593.

Crutchfield, J. P., Former, J. D., Packard, N. H. and Shaw, R. S., 1986, Chaos, Scient. Am., Vol. 255(6), pp. 38-49.

Dubounis, M., Rubio, M. A., and Berge, P., 1983, Experimental Evidence of Intermittencies Associated with a Subharmonic Bifurcation, Phys. Rev. Lett., Vol. 51, pp. 1446-1449.

Edwards, D. A., Bonilla, C. F., and Cichalle, M. I., 1948, Condensation of Water, Styrene, and Butadiene Vapors, Ind. and Eng. Chem., Vol. 40, pp. 1105-1122.

Fujita, H., Katoh, K., and Takahama, H., 1985, Falling Water Films over a Ring Attached to a Vertical Cylinder (Wave Charateristics Measured by a Capacitance Method), Bull. Japan Soc. Mech. Eng., Vol. 28, pp. 1401-1408.

Gill, L. E., Hewitt, G. F., and Lacey, P. M. C., 1964, Sampling Probe Studies of the Gas Core in Annular Two-phase Flow II.Studies of the Effect of Phase Flowrates on Phase and Velocity Distribution, Chem. Eng. Sci., Vol. 19, pp. 665-682.

Hagiwara, Y., Suzuki, K. and Sato, T., 1982, Studies on Thin Liquid Film of Annular-mist Two-phase Flow I.Wave Characteristics and Heat Transfer, Mem. Fac. of Eng. Kyoto Univ., Vol. 44(2), pp. 309-328.

Hagiwara, Y., Suzuki, K., Sato, T., and Chigusa, N., 1983, Simultaneous Measurements of Wall Shear Stress and Liquid Film Thickness in an Annular-mist Two-phase Flow in a Vertical Tube, Proc. ASME-JSME Thermal Eng. Joint Conf., Honolulu, Vol. 1, pp. 55-62.

Hagiwara, Y., Miwada, T., Suzuki, K., and Sato, T., 1984, A Study on Liquid Film Characteristics in Annular Two-phase Flow, in Multi-Phase Flow and Heat Transfer III, Part A: Fundamentals, ed. T. N. Veziroglu, and A. E. Bergles, pp. 249-263, Elsevier Publishers, Amsterdam.

Hagiwara, Y., Miwada, T., and Suzuki, K., 1985, Heat Transfer and Wave Structure in the Developing Region of Two-component Two-phase Annular Flow, PhysicoChemical Hydrodynamics, Vol. 6, pp. 141-156.

Hagiwara, Y., 1988, Experimental Studies on Chaotic Behaviour of Liquid Film Flow in Annular Two-phase Flows, PhysicoChemical Hydrodynamics, Vol. 10, pp. 135-147.

Hagiwara, Y., 1989a, Numerical Analysis of 3D Interfacial Shape and Velocity Distribution of Liquid FIlm, PhysicoChemical Hydrodynamics, Vol. 11, No. 1, pp. 49-62.

Hagiwara, Y., Esmaeilzadeh, E. and Suzuki, K., 1989b, Simultaneous Measurement of Liquid Film Thickness, Wall Shear Stress and Gas Flow Turbulence of Horizontal Wavy Two-phase Flow, Int. J. Multiphase Flow, Vol. 15, No. 3, pp. 421-431.

Hebbard, G. M., and Badger, W. L., 1933, Steam Film Heat Transfer Coefficient for Vertically Tubes, Trans. AIChE, Vol. 30, pp. 194-216.

Hewitt, G. F., 1978, Measurement of Two-phase Flow Parameters, pp. 111-116, Academic Press, London.

Ishii, M., and Gromles, M. A., 1975, Inception Criteria for Droplet Entrainment of Two-phase Concurrent Film Flow, AIChE J., Vol. 21, pp. 308-318.

Ishii, M., and Mishima, K., 1982, Liquid Transfer and Entrainment Correlation for Droplet-Annular Flow, Heat Transfer 1982 (Proc. 7th Int. Heat Trans. Conf., Munich), Vol. 5, pp. 307-312.

Kutateladze, S. S., 1963, Fundamentals of Heat Transfer, pp. 311-315, Edward Arnold, London, 1963.

Launder, B. E., and Spalding D. B., 1972, Mathematical Model of Turbulence, p. 91, Academic Press, London.

Levy, S., and Healzer J. M., 1981, Application of Mixing Length Theory to Wavy Turbulent Liquid-Gas Interface, J. Heat Transfer, Vol. 103, pp. 492-500.

Miya, M., Woodmansee, D. E., and Hanratty, T. J., 1971, A Model for Roll Waves in Gas-liquid Flow, Chem. Eng. Sci., Vol. 26, pp. 1915-1931.

Nusselt, W., 1916, Die Oberflächenkondensation des Wasserdampfes, Z. VDI, Vol. 60, pp. 541-546, 569-575.

Pletcher, P. H., and McManus, H. N., 1972, A Theory for Heat Transfer to Annular Two-phase, Two-component Flow, Int. J. Heat Mass Transfer, Vol. 15, pp. 2091-2096.

Spalding, D. B., 1975, GENMIX-A General Computer Program for Two-dimensional Parabolic Phenomena, Mech. Eng. Dep. Rep. HTS/75/17, Imperial College.

Suzuki, K., Hagiwara, Y., and Sato, T., 1983a, Heat Transfer and Flow Characteristics of Two-phase Two-component Annular Flow, Int. J. Heat Mass Transfer, Vol. 26, pp. 597-605.

Suzuki, K., Miwada, T., Hagiwara, Y., and Sato, T., 1983b, Analysis on Film Condensation Heat Transfer Taking Account of Interfacial Wave Effect (in Japanese), Proc. 20th National Heat Transfer Symp. of Japan, pp. 253-255.

Suzuki, K., 1986, Heat Transfer and Hydrodynamics of Two-phase Annular Flow, in Encyclopedia of Fluid Mechanics, Vol.3 (Gas-Liquid Flows), ed. N. P. Cheremisinoff, pp. 1356-1391, Gulf Pub. Co., Houston.

Suzuki, K., Hagiwara, Y. and Izumi, H., 1988, Numerical Study of Forced-convection Filmwise Condensation in a Vertical Tube Accounting for an Interfacial Wave Effect (in Japanese), Trans. Japan Soc. Mech. Eng., Vol. 54, Part B, pp. 2550-2555.

Takahama, H., Katoh, S., and Maeda, N., 1983, Longitudinal Flow Characteristics of Vertically Falling Liquid Films without Cocurrent Gas Flow (Effect of Liquid Viscosity) (in

Japanese) Trans. Japan Soc. Mech. Eng., Vol. 49, pp. 62-70.

Tsiklauri, G. V., Besfamiliny, P. V., and Baryshev, Yu. V., 1979, Experimental Study of Hydrodynamic Process for Wavy Water Film in a Cocurrent Air Flow, in Two-phase Momentum, Heat Mass Transfer, ed. F. Durst et al., Vol. 1, pp. 357-372, Hemisphere Pub. Co., New York.

Ueda, T., and Tanaka, T., 1973a, Studies of Liquid Film Flow in Two-phase Annular and Annular-mist Flow Regions. (Part 1. Downflow in a Vertical Tube) (in Japanese), Trans. Japan Soc. Mech. Eng., Vol. 39, pp. 2842-2852.

Ueda, T., and Nose, S., 1973b, Studies of Liquid Film Flow in Two-phase Annular and Annular-mist Flow Region. (Part 2. Upflow in a Vertical Tube) (in Japanese), Trans. Japan Soc. Mech. Eng., Vol. 39, pp. 2853-2862.

van Driest, E. R., 1956, On Turbulent Flow near a Wall, J. Aeronaut. Sci., Vol. 23, pp. 1007-1011.

Villeneuve, J. P., and Ouellet, Y., 1978, Laboratory Systems for Measuring Short-term Changes in Water Levels, Rev. Sci. Inst., Vol. 49, pp. 1425-1431.

CRITICAL POWER CHARACTERISTICS IN A ROD BUNDLE WITH NARROW GAP

K. ARAI, S. TSUNOYAMA and S. YOKOBORI
Toshiba Corporation
Toshiba Nuclear Engineering Laboratory
Kawasaki, Japan

K. YOSHIMURA
Toshiba Corporation
Isogo Engineering Center
Yokohama, Japan

ABSTRACT

It is important for the design of nuclear reactors to clarify, from various viewpoints, the thermal hydraulic behavior in a fuel rod bundle. This paper reports an analytical and experimental study on the critical power characteristics in a fuel rod bundle with a narrow gap under Boiling Water Reactor (BWR) conditions.

Some correlations have already been developed to predict critical heat flux (CHF), the local heat flux at which a rapid reduction in heat transfer coefficient occurs, for a tight lattice. These correlations are based on the so-called local condition hypothesis. This hypothesis is valid to describe high-pressure, high-flow phenomena. In BWR conditions, on the other hand, an integral approach is employed, since upstream history in a nonuniform heat flux profile is quite important. The upstream history strongly depends on flow regime and, thus, quality.

Therefore, a critical-quality-type correlation for a tight-spaced triangular lattice, based on the Biasi correlation has been studied. This correlation is a critical quality-boiling length (X_C-L_B) type correlation, based on the experimental boiling transition data derived from triangular arrays of rod clusters at the Bettis Atomic Power Laboratory of Westinghouse Corporation.

Further, the lattice tightening effect on the critical power characteristics was studied, using a simple-shaped experimental apparatus. Comparisons with the critical power data obtained are also presented for steady state conditions and transient conditions. From these comparisons, it is considered that the present correlation is applicable to the critical power prediction in a tight-spaced, triangular fuel lattice.

INTRODUCTION

Recently, studies on a rod bundle with narrow gap have been extensively performed in Europe and Japan (Oldekop et al., 1982; Frewer, 1984; Uotinen et al., 1981; Saji et al.,1988). This paper reports an investigation of the critical power characteristics of a triangular fuel bundle with narrow gap (smaller than 3.0mm).

From the thermal hydraulics design viewpoint, the prediction of boiling transition is primarily important. CHF correlations or critical power correlations are therefore required for triangular fuel bundles with tight spacing. Some CHF correlations have already been developed for triangular tight lattices. These are based on the so-called local conditions hypothesis, which is valid when the rapid deterioration of heat transfer occurs in a low-quality or subcooled region. In the BWR condition, however, boiling transition is primarily due to high-quality film dryout on the heated surface. Thus, an integral concept is valid, since the upstream history is quite important. The applicability of these CHF correlations to the BWR condition has not been clarified. A boiling transition correlation based on the integral concept was therefore studied which is valid to estimate the critical power of a closely spaced triangular fuel bundle.

Using boiling transition (BT) data, obtained with triangular arrays of fuel bundles at the Bettis Atomic Power Laboratory, the validity of the existing CHF correlation under BWR conditions was investigated. A critical quality-boiling-length-type correlation for triangular fuel lattices with narrow gap based on the Biasi correlation was also studied, based on the Bettis data.

To investigate the critical power dependence on the rod clearance in a BWR operating condition, critical power tests have been carried out on annulus tubes. The present correlation has been tested with these test data, both for steady state and transient conditions.

EXISTING CHF CORRELATION FOR A TIGHT LATTICE

Among the CHF correlations for tight lattices, the KfK (Kernforshungszentrum Karlsruhe) correlation has been extensively used (Dalle Donne, 1985). Its validity under the BWR condition was investigated. Most of its data base was obtained for higher pressures than in the BWR condition. The pressure of the BWR is about 7MPa while the KfK correlation is expected to be valid near 15MPa. The extrapolation to the lower pressure has not yet been tested.

The pressure dependence in the KfK correlation has been compared with the data base for the correlation. The test data were obtained with a triangular array of rod clusters consisting of twenty rods arranged as shown in Fig. 1. The rods were 7.1mm in outside diameter and had a 1.37m heated length. The rod pitch was 8.6mm.

Figure 2 shows the ratios for the predicted CHF to the measured CHF ($q"_{CHF,calc.}/q"_{CHF,exp.}$) versus mass velocity at 13.8MPa, 11.0MPa, 8.3MPa and 5.5MPa. It is clearly seen that the KfK correlation was conservative for the BWR condition and agreed reasonably well with test data near Pressurized Water Reactor (PWR) conditions. Consequently, it is concluded that the correlation is not appropriate for predicting boiling transition in BWR's.

CRITICAL POWER CORRELATION FOR TIGHT PITCH FUEL LATTICE

In fuel bundles, a boiling transition occurs in the high quality annular flow regime. Under these hydrodynamic conditions, the boiling transition is primarily attributed to dryout of the liquid film on the fuel rod surface. The film dryout is governed by evaporation, by entrainment from the film, and by deposition of droplets from the steam core.

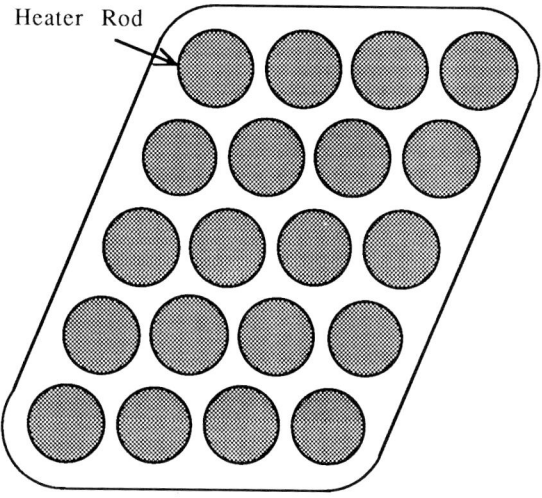

FIGURE 1. Bundle cross section (Letourneau et al., 1975)

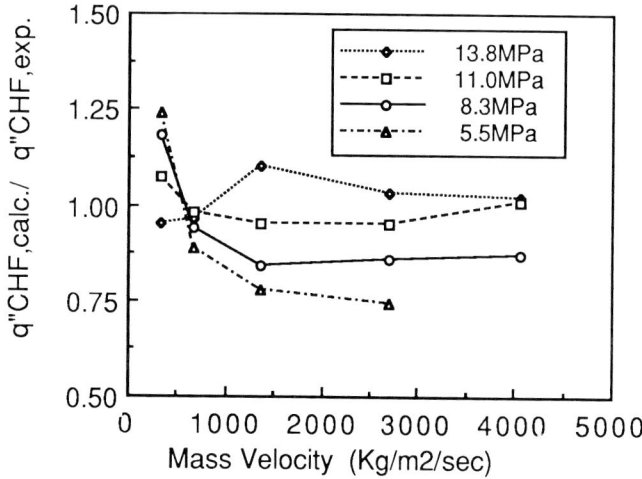

FIGURE 2. Ratio of predicted critical heat flux values to those measured as a function of mass velocity

It is well known that the integral concept is appropriate for the high quality boiling transition (Lahey and Moody, 1977). Some correlations, based on the integral concept, have been developed for predicting the boiling transition under BWR conditions (General Electric Co., 1973; Bertoletti et al., 1965). These correlations predict the critical power (i.e., the bundle power at the boiling transition point) in terms of critical quality and boiling length.

To estimate the critical power in a tight pitch fuel lattice under the BWR condition, critical-quality-type correlations for the triangular fuel bundle with narrow gap were studied. Experiments in tight pitch triangular fuel

TABLE 1. Summary of bundle geometry and experimental conditions

Number of Rods		20
Rod Arrangement		5x4 triangular array (Fig. 1)
Number of Rod Bundles		2
Rod Diameter	(mm)	6.4 and 7.1
Rod Pitch	(mm)	8.6
		(same for both bundle)
Pitch to Diameter Ratio		1.36 and 1.21
Infinite Array Hydraulic Diameter	(mm)	6.6 and 4.6
Heated Length	(m)	1.37
Heat Flux Distribution		uniform
Pressure	(MPa)	5.5 and 8.3
Mass Velocity	(Kg/m^2/sec)	340 - 4000
Inlet Subcooling	(KJ/Kg)	200 - 930
Average Exit Quality		0.06 - 0.70

lattices, conducted at Bettis Atomic Power Laboratory (Letourneau et al., 1975), were a part of the light water breeder reactor development program carried out with 20-rod bundles arranged in a 5x4 triangular array as shown in Fig. 1. The BT data were obtained with two rod arrays with different degrees of lattice tightness. The rods in the two arrays were of either 6.4mm or 7.1mm outside diameter, and the rod pitch was 8.6mm for the both arrays. The spacing between rods was maintained by four spacer grids. The fuel bundle geometry and the experimental conditions are summarized in Table 1.

Figure 3 shows the data obtained with the two different rod arrays, whose rod pitch-to-diameter ratios (P/D) were 1.36 and 1.21. It is observed that there is no significant difference between data for these two ratios, with similar inlet subcoolings. This indicates that the rod spacing effect on critical power may be captured by the critical quality type correlation.

As part of this study, the two versions of the critical quality type correlation are examined, versus the Bettis data (Letourneau et al., 1975) as shown in Fig. 3. The first is the CISE correlation (Bertoletti et al., 1965), and the second is the Biasi critical quality correlation (Biasi et al., 1967; Phillips et al., 1981). The Biasi correlation is based on a broad data base and has been used in TRAC-BD1 (Taylor et al., 1984), which is an advanced best estimate computer program for BWR transient analysis, developed at the Idaho National Engineering Laboratory.

Figures 4 and 5 show the comparisons of these correlations with the experimental data. From these figures, the following observations may be made;

1. Compared with the Biasi prediction, the CISE prediction shows a larger scatter;

2. For the CISE correlation, the deviations between the predicted and measured powers for the rod array whose P/D is 1.36, are larger than those for the other array.

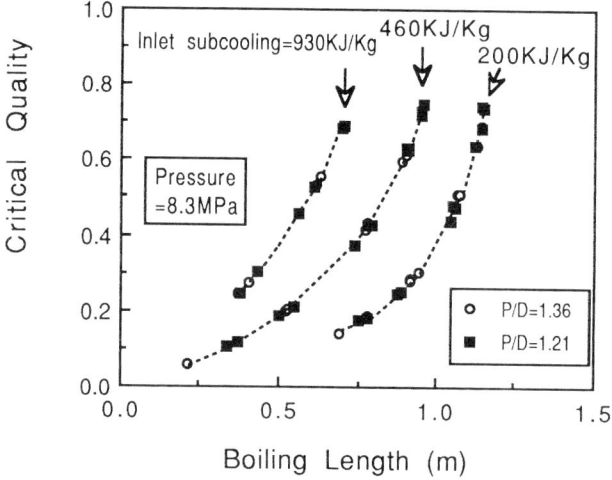

FIGURE 3 20-rod bundle test data (Letournean et al., 1975) compared on the basis of equilibrium boiling length

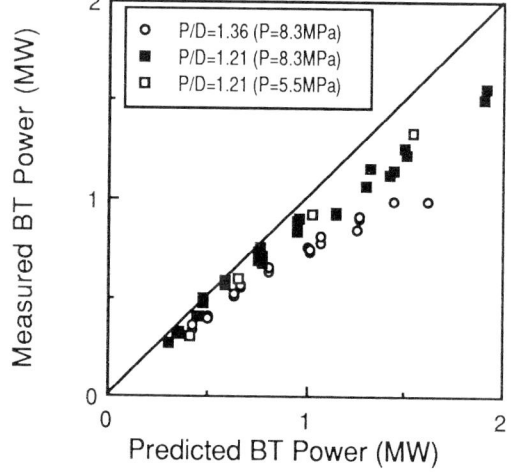

FIGURE 4. Comparison of measured BT power data with CISE (Bertoletti et al., 1965) prediction

This indicates that the CISE correlation cannot describe the rod spacing effect properly.

In the Biasi prediction (Fig. 5), the deviations for the two arrays are comparable. The Biasi correlation may capture the rod spacing effect rather well, even though it is inadequate to predict the critical quality for the tight pitch triangular fuel lattice.

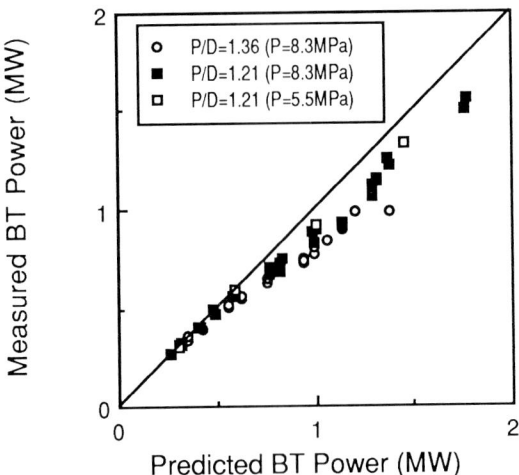

FIGURE 5. Comparison of measured BT power data with Biasi (Biasi et al., 1967; Phillips et al., 1981) prediction

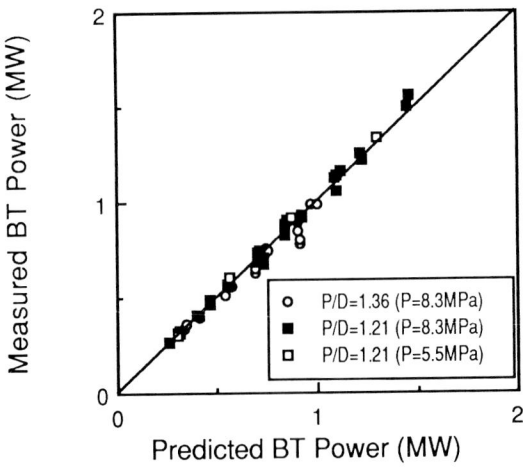

FIGURE 6. Comparison of measured BT power data with modified Biasi prediction

Based on the above observations, the Biasi correlation seemed preferable for this study, but it is necessary to modify the correlation for tight pitch triangular arrays. This modification was undertaken based on the Bettis data, and is compared with the original in Table 2. A parity plot is shown in Fig. 6.

It is seen that the modified correlation yields a better fit to the data, compared with the unmodified correlation shown in Fig. 5. The mean error in the predictions of the present correlation is -0.3%, and the standard deviation is 4.9%.

TABLE 2. Original Biasi critical quality correlation (Phillips et al, 1981) and modified correlation

$$A_1 = 1.0$$
$$B_1 = b_1 G^{m_1} D_h^{n_1} h_{fg} / h(p)$$
$$X_{c1} = \frac{A_1 L_B}{B_1 + L_B} \frac{P_h}{P_w} R_f^{-1/2}$$
$$A_2 = a_2 f(p)/G^{m_2}$$
$$B_2 = b_2 G^{m_3} D_h^{n_1} h_{fg}$$
$$X_{c2} = \frac{A_2 L_B}{B_2 + L_B} \frac{P_h}{P_w} R_f^{-1/2}$$
$$h(p) = -1.159 + 0.149 p \{\exp(0.019p)\} + 8.99p/(10+p^2)$$
$$f(p) = 0.7249 + 0.099 p \{\exp(-0.032p)\}$$
$$p = 10^{-5} P$$

Original Biasi

$100 < G < 300$; $X_c = X_{c1}$

$300 < G$; $X_c = \text{Max}(X_{c1}, X_{c2})$

$m_1 = 1.6$ $m_2 = 1/6$ $m_3 = 7/6$

$D_h > 0.01$; $n_1 = 1.4$ $b_1 = 1.048 \times 10^{-8}$

$a_2 = 1.468$ $b_2 = 5.707 \times 10^{-8}$

$D_h < 0.01$; $n_1 = 1.6$ $b_1 = 2.633 \times 10^{-8}$

$a_2 = 1.468$ $b_2 = 1.434 \times 10^{-7}$

Modified Biasi

$300 < G < 2000$; $X_c = X_{c1}$

$2000 < G$; $X_c = \text{Max}(X_{c1}, X_{c2})$

$m_1 = 1.7$ $m_2 = 0.15$ $m_3 = 1.27$

$n_1 = 1.5$ $b_1 = 1.320 \times 10^{-8}$

$a_2 = 1.413$ $b_2 = 7.132 \times 10^{-8}$

TEST OF THE CORRELATION WITH ANNULUS TUBE TEST DATA

To confirm the validity of the present critical power correlation, critical power tests were performed and compared with the predictions. The experiments were performed with an electrically heated rod enclosed by a thin-walled tube, which forms a coolant annulus (Fig. 7). The test section consisted of a heated inner rod with a 2.3m heated length, an unheated outer tube, and spacing elements. The geometrical details of the annulus tube, and the experimental conditions are listed in Table 3.

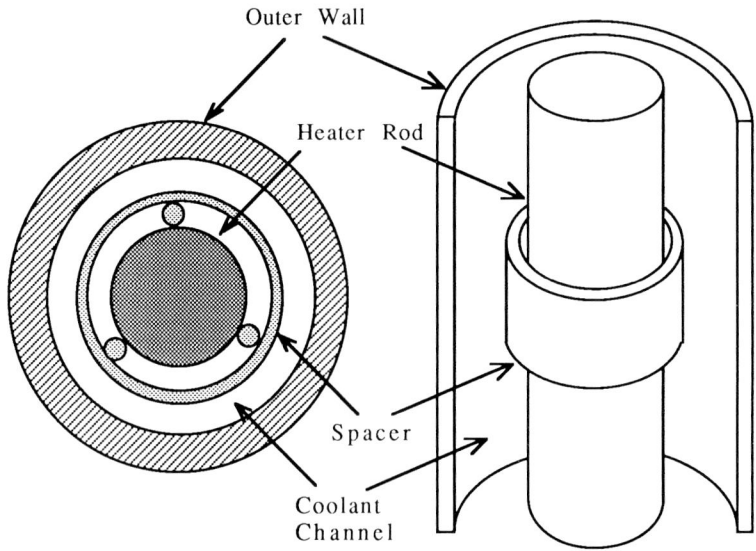

FIGURE 7. Schematic of the heated section of the annulus tube (Yokobori et al., 1987)

Three inner rods with different diameters were used, in order to simulate three triangular arrays with different rod spacings. The diameter of the inner rod was determined so that the annulus tube has the same heated equivalent diameter ($4A_x/P_h$) as that of the triangular array of the rod cluster which was to be modeled. Table 4 shows the relationship between the inner rod diameter of the annulus tube and the rod clearance (or the rod spacing) for the triangular, 12.2-mm diameter rod array. On the basis of equal heated equivalent diameters between the annulus tube and rod array, it is possible to determine the dependence between critical power and rod clearance from annulus tube test data. From comparisons of the present correlation with the annulus tube data, the rod clearance dependence of the correlation, derived from the Bettis rod bundle data, were verified. Details of the test facility and the test results were reported by Yokobori et al. (1987).

Critical Power Under Steady State Conditions

Figure 8 shows the critical quality data obtained on three annulus tubes. The converted clearance in the figure denotes the rod clearance of a triangular array, which is to be modeled by the annulus tube. The converted rod clearances correspond to the inner rod diameters of the annulus tubes, as shown in Table 4. The same trend can be seen as in the Bettis data; that is, the X_C-L_B lines are similar for three converted clearances, when the inlet subcoolings are the same.

Comparisons of the predictions with data taken from the annulus tubes are shown in Figs. 9 and 10. Figure 9 shows that the rod clearance effects are

TABLE 3. Geometrical details of annulus tube and experimental conditions

Inner Rod Diameter	(mm)	13.2, 13.6, 14.3
Inner Diameter of the Outer Tube	(mm)	17.0
Heated Length	(m)	2.3
Axial Peaking Factor		1.5
Pressure	(MPa)	7.0
Mass Velocity	(Kg/m^2/sec)	830 - 1390
Inlet Subcooling	(KJ/Kg)	23 - 70

TABLE 4. Relationship between inner rod diameter of the annulus and rod clearance for a triangular array of 12.2-mm diameter rods having identical heated equivalent diameters

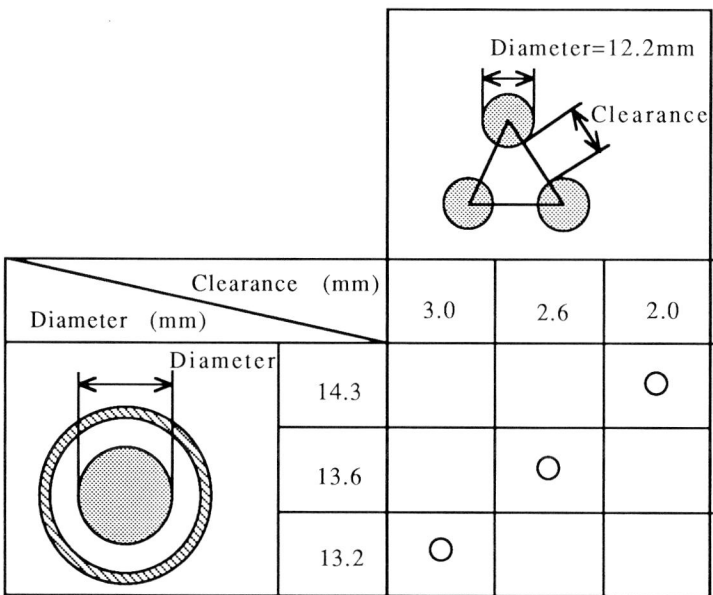

Diameter (mm) \ Clearance (mm)	3.0	2.6	2.0
14.3			O
13.6		O	
13.2	O		

well predicted by the present correlation. The slope of the critical power versus the rod clearance curve becomes smaller with increasing rod clearance, and this trend is well captured by the prediction which accurately represents the annular data obtained (Fig. 10).

Critical Power Under Transient Conditions

In the transient tests, the inlet mass flow decreased to about 60 % of the initial flow in two seconds, typical of a BWR flow coastdown under a recirculation pump trip condition (Fig. 11). The rod power, pressure and inlet subcooling were held constant throughout the mass flow transient. The times to boiling transition under the flow decay were measured at several power levels between 52KW and 68KW. The tests were conducted using two

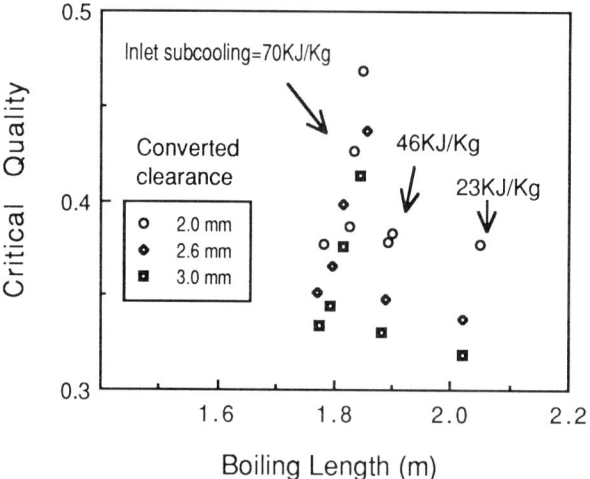

FIGURE 8. Annulus tube test data compared on the basis of equilibrium boiling length.

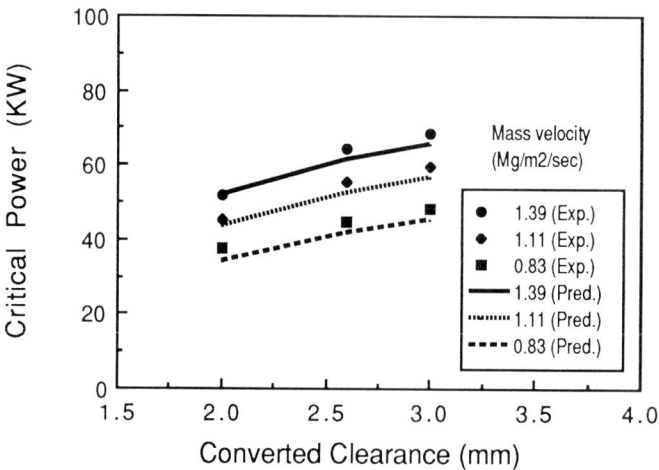

FIGURE 9. Comparison of measured critical power dependence on rod clearance with modified Biasi prediction

annulus tubes, which simulated triangular rod arrays with 2.6mm rod clearance and 2.0mm clearance. The transient test conditions are summarized in Table 5.

A one-dimensional, transient, thermal hydraulic model (General Electric Co., 1978) incorporating the modified correlation, was used to analyze the annulus tube data. Both transient flow and precalculated transient equilibrium boiling length were specified in order to obtain critical power predictions. In both cases, the time to boiling transition was conservatively predicted (Fig. 12).

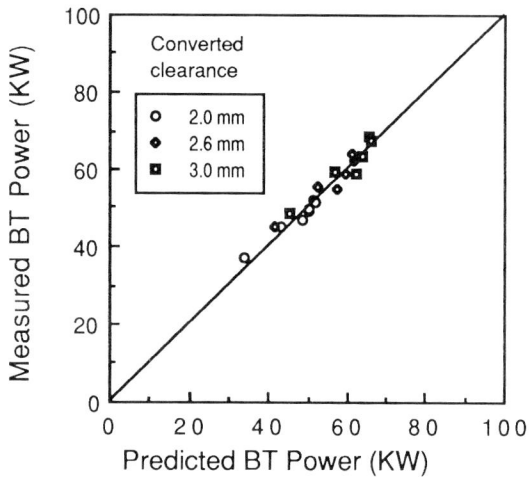

FIGURE 10. Comparison of annulus tube BT power data with modified Biasi prediction

TABLE 5. Transient test conditions

Inner Rod Diameter	(mm)	13.6 and 14.3
Pressure	(MPa)	7.0
Initial Mass Velocity	(Kg/m^2/sec)	1390
Inlet Subcooling	(KJ/Kg)	70
Rod Power	(KW)	52 - 68

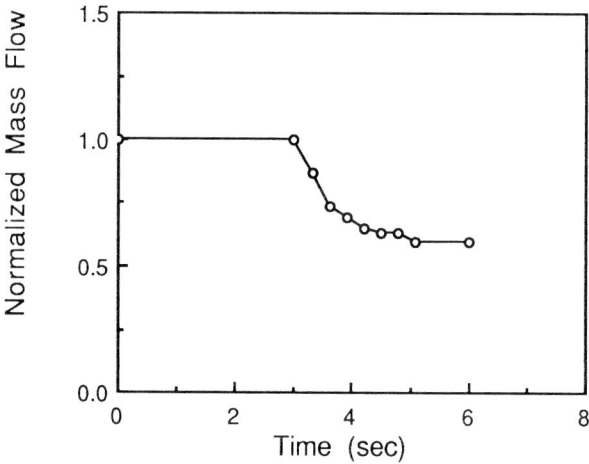

FIGURE 11. A typical flow decay curve

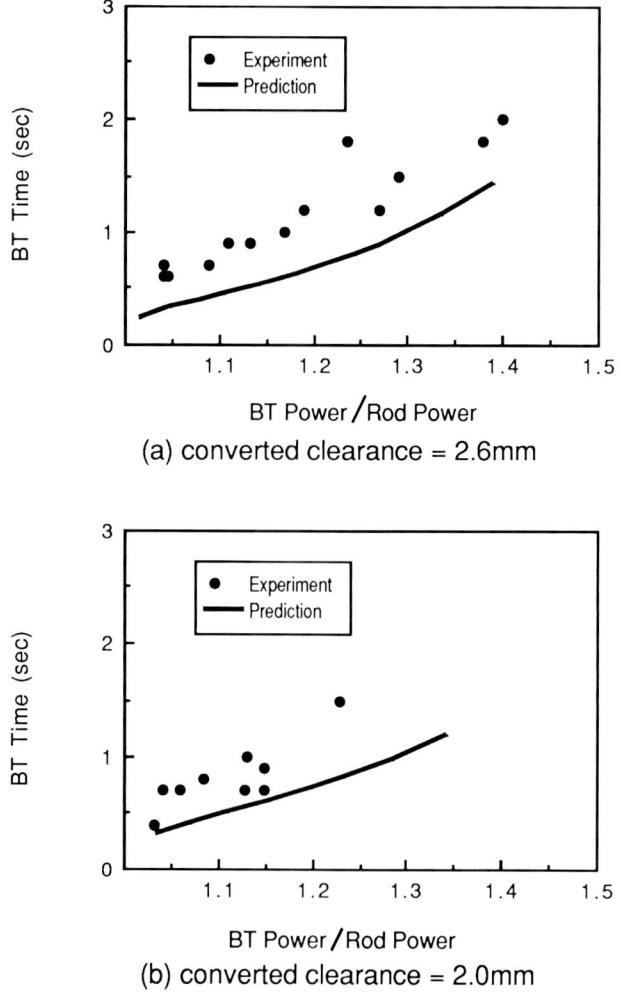

FIGURE 12. Measured versus predicted time to boiling transition

CONCLUSIONS

A critical power correlation for boiling transition was obtained by modifying the Biasi critical quality correlation, and compared both with existing tight-spaced triangular-pitch rod bundle data and new annulus tube data. The correlation predicts the steady state data base with mean and standard deviation of -0.3% and 4.9% respectively, and accurately predicts spacing and rod-clearance effects for the two geometries. Times to boiling transition were conservatively predicted when the correlation was incorporated in a one-dimensional, transient, thermal hydraulic model.

NOMENCLATURE

A	see TABLE 2
A_x	cross sectional flow area (m^2)
B	see TABLE 2
D_h	hydraulic diameter (m)
G	mass velocity (Kg/m^2/sec)
h_{fg}	latent heat of evaporation (J/Kg)
H_{sub}	inlet subcooling (J/Kg)
L_B	boiling length (m)
P	pressure (Pa)
P/D	pitch-to-diameter ratio
P_h	heated perimeter (m)
P_w	wetted perimeter (m)
q''_{CHF}	critical heat flux (W/m^2)
R_f	local peaking factor
X_c	critical quality

Subscripts

exp.	experimental
calc.	calculated

REFERENCES

Bertoletti, S., Gaspari, G. P., Lombardi, C., Peterlongo, M., Silvestri, M. and Tacconi, F. A. 1965, Heat Transfer Crisis with Steam-Water Mixtures, *Energia Nucleare*, 12, 3.

Biasi, L., Clerici, G. C., Garriba, S., Sala, R. and Tozzi, A. 1967, Studies on Burnout: Part 3, *Energia Nucleare*, 14, 9.

Dalle Donne, M. and Hame, W. 1985, Critical Heat Flux Correlation For Triangular Arrays of Rod Bundles with Tight Lattices, including The Spiral Spacer Effect, *Nuclear Technology*, Vol.71.

Frewer, H. 1984, KWU Advanced Nuclear Technology in Europe, KWU-symposium.

General Electric Company 1973, General Electric BWR Thermal Analysis Basis (GETAB): Data, Correlation and Design Application, NEDO-10958.

General Electric Company 1978, Qualification Of The One-Dimensional Core Transient Model For Boiling Water Reactors, Vol.1, NEDO-24154.

Lahey, R. T. and Moody, F. J. 1977, *The Thermal-Hydraulics of a Boiling Water Nuclear Reactor*, ANS.

Letourneau, B. W., Peterson, A. C., Coeling, K. J., Gavin, M. E. and Green, S. J. 1975, Critical Heat Flux and Pressure Drop Tests with Parallel Upflow of High Pressure Water in Bundles of Twenty 0.25- and 0.28-inch Diameter Rods, WAPD-TM-1013.

Oldekop, W., Berger, H. D. and Zeggel, W. 1982, General Features of Advanced Pressurized Water Reactors with Improved Fuel Utilization, *Nuclear Technology*, Vol.59.

Phillips, R. E., Shumway, R. W. and Chu, K. H. 1981, Improvements to the Prediction of Boiling Transition in BWR Transient Calculation, *Proceedings of the 20th ASME/AICHE National Heat Transfer Conference, Milwaukee, Wisconsin*.

Saji, E., Akiyama, Y., Kono, N., Namba, K., Hori, K., Umeoka, T. and Kono, T. 1988, Feasibility Studies On High Conversion Pressurized Water Reactors With Semitight Core Configurations, *Nuclear Technology*, Vol.80.

Taylor, D. D., Shumway, R. W., Singer, G. L. and Mohr, C. M. 1984, TRAC-BD1/MOD1: An Advanced Best Estimate Computer Program For Boiling Water Reactor Transient Analysis, Vol.1, NUREG/CR-3633.

Uotinen, V. O., Bloomfield, W. L., Haghi, M. A., Jones, H. M., Jones, J. H., Toops, E. C., Edlund, M. C. and Florian, R. J. 1981, Technical Feasibility of a Pressurized Water Reactor Design with a Low Water Volume Fraction Lattice, EPRI-NP-1833.

Yokobori, S., Kato, K., Nagasaka, H. and Yoshimura, K. 1987, Characteristics of Boiling Transition of Tight Lattice Rod Assembly, *Proceedings of the 24th National Heat Transfer Symposium of Japan*, C223, May (in Japanese).

THERMO-HYDRAULIC BEHAVIOR OF INVERTED ANNULAR FLOW (EFFECTS OF FLOW DIRECTION AND CHANNEL DIAMETER)

M. ARITOMI and A. INOUE
Research Laboratory for nuclear reactors
Tokyo Institute of Technology
2-12-1 Ohokayama, Meguro-ku
Tokyo 152, Japan

ABSTRACT

Inverted annular flow has been investigated experimentally in a vertical channel and in horizontal channels, with three different diameters under various heated wall temperature, inlet subcooling and inlet velocity conditions, using Freon 113 as a test fluid. This enables clarification of flow direction and flow channel diameter effects on the thermo-hydraulic behavior, and offers a data base for clues to the modeling of the thermo-hydraulics of two-phase flow in a thermal non-equilibrium state. The effects of flow direction and channel diameter on wall heat transfer coefficient and vapor film thickness are also shown. Furthermore, the relationship between wall shear stress and vapor film thickness is specified. The wall shear stress in inverted annular flow for a smooth interface between vapor film and liquid core jet is less than that in liquid single-phase flow, but becomes larger when significant waves develop.

INTRODUCTION

During the reflooding phase of a postulated loss-of-coolant accident in Light Water Reactors (LWRs), the cladding temperature may exceed the minimum film boiling temperature. When subcooled water refloods into the reactor core under bottom quenching conditions, such as during the reflooding phase in Pressurized Water Reactors (PWRs), superheated vapor flows up around the hot cladding and a subcooled liquid jet flows in the core space region among the rods. Since the positions of the liquid and vapor phases are exactly the opposite to that in normal annular flow, this flow pattern is called 'inverted annular flow'.

The most interesting aspect of this flow pattern is that a thermal non-equilibrium state, with both superheated vapor and a subcooled liquid jet, is readily formed through the cross section in a steady state. Furthermore, the shape of the interface is simpler than in other flow patterns such as bubbly flow and slug flow.

Transient two-phase thermo-hydraulics with thermal non-

equilibrium state is one phenomenon that is very difficult to analyze. The following two phenomena occur in the thermal non-equilibrium state:
(1) Though the local state of the flow is in a thermal equilibrium, the states of both phases obtained by averaging in a numerical control volume are found to be in thermal non-equilibrium.
(2) The local state of the flow is not in thermal equilibrium, so that it is necessary to consider a metastable state and relaxation.

For instance, subcooled boiling, superheated mist flow and two-phase condensation correspond to phenomena (1), while flash flow during blowdown and superheated liquid flow in accelerating state (such as choked flow) correspond to phenomena (2).

By considering the state of transient flow in terms of the transient time, which is treated as the transit time for positionally developing states of the flow, phenomena (2) occurs in states with a transient time faster than 10ms, while phenomena (1) occurs in slower transient states. Concerning safety analysis of LWRs, the transient phenomena (1) have a more significant effect on plant behavior than phenomena (2), because the period for which phenomena (2) appear is very short. It is obviously very important for the systematization of the numerical work of two-phase flow dynamics to clarify phenomena (2). From this point of view, the authors have been investigating the thermo-hydraulics of two-phase flow in thermal non-equilibrium states. They have first taken up inverted annular flow, in order to clarify the thermo-hydraulics involved, as well as to offer a data base to aid the modeling of two-phase flow in thermal non-equilibrium phenomena (1).

Most work on inverted annular flow has been performed at velocities lower than 0.1m/s, to simulate the reflooding phase of LWRs. Heat transfer is usually analyzed as film pool boiling. For instance, Berenson's correlation [1] is used in the RELAP-5 code [2], and Bailey's correlation which is a revision of Bromley's [3] is adopted in the TRAC code [4]. However, it is made clear in our previous work [5] and in Fung and Groeneveld's paper [6] that the heat transfer coefficient of inverted annular flow coincides with that of film pool boiling for flow rates lower than 0.1m/s, but is larger when the flow rate is higher than 0.5m/s.

The authors experimentally investigated inverted annular flow formed in a vertical channel with 9.8mm diameter, using Freon 113 as a test fluid. The wall heat transfer, wall shear stress, interfacial shear stress and superheated vapor film thickness were measured. Empirical correlations for the net vaporization rate from the interface, the heat transfer coefficient from the interface to liquid phase, the interfacial shear stress and the wall heat transfer coefficient were proposed [5]. Ishii [7] has reviewed many works on heat transfer and hydrodynamics as a post-CHF phenomenon, and modeled the flow regime transition from inverted annular flow to superheated dispersed flow, in terms of jet break-up length and droplet size. Although Elias and Chamble [8] concluded that void fraction at the quench front could be correlated with thermal equilibrium quality, and this correlation was also shown over a wider flow range by Fung and Groeneveld [9], it was made clear from previous work [5] that the void fraction in inverted annular flow could be well correlated with the equilibrium quality for only one flow rate, but could not be well correlated for different flow rates. Furthermore, from our previous work, it was supposed that the wall shear stress in inverted annular flow was smaller than that in liquid single phase, though the data were somewhat scattered.

In this paper, the wall heat transfer coefficient, wall shear stresses, and the vapor film thickness were investigated for inverted annular flow occurring in horizontal channels whose diameters were 3, 5 and 7mm. The effects of flow direction and channel diameter on these parameters are discussed, in comparison with the results [5] for vertical flow.

EXPERIMENTAL APPARATUS

Figure 1 is a schematic diagram of the forced-convection Freon 113 loop, operated at atmospheric pressure. The loop was devised to suppress flow fluctuations, by adopting a circulation pump with sufficient head and adequate throttling to the entrance of the test section, to create stable inverted annular flow. The vapor generated in the test section was separated in a tank located downstream of the test section, and returned to the liquid phase by a condenser attached in the tank. A precooler and a preheater were provided in the tank, to regulate the temperature of the circulating fluid. The flow rate was measured using an orifice flowmeter and was regulated by a flow control valve. A bypass line was installed parallel to the test section, to facilitate the formation of inverted annular flow. Fluid circulated through the bypass line while the required temperature and flow rate were obtained, allowing the test section to be preheated.

The test section, whose geometry and dimensions are shown in Fig. 2, consisted of a 200mm long stainless steel cylinder of 60mm O.D. This was installed horizontally in the loop; that is, the flow direction was horizontal. A microheater was fastened around a U-shaped groove on the outside of the cylinder. Metallic cement was used to attach the heater to the cylinder to enhance heat transfer between the two. A flow channel with fixed inner diameter was machined through the center of the cylinder, using an electric spark machine. Three test sections, whose flow diameters were 3mm, 5mm and 7mm, were used to investigate the effect of flow diameter on the thermo-hydraulic behavior of inverted annular flow. To ensure the development of stable, inverted annular flow, a cylindrical spot heater was inserted into a well drilled into one side at the inlet end of the test section.

Several thermocouples (0.5mm O.D.) were inserted in small holes (0.6mm O.D.) manufactured by the electric spark machine, to measure the temperature distribution in the cylinder and to obtain the wall temperature and wall heat

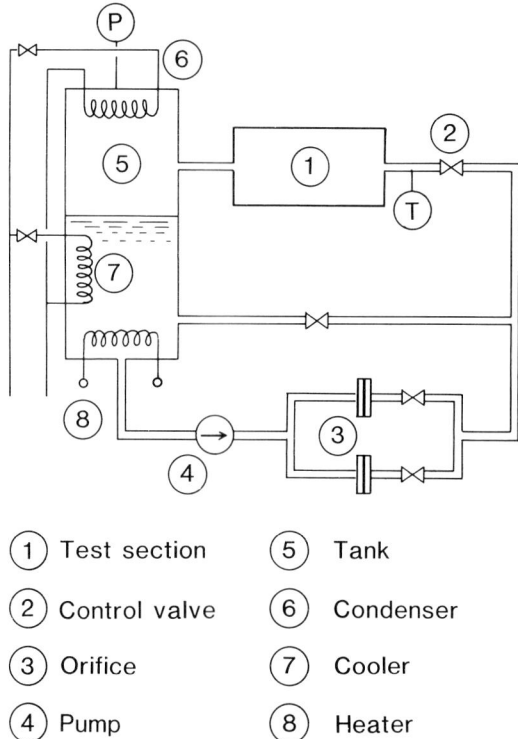

Fig. 1 Schematic diagram of forced-convection Freon 113 loop

① Test section
② Control valve
③ Orifice
④ Pump
⑤ Tank
⑥ Condenser
⑦ Cooler
⑧ Heater

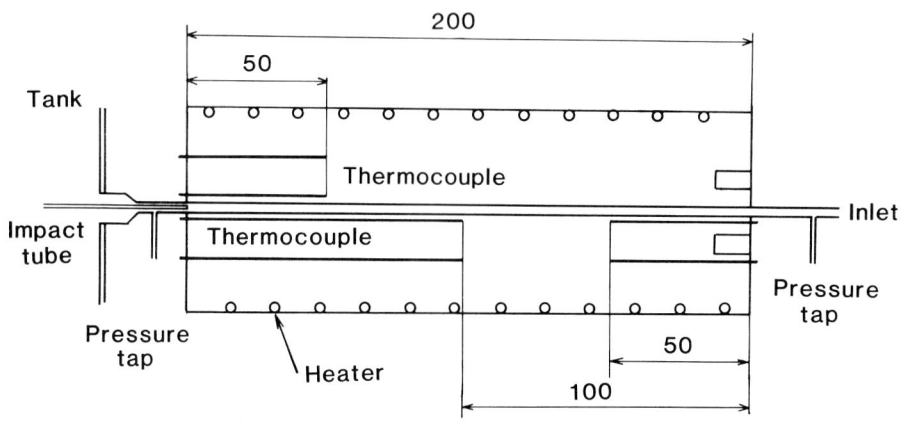

Fig. 2 Test section

Table 1 Experimental conditions

Test fluid	Freon 113
Pressure	Atmospheric pressure
Inlet velocity	0.5 - 2 m/s
Heat flux	0 - 170 kW/m^2
Inlet subcooling	5 - 20 K
Channel diameter	
Horizontal flow	3, 5 and 7 mm
Vertical flow	9.8 mm

flux, as shown in the figure. The pressure drop in the test section was measured between the pressure taps at the entrance and exit of the test section, using a differential pressure transducer. Since it was clear that the velocity profile in a fast liquid core jet during horizontal inverted annular flow was flatter than that during vertical flow because of less gravitational effects on the flow, the liquid jet exit velocity was measured using an impact tube, inserted at the center of the flow channel and positioned at the exit of the test section. The impact tube was calibrated with a single-phase flow, and was corrected by assuming that liquid single phase flow has a turbulent universal velocity profile. From the results, superheated vapor film thickness and void fraction were estimated, neglecting the change of liquid mass due to evaporation, since stable inverted annular flow is formed only in a subcooled state in thermal equilibrium, and the vaporization rate is fairly small.

Initially, fluid circulates through the bypass line, while the temperature and velocity are regulated and the test section is preheated to about 250°C with the microheater and the hot spot is observed with a thermocouple attached nearby, and is heated to 300°C. After temperatures are stabilized, the bypass is closed allowing the fluid to flow through the test section. Since the wall temperature of the test section and the hot spot temperature decrease slowly as the fluid begins to flow, the wall temperature is maintained at the required value by controlling the input of the microheater. The hot spot temperature is kept at 300°C by regulating the spot heater power. The experimental conditions of this work are summarized in Table 1. Conditions of our previous work [5] concerning vertical inverted annular flow are also shown in Table 1, to enable the results of the different flow directions to be compared.

RESULTS AND DISCUSSION

Three test sections with flow channel diameters of 3mm, 5mm and 7mm were used to investigate the effect of diameter on the thermo-hydraulic behavior of inverted annular flow.

Heat Transfer

The temperature distribution in heated cylinders was measured, to provide data for the evaluation of wall temperatures and wall heat transfer coefficients in inverted annular flow, by solving a steady-state thermal conduction equation. Using the measured wall temperature T_w and heat flux $q"$, the heat transfer coefficient h is defined by

$$h = \frac{q"}{T_w - T_{sat}}, \qquad (1)$$

where T_{sat} is the saturated temperature. All calculated heat transfer coefficients are given in this paper as Nusselt numbers Nu, which are defined by

$$Nu_d = \frac{h\,d}{\lambda_g}, \qquad (2)$$

where λ_g is the superheated vapor thermal conductivity for an average temperature T_g, and d is the flow channel diameter.

$$T_g = \frac{1}{2}(T_w + T_{sat}). \qquad (3)$$

Effect of flow rate. Experiments to investigate the effects of inlet flow rate on heat transfer coefficient were performed, under constant wall temperature and inlet subcooling conditions. The experimental results with constant wall superheating and thermal equilibrium quality are arranged by inlet Reynolds number Re_{in}, and are shown in Fig. 3. In the case of horizontal inverted annular flow, the Nusselt number is in proportion to the 0.35th power of the Reynolds number, regardless of the channel diameter. For the same inlet Reynolds number, the Nusselt number increases with a decrease in the channel diameter. The maximum inlet velocities are equal for all channel diameters at 2m/s, and the minimum inlet velocities are also the same at 0.5m/s. Therefore, the Nusselt number also increases for the same velocity, with a decrease in channel diameter. Since the Nusselt number of a narrow channel is larger than that of a wide channel, the heat transfer coefficient for a narrow channel has a higher value.

The Nusselt numbers of vertical inverted annular flow are shown in Fig. 3, but vary little with inlet Reynolds number. The following points were noted previously [5] for vertical inverted annular flow, and clarify the reasons for this small variation.
(1) The liquid-core jet is separated from the heated wall by a compressible superheated vapor film and flows against

INVERTED ANNULAR FLOW

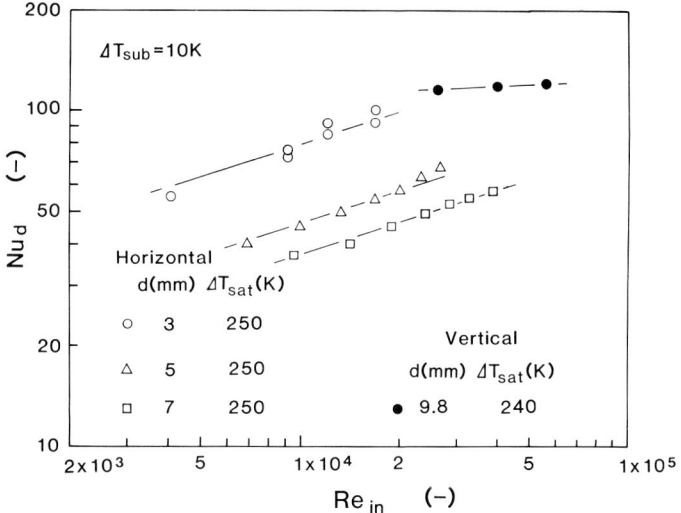

Fig. 3 Effects of inlet Reynolds number on Nusselt number

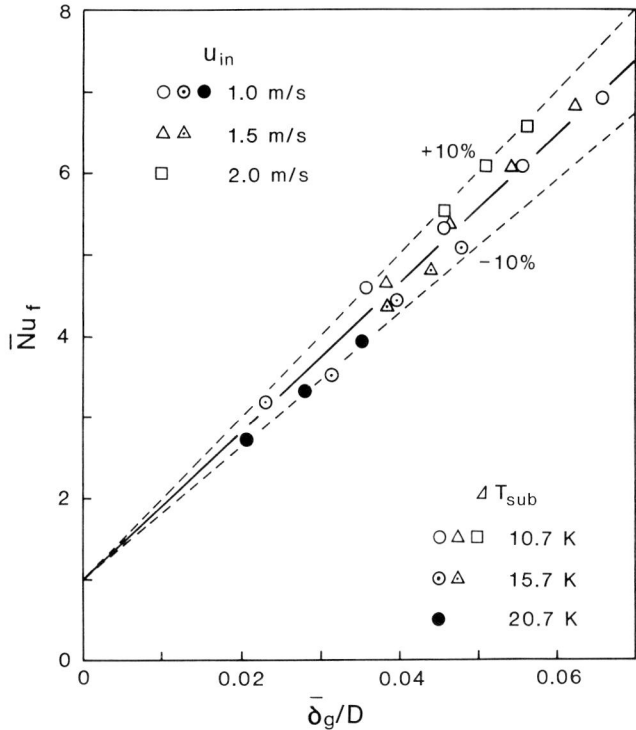

Fig. 4 Relationship between film Nusselt number and dimensionless vapor film thickness

the force of gravity, so the flow is unsteady. Consequently, the interface between the vapor film and the liquid jet becomes somewhat wavy and rough, as the film is thickening.
(2) The turbulence in the vapor film is governed by the interfacial roughness, which depends only on the vapor film thickness.
(3) As a result, the film Nusselt number Nu_f using vapor film thickness as the reference length, depends only on the vapor film thickness, as shown in Fig.4.
(4) The average vapor film velocity estimated analytically is not influenced by changes in the liquid jet velocity, and scarcely varies under all the experimental conditions. This is because an increase in velocity enhances the heat transfer from the interface to the subcooled liquid jet and reduces the vaporization rate.

For horizontal inverted annular flow, the following points are noted.
(1) The interface between the vapor film and liquid jet is smoother than in vertical flow for the same film thickness, because the gravitational force does not affect the flow direction.
(2) The vapor film velocity is affected by the liquid jet velocity, which is almost in proportion to the inlet velocity.
(3) With decreasing channel diameter, the vapor film thickness decreases (as presented later), so that the heat flux required for the formation of stable inverted annular flow increases.
(4) An increase in heat flux enhances the vaporization rate and the vapor film velocity.

From the above considerations, it is seen that heat transfer from the heated wall in horizontal inverted annular flow is governed by the vapor film velocity which is dependent on liquid jet velocity, so that the Nusselt number is in proportion to the inlet Reynolds number. Although the exponent of Reynolds number to Nusselt number is 0.8 if the heat transfer is governed by the vapor film velocity, in this case the power is 0.35. The reason is that the liquid jet velocity does not directly have an effect on vapor film velocity, but rather affects it indirectly through the change in interfacial velocity. Also, the turbulence in the vapor film is influenced by interfacial roughness, though the effect is smaller than that in vertical flow. An increase in the Nusselt number with a decrease in channel diameter is caused by higher vapor film velocity as the vaporization rate rises and the vapor film thickness becomes thinner.

Effect of thermal equilibrium quality. Experiments to investigate the effects of inlet subcooling on heat transfer coefficient were performed under constant wall temperature and inlet velocity conditions. Thermal equilibrium quality X_{eq} expresses the positional state. The experimental results are illustrated in Fig. 5.

Stable inverted annular flow is formed under subcooled conditions, so that thermal equilibrium quality is negative under all experimental conditions.   As the saturated condition is approached, the Nusselt number decreases; however, the rate of decrease becomes smaller with decreasing channel diameter, and the Nusselt number is independent of the quality at the 3mm diameter.   Since the heat flux on the wall increases with smaller diameters, the ratio of the heat flux to the heat transfer from the interface to the subcooled liquid jet decreases. Therefore, the effect of liquid jet subcooling on the vaporization rate becomes less pronounced as the diameter decreases.   In other words, the vapor film velocity varies little with changes in liquid jet subcooling as the diameter decreases.   Thus, Nusselt number for channel diameters less than 3mm should be constant for changes in inlet subcooling and local quality.

In order to consider the effect of different flow directions on the relationship between the Nusselt number and the thermal equilibrium quality, the Nusselt number of vertical flow is compared with those numbers of horizontal flow in the figure.   The Nusselt number has the same decreasing tendency as in horizontal flow when the saturated condition is approached.   Although the details will be discussed later, consideringthe change of the exponent of thermal equilibrium quality in the different horizontal channels from 3mm to 7mm diameter, the exponent for the vertical flow with 9.8mm diameter may be said to be in agreement with that of horizontal flow with the same diameter.

Effect of wall superheating.   Experiments to investigate the effects of wall temperature on heat transfer coefficient were performed under constant inlet velocity and inlet subcooling conditions.   The experimental results with almost constant thermal equilibrium quality are shown in Fig. 6.   The Nusselt number decreases with increasing wall superheating under all conditions.

The slope of Nusselt number to wall superheating, however, decreases with increasing channel diameter.   For narrower channels, a higher heat flux on the wall is required for the formation of inverted annular flow, and the ratio of the heat flux to the heat transfer between the interface and the subcooled liquid is lower.   Therefore, a small increase in the heat flux can cause a large rise in wall temperature, so that the decrease in the Nusselt number with increasing wall superheating is more pronounced.

On the other hand, in vertical inverted annular flow, the relationship between the Nusselt number and wall superheating is the same as that in the horizontal flow. Although the details will also be discussed later, considering the way the exponent of the superheating changes from 3mm to 7mm diameter in the horizontal flow, the exponent for the vertical flow with 9.8mm diameter may be taken to be in agreement with that of horizontal flow of the same diameter.

Summary of heat transfer.   In order to investigate further the effects of channel diameter and flow direction on heat

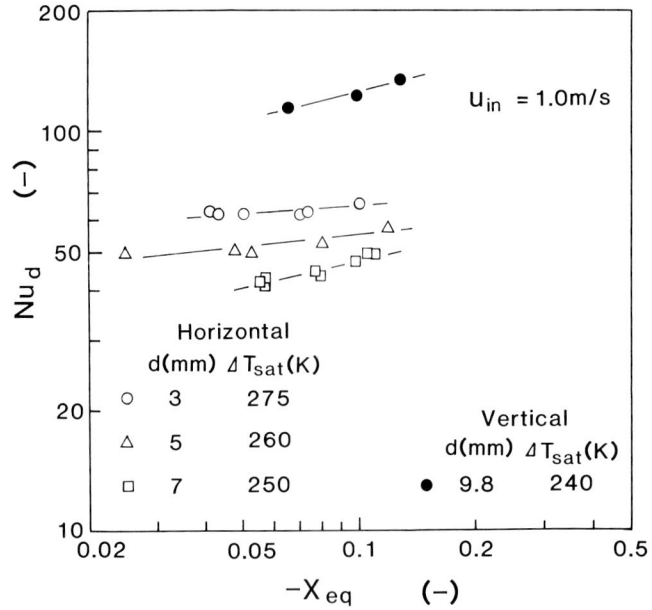

Fig. 5 Effects of thermal equilibrium quality on Nusselt number

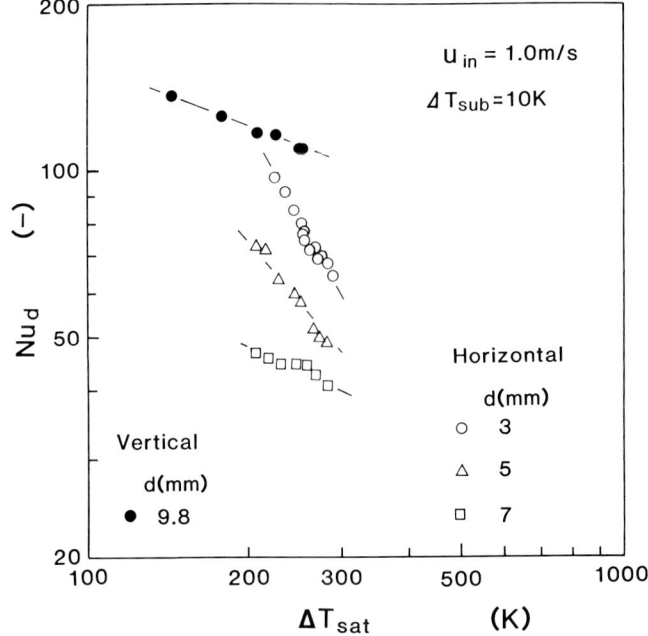

Fig. 6 Effects of wall superheating on Nusselt number

transfer, the effects of inlet Reynolds number, thermal equilibrium quality and wall superheating on the Nusselt number are determined from Figs. 3, 5 and 6. From these data, an empirical correlation regarding horizontal inverted annular flow is obtained as

$$Nu_d = 4.6 \, Re_{in}^{0.35} \left(\frac{L}{Cp \, \Delta T_{sat}} \right)^{\frac{14}{d'}} \left(-10 X_{eq} \right)^{0.16 d^* - 1.5}, \qquad (4)$$

where d^* is a dimensionless diameter defined by

$$d^* = d \sqrt{\frac{(\rho_\ell - \rho_g) g}{\sigma}}. \qquad (5)$$

The correlation is compared with all the experimental results and is shown in Fig. 7. It is seen from the figure that the correlation coincides with the experimental results, within a 20% error. Furthermore, in order to determine the differences in the relationships between quality, superheating and Nusselt number in the vertical and horizontal flow situations, the exponents of thermal equilibrium quality and wall superheating are calculated by substituting the channel diameter (9.8mm) into the correlation. The calculated exponent of the quality is 0.34, the result obtained from Fig. 5 is 0.3, and the relative error is therefore 5.9%.

Next, the calculated exponent of the superheating is 0.46, the result obtained from Fig. 6 is 0.4, and the relative error is 7.0%. Since heat flux was selected as an experimental parameter in the case of vertical flow, little data with constant thermal equilibrium quality and/or wall superheating was taken, as shown in Figs. 3, 5 and 6. Therefore, taking into account that the reliability of the exponents obtained from the figures is not very high, it is considered that the effects of thermal equilibrium quality and wall superheating on the Nusselt number are almost equal, in both vertical and horizontal flow. The effects of liquid jet velocity on the Nusselt number are different, as shown in Fig. 3; that is, in the vertical flow case the flow has no effect, but in the horizontal flow case the flow does have an effect.

Vapor Film Thickness

It was determined that the film Nusselt number in vertical inverted annular flow depended only on vapor film thickness [5], which is a very important parameter for modeling momentum, mass and energy transfer through the interface. However, it was difficult to measure the internal thickness of the channel in the test section used.

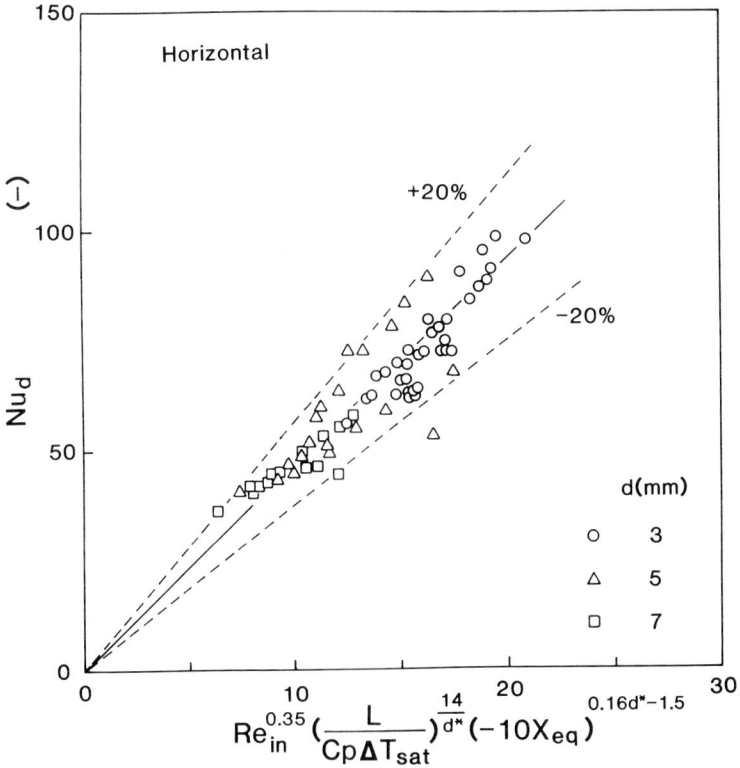

Fig. 7 Comparison between Nusselt number experiments and Eq. (4)

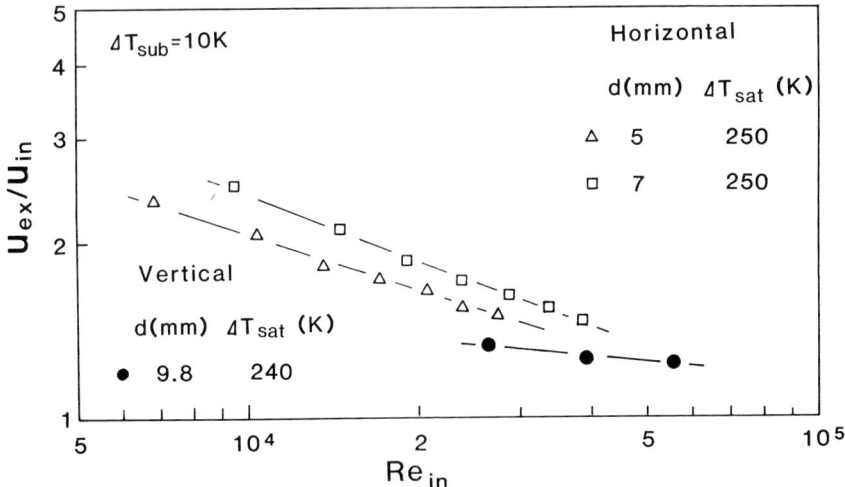

Fig. 8 Effects of inlet Reynolds number on velocity ratio

In order to establish a data base regarding thickness, and to evaluate the acceleration loss component of the pressure drop, liquid jet exit velocities were measured under various inlet velocity, inlet subcooling and wall superheating conditions. The thicknesses were obtained by neglecting the change of mass due to vaporization. This can be justified by the fact that the mass vaporization rate is very small in inverted annular flow. However, since it is clear that the correlation of the ratio of liquid jet exit velocity to inlet velocity for experimental parameters is better than that of vapor film thickness, the velocity ratio is adopted in this paper. The relationship between the velocity ratio and thickness is given by

$$\frac{\delta_g}{d} = \frac{1}{2}\left\{1 - \left(\frac{u_{in}}{u_{ex}}\right)^{0.5}\right\}, \qquad (6)$$

where d is the channel diameter, δ_g the vapor film thickness and u_{in} and u_{ex} the inlet and outlet liquid velocities. For a 3mm channel diameter, the exit flow pattern was not inverted annular flow, but rather liquid slug flow or superheated dispersed flow under most experimental conditions. Therefore, the experimental results are presented only for channel diameters 5mm and 7mm.

Effects of flow rate. The experiments to determine the effects of inlet velocity on vapor film thickness were performed in the same manner as the ones concerning heat transfer. The experimental results showing the effects of channel diameter are given in Fig. 8. Furthermore, the results in the vertical flow case are also shown in the figure, to enable the differences between the flow directions to be seen. As for the relationship between velocity ratio and inlet Reynolds number, the velocity ratio decreases with increasing inlet Reynolds number, regardless of the channel diameter and flow direction. The velocity ratio with a channel diameter of 7mm is slightly larger than that of 5mm. However, the relationship between the velocity ratio and the inlet velocity becomes equal for both the diameters, as shown in Fig. 9 in the horizontal flow case.

On the other hand, in vertical flow, vapor film thickness is thinner than that in horizontal flow, and the effect of inlet Reynolds number on the ratio is slightly smaller. The interface between the vapor film and the liquid jet in vertical flow is wavier for the same film thickness than that in the horizontal flow due to the force of gravity. Therefore, the interfacial shear stress is large. For vertical flow, both the vaporization rate from the interface and the interfacial shear stress are governed only by the interfacial turbulence, which is a function of vapor film thickness. Therefore, the thermo-hydraulic behavior of the vapor film scarcely changes, even if the velocity of the core liquid jet varies. On the other hand,

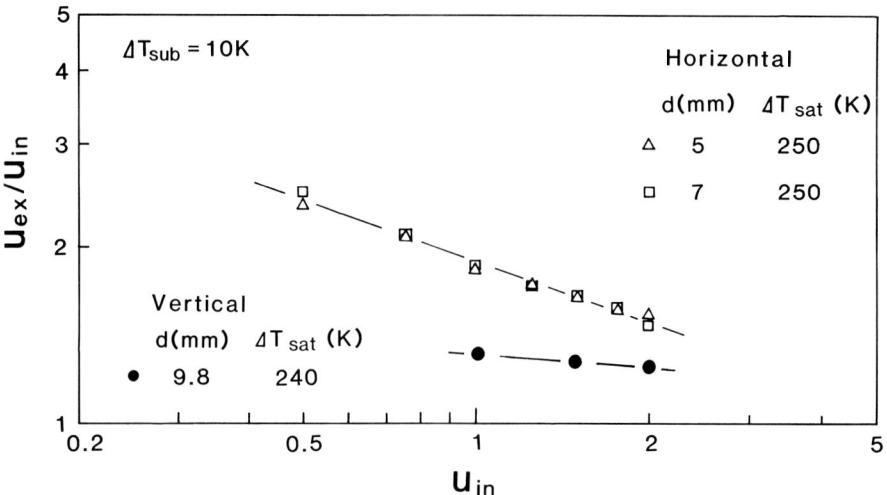

Fig. 9 Relationship between velocity ratio and inlet velocity

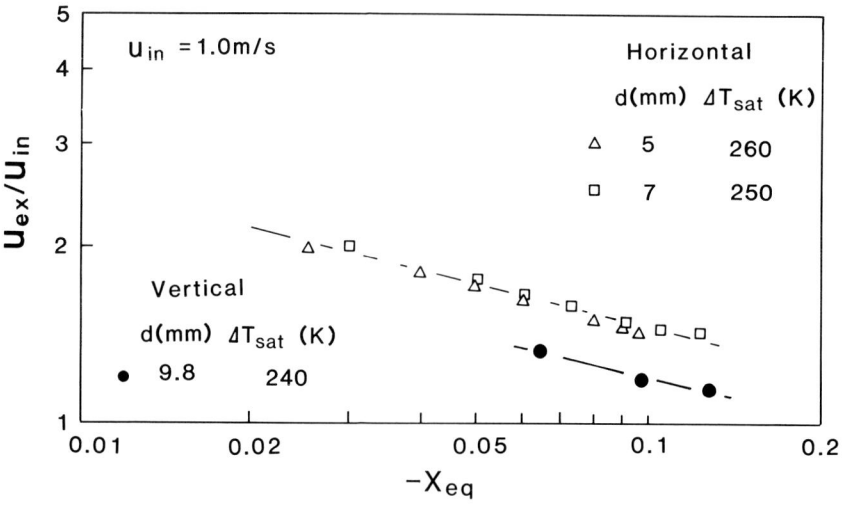

Fig. 10 Effects of thermal equilibrium quality on velocity ratio

since the vapor film velocity in horizontal flow (with the smoother interface) rises with increasing liquid jet velocity, the thickness decreases with increasing inlet Reynolds numbers.

Effect of thermal equilibrium quality. Experiments to investigate the effects of thermal equilibrium quality on vapor film thickness were performed in the same manner as those concerning heat transfer, and the experimental results are shown for the effect of channel diameter in Fig. 10. The results in vertical flow cases are also shown in the figure, to enable the differences between the flow directions to be seen.   When approaching the saturated condition, the velocity ratio increases, regardless of channel diameter and flow direction.   Furthermore, the effect of the quality on the velocity ratio in vertical flow is almost in agreement with that in horizontal flow, having the same tendency as the Nusselt number.

Effect wall superheating.   Experiments to investigate the effects of wall superheating on vapor film thickness were performed similar to those concerning heat transfer.   The experimental results are shown for the effect of channel diameter in Fig. 11, along with the results for vertical flow.   Although the data shown include results in which the thermal equilibrium quality varies slightly, the velocity ratio increases with an increase in wall superheating, regardless of channel diameter and flow direction. Correcting for quality the data shown in the figure, the effect of wall superheating on the ratio becomes smaller. From the figure, it is clear that changing the wall superheating slightly influences the velocity ratio regardless of channel diameter and flow direction, and that the slope of the velocity ratio versus the quality are almost in agreement in both flow situations.

Summary of vapor film thickness.   The results shown in Figs. 8, 9, 10 and 11 are rearranged in order to summarize the effects of each parameter on the vapor film thickness. The following emprical correlation is obtained:

$$\frac{u_{ex}}{u_{in}} = 0.154 \left(\frac{Re_{in}}{d^* \times 10^5} \right)^{-0.35} \left(-X_{eq} \right)^{-0.25} \left(\frac{Cp \Delta T_{sat}}{L} \right)^{0.2} , \quad (7)$$

where d^* is the dimensionless channel diameter defined by Eq. (5).   The correlation is compared with all the data, and the result is shown in Fig. 12. This coincides with the experiments, within an error of 10%.   The effects of different channel diameters on the dimensionless vapor film thickness is seen only with varying inlet Reynolds numbers, but not with varying flow rate, thermal equilibrium quality, or wall superheating.   On the other hand, the effects of the different flow directions on the same parameter appear only for the flow rate, but do not appear for the other

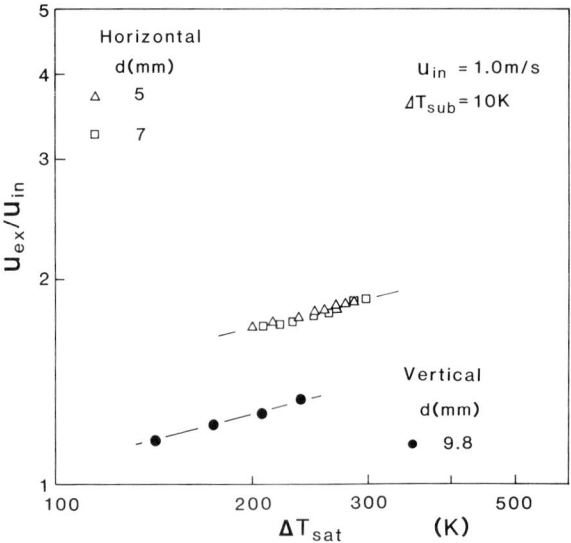

Fig. 11 Effects of wall superheating on velocity ratio

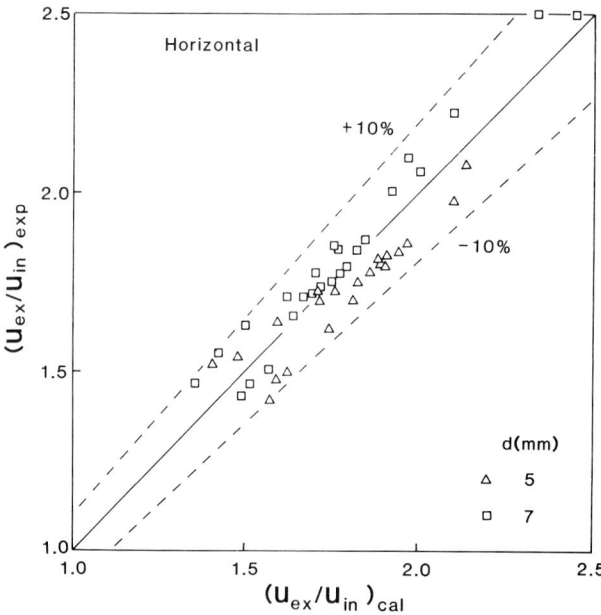

Fig. 12 Comparison between velocity ratio experiments and Eq. (7)

parameters in the same tendency as the Nusselt number.

Wall Shear Stress

Although the wall shear stress could not be discussed quantitatively, previous work on the vertical inverted annular flow [5] estimated qualitatively that the wall shear stress is smaller than that in liquid single-phase flow. Friction loss is moderately smaller than acceleration loss and gravitation loss in a vertical flow case. Liquid jet exit velocity could be measured with sufficient accuracy, but longitudinal void fraction distribution was estimated by the values obtained from liquid jet velocities measured at three points. Therefore, the small estimation error of the longitudinal void fraction distribution caused a large friction loss calculating error. In the case of horizontal flow, pressure drop is composed only of frictional loss and acceleration loss. Then, if liquid jet exit velocity is measured with adequate accuracy, the friction loss is also obtained with adequate accuracy. Therefore, to reduce errors, the horizontal flow situation was studied. The wall shear stress was obtained from the measured differential pressure between pressure taps by subtracting the liquid acceleration loss obtained in the above section, by correcting the pressure drop between the pressure tap and the heated section, and by neglecting vapor momentum based on the discussion above.

The data concerning wall shear stress were arranged for the friction multiplier ϕ_l^2 which is defined as the wall shear stress ratio of inverted annular flow to liquid single-phase flow, in order to eliminate the difference of channel diameter. Figure 13 shows the relationship between the friction multiplier and the average thermal equilibrium quality. It is seen from the figure that this arrangement expresses well the effect of inlet subcooling, but does not express the effects of inlet velocity and wall superheating, because these data are crowded near the quality of -0.05.

The relationship between the friction multiplier and the dimensionless vapor film thickness was also investigated, with results shown in Fig. 14. Although arranged data are scattered, compared to those of the Nusselt number and the velocity ratio, the effects of inlet velocity, local quality and wall superheating can be expressed with one correlation. Furthermore, the difference in channel diameter is made clear, and the friction multiplier increases with a decrease in diameter, even though the dimensionless vapor film thickness does not change with varying channel diameters. The reason that vapor film velocity increases with a decrease in channel diameter is because of an increase in the vaporization rate, due to an increase in heat flux. It is seen from the figure that the friction multiplier increases with an increase in vapor film thickness, and that the friction multiplier is less than unity for a thinner vapor film thickness, but is larger than unity for a thickness of more than 0.4mm. Thus, the multiplier is not always less than 1.0.

The results in vertical flow are shown in Fig. 4, together with those in horizontal flow.   The dimensionless

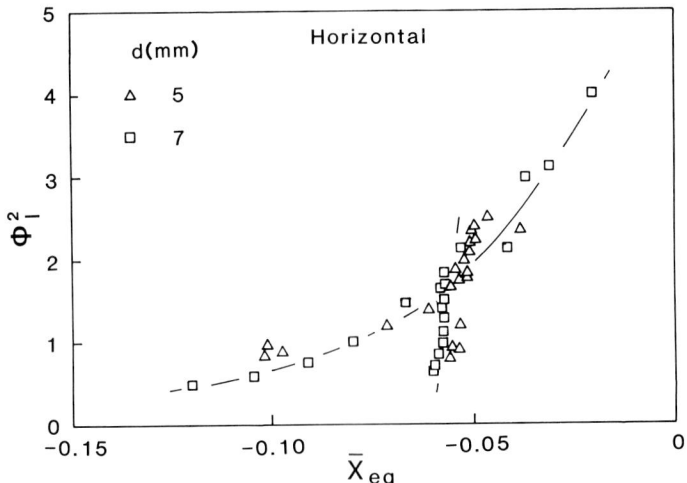

Fig. 13 Relationship between friction multiplier of inverted annular flow and thermal equilibrium quality

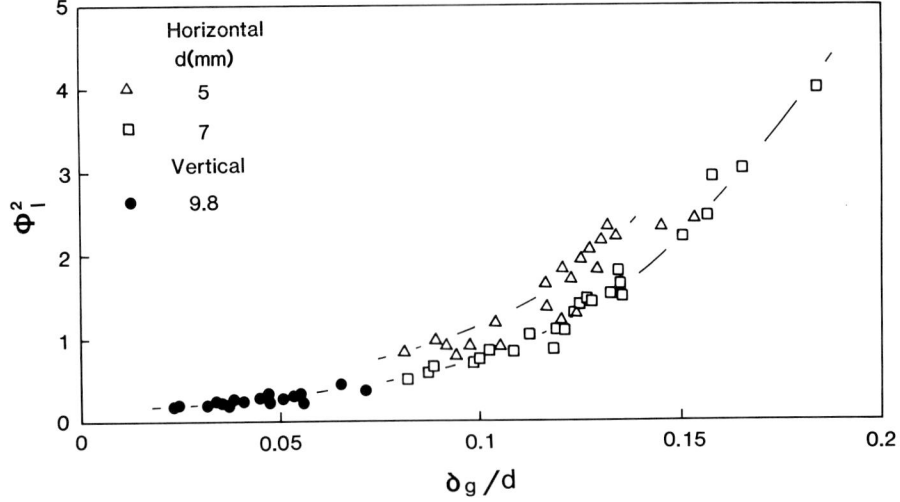

Fig. 14 Relationship between friction multiplier of inverted annular flow and dimensionless vapor film thickness

film thickness is very small in vertical flow, so the friction multiplier becomes smaller than unity. By extrapolating the results in horizontal flow, it is seen that the friction multiplier with the same diameter in vertical flow is larger than that in horizontal flow. This is because the interfacial wave and shear stress in vertical flow are larger than those in horizontal flow, due to the effect of gravity.

CONSIDERATION OF THERMO-HYDRAULICS

The thermo-hydraulic behavior of inverted annular flow is governed by the vapor film thickness and its turbulence, which increases with thickening vapor film. As channel diameter decreases, vapor film becomes thinner, so the heat transfer ratio of a wall to a liquid jet increases, as does the proportion of vaporization rate to heat input. Consequently, as the diameter decreases the vapor film velocity increases, so the heat transfer coefficient and wall shear stress increase.

On the other hand, in vertical flow, a liquid jet is more unstable than in horizontal flow, because of the effect of gravity. Therefore, vapor film becomes thinner and the interface has more waves. Consequently, the heat transfer rate is larger than that in horizontal flow. However, the vaporization rate is smaller in comparison with the wall heat flux, because the heat transfer from the interface to a liquid jet is larger. This is because of the wavier interface, so the vapor film velocity does not become larger than that in horizontal flow. Thus, wall shear stress is not much larger in vertical flow than in horizontal flows.

The effects of equilibrium quality and wall superheating on heat transfer coefficient and vapor film thickness are identical in both flow directions. However, the effect of liquid jet velocity appears in horizontal flow but not in vertical flow, because of a more unstable liquid jet.

CONCLUSIONS

To understand the effect of flow channel diameter, the thermo-hydraulic behavior of inverted annular flow was investigated experimentally, in three horizontal flow channels with different diameters. Furthermore, the experimental results are compared with that of vertical inverted annular flow, to show the effects of different flow directions on thermo-hydraulic behavior. The following conclusions can be drawn from this study.
(1) The vapor film thickness increases with increasing wall superheating and thermal equilibrium quality, regardless of different channel diameters and flow directions. These effects are identical regardless of different channel diameters. As the liquid jet velocity increases, the film becomes thinner in horizontal flow, but varies little in vertical flow.
(2) The heat transfer coefficient decreases with increasing superheating and quality, regardless of different

channel diameters and flow directions. The effect of the superheating on the heat transfer coefficient becomes smaller with increasing channel diameters, and the effect of quality becomes smaller with decreasing channel diameter.
(3) The heat transfer coefficient increases with increasing liquid jet velocity only in horizontal flow, and the effects of the flow rate on the heat transfer coefficient are identical, regardless of different channel diameters. However, changing the liquid jet velocity has little influence on the heat transfer coefficient in vertical flow.
(4) The effects of flow rate, superheating and quality on wall shear stress can be correlated for one channel diameter with dimensionless vapor film thickness, channel diameter and flow direction. The wall shear stress increases with an increase in the dimensionless thickness and with a decrease in the channel diameter.
(5) The wall shear stress in inverted annular flow with a smooth interface having a thin vapor film, is less than that in liquid single-phase flow, but increases as the interface becomes wavy.
(6) Assuming inverted annular flow in the same channel diameter, the wall shear stress in vertical flow is larger than that in horizontal flow, because of the wavier interface.

REFERENCES

1. Berenson, P. J., 'Film-Boiling Heat Transfer from a Horizontal Surface', Trans. ASME, J. Heat Transf. Vol.83, pp. 351-357, 1961.
2. 'RELAP/MOD2 Code Manual Volume 1: Code Structure, System Models and Solution Methods', NUREG/CR-4312, 1895.
3. Bromley, L., 'Heat Transfer in Stable Film Boiling', Chem. Eng. Prog., Vol.46, pp. 221-227, 1950.
4. 'TRAC-P1A: An Advanced Best-Estimate Computer Program for PWR LOCA Analysis', NUREG/CR-0665, 1979.
5. Aritomi, M., Inoue, A., Aoki, S. and Hanawa, K., 'Thermal and Hydraulic Behavior of Inverted Annular Flow', 2nd Int. Top. Meet. Nuclear Power Plant Thermal Hydraulics and Operation, Tokyo, pp. 1.34-42, 1986.
6. Fung, K. K. and Groeneveld, D. C., 'Subcooled and Low Quality Flow Film Boiling of Water at Atmospheric Pressure', Nucl. Eng. Design, Vol.55, pp. 51-57, 1979.
7. Ishii, M. and Jarlais, G. De, 'Flow Regime Transition and Interfacial Characteristics of Inverted Annular Flow', Proc. Japan-US Seminar on Two-Phase Flow Dynamics, pp. B.7.1-9, 1984.
8. Ellias, E. and Chamble, P., 'Inverted-Annular Film Boiling Heat Transfer from Vertical Surface', Nucl. Eng. Design, Vol.64, pp. 249-257, 1981.
9. Fung, K. K. and Groeneveld, D. C., 'Measurement of Void Fraction in Steady State Subcooled and Low Quality Film Boiling', Int. J. Multiphase Flow, Vol.6, pp. 357-361, 1980.